ARMY SCIENCE: THE NEW FRONTIERS

MILITARY AND CIVILIAN APPLICATIONS

EDITED BY

Daphne Kamely **Kenneth A. Bannister**

Robert M. Sasmor

Borg Biomedical Books
PO Box 986, Saratoga, WY 82331

Army Science: The New Frontiers
Military and Civilian Applications

Library of Congress Cataloging-in-Publication Data

Army science : the new frontiers : military and civilian applications
[edited by] Daphne Kamely, Kenneth A. Bannister, Robert M. Sasmor.
p. 744 cm.
ISBN 0-943255-36-8
1. Military research — United States.
2. United States. Army.
I. Kamely, Daphne
II. Bannister, Kenneth A.
III. Sasmor, Robert. M.
U393.5.A76 1993
355' .07' 0973 — dc20
for Library of Congress

93-9121
CIP

ISBN 0-943255-36-8

Table of Contents

Summaries

Advanced Electronics

Summaries

Advanced Materials Science

Summaries

Battlefield Environment

Summaries

Foreword

The 18th Army Science Conference was held in Orlando, Florida on June 22-25, 1992. This biennial conference was inaugurated in 1957 to provide a forum for the presentation, discussion, and recognition of significant accomplishments by U.S. Army scientists and engineers in their efforts to support the combat soldier of tomorrow.

The focal point of the 1992 conference was Army Key Emerging Technologies, including systemic issues and supporting capabilities. Over 80 scientific papers were presented, featuring innovative advances in Army research and development. Keynote and plenary addresses by nationally and internationally recognized scientists and engineers supplemented the scientific presentations throughout the conference. The conference also hosted exhibits to highlight the latest state-of-the-art research and development underway at Army laboratories and Research, Development and Engineering Centers (RDECs), and to provide an atmosphere for interaction and discussion.

This year, for the first time in the history of the Army Science Conference, participants from academia, industry, and private institutions, both from this country and abroad, were invited to attend. Over 500 scientists, engineers, and representatives from the public and private sectors attended and interacted with Army representatives and other conference participants in an effort to promote Army partnerships with academia and industry. The conference also was successful in paving the way for growth in industrial technology programs of interest to the Army and in strengthening the link with industry needed for the transfer of technologies from Army laboratories and RDECs to the private sector.

This book contains a selection of technical papers which were presented by Army scientists and engineers at the 1992 Army Science Conference. An independent panel of experts thoroughly reviewed all of the papers and selected for publication only the most original and innovative among them. There are a total of fifteen chapters, corresponding to the fifteen technical sessions of the conference. Where possible, each chapter is introduced with a paper by the keynote speaker who was invited to introduce the corresponding session at the conference.

Overall, we are pleased to report that the 1992 Army Science Conference proved to be a great success in encouraging and fostering collaborations among scientists and engineers from the Army laboratories, RDECs, academia, industry, and other government laboratories. This book is therefore published with the intent of sustaining, strengthening, and expanding this spirit of collaboration between the Army and its technological partners to support our most important customer – The Combat Soldier.

Daphne Kamely
Director of Research
Office of the Assistant Sectetary of the Army
for Research, Development and Acquisition
The Pentagon
Washington, DC 20310-0103

Robert M. Sasmor
Office of the Deputy Chief of Staff
for Research, Development and Engineering,
U.S. Army Materiel Command
Alexandria, VA 22333-0001

Kenneth A. Bannister
U.S. Army Research Laboratory
Aberdeen Proving Ground, MD 21005-5066

Acknowledgments

The 1992 Army Science Conference organizing committee wishes to thank Mr. George T. Singley III, Deputy Assistant Secretary of the Army for Research and Technology, for hosting the conference and Mr. Darold Griffin, Principal Assistant Deputy for Research, Development, and Acquisition, U.S. Army Materiel Command, for financial and administrative support. The committee especially thanks Mr. Richard L. Eichholtz of the U.S. Army Toxic and Hazardous Materials Agency, for his untiring efforts and expert administrative skills leading up to and during the conference. Thanks are also due to Dr. Robert A. Fifer of the U.S. Army Research Laboratory, Aberdeen Proving Ground, for his help during the planning and execution phases of the Conference, as well as in initiating the groundwork for this book.

Editorial Note

More than 80 presentations were given at the 1992 Army Science Conference. It was intended that only select papers, representing the wide range of research being conducted by the U.S. Army, be published. Independent experts from different scientific disciplines in the private sector reviewed all the papers presented and recommended those for publication. However, due to the tremendous amount of material selected, it was necessary to run some papers in summary format to keep within reasonable book length limits. By including these summaries, we wish to provide our readers with an introduction to important research topics, and the ongoing research it represents, as well as the names and affiliations of the persons leading the research to contact for further information.

—*The Editors*

The New World

Army Key Emerging Technologies:

Overview of the 18th Army Science Conference

Kenneth A. Bannister
U.S. Army Research Laboratory
Aberdeen Proving Ground, MD 21005-5066

This Book is based on the 18th Army Science Conference, held June 22-25, 1992, in Orlando, FL. The Conference provided a forum for active discussions of the Army's key emerging technologies. The Conference was unique because for the first time attendance was open to representatives from academe, industry, and private research and development (R&D) institutions. In these times of declining manpower and research dollars, the Army recognizes the need to use major conferences of this type as a vehicle to strengthen existing research and development partnerships with universities and industry and begin to foster new partnerships where possible. A second important goal of the Conference was to promote the growth of industrial programs of interest to the Army, as well as promote technology transfer from the Army's laboratories and R&D centers to the private sector. In order to provide an atmosphere for Army, academic, and industry representatives to interact and discuss work of common interest, the Conference hosted a large number of exhibits to highlight the latest state-of-the-art R&D underway at Army laboratories and centers. During the sixteen technical sessions of the Conference, over eighty scientific papers were presented by Army in-house researchers. Invited keynote and plenary addresses by nationally and internationally recognized scientists and engineers supplemented the scientific presentations.

In this Book, selected sessions and papers from the Conference are highlighted. The original sixteen session topics of the Conference have been combined into five topics. Because of this, some papers appear under a slightly different heading. In an attempt to maintain a uniform style, some papers have been edited and appear in summary form. In selecting which Conference papers to publish, the goal was to end up with a book of reasonable size, truly representing the scientific tenor of the Conference. Some very hard decisions were made in selecting which papers to publish. The scheme developed therefore was to form a selection board. The selection board membership was comprised of Department of the Army staff, Office of the Secretary of the Army for Research, Development and Acquisition staff, the publisher, and four prominent outside reviewers. The board selected papers as follows: (1) all those papers formally acknowledged as "best papers" at the Conference; (2) the available Conference keynote papers; and (3) other Conference papers. Emphasis was placed on uniqueness and thoroughness of treatment of subject matter in selecting the latter. With so many excellent papers available, the board truly regrets any omissions. The stringent size limitations on the Book simply made it impossible to include all of the Conference papers in their entirety.

No one who attended the previous Army Science Conference in June 1990 could have predicted the astonishing historical events that have occurred between that time and the most recent Conference. Who could have predicted the calamitous breakup of the Soviet Union? Who could have predicted the sudden outbreak of hostilities with Iraq, or the subsequent lopsided victories over that country's forces by the Western Allies during Desert Storm? The eloquent words of Dr. Edward Teller, now 85, help us understand these remarkable events and give his unique insights on what technologies we need to emphasize in order to maintain our leading edge. Dr. Teller recommends building on the successes of weapons such as the Stinger missile to continue to improve the accuracy of our weapon systems. Secondly, he advises us to greatly improve our satellite communication and surveillance capabilities, and related computer technologies, for both military and peacetime purposes. Dr. Keyworth cautions us that the old Soviet military apparatus has not gone away entirely; it is merely regrouping for the time when Russia resolves its present organizational problems and becomes a major economic power. He points out that our nation is undergoing some fundamental changes, partly due to external causes and partly due to lingering internal infrastructure problems. He says that we are moving away from an economy based on "economies of scale" to one based on information exploitation. He cites putting low-cost powerful personal computers in the hands of millions of users, at the expense of large computer mainframes, as the most visible sign of this fundamental transition. Dr. Keyworth goes further to claim that our military, including even the Defense Advanced Research Projects Agency, appears to be lagging behind industry during this transition. Keyworth concludes his paper by echoing and expanding on some of Dr. Teller's recommendations for developing more precise "smart weapons" and the need to exploit space-based surveillance systems. Finally, Dr. Donald Langenberg of the University of

Maryland offers valuable perspectives on strategies which he and the rest of the leadership of his large university have developed to cope with shortfalls in funds and resources. Dr. Langenberg points out the strong ties that have existed between the military and academe, dating from the days of the 1862 Morrill Land Grant Act and greatly augmented during and after World War II. He underscores the need to strengthen this relationship and use it as a means to build additional cost-sharing R&D partnerships and foster technology transfer to private industry.

The fifteen papers in the Biosciences section deal with three biotechnology problems of major concern to the Army: the environment, protection of soldiers against biological/chemical attack, and protection of soldiers against eye injuries due to lasers. The keynote papers by Kornguth, Anfinsen, and Chakrabarty, and the papers by Harvey, Dillon, and Wentsel focus on environmental science issues. Kornguth discusses biopolymers which as biomaterials could potentially be applied either to degrading or synthesizing energetic compounds and low-density/high-strength materials. Kornguth argues that these materials could have application in sensor development related to chemical-biological defense and detection of concealed ordnance, optoelectronic switching and encoding devices, and application to environmental remediation and demilitarization processes. Anfinsen's research on hypothermic bacteria has potential use in decomposing mustard gas. Chakrabarty discusses the fundamental question of how microorganisms evolve genetically for survival under stressed environments. This question directly bears on biodegradation of synthetic chlorinated aromatic compounds. Agricultural, industrial, and domestic activities introduce herbicides, pesticides, chlorinated solvents, and fluorocarbons, many of which are synthetic compounds, into the environment. Microorganisms traditionally are expected to degrade such chemical wastes in the environment; however, new synthetic compounds resist breakdown by microbial processes. Harvey discusses research on use of biodegradation as an alternative to incineration of chemical agents such as mustard gas. Careful attention must be paid to treatment of the various chemical byproducts, however, to make certain no toxic products remain. Dillon discusses the first development of a general framework for assessing environmental risks posed by contaminated sediments at Army facilities. Wentsel points out the critical need for soil test methods and requisite criteria for environmental clean-up at Department of Defense (DoD) facilities. A change of one order-of-magnitude in such criteria can lead to unacceptably high clean-up costs. Wentsel has developed a method to address the transport dynamics and transformation of chemicals in soil and the soil fauna microcosm. This method will provide soil community and ecosystem level data on soil systems which can be used to help develop reasonable clean-up criteria for DoD sites.

In biomedical research, the excellent papers by Smith, Reid, Hunt, Sjogren, Lebeda, Sadoff, Brimfield, and Friedlander deal with extremely important research issues of chemical and biological defense. The Army's concerns here were greatly heightened during the recent Desert Shield/ Desert Storm operation because of the threat that these dreadfully toxic agents might be used against our soldiers. Schuschereba reports research on treatment of eye injuries due to exposure to battlefield lasers. The experience from past wars shows that care of the blinded requires large numbers of personnel. No treatment for laser-induced eye injuries exists at the present time. The major focus of this research, therefore, is to develop an understanding of the nature of such eye injuries and then to devise means of treatment.

Research on Advanced Electronics is crucial for support of the Army's thrusts in surveillance, communications, sensors, detection, computer- and electronics-based simulation for training and readiness, and laser-based weapons. The thirteen papers and summaries contained here concentrate for the most part on ongoing research work at Army laboratories and research, development and engineering centers on basic physics, fabrication techniques, and determining the performance characteristics of new semiconductor and opto-electronic devices. The Army needs these devices to build larger integrated components and systems to support the thrust areas just mentioned. At the Conference, keynote speaker Koch compared and contrasted optoelectronic integrated circuit (OEIC) and photonic integrated circuit (PIC) devices, and gave an excellent overview describing many of the advantages of these technologies. Sooy discussed laser weaponry and summarized the history of attempts at using high-power visible light in battle. It appears that laser weapon technology must continue maturing and is still substantially more costly than competing gun and missile systems. Several DoD laser weapon development projects are currrently directed toward future battlefield applications.

Advanced Computing offers great promise to the Army. For the first time, the means to obtain solutions to very complex, real-time battlefield simulation problems appears to be in hand. This is certainly true for deterministic problems such as predicting dynamic motions of Army vehicles. Four technologies have now matured to the point where real-time operator-in-the-loop simulations are possible: (1) recursive dynamics formulations for mechanical systems, robustly implemented in the form of computer algorithms and software; (2) modest-cost parallel computers which provide vast computational power; (3) high-speed computer graphics systems motivated by Air

Force work on aircraft flight simulators; and (4) precision motion base control, such as multi-degree-of-freedom mechanical hardware to simulate vehicle acceleration environments. This simulation capability gives the Army a unique opportunity for virtual prototyping of vehicle design, realistic treatment of human and environmental factors on driver performance and driver-vehicle interactions, and for numerous other important military vehicle applications. Such modeling can save the Army millions of dollars by reducing or eliminating expensive, time-consuming, and sometimes dangerous driver-vehicle field tests.

It is ironic that while much of the progress in modeling dynamic systems is directly applicable to robotics, Brockett points out in his cogent keynote paper that progress on full exploitation of computers in robotics nevertheless remains slow. The slow progress is for reasons not unlike those that have hampered progress in other areas of the field of artificial intelligence, for example computer vision and language understanding. Namely, (1) robotics problems have turned out to be much harder than everyone initially thought, and (2) robotics depends on a blend of technologies, some of which are not nearly as far advanced as needed. Brockett sees the robotics problem as one requiring highly integrated computing, communication and control, especially if vision systems are to be involved. Unfortunately, progress in integrating these three areas is still very much in its infancy. He also mentions the critical need for robust, portable robotic motion control languages, such as the vendor-independent computer languages FORTRAN, C and Pascal, already available for a wide variety of computing platforms. Allied with the development of control languages, there is a need to develop symbol-to-analog signal transducer capabilities. Motion control for a robot is the central problem, since the robot's motion is the robot's reason for being. Furthermore, there is the problem of redundancy; that is, if multiple paths exist, then due consideration needs to be given to the consequences of following the various paths. Lastly, it has been found that using a computer to generate robot motion descriptions is a very much more difficult computing task than other routine numerical computing and word-processing tasks.

Brown and Shires report on progress in robot vision. They discuss challenging machine vision problems relating to autonomous navigation and remotely-driven (teleoperable) combat vehicles, respectively. Brown addresses methods for solving scene segmentation and definition problems for guiding vehicles along previously unmarked pathways. Shires devised techniques for imagery data compression needed for real-time control of teleoperated vehicles which view the battlefield with currently available forward-looking infrared (FLIR) sensor technology.

In another important area of advanced computing, artificial intelligence (AI), Hoock reported on a neural network-based technique, involving learning, to reduce effect of propagation losses and battlefield smoke and dust obscuration effects on the performance of automatic classification of manmade objects on the battlefield. Gupta describes research on complex semiconductor hardware components that could be used to build a novel neural network-based processing system for improved target recognition.

In the more traditional area of computer modeling, Hursh describes important extensions and improvements of the Army's Unit Resiliency Analysis (AURA) model for predicting unit effectiveness. Hursh has inserted modeling in AURA of the effects of health hazards that tend to degrade soldier performance such as infectious disease, CBW effects, stress and fatigue, environmental extremes (heat stress and dehydration), and traumatic and accidental injury. Parallel with this, he has inserted modeling of the effects of medical interventions which can reduce the degradation effects of all or some of the health hazards. These are extremely realistic capabilities never previously available in AURA. Norvelle describes an efficient new technique for updating DEM data using iterative orthophoto refinements (IORs). Cartland presents results of an interesting computational-experimental approach for studying simple atom-molecule reaction kinetics relevant to combustion and atmospheric chemistry. The computational model permits sufficient flexibility in modeling known and conjectured details of the chemical reactions, reaction rates, and intermediate products of reaction, so that when the model is coupled with judiciously planned experiments a much more complete portrayal of the overall behavior of the reactions can be deduced than by either experiment or modeling alone. Weinacht discusses an important aspect of projectile aerodynamics which can now be captured, owing to recent revolutionary advances in computational fluid dynamics (CFD) modeling of projectile aerodynamics.

The inescapable importance of materials science and engineering in all phases of the Army's activities is well illustrated in the section on Advanced Materials. Keynote speakers Eager and Wilkins pointed out the remarkable sophistication that this field has attained. In just the last twenty years we have moved from the traditional mode of "designing with materials" to one of "designing of materials". By this it is meant new products and systems are conceived and then materials with the required properties are developed. This is certainly true for many advanced composite materials applications. Both Eager and WIlkins stressed the importance of developing cost-effective manufacturing and fabrication processes for new materials. Merely "inventing" a material no longer guarantees success; the material must be brought to market in the form of useful products.

The remaining papers in the Advanced Materials section cover a wide range of topics. Beatty discusses recent work directed at gaining better understanding of the effects of adiabatic shear localization and microstructure on penetration failures of VAR4340 steel armor plates. Beaudry reports on microreticular resins which offer utility for chemical defense. His work focuses on characterizing the adsorption properties of these materials. McClane's paper discusses a new processing technique for etching gallium arsenide wafers, reactive ion plasma etching, which has potential use to the Army in fabricating microwave integrated circuits, detectors, and smart munitions. The last paper in this section, by Ruzga, deals with research on fabricating oxide/non-oxide ceramic materials using cotton fibers infiltrated with sol-gel solutions of silicon carbide. The research offers innovative low-cost alternatives to present methodologies for fabricating complex SiC structural components such as vehicle panels, machine parts, ballistic surfaces, helmets, and high temperature insulators.

The last section of the Book, Battlefield Environment, contains several related papers on major weapon system and propulsion technologies. Keynote speaker Hardison spoke to one of the most difficult problems in modern battle damage assessment, that of predicting the damage potential of modern conventional munitions. Bishop also reports on a damage assessment problem of considerable interest to the Army—the protection of trucks against land mine attack. His paper discusses the use of the AUGUST CFD computer code applied to modeling the damage response of a 5-ton M925A1 cargo truck to the detonation of an M15 AT land mine, which contains about 20lbs of TNT.

Lieb, Fisher, and Bulusu report advances in the understanding of the sensitivity and vulnerability of energetic materials. The first paper is concerned with establishing cause-and-effect relationships between mechanical damage and fracture of propellant grains and subsequent burning behavior of the grains in the gun chamber. It is known that uncontrolled breakup of grains can lead to damaging pressure waves in guns. In a closely related area of work, Fisher has used a hydrocode and experimentally-derived shockwave sensitivity and equation-of-state data to study the vulnerability of a solid propellant rocket motor to fragment impact. This research has led to intelligent redesign of the rocket motor to substantially improve its vulnerability resistance to threat stimuli. Busulu reports research on thermal decomposition of RDX and HMX. Improved knowledge on the thermal behavior of RDX and HMX is needed to improve the vulnerability properties of energetic formulations using these materials.

Fair gives an excellent state-of-the-art review of the potential benefits of electric devices and electric propulsion concepts, including electrothermal

(ET) and electromagnetic (EM) guns, to the Army. While basic research on EM guns progresses, ET gun concepts are considered the near-term electric gun solutions for achieving hypervelocity launch. Considerable work is underway on ET technology in the Army R&D community. For example, Juhasz explores the use of plasmas as an alternative means of igniting conventional gun propellants.

Bryzik discusses results of research done on advanced low heat rejection (LHR) diesel engine concepts, which show definite promise as powerplants for smaller, lighter, and more mobile Army vehicles.

Beyer, Sausa, Johnson, and Willingham all report research related to laser energy applications. Beyer reports on successful laser ignition of condensed (liquid and solid) gun propellants. The long-term goal here is to replace or augment liquid and solid propellant igniter systems in guns, howitzers and tanks. Sausa, in the paper awarded best at the 18th Army Science Conference, describes a novel ultra-sensitive laser-based detector for trace nitrogen dioxide-containing compounds. This technology can be used to detect or monitor the presence of energetic materials on the battlefield, such as propellants and explosives. Army applications include mine countermeasures, anti-terrorist operations, and possibly detection of CW agents. Johnson describes developments to solve some of the problems associated with laser satellite networking. This offers improved communications to support Army strategic and tactical missions. Willingham has demonstrated an optical limiting effect, resulting from ultrafast refractive index changes in specialized materials which can provide eye protection against incident laser energy.

The papers by Gillespie, Morrison, and Welsh deal with battlefield environment and sensing problems. Melloh and King also touch on related work concerning remote sensing capabilities for solving Earth surface and subsurface imagery problems. Morrison describes a new, highly military CW agent filters. Welsh reports progress on development of an integrated, physics-based, scene generation process for identifying and quantifying worldwide environmental parameters affecting smart weapons system performance. Of particular interest here is the capability to perform site-to-site and area-to-area comparisons for electromagnetic and electro-optical weapons systems. Because the work draws on many sources of quantitative data that must be stored and rapidly recalled in many different ways, Welsh makes use of novel relational database management systems approaches.

The Army has a need for high-power electrical impulse generators that can radiate electromagnetic pulses with narrow pulsewidths and picosecond rise times for use in sensors, battlefield synthetic aperture radars, and smart high-power microwave jammers. Kim describes the development of a gallium arsenide switching device which has successfully been used to generate picosecond pulses at 40 megawatts of power.

This Book is unique in that it provides a timely and very detailed overview of large portions of the Army's ongoing in-house research efforts across a broad range of technologies. Every effort has been made to publish this Book soon after the 18th Army Science Conference so that the information is current. Consequently, the Book helps fulfill the main objective of the Conference: to promote technology transfer to users outside the DoD.

The main goals in this Overview were: (1) to try to capture in a few sentences the main points of each section; and (2) to give a sense of the continuity and connectivity of the Army's many and varied research and development efforts. The Overview also was done for the benefit of readers interested in an overview of the Book's contents for purposes of selecting individual papers for further study.

In conclusion, the organizing committee of the 18th Army Science Conference, the editors, the publisher, and the paper selection board are acknowledged for their outstanding contributions to the successful completion of this Book.

The New World

Edward Teller
Director Emeritus
Lawrence Livermore National Laboratory
Livermore, CA 94550

I will try to talk simply and briefly about a few things that I consider really important. I hope that when I finish, you may have questions, any questions, short ones so that I can give long answers, because I do want to talk about what you want to hear.

I have to start by doing something that for a completely understandable reason not everybody is doing. That is to congratulate you for a tremendous success. I mean the war in the Gulf. I do not know when there has been such a war except perhaps when Cortez conquered Mexico. The stakes have been extremely high, and the remarkable thing is that there are so many people who said, "We cannot do it! It will mean a hundred thousand lives." They were off by a factor of a thousand, and those were some of the people who claim to be the wisest and know the most about military affairs. Why does anybody continue to listen to them?

Now, one other thing. We have a President whom I do not know well at all, and yet from a distance, I cannot help but admire him. He is a man of compromise. In that one case when it was the question to stop aggression, that is the one case where compromise would certainly have hurt, and where probably almost everybody would have compromised. And, he did not. I like a man who knows precisely the one case on which one cannot compromise.

I want to tell you a story which seems irrelevant. I hope you will catch the distant connection. It is an old story. A mother whose son was swept away by a tidal wave prayed to God, "He was a good son. Why did you take him away? I want my son back. Give my son back." So God performed a miracle. He sent back the tidal wave, and deposited the son hale and hearty and smiling. Then the mother raised her hands to Heaven and said, "But, Lord, where did you leave his hat?"

What are we complaining about? The United States has overcome a powerful enemy, the Soviet Union. They were so frightened by our success that they vanished. That is the miracle, the reversal of the tidal wave.

Now, of course, people say that the Soviets never really amounted to anything. I am Hungarian. Tell it to the Hungarians that the Soviets never amounted to anything. Four-hundred million people in the former Soviet Union, and in the former satellites, are now free to make their own mistakes. And they will make them, plenty of them. That is natural because you know what Democracy is? I know it because I learned it slowly and painstakingly. Democracy is the art of making many small mistakes, which is incomparably better then making a very few giant mistakes. We are throwing four hundred million new people into the difficult art of Democracy. And they need help, because this making and balancing of small mistakes is something that I greatly prefer and I want to explain to you why.

I claim that the American Revolution was the only revolution that really succeeded, that truly accomplished all its aims. The revolutionaries say, "We'll change the world. We'll look under the hood and find whatever is broken. We'll do it better." The Federalist papers make no such claims. They say we are poorly governed, and government is always poor; therefore, do not trust anybody. When we come to power, do not trust us either. Have checks and balances. Be inefficient, but avoid the mistakes of a big government. This is a complicated world into which we are now entering.

And now I would like to jump into an entirely different phase of the problems that we are facing. The Cold War, at least as it was, is over. We need not beat the Soviets. We need not outbuild the Soviets. They outbuilt us. We were ahead of them in research. We were flexible in electronics, in accuracy, in computing machines, in the kinds of things that made it possible to defeat a big and tough army in a few days with a loss of a hundred lives.

The situation is now different. We all know, and I think we all understand, that in the new situation, we do not need to spend as much money on the Armed Forces. Routine activities can be reduced. I just heard that one-third of our men in Europe have already been withdrawn, another third will be brought home in another year. Considering all things, I think this is the time for unlimited opportunity for our Armed Forces because our problem is not one big enemy; there are many dangerous trouble spots, 20 nations with missiles that can initiate sudden warfare. Many, many nations look to us for help. The days of "big bang" are clearly beginning to be replaced by the days of accuracy.

I have tried to check on figures concerning a spectacular success that we achieved while the Soviet Union was still in its full power. That was Afghanistan and the Stinger Missile. The courageous defendants of their homeland in Afghanistan were defeated by bombs from low-flying Russian airplanes and helicopters. But then we delivered to them a thousand products of our research, early products of accuracy: the Stinger missiles. The situation changed, low-flying missiles against low-flying aircraft. The incredible point is that the total cost to the United States was estimated at $100 million, a factor three-thousand less than one-year's U.S. military expenditure. That monetarily small effort defeated the Soviet aircraft within a few hundred miles of the Soviet boundary.

The Soviet leadership realized that they could not compete with American technology. But, they believed that they could compete with America, because, after all, America is a democracy and can never make up its mind. And we have a President who will compromise. But on the matter of aggression, the President did make up his mind. With a minimal expenditure of money, and an absolutely minimal expenditure in lives, he stopped a dangerous dictator. Without our intervention, Saddam Hussein would now control all of the Middle East oil, and all the political power that is derived from the oil. Furthermore, he was on his way toward building an atom bomb.

I claim that our success frightened the Soviet leadership. It was no longer monolithic; it could no longer withstand the old demand of the people in the Soviet Union for freedom. I honor that demand of the people, but we all have to understand that we, and in particularly you in the Armed Forces, have contributed to that spectacular change. The Soviet Union made a tremendous improvement. It went out of existence.

Now what next? Our Armed Forces are needed in a completely new situation. International cooperation will count for a lot. Routine operations are no longer needed as much as they used to be. Stockpiles need not be so high. But one thing will continue to count: research. Research must continue on the various ways of how weapons can be used, how they can be used rightly, and whether they have been used rightly in the recent past. I do not know precisely where to draw the line between research and development. I therefore do not know whether I should guess that the U.S. Army has spent one percent or three percent of its budget on research. But I can say that the fruits of research depend on your decisions as never before.

Because you can and must reduce the routine expenditures, you have the opportunity to double and triple the actual effort on research. I know that I cannot guess what research will produce. Nobody can. One cannot plan what the most clever people in the world will do. Believe it or not, no matter how stupid they are in politics, scientists in their own way are clever.

For the relatively near future, our development seems to be necessary and possible. I have hoped for a long time that defensive weapons will become effective. Within 10 years, we will have come far in accuracy, far in electronics, and very far in computers. Apparatus similar to the Stinger missiles will have been improved many fold. Accuracy was the secret of the surprise and success in the Persian Gulf. Accuracy will play a decisive role in defense.

In looking at what research will do next, I would put all kinds of emphasis on space. In many ways, space should be used differently from the way it has been used until now. So far, the most interesting space objects have been in synchronous orbit 25,000 miles away. I like small, inexpensive low-flying satellites. I was told they can be made in reasonable quantities at a few hundred pounds weight for $10 million each. I hope that they can be made even smaller, and even less expensive, maybe even $1 million each. The biggest hope today in the strategic defensive initiative is the brilliant pebbles. The great hope is in the plan for Global Protection Against Nuclear Strikes, which is the peacekeeping mission that we are all interested in. But, beyond that, one can make these small satellites to do many peaceful things, such as getting better information over the whole globe. With a revisit time of a few minutes, we could extend weather predictions from one uncertain week to two uncertain weeks, and eventually to one certain week. We could really see not only how much carbon dioxide or sulfur dioxide is in the air, but what it does to the climate. We could, I hope, find out about pollution even before we regulate it. We did regulate DDT and did a tremendous favor to the mosquitoes. Malaria was on its way out, and recently malaria has killed tens of millions of people due to thoughtless environmentalists who insisted on premature regulations. I believe by getting good information on the weather, its composition and pollution, we could replace thoughtless environmentalism with careful environmentalism. At the same time, we can find out what is badly needed for the military.

I love small satellites, for the same reason I love Democracy. With small satellites you make small mistakes and you can correct them. With big satellites, you make billion dollar mistakes which are much harder to correct. And, I must say that I see now a wonderful trend in the new NASA administration on matters clearly working in this direction, carefully, conservatively, not changing things too fast, but having it in mind what miracles one can perform with small satellites. But that is only one reason for small satellites. There is an additional reason. Small satellites, close by, need not only look, they can illuminate and look. The looking gets more difficult with "one over R-squared" at bigger distances. Illuminating and looking, as you all know from the radar equation, gets even more difficult as "one over R to the fourth." From synchronous orbit you cannot illuminate. From three hundred miles up, you can. With radar, with synthetic aperture radar, with lasers, you can gather much more data than you can ever evaluate or even understand. So why do you get them? Because in addition to everything else, we have computing machines that can select, with the help of adjustable programs, the interesting things. They need not look just for phenomena, but look just for the changes. And even further, when we have human reasons to be worried about what is going on in Inagu, in Equatorial Africa, or what is going on anywhere. If there is a volcanic eruption and we want to know how much non-manmade carbon dioxide is going into the air, the computing machines can be instructed to select specific data out of the infinity of data collected.

I am not telling you that the computers are more clever than people. I am not telling you even that people are more clever than computers. I am telling you that people and computers together are much better than either alone. We can put into each of the small satellites extremely powerful computers which can be flexibly programmed and reprogrammed depending on the variable circumstances. What that will do for a commander in the field is very hard to imagine because it will be more successful than any of us can guess.

I would like to mention to you just one of the many problems that will arise in connection with such a development. There will be a new world. We must cooperate. What about secrecy? We cooperate, let everybody know everything? I say, not quite yes, but almost yes. Because a democracy is not good at playing at secrecy. We are very good at keeping secrets from our own people to confuse them, but not as good at keeping secrets from our opponents. It is better to share our secrets, share all information and to work together. But some secrets are important and are easy to keep for a short time. A commander in the field can get the information more rapidly than anybody else. I am not telling you in a simplistic way to forget about secrecy. I am telling you to reform secrecy. Do not keep anything secret that cannot be kept secret. Think carefully what you really need to keep secret. In the long term, in the name of international cooperation, it is better find the ways of openness.

I talk about the problems I see, and about the direction in which I would look for solutions. The biggest problem to which I alluded and which I want to name outright here today—the biggest problem today in America is fear. It is the fear of progress, an incredible delusion that comes from the widespread publicity that technology is dehumanizing. It is this fear that is discouraging our young people from going into research. You need research. You can and must sponsor research. Together, and very much including the Armed Forces, we must find ways to give hope to the American people, to share with them the pride that we can rightly take in a war won without practically any loss of life. A great step has been made toward international stability, a step toward the mutual understanding with the Russians that we achieved in midcentury with the Germans and the Japanese.

Your work has contributed mightily toward a situation of the next century in which the world will be safe. Now wait a moment. Safe? Boring? I think that we might find ways that big wars may no longer be a danger. But to be really safe? No, because knowledge can be used and can be misused. Knowledge rightly used will produce decisions, countless decisions, many of which will still be mistakes. When we really begin to understand biology and can control inherited properties, we will be faced with a flood of new problems. Man has been described as a problem-solving animal. I contradict that description. The best description of man and woman is that of a problem-creating animal. Let's replace the great insoluble problems by a multitude of small, solvable problems.

And now I will end with one more comment—a pair of definitions. Obviously, I have talked to you as an optimist. I want to tell you what a pessimist is and what an optimist is. A pessimist is a person who is always right, but who does not get any enjoyment out of it. An optimist is somebody who imagines that the future is uncertain. I claim that it is my duty and your duty to be optimists. Because if you imagine that the future is uncertain, you will go out and do something about that uncertain future.

Discussion

Question: Do you offer any advice to the United States with regard to enhancement or nurturement of research in the future?

Professor Teller: Obviously I believe in a field of technology. A field of science needs great vision. I will tell you about something that happened more than 500 years ago in China. There was a fight, quite reminiscent of the fight between engineers and environmentalists today. That fight was between the Neo-Confucian administrators, the Mandarins, and the technological people who were the servants of the Emperor. Almost all of them were eunuchs who were bought and employed in this inhuman way by the Emperor. But, they led the Chinese fleet to explore the Indian Ocean and they landed in Africa before the Portuguese. They also criticized the Neo-Confucian administrators and there was a political battle that the Confucians won. The Ming Emperor made a decision: "No more explanations. Anything that's interesting can be found in this celestial empire—Chinese porcelain, Chinese philosophy—we need not go abroad to the barbarians."

That was the beginning of China's decline. That is how China degenerated from a leading technological society into an underdeveloped country. I think we are running a similar risk today. Yes, I believe research and technology must win.

Otherwise, America will follow China into the hovelss of the underdeveloped world.

Question: In the last few days, we have heard various speakers talk about our competitiveness, the advantage of Japan, and the role of Europe. Where do you see a role for the United States in this changing world market?

Professor Teller: I don't think that the Japanese are unfair. I think we are stupid. We are not competitive. Why? In the hard subjects in our good universities, universities that are probably the best in the world, half the students are foreigners at the graduate level. I don't protest. I'm a foreigner. The better half are foreigners. I still don't protest. Some of these foreigners stay here and help. Some go home and make international cooperation easier. The trouble is that the poorer half is American. They are not the poorer half because Americans are stupid, but, because our young people hear from their parents, from the press, even from their teachers, that technology is dehumanizing. They ask, "Why should I work hard to be dehumanized?"

I claim that science is the humanizing factor. Technology is the one thing that makes a tremendous difference between us humans and all other forms of life. I have not said that technology is good, but I do want to say that knowledge is good. Technology can not only be used, it can also be misused.

When I came to this country almost 60 years ago, science and technology were on a high pedestal. Now they are not. And that is the reason why we are not competitive. I think that not only students, but also the administrators in our big companies no longer really believe in technology, no longer really believe in its importance. This is a black and white statement, and I am sure that I have just now exaggerated. But, there is some truth to that statement. At any rate, there is much less respect for technology today than there was in the beginning of the century. We are now living on the capital which was accumulated when we paid proper respect to technology.

I think, that this point—this change in public attitude—is more important than any legislation, than any taxation, than any economic factor. I believe that our Armed Forces, perhaps not particularly our Army, but to some extent our Army, has been ahead on technology. That is, I think, where the emphasis must be. Not with any plan, but with care, interest, appreciation, and publication of the results, along with an involvement of the public. We must find a way to get back to the old American tradition to have the greatest respect for progress and those many things that work and that make it possible to influence the world around us. And then, I have the confidence that on the whole we are going to use our influence in the right way.

Technology and Future Force Structure

G. A. Keyworth, II
Distinguished Fellow
The Hudson Institute
Washington, DC

I could begin by offering a perspective of the tortuous steps we are taking to define our future military role, missions, and force structure. And I would be just another tired voice commenting from the past—with a mixture of nostalgia, voyeurism, and perhaps even a little envy. Instead, let me direct your attention to Russia, and ask you to look beyond the current chaos of political and economic change to the steps being taken by *their* military. The value of this exercise will, I think, become self-evident, and has nothing to do with any residual fear that the Cold War is anything but over. It has to do with what we can learn from their actions.

Few can doubt that Mikhael Gorbachev will be an important figure in history. Armchair analysts comment on his failure to separate economic and political reform, on his failure to appreciate the obstacles in converting from a centrally-planned to a market economy, and on how, like the hero in a Greek tragedy, he was simply overwhelmed by the progress he created. Others say he simply had no alternatives. To me, he displayed the essence of courage; he looked beyond the present and played to the future. To the people of the Soviet Union, he offered something new: hope for a better generation. For us, he removed an almost unspoken dread, that a failed Soviet Union would lash out with its one strength, its military, and, like a wounded bear, wreak irrational damage.

Today's Russian military general staff, whose ranks look virtually indistinguishable from what would have been the Soviet general staff, is in a dilemma. They are still a superpower army, steeped in counter-force deterrence, but whose current mission is merely to defend a country that is hardly in demand. In short, you have a first-class military with a second-class mission.

Apparently their leaders recognize that they are strong enough for the near and mid-term tasks because they are reducing drastically their procurement budget, thereby placing at risk the vaunted Soviet military complex, and investing heavily in military R&D, which they can hardly afford. After all, during our own military buildup in the early eighties, as well as during the current downturn, the priority of long-term R&D largely succumbed to the pressure for readiness. The question I ask is might they see something we do not?

One difference between us is that we seek to preserve greatness and they seek to restore greatness. They feel secure, militarily, in the near-term; in the mid-term, they face management problems; and, in the long-term, wherein lies their one source of hope, they seek a sophisticated military, capable of supporting the needs of an economically strong Russia. In contrast, let me observe that we face management problems in *both* the near and long-term; we are preoccupied with preserving our military-industrial complex; and we are paying little real attention to the long-term.

I believe the Russians have correctly identified what is in their best interest; I believe we have not. There is ongoing in this country a fundamental economic transition that is not reflected in the military. That is the move from a classical industrial economy, where economies of scale predominate, to an economy where exploitation of information technology is defeating those economies of scale. Some call it massive decentralization. I see it as George Gilder's "World of the Microcosm," or the realm where the microprocessor replaces the blast furnace. And what I see as its clearest manifestation is the emergence of distributed computing, with the ubiquitous PC in the vanguard.

Just look at a few facts, facts that are often overlooked in the debate over industrial policy, a debate that was led, first by steel industries, and most recently by the automobile industry. In 1980, there were about one million computer users in the United States. In 1990, there were more than sixty million. In 1980, American companies produced about 68 percent of the world's computer products; in 1990, they produced about 72 percent. In the decline of the mainframe and the ascendency of the PC, we adapted well. We both produced and consumed this advance in technology as our climate for entrepreneurialism and the ready acceptance of change paid off. We entered the age of the information economy in growth, not decline.

The military is behind in this transition. It is behind because the military-industrial complex lags in information technology behind the commercial sector; it is behind because our view of requirements is based far more in preserving the status-quo than in assuring our future; and it is behind because our best minds are writing "shrink-wrap" software, not target-recognition software. Where DARPA was once the visionary leader in computer technology, it is now a relative laggard, looking to massively-parallel centralized database servers while the world moves on to the bigger problem of matching broadband digital data transmission to the ultimate analogue interface—the human user. Where defense industries were leaders in the race to develop the technology for next-generation force-multipliers, the upstart MicroSoft wins hands down. The challenge is not how to reverse this trend, but how to take advantage of it.

This is one of the more challenging aspects of the whole "new world order" that we're immersed in. It calls for clarifying the kind of threat, setting aside accumulated conventional wisdom, and drawing upon our strengths.

A few years ago a Soviet military strategist, Marshall Nikolai Ogarkhov, took a far-reaching look at the way future wars would be fought. He described modern warfare as a succession of phases. The first was dominated by nuclear deterrence, the post-World War II condition. At a time when such views were hardly mainstream, he foresaw the weakening of nuclear deterrence, simply because, in his view, it would lose its utility. He was right, and that is one reason we ought to pay close attention to what he suggested would follow. Another is that a group photograph of the same Russian general staff that is today eschewing procurement, in the name of adequacy, for an altogether different kind of future force structure, looks like a reunion of ex-Ogarkhov students.

In his second phase, advanced conventional munitions (ACMs), or smart munitions, would dominate force structures and doctrine. We have entered that phase now, a time when we can achieve precise military results with low-yield weapons. It is a development that can give substantial advantages to technologically advanced forces.

Finally, Marshall Ogarkhov envisions a time when we will be able to achieve the same degree of precision, but achieve it with weapons that act at the speed of light over great distances such as lasers and other so called directed-energy weapons.

What is common for both of the two post-deterrence phases is the central, even enabling, role that precise observation and targeting plays. Moreover, such observation and targeting must be as flexible and dispersed as the weapons themselves could be. That equates to targeting-quality reconnaissance on a global scale. Strategic Defense Initiative (SDI), of course, first introduced us to the notion of real-time targeting on a global scale, though the requirements were narrower. SDI's targets are highly visible, come from predictable sources, and follow courses that can be readily calculated.

But the dilemma we run into as we enter the post-deterrence phase of smart munitions is that all the precision in the world is without meaning when the potential threat is unknown and unidentified. Smart munitions are a misnomer. They are not so much smart as they are extraordinarily obedient. They go where they are told. And it is people's jobs to tell them where to go.

My central point here is that reconnaissance and surveillance played a crucial role, often unheralded, in the past. In the future, they will contain much of that often-elusive measure of the successful application of technology, the force multiplier. And I believe they hold the key to future stability in this new world we find ourselves in.

One concept that marries the new kind of global threat we envision with the technology that derives from the breakthrough in distributed computing I spoke of earlier is "distributed surveillance." This is a space-based system that offers the means to observe, in real time, the entire globe with sufficient resolution to identify and evaluate potentially serious threats before they become irreversible.

What is important is to be able to understand and observe these threats, which are not only diverse but which are also likely to be poorly defined. Observation in this case requires a high vantage point, which points to space-based systems. Although the magnitude of the threat is smaller than under nuclear deterrence, the probability of use is greater. This places less demand on the development of weapons systems and more on the means to evaluate these threats, to support decisions to act, and to develop international consensus.

To use one of the emerging cliches of the 1990s, distributed surveillance is a technology "paradigm shift" that is perhaps best understood in an analogy, which is the shift I referred to earlier that has taken place in the world of computing. That shift has seen what looked like an uninterruptible thirty-year progression to larger and larger computers suddenly, violently, overturned by the introduction of the lowly personal computer. The PC, whose primary attribute was that it put computing independence, and eventually prodigious computing power, in the hands of tens of millions of individuals, has given us what is called distributed processing. Now the once dominant mainframes are struggling for diminished shares of the computing market, and even looking for new roles as complements to PCs.

I will not go into technical detail about distributed surveillance, but I will define it sufficiently, I hope, to distinguish it from merely small satellites. As I am sure you are well aware, it is possible to make inexpensive, small satellites with very good sensing capabilities, simply by replacing steel with silicon. It should be possible to put enough of them into space to provide continuous coverage of any place on earth. In fact, it should be possible to provide, eventually, continuous coverage of many places on earth at the same time, and instantaneous coverage of many places on earth almost at will. A basic enabling feature of this system is that all the satellites are networked, not unlike the way PCs scattered all over the world can be networked.

There are a number of new capabilities that such a network provides, but the critical feature is that there can be many people, from the soldier in the field to the commanders-in-chief in headquarters to the President himself, who can use such a system. We can envision a capability in which a person on one side of the globe can dial up the coordinates of a location on the other side and get current or very recent views. They can even get continuous views of the same spot, because the network would permit one satellite, staring at a spot as it passes over, to hand off the assignment to the next satellite as it comes into view, without interruption of viewing.

Without belaboring the obvious, this describes a system vastly different from what we have been using to date, a system that could have aptly been called mainframes in the sky. And like PCs, the new system takes advantage of the kinds of technology that the U.S. is very good at: microprocessors, telecommunications, and software. In spite of all the legitimate concerns about the competitiveness of our industries, we Americans are simply very good at those things, and it makes sense to apply them to our most pressing national security needs.

The transition to distributed surveillance will not be easy. Radical changes in direction rarely are, and there are entrenched attitudes and infrastructures that have to be considered. Yet there is a sense of inevitability to this transition, so whether it happens sooner or later, it behooves us to think now about how we would develop and use that capability and how it can improve our ability to control the stability of a world of increasingly fractionated power.

A democracy has a difficult time taking preventive action against a threat. Reaction yes, but prevention, rarely. Yet, we can see a time when a U.S. President will no longer have adequate means of dealing with nuclear or chemical or biological threats from upstart nations. How will that President make the critical decision to take action before absorbing precisely the kind of attack that struck fear into generations of people during the time of superpower nuclear domination?

The answer is threefold: one, to know with far better confidence than we do today when and where threats have crossed the threshold to become unacceptable; two, to be able to use those post-deterrence precision weapons to take them, and only them, out of action; and three, to be able to make the case for doing so unambiguous enough to convince a democracy that a particular instance is moral because it is necessary.

We will encounter growing discussions of this issue. Weapons proliferation, no matter how well we can slow it down, will eventually force us to reassess our role vis-a-vis these dangerously armed nations. In the past they could harass each other or make political hay with hit-and-run attacks causing surprise but minimal damage. But once we are at risk of unacceptable damage from the same kind of political forces, we are going to face a wrenching shock.

The noted military strategist General Glenn Kent called attention to the natural evolution of this kind of defense by suggesting that future threats will require what might be a fourth phase of SDI that is a complement to SDI's familar three phases of defense, which are boost-phase, mid-course, and terminal defenses. He says we may need what he calls "pre-boost phase intercept." It is a clever term, designed to grab attention, but simply put, he is stressing that strategy may soon require us to attack an opponent's weapons while they are on the ground, but when our knowledge tells us they have crossed that point where they become too dangerous to be allowed to remain. This is just one manifestation of how the new world order challenges our conventional thinking, as well as the conventional wisdom.

To those of us involved in research and development, such a picture as I have tried to sketch here also means some rethinking of how we work. It certainly requires closer affinity with progress in commercial technology; it certainly means more merging of disciplines; and it certainly means more teamwork. The sometimes-comforting cloak of classification is wearing thin, just as the search for better means to control the transfer of technology is beginning to look like the quest for the Holy Grail. In short, we have to adapt to a more open, less sheltered, and more complex environment, often with people who do not even know our business. Most intolerable of all, certainly to those of us from Los Alamos, is that many of those people who do not know our business do not even care.

In spite of all this change and uncertainty, and in spite of all the trials of fiscal conflicts and competition, there truly is a renewed excitement in science resulting from the ongoing explosion of information technology. Applying that same explosion of technology to define wisely tomorrow's force-multipliers is not a simple task. Clearly, my bet is that the single greatest impact will come from better, more accessible, and more timely reconnaissance and surveillance. Furthermore, I think the broader challenge of capturing the nation's excellence in information technology for defense will instill a new excitement in defense R&D. And finally, I even think there are signs of movement toward a healthier balance in priorities for research and procurement.

A New Alliance for Progress

Donald N. Langenberg
Chancellor
University of Maryland System
College Park, MD

You have gathered here to hear and talk about emerging technologies, new technologies. That is not my subject. My subject is the emerging new world. I am convinced that we all are living in a once-in-a-lifetime period, one in which the world we knew is irretrievably gone, and is being succeeded by a new and unfamiliar world. You are all deeply immersed in that transition on the technology front. But that is only part of what is happening. The social, economic, and political changes of the past several decades have accelerated enormously during the past several years. All of us, each in our own institutions and communities, face an unprecedented challenge to adapt to the new world, and to adapt quickly, or risk losing what most of us have spent our lives building and protecting.

In his new book, *Head to Head, MIT* economist Lester Thurow writes:

> *"In 1945 there were two military superpowers (the United States and the Soviet Union) and one economic superpower (the United States). In 1992 there is only one military superpower (the United States) and three economic superpowers (the United States, Japan, and Europe). Without a pause, the contest has shifted from being a military contest to being an economic contest."*

In other words, it is a whole new ball game. It is not just that we are moving to a different playing field, the rules are changing as well. The key question before us is quite simple: how do we prepare for the new game? The answer, however, is far from simple, and has profound implications for the future role of both academe and the military in our society.

What I hope to suggest to you today is that we can and should work together to find the answer. We can proceed separately and come up with solutions that promise short-term results. In the long-run, however, our nation can win on the economic playing field only if we devise a unified strategy.

The reason for this is obvious: the federal government has limited resources. If it invests those resources primarily in maintaining military superiority, we are at a disadvantage with those nations that invest primarily in economic development. The myth of recent decades has been that we can afford to do both. One needs only to look at the Federal deficit to realize just how big a myth that is.

As you know, the Budget Enforcement Act of 1990 erected a wall between the military budget and the domestic budget. I believe that wall should and will be breached. But once it is, there will not be a torrent of cash flowing from the armed services to the nation's domestic sector, including our universities. What we will find when the dust settles from the crumbling of the budget wall is that the so-called "peace dividend" was consumed before we ever saw it, by interest payments on the national debt, by the savings and loan debacle, and by uncontrolled spending on entitlements. You and I do not need to be concerned over who gets the larger share of the pie. The pie has already been eaten.

When it comes right down to it, both higher education and the military are intimately tied to the health of the nation's economy, and so we both have a vested interest in supporting America's economic development. Higher education's dependency on the economy has been graphically demonstrated in the last two years. Public universities across the country are reeling from cutbacks in state support resulting from the extended recession and the drop in tax revenues. Unlike the federal government, the states can't print money to make up the shortfall.

In Maryland, for example, the University of Maryland System has suffered seven successive budget cuts in two years. Our state funds have dropped by $125 million (or about 20 percent), and that's without any adjustment for inflation or increased workload. As a result, we have had to eliminate programs, both academic and administrative; lay off and furlough employees; and greatly restrict purchases, travel and other expenditures.

It has not been easy, but there is an up side. I think the University of Maryland System is now better prepared to face the new realities of the 1990s. Those new realities include long-term budget constraints, increased demands for accountability, as well as changing societal conditions and expectations for delivery of our services. In the words of Robert Zemsky of the Pew Higher Education Research Program, we have reached the "end of sanctuary." That is to say, higher education can no longer afford to ignore contemporary political, economic, and social realities. And, of course, neither can the military.

Therefore, let me tell you just a bit of what we are doing to adjust the way we do business at the University of Maryland System, which comprises 11 campuses and three research and service components. Six months ago, I delivered to my Board of Regents a report called "Achieving the Vision in Hard Times." The crux of the report takes the form of ten propositions. I want to focus on five of them that are particularly germane to today's discussion.

The first of these propositions underscores the centrality of undergraduate education to the mission of the University of Maryland System and its degree-granting institutions. It calls upon the Regents to establish undergraduate education as a priority for enhancement. However important graduate and professional education, research, and technology transfer may be as functions of American higher education, the stature and reputation of most universities will depend very strongly on the quality of the learning environment they create for their undergraduate students. It is in this arena, in particular, that higher education must win back the hearts and minds of the American public. I mention this proposition because it underscores the more general principle that, in order to survive periods of crisis, organizations such as ours must focus on their mission and on their service to society.

The second proposition I want to discuss is the need to develop performance enhancement mechanisms, with the dual goals of increasing productivity and responsiveness while reducing bureaucracy and its attendant costs. The rapid growth in recent years of administrative costs in American universities is well documented. To reduce those costs, I have called upon the University of Maryland System to take a number of steps, including:

- develop a system for identifying and assessing the relative values and costs of all administrative functions;
- flatten administrative hierarchies and empower employees at all levels to make and carry out decisions;
- achieve greater efficiency through sharing and consolidating administrative services;
- eliminate some services that can be shown to be peripheral to institutional missions;
- contract for services in the private sector where doing so would be more cost-effective;
- work to reduce regulatory constraints and reporting requirements that provide the appearance of accountability without the substance.

The two most radical propositions in my report are seemingly the most straightforward and innocuous. One states that among all of our goals, quality is preeminent. The other states that we will explore the application of Total Quality Management (TQM) to our institutions.

Why are these the most radical? Has the University of Maryland System lacked a commitment to quality? Of course not! But like most institutions of higher education in this country, we have not found a way to operationalize and focus that commitment. Unless we establish quality as the *sine qua non* of our enterprise, and unless we adopt the requisite management practices, then American higher education will lose its competitive advantage as surely and as quickly as this nation's home electronics industry did.

One of the most controversial implications of making quality preeminent is that in times of budget constraints other goals may have to be compromised. This is especially true for public higher education, which has a historical and a moral obligation to provide broad access to students from all socio-economic backgrounds. However, it is my contention that access to mediocrity is of little value. The current budget crisis may force us to make the difficult choice of providing better programs to fewer people. (Please understand, limiting access does not mean limiting diversity among the student population. On the contrary, such diversity is one characteristic of the quality we are striving to achieve.)

The other proposition I want to mention states that the University of Maryland System will "foster the optimal application of modern information technology in both the academic and administrative functions." I bring this to your attention because it is the mechanism by which we will accomplish many of the other propositions, especially given resource constraints. Furthermore, I am convinced that information technology will not only change how we do things but what we do. It holds perhaps the greatest promise for reaching citizens in new ways and serving them better at lower cost. Information technology is, in fact, forcing us to rethink the fundamental business of higher education. Why? Because information is the basic working material of our business. It can be argued that the last fundamental change in our capability for handling information, the invention of printing, literally made possible the university as we now know it. Now, half a millennium later, we are in the midst of an equally fundamental change in our information technology. To me it seems evident that this one will transform our educational institutions just as profoundly as did Gutenberg's.

These are just half of the propositions I have put forward for changing the way we do business at the University of Maryland System. I offer them only as examples of the kinds of strategies the military, in general, and Army science, in particular, might consider as you come to grips with your new realities including, the dissolution of the "evil empire," the success of the technological war in the Persian Gulf, and the downsizing of the military. In short, I suggest you focus on your mission, streamline your administration, improve the quality of your services, and embrace new technologies in delivering those services.

I also encourage you to join with academe in a new level and new kind of partnership. We have, of course, worked together before. As far back as 1862, the Morrill Land Grant Act gave federal land to each state to endow a college for "agriculture, the mechanical arts and *military tactics*." More recently, the end of World War II saw the bill provide free tuition for veterans in American universities. Not long thereafter the federal government increased funding of university and military research, as well as science education, in response to the Soviet launch of Sputnik. The University of Maryland System has been a particularly close partner with the military through our University College programs on bases around the world. We hope to continue and expand on that tradition. Perhaps the turn of this century will mark a new era of cooperation between education and the military. Our businesses, after all, overlap substantially, at least in peacetime. We are both heavily engaged in research and in education—you call it training—and those are precisely the areas that will give America a competitive edge in the global economy.

Our two sectors receive the lion's share of federal R&D funds. Therefore, we have an interest in and, I believe, an obligation to join in a strategic alliance with each other and with industry to make the best use of those funds. We are seeing evidence of a new receptivity to such alliances on the part of the Bush administration, thanks in great part to science and technology advisor D. Allan Bromley. The White House has called for elimination of the fee the government charges defense contractors that sell commercial products based on technology developed with Pentagon funds. The director of the National Science Foundation, Walter Massey, is calling for science funded by NSF to become more immediately relevant to the improvement of our standard of living. He is also calling for greater "intellectual integration" and "organizational integration," suggesting that there must be increased cross-sector and cross-disciplinary cooperation.

On Capitol Hill as well, there are a number of proposals for cooperative approaches among the military, industry, and academe. For example, at a recent meeting of the American Association for the Advancement of Science, Representative George Brown, chairman of the House Science, Space, and Technology Committee, called for such a partnership in overhauling the weapons laboratories and shifting federal support from military to civilian R&D. The Defense Conversion bill currently being considered in the House calls for $1 billion plus in three areas titled "growth enterprises," "work force for growth," and "communities for growth." That sounds to me like the business I am in. I am not worried, though. There is plenty of room there for all of us; the more the merrier.

So far, however, these proposals are piecemeal, ad hoc, and reactive. It is time we developed a much more strategic approach to industrial competitiveness. Universities and the military are in a unique position, given our vast intellectual and technical resources, to help identify the priorities for concerted action.

The National Academies of Sciences and Engineering in their report *The Government Role in Civilian Technology* call for U.S. science and technology policy to go beyond support for basic research to include pre-commercial R&D. Lester Thurow, to whom I referred at the beginning of my remarks, has suggested that the emphasis should shift even further, toward process technologies and infrastructure. (I would remind you that the United States has the oldest and most obsolescent infrastructure in the industrial world.) Thurow argues it was not an inability to develop VCRs and fax machines, but an inability to produce them efficiently, that handed the profits on these American inventions to the Japanese.

Thurow also points out that competitive advantage in the 21st century will have little to do with availability of natural resources, investment capital, or high technology, since all of these have been globalized. The distinguishing characteristic in a nation's economic competitiveness will be education, particularly technical training in and beyond high school. Here, too, there is great potential for university/military cooperation, since between us we receive the vast majority of federal funding for post-secondary training.

Edmund Burke noted that "education is the cheap defense of nations." I would amend that to read "the cheaper defense of nations." Quality education is not cheap, but it is far cheaper than war, and it is cheaper still if it is coordinated among the various sectors of our society.

Let me conclude my remarks by taking note of a boat launching in Japan recently, a launching in some ways just as significant as that of Sputnik. The boat is Yamato I, a prototype vessel powered by a magneto-hydrodynamic (or MHD) propulsion system. MHD is said to be smoother, quieter, more energy efficient, and safer than the screw propeller, which has dominated ship design for more than a century. There are two points I think Yamato I drives home. The first is that it was developed by a Japanese research consortium of universities and industry with support from the government. The success of this joint venture simply reinforces my assertion that we must work together. American scientists, like American society, have tended to emphasize independence to the exclusion of interdependence. As a result we are particularly good at breakthrough science and rather poor at putting the pieces together into breakthrough products.

The second point I want to make about Yamato I is that the science and technology behind it is an interesting mixture of old and new. The basic physics of magnetohydrodynamic propulsion (the Lorentz force) have been familiar for more than a century. MHD propulsion was seriously explored here in the U.S. by Westinghouse in the 1960s. Westinghouse gave up because of the weight associated with creating the required magnetic field using conventional magnets, and the complexity of creating it using low-temperature superconducting magnets. The Japanese did not give up. With the vision and dogged persistence that seems to characterize them they soldiered on. Yamato I is the result. What it will teach them about MHD ship propulsion remains to be seen (or perhaps guessed), but it seems clear that if it is successful, it and the potential of high temperature superconducting materials may position them at the leading edge of yet another commercially successful technology.

Such foresight and perseverance have not been characteristic of American industry, nor of higher education and the military, in recent decades. That must change. Some oxen will have to be gored, some sacred cows will have to be slaughtered. But that is how you make hamburgers. And there is a big worldwide demand for Big Macs.

In our respective enterprises, we must move beyond the limited perspectives of the battlefield and the academic research lab. We must target our efforts to the areas of greatest societal need. In addition to economic competitiveness, I would count among those areas: health care (with an emphasis on preventive care), environmental degradation (with an emphasis on water quality), energy production and conservation (with an emphasis on renewable resources), elementary and secondary education (with an emphasis on "techno-literacy," for want of a better word), and social justice (with an emphasis on inner cities).

That is an ambitious agenda. However, I firmly believe a new alliance between academe and the armed services can have a significant impact on all these issues. Furthermore, *unless* we form such an alliance and *unless* we attend to social priorities, we may well all become irrelevant anachronisms in the new world of the 21st century.

BioSciences

Summaries

1

Molecular Recognition and Imaging in Biotechnology and Neuroscience

Steven Kornguth
Department of Neurology and Biomolecular Chemistry
University of Wisconsin-Madison
1500 Highland Avenue
Madison, WI 53706

Biotechnology provides the biomaterials that have major applications in the development of sensors, electronic and optical switching and encoding devices, and environmental and demilitarization processes. The advantages of such materials include their nanometer dimensions, high-affinity binding properties, self-assembly and electro-optical conductive properties, and the potential for their inexpensive synthesis in bioreactors. This report considers the role that specific biopolymers/biomimetics may play in particular militarily relevant devices. Recognition, catalytic, adhesion, and electron/phonon conductive properties are reviewed. The stabilization of protein structure in the living cells and the control of the spatial organization of cells during biologic development are discussed and related to the construction of devices. The applications of biomaterials to neural modeling, pattern recognition, and information processing are considered using examples of near-microscopic neural imaging, functional neural imaging in humans, and biopolymer modified coupled optical waveguides. The final portion of the review relates the novel biomaterials to specific military and industrial critical technologies.

Introduction

This chapter will focus on one major intersection of the cutting edges of biotechnology and neuroscience. This intersection involves the basis of pattern recognition. Pattern recognition in biotechnology and neuroscience involves the analysis of molecular or cellular surfaces and the consequences emerging from such edge analysis at the interactive surfaces.

In the defense arena, surface and edge detection at the macro-scale is a familiar effort. For example, objects are imaged as heat emitting or radiofrequency reflecting entities; protection can be had from such imaging systems by increasing the number and spatial distribution of edges imaged and thereby confound the imager by masking the target (one example, chaff release from craft) or by altering the reflective properties of the surface with stealth materials.

Recognition, Catalytic, Adhesion, and Electron Transfer Properties

Biologic or polymeric recognition systems occur at a smaller scale, micro- to nanometer scale (10Å level). The processes by which biologic cells form organs and exhibit appropriate intercellular communication involve interactions of chemicals at the surfaces of the cells. The pathways along which nerve cells grow and form functional communication points is controlled by chemicals acting at the growing edges/surfaces of the cells. These edge or surface effects, demonstrated *in vivo*, can be mimicked by chemically or physically modifying matrix surfaces *in vitro*. The Dulcey group[1] at the Naval Research Laboratory (NRL) has shown that neural cells can be deposited in an organized manner on electrically conductive or resistive surfaces, after the surfaces were modified physically by etching or modified chemically with a biopolymer. Such studies involving the control of cell-cell or cell-matrix interactions have implications for the preparation of sensors or high-density information storage disks.

The molecules on the surfaces of cells govern not only the adhesion properties of cells but also control the communication between adjacent cells. The mechanisms by which nerve cells communicate involve proteins (receptors) on the surfaces of the cells that respond selectively to chemical signals (neurotransmitters). Chemicals that modify the receptor properties of the nerve cells have applications as nerve agents (the fluorophosphates) or disorienting agents, as well as therapeutic compounds in the treatment of neurologic and psychiatric disease. The utility of fiber-optic waveguides, which are coated with transmitter or receptor binding biopolymers, as sensors for chemical or biologic agents has been shown in the laboratories of Valdes at the Edgewood Research, Development and Engineering Center (ERDEC) at Aberdeen Proving Ground, MD, and Ligler at the NRL. Valdes has demonstrated that the enzyme acetylcholinesterase can be coupled to quartz optical fibers; the immobilized enzyme retains its affinity for drugs that bind the free enzyme, and the system provides a biosensor for edrophonium, neostigmine, and organic phosphates in the aqueous phase.[2] Ogert and colleagues[3] have

Table 1
Advantageous Properties of Biopolymers and Biomaterials

Self assembly of protein-lipids
Small dimensions—20 to 100Å diameter
Low density—Mass/volume = 1.13
Highly selective, high-affinity ligand binding—K_d= 10^{-8} to 10^{-14}
Conformation of macromolecules known
Light-absorbing properties known
Electron transfer properties known
Piezoelectric properties characterized

shown that anti-botulinum toxin A antibodies, when attached to a quartz fiber, can be used to detect and quantify the amount of toxin in solution.

At the tens of angstroms level of organization (10^{-9}m), the realm of the biopolymer is entered, with applications in the areas of new biotechnology and biomaterials. Biopolymers function as catalysts (enzymes that synthesize or degrade compounds); recognition molecules with high affinity and specificity (for example, antibodies, receptors); regulators of ion flux, metal binders (for example, metallothionines that can bind strategic metals); electron/phonon transfer materials (porphyrins, phthalocyanines, membrane-associated rhodopsin); and structural/support elements (for example, microfilaments, chitin, tubulin). At this nanometer level, it is the surface properties of proteins and biomaterials that affect their functions. As a result of the interaction of two materials at polymer-matrix or cell-matrix interfaces, the chemical or physical properties of the system may change. Such changes in properties provide the basis for the construction of sensors, electric/optical switching devices, obscurants, and decontamination and demilitarization materials.

The advantageous properties of biopolymers that contribute to their utility as materials in defense systems are shown in Table 1. Table 2 identifies specific applications for biomaterials that are currently available. A correlation of specific applications with critical military technologies is shown in Table 3. These applications are relevant to all science and technology thrusts that are identified in Table 4.

Protein Structure and Function

All of the sensor studies are concerned with interactions between chemical agents and biopolymers. These biopolymers, with dimensions in the 20- to

Table 2
Specific Applications for Biomolecules

Phthalocyanines and Porphyrins	organometallic electron transfer photochemical reaction center
Rhodopsin (bacterial or mammalian)	electron transfer
Antibodies/receptors/enzymes	high-affinity, specificity ligand binding $K_d = 10^{-8}$ to 10^{-14}
Polynucleotides	genetic engineering, electron/phonon transfer
Lipids/polysaccharides/crown ethers	encapsulation, films

Table 3
Applications of Biomaterials to Critical Military Technologies

Nanodimension electron transfer or switch elements[1]
Porphyrin or phthalocyanine containing proteins/peptides
Rhodopsin
Phthalocyanines—stacked or planar
Polynucleotides
Lipids as matrices for metal deposition—masks

Sensors[2]
On optical fibers or silicon wafers or gallium arsenide for chemical/biologic threat agents—antibodies, receptors, nucleic acid fragments

Environmental monitoring for
Strategic metals, signature elements—metal binding proteins (metallothionine)
Pesticides—antibodies, receptors
Degradation products of energetic, toxic materials

Acoustic wave detector for submerged vessels
Piezoelectrically active proteins, Polyamides, Magnetosomes

[1] These biomaterials have the following military applications: electronic devices (nanoscale electronic and opto-electronic); computers; data storage; sensors; signal processing; imaging; materials and processes (improved structure, signature reduction, electronics); energy storage; design automation (CAD, electromagnetics); and man/system interfaces.

[2] These biomaterials have the following military applications: sensors (active and passive, imaging); communications networking; environmental effects; materials and processes (improved structure, signature reduction, electronics); and man/system interfaces.

100-Å range, are typically comprised of several hundred amino acids that are covalently coupled by amide bonds. Although several folding patterns of the biopolymer may retain normal function, there is an abundance of inappropri-

Table 4
Science and Technology Thrusts

Global surveillance and communications
Precision strike
Air superiority and defense
Sea control and undersea superiority
Advanced land combat
Synthetic environments
Technology for affordability

ate folding patterns. An inappropriately folded protein will not exhibit normal function, and its utility as a catalyst to degrade agents or as a sensor is then lost. The processes controlling the intracellular folding of proteins involve pattern recognition on the molecular level. The functional properties of proteins, be they enzymes, antibodies, receptors/transducers, or electron transfer materials, are dependent upon retaining an appropriate three-dimensional structure.

The question then is, "What determines the three-dimensional shape of proteins and other biopolymers?" The landmark and elegant studies of Anfinsen in the 1950s demonstrated for the first time that some proteins such as ribonuclease have the intrinsic capacity to fold correctly, even after being unfolded.[4]

The ability of a denatured protein to refold correctly depends on the kinetics of the folding process, the presence of "sticky" intramolecular regions that can lead to inappropriate foldings, and the presence of other proteins in the environment that may influence the folding process. In a very large number of instances, cells utilize a set of proteins called "chaperones" to stabilize the proteins in a metastable state.[5,6] This stabilization reduces the frequency of adhesions between inappropriate portions of the unfolded protein surface.[7] The chaperone therefore prevents incorrect folding of the protein in the cell. The chaperones themselves recognize particular regions on the unfolded protein; in this process they engage in pattern recognition on a molecular level. When the chaperone-bound proteins come into the presence of a molecule that induces proper folding of the nascent protein, the appropriate functional shape of the protein is assumed, and the chaperones release the correctly folded stable protein. Chaperone-like molecules have potential applications in the stabilization of functional proteins in the field and may lead to increased shelf life and field hardened kits.

The role of chaperones in maintaining metastable states of biopolymers has analogies in developmental neurobiology. The human brain has approximately 10 billion nerve cells. Although these cells are not connected in a one-to-one predetermined manner, there is great specificity in the spatial address of nerve cells. For example, the nerve cells detecting a precise visual field in one eye have their thalamic communication output very close in space to the output of nerve cells in the other eye seeing the same visual field. For this type of register to occur, the growing axons from each eye must grow to the target cell in the thalamus that is a centimeter or more distant from the retinal cell bodies; a centimeter distance is the equivalent of 500 to 1000 cell body diameters.

Spatial Organization of Biologic Cells

An area of intense investigation involves the molecular basis of the address system; the questions addressed include, 1) How does a nerve cell recognize the final spatial location for its cell body? and, 2) How does the axonal tip of the neuron recognize target cells involved in synaptic contact? In general, there are two mechanisms used in development of the nervous system: one involves the recognition of target cells and is based on signals of attraction (taxis); the second prevents the formation of inappropriate connections and is based on restrictive/repulsive forces. The molecules preventing inappropriate connections would be functionally similar to "chaperones." Such molecules maintain the growing nerve process in a metastable state until a most appropriate target is encountered. Among the classes of protein molecules subserving this function are the cadherins, the immunoglobulin superfamily, the selectins, the integrins,[8] and the 90-kDa chaperones that maintain class I major histocompatibility complex antigens in a metastable state.[9] The principle of masking a cell surface to maintain the cell in a metastable state for a period of time, until a useful functional connection can be made, has applications in neural regeneration, sensor design, and networking systems.

Near-Microscopic Neural Imaging

A natural engineering outgrowth of edge analysis and pattern recognition is the brain imaging technology of the 1990s. In this case, edge analysis is defined electronically in terms of the pixel elements. With a 9.4-T magnetic resonance instrument, it is possible to achieve an imaging resolution of 100 x 100μm in the X,Y plane and 200μm in the Z plane (Figure 1) of living rat or

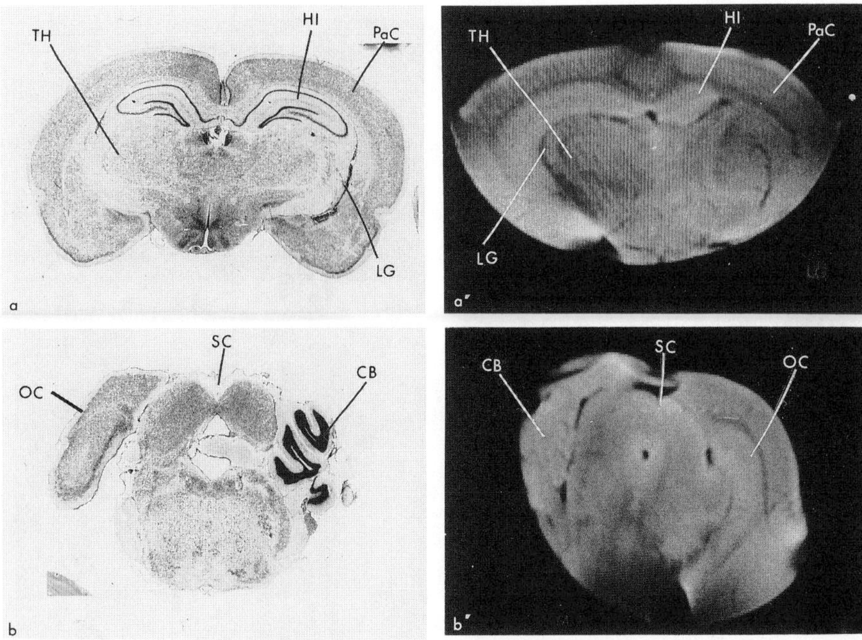

Figure 1. Comparison of magnetic resonance images (MRI) of intact adult rat brain with thionine stained section of rat brain cut through the same plane that was imaged. Figures a,a' are coronal sections through the frontal-parietal region and b,b' are through the cerebellar-occipital-collicular region. MRI images a',b' were acquired with a resolution of 95 x 95µm in the X,Y plane and 500µm in the Z plane. Echo time was 16msec and repetition time was three seconds; these are proton spin density images. Brain structures revealed include thalamus (TH), frontal-parietal cortex (PaC), cerebellum (CB), superior colliculus (SC), and occipital cortex (OC). Images and micrographs are 7x. Images were obtained at the National Magnetic Resonance Facility, University of Wisconsin-Madison, by Markley, Anderson, and Kornguth (unpublished).

mouse brain. These dimensions are the pixel elements. These images permit anatomic structure to be identified at a resolution of 5- to 10-cell diameters.[10,11] We have the capability to visualize non-invasively, at 1.5T, the precise structure of the living human brain at a resolution of 600 x 600µm in the X,Y plane and 1.5mm in the Z plane.

Functional Neural Imaging in Humans

Another imaging technique, positron emission tomography (PET), permits the study of functional and metabolic features of human and animal brain. The resolution in PET is 4 x 4mm in the X,Y plane and 4 mm in the Z plane.

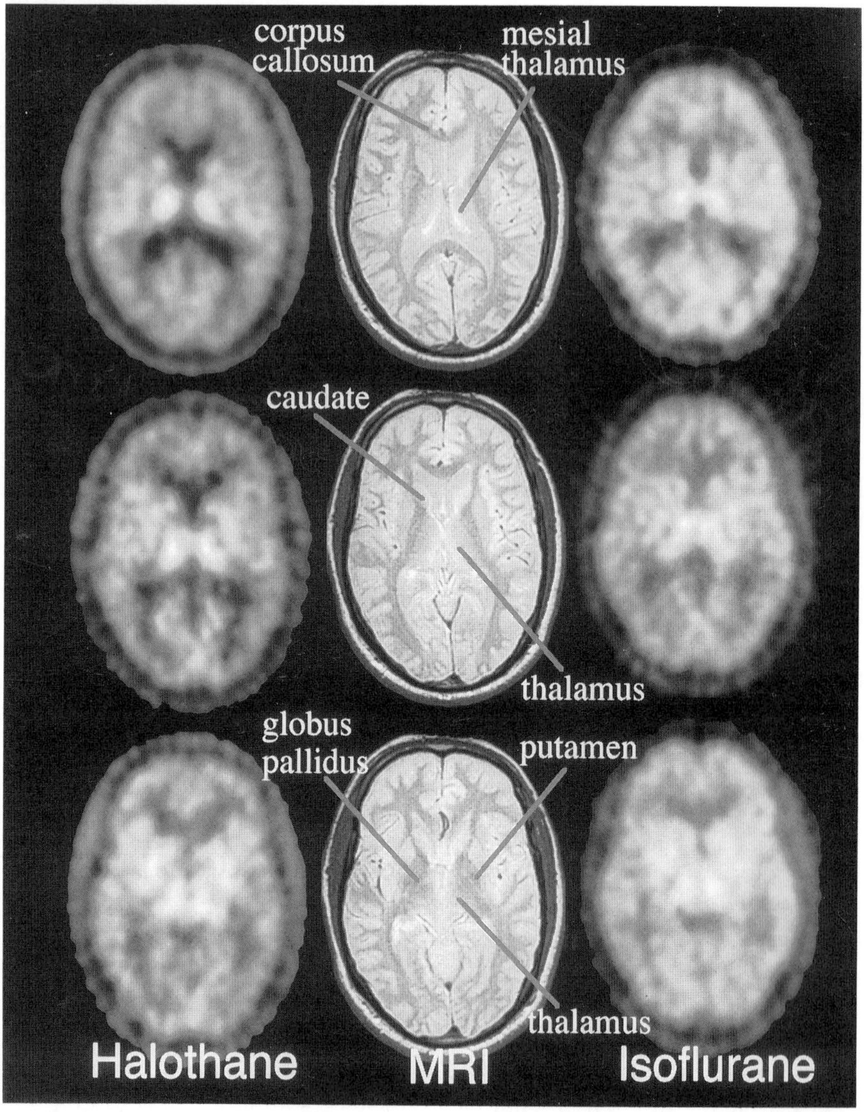

Figure 2, Color Plate 1. Positron emission tomographs (PET) and MRI of human brain exposed to ^{18}F-labeled halothane and isoflurane. The top set of three images are in horizontal plane at most apical region imaged, whereas the bottom set of three are the most basal regions imaged. High concentrations of ^{18}F are indicated in PET images, left and right columns, by yellow to white colors; the lowest concentrations of ^{18}F are indicated by dark red to black. The bright regions in MRI are gray matter whereas the darker blue regions are the myelinated areas. Highest concentration of anesthetics in PET images are localized to globus pallidus, putamen, thalamus, and caudate nucleus, as indicated in the MRI. These images were prepared by Nickles and Satter of the University of Wisconsin-Madison, Medical Physics Department (unpublished data).

In Figure 2, for instance, (courtesy of Satter and Nickles, University of Wisconsin-Madison), we see the selective binding by human brain of fluorine 18 (^{18}F)–labeled halothane and isoflurane, which are volatile anesthetics. ^{18}F is a positron emitting nuclide. By comparing PET and magnetic resonance imaging (MRI) images (Figure 2), it is possible to determine the anatomic location of the anesthetic binding sites. By attaching the ^{18}F to a variety of organic chemicals, we can measure (1) the flow of blood in the living alert human brain (with fluoro-methane), or (2) the cellular activity of nerve cells (with fluoro-deoxyglucose). Because increased firing of neurons is accompanied by an increased uptake of glucose by these cells, we may determine their physiologic activity by measuring their glucose uptake. ^{18}F-labeled compounds may also be used to map the distribution of receptors for neurotransmitters in the brain. The technology used to map the density of neurotransmitter receptors and uptake sites in the living, alert subject by PET has been reviewed recently;[12] it may have applications in evaluating human performance characteristics in the coming decades.

Novel computer software packages provide the capability of overlaying MRI, PET, and magneto-encephalographically derived images.[13] This permits a comparison of the structural features of human brain with their functional or metabolic features (Figure 3). These software packages were developed by Thomas Rock Mackie, Stuart Swerdloff, James Holden, and Mark Gehring, of the Medical Physics Department, University of Wisconsin-Madison. The overlayed images provide a sense of virtual reality and permit high resolution analysis of the brain regions that bind selectively the ^{18}F derivatives of glucose or neurotransmitter agonists/antagonists.

Information processing in the brain may now be imaged on-line by using more invasive techniques. An excellent demonstration of the pattern of electrical activity in the visual cortex, as the eye is stimulated visually, may be seen from the studies of Daniel Ts'o and colleagues.[14] Charge coupled devices permit the direct visualization, on the brain surface, of functionally meaningful information processing by the visual system of the brain.[14]

Sensors Utilizing Coupled Optical Waveguides and Biopolymers

We now turn our attention to the role that miniaturized optical waveguides may play in the construction of sensors or information storage elements. One model utilizes antibodies or polynucleotide fragments for the rapid detection of antigens. This model is based on the periodic coupling observed between parallel optical waveguides or between nonblocking intersecting waveguides.

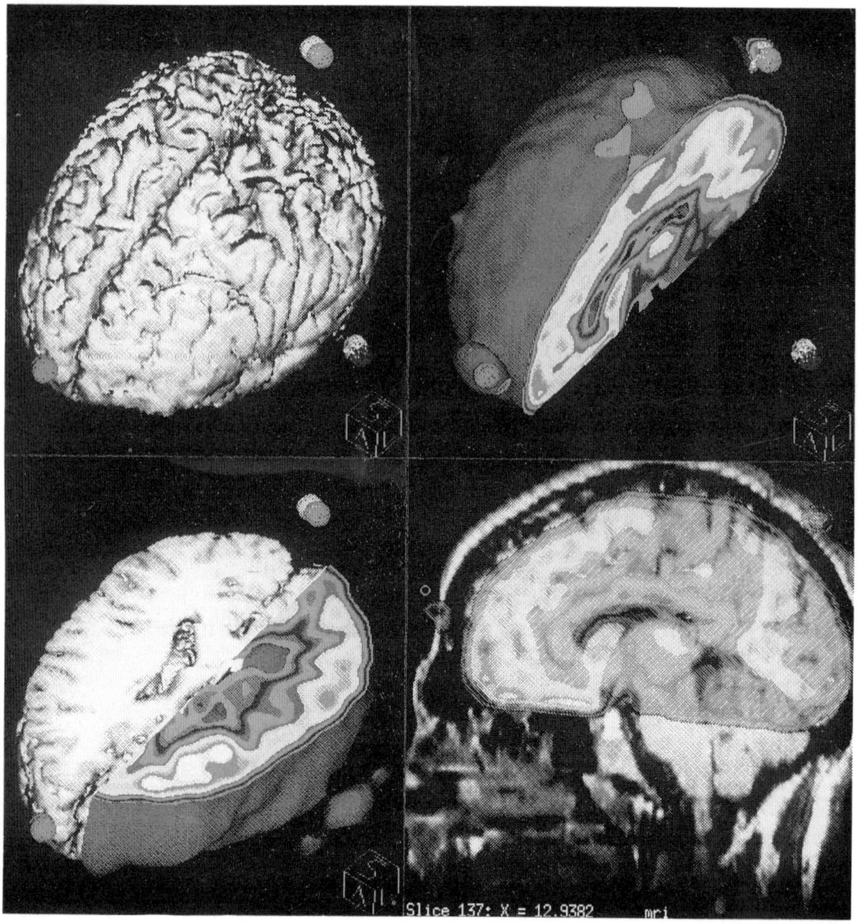

Figure 3, Color Plate 2. Computer generated overlap MRI and PET images of an intact brain from an alert human subject following injection of ^{18}F deoxyglucose. Left upper quadrant shows MRI of human brain. Sulci, gyri, and major blood vessels can be seen. Red and yellow circles surrounding brain are water capsules placed over fiducials. Anterior region of brain (frontal cortex) is at lower left portion of this image. PET image of entire right hemisphere of brain, imaged in sagittal plane, is shown in upper right quadrant. Cut-surface represents the mid-line image. The red to white colors indicate the highest level of ^{18}F. Fiducial markers may be discerned. Lower left quadrant shows a horizontal slice projection revealing MRI (right portion of brain) and PET (left portion of brain). Lower right image is a sagittal section through midline. Corpus callosum appears as a pink region, ventricles are dark black, frontal cortex is yellow, and occipital cortex is blue. Overlap images were constructed by Gehring, Swerdloff, Holden, and Mackie of the University of Wisconsin-Madison, Medical Physics Department (unpublished data).

An example of such a system is shown in Figure 4; in this figure the antibodies (Y) are attached to fluorophore-labeled antigens and to one of a pair of coupled parallel optical waveguides. The fluorophore-labeled antigen modu-

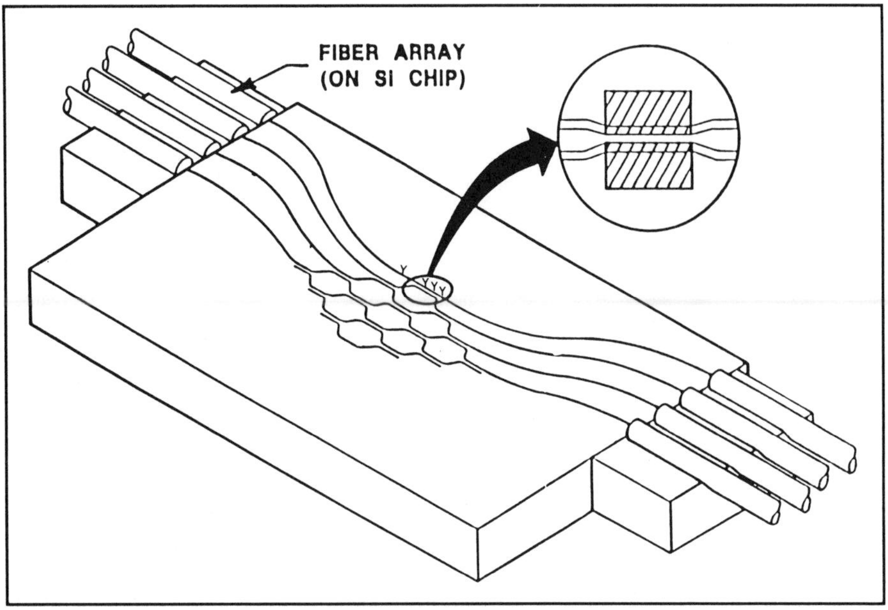

Figure 4. Illustration of a 4 x 4 directional coupler that is modified on one channel with an antibody (Y) and fluorescent labeled antigen. Matrix serves as a sensor for presence of non-fluorescent antigen in environment. Periodic coupling between light channels is a function of gap distance and length of coupled region. Displacement of fluorescent-labeled antigen from antibody provides a change in emerging light signal, the basis of the sensor.

lates the light passing through the adjacent waveguide. When the sensor is exposed to antigen in the environment, the fluorophore-labeled antigen is released, and the light intensity passing through the guide changes, thereby generating a signal. Because of the low optical extinction of nonlabeled antigens, the antibody-based immunodetection system will likely require the displacement of a fluorophore-tagged antigen if a sufficient signal for detection is to be generated; the extinction coefficient of the nucleic acids may provide sufficient signal to preclude the necessity of a fluor-labeled marker.

The construction of the sensor just described has applications for the detection of chemical or biologic agents, industrial pollutants and pesticides, or medically important antigens such as bacterial or viral materials. These sensors may provide rapid analysis of antigens in the environment chemical-biologic defense.

In a somewhat more versatile system, a 4 x 4 integrated optical crossbar switch array that is coated with antibodies or DNA fragments (Y) may be used as a sensor or an optical information address system (Figure 5). The system is based on an integrated-optical crossbar switch array described by McCaughan

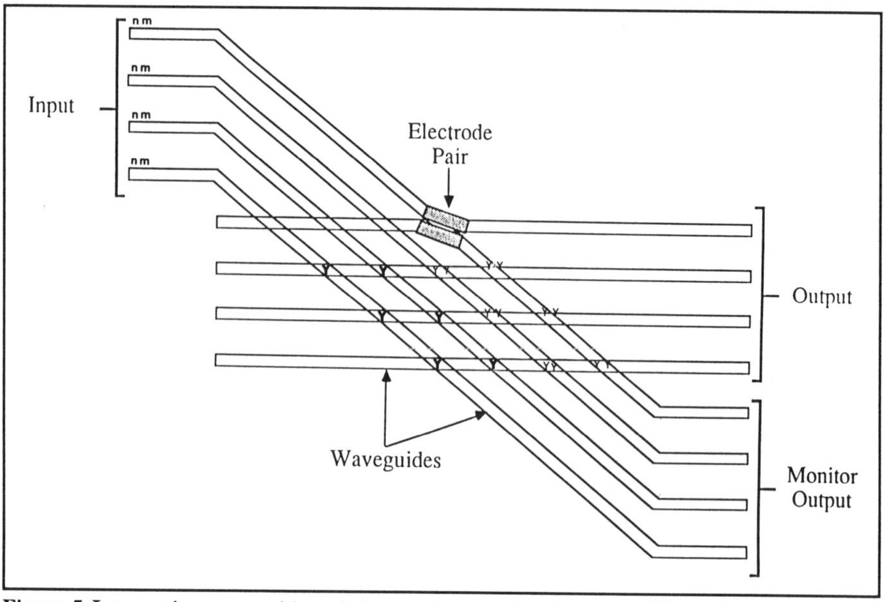

Figure 5. Intersecting waveguide switch array that can function as an addressable memory unit. Light input is at upper left; input wavelengths (nm) must be in a narrow range and not generate interference. Y units represent antibody coupled to the fluorescent antigen, or nucleic acid fragments. Electrode pair at upper waveguide permits regulation of the fraction of light passing through each of the emerging waveguides. Displacement of fluorescent-labeled antigen or the binding of a complementary nucleic acid fragment will generate a change (delta) in the light signal output. Design of intersecting waveguide is derived from studies in Reference 14.

and Bogert.[15] The presence or absence of fluorophore-labeled antigen, bound to the antibodies (Y), at the intersecting nodes of the crossbar guides, will affect the light reaching the output sensors. The electrodes at the crossbar points permit the tuning of the ratio of light traveling along each of the two bars after the crosspoint. Such tuning can compensate for the absorption of light by the bound antibody and for the adsorbed antigen that is labeled with a fluorescent tag. The address of the antigen fragment released may be identified by the sensing output unit. A nucleic acid fragment may be substituted for the antibody in the development of alternative sensors. This approach has applications for information storage and recall.

Biopolymers in Information Storage

As indicated earlier, selected biopolymers have utility for information storage because of their small size, their ability to self-assemble when inserted

Table 5
Defense-Related Technologies

Computers and software (neural nets)
Sensors (active and passive, imaging)*
Communications networking
Electronic devices (nanoscale electronic and optoelectronic)*, computers, data storage, sensors, signal processing, imaging
Environmental effects
Materials and processes. (improved structure, signature reduction, electronics)*
Energy storage*
Propulsion and energy conversion
Design automation (CAD, electromagnetics)*
Man/system interfaces

* These technologies are related to specific biomaterials in Table 3.

into membranes, and their electron/phonon conducting properties. Functionally active biopolymers, such as bacterial rhodopsin, have been attached to electro-conductive or optico-conductive surfaces,[16] and their information storage capabilities have been demonstrated. The pixel in the Miyasaka[16] experiment was limited by the size of the electric leads to the conductive film. In principle, the pixel size could be reduced to the submicron level as the address and readout components approximate the bacteriorhodopsin dimensions.

The technology needed to achieve high density packing of biopolymers has been advanced by scientists at Affymax Research Institute, Palo Alto, CA.[17] These researchers activated small regions of glass by photolithography, and the activated regions were then reacted with linker molecules and biopolymers. The utilization of light as an activating source permits the minimum spacing between biopolymers on the activated surface to approach 500nm (0.5μm). Because the wavelength of light limits the packing density of biopolymers, alternative mechanisms such as X-ray or electron beam activation are required to achieve high density packing of biopolymers.

The technologies required for the preparation of sensors or data storage elements, which contain densely packed biopolymers (intermolecular distances approximately 100 to 300Å) as the pixel elements, are now emerging. The characterization of the densely packed surface and the development of novel mechanisms required for addressing and reading the pixel elements are likely to be at the front edge of these new technologies.

These studies indicate the utility of biomaterials and imaging systems (edge analysis) in a variety of leading edge technologies. The opportunities are most evident in the area of sensors, electronic devices, materials and processes, design automation [these are defense related technologies (Table 5)], and pharmaceuticals. The direct relationship of these opportunities to economic competitiveness in the global arena suggests that biomaterials and imaging will receive intense investment in the industrialized nations.

References

1. C. Dulcey *et al.*, *Science* **252**, 551 (1991).
2. K. Rogers *et al.*, *Fund. Appl. Toxicol.* **16**, 810 (1991).
3. R. Ogert *et al.*, *Analyt. Biochem.* **205**, 306 (1992).
4. C.B. Anfinsen, *Science* **181**, 223 (1973).
5. J. Rothman, *Cell* **59**, 591 (1989).
6. E.A. Craig, *Crit. Rev Biochem.* **18**, 239 (1985).
7. K. Liberek *et al.*, *J. Biol. Chem.* **266**, 14491 (1991).
8. R. Hynes and A. Lander, *Cell* **68**, 303 (1992).
9. F. Hochstenbach *et al.*, *Proc. Natl. Acad. Sci.* **89**, 4734 (1992).
10. S. Kornguth *et al.*, *Am. J. Neuroradiol.* **11**, 313 (1990).
11. S. Kornguth *et al.*, *Teratology* 45, 383 (1992).
12. S. Gilman, *New England J. Med.*, **326**, 1608 (1992).
13. M. Gehring *et al.*, *Int. J. Radiat. Oncol. Biol. Phys.* **21**, 491 (1991).
14. D. Ts'o D *et al.*, *Science* **249**, 417 (1990).
15. L. McCaughan and G.A. Bogert, *Appl. Phys. Lett.* **47**, 348 (1985).
16. T. Miyasaka, K. Koyama, and I. Itoh, *Science* **255**, 342 (1992).
17. S. Fodor *et al.*, *Science* **251**, 767 (1991).

2

Protein Folding:

Submolecular Initiating Sites, Enzymatic Monitoring, and Stability of Proteins From Hyperthermophilic Bacteria

The folding of proteins appears to be entirely determined by their amino acid sequences. A number of experiments, particularly using immunologic methods, suggest that the rapidity and precision of the folding process may be due to local initiation of native substructures in the peptide chain, which then associate rapidly to form the final product. Our recent studies on enzymes from the hyperthermophilic bacterium *Pyrococcus furiosus* indicate that this folding process can occur even at temperatures around 100°C. The enzymes that are formed are themselves stable and maximally active at this temperature. Because of the unusually high activity of the thermophilic enzymes, we have recently begun studies on the use of these catalysts for the degradation of thiodiglycol, the first breakdown product of mustard gas. It is hoped that the purified enzyme system might be susceptible to attachment to an immobilizing column through which the material could be passed in an on-line fashion.

Christian B. Anfinsen
Department of Biology
The Johns Hopkins University
Baltimore, MD 21218

Introduction

It is generally accepted that the three-dimensional structure of a native protein in its normal physiologic milieu (solvent, pH, ionic strength, presence of other components such as metal ions or prosthetic groups, temperature, and so on) is the one in which the Gibbs' free energy of the whole system is lowest. Studies on ribonuclease and on a large number of other proteins have established that the amino acid sequence alone, in a proper environment, determines the native conformation.

Proteins that contain disulfide bonds require the proper pairing of half-

Figure 1. Fractionation of antinuclease antibodies by successive absorption of multicomponent mixture of antibodies against native nuclease by passage through three columns to yield anti-$(99\text{-}126)_n$.

cystine residues. Pancreatic ribonuclease contains eight half-cystines, the pairing of which to form four disulfide cross-linkages can take place in 105 ways. During the *in vitro* renaturation of the fully reduced chain, most of these incorrect bonds are formed, and many hours are required for the slow rearrangement to yield the correct four pairs of SS bonds.[1] An enzymatic system has been discovered that could catalyze the rapid interchange of these incorrect cross-linkages. This enzyme,[2] called protein disulfide isomerase, attacks exposed SS bonds by means of its single SH group and apparently continues its action until the protein under treatment has reached its state of native minimum Gibbs' free energy, with SS bonds no longer exposed to solvent. The enzyme helps achieve the fully native conformation of ribonuclease in less than two minutes.

Experiment

Some proteins lacking SS bonds have been shown to form their fully native states in a much shorter time. Using fluorescence emission techniques, it was possible to show that the residue staphylococcal nuclease, which contains a sequence of 149 amino acids, was totally converted from the random coil to the native structure in less than 250 milliseconds.[3] It seemed essential that such a rapid folding process must involve a number of nucleation sites along the polypeptide chain that could independently assume a "native format" and subsequently interact with one another to achieve the native geometry. Using immunologic techniques,[4,5] it was possible to show that various peptide fragments of the protein chain did, indeed, "remember" their conformation as part of the native protein.

In these immunologic experiments, antibody was first prepared against the native nuclease. The antibody molecules specific for the numerous determinants on the molecule were first separated from the other molecules in the antiserum by absorption to nuclease immobilized on a Sepharose affinity chromatography column (Figure 1). A subfraction was then obtained by a second column bearing the attached peptide, residues 99 to 149, a terminal portion of the native protein containing two helical segments. This antibody fraction was, however, not suitable because it was a precipitating fraction and could not be used in spectrophotometric experiments—presumably because it

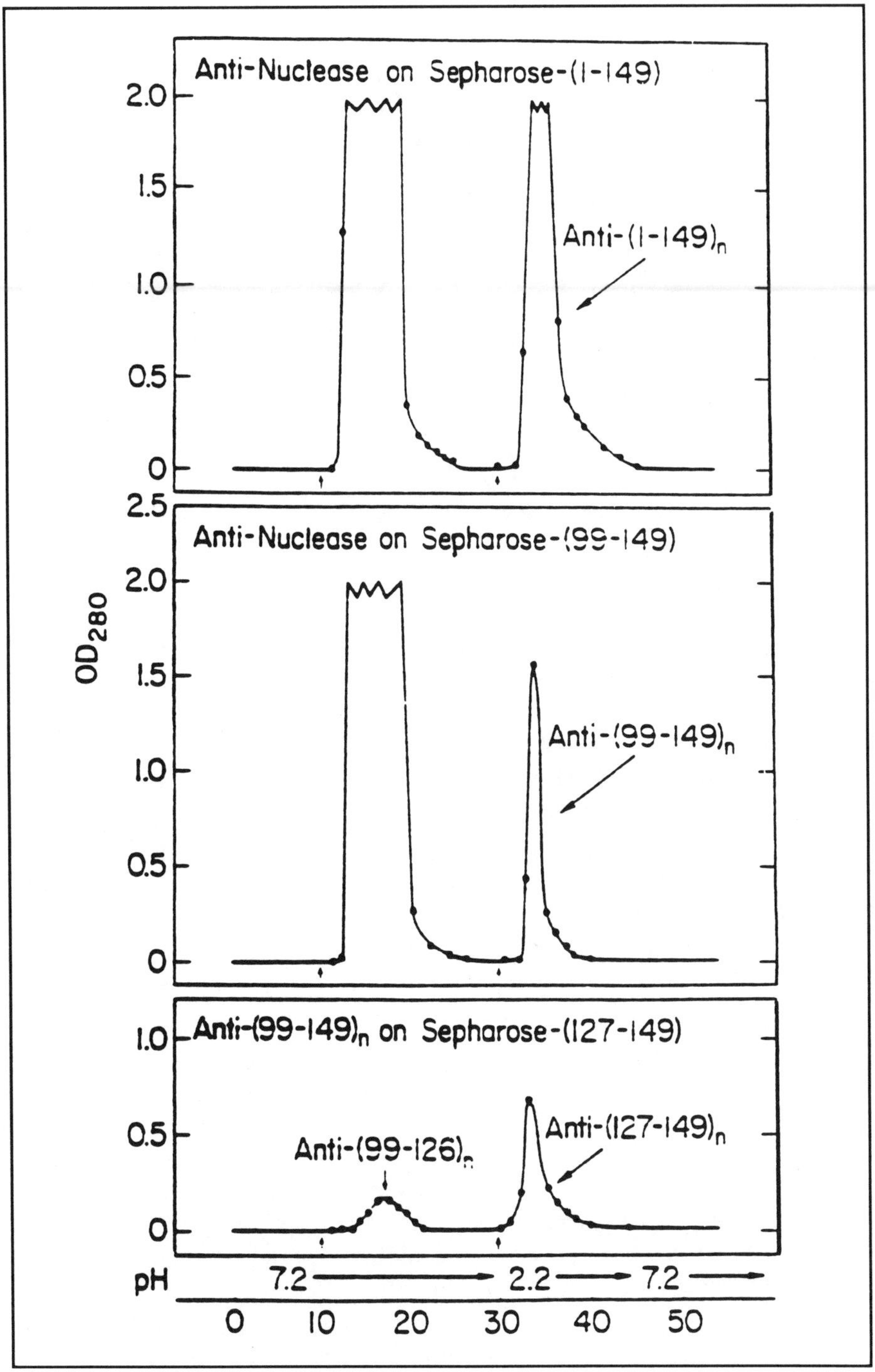
Anti-Nuclease on Sepharose-(1-149)
Anti-(1-149)$_n$
Anti-Nuclease on Sepharose-(99-149)
Anti-(99-149)$_n$
Anti-(99-149)$_n$ on Sepharose-(127-149)
Anti-(99-126)$_n$
Anti-(127-149)$_n$
OD$_{280}$
pH
7.2
2.2
7.2
0
10
20
30
40
50
2.0
1.5
1.0
0.5
0
2.5

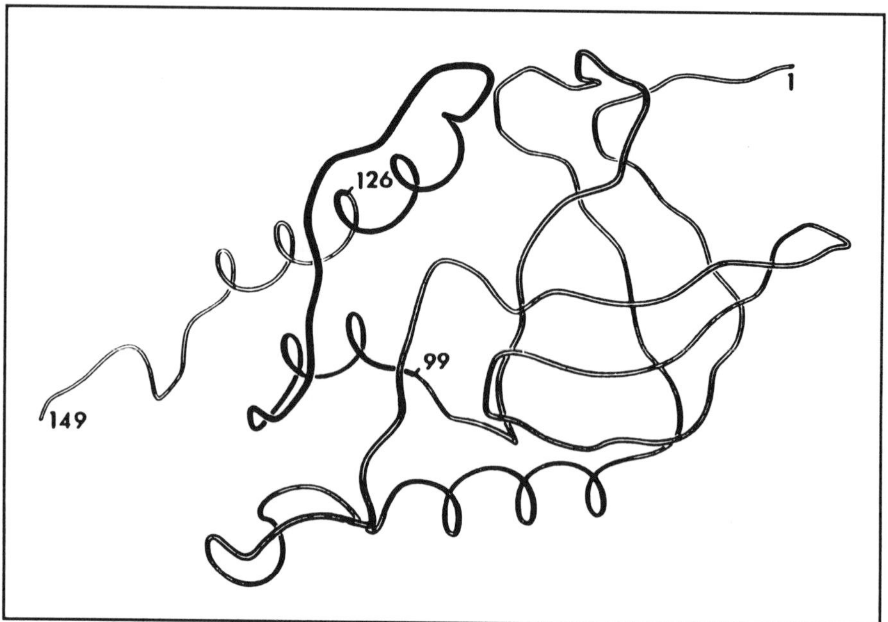

Figure 2. Schematic representation of staphylococcal nuclease indicating the location of the structure between 99 and 126 for which a monoclonal antibody was isolated by affinity chromatography.

contained antibodies against two or more antigenic sites. A subsequent affinity column bearing the ligand (residues 127 to 149) yielded monospecific and nonprecipitating antibody against (residues 99 to 126). This antibody bound to, and inactivated, native nuclease. However, peptide (residues 99 to 126) was capable of competing with native nuclease and thus released the active enzyme from its bound complex with antibody. It could be calculated from the stoichiometry of the competition between nuclease and peptide (residues 99 to 126) that the peptide must occupy the "native format" (as shown in Figure 2) about 0.05 percent of the time.

Future Studies

Several other enzyme systems that are required for correct protein folding are under investigation.[6] The peptidyl-prolyl-*cis-trans* isomerases[7] catalyze the rate-determining step in the folding of proteins containing "accessible" and "essential" *cis*-proline residues. So-called chaperonins[8] and "heat-shock" proteins appear to help keep protein chains in a more or less unfolded form until preliminary folding steps or cell attachments have taken place.

A new and intriguing aspect of protein folding is now under active study

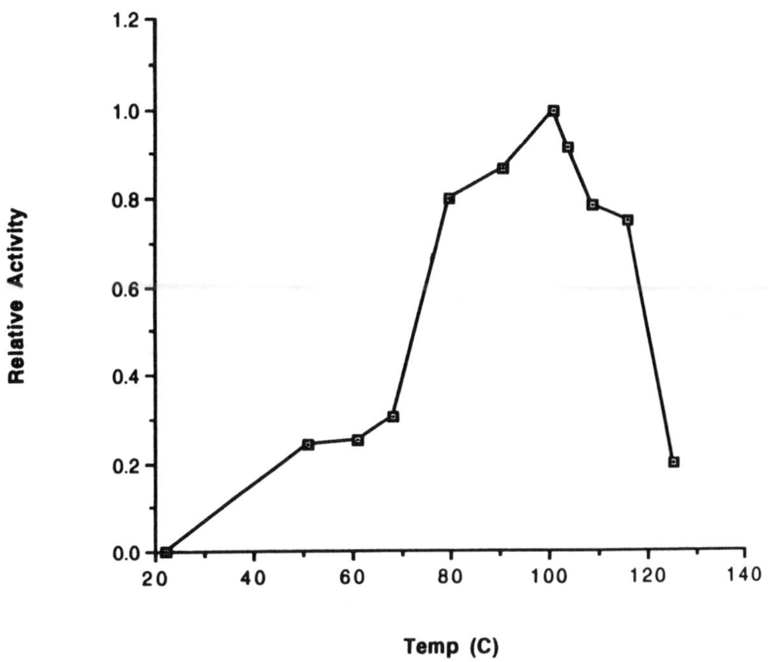

Figure 3. Temperature dependence of pyrococcal α-amylase activity.

in a number of laboratories. These studies arise from the availability of numerous strains of hyperthermophilic Archaebacteria that grow at temperatures in the neighborhood of 100°C and whose enzymes are optimally active only at temperatures near the boiling point of water. My own interests in this area are particularly tied to the possibilities of using such bacteria and their sonicates in the digestion of massive stores of chemical warfare agents such as mustard gas and its derivative, thiodiglycol (TDG).

Current studies on enzymes from hyperthermophilic bacteria suggest that some new stabilizing parameters must be involved to account for the thermostability of these proteins. An α-amylase[9] from *Pyrococcus furiosus* reaches its maximum activity at 100°C and is essentially inactive at temperatures below 60°C (Figures 3 and 4). This protein, with a molecular weight of 130,000kD, has a very low half-cystine content. Disulfide bond stabilization must be minimal. A somewhat elevated content of hydrophobic amino acids suggests that hydrophobic interactions might account for a considerable portion of the heat stability.

Interest in the bacterial degradation of TDG has recently been stimulated by the finding that certain soil bacteria can digest this substance during growth at moderate temperatures.[10] We have recently obtained evidence for degrada-

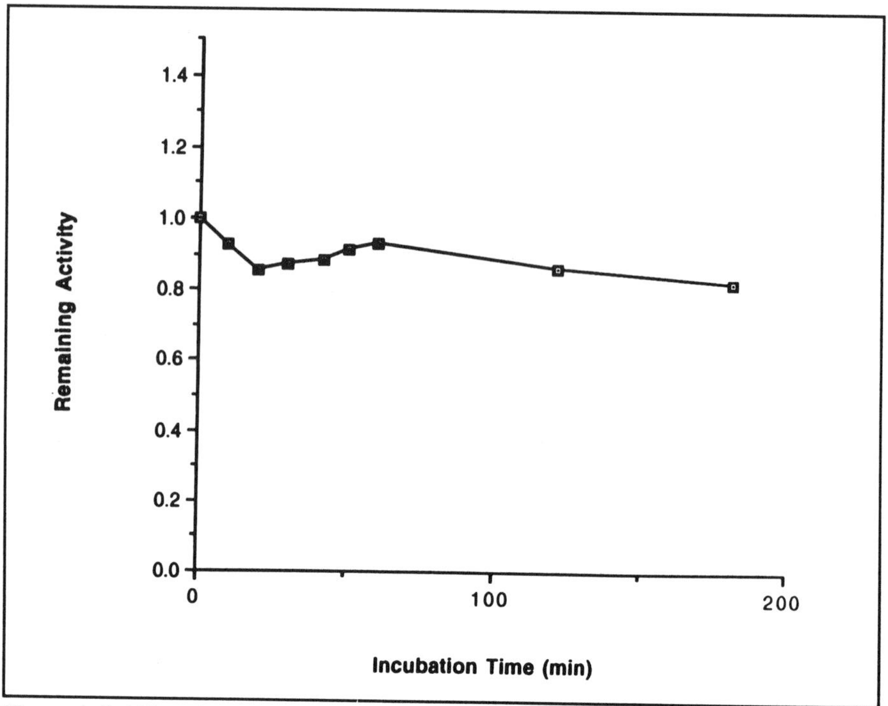

Figure 4. Stability of pyrococcal α-amylase activity at 100°C.

tion of TDG during growth of *Pyrococcus furiosus* and have begun studies on the possibilities of employing cell sonicates. Should such an attack be successful, it should be possible to prepare immobilized enzyme columns that might operate in an on-line manner at elevated temperatures.

References

1. C.B. Anfinsen, *Science* **181**, 223 (1973).
2. C.J. Epstein, R.F. Goldberger, and C.B. Anfinsen, *Cold Spring Harbor Symp. Quant. Biol.* **28**, 439 (1963).
3. A.N. Schechter, R.F. Chen, and C.B. Anfinsen, *Science* **167**, 886 (1970).
4. D.H. Sachs *et al.*, *J. Immunol.* **109**,1300, (1972).
5. D.H. Sachs *et al.*, *Proc. Natl. Acad. Sci. USA* **69**, 3790 (1972).
6. R. Jaenicke, *Biochemistry* **30**, 3147 (1991).
7. J.F. Brandts, H.R. Halverson, and M. Brennan, *Biochemistry* **14**, 4953 (1975).
8. R.J. Ellis, *Semin. Cell Biol.* **1**, 19 (1990).
9. K.A. Laderman *et al.*, *J. Biol. Chem.*, submitted.
10. S. Harvey *et al.*, in *Biotechnology: Bridging Research and Applications.* (Kluwer Academic Publishers, The Netherlands, 1991), pp.221-230.

3

Oral, Microencapsulated, CFA/II Vaccine Against *E. coli* Diarrheal Disease:

Preclinical Evaluation

Robert H. Reid
Edgar C. Boedeker
Charles E. McQueen
Catherine L. Wilhelmsen
Lee-Ying Tseng
David Davis
James A. Kodak
Keya Sau
U.S. Army Medical Research
and Development Command
Walter Reed Army
Institute of Research
Washington, DC 20307-5100

Hitesh R. Bhagat
School of Pharmacy
University of Maryland
Baltimore, MD 21202

During Operation Desert Shield, 57 percent of U. S. Troops studied had at least one episode of diarrhea. In 50 percent of the cases, a bacterial pathogen was found, with antibiotic-resistant, enterotoxigenic *E. coli* (ETEC) and *Shigella sonnei* being the most common pathogenic species found. The isolated ETEC were identified by their pilus colonization factor antigens (CFAs). Eleven percent expressed CFA/I, 31 percent expressed CFA/II, 32 percent expressed CFA/IV, and 26 percent expressed none of these antigens. Colony factor antigen (CFA/II) from enterotoxigenic *E. coli* (ETEC) was successfully incorporated into biodegradable polymer microspheres (CFA/II BPM) and found to be safe and immunogenic when administered intraduodenally to rabbits. Following vaccination, Peyer's patch cells responded by lymphocyte proliferation to *in vitro* challenge with CFA/II, indicating the CFA/II BPM to be immunogenic when administered intra-intestinally. B cells secreting specific anti-CFA/II antibodies were also found in spleens following vaccination. No pathologic changes were found following total necropsies of ten rabbits vaccinated with CFA/II BPM. As a potency test, high serum IgG antibody titers to CFA/II were produced following intramuscular administration of CFA/II BPM to additional rabbits. Sixty-three percent of the CFA/II BPM were between 5 and 10μm and all passed the safety test. CFA/II BPM oral vaccine is immunogenic and safe to begin a Phase I clinical study .

Introduction

During Operation Desert Shield, 57 percent of U. S. Troops studied had at least one episode of diarrhea. In 50 percent of the cases studied, a bacterial pathogen was found, with antibiotic-resistant, enterotoxigenic *E. coli* (ETEC) and *Shigella sonnei* being the most common pathogenic species found. The isolated ETEC were identified by their pilus colony factor antigens (CFAs). Eleven percent expressed CFA/I, 31 percent expressed CFA/II, 32 percent expressed CFA/IV, and 26 percent expressed none of these antigens.

The first step in pathogenesis is adherence to the small intestine epithelial cells by protein fimbrial (pilus) or fibrillar adhesins called CFAs. Three major CFAs have been recognized: CFA/I, CFA/II, and CFA/IV.[1]

Ten human volunteers who were immunized orally twice weekly for four weeks with CFA/II developed a poor antibody response and did not show any significant protection when challenged with pathogenic ETEC.[2] This disappointing response was attributed to adverse effects of gastric content, even at neutral pH, on fimbrial proteins.[3] When the vaccine was administered by inoculation directly into the duodenum, four of five immunized volunteers developed a significant rise in secretory IgA with CFA/II antibody.[2]

D and L-lactide and glycolide, as homo- and copolymers (PLGA), are biodegradable and permit slow and continued release of antigen, with a resultant adjuvant activity. These polymers have been shown to be safe in a variety of applications in human beings and in animals.[4-8] Delivery of antigens via microspheres composed of biodegradable, biocompatible lactide/glycolide polymers[5-8] may enhance the mucosal response by protecting the antigens from digestion and targeting them to lymphoid cells in Peyer's patches.[5-8] McQueen *et al.*[9] have shown that *E. coli* AF/R1 pili in PLGA microspheres, introduced intraduodenally in rabbits, protected them against diarrhea and weight loss when challenged with the parent rabbit diarrheagenic strain of *E. coli* (RDEC-1). Only one vaccinated rabbit of six lost weight, and only one had soft, pelleted stool. In contrast, all control unvaccinated animals became ill, lost weight, and shed soft pellets or unformed mucoid stool. Significant lymphocyte proliferation to AF/R1 from Peyer's patches and biliary IgA anti-AF/R1 antibody levels were seen.

Methods

To improve the CFA/II vaccine, it was incorporated into PLGA microspheres under Good Manufacturing Practices (GMP) in order to protect it from digestion and target it to the intestinal lymphoid system. The CFA/II biodegradable polymer microspheres (BPM) vaccine has undergone preclinical evaluation and has been found to be safe and immunogenic. Under Good Laboratory Practices (GLP) and GMP, *E. coli.* strain M424C1-06:H16, producing CFA/II, were cultured in 75-80 CFA agar plates (24 x 24 cm) for 24 hours, and then were harvested by scraping. The harvest was homogenized at slow speed for 30 minutes in a Sorvall omni mixer with overhead drive unit and cup immersed in an ice bath. The homogenate was centrifuged at 4°C at 16,500 x g for 30 minutes. The supernatant was saved and the pellet rehomogenized and centrifuged with the supernatants pooled. The supernatant pool was centrifuged at 50,000 x g for 45 minutes. The resulting supernatant was treated with ammonium sulfate at 20 percent saturation, stirred 30 minutes at 4°C, stored at 4°C for 16 hours, then centrifuged at 19,700 x g for 30 minutes. The supernatant was saved and treated with ammonium sulfate at 45 percent saturation, stirred 30 minutes at 4°C, stored at 4°C for 66 to 72 hours, then centrifuged at 19,700 x g for 45 minutes. The pellet was resuspended in about 100ml of phosphate buffered saline (PBS) containing 0.5 percent Formalin, held at 22° for 18 hours, then dialyzed for 45 to 50 hours against PBS at 4°C using a total of 12 liters in two-liter amounts. The dialysis was terminated when the PBS contained less then 0.03 percent Formalin using Nessler's reagent and fuchsin sulfuric acid reagent. The final product, formalinized CFA/II, contained 1mgm protein/ml PBS, was sterile, and passed the general safety test. It contained 90 percent CS3 and 10 percent CS1 by polyacrylamide SDS gel analysis.[10] The CFA/II was desalted and freeze-dried.

Solvent extraction technique was used to encapsulate the freeze-dried CFA/II into poly(lactide-co-glycolide)(Medisorb Techologies International, viscosity 0.73dl/g) microspheres in the one to 10µm size range to achieve theoretical antigen loading of one percent by weight. The freeze-dried antigen-sugar and matrix was dispersed in an acetonitrile solution of the polymer and then emulsified to achieve desired droplet size. Microspheres were solidified and recovered by using heptane as extracting solvent. Four microsphere batches were pooled and vacuum-dried to remove traces of solvent.

Using a precalibrated eyepiece micrometer, the diameter of 150 randomly chosen microspheres was determined, and the microsphere size distribution was calculated. The samples were examined with an Hitachi S-450 scanning electron microscope operated at 15 to 20KV. Protein content was determined using the microbicinchonic acid (BCA) method. Determinations were made of moisture, acetonitrile, and heptane contents. Tests were conducted to determine release of CFA/II from microspheres and the general safety of the CFA/II microspheres.

Two naive rabbits were immunized with CFA/II microsphere vaccine at

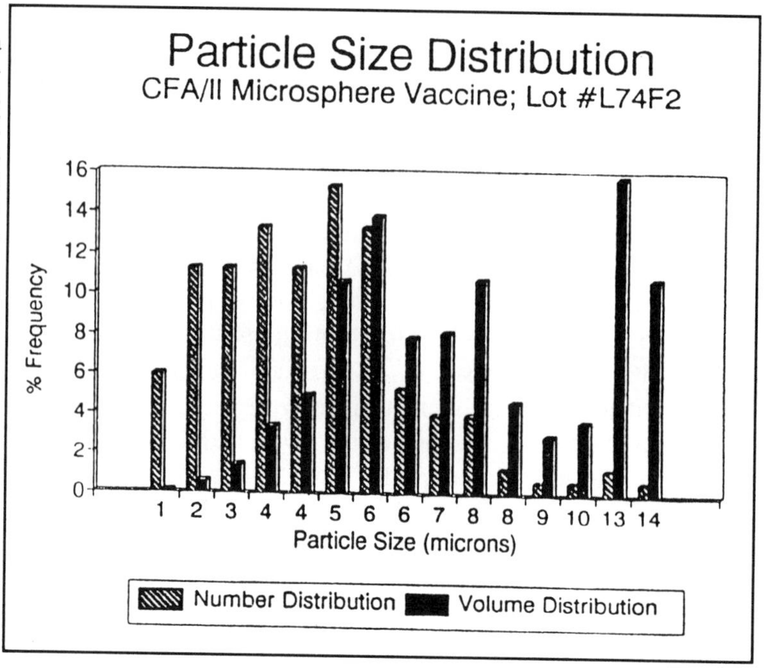

Figure 1. Results of a size frequency analysis on CFA/II MS Vaccine Lot L74F2 expressed as percent frequency versus particle size (diameter) in micrometer.

25μg protein in two different sites intramuscularly on Day 0. Sera were obtained from all animals before immunization on Day 0 and Days 7 and 14. The sera were tested by ELISA for IgG antibodies to CFA/II antigen and to individual *coli* surface (CS) proteins CS3 and CS1.

Five rabbits were vaccinated intraduodenally with CFA/II microspheres containing either 25 or 50μg of protein on Days 0 and 7. The rabbits were sacrificed by chemical euthanasia on Day 14 and subjected to complete necropsy. The Peyer's patches were removed, and the rabbit cells were suspended in a concentrated modified medium. The cells were placed in 2ml of media at a concentration of 2.5 x 10^6 cells/ml for each well of a 24-well plate. These cells were challenged separately with Bovine Serum Albumin (BSA) and the CFA/II antigen at doses of 500-, 50- and 5ng/ml in triplicate wells. The plates were incubated at 37°C with five percent CO_2. Spleen cells were obtained from immunized rabbits on Day 14 following intraduodenal immunization with CFA/II microsphere vaccine. The cells were placed in 96-well round bottom microculture plates at a final concentration of 6 x 10^5 cells per well and incubated for zero, one, two, three, four, and five days at 37°C with $5CO_2$. Ninety-six-well flat bottom microculture plates were coated with 3μg/ml of CFA/II antigen overnight and blocked with PBS with 0.05 percent polysorbate 60R. On the harvest days, the cells were gently flushed out of the

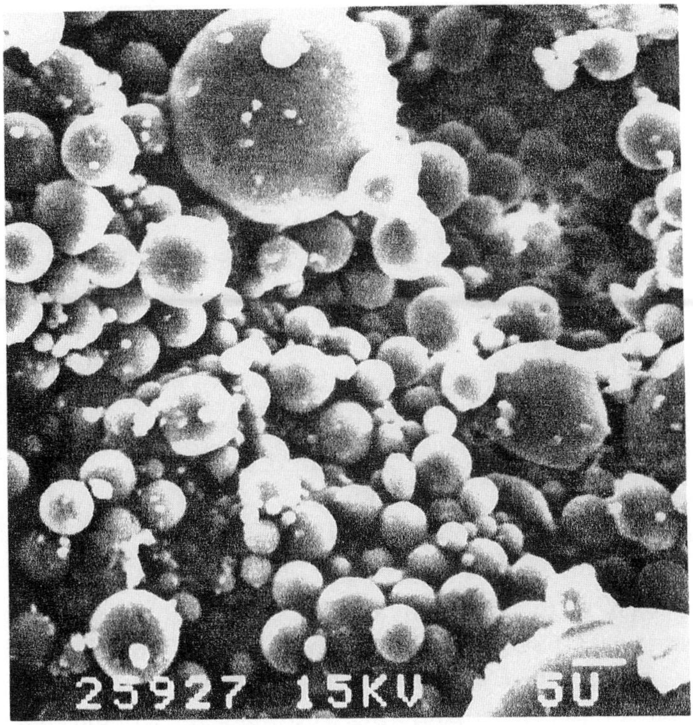

Figure 2. Scanning electron micrograph of the microspheres of Lot L74F2.

wells of the round bottom plates and transferred to the corresponding wells in the antigen-coated, 96-well flat bottom microculture plates to be tested for the presence of antibody secreting cells using a modified ELISA technique in which insoluble reagent-substrate product precipitates out of solution to form spots when developed in an agarose-substrate mixture, in reaction to antibody secreted from B cells.

Results

The results of size frequency analysis on CFA/II microsphere vaccine Lot L74F2 are shown in Figure 1. The average number frequency diameter is 4.5μm. The average volume frequency diameter is 9.4μm. The percent volume between diameters of 5 to10μm is 63 percent, and the percent volume less than 10μm diameter is 77 percent.

The microspheres of Lot L74F2 are seen in Figure 2, which is a scanning electron photomicrograph. They appear as smooth and nonporous, and nearly all the microspheres are less than 10μm as compared to the 5μm bar.

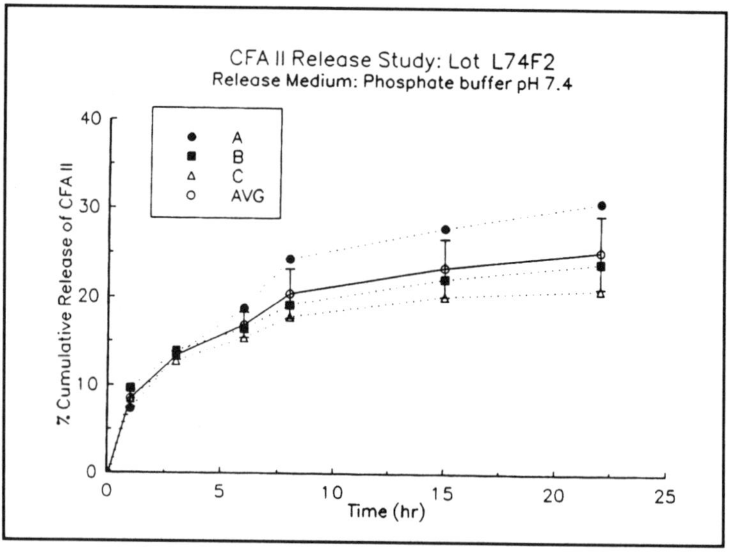

Figure 3. Results of a release study on Lot L74F2 expressed as percent release of CFA/II against time in hours for three samples.

The protein loads of the four individual batches are the following: K62A8, 1.16 percent ± 0.10 SD; K63A8, 1.023 percent ± 0.17 SD; K64A8, 1.232 percent ± 0.13 SD; and K65A8, 0.966 percent ± 0.128 SD. The mean average protein load is 1.16 percent ± 0.15 SD. The protein load of the CFA/II microsphere vaccine in the final dose vial is the following: Lot L74F2, 1.175 percent ± 0.17 SD.

Using the Karl Fischer titrimeter method, the CFA/II microsphere vaccine (Lot 74F2) percent water content was found to be 2.154 percent, using triplicate samples.

The acetonitrile residuals of the four individual CFA/II microsphere batches are the following: K62A8, <0.1 percent; K62A8, <0.1 percent, K64A8, <0.1 percent; and K65A8, <0.1 percent. The acetonitrile residual of the CFA/II microsphere vaccine in the final dose vial is the following: Lot L74F2, 0.07 ±0.05 percent. The heptane residuals of the four individual CFA/II microsphere batches are the following: K62A8, 1.9 percent; K63A8, 1.4 percent; K64A8, 1.6 percent and K65A8, 1.6 percent. Following pooling in heptane and subsequent drying, the heptane residual of the CFA/II microsphere vaccine in the final dose vial is the following: Lot L74F2, 1.6 ± 0.1 percent.

Twenty-two colonies grew on the blood agar plates after 48 hours of culture. Twenty-one were identified as coagulase negative staphylococci and one as a *Micrococcus* species, all of which are considered to be nonpathogenic to humans. Three fungal colonies grew on the fungal culture plates, and each was identified as *A. glaucus.*

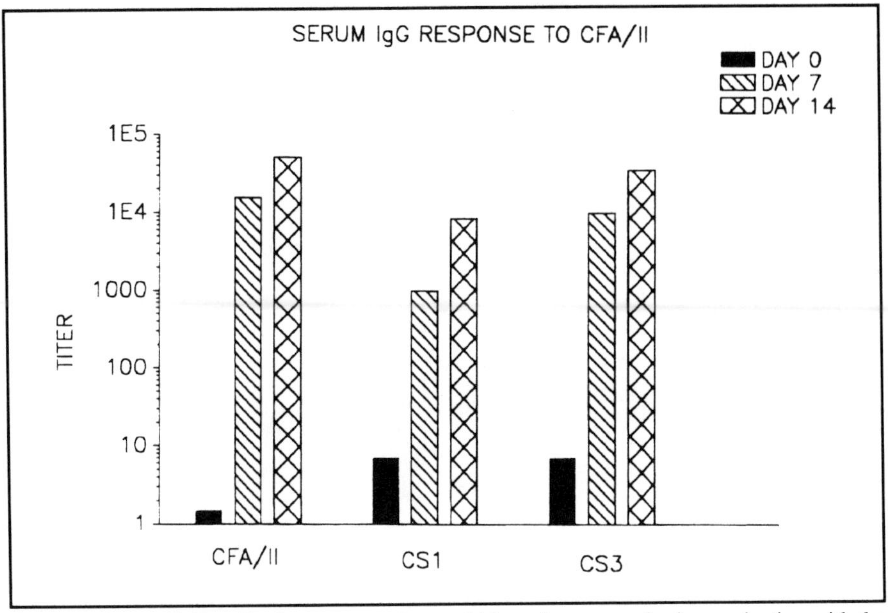

Figure 4. Serum IgG antibody response of two rabbits to intramuscular immunization with the CFA/II MS Lot L74F2.

The results of the release study on Lot L74F2 are shown in Figure 3. The results are plotted as percent release of CFA/II against time in hours for three samples: A, 33.12mgm; B, 29.50mgm; C, 24.20mgm at one-, three-, six-, eight-, 12-, and 22-hour intervals. An average of eight percent of CFA/II is released at one hour, rising to 20 percent at eight hours, then a slower release to 25 percent at 22 hours. The average represents the mean ±1 SD.

The CFA/II MS were administered intraperitoneally to two mice and by gastric lavage to two guinea pigs as a general safety test. The animals were weighed both before and seven days following immunization, and were monitored for any signs of toxicity. None of the animals displayed any signs of toxicity for seven days. The mice gained an average of 2.3g, and the guinea pigs gained an average of 43g.

The serum IgG antibody titers to CFA/II microsphere vaccine Lot L74F2, CS1, and CS3 proteins following (25μg) protein intramuscular immunization on Day 0 in two rabbits were determined using ELISA; the results are shown in Figure 4. The results were determined on serial dilutions of sera and are expressed as mean antibody titers against the different antigens at zero, seven, and 14 days. Antibody titers greater than 1000 were seen at seven days to both CS1 and CS3 protein, which rose to greater than 10,000 by Day 14. Rabbit 109 developed an antibody titer to CFA/II of 1,000 by Day 7 rising to

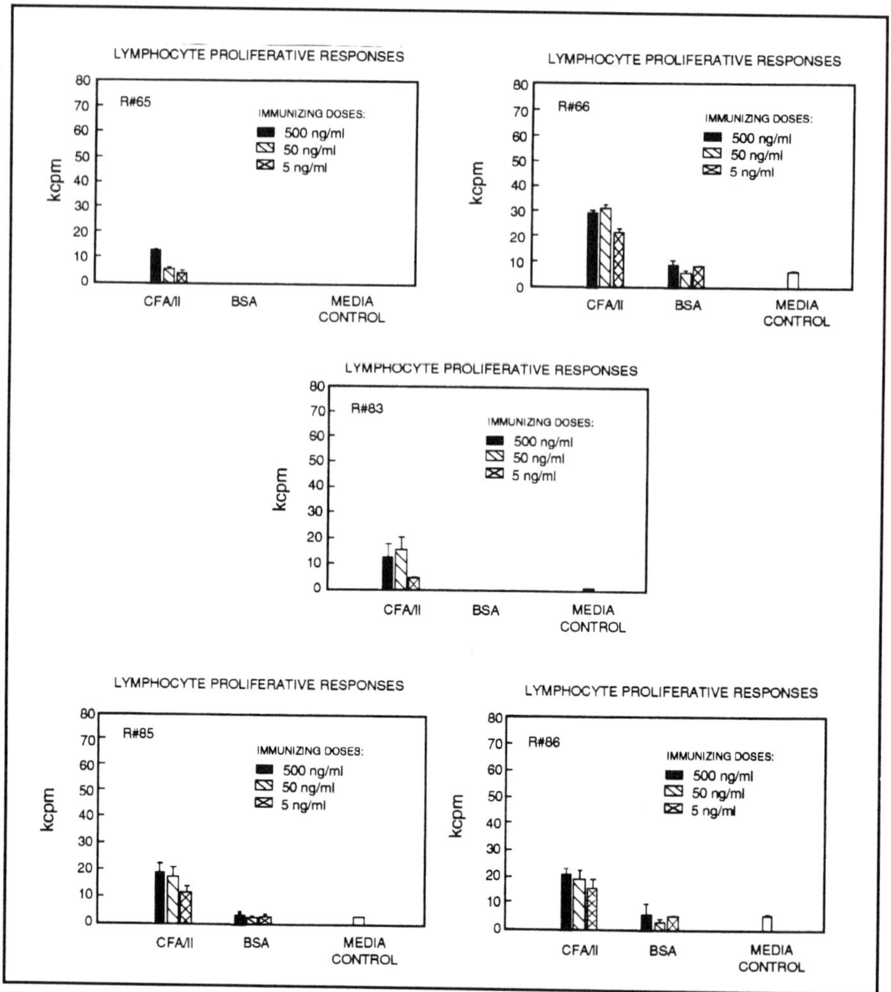

Figure 5. Lymphocyte proliferative responses from Peyer's patch cells of five rabbits immunized intraduodenally with 50μm of CFA/II in microspheres.

3,000 by Day 14. Rabbit 108 had a log higher rise at Day 7 and 2 log higher rise at Day 14, being 3 x 10^4 at Day 7 and going to 1 x 10^5 at Day 14 (data not shown).

The lymphocyte proliferative responses for Peyer's patch cells of rabbits 65, 66, 83, 85, and 86 immunized intraduodenally with 50μg protein of CFA/II microsphere vaccine 14 and seven days earlier are seen in Figure 5. The cells are challenged *in vitro* with CFA/II or BSA at 500-, 50-, or 5ng/ml of media in triplicate. The lymphocyte transformation was determined by tritiated thymidine incorporation. The uptake of tritiated thymidine in kcpm is expressed as mean ± 1 SD. The results are expressed as kcpm against antigen

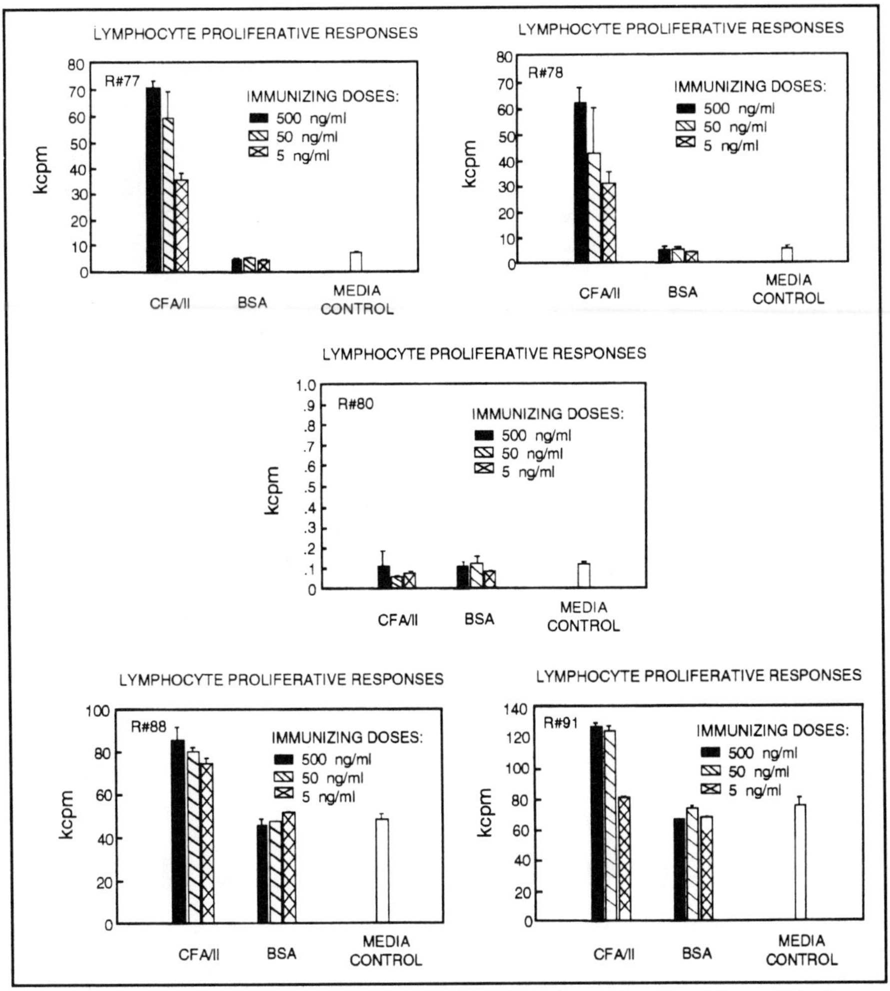

Figure 6. Lymphocyte proliferative responses from Peyer's patch cells of five rabbits immunized intraduodenally with 25μg of CFA/II in microspheres.

dose. No response to BSA or media control is seen in any of the five rabbits. All rabbits responded by lymphocyte transformation in a dose-dependent manner to the CFA/II. The highest dose responses were three to 10 times the media/control and are highly significant, with a p value of <0.006. Using the paired Student's *t*-test, the p values of 500μg/ml dose of CFA/II compared with media control are: 65, p = 0.002; 66, p = 0.002; 83, p = 0.006; 85, p = 0.002; and 86, p = 0.0002.

The lymphocyte proliferative responses from Peyer's patch cells of rabbits 77, 78, 80, 88, and 91 immunized intraduodenally with 25μg protein of CFA/II microsphere vaccine 14 and seven days earlier are seen in Figure 6.

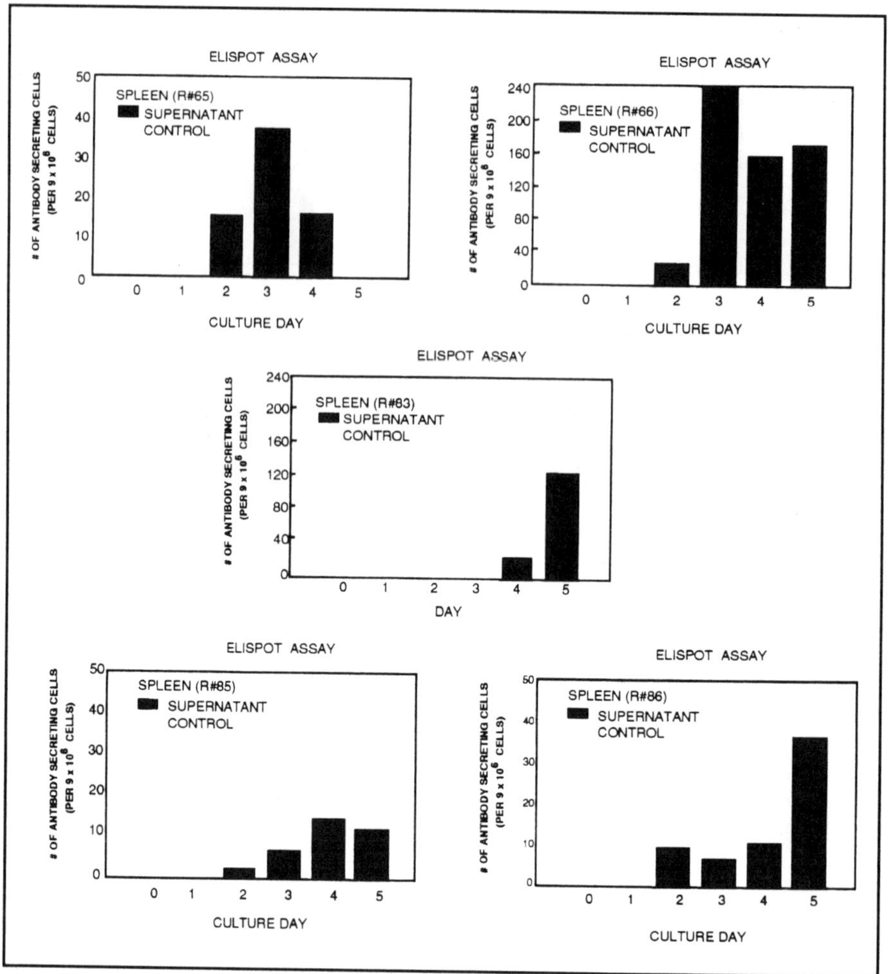

Figure 7. Results of an ELISPOT assay of spleen cells from five rabbits immunized intraduodenally with the CFA/II MS vaccine, expressed as number of antibody-secreting cells found on each day of culture.

The cells are again challenged *in vitro* with CFA/II or BSA at 500-, 50-, or 5ng/ml of media in triplicate. The lymphocyte transformation was determined by tritiated thymidine incorporation. The uptake of tritiated thymidine in kcpm is expressed as mean ± 1 SD. Rabbit 80 gave no response, probably due to a poor Peyer's patch cell population, which did not respond well to Concanavalin A mitogenic stimulation either. The remaining four rabbits gave positive responses, with the high CFA/II dose response being two to eight times the media control and highly significant, with p values of <0.009.

Again, no response was seen to BSA as compared with the media control. Using the paired Student's *t*-test, the p values of 500ng/ml dose of CFA/II compared to media control are: 77, p = 0.001; 78, p = 0.0015; 80, p = insignificant; 88, p = 0.0093; and 91, p = 0.001.

The results of the ELISPOT assay of spleen cells from rabbits 65, 66, 83, 85, and 86 immunized intraduodenally with 50µg protein of CFA/II microsphere vaccine 14 and seven days earlier are seen in Figure 7. These cells were placed into microculture and tested on Days 0, 1, 2, 3, 4, and 5 by ELISPOT for cells secreting antibodies specific for CFA/II antigen. The results are expressed as number of antibody secreting cells per 9 x 10^6 spleen cell against culture days. Positive responses were seen in all five rabbits on Days 2 through 5. Days of maximum responses occurred on Day 3 for rabbits 65 and 66; Day 4 for rabbit 85; and Day 5 for rabbits 83 and 86. The responses are highly significant, being seven to 115 times higher than the one to two cells seen on all days in four control rabbits (67, 69, 72, 89); (data not shown).

A consistent finding in the spleens of all rabbits—both the 25- and 50µg protein dose groups—was minimal to mild diffuse lymphocytic hyperplasia of the periarteriolar lymphatic sheaths (T-cell-dependent). Two of five rabbits of the 50µg dose group (83 and 86) also had mild lymphocytic hyperplasia of splenic follicular (B-cell-dependent) areas. The three rabbits in an untreated control group had histologically normal spleens.

Reactive hyperplasia of mesenteric lymph nodes was often seen in vaccinated rabbits. Two of five rabbits in the 25-µg human dose equivalent group (88 and 91) and four of five rabbits in the 50-µg dose group (65, 66, 83, 86) had diffuse, minimal to mild lymphocytic hyperplasia of paracortical (T-cell-dependent) areas of the mesenteric lymph nodes. Two rabbits in the 50-µg dose group (83 and 86) also had minimal to mild lymphocytic hyperplasia of cortical follicular (B-cell-dependent) areas. The mesenteric lymph nodes of the other vaccinated rabbits and of the untreated control rabbits were within normal limits. Incidental or background lesions found in one or more rabbits of all three groups were acute minimal to mild pnuemonia and foreign body microgranulomas of the cecal gut-associated lymphoid tissue.

Discussion

McQueen *et al.*[9] have found that the AF/R1 adhesin of rabbit diarrheagenic *E. coli* (RDEC-1) incorporated into biodegradable microspheres could function as a safe and effective oral intestinal vaccine in the rabbit diarrhea model. The AF/R1 was incorporated into poly (D,L-lactide-co-glycolide) microspheres and administered intraduodenally. Jarboe *et al.*[11] reported that Peyer's patch cells obtained from rabbits immunized intraduodenually with AF/R1 in microspheres responded with lymphocte proliferation upon *in vitro* challenge with AF/R1. This early response at 14 days gave a clear indication as to the immunogenicity of *E. coli* pili contained within the polymer microspheres.

CFA/II from enterotoxigenic *E. coli* given as an oral vaccine was found to be ineffective, as the fimbrial and fibrillar proteins were found to be rapidly degraded when treated with conditions mimicking those contained in the stomach.[3] The CFA/II was found to be immunogenic when given in high doses intra-intestinally—producing intestinal secretory IgA antibodies.[2]

The CFA/II vaccine has been incorporated into poly(D,L lactide-co-glycolide) microspheres under GMP and tested under GLP. The microspheres are spherical, smooth surfaced, and without pores. The majority (63 percent) are between five and10μm in diameter by volume, a size range that has been suggested to promote localization within the Peyer's patch in mice and perhaps to enhance local immunization.[5-8] The protein content is 1.174 percent, and 0.62 percent was the core loading of the AF/R1 microspheres that were effective.

The organic residuals are at a safe level, with the heptane exposure being 1.7mgm per vaccine dose and the acetonitrile exposure being 0.1mgm per vaccine dose. This is compared with the occupational maximum allowable exposure of 1800mgm/15 minutes for heptane and the human oral TDLO (any nonlethal toxicity) of 570mgm/Kg for acetonitrile.

The CFA/II protein was produced under aseptic conditions; however, the incorporation of protein into microspheres and microsphere pooling were done in a clean room, allowing entrance of microbes. The World Health Organization (WHO) Requirements for Typhoid Vaccine (live, atttenuated, Ty2la oral) allow 200 nonpathogenic bacteria as well as 20 fungi per dose. The CFA/II microsphere vaccine has only 22 nonpathogenic bacteria and three fungi per dose.

A general safety test [from the WHO Requirements for Typhoid Vaccine (live, attenuated, Ty21a oral)] was performed in which the microsphere vaccine was given by gastric lavage to guinea pigs. Both mice and guinea pigs demonstrated no toxicity and gained weight over the seven-day test, clearly indicating the innocuous nature of this vaccine by passing this safety test.

The CFA/II microsphere vaccine (Lot 74F2) is immunogenic, giving high titer serum IgG antibody responses as early as seven days following intramuscular injection in rabbits. This test will be used as a potency test for future lots of the CFA/II microsphere vaccine.

The CFA/II microsphere vaccine was also immunogenic following intraduodenal administration to rabbits. The highest lymphocyte proliferative responses from Peyer's patch cells were seen with the lower 25-μg dose. This is the human equivalent dose and suggests that higher doses of antigen in polymer microspheres may attenuate this immunologic reponse.

The antibody secreting B cells demonstrated in the rabbit spleen at 14 days is a clear indication that B cells have been immunized. They may represent resident B cells immunized in the spleen or B cells immunized at the level of the Peyer's patches that are migrating through the spleen to return to the intestinal mucosal lamina propria.[12] The delay of several days before secreted antibody is detected suggests either maturation is required of the B cells or that down-regulation may be present initially and lost with time in culture.

Further evidence of immunization by the CFA/II microsphere vaccine given intraduodenually is demonstrated by the lymphocytic hyperplasia in the spleen, particularly in the T-cell-dependent areas in all vaccinated rabbits. On the other hand, greater T-cell-dependent area lymphocytic hyperplasia in the mesenteric lymph nodes was seen in rabbits receiving the higher 50-µg dose, four of five compared to two of five with the lower dose, indicating a dose response. These changes are most likely due to the vaccine, because no similar changes were seen in three untreated control rabbits. Also, no abnormal pathologic changes attributable to the vaccine were seen.

The CFA/II BPM vaccine has undergone preclinical evaluation and has been found safe and immunogenic. This vaccine is ready for clinical Phase I safety testing following the Food and Drug Administration's approval of the Investigational New Drug Exemption (IND).

References

1. J.B. Kaper and M.M. Levine, *Vaccine* **6**, 197-199 (1988).
2. M. Levine *et al.*, in *Molecular Biology of Microbial Pathogenicity, Protein- Carbohydrate Interactions in Biological System*, D. Lark *et. al.*, Ed. (Academic Press, London, 1986), pp. 143-145.
3. M. Schmidt *et al., Gastroenterology* **88**, A1575 (1985).
4. D.L. Wise *et al.*, in *Drug Carriers in Biology and Medicine*, G. Gregoriades, Ed. (Academic Press, London, 1979), pp. 237-270.
5. J.H. Eldridge *et al., Curr. Top. Microbiol. Immunol.* **146**, 59-66 (1989).
6. J.H. Eldridge *et al., J. Controlled Release* **11**, 205 (1989).
7. J.H. Eldridge *et al., Mol. Immunol.* **28**, 287-294 (1991).
8. Z. Moldoveanu *et al., Curr. Top. Microbiol. Immunol.* **146**, 91-99 (1989).
9. C.E. McQueen *et al., Vaccine* (in press) (1992).
10. M.M. Levine *et al., Infect. Immun.* **44**, 409- 420 (1984).
11. D. Jarboe *et al., Abstracts of the Annual Meeting of the American Society of Microbiology*, (May 1990), p. 121.
12. J. Tseng, *J. Immunol.* **132**, 2730-2735 (1984).

4

Host Respiratory Protection For Lethal Staphylococcal Enterotoxin B

Robert E. Hunt
Anthony J. Johnson
Marti Jett
Jeenan Tseng
Jack L. Komisar
Catherine L. Wilhelmsen
U.S. Army Medical Research
and Development Command
Walter Reed Army Institute
of Research
Washington, DC 20307-5100

Louise Pitt
David Ruble
U.S. Army Medical Research
Institute of Infectious Diseases
Fort Detrick
Frederick, MD 21702-5012

For more than three decades the U.S. Army Medical Research and Development Command has conducted research on staphylococcal enterotoxin B (SEB). It is a well known natural disease threat and a feared biologic warfare (BW) agent. A major thrust of the research has been to develop an effective vaccine that fully protects against the lethal intoxication that can occur from respiratory tract exposure to a minute quantity of the toxin. A formalinized soluble toxoid vaccine provided partial protection; however, immediate hypersensitivity reactions were associated with the immunization when the host was challenged with SEB via the respiratory route. We are reporting a novel first time immunization scheme that protected rhesus monkeys against aerosolized SEB, a species in which SEB intoxication closely mimic humans in sensitivity. The immunization scheme involved biodegradable and biocompatible poly (D,L-lactide-co-glycolide) microsphere encapsulation of the formalized SEB toxoid and the administration of an intratracheal booster after the initial intramuscular or intratracheal vaccine dose. Complete protection against a respiratory lethal dose challenge of SEB in the rhesus monkey following immunization with the microsphere SEB toxoid vaccine strongly indicates that similar results can be achieved in humans.

Introduction

During the 1960s, Army scientists prepared, purified, assayed, and determined the effects of staphylococcal enterotoxin B (SEB) in primates.[1-3] A significant observation was a pattern of acute intoxication that progressed to a severe shocklike syndrome with high mortality following the administration of microgram (μg) quantities of purified SEB. Because of the toxin's overall biologic and physical properties, attention was directed to its potential usefulness as a biologic warfare agent in overt, covert, terrorist, and conventional warfare activities. It was the threat agent potential that led the U.S. Army Medical Research and Development Command to conduct biomedical research for development of defenses against the toxin's incapacitating and lethal effects.

SEB, along with six structurally and biologically related serotyped classes of enterotoxins (SEA, SEB, SEC1, SEC2, SEC3, SED, and SEE), have been major causes of food poisoning worldwide.[4,5] These enterotoxins are extracellular proteins produced by the ubiquitous bacterium *Staphylococcus aureus*. It was the incidence and misery of staphylococcal food poisoning that led scientists to develop vaccine candidates and attempt to immunize humans as a means of reducing the incidence of staphylococcal enterotoxin foodborne illness. Early attempts were hampered by unpurified enterotoxin and questionable serologic assays.[6] It was only after the enterotoxins were properly identified serologically and purified as distinct molecules that specific immunization regimens could be addressed. Bergdoll, as early as 1966, investigated the immunologic properties of purified SEB in rhesus monkeys after toxoiding the molecule with Formalin.[7] After alum-adsorption of the formalinized SEB toxoid, he was able to induce antibody formation in rabbits and increase the resistance of rhesus monkeys to oral SEB challenge about 100-fold. The results of the immunization experiment indicated that three-intramuscular injections of 62.5μg for the initial and first booster doses of SEB toxoid and 250μg for the second booster, all at five-week intervals, were needed for this protection. His results also indicated that the resistance to SEB was due, at least in part, to the presence of antitoxin antibodies in the plasma of resistant animals. His research and that of others have contributed valuable information on immunoprophylaxis through vaccination against staphylococcal enterotoxins and toward development of an effective toxoid vaccine against SEB. Yet there remains no effective vaccine that fully protects against these staphylococcal enterotoxins.

Beginning in February 1991, scientists from the Division of Pathology, Walter Reed Army Institute of Research and the U.S. Army Medical Research

Institute of Infectious Diseases have been testing the degree of protection afforded rhesus monkeys immunized with microspheres of a biocompatible, biodegradable poly (D,L-lactide-co-glycolide) polymer (DL-PLG) that encapsulates the SEB toxoid. The decision to encapsulate the SEB toxoid was made following failures of soluble and alum-adjuvanted SEB toxoids to provide appropriate protection against SEB respiratory exposure in rhesus monkeys.[8,9] Severe immediate hypersensitivity was an additional adverse response observed when immunized and SEB exposed monkeys were subsequently challenged by the respiratory route. Major factors for microsphere encapsulation of the SEB toxoid were to develop an oral vaccine against SEB that especially stimulated the production of secretory immunoglobulin A (sIgA) and to provide protection in immunized rhesus monkeys against SEB aerosol challenge. Additionally, it was theorized that effective production of anti-SEB sIgA antibodies would suppress the immediate hypersensitivity response encountered following immunization with soluble or alum-adsorbed SEB toxoid.

The microsphere formulation, SEB toxoid encapsulation, and biologic testing in mice had been previously accomplished through contract arrangements with a commercial firm.[10] After optimizing the formulation of microspheres composed of DL-PLG and SEB toxoid, immunization studies were conducted in mice.[11] It was shown that (a) three oral immunizations, (b) an initial parenteral immunization followed by an oral, intramuscular, or intratracheal booster immunization, or (c) an initial and booster immunizations administered intratracheally provided antibody responses in the plasma and bronchoalveolar lavage (BAL) fluid. Also, a strong IgA antibody response was induced in the bronchoalveolar fluid following the intratracheal administrations. This led to questions and the decision to immunize rhesus monkeys by various routes to determine the most effective method to induce protection against lethal SEB aerosol exposure. It is the unique results of these early SEB protection studies accomplished in rhesus monkeys that form the basis for this report.

Methods

The study consisted of 40 young, healthy rhesus monkeys: four controls and 36 immunized. Nine different immunization regimens employing three different routes of immunization were accomplished (Table 1).

The routes of immunization were oral (OR), intramuscular (IM), and intratracheal (IT). Each immunization regimen consisted of a primary immunization (OR, IM, or IT) followed 49 days later by a secondary booster

Table 1
Immunization Regimens

OR/OR	OR/IM	OR/IT
IM/OR	IM/IM	IM/IT
IT/OR	IT/IM	IT/IT

Primary/booster immunization regimen.
OR=Oral; IM=Intramuscular; IT=Intratracheal.

immunization (OR, IM, or IT). The microsphere vaccine was administered at a dose of 100μg in 0.5ml volume for the intramuscular and intratracheal immunizations and 1.0mg in 0.5ml volume for the oral immunizations. Operational and resource constraints allowed only ten monkeys, one from each immunization regimen and one control, to be studied as a group.

The aerosol SEB challenge was performed 80 days after the booster immunization. A standard lethal SEB challenge dose of 15 LD_{50} was scheduled for all rhesus monkeys; however, the actual challenge doses varied among study groups. The SEB challenge dose received by each rhesus is listed in Tables 2 and 3. The actual exposure doses were determined from samples taken from the exposure chamber during challenge.

To gain more information on the role of anti-SEB antibodies in protection and to obtain additional data on the reported incidence of immediate hypersensitivity, the immunoglobulin (Ig) isotype IgM, IgG, and IgA titers were measured in the plasma and bronchoalveolar lavage fluids of all monkeys using an ELISA test. The isotype titers for IgG and IgA are listed for each monkey (Tables 2 and 3). It was determined that IgG and IgA in the plasma and bronchoalveolar lavage fluid contributed to protection because the majority of the survivors had high levels of both isotype antibodies. The isotype IgM response was inconsistent and had no noted relationship to protection.

Results

All four rhesus monkeys (100 percent) administered the IM/IT immunization regimen were protected against the aerosol lethal dose challenge. In addition, two monkeys in the IT/IT and two in the OR/IT immunization regimens were protected. In contrast, no other immunization regimen protected more than one monkey, and there were no protected monkeys in the OR/OR, OR/IM, and IM/OR regimens. These data emphasize the importance of the intratracheal route as a booster immunization for protection against aerosol exposure.

Table 2
Protection: Monkeys Immunized With Microspheres

Day 98 Prebleed/Prebal Antibody Titers (SRI) Endpoint Titration-ELISA						
SURVIVORS Monkey/Immu*		**IgG Plasma**	**IgG BAL†**	**IgA Plasma**	**IgA BAL**	**Challenge LD_{50}**
89B161	IM/IT	800000	320	45	40	13.8
90B003	IM/IT	1000000	1280	1800	320	18.1
90B031	IM/IT	500000	160	250	20	7.0
90B078	IM/IT	1800000	1280	1000	320	22.2
90B009	IT/IT	1000000	320	380	160	17.6
90B055	IT/IT	900000	2560	1800	1280	7.4
89B064	OR/IT	200000	80	25	20	12.9
90B066	OR/IT	5000	<10	60	20	7.8
90B025	IT/IM	20000	10	25	20	7.1
90B030	IM/IM	1800000	80	280	<10	6.7
90B062†	OR/IM	900000	640	30000	<10	18.4
90B067	IT/OR	80000	40	400	20	7.4

Primary/booster immunization regimen.
OR=Oral; IM=Intramuscular; IT=Intratracheal.
Immu*=Immunization regimen; BAL=Bronchoalveolar lavage.
† Deleted from study due to illness and treatment prior to challenge.

There are several implications in this study that are relevant to protection against biologic threat agents delivered as an aerosol. First, the toxin was directed toward the respiratory tract, it was hypothesized that protection would correlate with the level of local IgA anti-SEB antibodies in bronchial fluid. As shown in the data in Tables 2 and 3, this was indeed the case. Monkeys with the highest IgA and IgG levels in BAL were protected, whereas those with the lowest levels were more susceptible to the lethal challenge. Second, it was hypothesized that the route of immunization would influence the level of respiratory antibodies and hence be a prime determinant of protection. Indeed, the question of which route of immunization best induces secretory immunity was central to this study. Although researchers have long suspected the importance of local immunization, there is a notable paucity of protection data comparing various immunization regimens, especially in

Table 3
Protection: Monkeys Immunized With Microspheres

Day 98 Prebleed/Prebal Antibody Titers (SRI)
Endpoint Titration-ELISA

NONSURVIVORS Monkey/Immu*		IgG Plasma	IgG BAL*	IgA Plasma	IgA BAL	Challenge LD_{50}
89B088	Control	25	<10	25	<10	13.0
90B001	Control	50	<10	50	<10	18.1
90B029	Control	25	<10	25	<10	6.8
90B050	Control	25	<10	25	<10	18.4
89B001	IT/IT	189000	<10	25	<10	13.1
90B075	IT/IT	700000	1280	400	640	21.8
90B008	OR/IT	28000	<10	50	<10	17.1
90B098	OR/IT	20000	<10	25	<10	21.9
89B116	IT/IM	200000	20	25	<10	13.8
90B021	IT/IM	22000	320	50	<10	18.2
90B103	IT/IM	40000	40	25	<10	20.9
89B210	IM/IM	850000	40	25	<10	13.2
90B010	IM/IM	450000	<10	50	<10	18.2
90B060	IM/IM	20000	<10	25	<10	18.6
89B034	OR/IM	100000	<10	25	<10	13.3
90B018	OR/IM	28000	<10	50	<10	18.1
90B043	OR/IM	10000	10	25	<10	7.6
89B060	OR/OR	25	<10	25	<10	13.1
90B014	OR/OR	875	<10	50	<10	18.5
90B027	OR/OR	2000	<10	170	<10	7.3
90B079	OR/OR	1000	<10	700	<10	22.4
89B096	IM/OR	30000	<10	50	<10	13.0
90B002	IM/OR	13000	<10	50	20	17.4
90B016	IM/OR	40000	<10	25	<10	7.2
90B110	IM/OR	20000	<10	25	<10	21.
89B079	IT/OR	25	<10	25	<10	13.2
90B023	IT/OR	110000	20	50	<10	18.5
90B092	IT/OR	9000	20	25	<10	18.7

Primary/booster immunization regimen.
OR=Oral; IM=Intramuscular; IT=Intratracheal; Immu*=Immunization regimen; BAL=Bronchoalveolar lavage.

primates. For example, based on what is known about lymphocyte trafficking following oral immunization, it may have been thought that oral immunization would be superior to respiratory (and certainly IM) immunization regimens. On the other hand, if systemic immunity alone were important for protection against the lethal event induced by this toxin, the IM/IM immunization regimen may have been the most effective regimen. The success of intratracheal immunization, especially as a booster in the IM/IT regimen, in generating protection against the highest lethal toxin challenges attempted in this study demonstrates that immunization via the respiratory tract may be the most promising regimen to induce antibody-mediated immunity against aerosolized biologic warfare agents such as staphylococcal enterotoxins.

Conclusions

In summary, the lives of rhesus monkeys were completely protected from a high dose ($15LD_{50}$) SEB aerosol lethal challenge by an initial intramuscular injection of 100μg of formalinized SEB toxoid encapsulated in biodegradable and biocompatible DL-PLG copolymer microspheres and a 49th day booster 100μg dose administered intratracheally. No evidence of hypersensitivity was noted following the aerosol challenge. Additionally, the data indicate that protection against SEB is provided, at least partially, by the production of IgG and IgA anti-SEB isotype antibodies in the plasma and bronchoalveolar lavage fluid. The immunization regimen, consisting of an initial intramuscular injection followed by a booster vaccine dose administered intratracheally, the latter being the mucosal site of SEB challenge, proved to be an important factor in protection.

Future endeavors will include aerosol rechallenge of surviving monkeys to evaluate the duration of protection and to further assess the potential for immediate hypersensitivity reactions to occur in immunized monkeys. Additionally, following the technical development and evaluation of DL-PLG microsphere SEB toxoid booster immunization by the intranasal (IN) route in preliminary mouse studies, a scientific study is planned to aerosol challenge and assess a group of monkeys immunized with an IM/IN immunization regimen. Complete characterization of respiratory route immunization to include safety test analysis of empty DL-PLG microspheres administered to the lungs of monkeys is planned. Formulations of DL-PLG copolymers constructed as resorbable sutures, resorbable suture clips, and as microsphere control release implants are approved for use in humans and currently have a safe history.[11] Continued success with novel selective immunization of the respiratory tract against lethal toxins and potentially other agents can have a major positive impact in protecting the health and fighting strength of soldiers.

References

1. W.R. Beisel, *Toxicon* **10**, 433-440 (1972).
2. G.J. Crawley *et al.*, *J. Infect. Diseases* **116**, 48-56 (1966).
3. L.F. Hodoval *et al.*, *Appl. Microbiol.* **15**, 187-192 (1968).
4. M.S. Berdgoll, Staphylococcal Intoxications, in *Foodborne Infections and Intoxications*, H. Riemann and F.L. Bryan, Eds. (New York, Academic Press, Inc., 1979), pp. 443-452.
5. M.S. Bergdoll, Enterotoxins, in *Microbial Toxins, III, Bacterial Protein Toxins,* T.C. Montie, S. Kadis, and S.J. Ajl, Eds. (Academic Press, Inc., New York, 1970), pp. 265-326.
6. G.M. Dack, E.O. Jordan, and O. Woolpert, *J. Preventive Med.* **5**, 151-159 (1931).
7. M.S. Bergdoll, *Infect. & Immun.* **116**, 191-196, (1966).
8. U.S. Army Medical Unit, *Annual Progress Report, Fort Detrick, MD*, pp. 101-102 and 243-250 (1968).
9. U.S. Army Medical Unit, *Annual Progress Report, Fort Detrick, MD*, pp 227-233 (1969).
10. Southern Research Institute, *Annual Report, Project 6160, Contract DAMD17-C-6162*, 1991.
11. J.H. Eldridge *et al.*, *Infect. & Immun.* **59** (9), 2978-2986 (1992).

5

Thermodynamic Analyses of Transmembrane Channel Formation by Botulinum Toxin

Frank J. Lebeda
Neurotoxicology Branch
U.S. Army Medical Research
Institute of Chemical Defense
Aberdeen Proving Ground,
MD 21010-5425

Mary K. Gentry
Biological Chemistry Branch
U.S. Army Medical Research
and Development Command
Walter Reed Army
Institute of Research
Washington, DC 20307-5100

Dallas C. Hack
Virology Division
U.S. Army Medical Research
and Development Command
Institute of Infectious Diseases
Fort Detrick
Frederick, MD 21702-5011

The translocation of botulinum neurotoxin (BoTX) into presynaptic terminals of neuromuscular junctions may involve its well-known capacity to form transmembrane (TM) channels. The present study predicts which peptide segments of this toxin are energetically favored to be inserted into a biological membrane, by calculating thermodynamic estimates of the free energies of transfer of amino acid residues from an aqueous to a nonpolar, lipid environment. Hydropathic analysis showed that the N-terminal half of the heavy (H) chain of BoTX (A, C1, and D serotypes) possessed three adjacent hydrophobic regions that are potentially TM segments. An alternative calculation of the free energy cost predicted that the two longest hydrophobic regions on the H-chain are energetically favored to be inserted and stabilized in a lipid environment. Amphipathic analysis revealed that this half of the H-chain has a single region that contains the requisite hydrophilic residues for channel formation. The amino acids in this region are also similar to those in putative TM regions in neuronal voltage- and agonist-gated channels. At least two H-chains of the BoTX serotypes form conducting ionic channels. This predicted channel-forming region may be involved in a process common to all the BoTX serotypes. and will be of strategic importance in designing and developing single pharmaceuticals or medical products that could counteract the effects of various BoTX serotypes.

Introduction

The peptide neurotoxins from *Clostridium botulinum* (botulinum toxin, BoTX) are the most lethal substances known. At the present time, seven immunologically distinct serotypes (A-G) have been identified and are the subjects of numerous and extensive reviews.[1-4] The BoTX-A serotype is the most toxic species and can now be massively cloned with low-cost genetic engineering techniques.[5] Although a pentavalent toxoid is available (for the A, B, C1, D, and E serotypes), clinically relevant prophylaxis takes months to develop. Immunizing personnel directly by injecting antibodies for the different serotypes are currently being developed, but despite this progress there are no effective antidotes that can be administered to exposed individuals who are either unvaccinated or who have insufficient titers of the circulating, neutralizing antibodies. Thus, because of their extreme potency, availability, and the lack of an adequate set of treatment compounds, development of medical countermeasures to BoTX is considered by the Medical Biological Defense Program as high priority in accordance with the seriousness of this agent's threat to U.S. military personnel.

These concerns raise additional issues. For example, other, but as yet uncharacterized, naturally occurring toxin serotypes may exist, and the present approach to counteract these new species with specific antibodies would require further developmental efforts and attendant high costs. Furthermore, site-directed mutagenesis of genetic material could conceivably produce an infinite number of novel toxins that could possibly circumvent the presently available set of neutralizing antibodies. In light of these problems, an alternative strategy to discover and develop new countermeasures forms the basis of the present research. This study's ultimate goal is to identify and develop single pharmaceuticals or other medical products for clinical care applications that can counteract various BoTX serotypes (rather than one for each serotype), regardless of their biological origin or source. To achieve this goal, a detailed knowledge of the toxin mechanism of action is required.

The primary toxic effect of all the BoTX serotypes (excluding the C2 and C3 serotypes[6]), is flaccid paralysis of skeletal muscles. The toxin prevents the release of acetylcholine in the periphery and blocks neurally induced muscle contraction[7,8] which can eventually cause respiratory failure and death. The toxin also inhibits the release or exocytosis of other neurally active substances in cholinergic and noncholinergic systems—but at concentrations that are much higher than those required to exert effects at the cholinergic peripheral synapse. This discriminating action is presumably due to the selective binding of the toxin to receptors that are predominantly localized at these presynaptic endings.[3]

Like diphtheria toxin (DiTX), which inhibits protein synthesis,[9] the clostridial neurotoxins (including tetanus; TeTX) block neurochemical transmission by a complex series of reactions that is related to their tripartite structure. The C- terminal half of the heavy (H) chain binds to the target site (peripheral presynaptic nerve terminal); the N-terminal half is involved in the endocytotic internalization process; and the light (L) chain exerts the toxic effect. Once internalized, the release (translocation) of the toxic L-chains from the endocytotic vesicles involves a pH-dependent process that may include the formation of ionic channels. Although the heavy chains of TeTX and DiTX (and their N-terminal halves) are capable of forming functional ionic channels in artificial membranes,[10-16] studies with DiTX indicate that hydrophobic segments of the light chain can also interact with the membrane[17] and take part in translocation.

If a process in this multireaction scheme is common to all of the serotypes, then it is possible that a single treatment drug could be found and developed for clinical use. Evidence from biochemical studies suggests that the first step in the scheme previously outlined differs for the various serotypes. At least two different serotype binding sites exist.[2] From electrophysiological data, the pattern of the final, toxic reaction also varies somewhat among the serotypes.[18] In contrast, the intermediate process, the toxin translocation step, is not as well characterized and may yield information on a reaction that the neurotoxin serotypes could have in common.

Because ionic channel formation by the toxins is believed to play a role in their internalization, we focused on identifying regions within the heavy chains of BoTX, TeTX, and DiTX that could be important in forming ionic channels. Our approach in localizing channel-forming regions was to use several different computer-assisted calculations to determine free energies of peptide insertion into biological membranes. The main assumption in this study was that channels are comprised of sequences containing both hydrophobic and hydrophilic residues.[19-22] Our aim was to determine where sufficiently long hydrophobic regions occurred and whether they included amphipathic segments. In conjunction with the presently known electrophysiological properties of these channels, our calculations support the conclusion that short segments located in the N-terminal half of the toxin heavy chains could participate in channel formation.

Methods

Amino acid sequences of the BoTX-A, C1, and D serotypes and TeTX were obtained from the National Biomedical Research Foundation (NBRF) database at the Advanced Scientific Computing Laboratory (National Cancer Institute, Frederick, MD). Peptide search, comparison, and analysis software included FASTA,[23] the Sequence Analysis Software Package (Genetics Computer Group, Inc.[24]), the IDEAS package,[25] and the programs supported by the Protein Identification Resource (NBRF, Georgetown University Medical Center, Washington, DC). Programs on personal microcomputers included the sequence alignment program MACAW (Center for Biotechnology Information[26]) and the MacVector analysis package (IBI, New Haven, CT).

The Kyte-Doolittle hydropathic index[27] was calculated using the authors' SOAP program and amino acid hydropathy scale (the free energy of transfer of an amino acid from a nonpolar medium to water). This index was calculated as a sum of hydrophobicity values of short peptide segments contained within a consecutively moving window. We wrote a second program (GES) to calculate this free energy of transfer (dG(tr)) using the amino acid polarity scale, window size, and method described by Engleman *et al.*[28]

Analysis of peptide amphipathicity used the MOMENT program, which uses the Fourier transform to calculate the sequence hydrophobic moment:

$$<\mu_H> = \{[\sum_{n=1}^{N} H_n \sin(\delta n)]^2 + [\sum_{n=1}^{N} H_n \cos(\delta n)]^2\}^{1/2} \qquad \text{(Equation 1)}$$

where N = window size, H^n = hydrophobicity of the nth residue, δ = an angle from 0-π radians, and a consensus scale for the free energy of transfer of an amino acid residue from a nonpolar to a polar medium.[19,20,29,30] Multimeric transmembrane (TM) segments were defined on the basis of values for the moment and the hydrophobicity of the sequence of amino acids within a window. Two different amino acid scales were used with MOMENT. The consensus hydrophobicity scale of Eisenberg was used initially. For comparison, we also developed a novel amino acid scale that is based on the frequency of an amino acid's occurrence in the (nonhelical) nontransmembrane and helical (transmembrane) segments of the photosynthetic reaction center (see Table II in reference 31). The values for *N* for each of these algorithms were selected on the basis of test runs using the amino acid sequence of the M-chain of the photosynthetic reaction center of *R. viridis*,[31] when all five of the known membrane-spanning regions of this peptide were detected.

Results

Sequence alignments. Blocks of similar sequences within the toxins examined in this study were determined with the MACAW program,[26] using

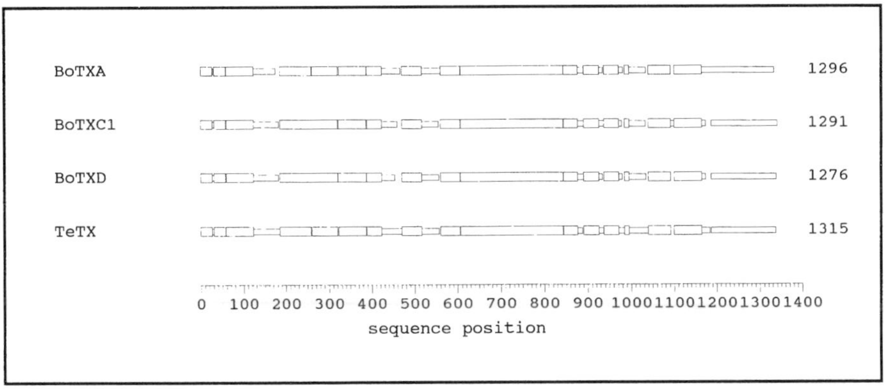

Figure 1. Schematic alignment of the primary structures of BoTX-A, C1, D and TeTX. The larger rectangles represent regions of similar residues. The N-terminal of the L-chain is on the left, and the C-terminal of the H-chain is on the right. The H-chain of BoTX-A starts at residue 448. Gaps were optimally inserted by MACAW.

a modified PAM-120 matrix.[32] Regions of amino acid similarity are represented as the wider blocks in Figure 1. Gaps were automatically inserted by MACAW to optimize the alignment. The GCG GAP program[33] was used to independently quantify the amount of similarity in the aligned peptide pairs (Table 1). Similarity in the single chain sequences of BoTX (A, C1, D) and TeTX ranged from 55.2 to 68.9 percent. The highest similarity scores (73, 69, and 61 percent) were obtained between BoTX-C1 and BoTX-D. The interchain comparison between the L- and the H-chains of TeTX and BoTX-A, respectively, showed about a 40 percent similarity (not shown). Because comparable percentages resulted when randomized sequences were used, data from the L- and H-chain comparisons were not considered to be significant. These results contrast with the conclusion that the two chains of TeTX are similar.[34] In contrast to the similarities in the primary structures between the BoTX serotypes and TeTX, only a short segment in the B-chain of DITX was similar to a region in the H-chains of the clostridial toxins. The importance of this region was revealed during the amphipathic analysis (see following text).

Hydropathic analysis. The KD algorithm and amino acid hydrophobicity scale developed by Kyte and Doolittle[27] were used in this portion of the study to obtain the first estimate of the number of possible TM segments in the clostridial toxin H-chains. The horizontal bars in Figure 2 show the extent of the three calculated hydrophobic regions in the N-terminal half of the BoTX type A H-chain. From the MACAW analysis, these regions, in general, were similar in the other toxins examined. For convenience, these three sequences will be referred to in this paper as toxin hydrophobic regions I (residues 596-610), II (624-643), and III (650-695).

Table 1
Comparison of Whole Toxin (W) and Light (L) and Heavy (H) Chains of Selected *botulinum* Toxin Serotypes and Tetanus Toxin

	BoTX-A(W)	BoTX-C1(W)	BoTX-D(W)	TeTX(W)
BoTX-A	1/1*	0.552/0.339	0.552/0.343	0.552/0.354
BoTX-C1	—	1/1	0.689/0.536	0.562/0.351
BoTX-D	—	—	1/1	0.553/0.341
TeTX	—	—	—	1/1
	BoTX-A(L) (1-448)†	BoTX-C1(L) (1-449)	BoTX-D(L) (1-445)	TeTX(L) (1-457)
BoTX-A(L)	1/1	0.526/0.345	0.544/0.341	0.528/0.321
BoTX-C1(L)	—	1/1	0.611/0.467	0.555/0.348
BoTX-D(L)	—	—	1/1	0.549/0.345
TeTX(L)	—	—	—	1/1
	BoTX-A(H) (449-1296)	BoTX-C1(H) (450-1291)	BoTX-D(H) (446-1276)	TeTX(H) (458-1315)
BoTX-A(H)	1/1	0.565/0.337	0.558/0.343	0.572/0.379
BoTX-C1(H)	—	1/1	0.730/0.572	0.566/0.354
BoTX-D(H)	—	—	1/1	0.561/0.345
TeTX(H)	—	—	—	1/1

*Values represent the fraction of similar/the fraction of identical amino acid residues for a given pair of aligned sequences.
†Values represent the range of amino acid positions.

The computational approach developed by Engleman, Steitz, and Goldman[28] was the second method used to estimate the number of TM segments. It is similar to the KD algorithm except that it uses a different amino acid scale and requires a larger window size. The results from the GES program shown in Figure 3 indicate that there are two potential membrane insertion regions within the BoTX-A H- chain. These regions have residues in common with hydrophobic region II (625- 630; labeled G1 in Figure 2) and the N-terminal half of hydrophobic region III (IIIa; 647-676; labeled G2 in Figure 2). Although the first region has too few residues to span a biological membrane, it was still considered to be an important area (see following text). As with the KD algorithm, similar results were obtained with the other toxins examined, which again emphasizes the similarities of the different BoTX serotypes when considering these TM potentially channel-forming segments.

Hydrophobic moment analysis. Transmembrane amphipathic regions were identified based on values of their sequence hydrophobic moment (see

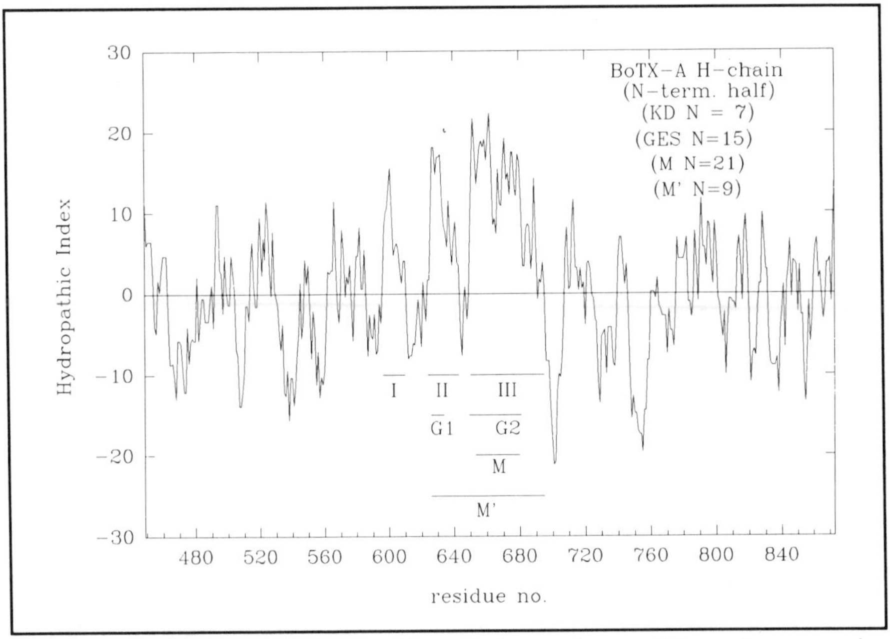

Figure 2. The Kyte-Doolittle hydropathic analysis reveals potential membrane-spanning sections on the N-terminal half of the H-chain in BoTX-A. Three relatively long hydrophobic regions are indicated by the horizontal lines labeled I, II and III. Window size (N) = 7. Regions calculated to be transmembrane by GES (G1 and G2) and MOMENT (M and M´) are also shown with their respective window sizes.

Equation 1) and hydrophobicity.[22] Both the consensus amino acid free energy scale derived by Eisenberg *et al.*[29] and the scale that was developed by us (see Methods) were used with MOMENT. Using the Eisenberg consensus scale, a single region (BoTX-A residues 625-695) was predicted by the MOMENT program to be amphipathic. This region corresponded to hydrophobic region IIIa (labeled M in Figure 2). In additional calculations, the region found by MOMENT using our novel amino scale also included hydrophobic region II (labeled M´ in Figure 2).

Although the low values of $<\mu_H>$ for BoTX could not be categorized as strongly reflecting either an amphipathic helix or ß-strand in these single regions, a helical arrangement for this segment was considered. One possible configuration of this amphipathic region was visualized by the helical wheel display.[35] Plots shown in Figure 3 illustrate the relative orientation of hydrophobic (boxed) and hydrophilic (unboxed) residues in regions II and IIIa (the N-terminal half of region III) of BoTX-A. These hydrophobic residues tend to be clustered within one quarter of a possible helical conformation, an arrangement that will be discussed next.

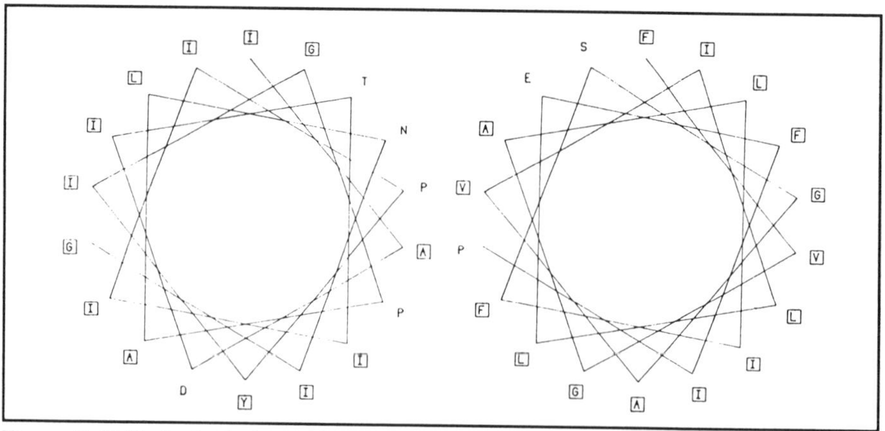

Figure 3. Helical wheel plots of BoTX-A show the relative orientation of the hydrophobic (boxed) and hydrophilic (unboxed) residues in region II (residues 627-644; left panel) and the N-terminal half of region III (652-669; right panel).

Additional sequence alignments. Interestingly, one of the two amphipathic regions in the DiTX heavy (B) chain (not shown) was found by MACAW to be similar to the single amphipathic sequence (region III) in both BoTX-A and TeTX. This alignment suggests that the channel-forming regions of these three different toxins may have some common structural features.

A final set of sequence comparisons was performed for which one result is illustrated in Figure 4. Amino acid residues in hydrophobic regions I and II of BoTX-A and TeTX are similar to those in the putative TM S2 and S3 segments of voltage-gated K^+ channels in vertebrate and invertebrate neurons.[36] In addition, there were significant matches between these toxin regions and the M2/M3 segments within the ß- and γ-subunits of peripheral nicotinic receptors[37] (not shown). The toxin hydrophobic regions, however, did not resemble the sequences that are considered to line the ionic conduction pathways and function as selectivity filters. Similarity between peptide toxins and membrane channels, nevertheless, lends plausible evidence in support of the hypothesis that amphipathic regions in these toxins are involved in the formation of ionic channels.

Discussion

Based on three different sets of calculations, a section containing the necessary hydrophobic and hydrophilic residues for channel construction was localized in the N-terminal portions of the H-chains of all the toxins examined.

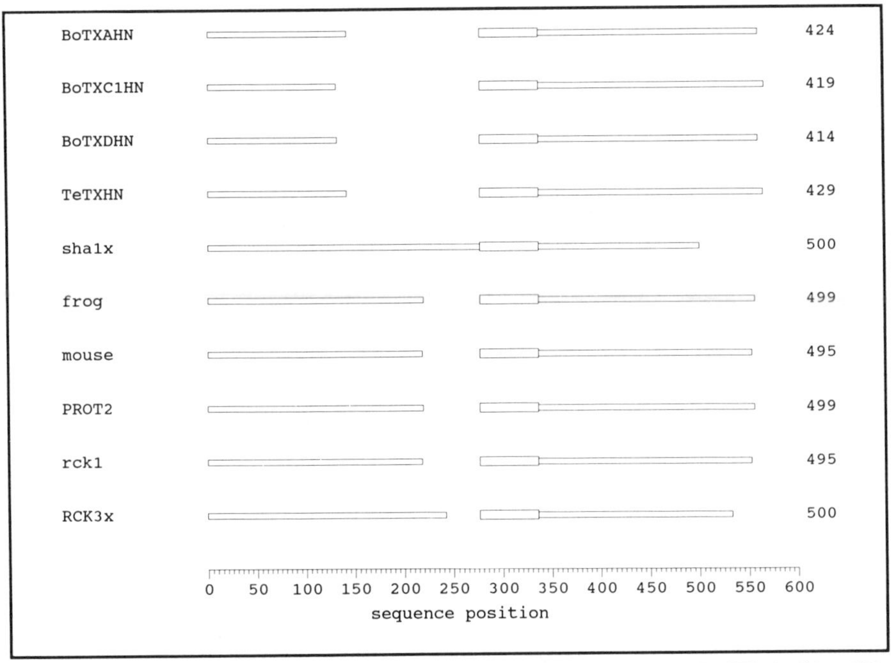

Figure 4. Sequences of amino acid residues in the hydrophobic regions of BoTX-A, C1 and D, and TeTX are similar to those of transmembrane segments S2 and S3 of peptides forming voltage-gated potassium channels from a wide variety of species. Only the N-terminal halves of the H-chains (HN) are illustrated. The residue positions are relative to their location on the H-chain.

Two regions within this amphipathic section could span biological membranes. It is not clear from these calculations, however, if these two regions are helices or ß-strands. Experiments with TeTX H-chain fragments[38] showed that the N-terminal portion of the H-chain is involved in a pH-dependent translocation process, and according to Binz *et al.*,[5] this region includes regions II and III as defined by the KD analysis.

Possible TM regions in the H-chains that were identified in this study lack the series of charged residues that exist in voltage-gated K^+ channels.[36] It is possible that the toxin segments serve only as structural components.[39] On the other hand, the importance of uncharged hydrophilic residues lining ionic channels has been emphasized by Eisenman and co-workers.[40] Because of the similarity of some residues with those in the TM regions of biologic channels, it is likely that these regions adopt similar secondary structures. Despite apparent similarities in peptide structure between toxins and ionic channels, the cautionary remarks of Doolittle[41] on spurious matches and inferences about ancestral relationships must also be considered.

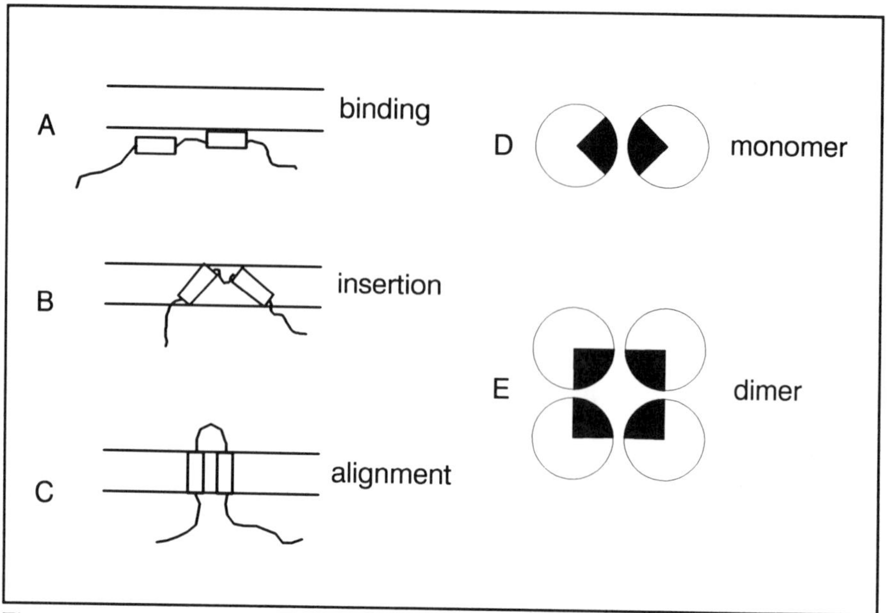

Figure 5. Two hydrophobic regions (rectangles) within the single amphipathic region are shown to bind (A), insert (B), and align (C) themselves starting on the *cis* side of the membrane. The resulting transmembrane topology predicts that both the N- and C- termini of the toxin H-chain are located on the *cis* side. Top view (D) shows how the hydrophilic residues (shaded areas) could be aligned. E: A dimer forms a membrane-embedded hydrophobic structure whose core contains hydrophilic residues.

The schematic diagrams in Figure 5 may be used as a starting point in portraying the formation of ionic channels by toxin H-chains. If only a relatively short portion of the H-chain traverses the membrane to form two antiparallel segments, it would be expected that both the N- and C-terminals would remain on the *cis* side of the membrane (Figure 5A-C). The single amphipathic region that was predicted was divided into two hydrophobic regions (II, IIIa). The steps for binding (5A), insertion (5B), and alignment (5C) are shown from a view lateral to the membrane (parallel lines). Further details in the alignment process are illustrated in the view normal to the membrane (Figure 5D). Shaded areas that correspond to the hydrophilic residues depicted in the helical wheel plots (Figure 3) are aligned toward the hydrophilic residues located within the opposite region.

From the electrophysiological evidence obtained with artificial membranes, the concentration-effect relation for DiTX and BoTX is a power function,[11,14] a result which suggests that there must be at least two toxin molecules involved in a conducting channel. The illustration of a dimeric arrangement in Figure 5E is based on the postulated topology of peptides that form voltage-[42,43] and agonist-gated channels.[44,45] These segments are depicted to have residues with hydrophilic side chains that line a core, interleaved between residues with hydrophobic side chains that face toward the lipid environment.[46]

Toxin insertion was viewed in our study as a spontaneous, thermodynamically favorable reaction, an approach that was prompted by the signal peptide hypothesis.[47] We have intentionally focused only on the toxin H-chains and have not incorporated other data that are available on the translocation process. For example, both chains of DiTX have been demonstrated to interact with lipids[17,48] or undergo structural alterations that occur with changes in pH.[48,49] Further, hydrophobic and other noncovalent interactions between the chains may also play a vital role. Misler[12] has suggested that the DiTX B-chain envelopes the A-chain either during synthesis or membrane insertion. Once this complex is inserted, the A-chain migrates to the *trans* side, leaving the B-chain within the membrane to serve as a residual structure that has properties of an ionic channel.

The main goal of this study was to localize regions involved in the ionic channels formed by the H-chains of BoTX and TeTX. Because the amino acid sequences in the H-chains of these toxins are more than 50 percent similar (see Table 1) and all are predicted to have hydrophobic and amphipathic TM segments in the same region (Figure 2), it is possible that the ionic channels formed will be structurally and functionally similar. We hypothesize that the toxin-induced ionic channels represent targets common to all of the diverse serotypes and that these structures serve as focal points for the development of therapeutic drugs. Rather than develop drugs that are unique for each serotype (because of their different antigenic, binding, and toxic properties), it is proposed that a single drug could be developed that would be an effective countermeasure against all of the serotypes that form ionic channels.

Acknowledgments

Staff support and allocation of computer time was generously provided by the National Cancer Institute's Advanced Scientific Computing Laboratory of the Frederick Cancer Research Facility.

References

1. L.L. Simpson, *Pharmacol. Rev.* **33**,155-188 (1981).
2. E. Habermann and F. Dreyer, *Curr. Topics Microbiol. Immunol.* **29**, 93-179 (1986).
3. J.L. Middlebrook, in *Botulinum Neurotoxin and Tetanus Toxin* (Academic Press, New York, 1989), pp. 95-119.

4. H. Niemann, in *Sourcebook of Bacterial Protein Toxins,* J.E. Alouf and J.H. Freer, Eds., (Academic Press, London, 1991), pp. 303-348.
5. T. Binz *et al., J. Biol.Chem.* **265**, 9153-9158 (1990).
6. K. Aktories and A. Hall, *Trends Pharmacol. Sci.* **10**, 415-418 (1989).
7. E.C. Dickson and E.J. Shevky, *J. Exp. Med.* **38**, 327-346 (1923).
8. A.C. Guyton and M.A. MacDonald, *Arch. Neurol. Psychiat.* **57**, 578-592 (1947).
9. R.J. Collier, in *ADP-ribosylation reactions*, D. Hayaishi and K. Ueda, Eds. (Academic Press, New York, 1982), pp. 575-592.
10. J.J. Donovan *et al., Proc. Natl. Acad. Sci. USA* **78**, 172-176 (1981).
11. B.L. Kagen, A. Finkelstein, and M. Colombini, *Proc. Natl. Acad. Sci. USA* **78**, 4950-4954 (1981).
12. S. Misler, *Biophys. J.* **45**, 107-109 (1984).
13. D.H. Hoch *et al., Proc. Natl. Acad. Sci. USA* **82**,1692-1696 (1985).
14. J.J. Donovan and J.L. Middlebrook, *Biochem.* **25**, 2872-2876 (1986).
15. R.O. Blaustein *et al., FEBS Let.* **226**,115-120 (1987).
16. C.C. Shone, P. Hambleton, and J. Melling, *Eur. J. Biochem.* **167**, 175-180 (1987).
17. C. Montecucco, G. Sciavo, and M. Tomasi, *Biochem. J.* **231**, 123-128 (1985).
18. J. Molgo *et al., J. Physiol. (Paris)* **84**, 152-166 (1990).
19. D. Eisenberg *et al., J. Mol. Biol.* **179**, 125-142 (1984a).
20. D. Eisenberg, R. Weiss, and T.C. Terwilliger, *Proc. Natl. Acad. Sci. USA* **81**, 140-144 (1984b).
21. J. Finer-Moore and R. Stroud, *Proc. Natl. Acad. Sci. USA* **81**,155-159 (1984).
22. F. Jähnig, in *Prediction of Protein Structure and the Principles of Protein Conformation,* G.D. Fasman, Ed. (Plenum Press, New York, 1989), pp. 707-717.
23. D.J. Lipman and W.R. Pearson, *Science* **227**, 1435-1441 (1985).
24. J. Devereux, P. Haeberli, and O. Smithies, *Nucl. Acids Res.* **12**, 387-395 (1984).
25. M. Kanehisa, J.W. Fickett, and W.B. Goad, *Nucl. Acids Res.* **12**, 149-158 (1984).
26. G.D. Schuler, S.F. Altschul, and D.J. Lipman, *Proteins Struct. Funct. Gen.* **8**,180-190 (1990).
27. J. Kyte and R.F. Doolittle, *J. Mol. Biol.* **157**, 105-132 (1982).
28. D.M. Engleman, T.A. Steitz, and D. Goldman, *Ann. Rev. Biophys. Biophys. Chem.* **15**, 321-353 (1986).
29. D. Eisenberg *et al., Faraday Symp. Chem. Soc.* **17**, 109-120 (1982).
30. D. Eisenberg, M. Wesson, and W. Wilcox, in *Prediction of Protein Structure and the Principles of Protein Conformation,* G.D. Fasman, Ed. (Plenum Press, New York, 1989), pp. 719-759.
31. H. Michel *et al., EMBO J.* **5**, 1149-1158 (1986).
32. M.O. Dayhoff, R.M. Schwartz, and B.C. Orcutt, in *Atlas of Protein Sequence and Structure,* M.O. Dayhoff, Ed. (National Biomedical Research Foundation, Washington, DC, 1978), Vol. 5, Suppl. 3, pp. 345-352.
33. S.B. Needleman and C.D. Wunsch, *J. Molec. Biol.* **48**, 443-453 (1970).
34. C.F. Taylor, P. Britton, and S. Van Heyningen, *Biochem. J.* **209**, 897 (1983).
35. M. Schiffer and A.B. Edmundson, *Biophys. J.* **7**, 121-135 (1967).
36. B.L. Temple *et al., Science* **237**, 770-775 (1987).
37. K. Imoto *et al., Nature* **335**, 645-648 (1988).
38. J.K. Rao and P. Boquet, *J. Biol. Chem.* **260**, 6827-6835 (1965).
39. H.R. Guy and F. Conti, *Trends Neurosci.* **13**, 201-206 (1990).
40. G. Eisenman, A. Oberhauser, and F. Bezanilla, in *21th Jerusalem Symposium on Quantum*

Chemistry, A. Pullman, J. Jortner, and B. Pullman, Eds. (Kluwer Academic, Dordrecht, Netherlands, 1989), pp. 27-50.

41. R.F. Doolittle, in *Prediction of Protein Structure and the Principles of Protein Conformation,* G.D. Fasman, Ed. (Plenum Press, New York, 1989), pp. 193- 301.

42. S. Oiki, W. Danho, and M. Montal, *Proc. Natl. Acad. Sci. USA* **85**, 2393-2397 (1988).

43. H.A. Hartmann *et al., Science* **251**, 942-944 (1991).

44. M. Montal, M.S. Montal, and J.M. Tomich, *Proc. Natl. Acad. Sci. USA* **87**, 6929-6933 (1990).

45. J.-C. Galzi *et al., Annu. Rev. Pharmacol.* **31**, 37-72 (1991).

46. M.S.P. Sansom, *Prog. Biophys. Mol. Biol.* **55**, 139-235 (1991).

47. G. Von Heijne, *Curr. Top. Mem. Transp.* **24**, 151-179 (1985).

48. J.X. Jiang, F.S. Abrams, and E. London, *Biochem.* **30**, 3857-3864 (1991).

49. V. Cabiaux *et al., J. Biol. Chem.* **264**, 4928-4938 (1989).

6

Acetylcholinesterase:

Cloning, Expression, Production, and X-Ray Crystallographic Structure

Jerald C. Sadoff
U.S. Army Medical Research and Development Command
Walter Reed Army Institute of Research
Washington, DC 20307-5100

Hermona Soreq
The Hebrew University
Jerusalem 91904
Israel

Joel Sussman
Israel Silman
The Weizmann Institute of Science
Rehovot 76100
Israel

Avigdor Shafferman
Israel Institute for Biological Research
Ness-Ziona 74050
Israel

Human acetylcholinesterase (HuAChE) is the target of all organophosphate chemical warfare nerve agents and may be developed as a prophylactic scavenger to prevent nervous system intoxication. To modify HuAChE so that it can rapidly hydrolyze nerve agents, it has been possible to clone, sequence, and express HuAChE at very high levels. Cloning was accomplished by identifying tissues that produce specific mRNA using oocyte injection, constructing libraries from these tissues, and probing them with a differential screening procedure based on presumed sequence analogy to *Torpedo* AChE. A series of eukaryotic multicistronic expression vectors were constructed and stably transfected human 293 cell lines produced. These cell lines produced >10 μg HuAChE/10^6 cells/24 hours,1000-fold higher than any other HuAChE-producing cell line. Procedures for purification and crystallization of *Torpedo* AChE were established. Unlike other serine esterases, its active site catalytic triad contains Glu instead of Asp, is of the opposite "handedness," and is located at the base of a 20-Å long, convoluted gorge that penetrates halfway into the enzyme, widens at the base, and is lined with 14 aromatic residues. Modeling of acetylcholine binding suggests that its quaternary ammonium ion binds to some of these aromatic residues. Site-directed mutagenesis confirmed that Ser^{203}, His^{447}, and Glu^{334} are involved in the catalytic triad of HuAChE and that Asp^{175} and Asp^{404} are involved in salt bridges critical for the folded state.

Introduction

Chemical warfare nerve agents are one of the most serious threats to U.S. military operations worldwide. Their use, or even the potential for their use, has enormous implications in terms of military operational capability, training, and equipment requirements as well as civilian political and psychological factors. Current strategies to protect military and civilian personnel against these airborne agents, which can kill a human being in less than a minute, are based on physical barrier protection and pharmacologic prophylaxis and treatment. Physical barriers such as gas masks and protective suits obviously impair operations and are subject to failure during battle. The pharmacologic prophylaxis and treatment drugs we currently have are only partially effective against the more potent agents and may have undesirable side effects if used at the wrong time or in an inappropriate dose.

Chemical warfare organophosphate nerve agents work by destroying the ability of the human enzyme acetylcholinesterase (HuAChE) to inactivate the neurotransmitter acetylcholine (ACh).[1] During normal synaptic transmission, within about 1msec, ACh is released from the presynaptic nerve cell, diffuses across the synaptic cleft, reversibly binds to the ACh receptor, and is hydrolyzed by HuAChE.[2] When HuAChE is poisoned by organophosphate nerve agents, ACh is not hydrolyzed and continues to stimulate nerves and muscles in an inappropriate manner, leading to respiratory paralysis, seizures, and rapid death.

Strategy

Recombinant produced HuAChE can be given prophylactically as a scavenger to bind nerve agents prior to their contact with HuAChE in the nervous system. HuAChE is the target of all nerve agents and is also the molecule with the highest affinity for the nerve agents. HuAChE should not be immunogenic because it is a natural human protein and should not have physiologic effects because it will not cross the blood-brain barrier. This concept has recently been validated in animal models where it has been shown that fetal bovine serum derived acetylcholinesterase (AChE) can act as an effective prophylactic scavenger against the powerful nerve agent Soman.[3] In order to use HuAChE as a prophylactic scavenger agent, it is necessary to clone, sequence, and express the enzyme at a high level.

It is also possible that HuAChE can be converted from an enzyme that is inactivated by nerve agents to an enzyme that can very rapidly hydrolyze them. Such an enzyme would be a nonpharmacologic prophylactic agent that could be given in much lower doses than HuAChE, with the capability of preventing nerve agent intoxication. In order to produce such a modified enzyme, it is necessary to understand the detailed structure of HuAChE and its interaction with AChE, nerve agents, and reactivating agents at an atomic level. This can be accomplished by X-ray crystallography and computer modeling studies linked to functional studies of mutant proteins produced by site-directed mutagenesis. These structure-function relationships can then be used to guide site-directed mutagenesis experiments in the construction of recombinant HuAChE capable of rapidly hydrolyzing organophosphate nerve agents. These studies may also be useful in the rational design of new treatment drugs that can displace nerve agents from HuAChE by determining contact residues and structural requirements for drugs of this nature.

Cloning of HuAChE

At the time these studies began, HuAChE could not be obtained in large enough quantities to determine protein sequence or produce specific antibodies. Furthermore, most cells make extremely low levels of HuAChE-specific messenger RNA. This made it difficult to construct cDNA libraries containing the full-length sequence, hard-to-screen expression libraries with weakly cross-reactive antibodies or to use the subtraction techniques that had been successfully used to clone other proteins under such circumstances.

Cloning of HuAChE was finally accomplished[4] by identifying tissues that expressed translatable HuAChE mRNA by microinjecting their mRNA into *Xenopus* oocytes and assaying for functional HuAChE. cDNA libraries from such tissues were then either constructed or purchased. A differential screening procedure using synthetic oligonucleotide probes was devised to find HuAChE-specific clones in these libraries based on the assumption that there would be similarities between the known sequence of *Torpedo* AChE[6] and HuAChE. A 5′-end labeled 35-mer oligodeoxynucleotide GTAChE, complimentary to the sequence that encodes the peptide Tyr-Met-Met-His-Trp-Lys-Asn-Gln-Phe-Asp-His-Tyr found in the carboxyl-terminal region of *Torpedo* AChE, was used for primary screening of the libraries. This probe was designed not to hybridize with cDNA for a similar human enzyme, butyrylcholinesterase (BuChE). False positives were eliminated by probing with OPSYNO, a mixed oligonucleotide probe designed to hybridize with the active site region of *Torpedo* AChE and presumably all cholinesterases.

This differential screening procedure resulted in the isolation of two similar 1.5kb clones designated NBG8A from a single library. Rescreening of several libraries with ^{32}P-labeled NBG8A resulted in the isolation of >25

Figure 1. Genomic DNA clone of HuAChE, comprising, clones of GNACHE, ABGACHE, and FEMA-CHE.

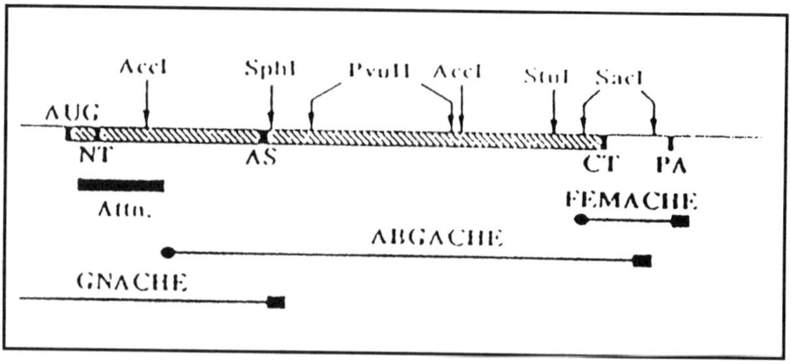

positive clones, all encoding nucleotides downstream from the expected initiator AUG codon. One 500-base pair clone, FEMACHE, included a complete 3´ nontranslated region ending with a polyadenylation site and a poly (A) tail. Another clone, designated ABGACHE, included most of the coding and 3´ nontranslated sequence. The location of these sequences can be seen in Figure 1.

To derive a complete HuAChE coding sequence, a 1/mer probe, 5´-[d(CG-CCC-ATG-GTA-CAC-GTC)]-3´, was designed according to the nucleotide sequence at the 5´ end of the clone NBG8A and used to screen a human genomic DNA library. The resultant genomic DNA clone, GNACHE (see Figure 1) included part of the coding region for HuAChE, intronic sequences, the absent upstream sequence encoding the amino terminus of HuAChE, and a putative signal sequence preceded by the AUG codon, which was imbedded in an appropriate consensus sequence for initiation of translation. The entire sequence derived from these overlapping clones is shown in Figure 2. Nucleotides are numbered at left, predicted amino acids (one letter symbols) are shown below the corresponding codons and are numbered at right. The active site sequence 231-240 is underlined. When a synthetic 2.2-kb mRNA transcript of the recombined sequence was microinjected into *Xenopus* oocytes, catalytically active HuAChE was produced.

Expression of Recombinant HuAChE

Once HuAChE was cloned and sequenced, attempts to develop an efficient mammalian cell system for its expression were begun. A multicassette mammalian expression vector for human recombinant acetylcholinesterase (rHuAChE) was designed and is shown in Figure 3. A-K shown in Figure 3 are unique restriction sites engineered into the vector to allow for maximum flexibility in construction.

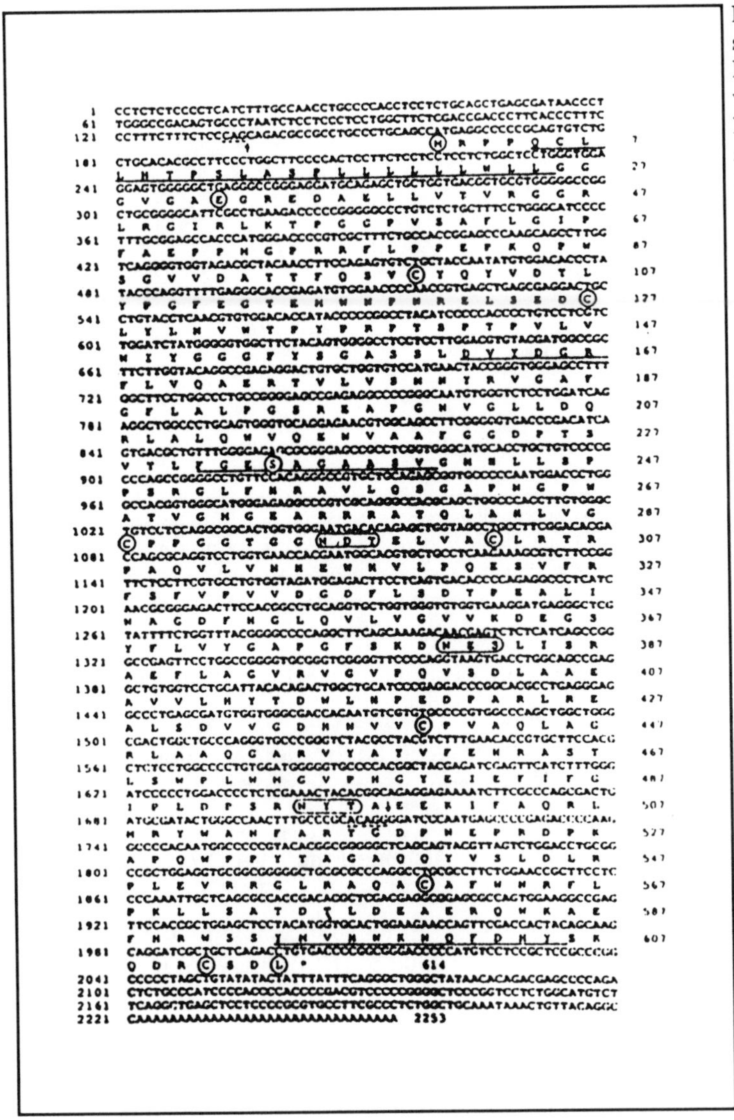

Figure 2. DNA sequence of HuAChE derived from overlapping clones shown in Figure 1.

The main genetic elements in this vector were: (a) HuAChE cassette, which includes the coding region of HuAChE, a SV_{40} poly A signal (from SV_{40}), and one of three efficient promoters of transcription: SV_{40} early, CMV early, or RSV-LTR; (b) Neo^r cassette, which includes the Neo^r gene, a poly A signal (from SV_{40}), and a weak transcription promoter—the mouse histocompatibility $H2L^d$ promoter. This cassette is used for selection of cells that have incorporated the vector, thereby making themselves resistant to the antibiotic G418 through expression of the Neo^r gene, which codes for C418 resistance;

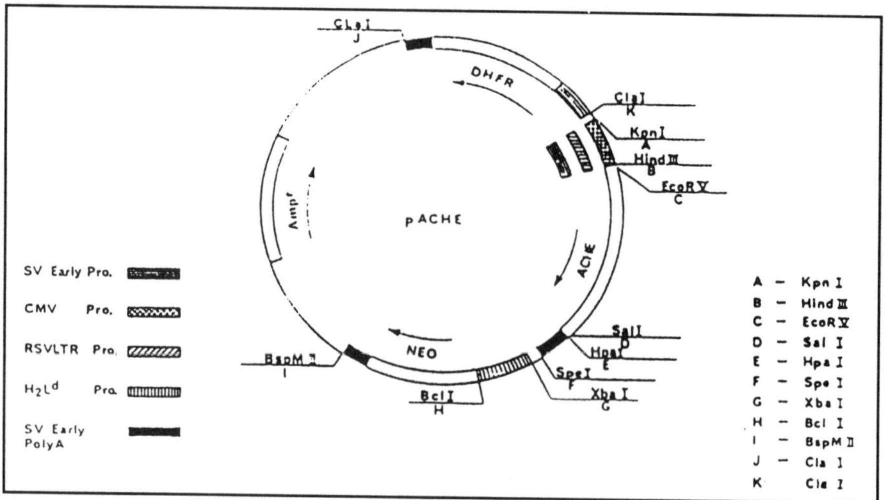

Figure 3. Design of multicassette mammalian expression vector of rHuAChe. Unique restriction sites (A-K) were engineered into a pGEM derivative to allow for maximal flexibility in construction/replacement of genetic element (such as *neo, dhfr, ache* enhancer-promoters, and poly A signals). Actual sites were selected by computer-aided sequence elimination procedure.

and (c) DHFR cassette, which includes the DHFR gene, an SV_{40} poly A signal, and the SV_{40} early promoter. This cassette was inserted in order to select for cells that have undergone gene duplication as evidenced by increased expression of the DHFR gene, which codes for resistance to the cytotoxic agent methotrexate.

Main considerations in the construction of the HuAChE cassette were:

1. Elimination of all sequences 5′ to the coding region, except for the translation initiation signals, which in rHuAChE appear to conform with known efficient signals.
2. Elimination of superfluous sequences 3' to the coding region.
3. Provision of sites upstream and downstream from the coding region, which allow insertion and shuffling of promoters and poly A signals.
4. Introduction of additional translation stop codons at the 3′ end.
5. Insertion of the minimal essential sequence of the polyadenylation signal of SV_{40} early poly A signal.

Over 40 plasmids were designed and constructed to produce expression plasmids with the elements shown in Figure 3. A transient transfection method was used for screening plasmids and cell lines for production of rHuAChE. A method for producing a medium with no intrinsic acetylcholinesterase activity was developed for this purpose. It was found in these experiments that the human embryonic kidney cell line 293, when transiently transfected with a number of our expression vectors, efficiently expressed catalytically active rHuAChE. Eighty percent of the enzymatic activity appeared in the cell growth medium as soluble rHuAChE in the monomeric, dimeric, and tetrameric globular hydrophilic forms of this enzyme.

Experiments were then performed to make stable cell lines producing high levels of rHuAChE. Plasmids utilizing the CMV and RSVLTR promoters were transfected into 293 cells. Transfected cells that had integrated the plasmid into their chromosome were selected by exposure to the neomycin analogue G418. Clones secreting rHuAChE at high levels were identified and expanded. The CMV promoter appeared superior in these experiments. From five independent transfection experiments, we obtained several recombinant 293 cloned cells that produced approximately 10μg/10^6 cells/24 hours. Generally, synthesis of over 1μg of foreign protein per 10^6 cells per 24 hours is considered to be a high level of expression,[7] and very efficient systems display synthesis of 5-10μg/10^6 cells/24 hours.[8] One of our cell lines, 293-R11, expressed 27μg/10^6 cells/24 hours. Therefore, our level of production is within or exceeds the range of maximal levels reported in other mammalian expression systems. To the best of our knowledge, these engineered cell lines have a production potential exceeding, by at least 1000-fold, any other HuAChE-producing cell line.

No increases in production by selection for gene duplication using MTXr-dhfr selection have been found. The copy number of chromosomally integrated cDNA, as revealed by Southern blotting of a number of stably transfected cell lines, was, however, related to rHuAChE production levels. We believe that in our stably transfected cell lines, 293 cells are expressing at the limit of their capability.

Using our expression system, we have been uniquely able to produce a large amount of highly purified rHuAChE. With this highly purified material we were able to accurately determine: Km (120uMK) and kcat values ((3.9 x 10^5) min^{-1}); substrate specificity (Acetylthiocholine : Butyrylthiocholine = 90:1, Acetylthiocholine : Propionylthiocholine = 2:1); sensitivity to various inhibitors (IC_{50} of BW264G51, which is a specific inhibitor of HuAChE = 0.008uM and the IC_{50} of BW284C51, which is a specific inhibitor of butyrylcholinesterase = 200uM; substrate inhibition (>1mM Acetylthiocholine); and a specific activity of 6000U/mg. Thus, our expressed rHuAChE has all the characteristics of a true AChE.

Amino acid sequencing of the NH_2 end of the purified protein revealed that the protein consisted of two molecular species differing at their amino end. The terminal 14 amino acid residues of one of these species corresponded to codons 32-44 of the HuAChE coding region, which suggests that the first

31 amino acid residues code for a signal peptide, which is subsequently removed. The sequence of the second species originates at a distance of two amino acid residues. An arginine residue occupies the +1 site of this species, while alanine and glyoine residues are located at -3 and -1 relative to the peptidase cleavage site, respectively. Thus, this minor cleavage site conforms with other known NH_2 processing sites of secreted proteins. To the best of our knowledge, this is the first demonstration of the existence of the previously postulated bidirected processing of a native protein in a eukaryotic system.

X-ray Crystallography of *Torpedo* AChE

Because HuAChE could not be obtained, efforts to crystallize AChE from *Torpedo californica* were carried out at the same time as our cloning and expression efforts. In *Torpedo*, a major form of AChE is a homodimer bound to the plasma membrane through covalently attached phosphatidylinositol (PI). This dimer can be selectively solubilized by a bacterial PI-specific phospholipase G.[9] Combining this mild procedure with affinity chromatography, we were able to obtain large amounts of pure, unnicked enzyme for crystallization. Extensive hanging drop experiments yielded conditions using ammonium sulfate in which single crystals grew within a few weeks to dimensions of 0.8mm by 0.4mm by 0.4mm. They belong to space group $P3_121$, contain one monomer per asymmetric unit, and diffract to 2.6Å resolution.[10]

The structure was solved by standard methods (see reference10). The sulfhydryl of Cys^{231} is the only free thiol in the enzyme and seemed to be a reasonable attachment site for a mercury derivative. We looked near the buried mercury atom for densities whose shape corresponded to the sequence CYS^{231}-Pro^{232}-Trp^{333}. When this sequence proved to fit the observed density satisfactorily, we built out from the tripeptide in both directions. The amino acid sequence could be fitted accurately to the electron density at all points. After crystallographic refinement, the model consisting of 526 residues and 71 water molecules had an R factor of 0.186.

The molecule has an ellipsoidal shape with dimensions approximately 45Å by 60Å by 65Å. It belongs to the class of alpha/beta proteins and consists of a 12-stranded central mixed beta sheet surrounded by 14 alpha helices. The overall structure is shown in the ribbon diagram in Figure 4 and the secondary structure cartoon in Figure 5 where the beta sheets are represented by gray arrows and the alpha helices by rods.

Early pharmacologic studies indicated that the active site of AChE contained two subsites, the "esteratic" and the "anionic" subsites,[11] which

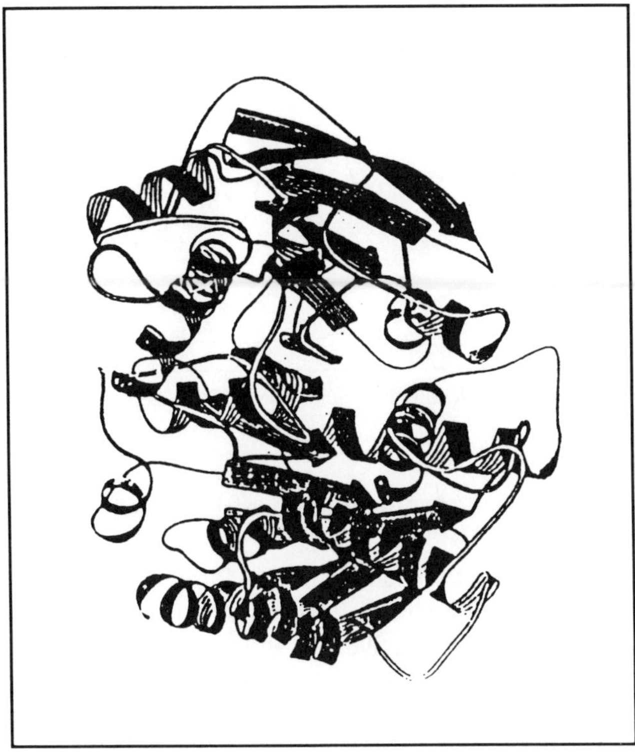

Figure 4. Ribbon diagram of structure of *Torpedo* AChE .

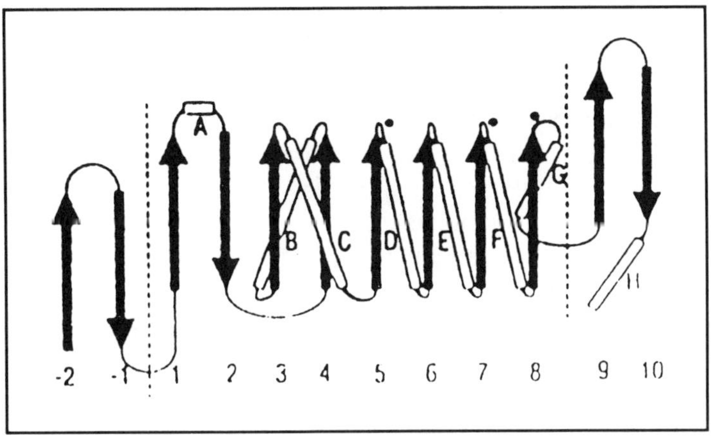

Figure 5. Secondary structure of *Torpedo* AChE showing beta sheets (arrows) and alpha helices (rods).

correspond to the catalytic machinery and the choline binding pocket, respectively. The esteratic subsite was believed to resemble the catalytic subsites of other serine hydrolases.[12] The ACh molecule is shown in Figure 6, spanning these esteratic and anionic sites of this widely accepted model. We have made

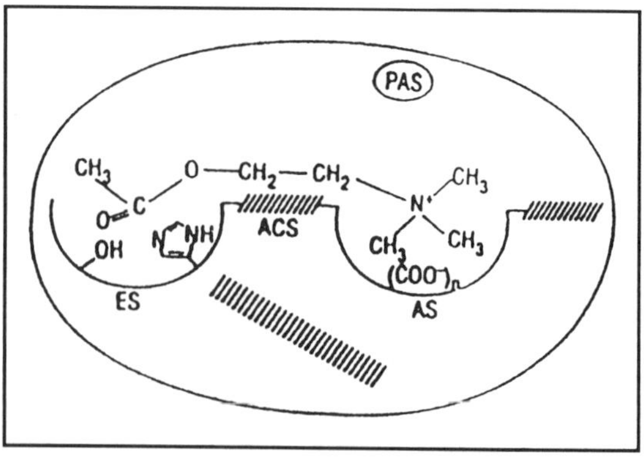

Figure 6. Schematic representation of the binding sites of AChe, based upon previous kinetic, spectroscopic, and chemical modification studies: ES, esteratic site; AS, anionic substrate binding site; ACS, active site-selective aromatic cation binding site; and PAS, peripheral anionic binding site or sites. Hatched areas represent putative hydrophobic binding regions. ACh molecule is shown spanning the esteratic and anionic sites of the catalytic center. Imidazole and hydroxyl side chains of His and Ser are shown within the esteratic site. Within the anionic site, $(COO^-)_n$ represents six to nine putative negative charges.

surprising discoveries concerning the nature of this model and each of these two subsites.

Organophosphate nerve agents poison AChE by forming a covalent bond to Ser^{200} in the esteratic active site of *Torpedo* AChE.[13] His^{440} has recently been implicated as part of the active site as well.[14] Our structure shows three residues, His^{440}, Ser^{200}, and Glu^{337}, that are all near one another and form a planar array that resembles the catalytic triad of chymotrypsin and other serine proteases.[15] The catalytic triad of *Torpedo* AChE, which we have discovered, differs from other serine proteases in that it contains Glu instead of the classical Asp in the triad and is of the opposite "handedness," that is, a mirror image of that seen in serine proteases. These new and unexpected findings predict that the "oxyanion hole",[15] which is formed by the amide NH of the active site Ser in the serine proteases (N-terminal), would differ in AChE. Molecular modeling studies suggest that the "oxyanion hole" of HuAChE would be formed by the main chain nitrogens of Ala^{204} and Gly^{122} interacting with the carbonyl oxygen and that of Gly^{121} interacting with the ester bond oxygen of ACh or its analogues.

The most remarkable feature of the three-dimensional structure is a narrow, deep, and convoluted gorge, about 20Å long, which penetrates halfway into the enzyme and widens out close to its base. This cavity has been named the "active site gorge" because it contains the AChE catalytic triad of amino acids centered around Ser^{200}. This gorge can be seen in Figure 7A. The O_γ atom of Ser^{200}, which is located approximately 4Å above the base of the gorge, can be seen from the surface of the enzyme. This is shown in Figure 7B.

Despite the structural complexity of the gorge and flexibility of ACh, manual docking experiments yielded a good fit of the extended all-trans conformation of ACh into the gorge. As shown in Figure 8, the acyl group of

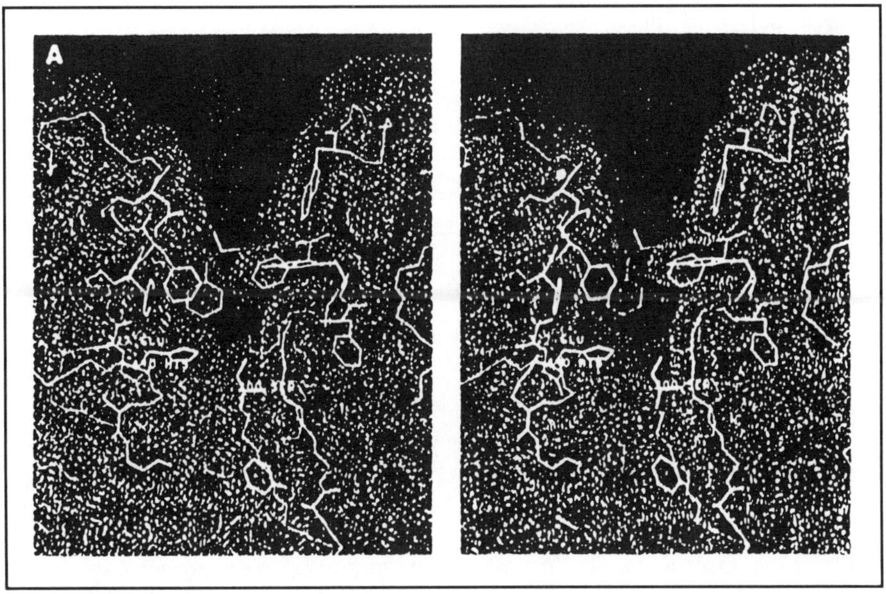

Figure 7a. Stereo van der Waals surface dot drawing of a thin slice of the AChE monomer showing the active site gorge containing the catalytic triad of amino acids centered around Ser^{200}.

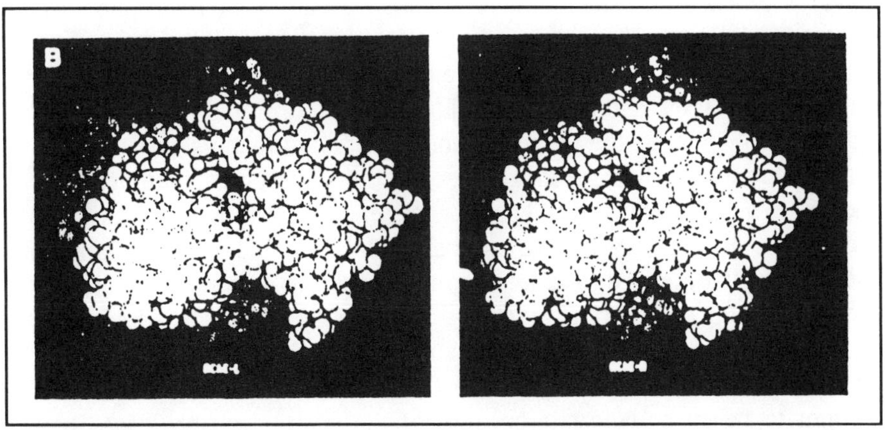

Figure 7b. Space filling stereo view of the AChE molecule looking down into the active site gorge. Ser^{200} appears as a black space in the middle of this view.

ACh is positioned to make a tetrahedral bond with O_γ of Ser^{200}, whereas the quaternary group of the choline moiety of ACh was within a van der Waals distance (approximately 3.5Å) of Trp^{84}. An affinity labeling study also recently identified Trp^{84} as part of the putative anionic (choline) binding site.[16]

Fourteen aromatic residues line approximately 40 percent of the surface

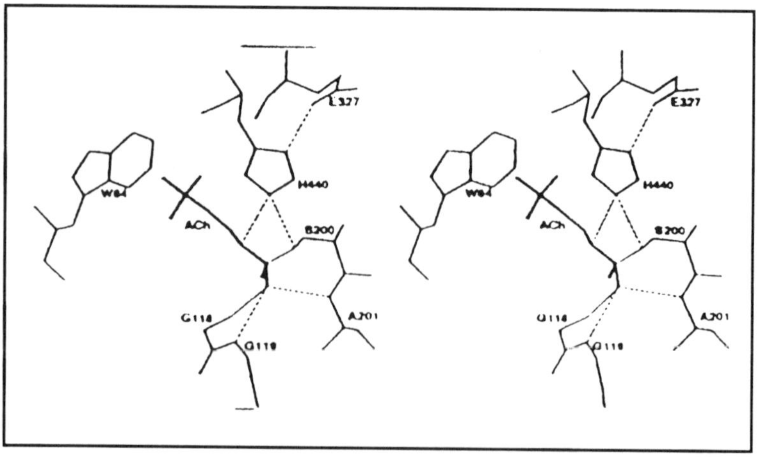

Figure 8. Acyl group of ACh is shown to form a tetrahedral bond with O_γ of Ser^{200} in the enzyme. Thin solid line represents bond produced in the transition state between Ser^{200} O_γ of AChE and the carbonyl carbon of ACh. Dotted lines indicate putative hydrogen bonds.

of the gorge. Only a few acidic residues are contained in the gorge; Asp^{285} and Glu^{273} at the very top; Asp^{72} which is hydrogen bonded to Tyr^{334} about halfway down, and Glu^{199} near the bottom. The high aromatic content of the walls and floor of the gorge, together with its dimensions, help explain why previous biochemical studies have revealed a variety of hydrophobic and anionic binding sites distinct from, or overlapping, the active site. The gorge is so deep and its aromatic surface so extensive that there are many ways and places for substrata, agonists, and inhibitors to bind. Certain ligands may be too bulky to penetrate the gorge but could partially block its entrance, while potent inhibitors such as the longer bisquaternary compounds might attach at one end to a peripheral site and at the other end to an aromatic residue in the wall of the gorge. Shorter bisquaternary inhibitors and oxime reactivators (which are the drugs we field for use in treatment of organophosphate nerve agent poisoning) might bind entirely within the gorge.

We believe the choline binding anionic subsite (see Figure 6) is clearly misnamed because it contains at most one formal negative charge. It was thought to be an anionic site because of the positive charge of ACh and numerous other potent ligands and the six to nine putative negative charges in the binding site.[17] AChE is one of the most active enzymes known, with an extremely fast-on reaction. If "electrostatic guidance" is not responsible for this high catalytic activity, how can it be explained? We postulate the existence of an "aromatic guidance mechanism." Recent data and theoretic considerations support a preferential interaction of quaternary nitrogens with the pi electrons of aromatic groups.[18] After initial binding of AChE to low affinity sites at the top of the gorge, the aromatic lining could then function by providing an array of low-affinity binding sites that would permit rapid diffusion to the active site.[19]

These X-ray studies have, therefore, demonstrated that (1) the previous concepts of the esteratic site needed to be modified in terms of its physical orientation and the existence of Glu and not Asp in the catalytic triad; (2) that the concepts of the anionic substrate binding site and peripheral anionic binding sites needed to be modified in favor of a deep aromatic lined gorge; and (3) that "electrostatic guidance" should be replaced with the concept of "aromatic guidance."

Site-Directed Mutagenesis of rHuAChE.

A replacement method that is of higher fidelity than methods involving replication by polymerases[19] was utilized for site-directed mutagenesis studies. HuAChE sequence variants that conserve the wild-type coding specificity but carry new unique restriction sites were created so that a synthetic DNA duplex carrying the mutated codon could replace the appropriate HuAChE fragment. Following transient transfection, specific activity of secreted enzyme was determined by directly measuring enzymatic activity and utilizing specific antibodies to determine rHuAChE protein mass. Expression levels were normalized by comparison to a *cat* reporter gene.

Evidence for the involvement of Ser^{203}, His^{447}, and Glu^{334} in the catalytic triad of rHuAChE was provided by substitution of these amino acids by alanine residues. Of 20 amino acid positions mutagenized thus far, these three were unique in abolishing detectable enzyme activity (<0.0003 of wild-type levels) in the presence of proper production, folding, and secretion. This is the first biochemical evidence for the involvement of a glutamate in a hydrolase triad and supports our X-ray crystal structure of *Torpedo* AChE. Attempts to convert rHuAChE to a Cys-His-Glu or Ser-His-Aps did not yield effective rHuAChE. Ser and Cys differ only by substitution of a sulfur in Cys for an oxygen in Ser, but this one atom difference eliminates all activity. Substitution of Asp^{74} by Gly or Asn generated an active enzyme with increased resistance to succinylcholine and dibucaine, thus mimicking in an rHuAChE molecule the phenotype of an atypical BuChE natural variant (Asp^{70} to Gly mutation). This demonstrates our ability to change the natural properties of rHuAChE with site-directed mutagenesis. Complete absence of secreted rHuAChE polypeptide was observed when Asp^{175} or Asp^{404} were substituted by Asn. These two aspartates play a role in the folded state of the polypeptide. This is supported by the X-ray structure that reveals the participation of these residues in salt bridges between neighboring secondary structure elements.

Conclusions

We have for the first time cloned, sequenced, expressed at high levels, and performed site-directed mutagenesis of HuAChE, which is the target of all organophosphate chemical nerve threat agents. We have also crystallized and solved the X-ray structure of *Torpedo* AChE and are currently working on crystallization of rHuAChE. These studies have provided us a much deeper understanding of the specific structure-function relationships of this enzyme at an atomic level and demonstrated our ability to manipulate this molecule and produce it at high levels. We have laid the foundation for use of rHuAChE in prophylaxis against nerve agent intoxication and the development of rHuAChE with hydrolytic activity for these organophosphate poisons.

References

1. P. Taylor in *Pharmacologic Basis of Therapeutics*, A.G. Goodman *et al.*, Eds. (MacMillan, New York, 1985), pp 110-129.
2. T.L. Rosenberry, *Adv. Enzymol. Relat. Areas Mol. Biol.* **43**, 103 (1975).
3. Y. Ashani *et al.*, *Biochemical Pharmacology* **41**, 37 (1991).
4. H. Soreq *et al.*, *Proc. Natl. Acad. Sci. USA* **87**, 9688 (1990).
5. Schumacher *et al.*, *Nature (London)* **319**, 407 (1986).
6. C. Prody *et al.*, *Proc. Natl. Acad. Sci. USA* **84**, 3555 (1987).
7. S.B. Yan, B.W. Grinell, F. Wold, *TIBS*. **14**, 264 (1989).
8. M.B. Hendricks, C.A. Luchette, M.J. Banker, *Biotechnology* **7**, 1271 (1989).
9. I. Silman and A.H. Futerman, *Eur. J. Biochem.* **179**, 11 (1987).
10. J.L. Sussman *et al.*, *Science* **253**, 872 (1991).
11. D. Nachmansohn and I.B. Wilson, *Adv. Enzymol.* **12**, 259 (1951).
12. T.L. Rosenberry, *Adv. Enzymol.* **43**, 103 (1975).
13. K. MacPhee-Quigley, P. Taylor, S. Taylor, *J. Biol. Chem.* **260**, 12185 (1985).
14. G. Gibney *et al.*, *Proc. Natl. Acad. Sci. USA* **87**, 7546 (1990).
15. T.A. Steitz and R.G. Shulman, *Annu. Rev. Biophys. Bioeng.* **11**, 419 (1982).
16. C. Weise *et al.*, *EMBO J.* **9**, 3885 (1990).
17. H.J. Nolte, T.L. Rosenberry, E. Neumann, *Biochemistry* **19**, 3705 (1980).
18. D.A. Dougherty and D.A. Stauffer, *Science* **250**, 1558 (1990).
19. L. Hedstrom *et al.*, *Methods in Enzymol.* **202**, 671 (1991).

7

Catalytic Antibodies as Biologic Scavengers for Organophosphorus Poisons

Alan A. Brimfield
David E. Lenz
Donald M. Maxwell
Clarence A. Broomfield
U.S. Army Medical Research and Development Command
Institute of Chemical Defense
Aberdeen Proving Ground, MD 21010-5425

Pharmacologic protection against organophosphorus (OP) anticholinesterases is probably near the optimum achievable. However, several drawbacks remain. Exposure to nerve agents still causes severe incapacitation even though lethality is reduced; timing for the administration of pre- and postexposure treatment is critical; some users suffer side effects from the pretreatment; and the treatment itself can be toxic. To combat these problems, mouse monoclonal antibodies capable of cleaving anticholinesterase agents to nontoxic fragments *in vivo* have been developed. Doing so required an analogue of the transition state assumed by OP compounds during hydrolysis. Such a structure required the design and synthesis of an unsymmetrically substituted, pentavalent phosphorus compound stable in an aqueous environment. To identify monoclonal antibodies arising from this hapten, an assay was developed in which protein from culture medium containing nascent hybridomas was trapped on a nitrocellulose membrane. This allowed the creation of conditions optimal for the measurement of catalytic activity. Application of these techniques yielded a mouse IgG2a antibody designated IIA12 ID10. Examination of this antibody using ^{14}C-soman revealed hydrolytic activity with a K_m of 0.33mM and a V_{max} of 25nmol/min/mg protein. This is the first catalytic antibody able to hydrolyze an OP anhydride and confirms the concept for an antibody capable of hydrolyzing nerve agents.

Introduction

Recent research has been reported on the use of injectable protein scavengers that can react with the toxic organophosphorus nerve agents on a one-to-one molar (stoichiometric) basis and remove them from circulation. When rhesus monkeys were given equine butyrylcholinesterase or fetal bovine serum acetylcholinesterase prior to soman, they were protected against 2 MLD_{50}[1] or up to 4 MLD_{50} of soman,[2] respectively, with no signs of toxicity or postexposure behavioral deficit. Equine butyrylcholinesterase also afforded protection in monkeys challenged with one MLD_{50} of sarin. These results have been reviewed recently.[3]

Although stoichiometric scavengers offered promising results, there were drawbacks to their use. Each molecule of scavenger reacted with only one molecule of nerve agent, thus requiring the administration of large doses of protein to achieve the protective effect. If protection against a recurring exposure to a chemical threat was desired, stoichiometric scavengers would have to be administered repeatedly, which is an additional disadvantage.

We have now begun work that capitalizes on the striking effectiveness of the biologic scavenger approach to protection by turning our attention to biologic scavengers that possess catalytic activity. Molecules of that type would render a toxic organophosphorus poison nontoxic by catalyzing its hydrolysis. Because they would repeatedly hydrolyze new substrate molecules, catalytic scavengers could provide protection with smaller doses of exogenous protein. Additionally, they would remain active for the length of their circulatory life and provide continuing protection against agent exposures. Although a recent report on the utility of using an enzymatic scavenger to protect mice[4] has been supportive of the concept of catalytic biologic scavengers, the human gene for this enzyme has not yet been expressed.

With the advent of this new concept, we felt that an examination of the potential of catalytic monoclonal antibodies as biological scavengers was well justified. The model on which our antibody was based was the hydrolysis of fluorophosphate and fluorophosphonate esters, which is known to proceed via a trigonal bipyramid transition state.[5] This report summarizes our efforts to date to produce a monoclonal catalytic antibody capable of hydrolyzing the highly toxic nerve agent soman.

Materials and Methods

Chemicals. Our transition state analogue (TSA) was 4-amino-3,3-dimethyl-2-butoxy-2-methoxy-2-methyl-1,3-dioxa (4,6-di-*tert*-butyl-2-benzo)

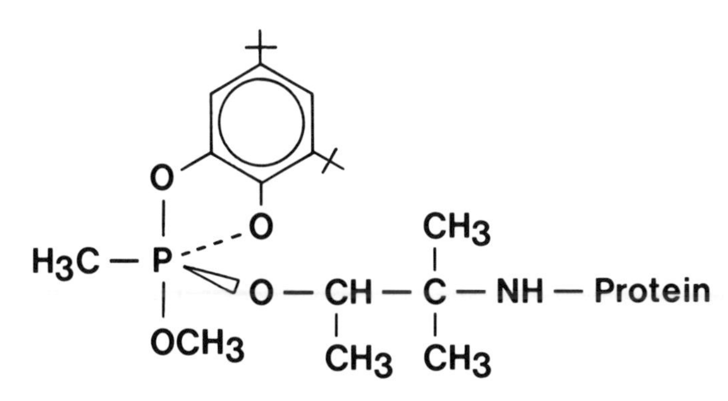

Figure 1. Structure of the transition state analogue.

phosphole synthesized under contract by R.M. Moriarty and his associates in the Department of Chemistry at the University of Illinois at Chicago[6] (Figure 1). The phosphole was conjugated to bovine serum albumin (BSA) through an amide linkage to yield an immunogen with an epitope density of 38mol/mol. An additional conjugate, 4-aminophenyl pinacolyl methyl-phosphonofluoridate diazotized to keyhole limpet hemocyanin (KLH-GD), was used in hapten affinity purification, and was obtained from Ashe-Stevens, Detroit, MI. Soman was obtained from the Chemical Research, Development and Engineering Center, Aberdeen Proving Ground, MD. ^{14}C-soman (pinacolyl[^{14}C]methylphosphonofluoridate; sp. act. 54.5mCi/mmol) with a purity greater than 90 percent by thin-layer chromatography was purchased from Los Alamos National Laboratory (Los Alamos, NM). Immunochemicals were purchased from Sigma Chemical Co. (St. Louis, MO). Other chemicals were of analytical grade and were commercially obtained.

Animals. Female BALB/c mice, 20-22g, were obtained from Charles River Laboratories (Wilmington, MA). They were housed in facilities approved by the American Association for the Accreditation of Laboratory Animal Care. They received Purina Lab Chow (Purina Mills Inc., Richmond, IN) and tap water *ad libitum.*

Immunization. Mice were immunized by IP injection with the transition state analogue conjugated to BSA carrier protein. They received 50μg (protein) of conjugate in saline emulsified in Freund's complete adjuvant in a final volume of 0.1ml. The animals were rested for two months and were boosted with 50μg of the same conjugate by the IP route in Freund's incomplete adjuvant. Three days prior to fusion, mice received a final boost of 50μg per animal, IP, in 0.85 percent saline. The state of immunization was monitored using ELISA on Immulon 2 microtiter plates (Dynatech Laboratories, Chantilly,

VA) coated with a TSA conjugate to human serum albumin, complementary to the BSA conjugate used for immunization. Detection employed the alkaline phosphatase system.

Fusion. Cell fusion for hybridoma generation was polyethylene glycol mediated (1300-1600 mol. wt.) with HAT selection according to a standard protocol.[7] The nonsecreting SP2/0-Ag-14 myeloma[8] employed as a fusion partner was a gift from Howard Meyer of our Institute.

Identifying catalytic monoclonals. Testing of the nascent hybridoma cultures to identify those producing antibody with the desired catalytic activity was carried out in a visually scored colorimetric screening assay based on the release of H^+ during soman hydrolysis and the resultant change in pH. It is a system developed in our laboratory by Little *et al.*,[9] modified for use with Multiscreen™ microtiter plates (Millipore Corp., Bedford, MA), and which have nitrocellulose membrane in the bottom of the wells in place of the plastic. Briefly, 200μl of tissue culture fluid was sterilely removed from the hybridoma culture plate and pipetted onto the membrane prewetted with saline. The sample was drawn through the membrane using a vacuum manifold, also supplied by Millipore, and the membrane was rinsed five times by filling the well with saline and drawing it through the membrane. Antibodies in the culture fluid are bound to the membrane, and potential sources of buffering action are washed away. A solution of 2×10^{-4} M soman in 0.85 percent saline containing 20μl/ml of methyl red (1mg/ml in ethanol) was added, and the plate was incubated at room temperature for several hours. A positive control was included consisting of a dilute solution of the organophosphorus acid anhydride hydrolase cloned from *Pseudomonas diminuta*[4] in fresh tissue culture medium. Fresh tissue culture medium alone served as a negative control. Positive wells were amber to rosy red in color, and negatives were lemon yellow.

Cloning. Cultures from positive wells were expanded and cloned by limit dilution.[10] Clones were reevaluated with the methyl red assay and ELISA. The most active clone, designated IIA12 ID10, was mass cultured for immunoglobulin production.

Immunoglobulin purification. Culture fluid was diluted 1:4 in Tris buffered saline (TBS) filtered through a 0.45 micron filter (Nalge Co., Baltimore, MD), and affinity purified by a modification of the method of Ey *et al.*[11] with KLH-GD replacing protein A as the immobilized ligand. Eluted immunoglobulin was dialyzed overnight against 4L of TBS, and concentrated against high molecular weight polyethylene glycol. Protein determination was by O.D. 280, using an extinction coefficient of 1.4/mg. For purposes of kinetic assay, some antibody preparations were exhaustively dialyzed against 0.85 percent saline.

Immunoglobulin typing. Antibody class and subclass determinations were done by Ouchterlony double diffusion assay using a commercial kit available from The Binding Site (San Diego, CA).

Kinetic analysis. The method used for our kinetic assays was described by Broomfield *et al.*[12] Briefly, enzyme and radiolabelled phosphate or phosphonate ester substrate are combined in a suitable low-volume container (microtiter plates were used) at the desired concentrations and incubated. Incubation is terminated by pipetting the contents of each reaction vessel into a tube containing 1ml of toluene and 1ml of distilled water. The tube is vortexed, the phases allowed to separate, and the phases are drawn off by Pasteur pipette and added to scintillation vials for quantification of the radioactivity in the aqueous and organic phases. In the assay presented here, four substrate concentrations were run in triplicate both with antibody (0.1ml, 39μg/mL in 0.85 percent saline) and without. The counts from the non-antibody containing wells were subtracted from the antibody catalyzed counts to correct for "nonenzymatic" hydrolysis. The data were analyzed by standard enzyme kinetic techniques using double reciprocal plots.

Results

The immunization regimen was successful. Sera from the immunized mice titered out to greater than a 1:100,000 dilution when they were tested on a microtiter plate coated with the HSA conjugate of the transition state analogue (Figure 2). No estimate of the fusion efficiency was possible because of bacterial contamination of the fusion which, was brought under control with antibiotics.

At two weeks post fusion, culture supernatant from the remaining fusion wells was tested for catalytic activity. Several produced positive results, but the cultures growing in two of the wells, designated II A3 and II A12, stood out. These were cloned immediately, and supernatant collection from expanded cultures was begun to initiate affinity purification. Additional assays for catalysis indicated that activity increased with purification on both the KLH-GD hapten and on protein A Sepharose, a protein isolated from *Staphylococcus aureus* that binds immunoglobulins independently of their specificity. The protein A indicated that the source of our activity was an antibody. Material originating with the II A12 culture and purified on each of these immunoadsorbants was tested for mouse immunoglobulin type and subclass. The results of this test were positive for the presence of mouse immunoglobulin, kappa light chain character (kappa is the most prominent species among

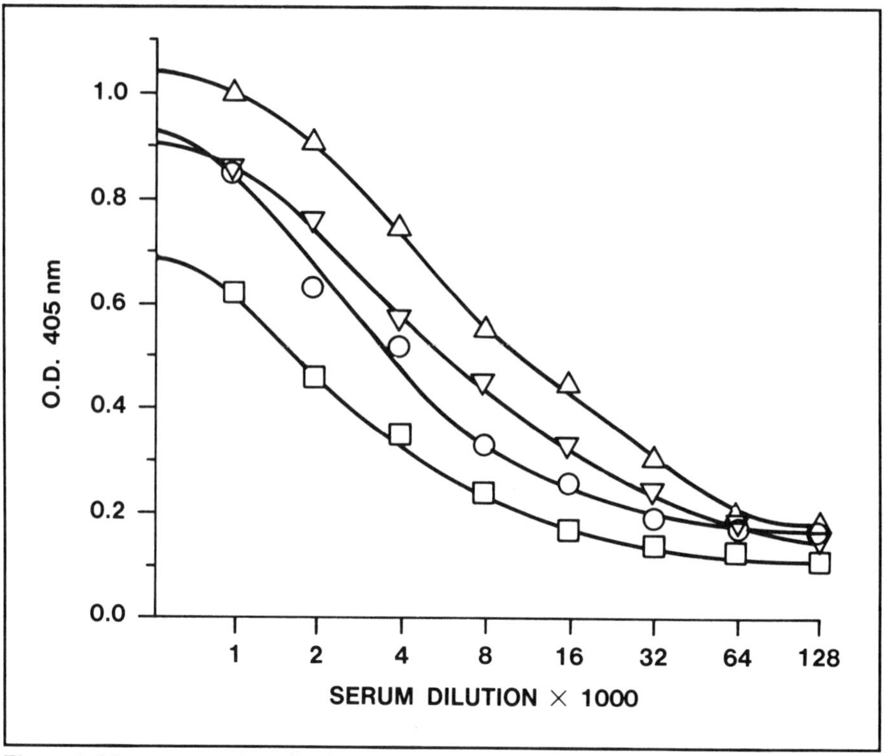

Figure 2. Mouse serum titer after immunization with the BSA-transmission state analogue.

the light chains produced by mice), and IgG2a. The outcome relative to type and subclass was the same for either method of purification.

The kinetic experiments were carried out with KLH-GD affinity purified material from one of the clones of the II A12 culture designated IIA12 ID10. The basic steps in the experiment are outlined in the section Materials and Methods. Additional details are found in the legend to Figure 3. Counts done prior to the start of the analysis showed that the substrate soman was 7.4 percent hydrolyzed. In the toluene-water system used to separate soman from the hydrolyzed product, pinacolyl methylphosphonic acid, the hydrolysis product, and its associated radioactivity are extracted into the aqueous phase. The results for the aqueous counts in the experimental and control samples were corrected to reflect this preexisting contamination. V_{max} equaled 25nmol min^{-1} mg^{-1} protein; Km was 3.3 x 10^{-4}M, and k_{cat}, defined as V_{max} divided by the total antibody concentration, was 4 min^{-1}.

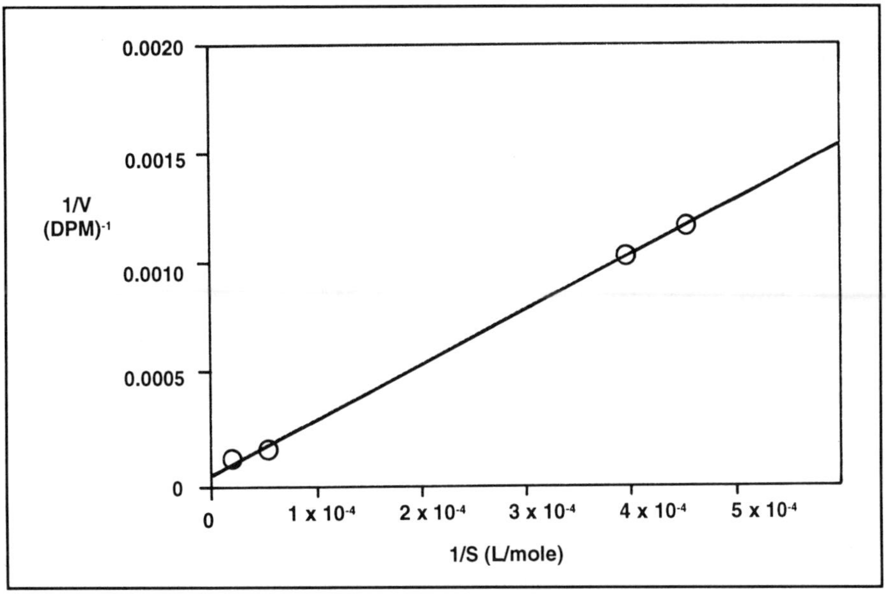

Figure 3. Double reciprocal plot of ^{14}C-soman hydrolysis by IIA12 ID10. Samples were run in triplicate. Other conditions were: temperature 26°C; 3.9µg of antibody protein per sample; molar soman concentrations, 5×10^{-4}, 1.8×10^{-4}, 2.5×10^{-5}, 2.2×10^{-5}, 120min incubations.

Discussion

Given our interest in proteins with catalytic activity, we felt that the recent development of catalytic monoclonal antibodies presented us with a unique opportunity. Immunoglobulins (antibodies) represent a totally different class of physiologically benign proteins that lend themselves to use *in vivo*. Antibodies in crude sera or in purified form have been used therapeutically for snake bite and to transmit passive immunity to individuals suffering from infectious disease.[13,14] Purified non-catalytic antibody preparations with targeted specificity have been shown to favorably alter the pharmacodynamics of toxic compounds.[15-17] The first catalytic monoclonal antibody was reported by Tremontano and his colleagues in 1986.[18] The technique depended on producing antibodies with specificity for a rationally designed structure representing the transition state for a modeled organic or enzymatic reaction. That first catalytic antibody hydrolyzed an aryl ester. Antibodies developed since then have catalyzed carbonate hydrolysis, amide hydrolysis, lactonization, aminolysis, and a Claisen rearrangement. The topic has been the focus of several recent reviews.[19,20]

The data reported here, although preliminary and subject to confirmation as greater quantities of IIA12 ID10 become available, suggest that a catalytic antibody capable of hydrolyzing the highly toxic nerve agent soman can be developed. The k_{cat} of 4 min^{-1} for our catalytic antibody compares favorably with turnover numbers reported for other catalytic antibodies.[19,20] Our results strongly support our selection of a trigonal bipyramid structure for a transition state analogue.

In an effort to minimize spontaneous base catalyzed hydrolysis there was no buffer included in this experiment. However, Broomfield *et al.*[12] showed that in the absence of buffer, hydrolysis was accelerated as the reaction proceeded due to the generation of hydrogen ions and a transition to acid catalyzed hydrolysis at the lower pH. We cannot be certain that the same phenomenon was not at work here. The outcome of this experiment under those circumstances would be worse than expected rather than better. Spontaneous hydrolysis would have increased, and the antibody catalyzed reaction would probably have slowed down. Therefore, we may see an increase in the k_{cat} when the reaction is repeated in a buffered system.

Due to a limited supply of the antibody, it was not possible to include a second set of control samples containing heat-inactivated antibody protein. This would have ensured that the activity observed was not attributable to a nonspecific function of the presence of protein in the experimental system. This is a valid concern. de Jong *et al.*[21] showed that serum proteins produce a finite increase in the hydrolysis of pure soman isomers in the absence of enzymatic activity. This effect will be addressed in future work.

The prospect of a catalytic antibody capable of catalyzing the hydrolysis of soman has the potential for practical application. Extensive research over the last several years has probably maximized the protection achievable against organophosphorus anticholinesterases using pharmacologic methods. Despite those advances, several drawbacks remain. Exposure to nerve agents still causes severe incapacitation even though lethality is reduced. Timing is critical for the administration of pre- and postexposure treatment to be successful. Finally, a portion of the population experiences undesirable side effects from the current pretreatment,[22,23] and, in the absence of agent exposure, the therapeutic compounds themselves can exhibit some toxicity.[24,25]

Work with the catalytic antibodies has not yet reached the point where we can actually perform *in vivo* tests of their protective potential. However, we are very optimistic as a result of the recent report of Broomfield's work[4] in which he successfully protected mice against the effects of twice the lethal dose of soman using a recombinant organophosphorus anhydride hydrolase from *Pseudomonas diminuta*. Our work is continuing with the use of additional protein conjugates that will allow us to evaluate the effectiveness of other transition state analogues.

References

1. C.A. Broomfield *et al.*, *J. Pharmacol Exp. Ther.* **259**, 633 (1991).
2. B.P. Doctor *et al.*, *FASEB J.* **5**, A1599 (1991).
3. D.M. Maxwell and B.P. Doctor, in *Chemical Warfare Agents*, S.M. Somani, Ed. (Academic Press, San Diego, CA, 1992), pp. 195-207.
4. C.A. Broomfield, *Pharmacol. Toxicol.* **70**, 65 (1992).
5. A. Fersht, *Enzyme Structure and Mechanism* (W.H. Freeman & Co., New York, 1985), pp. 236-237.
6. R.M. Moriarty, J. Hiratake, and K. Liu, *J. Am. Chem. Soc.* **112**, 8575 (1990).
7. J.W. Goding, *J. Immunol. Meth.* **39**, 285 (1980).
8. M. Schulman, C.D. Wilde, and G. Kohler, *Nature* **276**, 269 (1978).
9. J.S. Little *et al.*, *Biochem. Pharmacol.* **38**, 23 (1989).
10. D.P. Dumas *et al.*, *J. Biol. Chem.* **264**, 19659 (1989).
11. P.L. Ey, S.J. Prowse, and C.R. Jenkin, *Immunochem.* **15**, 429 (1978).
12. C.A. Broomfield, D.E. Lenz, and B. MacIver, *Arch. Toxicol.* **59**, 261 (1986).
13. J.B. Sullivan, *Med. Toxicol.* **1**, 47 (1986).
14. J-M. Sherrmann *et al.*, *Clin. Toxicol.* **27**, 1 (1989).
15. M.J. Hursting *et al.*, *Clin. Toxicol.* **27**, 53 (1989).
16. K.W. Hunter *et al.*, *J. Pharmacol. Exp. Ther.* **255**, 1183 (1990).
17. T.W. Smith *et al.*, *New Eng. J. Med.* **307**, 1357 (1982).
18. A. Tremontano, K. Janda, and R.A. Lerner, *Science* **234**, 1566 (1986).
19. A. Tremontano and D. Schloeder, *Meth. Enzymol.* **178**, 531 (1989).
20. K.M. Shokat and P.G. Schultz, *Ann. Rev. Immunol.* **8**, 335 (1990).
21. L.P.A. deJong, C. VanDijk, and H.P. Benschop., in *Enzymes Hydrolysing Organophosphorus Compounds*, E. Reiner, W.N. Aldridge, and F.C.G. Hoskin, Eds. (Ellis Horwood Publishers, Chichester, United Kingdom, 1989), pp. 65-78.
22. M.A. Dunn and F.R. Sidell, *J. Am. Med. Assoc.* **262**, 649 (1989).
23. J. Keeler, *Mil. Med.* **155**, 527 (1990).
24. V. Vojvodic, *Pharmacol. Clin.* **2**, 216 (1970).
25. G.R. Wenger, *J. Pharmacol. Exp. Ther.* **209**, 137 (1979).

8

Postexposure Prophylaxis Against Experimental Inhalation Anthrax

Arthur M. Friedlander
Susan L. Welkos
Margaret L.M. Pitt
John W. Ezzell
Patricia L. Worsham
Kenneth J. Rose
Bruce E. Ivins
John R. Lowe
Gerald B. Howe
Perry Mikesell
Wade B. Lawrence
U.S. Army Medical Research and Development Command Institute of Infectious Diseases
Fort Detrick
Frederick, MD 21702-5011

Inhalation anthrax is a rare disease that is almost invariably fatal. In this study, we determined whether a prolonged course of postexposure antibiotics with or without vaccination would protect when the drug was discontinued. Rhesus monkeys were exposed to a lethal aerosol dose of *Bacillus anthracis*. Beginning one day after exposure, groups of 10 animals were given penicillin, ciprofloxacin, doxycycline, doxycycline plus vaccination, vaccination alone, or saline. Antibiotics were administered for 30 days and then discontinued. Vaccine was given on Days 1 and 15 post exposure. Two animals died of causes other than anthrax and were not included in the statistical analysis. Nine of 10 controls and eight of 10 animals given vaccine alone died. There were no deaths due to anthrax in any antibiotic-treated animal while on therapy. Each antibiotic regimen provided significant protection upon discontinuance of the drug (penicillin, 7 of 10 survived, $p<0.02$; ciprofloxacin, 8 of 9 survived, $p<0.002$; doxycycline, 9 of 10 survived, $p<0.002$; doxycycline plus vaccination, 9 of 9 survived, $p<0.0002$). Thus, postexposure treatment of inhalation anthrax with antibiotics for 30 days beginning one day after exposure completely protected animals while on treatment and provided significant long-term protection upon discontinuance of therapy. Protection against rechallenge was provided by combining postexposure antibiotic treatment with vaccination.

Introduction

Anthrax is a zoonotic infection, mainly of herbivores, caused by *Bacillus anthracis*. Humans become infected by contact with infected animals or contaminated animal products. Anthrax begins by introduction of the spore through: (1) the skin, producing cutaneous anthrax; (2) the gastrointestinal tract, causing oropharyngeal or gastrointestinal anthrax; or (3) the respiratory tract, causing inhalation or mediastinal anthrax. The inhalation form of the disease was first recognized in the 19th century during the Industrial Revolution with the development of the textile and tanning industries and was one of the first occupational pulmonary diseases identified.[1] Workers in these industries were exposed to high concentrations of anthrax spores in the air, which were generated by the processing of contaminated wool, hair, or hides.[2] Presently, inhalation anthrax is extremely rare, with approximately 30 cases reported in this century, most often associated with industrial exposure to spores.[3,4] Inhalation anthrax has been almost uniformly fatal because of the difficulty in establishing the diagnosis and the rapid progression of the disease. Previous experimental studies demonstrated that treatment with penicillin for five or 10 days, beginning one day after aerosol exposure of rhesus monkeys, was protective during the drug therapy, but animals died when the antibiotic was discontinued.[5] Long-term protection was afforded only by combining penicillin therapy with postexposure vaccination. Recent events in the Gulf War have heightened the awareness of the possibility of using anthrax as a biologic weapon. For this reason, we determined whether a more prolonged course of antibiotic therapy with or without vaccination, begun after an aerosol exposure, could protect animals from lethal inhalation anthrax. We also determined if any long-term survivors were immune upon subsequent rechallenge.

Materials and Methods

Bacterial strain preparation, animals, and aerosol exposure. Spores of *B. anthracis* Vollum 1B were prepared and stored as previously described, except for the omission of centrifugation through Renografin gradients.[6] On the day of the experiment, spores were diluted in sterile water, heated at 60°C for 45 minutes, and then kept on ice. Rhesus monkeys (*Macaca mulatta*) of both sexes, weighing from 5.8 to 13kg, were anesthetized with tiletamine/zolazepam (3mg/kg, A. H. Robins Co., Richmond, VA) and exposed in a head-only 1cu ft plexiglass chamber contained within a Class III biologic safety cabinet to an aerosol generated with a Collison nebulizer for 10 minutes followed by a 5-minute air wash. The mass median diameter of the particles generated was 1.2μ, as determined by cascade impacter and aerodynamic particle sizer (Model 3310, TSI Incorporated, St. Paul, MN). The concentration of bacterial spores in the aerosol was determined during exposure for each monkey by sampling the aerosol with an all-glass impinger. The minute respiratory volume was measured on each anesthetized animal immediately before exposure. In the first challenge experiment, animals were exposed to an inhaled dose of $4.0 \pm 1.6 \times 10^5$ spores (mean ±SD), corresponding to approximately $8LD_{50}$ (Reference 5 and Welkos *et al.*, unpublished data, 1991). All survivors from the first experiment were rechallenged by aerosol with an inhaled dose of $2.6 \pm 1.4 \times 10^6$ spores corresponding to approximately $50LD_{50}$.

Experimental groups. The animals were distributed randomly by sex and weight in six experimental groups. (1) Controls: 10 animals were given saline as a control solution intramuscularly every 12 hours, beginning one day after exposure. (2) Penicillin: 10 animals were given procaine penicillin G intramuscularly at a dose of 180,000U (0.6ml) every 12 hours, beginning one day after exposure and continuing for 30 days. (3) Ciprofloxacin: 10 animals were given ciprofloxacin at a dose of 125mg (in 5ml H_20) by orogastric tube every 12 hours, beginning one day after exposure and continuing for 30 days. (4) Doxycycline: 10 animals were given doxycycline at a dose of 30mg (6ml) by orogastric tube every 12 hours, beginning one day after exposure and continuing for 30 days. (5) Doxycycline plus vaccination, 10 animals were given doxycycline at a dose of 30mg (6ml) by orogastric tube every 12 hours, beginning one day after exposure and continuing for 30 days. They were also given 0.5ml of the human anthrax vaccine (Michigan Department of Public Health, Lot No.FAV001) on Days 1 and 15 after the aerosol exposure. (6) Vaccination: 10 animals were given 0.5ml of the human anthrax vaccine on Day 1 and, if still alive, on Day 15 after the aerosol exposure. They were also given water by orogastric tube every 12 hours, beginning one day after exposure and continuing for 30 days. Animals given ciprofloxacin, doxycycline, or vaccination were anesthetized with tiletamine/zolazepam (3mg/kg) every 12 hours in order to give the medication by orogastric tube. After 30 days of treatment, antibiotics were discontinued, and the animals were given no additional therapy. Survivors were rechallenged 131 to 142 days after the initial exposure together with five additional new untreated controls.

Clinical, microbiologic, and pathologic studies. Daily blood cultures were obtained from the untreated controls and vaccination groups until death or for 14 days. In the antibiotic-treated groups, blood was cultured every other day until 80 percent of the controls died, then twice weekly until Day 30 when antibiotics were discontinued, then every other day until approximately Day 60, and then once a week until rechallenge. Blood from untreated animals was collected in an Isolator 1.5 (Du Pont Co., Wilmington, DE) and cultured in 10-fold dilutions in triplicate on trypticase soy agar. Blood from treated animals was collected in an Isolator 1.5 and was cultured undiluted and at a 1:100 dilution on trypticase soy agar. In addition, 1ml was cultured in a Bactec Peds Plus bottle (Becton Dickinson, Towson, MD). Blood obtained before and at various times after challenge was analyzed for IgG antibodies to the anthrax protective antigen by an ELISA with protein A-horseradish peroxidase (Boehringer Mannheim, Indianapolis, IN) and 2,2'-azinobis(3-ethylbenzthiazolinesulfonic acid) as substrate (Sigma, St. Louis, MO). Blood was also analyzed at various times for routine hematology and biochemistry.

All animals were observed at least twice daily until death. Moribund animals were killed by deep anesthesia (tiletamine/zolazepam, 6mg/kg) and exsanguination. All animals were autopsied. A diagnosis of anthrax was confirmed in all animals by isolating *B. anthracis* from the blood. Bacteria were identified by typical morphology, production of capsule when grown in 20 percent CO_2, and production of protective antigen as detected by a precipitin halo when they were grown on agar plates containing antibody to protective antigen. In some cases, organs were cultured quantitatively. In all deaths in which antemortem blood cultures were negative, cultures were obtained at autopsy of the blood, spleen, lung, liver, intrathoracic lymph nodes, and brain.

Antibiotic sensitivity testing and serum levels. Minimal inhibitory concentrations (MIC) of the *B. anthracis* Vollum 1B strain were determined in Mueller-Hinton broth dilutions by using an inoculum of 2.5 to 3.0 x 10^5/ml in tubes and in a microtiter format. The MIC was 0.08µg/ml for penicillin, 0.08µg/ml for ciprofloxacin, and 0.02µg/ml for doxycycline. The minimal bactericidal concentration was 0.32µg/ml for penicillin and 0.08µg/ml for ciprofloxacin.

Peak levels of antibiotics in the serum were determined one hour (ciprofloxacin) or two hours (penicillin and doxycycline) after a dose. Trough levels were measured 12 hours after a dose for all drugs.

Procaine penicillin G, penicillin G potassium used as a reference standard, and doxycycline monohydrate suspension were purchased from Pfizer, Inc. (New York, NY). Powdered doxycycline hyclate for a reference standard was a gift from Pfizer, Inc. (Groton, CT). Ciprofloxacin tablets and powdered ciprofloxacin for a reference standard were gifts of Miles Pharmaceuticals, Inc. (West Haven, CT).

The experiments were carried out under the guidance of the Veterinary Medicine Division in compliance with the Animal Welfare Act and other

Federal statutes and regulations relating to animals and experiments involving animals; the experiments adhered to principles stated in the Guide for the Care and Use of Laboratory Animals, NIH publication 86-23, 1985 edition. Our facilities are fully accredited by the American Association for Accreditation of Laboratory Animal Care.

Results

Description of the disease in untreated controls. Nine of the 10 untreated control animals exposed to an inhaled dose of $8LD_{50}$ died within three to eight days following the challenge (mean ± SE = 5.6±1.1 days). The animals were ill for one to four days before death, demonstrating decreased spontaneous activity, weakness, and anorexia. One animal had a single seizure on the day of death and was found to have meningitis on autopsy. Respiratory distress was observed in only one animal. Bacteremia at levels of 10 to 10^5 colony forming units (CFU) per millilter, was present for a mean of 1.8±0.9 days before death. Terminal bacteremias in eight of the nine animals that died varied from 10^4 to 10^9CFU per milliliter. The one animal with a low terminal bacteremia of 2 x 10^2CFU per milliliter had meningitis with 2 x 10^7CFU per gram of brain tissue. Five of nine animals had gross findings of mediastinitis and enlarged intrathoracic nodes with hemorrhagic lymphadenitis. A glistening edema in the mediastinum was also present in several animals. Meningitis was present in five of nine animals and was hemorrhagic in three of the cases. The one animal that survived never had a positive blood culture.

Antibiotic serum levels. The mean peak and trough serum antibiotic levels for each of the antibiotics did not vary significantly when measured on Days 5, 9, 20, and 30. The peak levels were at least 10 times the minimal inhibitory concentration (MIC) for all antibiotics, and the trough levels varied from one times the MIC for ciprofloxacin to 10 times for penicillin and doxycycline (D. Kelly *et al.*, manuscript submitted).

Effect of postexposure treatment on survival. Survival of the various treatment groups is shown in Table 1. Eight of 10 animals treated with vaccination alone died five to 10 days after challenge (mean, 7.0±1.5 days); which did not differ significantly from the untreated controls. The disease in the group given only vaccine was similar to that in the untreated controls. Those given vaccine alone were bacteremic for a mean of 1.9±0.7 days before death at the same low levels observed in the controls. The degree of terminal bacteremia and the findings on autopsy were also similar to those observed in the untreated controls. Of the two vaccinated animals that survived, one had persistently negative blood cultures, whereas the other had transiently positive

Table 1
Survival After Postexposure Treatment of Inhalation Anthrax

Group	Anthrax Deaths	p Value vs Control
Control untreated	9/10	
Vaccine alone	8/10	>0.1
Penicillin	3/10	<0.02
Ciprofloxacin	1/9*	<0.002
Doxycycline	1/10	<0.002
Doxycycline + vaccine	0/9†	<0.0002

* One animal died five days after exposure from aspiration pneumonia and had no evidence of anthrax on autopsy. This animal was excluded from analysis. A second animal died due to urethral obstruction 73 days after discontinuance of ciprofloxacin and had no evidence of anthrax at autopsy. This animal is included in the statistical analysis as a survivor.

† One animal died six days after discontinuing doxycycline, with no evidence of anthrax on autopsy. The cause of death remains unknown, and the animal was excluded from statistical analysis.

blood cultures on Days 5, 11, and 12 at low levels of 10 to 20CFU per milliliter. Blood cultures in this animal remained persistently negative after Day 12.

We observed significant protection against death in each of the antibiotic-treated groups. All animals in the penicillin group survived the 30 days while on treatment, during which their blood cultures remained negative. Upon discontinuation of the penicillin, three of 10 animals died of anthrax. Deaths occurred on Days 9, 12, and 20 after stopping the penicillin. In all three cases, death was due to anthrax.

One of the ciprofloxacin-treated animals died five days after exposure to anthrax from an aspiration pneumonia that developed 24 hours after the inadvertent introduction of drug into the trachea. All antemortem blood cultures were negative for anthrax, as were the postmortem cultures of blood, lung, liver, spleen, and brain. Pathologic analysis showed an aspiration pneumonia with no histologic evidence of anthrax in any organ. This animal was eliminated from analysis because no assessment of the effectiveness of antibiotic treatment on long-term survival was possible. All other ciprofloxacin-treated animals survived the 30 days of treatment with negative blood cultures. One animal died of anthrax with a positive terminal blood culture six days after stopping the antibiotic. This animal also had severe endometriosis on autopsy. Another animal in the ciprofloxacin group died 73 days after discontinuance of antibiotic. Autopsy cultures of blood and organs were negative for anthrax, and there was no other pathologic evidence of anthrax. This animal died of urethral obstruction due to rubbery plugs (concretions) in the proximal urethra and bladder. We considered this animal to be a survivor of the anthrax challenge.

In the group treated with doxycycline alone, all animals survived during therapy and had negative blood cultures. One animal died of anthrax 28 days after treatment was stopped.

None of the animals treated with doxycycline plus postexposure vaccination died of anthrax. One animal in this group died six days after discontinuation of doxycycline but had no evidence of anthrax on autopsy by either culture or histologic examination. Mild myocardial degeneration and necrosis were observed, but the exact cause of death could not be definitively determined. The effect of treatment on long-term survival from anthrax could not be evaluated in this animal, and it was eliminated from statistical analysis.

The gross and microscopic pathologic findings in the antibiotic-treated animals that died of anthrax after discontinuation of the drug did not differ from those observed in the untreated controls.

The animals that survived exposure to the aerosol challenge were examined for evidence of an immune response. Sera obtained 131 to 142 days after exposure were tested for antibody to the protective antigen component of the anthrax toxin. As seen in Table 2, none of the surviving animals treated with penicillin, ciprofloxacin, or doxycycline alone had a demonstrable immune response. In contrast, the exposed and surviving animals that had been given vaccine in addition to doxycycline therapy all developed a fourfold or greater rise in antibody to protective antigen. In addition, the two surviving animals that had been given vaccine alone also developed an antibody response to protective antigen. Thus, the only anthrax-challenged animals to develop an immune response were those that were vaccinated.

Resistance of surviving treated animals to rechallenge. Surviving animals were rechallenged with anthrax spores 131 to 142 days after the initial challenge. The dose was approximately $50LD_{50}$. As seen in Table 3, significant protection against rechallenge occurred only in the group that had been vaccinated and treated with doxycycline. There was no significant protection afforded by antibiotic treatment alone. Data not shown in Table 3 indicated that the two animals who were given vaccine alone without doxycycline and who survived the initial challenge also survived the rechallenge. Thus, the only animals to survive rechallenge were those that had been vaccinated. These surviving animals have now been observed for one year after the rechallenge and have remained free of disease.

Table 2
Serologic Response to Anthrax Protective Antigen in Survivors After Aerosol Challenge with Anthrax Spores

Group	≥ Fourfold Rise in Antibody Titer*
Control untreated	0/1†
Vaccine alone	2/2
Penicillin	0/6
Ciprofloxacin	0/7
Doxycycline	0/9
Doxycycline + vaccine	9/9

* Sera obtained 131 to 142 days after challenge were tested for antibody to anthrax protective antigen by ELISA as described in the Materials and Methods section.

† Number of animals positive/number of animals tested.

Discussion

The clinical and pathologic findings, including the duration of bacteremia preceding death, which we observed in rhesus monkeys after aerosol exposure to anthrax spores, are consistent with previously reported studies.[7,8] The observations are strikingly similar to those reported in cases of human inhalation anthrax,[3,9,10] a similarity which confirms the relevance of this experimental model. The most striking pathologic findings were the intrathoracic hemorrhagic lymphadenopathy, mediastinitis, and meningitis.

One of the control animals survived the initial aerosol challenge and appears not to have become infected because all blood cultures were negative, and an immune response to protective antigen never developed (see Table 2). There were two animals in the group given vaccine alone that also survived. One of these had a positive blood culture with low-level bacteremia. This suggests that postexposure vaccination alone may occasionally be protective. It is perhaps more likely that, on occasion, anthrax may not always be fatal in non-human primates. This is supported by the previous observations that two of four chimpanzees exposed to an aerosol of anthrax spores had transient low-grade bacteremias and survived.[11]

There were three deaths that were not due to anthrax. One of the ciprofloxacin-treated animals died of an aspiration pneumonia five days after starting treatment and had no evidence of anthrax at autopsy. One animal in the doxycycline plus vaccine group died six days after discontinuing the antibiotic on Day 36. There was no evidence of anthrax on autopsy, and the

Table 3
Resistance of Survivors of Postexposure Treatment to Rechallenge

Group	Anthrax Deaths	p Value vs. Control
Control*	4/5	
Penicillin	7/7	>0.1
Ciprofloxacin	6/7	>0.1
Doxycycline	9/9	>0.1
Doxycycline + vaccine	0/9	<0.005

* Controls consisted of five additional animals not previously exposed to anthrax.

cause of death could not be established. These two animals were removed from statistical analysis because the effect of the treatment regimen could not be adequately evaluated. The third animal was in the ciprofloxacin group and died 73 days after discontinuation of the antibiotic; this animal had no evidence of anthrax at autopsy. This animal died of urethral obstruction due to rubbery concretions. The relationship of this finding to the crystalluria induced by ciprofloxacin in nonhuman primates is unclear because the animal had not been taking ciprofloxacin for more than two months. We considered this animal to be a survivor of anthrax.

Critical to the rational treatment of inhalation anthrax is an understanding of the initial pathogenesis of the disease. The spore is the naturally infectious form of the organism. Studies by Ross[12] suggest that inhaled spores are rapidly phagocytosed by alveolar macrophages and transported by draining lymphatics to the regional lymph nodes where germination to the vegetative bacillus occurs. With release of bacilli from phagocytes, there is rapid extracellular bacterial proliferation throughout the node, with subsequent spread locally and to other organs. However, some of the inhaled spores remain dormant and do not germinate. Most important, from the perspective of therapy, these spores may remain within the lung for extended periods of time. Henderson *et al.* demonstrated that 42 days after inhalation of spores by rhesus monkeys, 15 to 20 percent of the initially retained dose of spores were still present in the lung.[5] The significance of this for treatment of anthrax was initially appreciated by Barnes[13] as early as 1947, who stated in reference to penicillin that spores may persist in the tissues and germinate after the blood level falls. The previous findings of Henderson *et al.*[5] on postexposure treatment of inhalation anthrax are consistent with this idea. They showed that animals treated for five or 10 days with penicillin remained free of disease while receiving the drug but died when the antibiotic was discontinued.

Our experiments clearly demonstrate that more prolonged antibiotic treatment for a 30-day period results in statistically significant long-term survival after discontinuation of treatment. Seven of 10 penicillin-treated, eight of nine ciprofloxacin-treated, and nine of 10 doxycycline-treated animals survived. This result supports the hypothesis that treatment with antibiotics alone will be successful if the treatment continues until the level of retained persistent spores falls to less than the infectious dose. However, the five deaths that were attributable to anthrax, particularly the animal in the doxycycline group who died 58 days after the initial aerosol exposure, directly support the concept that spores persist for prolonged periods in the host. The present data, taken together with the previously reported treatment failure with a short course of antibiotics,[5] suggest that an even more prolonged course of antibiotics might have prevented all deaths from anthrax in these experiments. Although the duration of postexposure antibiotic treatment required to prevent relapse is likely to be a function of the initial retained dose of spores, data are insufficient to determine accurately what the relationship is and how long treatment should be continued.

Our results did show that complete long-term survival, after discontinuation of antibiotics, occurred when postexposure antibiotic-treatment was combined with vaccination. This result confirms the previous reports of Henderson *et al.*[5] and Lincoln *et al.*[14] The survival of all nine animals in the doxycycline plus vaccine group did not differ statistically from that of the animals treated with antibiotics alone. Nevertheless, antibiotics plus vaccination appears to be the optimal approach to postexposure treatment of inhalation anthrax.

The serologic response to the initial aerosol challenge was determined by measuring antibodies to the anthrax protective antigen. As shown in Table 2, none of the animals treated with penicillin, ciprofloxacin, or doxycycline alone developed evidence of an immune response to anthrax. This suggests that the antibiotic treatment inhibited significant bacterial growth and prevented the infection from developing. The only animals that seroconverted after the aerosol challenge were those that had been vaccinated. Seroconversion does occur in humans who recover from anthrax.[15-18] Vaccination likewise induces serologic conversion in humans[16,19] and nonhuman primates (Ivins *et al.*, unpublished data, 1991).

The development of an immune response was found to predict resistance to rechallenge. As seen in Table 3, the only animals resistant to a second aerosol challenge with anthrax spores were those that had been vaccinated and had seroconverted. Antibiotics, which had protected animals against the initial infection, prevented the development of an effective immune response, leaving the animals susceptible to a subsequent infection. This agrees with a previous report in which surviving animals treated postexposure with antibiotics and by hyperimmunization with five doses of vaccine were protected upon rechallenge.[14] The protection afforded by vaccination before exposure is to be expected, based upon prior reports[20-22] as well as our experiments (Ivins *et al.*, unpublished data, 1991).

Thus, these results suggest that the optimal therapy for an unimmunized host exposed to an aerosol of anthrax spores consists of long-term suppressive antibiotic treatment together with vaccination to protect against both the initial challenge and any subsequent exposure.

References

1. F. M. LaForce, *Bull. N. Y. Acad. Med.* **54,** 956 (1978).
2. C. M. Dahlgren *et al., Am. J. Hyg.* **72,** 24 (1960).
3. P. S. Brachman, *Annals N.Y. Acad. Sci.* **353,** 83 (1980).
4. K. Vessal *et al., Clin. Radiol.* **26,** 471 (1975).
5. D. W. Henderson, S. Peacock, and F. C. Belton, *J. Hygiene* **54,** 28 (1956).
6. B. E. Ivins *et al., Infect. Immun.* **58,** 303 (1990).
7. C. A. Gleiser *et al., Brit. J. Exp. Pathol.* **44,** 416 (1963).
8. R. E. Lincoln *et al., J. Infect. Dis.* **115,** 481 (1965).
9. W. S. Greenfield, Eleventh Annual Report of the Medical Office of the Local Government Board, London (1881-1882), pp. 207-238.
10. W.S. Albrink *et al., Amer. J. Pathol.* **36,** 457 (1960).
11. W. S. Albrink and R. J. Goodlow, *Amer. J. Pathol.* **35,** 1055 (1959).
12. J. M. Ross, *J. Pathol. Bact.* **73,** 485 (1957).
13. J. M. Barnes, *J. Pathol. Bact.* **59,** 113 (1947).
14. R. E. Lincoln *et al., Antimicrobial Agents and Chemotherapy - 1964,* (American Society for Microbiology, 1965), pp.759-763.
15. T. M. Buchanan *et al., J. Immunol.* **107,** 1631 (1971).
16. P. C. B. Turnbull *et al., Infect. Immun.* **52,** 356 (1986).
17. T. Sirisanthana *et al., Am. J. Trop. Med. Hyg.* **39,** 575 (1988).
18. H. Winter and R. M. Pfisterer, *Schweiz. Med. Wochenschr.* **121,** 832 (1991).
19. A. Johnson-Winegar, *J. Clin. Microbiol.* **20,** 357 (1984).
20. F. C. Belton and R. E. Strange, *Brit. J. Exp. Pathol.* **35,** 144 (1954).
21. G. G. Wright, T. W. Green, and R. G. Kanode, *J. Immun.* **73,** 387 (1954).
22. H. M. Darlow, F. C. Belton, and D. W. Henderson, *Lancet* **II,** 476 (1956).

9

Environmental Pollution and Microbial Response to a Stressed Environment

A.M.Chakrabarty
Department of Microbiology
and Immunology
University of Illinois
College of Medicine
Chicago, IL 60612

The release of large amounts of agrochemicals, industrial chemicals, or chemicals used by the military as chemical warfare agents, explosives, and munitions has created massive pollution problems that require urgent solutions. Microbial recycling of natural compounds is a major route to pollution prevention. Unfortunately, synthetic compounds are only inefficiently degraded, if degraded at all, by natural microflora, and therefore evolution of degradative functions in microorganisms becomes of paramount importance. Not only do new genes need to be evolved, but their expression in the polluted environment needs also to be assured. It is therefore important to understand how a stressed environment influences microbial activities through evolutionary as well as sensory and metabolic processes. This chapter deals with two specific instances where either the evolutionary or the sensory/metabolic aspects of microorganisms are significantly affected by the environment.

Introduction

We live in an everchanging environment. With the introduction of herbicides and pesticides for enhanced agricultural productivity and the continued use of chlorinated solvents and fluorochlorocarbons in industrial and domestic activities, the level of pollution both in land and air is increasing, which is now a matter of great concern. Many of these pollutants are synthetic compounds that have only been manufactured during the last few decades and need to be removed from the environment. Microorganisms traditionally are responsible for the recycling of natural wastes and cannot readily handle synthetic compounds that are newly made. Such compounds are therefore quite recalcitrant to microbial attack and accumulate in the environment, creating problems of toxicity and health hazard. Such stressed environments, in turn, modulate microbial activities in a significant way. Two areas of microbial processes profoundly affected by the nature of the environment will be discussed: how the environment controls the evolutionary processes for biodegradative genes, and the extent of expression of some genes that allow microorganisms selective advantage in a stressed environment.

Environmental Modulation of Microbial Gene Evolution

As mentioned before, a large number of synthetic compounds have been released into the environment in the form of agricultural or industrial chemicals or for domestic use (paints, varnishes, fire retardants, and so on). Many of these chemicals are aromatic compounds containing chlorine substituents, rendering them recalcitrant to microbial attack. These compounds have, however, been released in massive amounts during the last several decades, so that natural microorganisms have been exposed to these chemicals for a fairly long time. It is thus pertinent to ask how microorganisms evolve genes for the degradation of synthetic chlorinated aromatic compounds. Indeed, many microorganisms can be isolated from nature that are capable of degrading simple chlorinated aromatic compounds such as chlorobenzoates, chlorophenoxyacetates, chlorophenols, and chlorobenzenes.[1,2] Because many of these compounds are converted to chlorocatechols before being metabolized further by soil microorganisms, an important question is whether the chlorocatechol genes evolved in these different strains would show organizational and sequence similarity with one another, as well as with the catechol degradative (*cat*) genes found in a variety of aromatic degrading microorganisms.[3] The pathways of catechol and chlorocatechol degradation, where the catechol or chlorocatechols are derived from different aromatic compounds, are shown in Figure 1. The enzymes involved in these degradative pathways and the genes encoding such enzymes are also shown. It should be noted that the chlorocatechol degradative genes (clcABD) evolved as part of a plasmid pAC27 in a strain of *Pseudomonas putida* isolated in the United States,[4] whereas the chlorocatechol degradative genes (*tfd*CDEF) evolved as part of a plasmid pJP4 in a strain of *Alcaligenes eutrophus* isolated in Australia.[5] Another *Pseudomonas* sp. strain P51 isolated in the Netherlands has been shown to harbor a plasmid pP51 encoding degradation of 1,2-dichloro-, 1,4-dichloro- and 1,2,4-trichlorobenzene,[6,7] the latter compound giving rise to the intermediate 3,4,6-trichlorocatechol. Essentially similar types of enzymes are involved in chlorocatechol degradation, specified by the three plasmids pAC27, pJP4, and pP51, that is, the pyrocatechase II type of enzymes encoded by *clc*A, *tfd*C, or *tcb*C, the cycloisomerase II (muconate lactonizing enzyme II) type of enzymes encoded by *clc*B, *tfd*D or *tcb*D and the hydrolase II type of enzymes encoded by *clc*D, *tfd*E or *tcb*E. It is important to note that these three plasmids were found in strains that were selected for three different compounds: 3-chlorobenzoate; 2,4-dichlorophenoxyacetate (2,4-D); and 1,2,4-trichlorobenzene, the metabolism of which resulted in three structurally different intermediate products: 3-chlorocatechol, 3,5-dichlorocatechol, and 3,4,6-trichlorocatechol (Figure 1). The three evolved pyrocatechase II enzymes, specified by *clc*A, *tfd*C, and *tcb*C reflected this difference by using different chlorocatechols as primary substrates.[7,8] This specialization of substrate specificities by the three pyrocatechase II enzymes is also reflected in subsequent steps,[8] suggesting that newer types of enzymes are produced by microorganisms in response to selection against different chlorinated compounds.

How did three chlorocatechol degradative pathways, having analogous enzymes with somewhat different substrate specificities, evolve in three different microorganisms isolated in three continents? Because natural microorganisms, particularly those of the genus *Pseudomonas*, can degrade the non-chlorinated parent compound catechol, how do catechol degradative enzymes or their genes compare with those of the evolved chlorocatechol degradative enzymes or their genes? Figure 2 shows the organization and nucleotide sequence identity of the catechol and chloro-catechol degradative genes cloned from the chromosome of *P. putida* (*cat*R-*cat*BC operon, reference 9), from the plasmid pAC27 harboring the *clc*R-*clc*ABD operon,[10] from the plasmid pJP4 harboring the *tfd*X-*tfd*CDEF operon,[11,12] and from the plasmid pP51 harboring the *tcb*R-*tcb*CDEF operon.[13] The organizational similarity among the four gene clusters is striking, as is the nucleotide sequence identity among the genes encoding isofunctional enzymes. It appears that natural evolution of the chlorocatechol genes, even when different chlorinated aromatic compounds were used as selecting agents, encompassed recruitment of genes analogous to the parental *cat* gene cluster, with divergence of individual genes that allowed for a different and broader substrate specificity.

We have also described[14] the chemostatic selection of a strain of *P. cepacia* (AC1100) capable of utilizing a recalcitrant compound such as 2,4,5-trichlorophenoxyacetic acid (2,4,5-T) as its sole source of carbon and energy. Because 2,4,5-T is normally a recalcitrant compound, the chemostatic selection involved continuous propagation for 10 months of a large number of microorganisms isolated from toxic dump sites with 2,4,5-T as the major source of carbon. Unlike natural evolution, where a variety of carbon sources

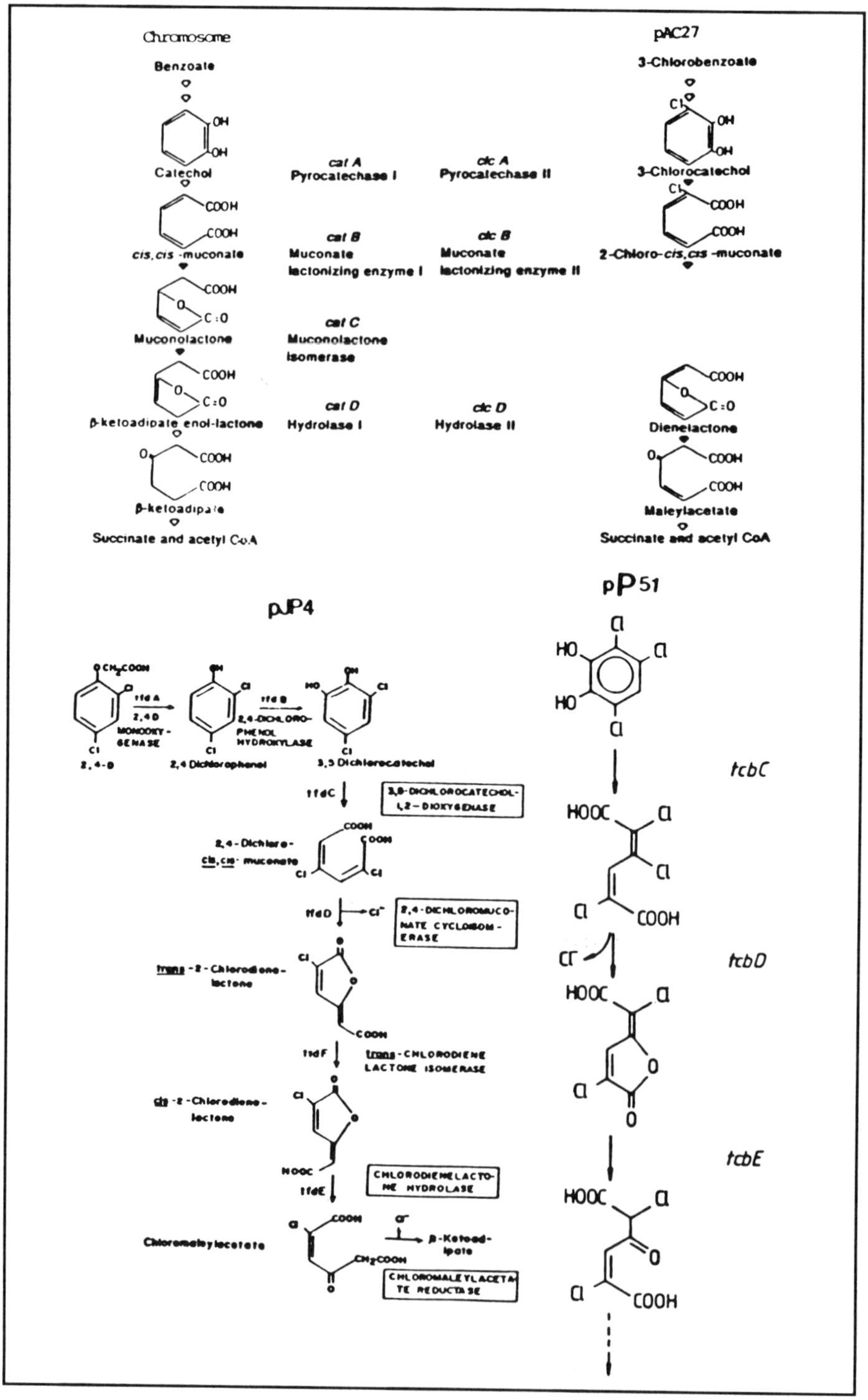
Chromosome
pAC27
Benzoate
3-Chlorobenzoate
Catechol
cat A
Pyrocatechase I
clc A
Pyrocatechase II
3-Chlorocatechol
cis,cis -muconate
cat B
Muconate
lactonizing enzyme I
clc B
Muconate
lactonizing enzyme II
2-Chloro-cis,cis -muconate
Muconolactone
cat C
Muconolactone
isomerase
β-ketoadipate enol-lactone
cat D
Hydrolase I
clc D
Hydrolase II
Dienelactone
β-ketoadipate
Maleylacetate
Succinate and acetyl CoA
Succinate and acetyl CoA
pJP4
pP51
tcbC
tcbD
tcbE

Figure 1. (opposite page) Pathways of degradation of catechol; 3-chlorocatechol; 3,5-dichlorocatechol; and 3,4,6-trichlorocatechol encoded by chromosomal genes and plasmids pAC27, pJP4, and pP51, respectively. Genes in each step of pathways are indicated, as are enzymes in first three pathways. Type II enzymes involved in 3-chlorocatechol and 3,5-dichlorocatechol utilization are also involved in 3,4,6-trichlorocatechol dissimilation.

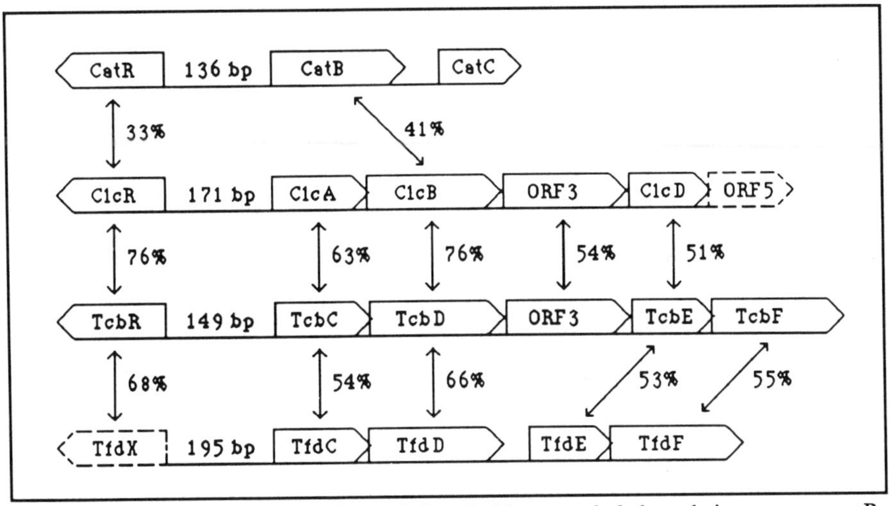

Figure 2. Genetic organization of catechol and chlorocatechol degradative operons *cat*R-*cat*BC, *clc*R-*clc*ABD, *tcb*R-*tcb*CDEF, and *tfd*X-*tfd*CDEF. Extent of amino acid sequence identity among two proteins encoded by two different operons is shown (arrows). Intervening distance between translational starts of regulatory gene and structural gene cluster shown in base pairs.

are available, the chemostatic selection provided a highly stressed environment with 2,4,5-T as the only major carbon source. A tentative pathway of 2,4,5-T degradation in *P. cepacia* AC1100 is shown in Figure 3A. Unlike the chlorinated catechols that are normally formed from chloroaromatic compounds, including 2,4-D (Figure 1), *P. cepacia* AC1100 metabolizes 2,4,5-T through a chlorohydroquinone intermediate. At least two of the 2,4,5-T degradative (*tft*) genes have been cloned,[15,16] both showing the presence in the upstream region of a repeated sequence RS1100, which shows little homology with *Pseudomonas* genomic digests.[17] The RSll00 has been shown to be a transposable element with outwardly directed promoters,[18] implying its potential role as a gene recruitment vector. The *tft* genes also exhibit little homology with genomic digests from a number of *Pseudomonas* species.[15,17] In addition, sequencing of one of the *tft* genes (Figure 3B) shows the presence of open reading frames demonstrating homology to glutathione reductase and glutathione S-transferase genes. Such genes are normally involved in xenobiotic degradation in eukaryotic or non-pseudomonad type of prokaryotic organisms,[19,20] which suggests that the *tft* genes recruited by *P. cepacia* AC1100

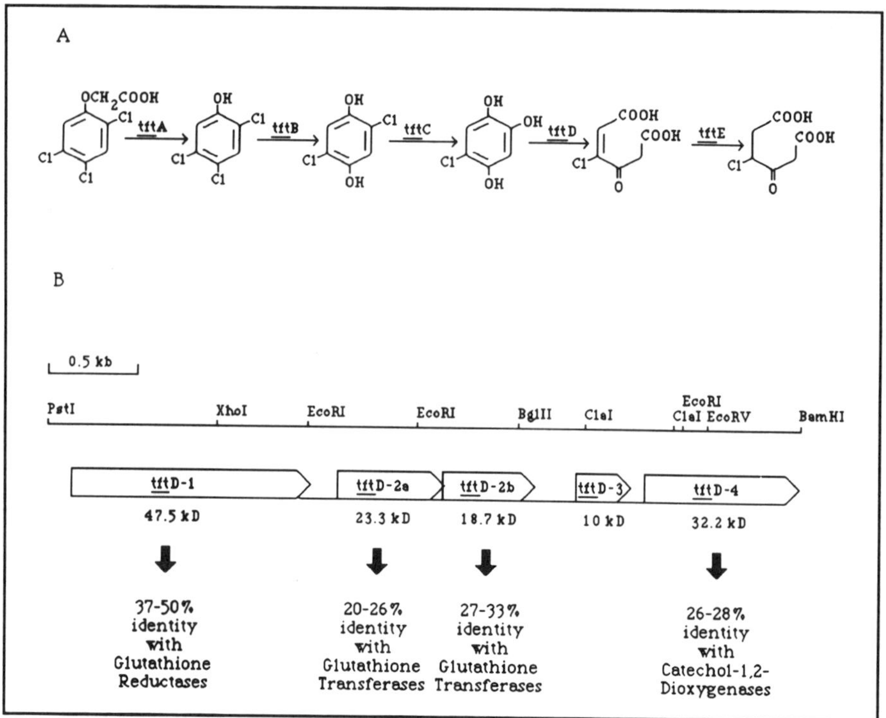

Figure 3, A. Tentative pathway for dissimilation of 2,4,5-T with gene designations for each step. **B.** A *Pst*I-*Bam*HI chromosomal fragment harboring the degradative gene cluster *chQ* (see reference 15) has been completely sequenced which shows identity to genes encoding glutathione reductases, glutathione S-transferases and catechol 1,2-dioxygenases.

may be of foreign ancestry. As contrasted with the extensive homology in evolved genes seen during natural evolution, lack of homology in the evolved *tft* genes with the chromosome of *P. cepacia* AC1100 appears to suggest that the directed evolution occurring over a short period of time under intense selection in a highly stressed chemostatic environment allows recruitment and facilitated expression of degradative genes without regard to any genetic homology with the host genome. The selective environment may therefore greatly influence the process of genetic evolution in microorganisms.

Environmental Regulation of Microbial Gene Expression

Microorganisms live in a changing environment. While release of toxic chemicals in the environment provides the microorganisms with opportunity to degrade new substrates by allowing them to evolve new genetic functions, such toxic compounds may also create problems of toxicity and may thus negatively affect the microorganism's survival and proliferation in that environment. Microorganisms may therefore have to switch on a completely new set of gene(s) to protect themselves in a highly stressed environment. An interesting example of how microorganisms react to an external environment for their survival is the infection of the lungs of cystic fibrosis (CF) patients by mucoid strains of *P. aeruginosa*.[21] The CF patients have deficiencies in the chloride channel, leading to accumulation of high levels of NaCl and a dehydrated, sticky mucus. Nonmucoid *P. aeruginosa* cells infecting the lungs of CF patients become heavily mucoid during proliferation in this environment. It is noteworthy that while *P. aeruginosa* causes other infections such as in burn, eye, or urinary tract, they are seldom mucoid, suggesting that mucoidy in *P. aeruginosa* is uniquely related to their proliferation in the CF lung environment. The mucoidy is due to the production of an exopolysaccharide termed alginate, which is a linear co-polymer of D-mannuronate and its C-5 epimer L-guluronate (Figure 4A). The mannuronate residues are usually epimerized to the guluronate residues either at the monomer or at the polymer level. Biosynthesis of alginate (Figure 4B) therefore involves the conversion of a glycolytic intermediate such as fructose 6-phosphate to guanosine diphosphate (GDP)-mannuronic acid by a series of enzymes such as phosphomannose isomerase (PMI, product of the *alg*A gene); phosphomannomutase (PMM, product of the *alg*C gene); GDP-mannose pyrophosphorylase (GMP, product of the *alg*A gene, which also encodes PMI); and GDP-mannose dehydrogenase (GMD, product of the *alg*D gene). GDP-mannuronic acid is then polymerized, epimerized, acetylated, and excreted as the exopolysaccharide alginate. The primary function of alginate appears to be the encapsulation of the infecting cells to protect them from the body's immune response as well as antibiotic therapy. The alginate-encapsulated mucoid cells are also believed to be better able to adhere to the epithelial cells,[22] thereby effectively colonizing the CF lung through formation of microcolonies. Indeed, *P. aeruginosa* is known for its ability to form biofilms on environmental surfaces, and alginic acid may be involved in biofilm formation as well as corrosion of metal surfaces.[23]

How does the environment of the CF lung trigger alginate formation? The alginate biosynthetic (*alg*) genes are normally silent in nonmucoid *P. aeruginosa*, as two critical *alg* promoters, *alg*C and *alg*D, are minimally active. Under specific environmental conditions such as high osmolarity, both the *alg*C and *alg*D promoters become activated.[24,25] The *alg*D gene maps at 34min of the *P. aeruginosa* chromosome at one end of the alginate biosynthetic gene cluster (Figure 4C), while the *alg*C gene maps at the 10min region near a cluster of alginate regulatory genes such as *alg*Rl, *alg*R2, and *alg*R3 (Figure 4C). Environmental factors such as nutrient limitation,[26] membrane

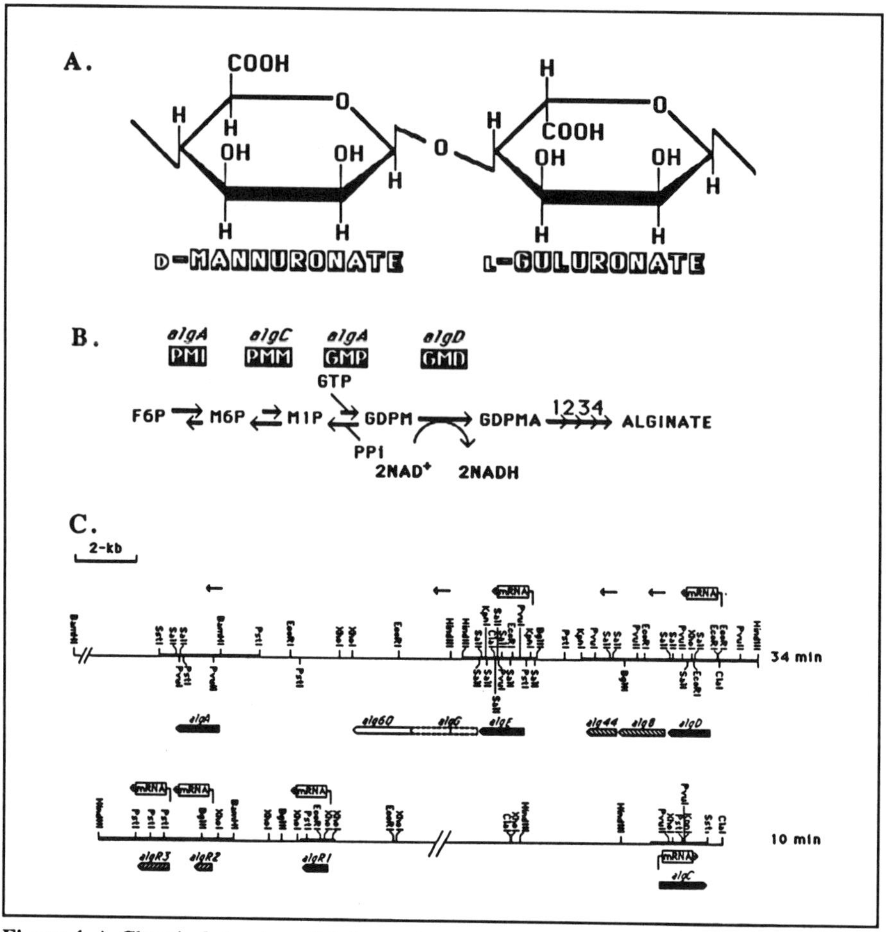

Figure 4. A. Chemical structure of alginate where D-mannuronate and L-guluronate residues are linked via ß-1,4 glycosidic bonds. **B.** Pathway of alginate synthesis. F6P, fructose 6-phosphate; M6P, mannose 6-phosphate; MlP, mannose l-phosphate; GDPM, GDP-mannose; GDPMA, GDP-mannuronic acid. Enzymes PMI, PMM, GMP, and GMD, along with genes encoding these enzymes (*alg*A, *alg*C, *alg*A, and *alg*D) are shown above individual steps. **C.** Organization of alginate gene clusters at 34min (top) and 10min (bottom) regions of *P. aeruginosa* chromosome; *alg*D is at upper right hand extreme. Gene *alg*C maps at 10 min with the regulatory genes *alg*Rl, *alg*R2, and *alg*R3.

perturbation caused by a dehydrating agent such as ethanol,[27] stimulate *alg*D promoter activation. This stimulation is sensed by two regulatory proteins, AlgRl and AlgR2. Bacteria, which scan the environment constantly, usually employ a pair of regulatory proteins used as a sensor and a response regulator.[28,29] This so-called "two-component system" involved in signal transduction usually allows a *trans*-membrane sensor protein to undergo autophosphorylation in the presence of appropriate environmental signals.

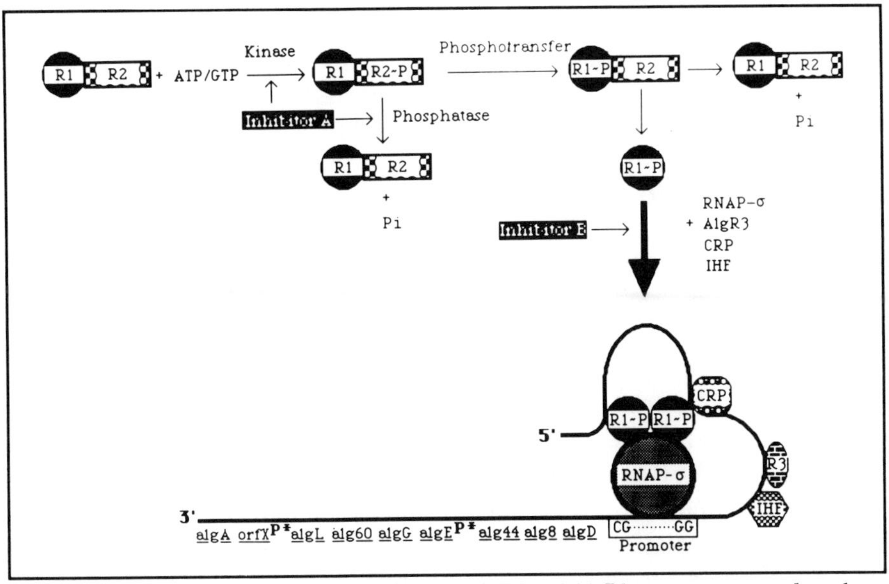

Figure 5. Phosphorylation/dephosphorylation of AlgR2 and AlgRl present as a complex where AlgR2 is initially phosphorylated in presence of ATP or GTP and phosphate is then transferred to AlgRl. AlgRl phosphate (designated Rl~P) then binds upstream of *alg*D gene cluster containing a number of alginate genes such as *alg*8, *alg*44, and so on. Other proteins such as AlgR3 and *P. aeruginosa* homologues of *E. coli* DNA-bending proteins such as CRP and IHF are also involved, resulting in looping out of the DNA to allow Rl~P molecules, bound at -382 and -458 regions, to make contact with RNA polymerase (complexed with appropriate sigma factor) bound at the promoter region.

The phosphorylation takes place in a histidine residue in the sensor kinase protein. This phosphate is then transferred to an aspartate residue of a DNA-binding response regulator protein, which in its phosphorylated form functionally binds the upstream region of a gene (or a set of genes), thereby initiating transcription. There are, of course, a number of variations of this theme.[29] AlgR2 has been demonstrated to be a protein kinase capable of undergoing autophosphorylation in the presence of either ATP or GTP.[30] It then transfers its phosphate to the response regulator AlgRl,[31] which is known to bind at -382 and -458 regions of *alg*D[32] and -91 region of the *alg*C[24] promoters. Several auxiliary proteins such as AlgR3 and *P. aeruginosa* homologues of two *E. coli* proteins CRP and IHF, which are DNA-bending proteins,[33,34] are believed to be required for functional transcription of the *alg*D promoter (Figure 5). Binding of phosphorylated AlgRl, in the presence of these DNA-bending proteins, allows looping of the intervening DNA to facilitate contact between phosphorylated AlgRl bound at -382 and -458 regions of the *alg*D gene and the RNA polymerase holoenzyme bound at the promoter region (Figure 5). This contact is essential for effective transcription by RNA polymerase and is possible only when the appropriate environmental signals are present to trigger the phosphorylation cascade mediated by AlgR2/AlgRl. The phosphorylation of AlgR2 and AlgRl as well as the DNA-binding by AlgRl can be inhibited by specific inhibitors (Figure 5), which may be potentially important in preventing environmental signal transduction mediated by AlgR2/AlgRl, and possibly by other members of the two component systems.[35] Thus, the environment plays a critical role in selectively turning on or switching off specific gene segments to modulate microbial activities.

Acknowledgments

This investigation was supported by grants from the National Institutes of Health (ES04050 and AI16790), the National Science Foundation (MCB-9100015), and the Cystic Fibrosis Foundation (P455). I thank all my students, postdoctoral associates, and faculty colleagues whose hard work, motivation, dedication, and advice made this work possible.

References

1. W. Reineke and H.-J. Knackmuss, *Annu. Rev. Microbiol.* **42**, 263-287 (1988).
2. R.K. Rothmel and A. Chakrabarty, *Pure Appl. Chem.* **62**, 769-779 (1990).
3. L.N. Ornston and D. Parke, *Biochem. Soc. Trans.* **4** (1976).
4. D.K. Chatterjee *et al., J. Bacteriol.* **146**, 639-646 (1981).
5. R.H. Don and J.M. Pemberton, *J. Bacteriol.* **145**, 681-686 (1981).
6. J.R. van der Meer *et al., J. Bacteriol.* **173**, 6-15 (1991).
7. J.R. van der Meer *et al., J. Bacteriol.* **173**, 2425-2434 (1991).
8. M. Schlomann, D.H. Pieper, and H.-J. Knackmuss, in *Pseudomonas: Biotransformations, Pathogenesis, and Evolving Biotechnology*, S. Silver *et al.*, Eds., (American Society for Microbiology, Washington, DC, 1990), pp. 185-196.
9. R.K. Rothmel *et al., J. Bacteriol.* **173**, 4717-4724 (1991).
10. R.K. Rothmel *et al.*, in *Biodeterioration and Biodegradation 8*, H.W. Rossmoore, Ed., (Elsevier Applied Science, New York, 1991), pp. 276-291.
11. E.J. Perkins *et al., J. Bacteriol.* **172**, 2351-2359 (1990).
12. D. Ghosal and I.-S. You, *Gene* **83**, 225-232 (1989).
13. J.R. van der Meer *et al., J. Bacteriol.* **173**, 3700-3708 (1991).
14. J.J. Kilbane *et al., Appl. Environ. Microbiol.* **44**, 72 (1982).
15. U.M.X. Sangodkar *et al., Gene* **71**, 267-277 (1988).
16. R.A. Haugland *et al., Gene* **100**, 65-73 (1991).
17. P.H. Tomasek *et al., Gene* **76**, 227-238 (1989).
18. R.A. Haugland, U.M.X. Sangodkar, and A. Chakrabarty, *Mol. Gen. Genet.* **220**, 222-228 (1990).
19. S.D. La Roche and T. Leisinger, *J. Bacteriol.* **172**, 164-171 (1990).
20. L. Xun, E. Topp, and C.S. Orser, *Biochem. Biophys. Res. Communs.* **182**, 361-366 (1992).

21. T.B. May *et al.*, *Clin. Microbiol. Rev.* **4**, 191-206 (1991).
22. P. Doig *et al.*, *Infect. Immun.* **55**, 1517-1522 (1987).
23. J.G. Jolley *et al.*, *Appl. Spectroscopy* **43**, 1062-1067 (1989).
24. N.A. Zielinski, A.M. Chakrabarty, and A. Berry, *J. Biol. Chem.* **266**, 9754-9763 (1991).
25. A. Berry, J.D. DeVault, and A.M. Chakrabarty, *J. Bacteriol.* **171**, 2312-2317 (1989).
26. J.D. DeVault *et al.*, *Bio/Technology* **7**, 352-357 (1989).
27. J.D. DeVault, K. Kimbara, and A.M. Chakrabarty, *Mol. Microbiol.* **4**, 737-745 (1990).
28. J.B. Stock, A.M. Stock, and J.M. Mottonen, *Nature* **344**, 395-400 (1990).
29. J.S. Parkinson and E.C. Kofoid, E.C. *Annu. Rev. Genet.* **26**, in press.
30. S. Roychoudhury, K. Sakai, and A.M. Chakrabarty, *Proc. Natl. Acad. Sci. U.S.A.* **89**, 2659-2663 (1992).
31. S. Roychoudhury *et al.*, *Gene* **112**, 45-51 (1992).
32. J. Kato and A.M. Chakrabarty, *Proc. Natl. Acad. Sci. U.S.A.* **88**, 1760-1764 (1991).
33. J.D. DeVault *et al.*, *Mol. Microbiol.* **5**: 2503-2509 (1991).
34. D. Wozniak, *Abstr. 92nd Annu. Meet. Am. Soc. Microbiol. D-232*, 134 (1992).
35. S. Roychoudhury *et al.*, *Proc. Natl. Acad. Sci. U.S.A.*, in press.

10

Biodegradation Of Chemical Warfare Agents:

Demilitarization Applications

Steven P. Harvey
Joseph J. DeFrank
U.S. Army Edgewood
Research, Development and
Engineering Center
Aberdeen Proving Ground,
MD 21010-5423

Although biodegradation has been considered as an alternative to incineration for the disposal of chemical agents,[1-3] the requisite organisms have not been available. Studies with sulfur mustard (2,2' dichlorodiethyl sulfide) made with soil samples collected from areas of Aberdeen Proving Ground (APG) demonstrate that this compound can be alkaline hydrolyzed and biodegraded by two bacterial strains. Because mustard quickly hydrolyzes to thiodiglycol in aqueous solution, it was necessary to degrade the hydrolysis product. Enrichment cultures were set up with thiodiglycol as the sole source of carbon. Two bacteria which utilize thiodiglycol as their sole source of carbon and energy were isolated from enrichments. Both pure and munitions-grade mustard were degraded by alkaline hydrolysis and subsequent biodegradation. At least 97 percent of the organic compounds in the medium were degraded. Microtoxicological tests detected no toxicity in the final culture medium. Small-scale fermentation studies on munitions-grade mustard from the U.S. chemical stockpile at APG demonstrated that mustard was degraded at the rate of 50µl/hour/l of culture. This rate is within the range feasible for biological degradation of the APG mustard stockpile. Cost is estimated at $50 million, less than one-fifth the current estimate for construction and operation of the APG mustard demilitarization incinerator. The fermentation equipment could subsequently be utilized for production of enzymes, antibodies, or other useful biotechnology products.

Introduction

By treaty agreements, the chemical stockpiles of both the U.S. and the former Soviet Union are currently slated for destruction. However, chemical demilitarization technology now centers on incineration, which has encountered public opposition due to environmental concerns and excessive cost, currently estimated at 50 times the cost of producing the chemicals.[1,4] There are several reasons why chemical demilitarization would be an ideal application of biodegradation. First, unlike other bioremediation scenarios where contaminants are found in soil or groundwater, in this case, the chemicals to be degraded are in an exactly known location and a relatively pure form (typically distilled). Consequently, it is possible to "feed" the chemicals to the bacteria in a precisely controlled manner. Second, biodegradation occurs at room temperature in aqueous solution, thus precluding any possibility, however remote, of a catastrophic vaporization of agent. Third, a major objection has been that so much money would be spent on incinerator construction that it would be necessary to subsequently use the incinerators for disposal of other hazardous wastes to amortize expenditures. This has brought criticism of the Army from surrounding communities. The fermentors that would be used for biologic demilitarization, on the other hand, could subsequently be very useful in the production of enzymes, antibodies or other biologic products. These fermentation processes have no adverse environmental impact.

The chemical stockpile at Aberdeen Proving Ground (APG) may be uniquely appropriate for biologic demilitarization. Mustard is the only agent stockpiled there, and it is stored in bulk. Consequently, there are no explosive components of munitions for disposed. A major criticism of previous demilitarization alternatives has been that they would only dispose of the chemicals, not the other equipment. The empty one-ton mustard storage containers can be decontaminated to the 3X level with sodium hydroxide, the contaminated wash from which could be fed directly into the fermentor medium. Decontamination to the 5X level can then be achieved in the decontamination/ detoxification incinerator, which is currently used for that purpose.

Methods and Results

Mustard hydrolysis has been studied extensively.[5-9] Briefly, the reaction proceeds through a series of sulfonium ion intermediates to thiodiglycol (Figure 1). The first (rate-limiting) step is the formation of a cyclic sulfonium ion, which then reacts with water to form the alcohol. Hydrolysis occurs most

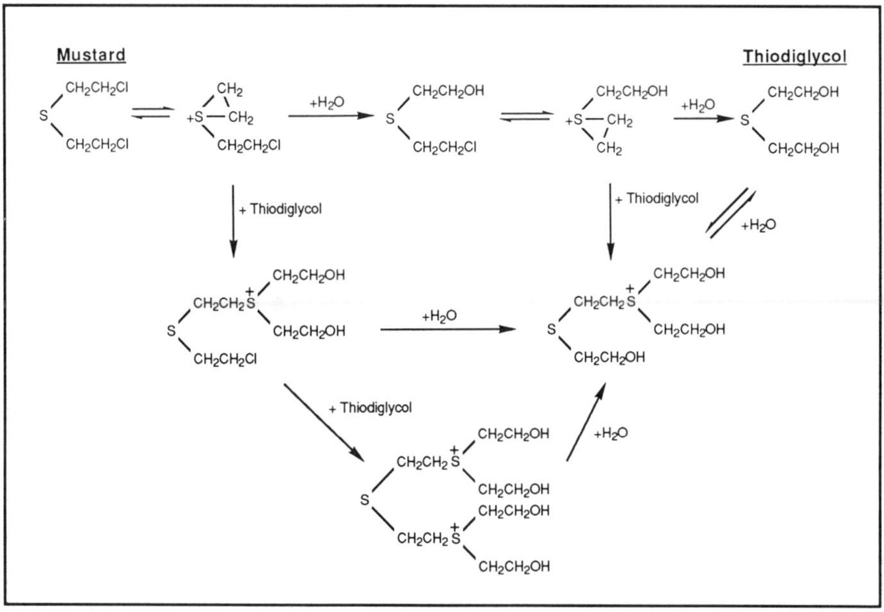

Figure 1. Hydrolysis of mustard. Mustard hydrolyzes through a series of sulfonium ion intermediates, finally to thiodiglycol. The first step, the formation of the cyclic intermediate, is rate-limiting. Thiodiglycol acts as a nucleophile to produce the branched intermediates. Adapted from reference 9.

rapidly at dilute mustard concentration, high pH, and low chloride ion concentration.

Isolation and identification of thiodiglycol-degrading bacteria. Our literature searches found no previous reports of thiodiglycol biodegradation. Enrichment cultures were set up in mineral salts medium using thiodiglycol as the sole carbon source. Soil samples containing indigenous bacteria were added, and cultures were incubated up to two weeks. Tubes that became turbid were diluted 1:500 in 10mM thiodiglycol minimal medium. After 10 to 15 such dilutions, bacteria were streaked on nutrient plates at least three times for single colonies and inoculated back into thiodiglycol minimal medium. Two different strains were isolated (SH18 and SH42) that grew on 10mM thiodiglycol as sole carbon source. Neither organism grew without an exogenously added carbon source, indicating CO_2 was not being utilized as a carbon source. Disappearance of thiodiglycol, with corresponding increase in optical density at 595nm ($O.D._{595}$) of the culture, was monitored by gas chromatography (GC).[10] Bacteria were identified by gas chromatographic fatty acid analysis (Microbial I.D. Inc., Newark, DE) and Biolog carbon source utilization plates with the corresponding Microstation System soft-

Strain	% Original Carbon Remaining	Compounds Detected
SH18	16	Thiodiglycol sulfoxide
SH42	2 - 3	Thiodiglycol, thiodiglycol sulfoxide, small amounts of ethers or thioethers

Figure 2. Results of ^{13}C NMR analysis of SH18 and SH42 grown on hydrolyzed mustard.

ware.[11] SH18 showed a fair (37.5 percent) identity with *Pseudomonas pickettii* by GC fatty acid analysis and was not identifiable with the Biolog plates. SH42 showed a very high (74.1 percent) identity with *Alcaligenes xylosoxidans ssp. xylosoxidans* by GC fatty acid analysis. When tested with the Biolog system, SH42 corresponded exactly (95 of 95 carbon sources) with the results most commonly seen with *A. xylosoxidans ssp. xylosoxidans*. Both bacteria grew optimally at 28 to 30°C and more slowly at temperatures up to 34°C. SH18 grew optimally on thiodiglycol between pH 7.0 and pH 8.6. Optimum growth of SH42 on thiodiglycol occurred between pH 8.4 and pH 9.4. All subsequent experiments with SH18 were conducted at pH 7.5 and with SH42, at pH 9.2.

Growth of SH18 and SH42 on hydrolyzed mustard as sole source of carbon. Bacteria were grown in hydrolyzed mustard (16mM) as sole carbon source. Resulting cultures were analyzed by ^{13}C nuclear magnetic resonance (NMR) to determine their postgrowth organic profiles (Figure 2).[12]

In the case of the SH18 culture, analysis detected only one remaining product, 2,2'-dihydroxydiethyl sulfoxide (thiodiglycol sulfoxide), which comprised approximately 16 percent of the original carbon from mustard. Thiodiglycol sulfoxide was not detected in a sterile control containing hydrolyzed 16mM mustard, consistent with a metabolic origin for the compound. Because SH18 will not utilize thiodiglycol sulfoxide as its sole source of carbon for growth, it may represent a metabolic dead-end product. In the case of the SH42 culture, a one-hour analysis detected no carbon-containing products in the culture medium. A subsequent 72-hour analysis of this same culture improved the signal-to-noise ratio, and it was seen that a maximum of approximately 3 percent of the carbon from mustard remained in the medium in the form of thiodiglycol, thiodiglycol sulfoxide, and small amounts of several compounds tentatively identified as ethers or thioethers.

Toxicity of hydrolyzed and biodegraded mustard. Because thiodiglycol is essentially nontoxic, the expectation was that the medium would be non-

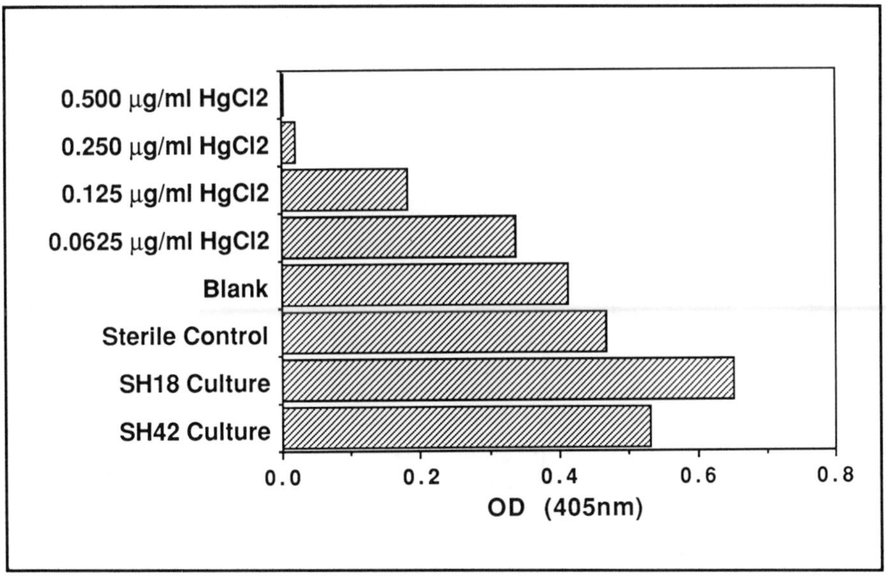

Figure 3. Results of Toxi-Chromotest. O.D.$_{405}$ is inversely proportional to toxicity. HgC_{12} is the standard toxic compound. The blank consists of test bacteria and buffer alone (no toxicity). Bacteria in wells of a microtiter plate are incubated with 100μl of the indicated concentration of HgC_{12}, hydrolyzed 16mM mustard, or postgrowth cultures of SH18 or SH42 bacteria grown on hydrolyzed mustard. HgC_{12} results are averages of two determinations; others are averages of eight determinations each.

toxic after hydrolysis prior to inoculation. However, the pathway of degradation is unknown, it is possible that toxic metabolites are formed during growth. In order to determine the toxicity, if any, of the culture, two commercial bacteria-based toxicity testing kits were used. Toxi-Chromotest (Orgenics, Inc., Columbia, MD) uses the chromosomally encoded ß-galactosidase enzyme of *E. coli* test bacteria to catalyze the cleavage of a chromogenic substrate. The intensity of color is inversely proportional to the degree of cell killing by the compound to which the bacteria are exposed. Results show no detectable toxicity from either the sterile medium containing hydrolyzed mustard only or the postgrowth culture medium with either SH18 or SH42. In all cases the O.D.405 is greater than that of the blank containing only *E. coli* cells and buffer (Figure 3).

The second test, SOS-Chromotest (Orgenics, Inc.), specifically detects genotoxicity with *E. coli* that have an SOS promoter cloned upstream of the ß-galactosidase gene. Genetic damage activates the SOS promoter, stimulating synthesis of ß-galactosidase. The chromogenic substrate is therefore cleaved in proportion to the degree of genotoxicity. This test was of particular

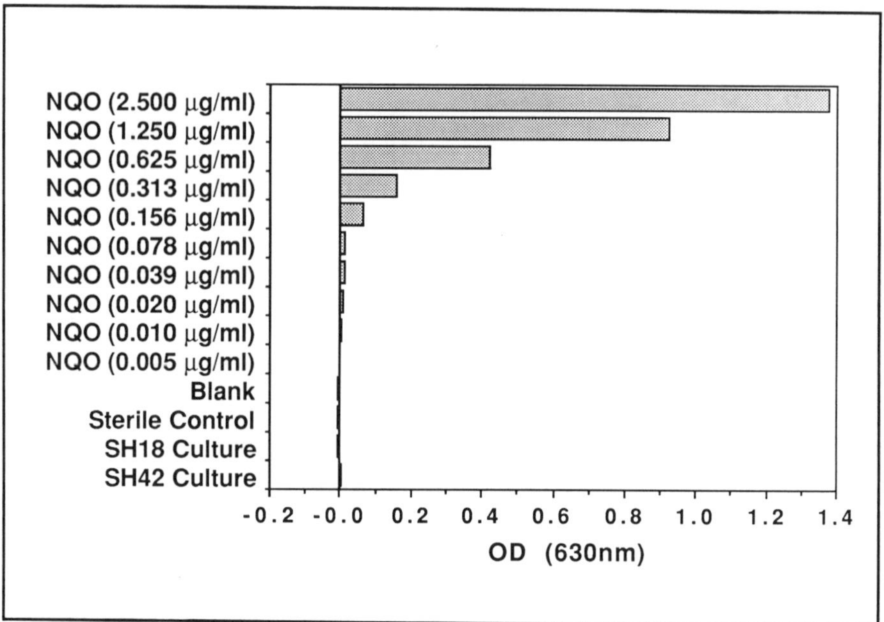

Figure 4. Results of SOS-Chromotest. O.D.$_{405}$ is proportional to toxicity. 4-Nitroquinoline is the standard genotoxic solution. Procedures are the same as described for Figure 3.

interest because of mustard's alkylating effects on DNA. Results (Figure 4) show that the sterile control and SH18 and SH42 cultures all exhibit no detectable genotoxicity (no increase in O.D.$_{405}$).

Fermentation study with munitions-grade mustard from the U.S. chemical stockpile. Although the initial studies demonstrated that a very large percentage of the carbon in the mustard hydrolysis medium could be consumed (at least 97 percent), it was also of interest to attempt the degradation on a larger scale. For that purpose, a fermentation study using munitions-grade mustard from the chemical stockpile was conducted. A sample was obtained from the circa 1940s chemical stockpile at APG. The mustard was dark brown to black in color (highly purified mustard is colorless) and contained some insoluble particulate material. Presumably, the color and particulate material were remnants of the one-ton steel container in which the mustard had been stored for approximately 50 years. One hundred twenty-five milliliters of this mustard was hydrolyzed in a slight molar excess of NaOH. Mustard was measured by volume, and no impurities were accounted for in the stoichiometry. The solution was stirred overnight, and undissolved particulate material was allowed to settle prior to decanting.

For fermentation, this solution was diluted 1:1 in dH_2O and 3g/l KH_2PO_4;

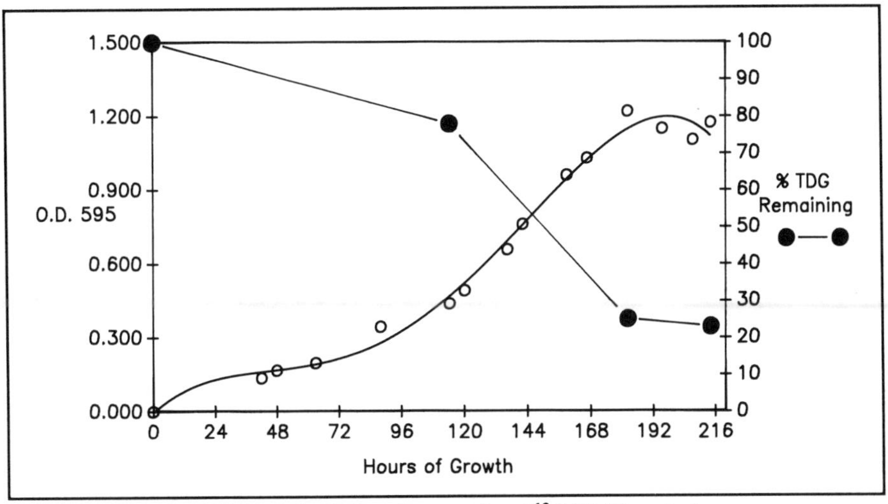

Figure 5. Growth of SH42 on hydrolyzed mustard and ^{13}C NMR analysis.

1g/l $(NH_4)_2SO_4$, 0.1g/l yeast extract, and 10ml/l Wolin Salts solution were added. pH was adjusted to 9.2 with a few drops of H_2SO_4. Assuming pure mustard and complete hydrolysis, the thiodiglycol concentration was 50mM. This medium was filter-sterilized and added to a sterile 1.25l fermentation vessel on a New Brunswick BioFlo III fermentor. One hundred milliliters of a midlog culture of SH42 growing on thiodiglycol as sole carbon source was added as seed culture. Fermentation was conducted at pH 9.2 at 30°C with constant aeration. Change in the concentration of thiodiglycol and possible appearance of metabolites in the medium were monitored by ^{13}C NMR at various times during growth. Results are shown in Figure 5. During the log phase of growth, mustard consumption was approximately 50µl/hr/l of culture.

^{13}C NMR results from fermentation study. The NaOH hydrolysis step yielded a relatively pure thiodiglycol product. There are several minor impurity peaks at 33ppm, 34ppm, and 72ppm. Approximately 77 percent of the thiodiglycol was consumed during the fermentation. An apparent minor degradation product peak also developed around 39.5ppm.

Conclusions

In summary, these are the first data to show that mustard can be (1) alkaline hydrolyzed and biodegraded and (2) utilized as the sole source of

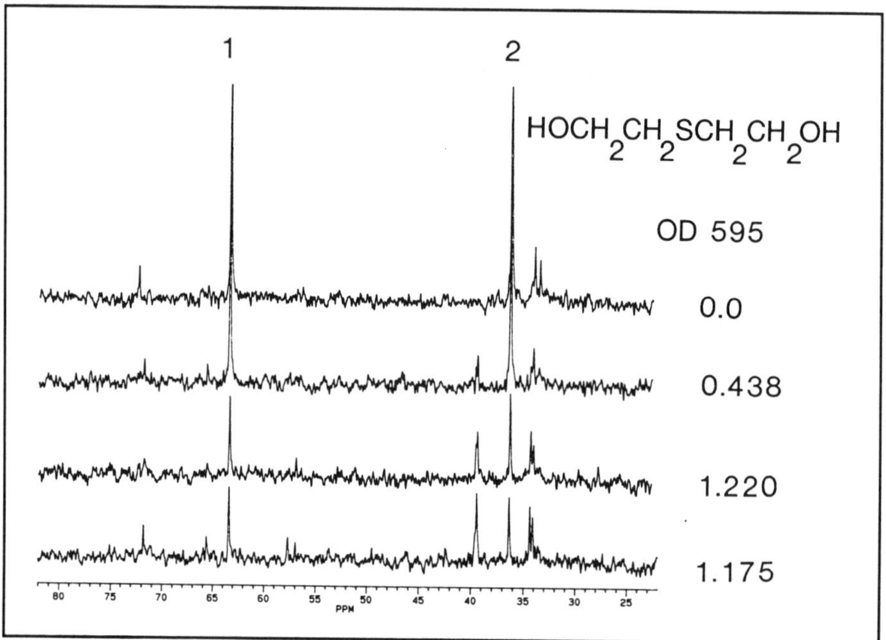

Figure 6. ^{13}C NMR results from fermentation study.

carbon by two different strains of bacteria. NMR analysis of the resulting cultures shows, in the case of SH18, approximately 16 percent of the original carbon from mustard remaining in the form of thiodiglycol sulfoxide. In the case of SH42, a maximum of only 3 percent of the original carbon from mustard remains in the form of (primarily) thiodiglycol, thiodiglycol sulfoxide, and small amounts of several possible ether or thioether compounds.

Two different microtoxicologic tests detected no toxicity from either culture. The tests show that hydrolysis alone will detoxify the aqueous mustard solutions (microtoxicology results from sterile controls). However, it is an important practical consideration that although thiodiglycol is essentially nontoxic, it is also a mustard precursor (chlorination yields mustard).[9] This is the rationale for U.S. export controls on thiodiglycol. Therefore, hydrolysis alone will not convert mustard to an acceptable product for disposal.

Munitions-grade mustard from the U.S. chemical stockpile was also degraded by SH42 in a one liter fermentation. The rate of degradation during the logarithmic growth phase was approximately 50µl/hr/l of culture. Refinements of the growth conditions would be expected to yield significant improvements in this figure. Even at this level, the degradation rate is in the range required to be able to degrade the entire APG stockpile in approximately three

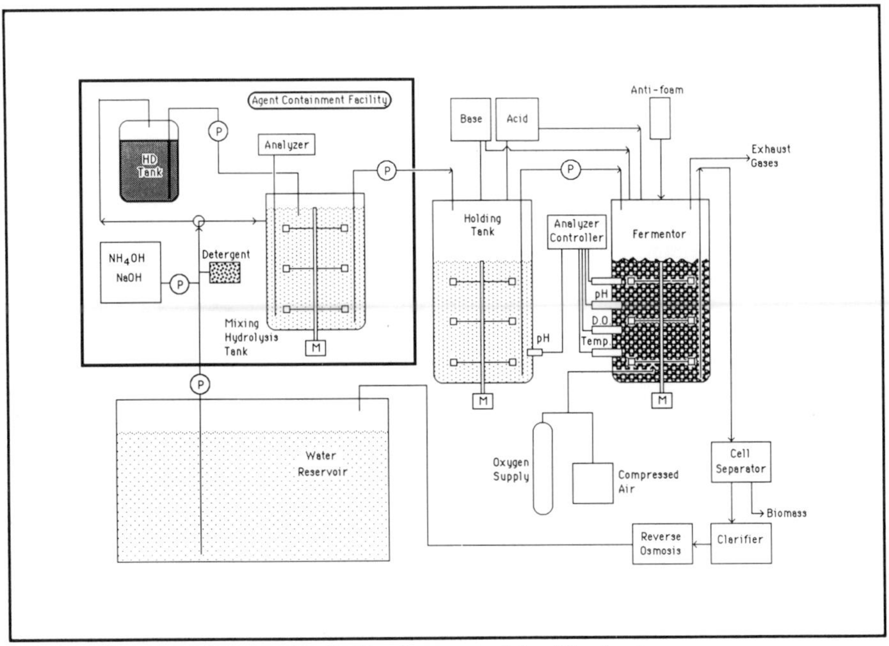

Figure 7. Proposed fermentation facility for mustard demilitarization.

to four years (the time required would depend on the final degradation rate achieved and the total fermentation capacity).

Proposed fermentation facility. The possible configuration of a mustard demilitarization facility (partial) based on the chemical/biologic technology described in this paper is shown in Figure 7. A relatively small agent containment facility would be the site for the removal of the mustard from the bulk container and its hydrolysis to thiodiglycol. It is estimated that each ton container of mustard would be diluted into 15 to 30,000 l of alkaline water. Once complete hydrolysis is confirmed, the thiodiglycol would be pumped to a holding tank within the fermentation facility. At this point, the pH of the solution would be adjusted to the optimum for growth of the microorganisms and any trace nutrients added. This solution would be fed into the fermentors in a continuous or sequential manner. It is anticipated that the fermentation plant will have four to eight fermentors running at any one time and at different stages in the process. Upon the completion of the mineralization process, the bulk of the culture would be harvested by separation of the biomass and removal of salts and other soluble materials by ultrafiltration/ reverse osmosis. If appropriate, 1 to 2,000 l of culture would remain in the fermentor to act as the inoculum for the next fermentation run. The bulk of the

Table 1
Approximate Cost of Startup for the Proposed Fermentation Facility

Capital equipment (Fermentors, tanks, pumps, purification equipment, and so on)	$25,000,000
Building	$5,000,000
Operating expenses (4 years) (Salaries, materials/supplies, utilities)	$20,000,000
TOTAL	$50,000,000

water used in this process would be recycled, thus greatly reducing the potential impact on waste water treatment facilities. Although an exact cost for this process is difficult to determine at this time, a rough estimate based on conventional fermentation facilities appears in Table 1.

Once the facility has been completed (two to three years), the actual demilitarization process would take three to four years.

Acknowledgments

We thank William T. Beaudry, Linda L. Szafraniec (CRDEC Chemical Division) and Adam N. Stroup (CRDEC Biotechnology Division) for ^{13}C NMR analysis of many samples, Jude J. Height (CRDEC Analytical Division) for assistance with the mustard work, and James J. Valdes (CRDEC) and Ananda M. Chakrabarty (University of Illinois, Chicago) for advice on the project.

References and Notes

1. L.R. Ember, *Chem. Eng. News* **68**, 9 (1990).
2. F.P. Ward, in *Biotechnology and Biodegradation*, D. Kamely, A. Chakrabarty, and G.S. Omenn, Eds. (Gulf Publishing Co., Houston, 1990), pp. 147-154.
3. A. Picardi, P. Johnston, and R. Stringer, Alternative Technologies for the Detoxification of Chemical Weapons: An Information Document, (Greenpeace International, Washington DC, 1991), pp. 20-39 .
4. T.M. Beardsley, *Sci. Am.* **263**, 48 (1990).
5. P.D. Bartlett and C.G. Swain, *J. Am. Chem. Soc.* **71**, 1406 (1949).
6. Y.C. Yang *et al.*, *Thermochim. Acta* **114**, 313 (1987).
7. S.P. McManus *et al.*, *J. Am. Chem. Soc.* **107**, 3393 (1985).

8. Y.C. Yang, J.R. Ward, and T. Luteran, *J. Org. Chem.* **51**, 2756 (1986).
9. Y.C. Yang *et al.*, *J. Org. Chem.* **53**, 3293 (1988).

10. Gas chromatography (GC) with flame photometric detector was performed as follows: One microliter of aqueous culture was directly injected into a Hewlett-Packard 5840A gas chromatograph equipped with a flame photometric detector with a sulfur filter to which was connected a Hewlett-Packard 5840A GC terminal. The GC was equipped with a six-foot coiled glass column (I.D. 2mm) packed with 100/120 mesh Gas Chrome Q support phase, which was coated with a 3 percent load of OV-101 as the stationary phase. Injection port temperature was set at 230°C, and detector temperature was set at 240°C. Oven temperature was programmed between 60°C and 230°C at a rate of 30°C/minute. The final temperature was held for 15 minutes prior to integration. Helium flowing at 90ml/minute was used as carrier gas. Detector gases were hydrogen, air, and oxygen flowing at 220, 110, and 60ml/minute respectively. Slope sensitivity was set at 0.04, and chart speed was set at 0.5cm/minute.
11. Biolog plates were purchased from Biolog, Inc., Hayward, CA. Thiodiglycol was purchased from Sigma, St. Louis, MO, and had a purity of greater than 99 percent. Vacuum-distilled mustard from our stock was used. Warning: Mustard is a dangerous vesicant and must be handled in a closed system or in a hood with a minimum velocity of 100ft/minute.
12. NMR analysis was according to the following procedure: Samples were monitored using ^{13}C NMR spectroscopy. The spectra were recorded using a Varian VXR-400S Superconducting FTNMR System operating at 100MHz for ^{13}C. In all cases, spectra were obtained at a probe temperature of 21°C in an unlocked mode. Tetramethylsilane (TMS) in chloroform was used as the external standard. For each sample analyzed, 592 to 1192 transients were acquired in absolute intensity mode. The number of transients was dependent on the signal-to-noise required or desired. The spectra were obtained using a pulse width of 8.8μsec (57°), a sweep width of 25KHz, an acquisition time of 1.0sec, and a pulse delay of 2.0sec. WALTZ decoupling was used for full proton decoupling. Quantitative data were obtained by digital integration of the peak areas. A sterile control was run under identical conditions prior to each sample analyzed. The area of the SCH_2 resonance (δ 32.5ppm) in the sample was divided by the area of the SCH_2 resonance in the control sample to obtain a value for the amount of thiodiglycol consumed. Similarly, the area of the OCH_2 resonance (δ 59.5ppm) in the sample was divided by the area of the OCH_2 resonance in the control sample. The two values were averaged to obtain the percentage of thiodiglycol consumed at time, t. GC-mass spectroscopy was used for confirmation of NMR results.

A Framework for the Environmental Risk Assessment of Contaminated Sediments at DOD Installations

Tom M. Dillon
USAE Waterways Experiment Station
Vicksburg, MS 39180-6199

Over 17,000 sites at more than 1,800 Department of Defense (DoD) installations are being considered for remedial action because of chemicals released into the environment. Many of these sites contain heavily contaminated sediments. A framework for the evaluation of environmental risks associated with these contaminated sediments has been devised. The framework is driven by the needs of the decision-maker and places heavy emphasis on the initial development of assessment and measurement endpoints.

This framework is similar in structure to traditional human health risk assessments, but with some notable exceptions. First, both human and nonhuman target receptors are considered. To use both target receptors is significant because Army installations provide critical habitats to commercially and ecologically important animals as well as threatened and endangered species. Second, an effects-based sediment bioassay approach is used to evaluate the complex mixtures of contaminants normally found in sediments. This approach is crucial for DoD installations where a variety of conventional pollutants and military-unique compounds exist (munitions, lethal or incapacitating agents, smoke and incendiary material, and so forth). The effects-based sediment bioassays represent a significant improvement over the more traditional chemical-by-chemical approach. Reference sediments are used to evaluate incremental or comparative environmental risks rather than absolute risks.

Qualitative and quantitative methods are explicitly used to address the uncertainties associated with estimates of environmental risk being evaluated. Finally, this framework can be used to compare alternative remedial actions as well as establish technically sound clean-up levels for contaminated sediments at DoD installations.

Cloning, Characterization, and Expression of Animal Toxin Genes for Vaccine Development

Leonard A. Smith
Department of Immunology and Molecular Biology
U.S. Army Medical Research and Development Command
Institute of Infectious Diseases
Fort Detrick
Frederick, MD 21702-5011

Gene libraries have been constructed from the messenger ribonucleic acid (mRNA) isolated from venom glands of poisonous animals such as snakes and scorpions. The gene banks thus created contain recombinant clones harboring DNA sequences encoding toxins with various pharmacological activities, ranging from myonecrosis-inducing to those affecting neuronal transmission. A number of these clones have been isolated and characterized, and gene expression has been attempted with limited success in *Escherichia coli*, baculovirus, and one mammalian cell expression system.

Safety and Kinetics of an Equine F(ab)$_2$ Heptavalent Botulinum Antitoxin in Human Volunteers

Maria H. Sjogren
United States Army Medical Research and Development Command
Division of Medicine
Fort Detrick
Frederick, MD 21702

Kelly McKee
M.A. Clayton
David Franz
R.M. Condie
C.J. Peters
United States Army Medical Research and Development Command
Institute of Infectious Diseases
Fort Detrick
Frederick, MD 21702

A despeciated equine botulinum antitoxin was prepared at the University of Minnesota by immunizing a single horse with *C. botulinum* toxoids and toxins derived from types A, B, C, D, E, F, and G. Plasma obtained from the immunized animal was purified and pepsin treated to yield predominantly F(ab´)2 antibody fragments. Five healthy male volunteers received 10ml of antitoxin by slow intravenous injection. Antitoxin levels for type A were assayed with a classical mouse neutralization assay in sequential sera collected for 21 days after infusion. No local or systemic side effects were observed in the five volunteers during the 4-month follow-up. Anti-type A toxin levels were detected in the first sample after infusion, which was taken at 2 hours (antitoxin range: 6.45 to 2.56IU/ml). The antitoxin was still detectable at Day 21 (values ranged between 0.02 and 0.03IU/ml).

Primate studies have shown that an antitoxin level of at least 0.02 in serum is protective against challenge with botulinum toxin. This despeciated equine heptavalent botulinum antitoxin is safe in humans and provides serum antitoxin levels of expected beneficial value.

Basic Fibroblast Growth Factor Accelerates Repair and Prevents Degeneration in Retinal Laser Lesions

Steven T. Schuschereba
Phillip D. Bowman
David J. Lund
Julie A. Quong
Joseph A. Vargas
Ronald E. Ferrando
Janet M. Yang
U.S. Army Medical Research and
Development Command
Letterman Army
Institute of Research
Presidio of San Francisco, CA 94129-6800

Ocular injury caused by exposure to battlefield lasers is anticipated to be a major problem for military medicine in future conflicts, but no treatment for this type of injury is currently available. Treatment is being devised for laser-injured eye tissue and a beneficial effect of basic fibroblast growth factor (bFGF) has been found both to accelerate repair and to ameliorate degeneration after laser injury to the eye.

Multiline argon laser irradiation (454-514nm) produced lesions at or slightly below subretinal hemorrhaging levels in pigmented New Zealand Red rabbits. Within 30 minutes after irradiation, eyes were intravitreally injected directly above the lesions with 10μl vehicle or 10μg/10μl bFGF. On subsequent days, funduscopy showed that bFGF-treated lesions were smaller in diameter, and fluorescein angiography showed that bFGF-treated lesions were less leaky than lesions in the control eyes. After four days of treatment, animals were terminated and eyes examined histologically. Treated eyes exhibited a reduction in lesion diameter and in the lesion periphery, decreased loss of photoreceptors, and greater numbers of pigmented epithelial cells. By bromodeoxyuridine (BrDU) immunocytochemistry, increased proliferation occurred in fibroblasts, retinal pigmented epithelial (RPE) cells, and inner retinal glial cells. These results indicate that bFGF both accelerated ocular tissue repair and more important, prevented photoreceptor degeneration. For the first time, a beneficial treatment for retinal laser injury has been demonstrated by using a neural-derived growth factor.

Novel Techniques to Determine Ecological Effects at U.S. Army Sites

Randall Wentsel
Ronald Checkal
U.S. Army Edgewood Research,
Development and Engineering Center
Aberdeen Proving Ground, MD 21010-5423

Robert Parmelee
Rutgers University,
New Brunswick, NJ 08903

To support U.S. Army Installation restoration and base closure programs, Ecological Risk Assessment (ERA) is required to assist in the determination of the appropriate cleanup levels. The increase in a cleanup criterion by an order of magnitude can cost millions of dollars. Therefore, technically sound environmental data that can be defended to the scientific and regulatory community are critical to reach informed decisions about U.S. Army sites.

Novel techniques developed at the U.S. Army Edgewood Research, Development and Engineering Center, which can support ERA, are under way. Two technologies have been developed. The transport and transformation of chemical contaminants in soil have been studied using intact soil columns. To assess ecological effects, research was conducted to develop a soil fauna microcosm. Data are available that illustrate the utility of the intact soil column system for monitoring and studying the movement of munition compounds or other chemicals in soils. The soil fauna microcosm uses nematode and arthropod populations to measure the impact of chemicals on soil communities.

Advanced Electronics

11

Metal Diffusion in Electronic Materials:

Accurate Measurement Using Secondary Ion Mass Spectrometry and Scanning Auger Microscopy

Stephen N. Schauer
Geo-Centers, Inc.
Fort Monmouth, NJ 07703-5601

Richard T. Lareau
Howard S. Lee
U.S. Army Research Laboratory
Electronics and Power
Sources Directorate
Fort Monmouth, NJ 07703-5601

Accurate chemical and structural analyses are critical in the development of devices using sub-micron patterning and sub-ppm elemental doping/control. Secondary ion mass spectrometry (SIMS) and scanning Auger microscopy (SAM) can be used to measure the three-dimensional distribution of elements in metal/semiconductor structures, but several factors often obscure true diffusion profiles. When going from a region of high to low concentration (such as a metal ohmic contact on a GaAs device structure), the knock-on effect of the primary ion beam and the typically rough metal interfaces can cause the observed metal penetration to be greater than the true diffusion depth. This can be solved by using the back-side depth profiling technique, which effectively reverses the concentration gradient by sputtering through the substrate into the top contact metal. Interface roughness or spiking of metals through interfaces can also broaden measured interfaces, because SIMS depth profiling typically averages the signal over the analyzed region. By using image depth profiling techniques, such as selected area depth profiling, we have deconvolved these effects. For the first time, back-side imaging depth profiling was investigated. SAM was utilized to examine surface topography (in the secondary electron microscopy mode) and chemical distributions in multiphase metal structures.

Introduction

Secondary ion mass spectrometry (SIMS) and scanning Auger microscopy (SAM) are two analytic techniques that permit measurements of three-dimensional elemental distributions.[1-3] SIMS is capable of measuring concentrations varying from values as high as one percent to as low as the part-per-billion (ppb) level. The spatial resolution can resolve sub-micron features, and in optimum circumstances depth resolution is better than 50Å. SAM has the advantages of excellent spatial resolution and semiquantitative information without standards at concentrations as low as 0.1 percent and excels in surface specificity.

Conventional ohmic contacts are typically alloyed in order to achieve low contact resistance between the contact metal and the semiconductor. These contact metals must not penetrate through the active device regions, which are typically less than 100nm thick; hence, the necessity to include a thicker buffer or cap layer. The ability to accurately determine even trace amounts of diffusion is necessary for fundamental studies of contact formation, in order to develop devices for future applications, as well as failure analysis of devices presently in application.

We often encounter lateral inhomogeneities when analyzing device structures. Until recently, depth profiling using SIMS or Auger electron spectroscopy (AES) averaged the signal over the entire analyzed area, so spatial information was very limited. We are now able to collect data in an image format, pixel by pixel, and deconvolve concentration gradients in a postanalysis data processing system.

SIMS and SAM both rely on an ion beam to sputter the sample in order to measure elemental concentrations as a function of depth. This ion beam causes mixing of the sample at least to the depth of penetration, typically on the order of 50Å. When going from a high concentration to a low concentration, this depth resolution is further limited by knock-on effects, memory effects from the sample chamber surfaces contaminated during sputtering in the high concentration region, and crater edge effects, where the element of interest is being sputtered from the crater walls and redeposited in the sample crater. Depth resolution is further limited by any surface topography or roughness, which cause different depths to be analyzed simultaneously, resulting in a signal from an average depth, instead of a more specific depth. Metal interfaces are typically rough, and alloys are often inhomogeneous in the lateral and depth directions. Roughening can be enhanced by the ion beam, particularly in metals. An innovative technique was pioneered at the Electronics and Power Sources Directorate (EPSD) to circumvent this problem[4,5] and since that time has become internationally recognized as an effective analytic method.[6-8] This technique is referred to as SIMS backside depth profiling, indicating that the sample is profiled from the back surface instead of the front. Because semiconductor wafers are too thick to sputter through, the semiconductor substrate must be carefully thinned (wafers are typically ~600μm thick; a realistic sputter depth is <5μm).

Metal contacts on semiconductor substrates are excellent examples of the benefits of backside depth profiling. Conventional, frontside depth profiling will show an apparent depth of penetration of the contact metal much greater than the actual diffusion length. Backside profiling can deconvolve true diffusion from ion beam induced artifacts.

We discuss several studies that indicate the potential these analytical techniques have for analyzing semiconductor device structures, and also illustrate the three-dimensional analyses that these techniques are capable of obtaining. The first two studies involved extensive materials investigations on ohmic contacts to both n+ and p+ III-V electronic device structures. The first study involves n+ and p+ contacts being studied for use in heterojunction bipolar transistor (HBT) device structures, where again the metal penetration depths into the active device areas are of importance. The second contact structure to be studied incorporated a quasi-epitaxial layer of Ge followed by a thicker Au outer contact. In this case, investigations centered around understanding the role of Ge and Au in forming a nonalloyed ohmic contact, and the elemental interaction between the metal atoms and substrate Ga and As atoms, in addition to the metal penetration depths achievable. The third study involved an ohmic contact to a novel device, the dual-channel high electron mobility transistor (DCHEMT) fabricated at EPSD. The success of the final device design was achieved after changes were suggested as a result of backside depth profiles from the initial device structures. The final example involved a failure analysis study at EPSD on a fully processed active device typically implemented in critical military systems, such as the Phoenix Missile. The success of this study was only possible with application of the techniques and capabilities detailed here.

Experimental Procedure

A Cameca IMS-3f secondary ion mass spectrometer was used in these studies. This double-focusing mass spectrometer is capable of providing a chemical image with sub-micron spatial resolution when used in a microprobe mode, or rapid analyses when used as an ion microscope. The primary ion beams used were O_2^+, to provide high sensitivity for electropositive elements, and Cs^+, to provide high sensitivity for electronegative elements and sub-micron ion imaging.

SAM and secondary electron microscopy (SEM) were performed with a Perkin-Elmer PHI 660 instrument. These techniques provided elemental and topographic maps, with approximate quantification. In this case, the sputter ion beam used was Ar^+, which provided good surface removal rates and the possibility of monitoring carbon and oxygen.

Samples analyzed with the backside SIMS technique were prepared following the procedure developed at EPSD.[4,5] The metal contacts were grown on GaAs substrates that had 500Å marker layers alternately doped with silicon at concentrations of $5x10^{18}$ and $5x10^{16}/cm^3$ to provide an accurate depth calibration to correlate metal penetration depths. Beneath the marker layers was a buffer layer of GaAs, followed by a chemical etch stop of $Al_{0.3}Ga_{0.7}As$. Samples were mounted on a glass slide with crystal bond, then mechanically lapped to a thickness of ~30 microns. Samples were chemically etched to the AlGaAs etch stop in a H_2O_2 solution buffered with ammonium hydroxide to a pH of 8.4 , which selectively etches GaAs 20 times faster than AlGaAs.[9] The etching was done in a jet etching system to provide uniform etching; this system permitted both continuous rotation of the etched sample and a continuously pivoting and recycling jet spray system. The endpoint was monitored by the surface morphology; the polishing scratches disappeared when the $Al_{0.3}Ga_{0.7}As$ layer was reached. Past experience with thinner samples showed a change in color when the etch stop layer was reached, due to a change in the index of refraction. This process is shown schematically in Figure 1. The etch stop was reached faster at the edges of the sample, and etching was stopped when enough area was exposed for an analysis. Because the samples were mounted on a glass slide, they were electrically isolated, which would cause a charge buildup during analysis. The samples were coated with ~300Å of gold to provide a conductive path from the sample to the sample holder.

Results

Excellent data can often be obtained on contacts with relatively routine analytical techniques. GeMoW contacts are being studied as ohmic contacts to p+ and n+ GaAs.[10] These contacts have been demonstrated to form ohmic contacts to both n- and p-type GaAs over a limited temperature range. The mechanism of this contact formation and degradation at higher annealing

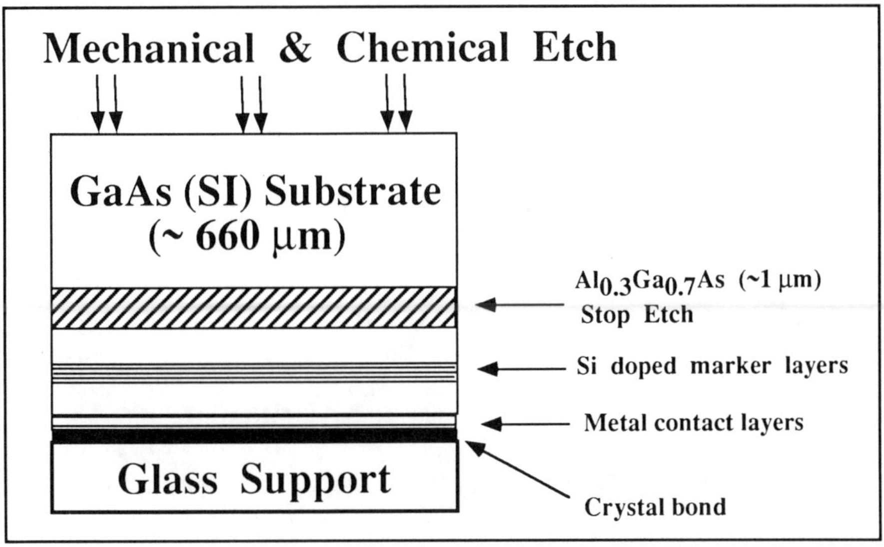

Figure 1. Schematic of sample preparation for the backside depth profiling technique.

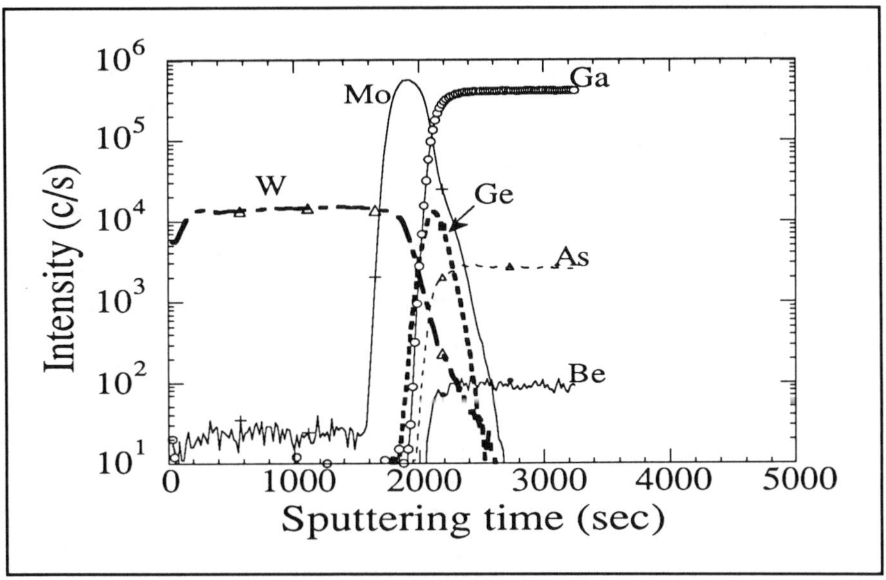

Figure 2. SIMS depth profile of an as-deposited GeMoW ohmic contact.

temperatures is currently being studied. Figure 2 shows the elemental distribution of W, Mo, Ge, Ga and As as a function of depth in an as-grown contact area on Be-doped GaAs. This same structure is shown in Figure 3 after

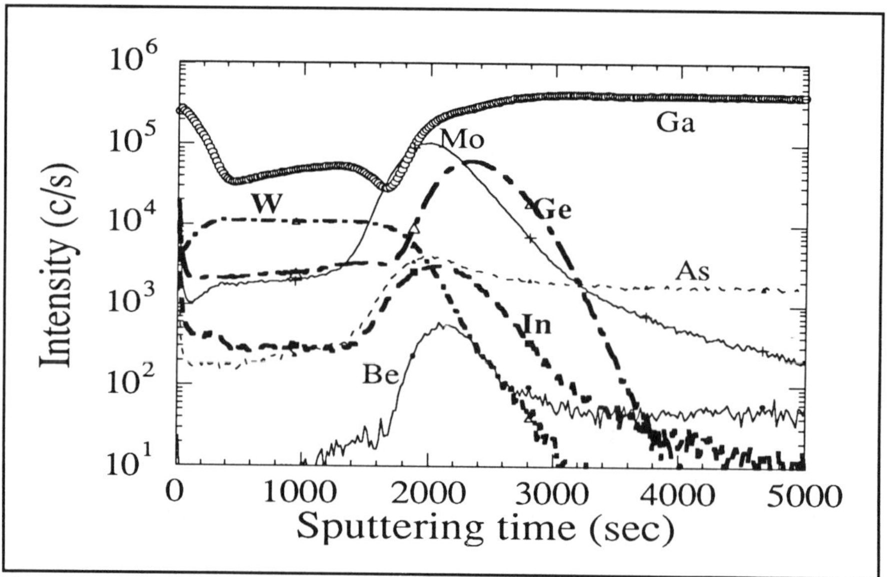

Figure 3. SIMS depth profile of an annealed GeMoW ohmic contact.

annealing in an InAs overpressure at 780°C. It is readily apparent that considerable diffusion of Be, Ge, Mo, Ga, As, and In occurs when these contacts are annealed.

SIMS quantification is notoriously difficult, because ion yields can vary by orders of magnitude, depending on the sample matrix and analyzed element. The yield of Auger electrons is less variable, so semiquantitative data is possible without standards. The SIMS signal intensities of Ga, As, Be, and In in the metal contact region were extremely high in the annealed sample; therefore, these same structures were analyzed using AES, in an attempt to measure the diffusion of Be, Ga, As, Ge, Mo, In and W. The As moved significantly into the Ge and Mo, but was below the detection limit for the W region, indicating the concentration was <0.1 percent. The concentrations of Ga, Be, and In were below the detection limits, putting an upper limit to the diffusion of these elements at <0.1 percent.

Many structures contain important details that can be very complicated to analyze. One example of this is an Au-Ge ohmic contact to n-GaAs grown in a UHV Molecular Beam Epitaxy (MBE) system.[11] The Ge was epitaxial with the GaAs, and the Au was highly oriented. Ohmic behavior of the contacts was observed after annealing at 320°C for three hours. For these conditions, specific contact resistivity and contact measurements on transmission line method (TLM) patterns yielded average values as low as $7 \times 10^{-6}\Omega$-

Figure 4. SEM picture of the surface of an AuGe contact on a GaAs substrate.

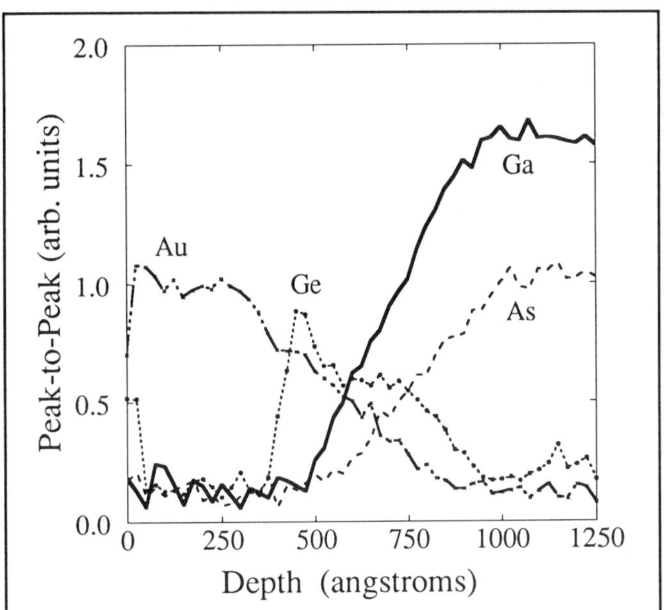

Figure 5. SAM depth profile of an as-grown AuGe contact.

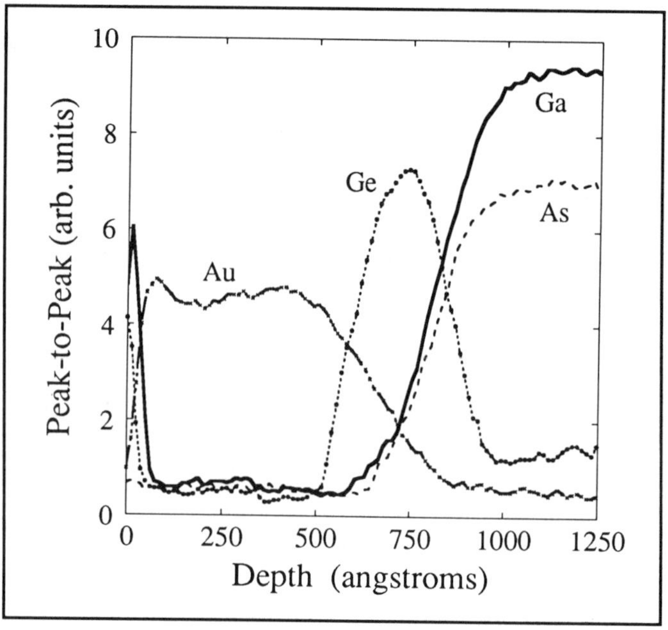

Figure 6. SAM depth profile of an annealed AuGe contact in light regions.

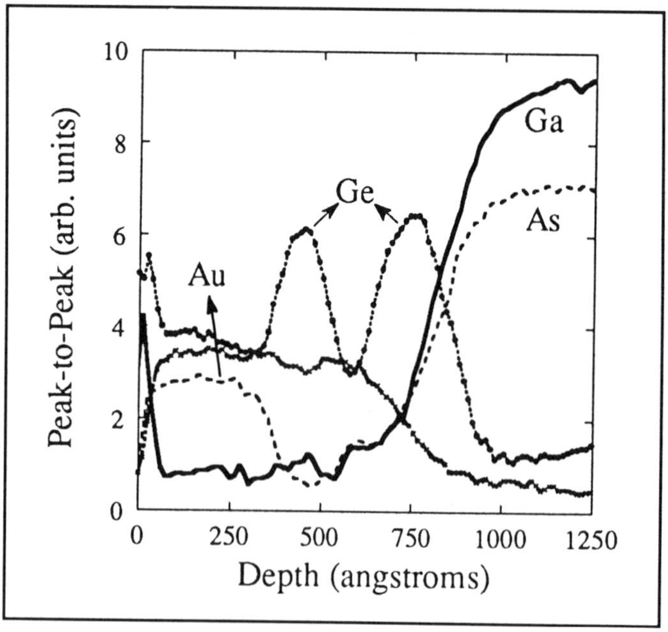

Figure 7. SAM depth profile of an annealed AuGe contact in dark regions.

cm^2 and 0.28Ω-mm, respectively. Figure 4 is a SEM picture of the surface of an annealed AuGe contact on GaAs. Nonuniformities are evident across the surface. Figure 5 is an Auger depth profile of the as-grown sample. Figures 6

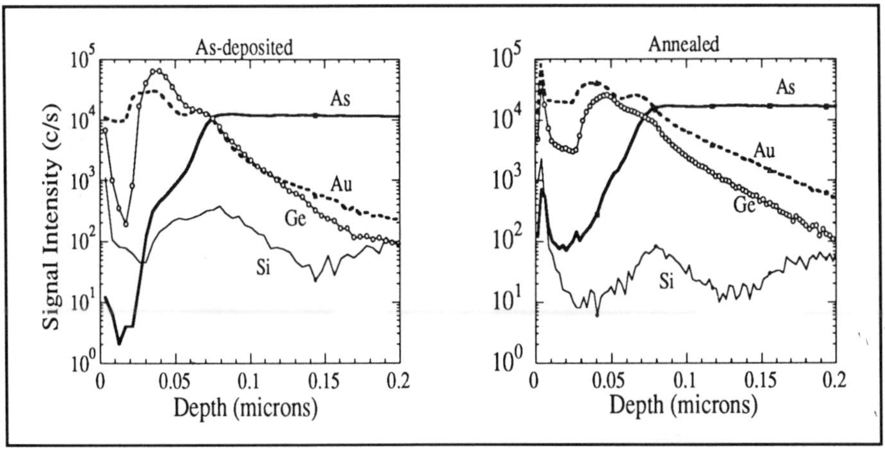

Figure 8. Frontside SIMS profiles comparing an as-grown and an annealed AuGe contact.

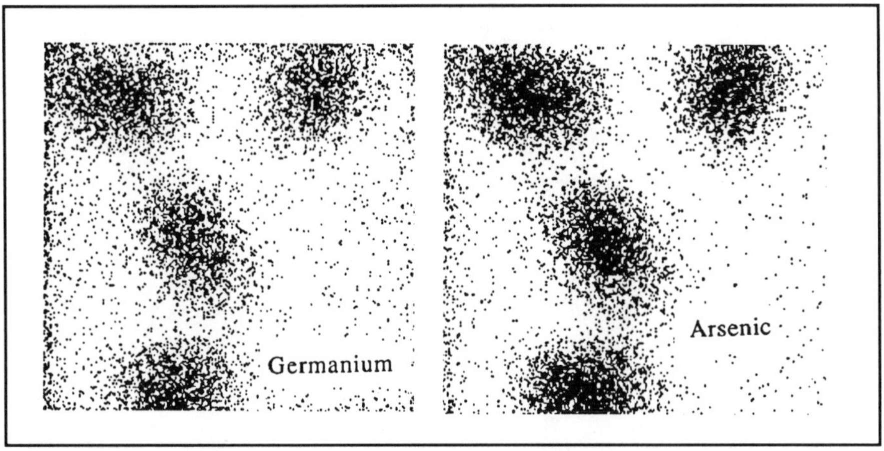

Figure 9. Ge and As ion images obtained during the depth profile in Figure 8 (annealed).

and 7 are depth profiles of the annealed sample measured in the dark and light regions, respectively. These two figures are from the same crater, and the data were extracted by image processing after the analysis. It is clear from the two distinctly different profiles that diffusion is occurring nonuniformly in a lateral direction, possibly forming nucleation sites.

Figure 8 compares SIMS front-side depth profiles of the as-grown and annealed samples. Considerable out-diffusion of both As and Ge is revealed in these profiles. The gold and germanium appear to have diffused considerably into the GaAs substrate in both samples. It is unlikely that diffusion will occur in the as-grown sample, so the apparent diffusion is due to ion beam mixing and interface roughness. Figure 9 presents ion images of the annealed sample. Similar distributions are observed here as in the previous SEM image

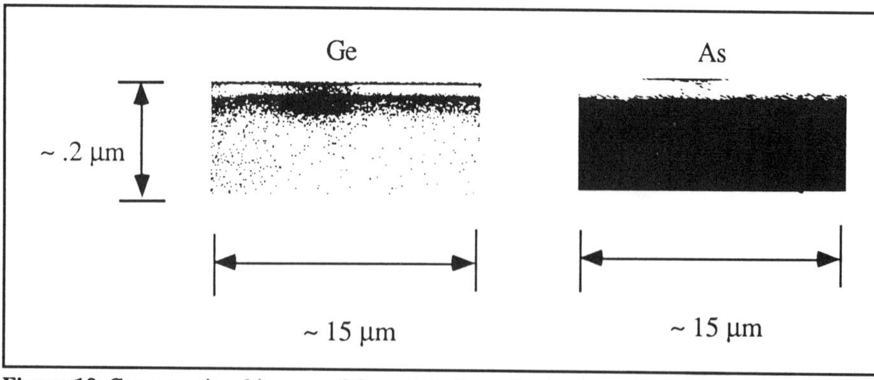

Figure 10. Cross-sectional images of Ge and As from the depth profile in Figure 8 (annealed).

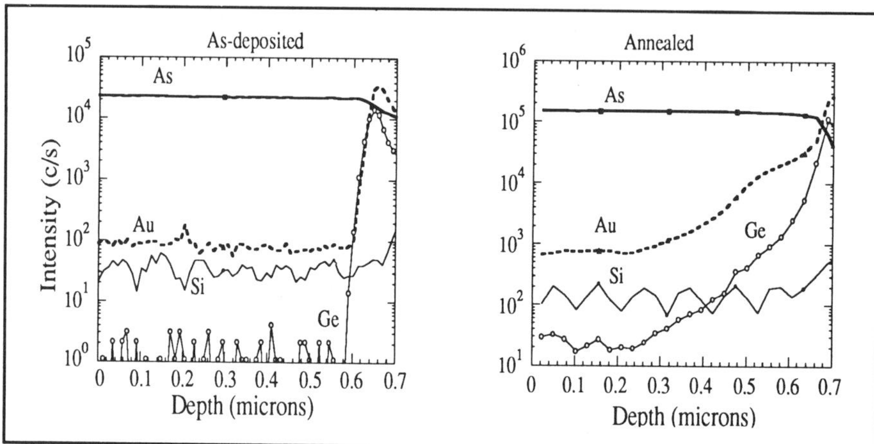

Figure 11. Backside SIMS depth profile comparing an as-grown and an annealed AuGe contact.

(see Figure 4). Again using image processing techniques, a cross-sectional image of the annealed sample is shown in Figure 10. Note both lateral and in-depth elemental distributions are not uniform, possibly indicating metal "spiking" or "pin-hole" formation. This is essentially a pictorial format of Figure 8. The ability to present the data in several different formats often gives a better qualitative understanding of the material interactions taking place.

To determine the actual amount of diffusion, the samples were analyzed using the backside depth profiling technique. The depth profiles were collected using the imaging capability, in order to determine if any diffusion was laterally nonuniform. Figure 11 shows the backside depth profiles of the as-grown and annealed samples. The metal penetration depth, calibrated to the marker layers, is approximately 4000Å. Lateral images of the annealed sample are shown in Figure 12. It is apparent that the Au and Ge did not diffuse uniformly into the GaAs substrate. The selected area depth profiles of the

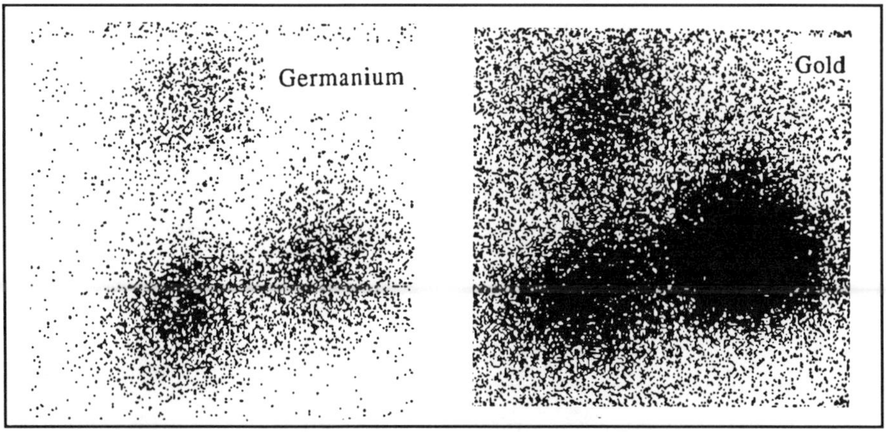

Figure 12. Ion images of Ge and Au from the backside depth profile of the annealed AuGe contact in Figure 11.

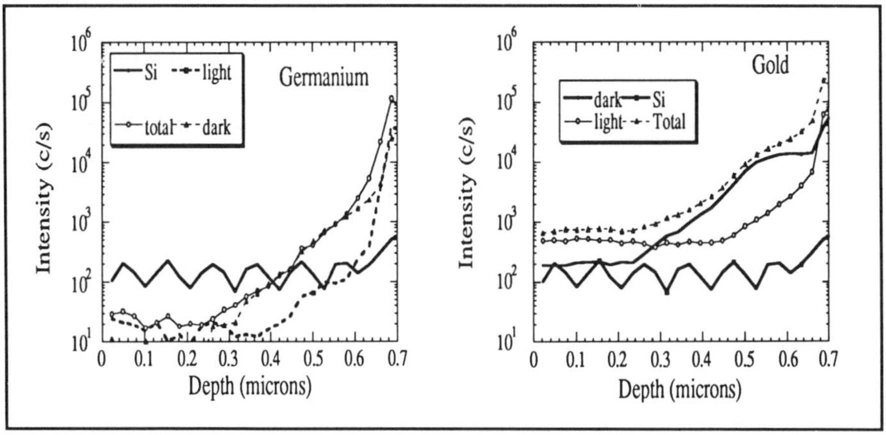

Figure 13. Selected area profiles of Ge and Au correlating the light (low concentration), dark (high concentration), and total regions of Figure 12, along with the Si marker layers used to calibrate the depth.

areas with high, low, and total Ge and Au concentrations are compared in Figure 13. It is significant that all three profiles, for both elements, were obtained simultaneously in the same crater, and deconvolved during the postanalysis data reduction. The areas to compare were determined from the images in Figure 12. Notice that almost all the diffusion occurs in the localized region. This would be impossible to determine without imaging capability. This study is the first application of the backside imaging depth profiling technique.

The backside SIMS depth profiling technique has also been applied for diagnosing prototype electronic devices. Figure 14 is a schematic of the EPSD

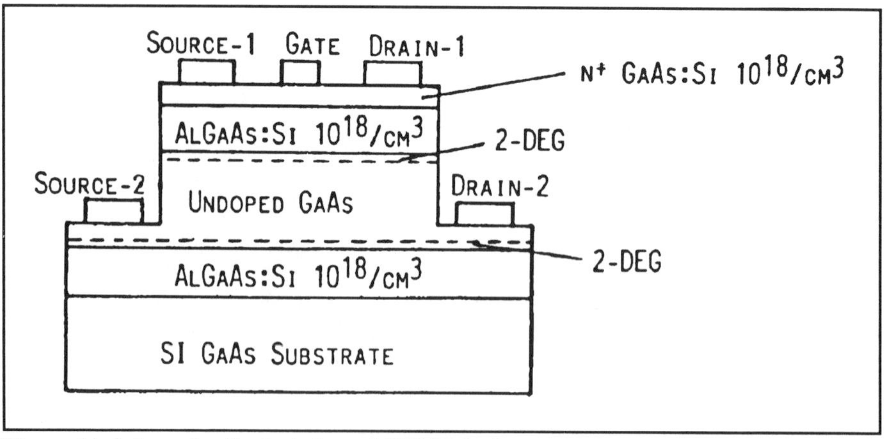

Figure 14. Schematic of a dual-channel HEMT device.

novel design for a DCHEMT. This device is the result of an attempt to produce an ultrahigh-speed inverter, as conceived by Iafrate *et al,*[12] and fabricated and tested at EPSD.[13] It employs a dual-channel structure in which a two-dimensional electron gas (2DEG) is formed on either side of a 50nm GaAs quantum well sandwiched between two n-AlGa As layers with appropriate spacers. Under zero bias, band bending forces electrons from the upper channel into the lower channel, allowing conduction only through the lower channel. Under forward bias, electrons are drawn into the upper channel from below, which reverses the effect. Biasing allows current-steering behavior to be obtained from a single device. This operation is possible due to the ability to contact each channel individually.

Figure 15 shows the SIMS frontside depth profile that confirms the initial electrical measurements and reveals the metal contact penetration into this 2DEG region and presence of severe Al diffusion. Some of these results could be explained as SIMS artifacts, particularly when rough interfaces between each metal and contacting semiconductor surfaces are concerned. From previous work involving similar ohmic contact structures and backside sputter depth profiling, it is clear that significant metal penetration from the contacts to the active device layers does occur, dictating the use of thicker n+ GaAs contacting layers to be included in the electronic device layer designs.[4] Figure 16 shows the backside depth profile of the same device structure with a thicker (~3000Å) n+ GaAs contacting layer. The 2DEG region has little diffusion and is not affected by the contact metal penetration. The frontside depth profile of this sample, which has the thicker n+ GaAs contacting layer, still indicates metal penetration deep into this region. The backside profile

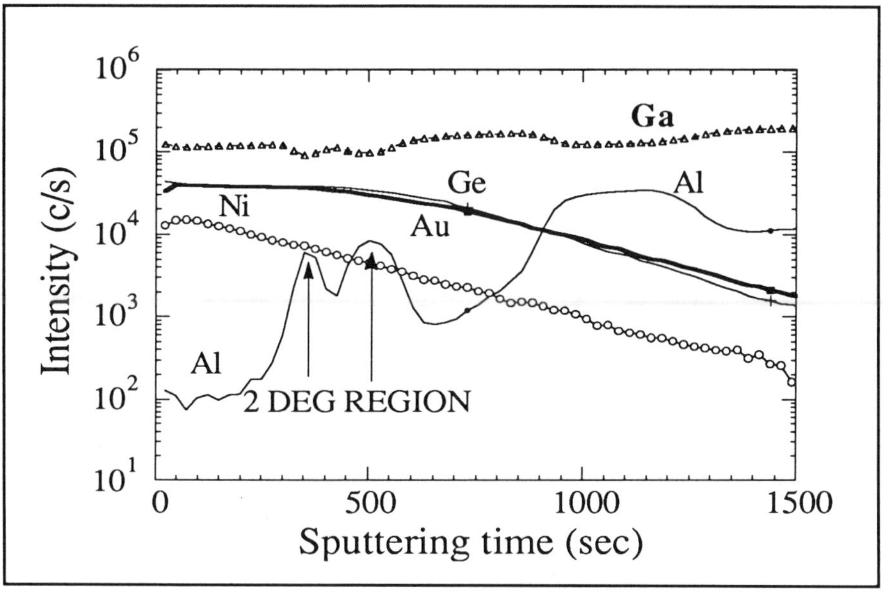

Figure 15. SIMS frontside depth profile of a dual-channel HEMT device.

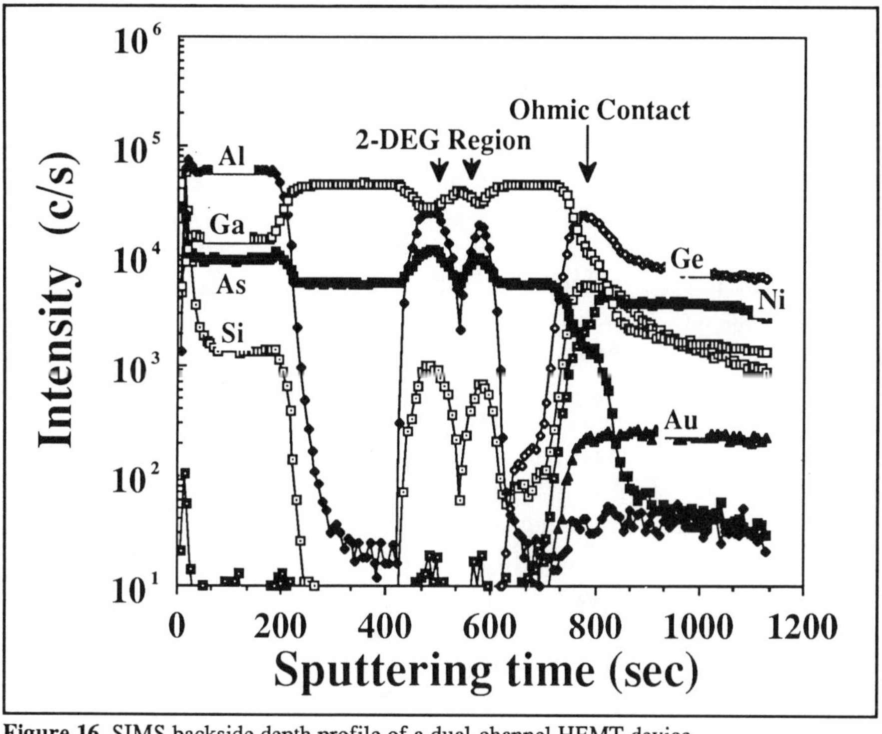

Figure 16. SIMS backside depth profile of a dual-channel HEMT device.

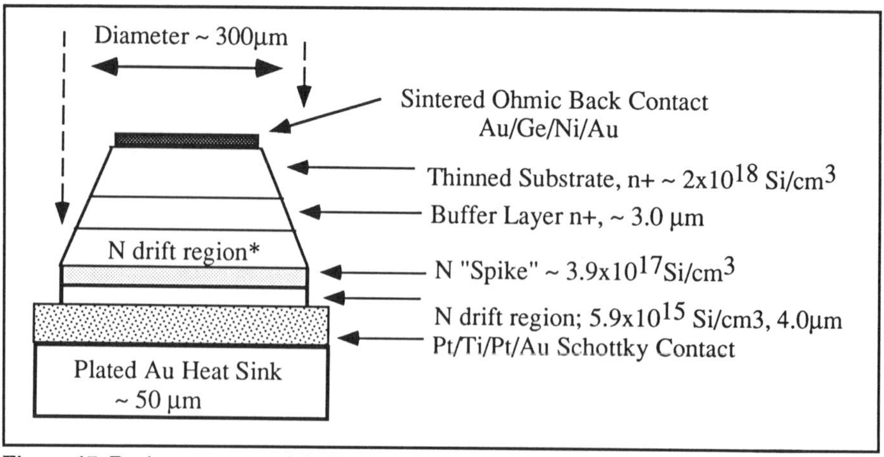

Figure 17. Design structure of the Phoenix Missile/IMPATT diode.

demonstrates that the contact metals do not actually diffuse into the device structure. The backside sputtering approach avoids sputtering effects due to surface roughness between the MBE layers and contact metallization layer(s), while at the same time providing the desired reverse concentration gradient of the penetrating metal layers.

We have also encountered samples that are of critical military importance and are well suited to the backside sputter depth profiling method. IMPATT diodes, which were being used for the Phoenix Missile System, were believed to be failing due to a diode "burnout" phenomena, whereby the sidewall of the mesa structure would blister and permit gold to flow from top to bottom contact, hence, shorting the electrical path. Although most of this work was performed by secondary electron microscopy at EPSD, the low level doping and metal penetration profiles provided additional and desirable information easily obtainable from SIMS. The device structure of the IMPATT diode required little sample preparation to measure a depth profile from the back of the contact structure because this diode is contacted from each side. Simple chemical etching of the top ohmic contact bonding pad allowed profiling through the entire III-V semiconductor structure as well as the metal contact structure from low to high concentration gradient. Presented in Figure 17 is a schematic of the intended IMPATT diode structure. The backside depth profile is presented in Figure 18.

The backside depth profile reveals two important features: (1) the Ti from the lower contact is penetrating very deep into the device layer; (2) the doping layer(s) consisting of Si at $5.9x10^{15}/cm^3$ level was quite higher than the original design structure, and the "spike" area was proceeded by a Si diffusion front. As in the previous studies, the additional information gained from

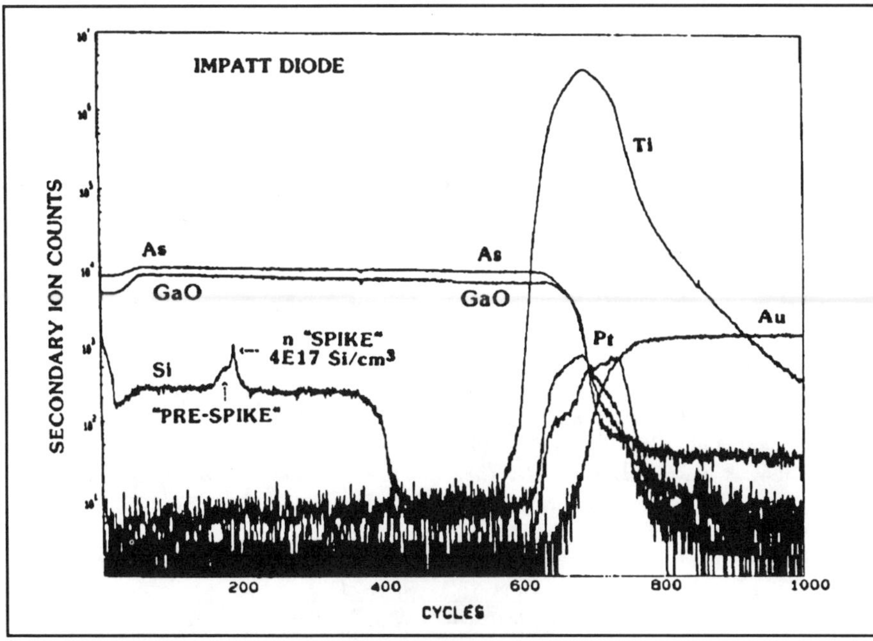

Figure 18. SIMS backside depth profile of a failed IMPATT diode.

performing backside depth profiling and SEM is invaluable in designing and fabricating electronic devices.

Discussion

The accurate measurement of elemental distributions in semiconductor devices is essential in developing and diagnosing advanced designs. Without sensitive analytic techniques, scientists and engineers are often operating blindly when trying to explain and understand device properties without accurate correlation to material properties and behavior.

Future work in this area includes expanding the backside sputter technique to Auger electron spectroscopy and Rutherford backscattering spectroscopy.

Acknowledgments

We would like to thank Captain Ken Merkel of Wright Laboratory (WPAFB) for his ongoing collaboration on novel contact structures through the Tri-Service Reliance Program.

References

1. R.G. Wilson, F.A. Stevie, and C.W. Magee, *Secondary Ion Mass Spectrometry* (John Wiley and Sons, New York, 1989).
2. A. Benninghoven, F.G. Rudenauer, and H.W. Werner, *Secondary Ion Mass Spectrometry* (John Wiley and Sons, New York, 1987).
3. D. Briggs and M.P. Seah, Eds., *Practical Surface Analysis by Auger and X-ray Photoelectron Spectroscopy,* (John Wiley and Sons, New York, 1983).
4. J.R. Shappirio *et al., J. Vac. Sci. Technol.* **A5**, 1503 (1987).
5. R.T. Lareau, in *Secondary Ion Mass Spectrometry, SIMS VI*, A. Benninghoven, A.M. Huber, and H.W. Werner, Eds. (John Wiley and Sons, New York, 1988), pp. 437-440.
6. J.A. Jackman *et al., Surf. Interface Analysis* **15**, 451 (1990).
7. C.J. Palmstrom *et al., J. Appl. Phys.* **67**, 334 (1990).
8. J. Herniman, J.S. Yu, and A.E. Staton-Bevan, *App. Surf. Sci.* **52**, 289 (1991).
9. K. Kenefick, *J. Electrochem. Soc.* **129**, 2380 (1982).
10. K. Merkel *et al., IEEE Proceedings,* 1991.
11. H.S. Lee *et al.*, in *Advanced III-V Compound Semiconductor Growth, Processing and Devices,* S.J. Pearton, J.M. Zavada, and D.K. Sadana, Eds. (Mater. Res. Soc. Proc. **240,** Pittsburgh, PA), in press.
12. G.J. Iafrate *et al.*, "Dual Channel High Electron Mobility Field Effect Transistor", S.N. 228795 , July 29, 1988.
13. R. Khanna *et al., IEEE Elect. Device Letters* **10**, 531 (1989).

12

Observation of rf-Induced Failures in Silicon Bipolar Junctions Using Synchrotron X-Ray Topographic Imaging

C. Fazi
U.S. Army Research Laboratory
Signatures, Sensors, Signal and Information Processing Directorate
2800 Powder Mill Road
Adelphi, MD 20783

M. Dudley
S. Wang
D. Gordon-Smith
Department of Materials Science & Engineering
State University of New York at Stony Brook
Stony Brook, NY 11794

The U.S. Army is concerned with electronics systems survivability in high transient rf environments such as high-power microwaves (HPM) and nuclear-generated electromagnetic pulses (EMP). Examining damage induced in semiconductor devices by transient electromagnetic fields is fundamental to assessing the survivability and hardening of such devices. Synchrotron white-beam x-ray diffraction topography, in both transmission and grazing Bragg-Laue geometries, has been used to investigate damage induced during the electromagnetic breakdown of *pn* junctions on silicon. Rf-induced filamentation effects are reported for the first time to give rise to characteristic contrast features on grazing Bragg-Laue topographs. Preliminary results show that filaments always occur at the interface between the junction and the metallic contact, indicating that discontinuous changes in topology as a function of depth increase susceptibility. Similar filamentation in *pn* junctions of planar technology was observed for both high-power microwave and reverse-biased video-pulse-type transients. These new observations explain why often it is so difficult to compare experimental failure data with the semiconductor failure models. Comparing results with optical microscopic studies show that optical microscopes reveal only surface damage, while the x-ray technique, capable of yielding subsurface information, provides reliable data on both the true location and extent of crystal damage.

Introduction

The U.S. Army is concerned about the survivability of military systems employing semiconductor devices in high transient radio frequency (rf) environments such as high-power microwaves (HPM) and nuclear-generated electromagnetic pulses (EMP). Understanding the nature and causes of the damage induced in electronic semiconductor devices by transient electromagnetic fields is an issue of central importance to assessing the long-term survivability of such devices and to hardening them. This damage is associated with drastic changes in the electrical characteristics of the devices, which occur during uncontrolled breakdown processes. Apart from detailed electrical characterization of the devices before and after breakdown, previous investigations of the nature of the resulting damage have been largely limited to optical microscopy and scanning electron microscopy (SEM) studies (for example, see Ricketts *et al.*[1]). By nature, optical and SEM studies can reveal modifications only in the surface morphology of the metallization and are not sensitive to subsurface damage. SEM of cross-sectionally beveled specimens can reveal some information (for example, distribution of chemical species), but it cannot give a reliable indication of the distribution of damage throughout the epilayer region. Transmission electron microscopy is generally not routinely applicable in such studies for two reasons: (1) the technique is destructive by nature, because thin specimens must be prepared (potentially removing some of the damaged regions distributed throughout the thickness of the device); and (2) the technique has a very small field of view, making it very difficult to prepare specimens specifically containing the localized damaged areas. In order to reveal such subsurface damage, a nondestructive technique capable of high strain resolution is desirable. X-ray diffraction topography provides a potential solution, although in either transmission or conventional Bragg geometries, sensitivity to distortions located only a few micrometers below the surface of a crystal is very limited. However, employment of grazing Bragg-Laue geometries enhances the near-surface sensitivity of the technique. The versatility of the technique in these geometries is greatly enhanced by the application of synchrotron white radiation, with its tunability and high brightness (see Yao *et al.*[2]). The use of synchrotron x-ray topography to image the damage created during rf- induced failures in semiconductors has already been proved to be an important diagnostic tool in understanding rf hardening techniques and failures in semiconductors (see Dudley *et al.*[3]). Unlike conventional x-ray sources, the synchrotron radiation source can produce area-filling beams of highly collimated white x-rays. The large beam area, coupled with high intensity, make it possible to record many images simultaneously (comprising a Laue pattern), with exposure times of just a few seconds. Transmission geometries enable bulk strain distributions in the crystal to be revealed, while the use of grazing-incidence Bragg-Laue geometries reveals detailed information of local strain distributions as a function of depth. This paper is a preliminary report of results of a synchrotron x-ray diffraction topographic investigation of damage deliberately induced in *pn* junctions on silicon by high-power microwave sources and video pulses of 3μs and 200ns duration. Information obtained from x-ray topographs is compared with that obtained from optical micrographs. Observed differences are discussed, and the superior nature of the information afforded by the x-ray technique is demonstrated.

Experimental Results

Specimens used in these experiments were fabricated at the Signatures, Sensors, Signal and Information Processing Directorate (formerly the U.S. Army Harry Diamond Laboratories). Samples consisted of arrays of diodes of different shapes and sizes on two-inch diameter (111) silicon wafers. The wafers consist of a 12-μm-thick *N*-type epilayer on a 0.01-in. thick N^+-type substrate, with an epilayer doping level of 10^{15} impurities per cubic centimeter. The use of wafers rather than individually packaged devices facilitated the testing and subsequent simultaneous imaging of statistically significant numbers of devices that had undergone identical processing. A cross-sectional view of a *pn* junction is given in Figure 1. We induced breakdown in selected 300-μm-diameter circular junctions by direct injection of a single 3-μs-duration or 200-ns-duration unipolar pulse of reverse polarity. This was achieved with the use of a coaxial probe technique developed by Fazi.[4] The amplitude was increased until permanent changes in the breakdown voltage were observed on a curve tracer. HPM breakdown was induced in selected triangular junctions of the same surface area as the circular ones. The purpose in switching topology for the HPM test was to induce failures with the lowest possible pulse amplitude, because high electric field at the corners would induce failures there. It did not; instead, it was surprising to observe that the induced failures for both topologies, driven by HPM and video-pulse-type waveforms (narrowband and wideband) occurred only between the metallic interconnect and the *pn* junction, as indicated in Figure 1 by an arrow showing the failure location. For discrete devices, the failure location has been observed to occur randomly around the P^+ well, and the failure levels are proportional to the junction volume. If the failure is attributed only to an edge

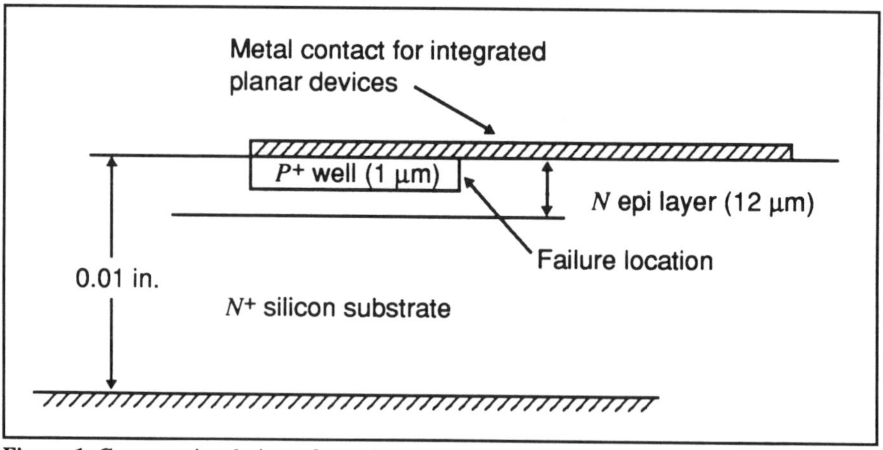

Figure 1. Cross-sectional view of a *pn* junction used in integrated planar devices. Contacting P^+ well with a metal line induces failures at this interface (arrow indicates failure location). Discrete semiconductor devices usually do not rely on this type of contact; hence, they may be less susceptible to failure.

effect, the modeling is no longer a three-dimensional issue. The correct physics must be taken into account when setting the boundary conditions for the damage modeling codes. Based on this preliminary observation, one may wonder if planar technologies (including very-high-speed integrated circuits and microwave monolithic integrated circuits) may be inherently more vulnerable to rf burnout than discrete vertically structured devices.

We carried out white-beam transmission and variable-penetration-depth topographic imaging in grazing Bragg-Laue geometries using the white-beam camera on the Stony Brook Synchrotron Topography Beamline (X-19C) at the National Synchrotron Light Source (NSLS), Brookhaven National Laboratory. The variable-penetration-depth grazing Bragg-Laue imaging was carried out using methodologies that can be found in the literature.[2,5-8] Figures 2 and 3 show optical and x-ray topographic images recorded in the grazing Bragg-Laue geometry from two 300-μm-diameter circular diodes following breakdown induced by pulses of 200ns and 3us duration, respectively. Note that several features are visible in the second optical picture that can be interpreted as failures. Only after comparing with the x-ray images can one determine that the feature at the interface edge is due to a damaging filament (indicated by an arrow). Figure 4 shows similar optical and x-ray topographic images recorded from a triangular diode of equivalent surface area following HPM-induced breakdown. The purpose of changing to a triangular topology is to attempt to induce failures at the sharp corners due to high field-induced breakdown. Again, the failure occurred at the metal-to-junction interface (indicated by an arrow).

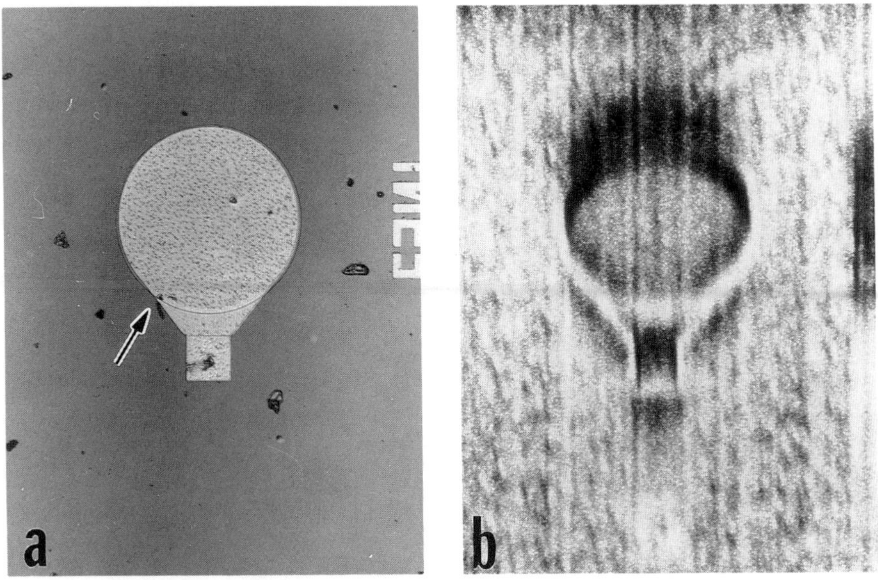

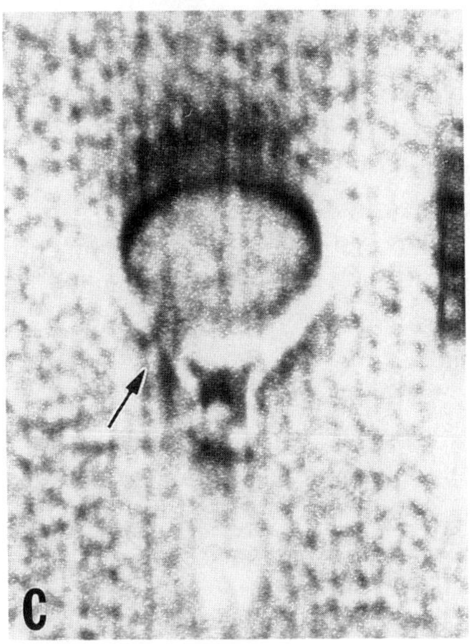

Figure 2. (**a**), Optical image, and (**b**), x-ray topographic image for an unfailed 300-μm *pn* junction. Image (**c**) is an x-ray topographic image recorded in the grazing Bragg-Laue geometry following breakdown induced in reverse-bias using a pulse of 3μs duration. Arrows point to damaging filaments. In this example, failure is evident in x-ray image, but not in optical image.

Conclusion

Based on the limited data taken so far on planar epitaxial devices, rf transients of either narrowband or wideband type may induce filamentary failures at the interface where the metallic line contact and the junction meet, as indicated in Figures 2(a) and 3(a) by the arrows. This poses a dilemma when

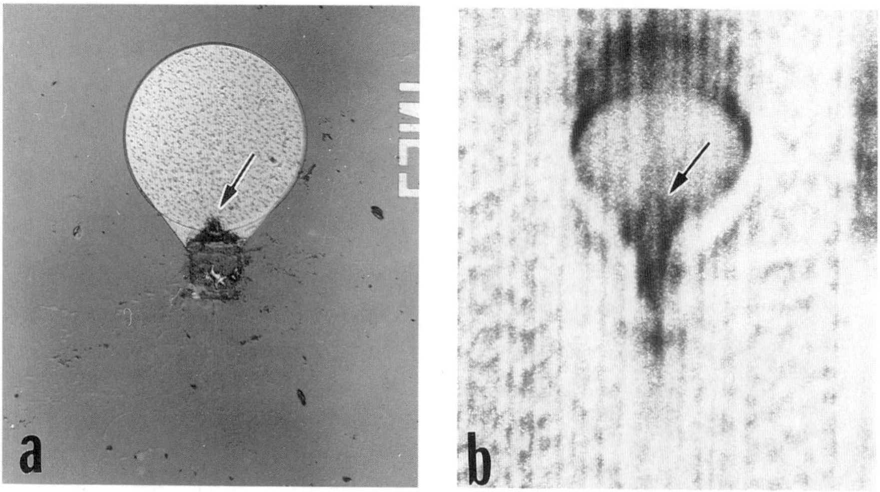

Figure 3. (**a**) Optical image, and (**b**) x-ray topographic image recorded in the grazing Bragg-Laue geometry following breakdown induced in reverse bias using a pulse of 200ns duration. Arrows point to damaging filaments. In this example, failure is evident in both types of image.

one needs to estimate the rf failure levels for a certain device, or the requirements of a Zener protection circuit for a VLSI circuit. Did the junction or transistor itself fail, or is the device shunted by a shorted path caused by a filamentary breakdown at the aforementioned interface, or both?

When failures occur only at this interface, improving either the junction topology and size or the metallic contact alone will not necessarily reduce the susceptibility of the device to rf transient burnout. Though efforts can be made to avoid sharp corners on the surface junction topology, it is also necessary to consider the effects of similar topologies from a cross-sectional point of view (which may be easily forgotten, because this view is not readily observable).

In light of our results, it would not be surprising that two equal devices, one being a planar device and the other a vertically structured device (for example, a 1N4148 silicon diode and equivalent size planar device), would fail at different rf power levels. This suggests important limitations on any rf burnout model or code, and that rf failure data for planar integrated devices in component reliability databases may be suspect.

Another important observation made when the x-ray results were reviewed is that surface evidence of failure, as seen optically, may not correspond with the extent and location of crystal damage. Some devices actually failed that showed little or no surface evidence, yet they had catastrophic crystal damage that only x-ray topographic imaging could reveal. Figures 2 and 3 show the comparison of optical micrograph and x-ray synchrotron

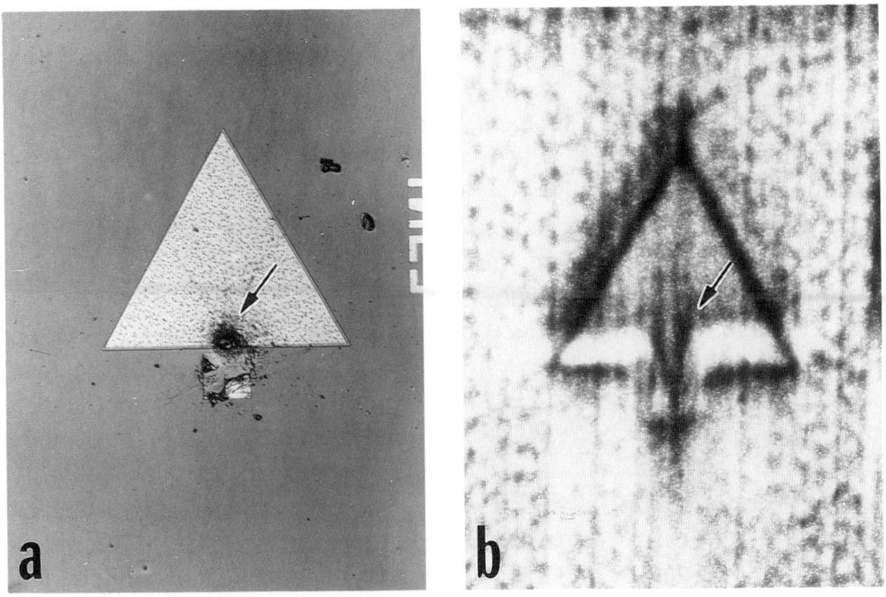

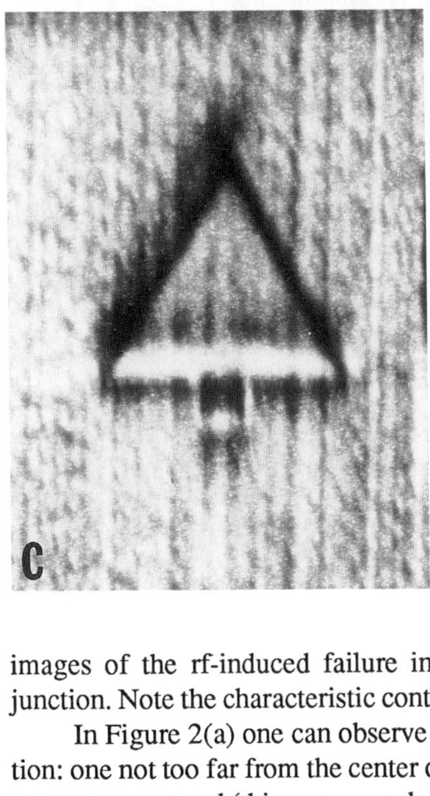

Figure 4. (**a**) Optical image, and (**b**) x-ray topographic image recorded through the use of grazing Bragg-Laue geometry following high-power-microwave-induced breakdown. Arrows point to damaging filaments. Image (**c**) is an x-ray topographic image for an unfailed triangular junction.

images of the rf-induced failure in a 300-μm-diameter silicon planar *pn* junction. Note the characteristic contrast feature associated with the filament.

In Figure 2(a) one can observe three spotlike features on the metallization: one not too far from the center of the circular anode, a second one in the square contact pad (this corresponds to the position of the probe contact), and a third one at the metal junction interface, as shown by the arrow. Of the two spots observed within the junction region, the one near the center is only a surface defect on the metallization, whereas the other (indicated by the arrow) is the visible portion of a filamentation channel that has a region of severe lattice damage associated with it just below the surface. Employing the grazing Bragg-Laue geometry, with its variable penetration depth, enables one to conclude that the maximum lattice distortion occurs closer to the surface than to the internal interface between the substrate and the epitaxial region of the junction (see Dudley *et al.*[3]).

Damage induced by HPM waveforms shows a similar pattern. Figure 4(a) reveals what appears to be a filament at the interface between the metal contact and the diode itself. Comparison between the x-ray topographic contrast features associated with this filamentary feature (on Figure 4c) and those observed from filaments induced in reverse bias reveals this similarity.

Although the synchrotron has not previously been used to probe active semiconductor devices, our research shows that the synchrotron is a unique and powerful diagnostic tool to image failures in semiconductor devices. This research tool is unique in spotting failures that are not visible to ordinary surface imaging methods. Based on these capabilities, it is now possible to better assess the susceptibility of different semiconductor technologies to rf-induced failures and understand how to improve the hardening of electronic systems. This work is continuing and is also probing new technologies such as silicon carbide (SiC) and GaAs, thus contributing to our understanding of the nature and causes of the damage mechanisms for the primary semiconductor technologies—an essential requirement for assessing the long-term survivability of battlefield assets employing electronics. Thus, this work is directly relevant to Army concerns for the survivability of future military systems employing semiconductor devices in high transient rf environments, such as HPM and EMP.

Acknowledgments

The support of the Signatures, Sensors, Signal and Information Processing Directorate (formerly the U.S. Army Harry Diamond Laboratories) Semiconductor Engineering and Materials Technology Facility is acknowledged for the silicon wafer processing, without which this study would not have been possible.

Synchrotron topography was carried out at the National Synchrotron Light Source, which is funded by the Department of Energy. D. G.-S. was on sabattical leave from the University of Warwick, Great Britain, during the course of this research.

References

1. L.W. Ricketts, J.E. Bridges, and J. Miletta, *EMP Radiation and Protective Techniques*, (New York, Wiley Interscience, 1976), pp 78-79.
2. G.-D. Yao, M. Dudley, and J. Wu, *J. X-ray Science Tech.* **2**, 195 (1990).
3. M. Dudley *et al.*, *Government Microcircuit Applications Conference Digest of Papers*, **Vol. XVI**, 651 (1990).
4. C. Fazi, Ph.D. Thesis, Johns Hopkins University, 1990.
5. M. Dudley, J. Wu, and G.-D. Yao, *Nucl. Instr. & Meth.* **B40/41**, 388 (1989).
6. M. Dudley *et al.*, *Mat. Res.Symp. Proc.* **160**, 469 (1990).
7. M. Dudley *et al.*, *Mat. Res. Symp. Proc.* **163**, 1031 (1990).
8. M. Dudley *et al.*, *Mater. Sci. & Engin.* **B10**, 75 (1991).

13

Strained Quantum Well Field Effect Lasers for Optoelectronic Integration

Paul W. Cooke
Geo Centers, Inc.
Fort Monmouth, NJ 07703-5601

Wayne Chang
U.S. Army Research Laboratory
Electronics and Power Sources Directorate
Fort Monmouth, NJ 07703-5601

G.W. Taylor
P.A. Evaldsson
AT&T Bell Laboratories
Holmdel, NJ 07733-3030

Optoelectronic integration would be desirable for both optical interconnected circuits and high frequency analogue optical links. A unique approach to fabricating both lasers and field effect transistors from the same epitaxial layer sequence is presented. The commonality of processing steps enables both types of devices to be concurrently fabricated, thereby simplifying the problem of integration. The authors present discrete device results and discuss the prospects for future circuit implementation.

Introduction

Considerable emphasis is currently being placed upon the ability of high-speed optical links to replace existing electrical coaxial cables in systems such as phased array radars.[1] Such applications place considerable demands upon the devices employed to perform the optoelectronic conversion at both the transmitter and receiver ends of the link. In particular, for the transmitter elements, high-speed and high-efficiency semiconductor lasers would be desirable. Future systems have the potential to require large numbers of such devices. From both cost and complexity considerations, integration of lasers and either other optical elements or drive electronics would be advantageous for these applications.

More esoteric functions can be envisaged for integrated circuits containing many optical transmitter and receiver units. Optical preprocessing has been proposed as a method to speed up and enhance many complicated digital processing tasks.[2] Examples of such systems include pattern recognition and target identification. In these cases, the input data are intrinsically parallel in nature, so that filtering or preprocessing with the input still in parallel form would be beneficial. With suitably adaptable circuits, arrays of "smart pixels" could be envisaged that use in-plane electronic processing and plane-to-plane multiple optical interconnects. Vertically stacking several planes would then provide the potential for a full parallel processing arrangement. In an ideal situation, even the output of the system could be in optical form.

Several different approaches have been tried to integrate lasers with either Field Effect Transistors (FETs) or bipolar devices.The most common approach is to employ a multiple or stacked growth technique,[3] where the laser structures are grown either above or below the transistor structures. This inherently creates problems of planarization and interconnection. An alternative approach is to use a common epitaxial layer sequence that enables fabrication of the different devices from the same basic structure. Appropriate choice of processing steps then ensures that both lasers and drive or logic electronics can be integrated at will.

Recently, the Heterojunction Field Effect Laser (HFEL) was demonstrated in the AlGaAs/GaAs material system.[4] The epitaxial layer sequence for the HFEL is totally compatible with that of the Heterojunction Field Effect Transistor (HFET).[5] Both the HFET and HFEL are formed by placing a heavily doped charge sheet adjacent to an isotype heterojunction. With the correct choice of dopings, an inversion channel is formed at the interface. Modulation of the inversion charge is achieved by using an ohmic gate contact. HFEL operation is obtained by configuring the doping profile to produce the inversion charge in the multiquantum well (MQW) active region of a suitable optical waveguiding structure. Due to the use of the ohmic contact, the gate contact of the HFET can be spatially removed from the channel to enable a waveguiding region to be incorporated. For the N-channel configuration, biasing the gate-source diode beyond breakdown provides hole injection from the gate contact and electron injection into the channel for lasing. As such, the HFEL effectively functions as a lateral current injection laser. Inherent compatibility of both devices ensures that they can be fabricated alongside each other with the appropriate choice of processing steps. Thus, they naturally lend themselves to the prospect of optoelectronic integration.

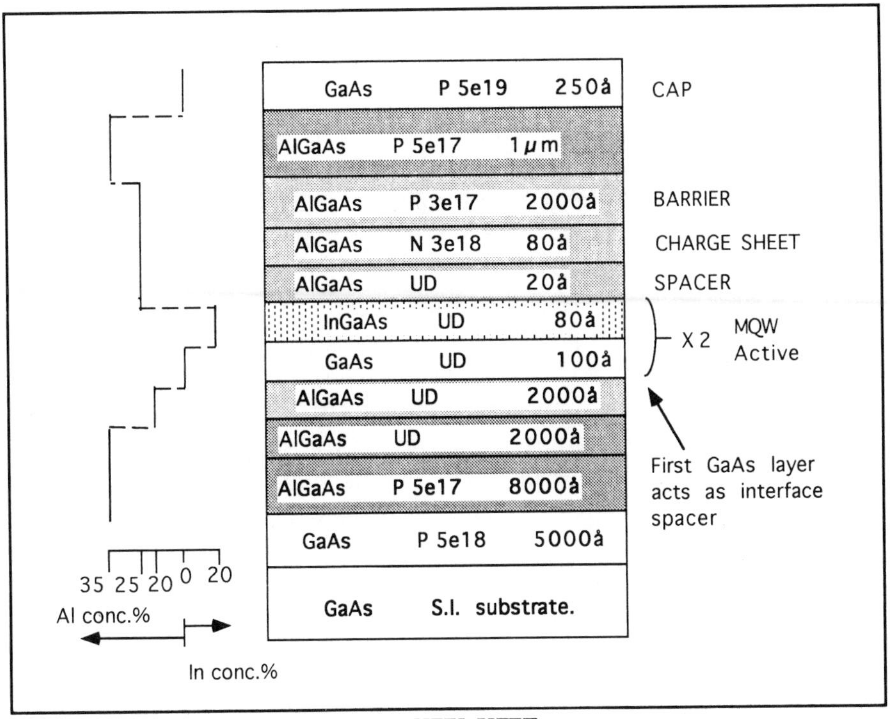

Figure 1. Epitaxial layer sequence for the HFEL/HFET.

Method

In this paper, results of fabrication of the HFEL and HFET using the strained InGaAs/AlGaAs material system are presented. Compared with the lattice-matched GaAs/AlGaAs material system, incorporation of coherently strained InGaAs quantum wells offers several advantages. The incorporated strain significantly modifies the valence band structure of the quantum well and reduces the in-plane hole mass. The lifting of the zone center degeneracy reduces the density of states function and has been demonstrated to produce lower threshold current densities and faster small signal response in conventional diode lasers.[6] The smaller bandgap also means the GaAs substrate is transparent at the emission wavelengths, an important consideration for the optical preprocessing applications just envisaged. From the HFET perspective, the use of strained MQW structures has the potential to increase the maximum channel charge that can be obtained prior to parasitic parallel conduction occurring in the immediate gate ("barrier") region.

Results and Discussion

The basic growth structure used for these experiments is shown in Figure 1. The entire structure was grown using Molecular Beam Epitaxy (MBE) on a semi-insulating GaAs substrate. The growth temperature for the AlGaAs and GaAs layers was 580°C, and for the InGaAs quantum well regions 525°C. The growth temperatures for the AlGaAs layers are much lower than those conventionally considered necessary for optimum laser performance.[7] The use of the strained MQW region with GaAs barriers means that the usually poor lower AlGaAs/GaAs interface can be spatially separated from the MQW active region. Thus, nonradiative parasitic interface recombination can be limited. Crucial to maintaining high luminescence efficiency is the correct choice of the binary (GaAs) spacer thickness. Typically, a 100 to 200Å GaAs spacer has been found to be necessary. To facilitate the change in growth temperatures for the InGaAs layers, a two-minute growth interruption is employed. The use of lower overall growth temperatures is also beneficial in terms of dopant stability, because the HFET voltage threshold is determined exclusively from the dopant profile within the structure (unlike, for example, conventional modulation doped FET [MODFET] devices that rely on gate recess procedures). Control over the FET parameters, such as the voltage threshold, becomes crucial when larger scale logic circuits are considered.

The detailed structure consists of a heavily doped charge sheet (N+) grown above the strained InGaAs MQW region. A small AlGaAs spacer region is incorporated in a similar manner to MODFET structures. The charge sheet is depleted both at equilibrium and under bias. The region above the charge sheet is p-doped and designated the barrier region. HFET operation can be viewed as the tradeoff of barrier depletion charge and inversion channel charge. The surrounding layers form an optical waveguide in a similar manner to a conventional diode laser. The lower waveguide layer also acts as a collector or backgate contact for the HFET/HFEL. Key to operation of both the HFEL and HFET is the use of a highly P+ doped cap layer as the gate or emitter contact.

A schematic device cross-section is shown in Figure 2. Fabrication of the HFEL proceeds by initially defining a refractory gate (stripe) contact. Lateral waveguiding is obtained by dry etching down to within 0.2μm of the active region. Using the refractory contact as an implant mask, Si ion implantation is used to form source and drain contacts. Depending upon the exact configuration desired, the HFEL can be single- or dual-side sourced. The devices

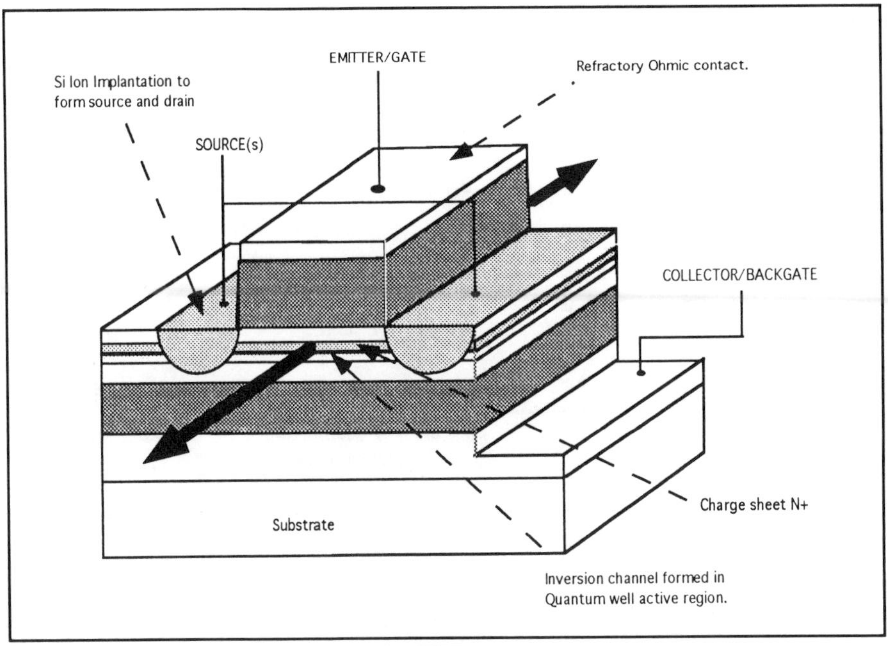

Figure 2. Schematic cross-section HFEL/HFET.

tested here were single sourced for simplicity. Activation of the implants is achieved using a rapid thermal anneal at 950°C for 10 seconds. The backgate or collector contact is formed by etching down below the MQW region. Final bond pad metalization steps using polyimide isolation layers are used to enable contacting the device. Thinning and cleaving are performed in an identical manner to diode lasers to produce facets. HFET fabrication is similar, except for the gate (stripe) dimensions and final thinning step.

Operation of the HFEL is achieved by biasing the emitter contact with respect to the source contact. Typical pulsed (200nsec,1kHz repetition) light output characteristics are shown in Figure 3. Threshold currents of 34mA, 32mA, and 22mA were obtained on 15-, 10-, and 5-µm wide devices. The corresponding threshold current densities were 500,710 and 970A/cm^2 respectively. Typical slope efficiencies were 0.25W/A per facet. Continuous operation of these devices was obtained with threshold currents of 40mA, 38mA, and 29mA. Maximum continuous wave (CW) output was limited due to the device mounting arrangement and higher than expected series resistance(15Ω). This in turn was attributed to a higher than expected source resistance term resulting from non-optimization of the implant conditions. With many lateral current injection lasers, there is a possibility of filamentary

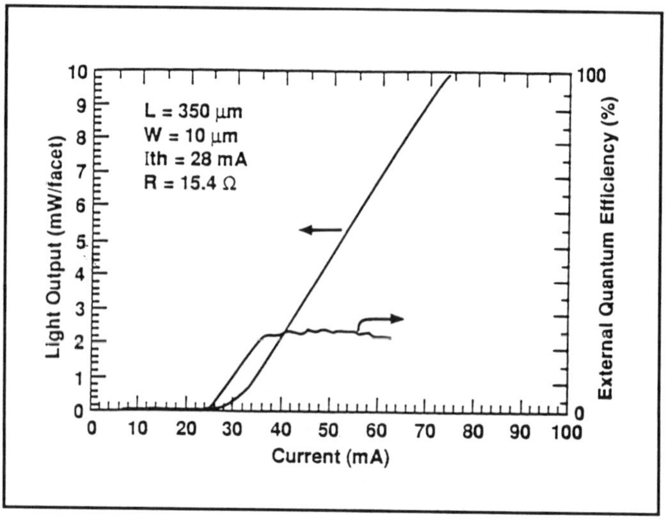

Figure 3. Pulsed light output characteristics.

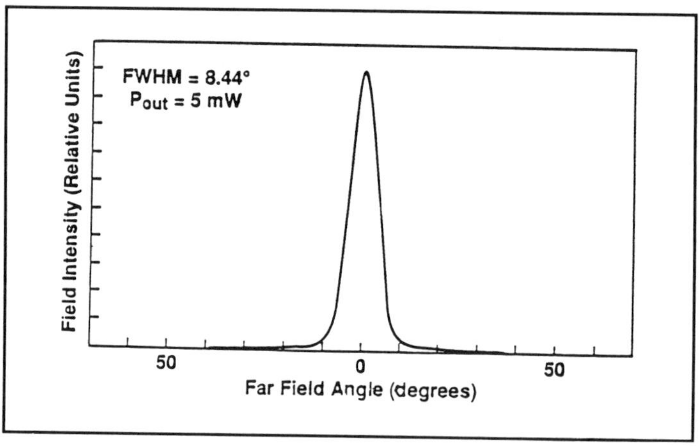

Figure 4. Measured farfield pattern, parallel to junction.

action from non-uniform current injection across the stripe. To confirm whether this was happening, the far field pattern of a 10-μm wide HFEL was recorded. This is shown in Figure 4. From the observed symmetry, uniform current injection across the stripe can be inferred. The wider 15-μm HFEL's exhibited multimode operation, as could be expected from the dimensions of the lateral waveguide structure.

The wavelength of emission of the HFEL was measured using an Anritsu spectrum analyzer. For CW operation, emission wavelengths were 1.02μm for 10-μm wide stripes. The output is shown in Figure 5.

Small signal response of the HFEL was measured using a high-speed

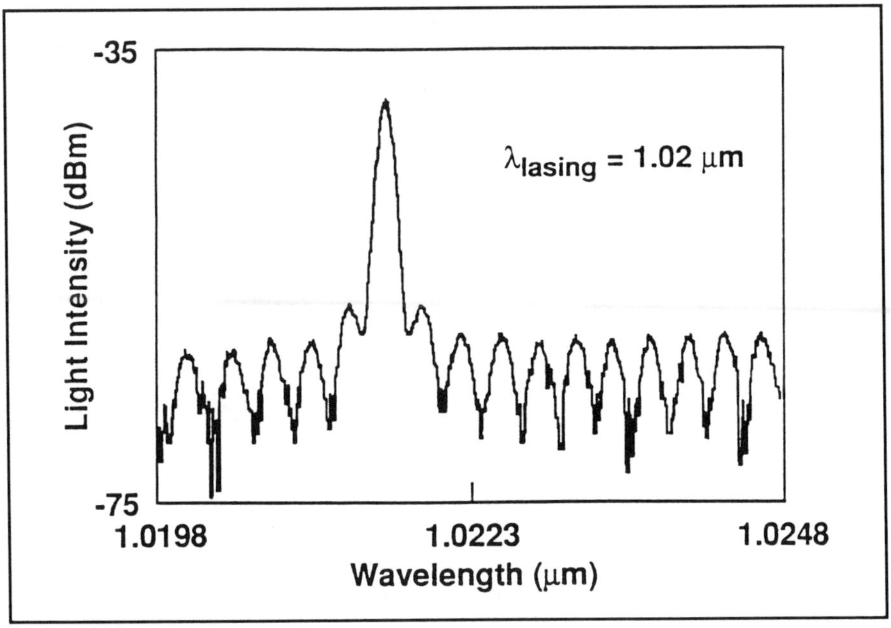

Figure 5. CW output spectrum.

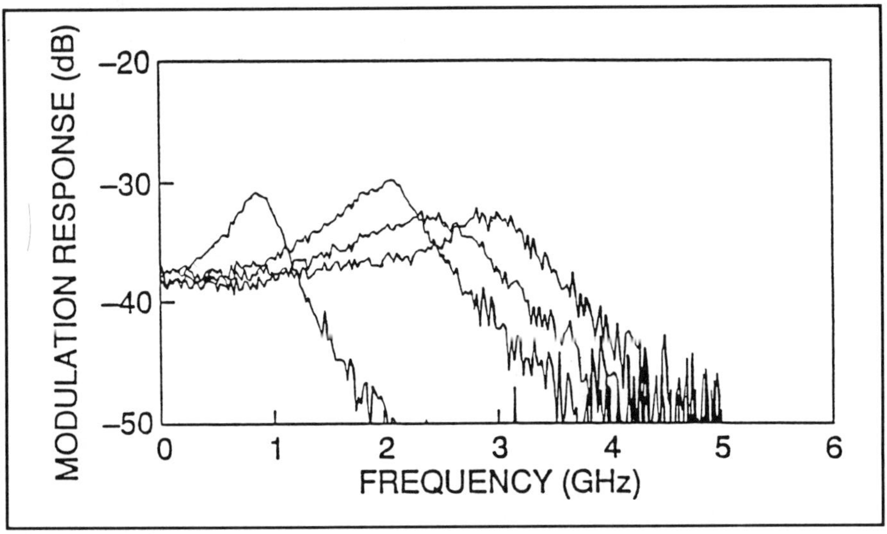

Figure 6. HFEL small signal response.

InGaAs photodiode and a network analyzer. Due to the poor response of this particular photodiode, the measured signal to noise was not particularly good. The expected increase in cutoff frequency with bias level was observed and is

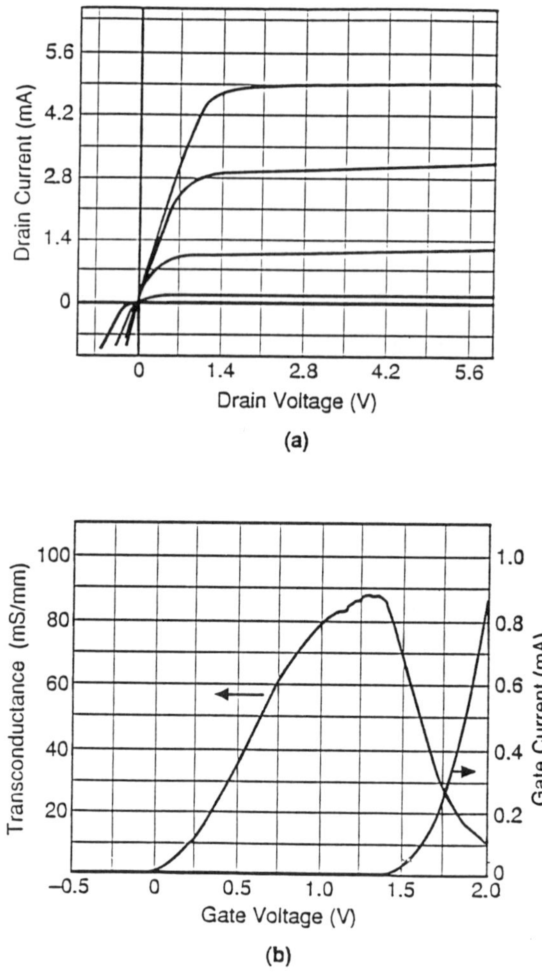

Figure 7. **(a)** HFET drain output characteristics; **(b)** HFET transconductance and gate current as a function of gate bias.

shown in Figure 6. Clear relaxation oscillations can be seen in the output. The maximum 3dB cutoff frequency obtained was 4GHz at 6mW output (approximately twice threshold). Further increases in modulation bandwidth were limited due to the falling CW output as a result of ohmic heating.

The measured DC drain output curves for the HFET fabricated from this material are shown in Figure 7a. Figure 7b shows the measured transconductance as a function of gate bias. Peak normalized transconductance is 87ms/mm for a nominal 2-μm long, 50-μm wide gate. The HFET threshold voltage was -0.3V. The maximum drain output current density obtained was 120mA/mm.

The DC compatibility of the two devices can be determined using the HFET drain output and HFEL threshold currents. Simple scaling of the present results suggests the capability to run the smaller 5-μm HFEL to just over threshold, assuming equal stripe length and gate width devices. Both the HFET and HFEL have considerable scope for optimization, though. The HFET drain output can be increased by increasing the charge sheet doping and barrier region doping to ensure a greater maximum inversion charge. Both values necessarily have to change to ensure an accurate threshold voltage is maintained for the HFET. The limit to this process is determined by two factors. First, practical considerations regarding achieving high Si doping levels using MBE; and second, the available conduction band offset obtained at the quantum well/barrier interface. This directly determines the available number of states that can be filled prior to thermionic emission into the barrier becoming a problem. The strained InGaAs/AlGaAs material system is clearly advantageous in this respect when compared with the lattice-matched GaAs/AlGaAs system. For comparison, strained MODFET, 550mA/mm channel currents have been reported for single channel structures.[8]

The need to make low resistance contact to the channel of the HFET means the etch depth and implant profile are not necessarily optimized for lateral waveguiding by the HFEL. The effect of optical losses can clearly be seen from the increasing threshold current density with decreasing stripe width. Refinements to this part of the processing sequence should enable lower threshold currents to be realized. The HFEL output efficiency can be improved by modifying the waveguide structure to provide for tighter carrier confinement within the active MQW region. Higher MQW barriers and graded optical confinement regions could be incorporated relatively easily. The operation of the HFET does impose some constraints upon the surrounding AlGaAs layers, though. Increasing the Al content of the barrier and collector layer immediately below the MQW active region would increase the confinement potential. However, due to the nature of Si donors in AlGaAs, Al mole fractions above 25 percent have a tendency to be resistive as a result of deep donor formation. Thus, such layers would further compound the source resistance problem for both the HFEL and HFET. Improvements in the small signal modulation bandwidth will first require the series resistance, and in particular the source resistance, to be lowered such that higher output powers can be acheived (comparable with the observed pulsed response). Thereafter, reduction in threshold and improvements in efficiency should significantly enhance the small signal response in an analogous manner to conventional diode lasers.[9]

Conclusions

A FET and lateral current injection laser have been grown and fabricated using a single epitaxial layer sequence. Initial results have established the concurrent operation of both devices using identical processing steps. Future integration should be possible to fulfill the potential of both mulitple optical microwave links and smart pixcl applications for optical preprocessing.

Acknowledgments

The authors thank the following persons: B. Tell and K. F Brown Goebler for assistance with the ion implantation steps; C. A Burrus for the device bonding; and P.A. Kiely and M.G. Young for assistance with the device testing.

References

1. K.Y Lau, *Proceedings of Device Research Symposium* (Charlotte, NC, 1991), p. 365.
2. S.R. Forrest, *Tech. Digest* (OSA Annual Meeting, San Jose, CA, 1991).
3. S.D. Offsey *et al., Elec. Lett.* **26**, 350 (1990).
4. G.W.Taylor *et al., Appl. Phys. Lett.* **58**, 666 (1991).
5. G.W. Taylor *et al., IEEE Elec. Dev. Lett.* **10**, 88 (1989).
6. R. Nagarakan *et al., Appl. Phys. Lett.* **58**(21), 2326 (1991).
7. W.T. Tsang, in *Physics and Technology of Molecular Beam Epitaxy*, E. H. C. Parker, Ed. (Plenum, New York, 1985).
8. L.D. Nguyen *et al., IEEE Trans. Elec. Dev.* **36**(5), 833 (1989).
9. G.P. Agrawal and N.K. Dutta, *Long Wavelength Semiconductor Lasers* (Van Nostrand, New York, 1986).

14

A Novel High-Contrast Optical Modulator in a GaAs-AlAs Structure

Mitra Dutta
Hongen Shen
Jagadeesh Pamulapati
Peter G. Newman
Wayne H. Chang
U.S. Army Research Laboratory
Electronics and Power Sources Directorate
Fort Monmouth, NJ 07703

The quantum confined Stark effect has given rise to several innovations in optoelectronic modulators. Such modulators have many potential applications in communications and special purpose computer systems. These devices are based on the shift of the absorption resonance as well as the change in its strength as an electric field is applied. Improved modulators would necessarily be required to operate at lower electric fields and to have better contrast. We present a different physical mechanism that shows promise as a future high-contrast optical modulator. We present experimental data on structures involving gallium arsenide-aluminum arsenide (GaAs-AlAs) multiple quantum wells where there is coupling between the Γ and X confined levels in the GaAs and the AlAs layers. Our data demonstrate that the electrons that are delocalized when the confined level is nearly degenerate with the χ-valley can be localized with the application of an external field. A blue shift of the luminescence wavelength and a dramatic increase in the oscillator strength are observed. In structures where there is a built-in field, the electric field can be modulated by the light intensity on the sample. An optical modulator with lower operating voltages and higher contrast will impact many potential Army applications where spatial light modulators and arrays are necessary.

Introduction

The performance advantages of an integrated optoelectronic approach stem from a massively parallel processing and interconnect function, wideband information processing and storage, increased switching speed, and immunity to electromagnetic radiation. To meet the challenge of future systems, it is important to develop integrated optics and optoelectronic devices so as to take full advantage of the photonics technology. This technology, including sources, sensors, modulators, switches, optical waveguides, and fibers will have a strong impact on future defense systems. The greatest impact will be felt in the fields of communications, computing, and sensors. Enhanced performance of any of the elements of sources, switches, modulators, interconnects, or sensors will increase the utility of optical techniques for signal processing applications.

Heterostructures and multiple quantum wells (MQWs) in III-V semiconductors promise fresh approaches and new options for high-performance electronic and optical devices required for future high-speed transfer, processing, and control of information. Gallium arsenide (GaAs)-based MQWs and hetero-structures have opened up new areas of science and engineering and have excited a large amount of research. With molecular beam epitaxy (MBE) and metallorganic chemical vapor deposition (MOCVD) techniques, it is now possible to prepare materials techniques with a synthetic periodicity superimposed on the characteristic crystal lattice structure. The superlattice and MQW materials have unique optical and electrical properties that are intrinsically different than those of bulk compounds. Bandgap, lattice constants, and refractive constants may be tailored in III-V heterolayer to confine carriers and radiation to impose built-in fields without the penalty of surface defects, and to enhance mobilities, hot carrier effects, exciton absorption, and optical nonlinearities. A number of devices not possible to be used before in bulk semiconductors have been developed, making use of the new physics associated with these materials and device structures. Low threshold current quantum well and wire lasers, tailored bandgap detectors, modulators, and switches are being researched and developed. Most modulators are based on the quantum confined Stark effect (QCSE), the most well known of these being the logic device, the self electrooptic device (SEED).

The QCSE is a phenomenon that arises when an electric field is applied across the planes of heterostructure superlattices. In a quantum well at zero electric field, the electron and hole energy levels are defined by the well width, and the electrons and holes are strongly confined in the well material layer.

However, when an electric field is applied, the electrons and holes are moved apart, and their energies are altered. This has the effect of shifting the absorption resonance to lower energy as well as modulating the strength of the absorption. This shift and decrease in strength of the absorption then provides for the optical modulation of the transmission as a function of applied electric field of any light appropriate wavelength that is incident on the heterostructure. The modulator concept that we will present is based on a different mechanism, the coupling between the Γ and X valleys in a GaAs-aluminum arsenide (AlAs) MQW that promises to provide better contrast and operate at lower voltages than the QCSE devices.

Type II quantum well structures have been studied[1] extensively in the past several years for their interesting physics and possible device applications. In these structures, where GaAs-AlAs short period superlattices (SPS) are archetypical, the electrons and holes are confined in adjacent layers of the materials, because the direct gap GaAs layers are narrow enough so that the confined electron level in the GaAs is higher in energy than the X-valley in the indirect gap AlAs barrier. However, even when the well layers are thick, with an indirect gap barrier, effects[2-5] due to the coupling between the Γ- and X-levels manifest themselves. When the allowed electron energy level is close to the X-valley in the barrier, the electron appears to be no longer confined, but delocalized. This has been seen previously from our measurements of exciton binding energy[2] which was found to be anomalously low when the electron confined level is close to the X-valley, indicating less confinement. Further evidence for this was obtained in photoluminescence excitation experiments[3] where charge transfer was observed between asymmetric wells coupled via the X-valley. In some samples, the X-valley was seen to be the relevant barrier for transport.[4] In asymmetric wells coupled via the X-valley, the application of electric field[5] showed a clear difference in the subsequent photoluminescence signal between samples that were coupled via the X-valley versus those that were not. In this paper, we report experimental results on samples where the reconfinement of the delocalized electron by the application of electric field is seen more dramatically.

Experimental Data

The samples were grown by MBE in a Varian Gen II machine. The electric field experiments on the asymmetric coupled wells were done on GaAs-AlAs MQW, with 18 narrow wells and one wide well. The wide well

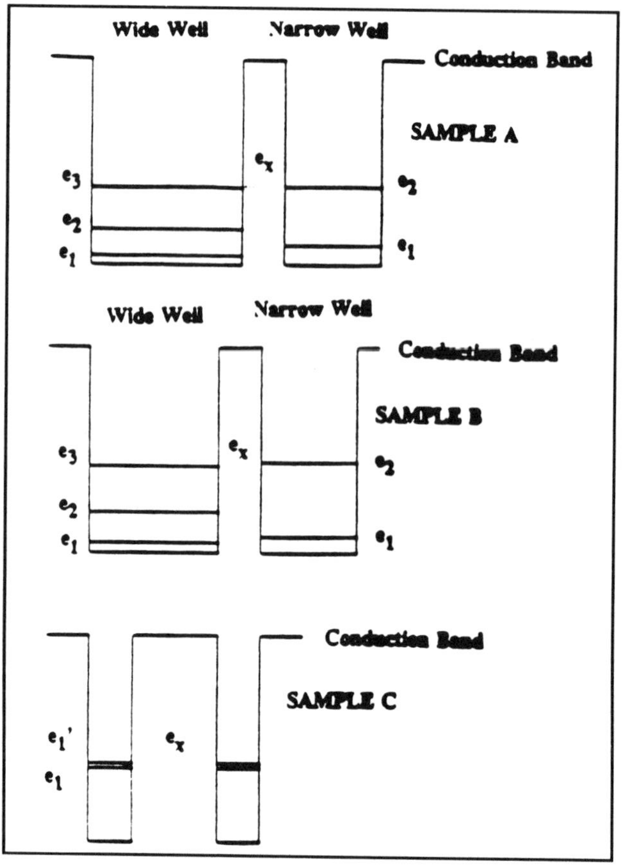

Figure 1. Energy band structures for GaAs-AlAs quantum wells.

is used to monitor the charge transfer and the coupling between the wells via the X-valley. Sample structures with electron levels are shown in Figure 1. Sample A has GaAs layers such that the third confined level in the wide well and the second confined level in the narrow well are both degenerate with the X-valley. In sample B, the wide and narrow wells are both made more narrow so that the confined levels are pushed up, and the X-valley is no longer degenerate with either electron level. In sample C, where the effect of reconfinement by electric field is seen more dramatically, the well width is such that the first electron level e_1 is nearly degenerate with the X-valley, and the e_1 level thus nearly delocalized. Schottky barriers of 80Å of Ni-Au or 50Å of Ti provided a transparent contact while an ohmic contact was made with the n^+ substrate. Preliminary experiments also have been performed with this structure C embedded in the intrinsic region of a surface-intrinsic -n^+ (s-i-n^+) structure,[6] where the large built-in field can be modified by the intensity of the

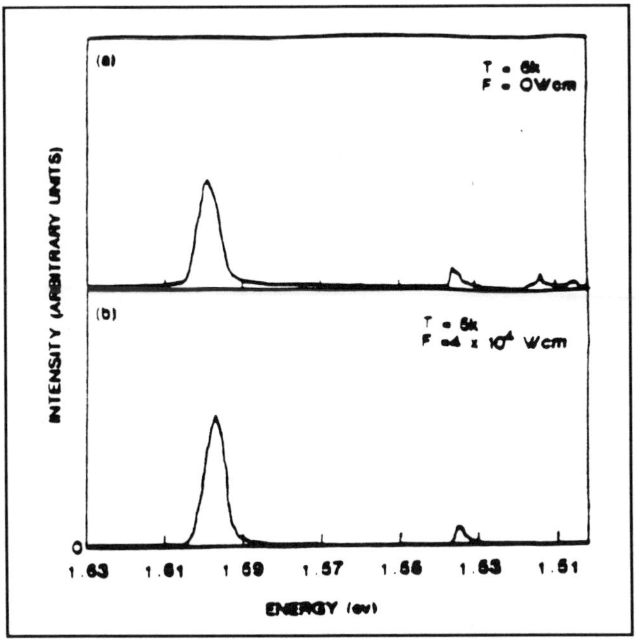

Figure 2. PL spectra for sample B; e_3 of wide well and e_2 of narrow well are not coupled with e_X; (a) F=0 V/cm (b) F=4x10^4 V/cm applied field.

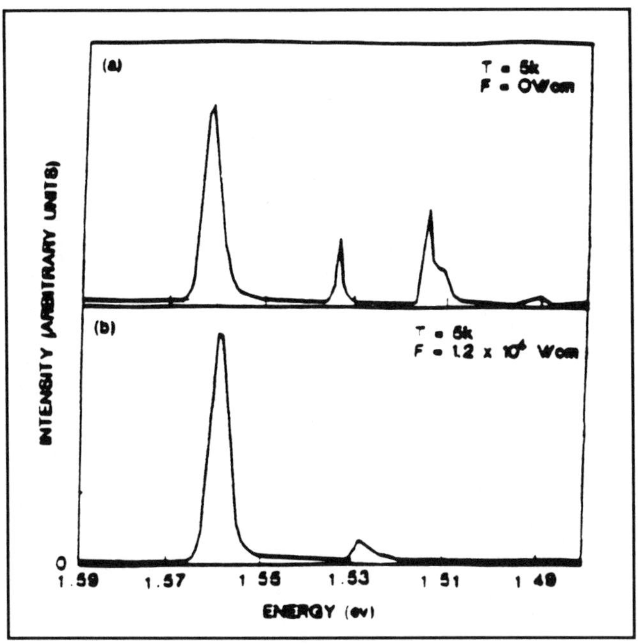

Figure 3. PL spectra for sample A; e_3 of wide well and e_2 of narrow well are coupled with e_X; (a) F=0 V/cm (b) F=1.2x10^4 V/cm applied field.

light on the sample, and an externally applied electric field is no longer necessary.

Figure 4. PL spectra for sample C; e_1 is coupled with e_X; (a) F=0 V/cm (b) F=2.5x10^4 V/cm applied field.

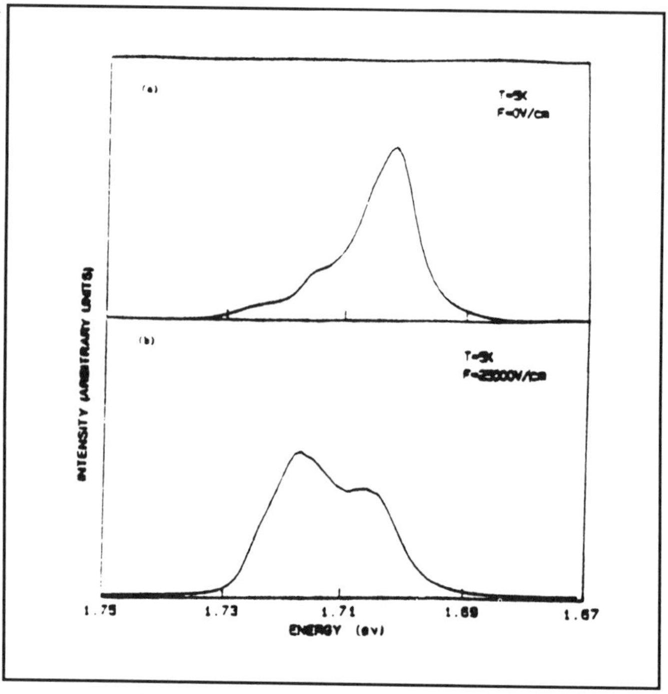

All of our photoluminescence (PL) experiments are performed with the 5145Å line of an Ar^+ laser at 5K. The laser power is about 2mW. The results of the application of the electric field on the asymmetric wells are summarized in Figures 2 and 3. Figures 2 and 3 show the PL spectra from samples B and A, respectively. Three features are seen from each of the samples, corresponding to signal from the narrow wells, the wide well, and the bulk GaAs material. In sample B (Figure 2), where the wells are not coupled, application of electric field ionizes the bulk GaAs exciton, and this feature disappears. The other two features from the wide and narrow wells show a red shift due to the quantum-confined Stark effect. It is important to notice that the ratio of the intensities of these two features remains the same, showing that there is no coupling between the wells. This can be contrasted with sample A (Figure 3), where there is a significant difference in the ratio of the photoluminescence intensities of the features with the application of a field; the intensity of the signal from the wide well drops more rapidly than that from the narrow wells. This is consistent with the interpretation that some of the intensity of the wide-well feature is due to electrons excited in the narrow wells, which transfer into the wide well via the coupling of the X-valley in the AlAs barriers. When the coupling is removed because of the application of electric field, electrons no longer transfer into the wide well, and the intensity of this feature is reduced. This effect of the confinement due to the application of an electric field is seen more convincingly in sample C.

The PL spectra from sample C are shown in Figure 4. Due to monolayer fluctuation, there are actually two such levels e_1 and e'_1, where e'_1 is one monolayer less and hence of higher energy and closer to the X-valley than e_1. The zero field spectrum shows three features, of which the low energy feature at 1.702eV is the strongest. We attribute this to the e_1-hh_1 transition where the e_1 level is confined more strongly and is at a lower energy than the X-valley, and where hh_1 is the first heavy hole level in the GaAs well. The next feature is at an energy 12meV higher and is the e'_1-hh_1 transition where the e'_1 electron state is nearly degenerate with the X-valley. Finally, the very weak feature at 1.725eV is the indirect X_1-hh_1 type II transition, where the electron is the X_1-electron in the AlAs. This transition is made possible by several mechanisms: (1) an overlap of the wavefunctions at the heterointerface, (2) a breakdown of the κ-vector conservation by super-periodicity and disorder at the interface, and (3) possible phonon assistance. Monolayer fluctuation between a 40Å GaAs well and a 37.5Å well will give rise to features separated by 17meV. However, the coupling between the Γ and X states will produce an anticrossing effect and will tend to depress in energy the e'_1 level, thus bringing it closer to the e'_1 level and to the experimentally observed 12meV between them. The electric field causes a localization of the delocalized electron, and hence there is an increase in the oscillator strength as the wavefunction of the e'_1 electron is confined more strongly in the well. This is analogous to Wannier-Stark ladders[7-10] observed when an electric field is applied to superlattice minibands. As the electric field is applied, the reduction in coupling between the Γ and X levels removes the anticrossing effect, resulting in a blue shift in the energy of the e'_1-hh_1 transition. From this blue shift, we have deduced the mixing potential between the Γ - X level to be about 3meV, which is in agreement with the results in references 11 and 12.

It should be pointed out that the nonvertical transitions due to electrons and holes in the adjacent wells and next adjacent wells cannot be observed because of the large periodicity in our structure and the heavy effective mass of the AlAs electron. Other researchers have observed comparable behavior in Type II short period superlattices,[10] but not in Type I GaAs-AlAs structures.

We obtained similar results with the structure of sample C embedded in the intrinsic region of the s-i-n^+ structure, where no external field need be applied, and a variation of the built-in electric field is performed by the change in intensity of the incident light. Strong dependence of the intensity of the phonon sidebands is seen as a function of electric field, and more experiments with different structures are under way to understand this phenomenon.

Conclusion

From the electric field results, it can be seen how dramatically the change is in both the shift and strength of the transition at a lower electric field. These initial results compare very favorably with the QCSE based structures.

We present experimental data on GaAs-AlAs multiple quantum wells, where the confined electron level is initially delocalized due to mixing between the Γ and X electron levels. The application of an electric field reduces this coupling and reconfines the electron in the GaAs layer. Two phenomon are associated with this: (1) an increase in the oscillator strength, and (2) a blue shift due to the reduction of the anticrossing between the Γ and X states; the reduction in the anticrossing is consistent with the value of the mixing potential, as measured by other research groups. This phenomenon can be utilized to produce a higher contrast, lower operating voltage optical modulator.

References

1. B.A. Wilson, *IEEE J. Quant. Electron* **QE-24**, 1763 (1988).
2. D.D. Smith *et al., Phys. Rev.* **B40,** 1407 (1989).
3. M. Dutta *et al., Phys. Rev.* **B 42,** 1474 (1990).
4. M. Dutta, K.K. Choi, and P.G. Newman, *Appl. Phys. Lett.* **55,** 2429 (1989).
5. M. Dutta *et al., Surface Science,* submitted.
6. H. Shen *et al., Appl. Phys. Lett.* **57,** 2118 (1990).
7. E.E. Mendez, F. Agullo-Rueda, and T.M. Hong, *Phys. Rev. Lett.* **53,** 2632 (1988).
8. J. Bleuse *et al., Appl. Phys. Lett.* **53,** 2632 (1988).
9. P. Voisin *et al., Phys. Rev. Lett.* **61,** 1639 (1988).
10. A.J. Shields *et al., Phys. Rev.* **B42,** 5879 (1990).
11. M.H. Meynadier *et al., Phys. Rev. Lett.* **60,** 1338 (1988).
12. N.J. Pulsford *et al., Phys. Rev. Lett.* **63,** 2284 (1989).

15

Physics and Device Applications of Silicon Microclusters

James F. Harvey
Hongen Shen
Robert A. Lux
Weimin Zhou
David C. Morton
Melanie W. Cole
Mitra Dutta
Madan Dubey
Richard C. Piekarz
Richard T. Lareau,
Armand M. Balekdjian
Jagadeesh Pamulapati
Clarence G. Thornton
U.S. Army Research Laboratory
Electronics and Power Sources Directorate
Fort Monmouth, NJ 07703

Charles M. Bowden
U.S. Army Missile Command
Research, Development and Engineering Center,
Redstone Arsenal, AL 35898

Raphael Tsu
University of North Carolina at Charlotte
Charlotte, NC 28223

Quantum confined silicon microclusters offer the possibility for novel high performance devices for Army mm-wave, high-speed digital and optical signal processing, and opto-electronics systems—all compatible with established Si technology. Si microclusters formed by two different fabrication techniques have been investigated to determine the physical properties applicable to device formulation. The two processes are the electrochemical etching of silicon wafers to form porous silicon and the recrystallization of a silicon microparticle film. The porous silicon exhibits bright photoluminescence at room temperature, with the promise of light emitting diode and other optoelectronic devices based on silicon. The recrystallized film has demonstrated resonant tunneling through the quantum confined states of the microparticles, with the promise of fast silicon tunneling devices. Structural, analytical, optical, and electrical techniques have been applied to these new materials, and results are discussed. The first Raman spectra and the first determination of the pressure dependence of photoluminescence, as well as a first unambiguous transmission electron microscopy identification of the crystalline planes of the porous Si microparticles, are presented. The potential for device applications is evaluated based on results from theory, experiment, and analysis. Providing these advanced devices in silicon integrated circuits at modest cost would have a major impact on the capability of Army electronics systems.

Introduction

The recent observations of bright, visible photoluminescence (PL) and resonant tunneling in microparticles of silicon have generated an exciting new field of research, with appealing attributes for novel devices of major impact on Army electronics and optics programs. After reviewing the stimulating discovery of PL from porous silicon (at Malvern, United Kingdom)[1] and the observations of resonant tunneling in silicon microcrystallites (at the University of North Carolina at Charlotte, supported by the U.S. Army Research Office (ARO)[2] a new program was initiated at the Electronics and Power Sources Directorate (EPSD) of the U.S. Army Research Laboratory. Its purpose was to exploit the rapidly developing prospects for new materials based on ultrasmall crystalline semiconductor clusters or structures formed without special epitaxial or lithographic procedures. It was apparent that these initial experiments opened the door to exciting new classes of devices based on new physics phenomena occurring in these novel materials, and promising opto-electronic, optical, and fast tunneling devices fully compatible with silicon integrated circuit (IC) technology. A coordinating group was quickly formed at EPSD to integrate experimental and theoretical approaches to this research across organizational and functional lines. Within several months EPSD researchers had reported the first Raman spectra from porous silicon, the first measurement of pressure dependence, among the first identifications by transmission electron microscope (TEM) of the crystalline planes in microcrystallites of silicon with sizes of several nanometers, and produced extremely small silicon microstructures using two promising alternative techniques. The EPSD program was leveraged by cooperative experiments with the nonlinear optics program at the Research, Development, and Engineering Center of the U.S. Army Missile Command (MICOM) and by capitalizing on ARO sponsored research at universities.

Bulk crystalline silicon has an indirect bandgap and displays only a faint infrared (IR) PL at room temperature. Porous silicon was found to demonstrate very bright visible PL at room temperature. Thus, porous Si offered the immediate promise for silicon IC based opto-electronic devices and because of this quickly fostered an extremely active research field. Because of the special promise of porous Si, the relatively easy fabrication conditions required to observe its bright PL, and the potential for benefiting from the ideas and research of so many other groups, the EPSD program has had the greatest initial success with porous silicon. As a result, most of the results reported in this paper deal with porous silicon. However, it should be remembered that this is a project with very general scope and wide implications.

The goal of these initial experiments with porous silicon and tunneling in silicon microclusters is to establish the unique transport and optical phenomena governing the behavior of materials composed of these ultrasmall silicon particles. Efforts are in progress in parallel with this research to formulate device concepts utilizing these phenomena. Initially, electroluminescent (EL) films and light emitting diode (LED) structures have been examined. Also, in parallel, promising new fabrication techniques for silicon microstructures, both in-house and in conjunction with ARO funded university research, are being assessed and developed. These new materials will require new approaches to device design, and the current research has focused on establishing the most advantageous approaches.

Luminescence From Porous Silicon

Porous silicon is made by electrochemically etching a silicon wafer in a solution of hydrofluoric acid (HF). As a result, many very small pores are formed in the silicon. As the etching process progresses, very small particles of crystalline silicon remain between the pores. The exact structure of the porous silicon and the mechanisms responsible for the bright PL are poorly understood.

The strong, visible PL from porous silicon is in sharp contrast to the weak PL in the IR from bulk crystalline Si. Silicon's indirect bandgap of 1.1eV (Γ point of the valence band to the Δ point of the conduction band) is mainly responsible for the weak IR PL. Optical transitions across this bandgap require a change in momentum of nearly the full Brillouin zone. Three important questions need to be addressed in order to explain the PL from porous Si. First, what happens to the 1.1eV bandgap of bulk silicon? Canham's[1] original report of PL from porous Si attributed its visible wavelength to a blueshift in the silicon PL as a result of quantum confinement in microcrystalline silicon structures with diameters on the order of 3nm. Second, what mechanism is responsible for the enhancement in the optical transitions across the bandgap? Although quantum confinement results in some uncertainty in the momentum (by the Heisenberg principle), this uncertainty is not by itself sufficient to significantly relax the momentum conservation requirement. Third, what is the role of localized states and their relation to the recombination process?

Other investigations have provided support of the quantum confinement model of the PL mechanism for porous Si, in which the silicon bandgap is widened due to the confinement of the carriers in an extremely small particle

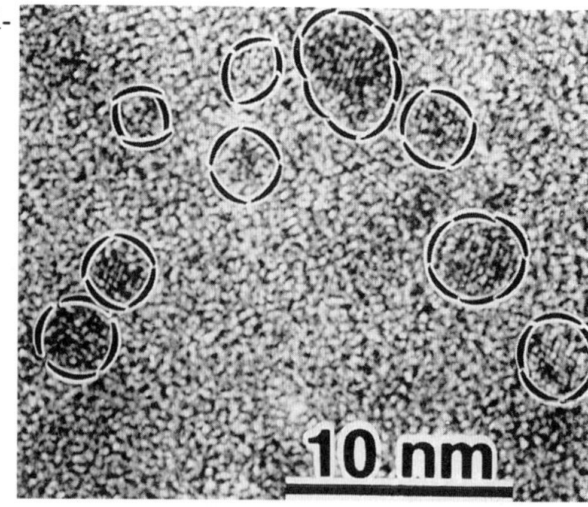

Figure 1. TEM Micrograph showing ultrafine crystallites in porous silicon.

in the same manner that the energy level spacing is increased for a particle in a box when the box dimensions decrease.[1,3,4] Alternative mechanisms for the visible luminescence have also been raised. Some reports suggest amorphous silicon could be the source of the PL.[5,6] Others have proposed optical transitions in individual molecules present at the silicon cluster surface or in the region between clusters, such as the siloxene molecule.[7,8] So far, experimental results have been ambiguous or conflicting with regard to a clear-cut determination of the PL mechanism. This portion of our paper discusses the results of structural, analytical, optical, and cathodeluminescence (CL) measurements on porous silicon. Like previous experimental work, these results are not by themselves definitive in identifying the PL mechanism. However, they do provide important additional information relevant to the three questions previously stated, and they provide important clues to guide efforts at device design.

Structural and chemical analysis. Early reports of PL and optical absorption in porous silicon attributed the results to quantum confinement effects in needle-like crystalline structures with diameters of 2 to 9nm.[1,3] TEM micrographs of the porous silicon film taken at EPSD[9] clearly show the crystalline planes in microcrystallites of silicon, as shown in Figures 1 and 2. As shown in Figure 2, the interplanar spacing is 0.31nm, which corresponds to the spacing for the (111) orientation of bulk crystalline silicon. The microcrystallites seen in these micrographs are not needlelike but rather of an irregular shape. Similar TEM micrographs were independently taken by Lehmann and coworkers.[10] The lattice planes in the crystallites shown in Figure 1 are rotated by a few degrees from each other in the plane of the micrograph. It is expected that other crystallites may be rotated by a few

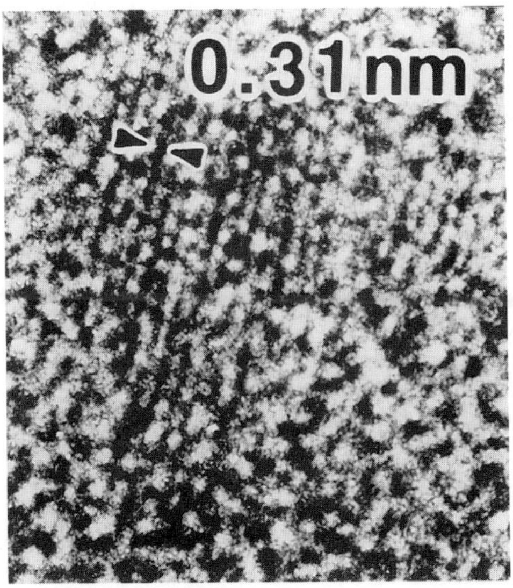

Figure 2. TEM image of a silicon crystallite from porous Si. The marker indicates the interplanar spacing, 0.31nm.

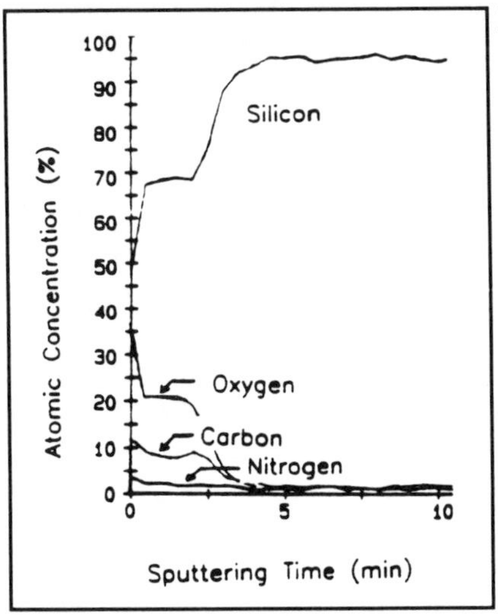

Figure 3. Auger analysis of porous silicon film.

degrees out of the micrograph plane, in which case their crystal planes would not appear in Figure 1, and they would appear as the amorphous-looking region between crystallites in this figure. Initial observations have not revealed the typical ring structure in transmission electron diffraction that is the signature of amorphous silicon.

Results of Auger analysis on the sample shown in Figures 1 and 2 are shown in Figure 3. Silicon, oxygen, carbon, and nitrogen atoms are seen in this analysis. Secondary ion mass spectroscopy (SIMS) analysis on the same sample also shows the presence of hydrogen atoms, which do not have an Auger signature, and a small amount of fluorine.[11] The composition of this film is about 70 percent (atomic concentration) of silicon, 20 percent oxygen, and 10 percent carbon and nitrogen, excluding the hydrogen concentration, which Auger cannot measure. The results of independent SIMS analysis elsewhere[12] confirm the qualitative nature of these measurements, and demonstrate a large variation sample-to-sample as the result of different environmental conditions, as well as long-time scales for interactions with the environment.

Electron paramagnetic resonance (EPR) measurements[13] on this sample indicated that the surface of the silicon microcrystallites was well passivated and free of dangling bonds, which was attributed to bonding with hydrogen or SiO_2. These conclusions were independently confirmed elsewhere by Fourier transform infrared (FTIR) measurements,[14-16] which noted that the strength of the PL correlated with SiH_2 and that the Si-H bonds were replaced by Si-O bonds over time.

Correlated Raman and PL measurements. For these measurements, both Raman and PL spectra were taken from the same spot on the sample and at nearly the same time. The average sizes of the silicon crystallites varied from sample spot to sample spot. At each sample spot, three spectra, parallel polarized Raman, cross-polarized Raman, and unpolarized PL, were taken. Reference Raman spectra were taken from the polished silicon surface that had not been exposed to the etching process. In contrast to the reference Raman spectra from the crystalline silicon surface, the Raman spectra from spots on the porous Si were unpolarized.[17-19]

Shown in Figure 4 are Raman and PL spectra from two spots (A and C) on a typical sample. The Raman spectra consist of sharp peaks due to crystalline silicon modes near the peak for the bulk crystalline Si reference, marked "c-Si" in the figure. The line widths of the porous Si peaks are about $5cm^{-1}$, compared with $3cm^{-1}$ for the reference c-Si peak. Peak C also displays the splitting of the TO and LO phonon modes. The PL spectra show the broad peaks at energies between 1.55eV and 2eV typical of porous Si PL.

The downshift of the porous Si Raman peaks from the c-Si peak and the LO-TO splitting can be used to estimate the size of the silicon microcrystallites. The electron-phonon interaction in a finite sized particle has a maximum corresponding to a phonon momentum $q=\pi/l$, where l is the confinement dimension of the particle. The phonon momentum of the Raman peak from a finite sized particle will therefore be shifted by π/l from the center of the Brillouin zone. From the experimental downshift in the energy of the

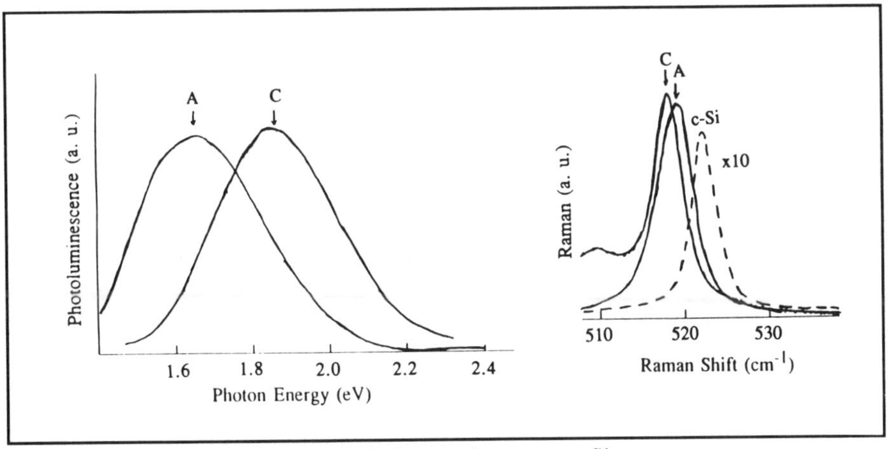

Figure 4. Correlation of Raman and PL Spectra from porous Si.

Raman peaks, the particle size can thus be estimated, using the known phonon dispersion in silicon. The phonon dispersion is different for the LO and TO phonons, so the experimental phonon splitting can also be used to estimate particle size. Our Raman data results in size estimates between 2 and 3nm for the porous Si microparticles.

Figure 4 also illustrates the correlation between PL and Raman spectra taken at different spots on a sample, which is typical of our data. The PL spectra from spots that have the largest Raman downshifts, and hence the smallest average particle size, have peaks at the largest PL photon energies. This indicates that quantum confinement due to the smaller particle size is increasing the spacing of the energy levels involved in the PL recombination.

Meausurements of the dielectric constants. Angle resolved ellipsometry measurements were made on the porous Si samples by fitting reflectance data to known formulas for thin films and extracting the index of refraction and film depth. Our data resulted in values for the index of refraction of the porous silicon film between 1.5 and 1.9, and therefore dielectric constants between 2.3 and 3.6. These values are much smaller than the dielectric constant for bulk crystalline Si at this excitation wavelength, 17.8. The reduction is due to two principal causes. This material is porous, with the voids contributing a dielectric constant of 1 to the mixture. Second, quantum confinement itself will reduce the dielectric constant[20] by increasing the critical optical transition energies. Based on our estimates of our porosities and on a simple estimation of the dielectric constant of a mixture of silicon particles and voids, the measured dielectric constants appear too small to be the result solely of the voids. Large void fractions, on the order of 90 percent, are not supported by

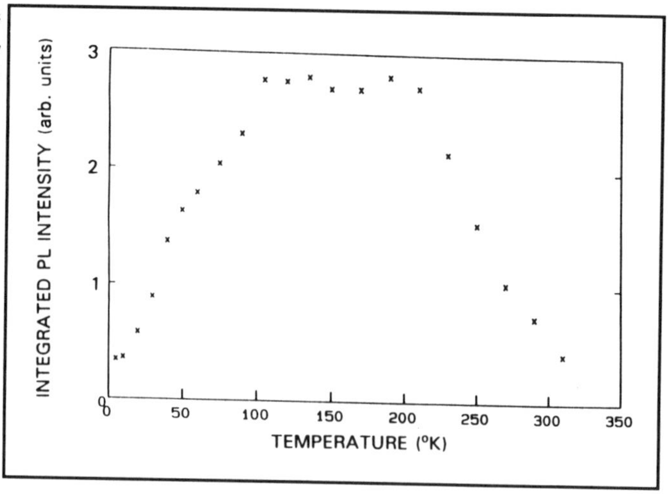

Figure 5. Temperature dependence of the PL intensity from porous silicon.

our structural studies. These measurements demonstrate that the quantum confinement effects can have a significant effect in the reduction of the dielectric constant of porous silicon[18,19] and that the dielectric constant within the microcrystallites themselves is reduced.

Temperature dependence of the PL. The temperature dependence of the integrated intensity of the PL is shown in Figure 5.[21] At low temperatures, the intensity rises with temperature, with a behavior typical of phonon-assisted recombination in semiconductors. Near room temperature, the intensity decreases. The temperature dependence of the porous Si is very similar to that reported for selenium doped Si.[22] In this report, the falloff in PL intensity near room temperature is attributed to the thermal dissociation of bound excitons. This suggests that the temperature behavior of the porous Si PL near room temperature is due to a similar mechanism. If the temperature dependence of the log PL intensity is plotted against inverse temperature, the relation is nearly linear near room temperature, and a thermal dissociation energy of 100meV can be extracted. The binding energy of the excitons is inversely proportional to the square of the dielectric constant inside the silicon microparticle. As a result of the quantum confinement on the dielectric constant, as discussed previously, relatively shallow defects or impurities in bulk Si would have their binding energies increased when confined in microparticles of these small dimensions. The increased binding energy would then be consistent with this measured dissociation energy. The temperature dependence reported here is not consistent with a recent report of the temperature dependence of the PL of siloxene,[23] which has been proposed as

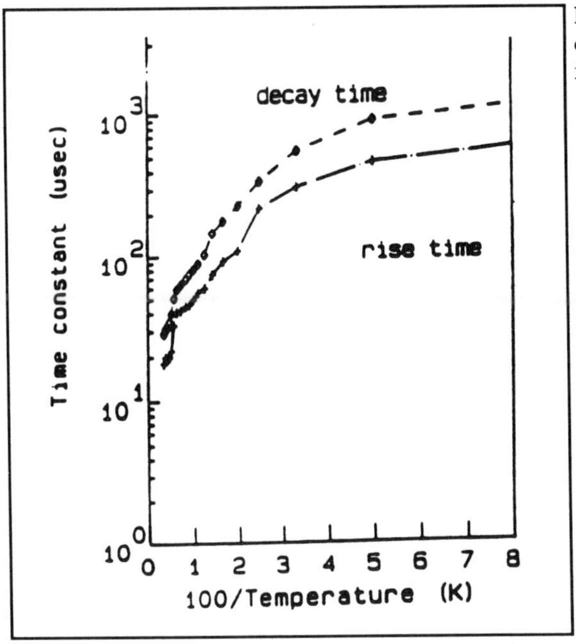

Figure 6. Temperature dependence of the PL time constants for porous silicon.

a possible cause of the bright PL from porous Si. Zheng and coworkers reported a steady increase in the PL from siloxene as the temperature was decreased to 10°K, which is contrary to these observations for porous silicon. Our temperature dependence measurements thus raise serious questions about a model based on PL originating from siloxene.

The temperature dependence of the PL time constants for porous Si is shown in Figure 6.[21] The time constants at low temperatures are very long (nearly 1 millisecond), decreasing by almost two orders of magnitude at room temperature. The shape of their temperature dependence and their magnitudes are very similar, again, to the PL lifetimes reported by Bradfield and coworkers[22] for Se doped Si, which again suggests the similarity in PL processes. The PL time constants are due to both radiative and nonradiative channels. At low temperatures, there are fewer nonradiative paths, phonons are frozen out, and electrons preferentially occupy a metastable excited state. As a result, the overall time constants are longer. Our results are thus consistent with the temperature dependence of transitions involving trapping or localized states.

Excitation power dependence of the PL. For these measurements, the PL was detected using a lock-in amplifier system with an acoustic modulator to chop the beam. Use of the acoustic modulator enabled the use of different "duty cycles," in which the peak power and the chopping period would remain

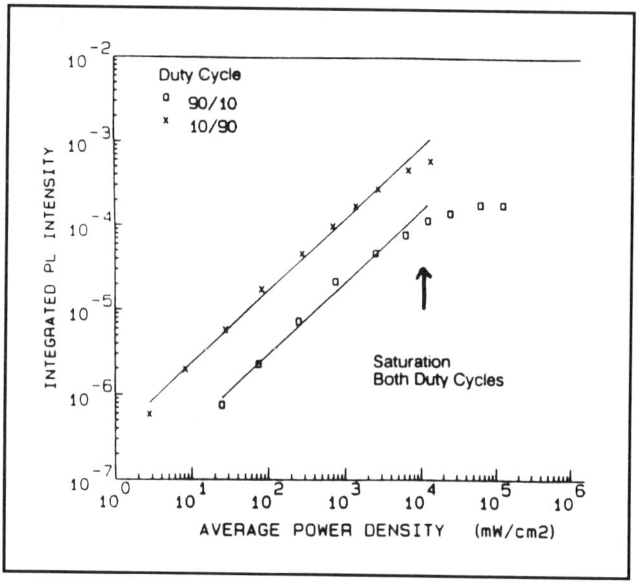

Figure 7. Excitation power dependence of the PL from porous silicon.

the same, but the portion of the period the excitation was on could be varied. A "10/90" duty cycle had the beam on only 10 percent of the time and thus had a lower average power than a 90/10 duty cycle at the same peak power.

The excitation power dependence of the PL is shown in the log-log plot in Figure 7. The slope of 1 on the log-log plot indicates a linear power dependence and that the excitation process is a one photon–one electron process. This linear dependence holds over nearly four orders of magnitude of excitation power, finally beginning to saturate at the higher powers. In order to investigate this saturation effect, different duty cycles were used for the excitation. The results for the 10/90 and 90/10 duty cycles are shown in Figure 7 plotted against the average excitation power. Figure 7 demonstrates that the saturation effect, occurs at nearly the same average power for different duty cycles, and therefore for different peak powers. This result indicates that the saturation is due to temperature increases in the illuminated spot on the porous Si sample as a result of increasing absorbed power, an effect that would depend on the average excitation power rather than the peak power. As shown in Figure 5, near room temperature an increase in temperature will cause a decrease in PL intensity. This also suggests that the one photon–one electron excitation process itself will be true at higher excitation powers.[19,24]

PL excitation spectrum. The PL excitation spectrum for the red peak of porous silicon is shown in Figure 8. This spectrum was taken using a xenon arc lamp and monochrometer. The energy of the PL peak is about 1.55eV, and

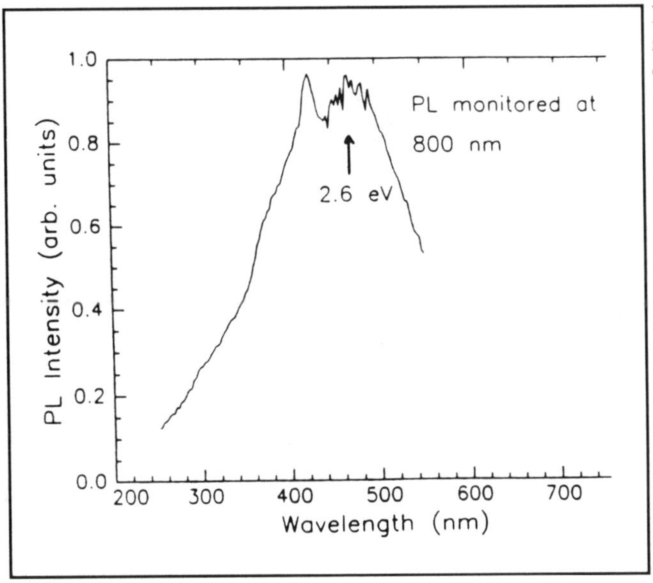

Figure 8. PL excitation spectrum of the red peak of porous Si.

this peak energy is quite insensitive to the excitation energy. For the excitation spectrum the PL is monitored at the 1.55eV peak and the PL intensity recorded as a function of the excitation wavelength. Figure 8 shows that the excitation spectrum is a fairly narrow feature peaked near 2.6eV. The 1.55eV PL is consistent with the indirect bandgap of Si, widened by quantum confinement. Because the direct bandgap energies of bulk Si are 3.3eV or greater, the 2.6eV excitation peak is probably not associated with these transitions. This suggests that it may be due to excitation to an excited level of a quantum confined state.[19,25]

Pressure dependence of the PL. The first measurements of the pressure dependence of the PL from porous silicon were made at EPSD.[26] The pressure dependence of optical transitions has been extensively used to determine the band structure of semiconductor material. The purpose of these measurements was to determine important information about the electronic structure of porous silicon. Figure 9 shows the pressure dependence of the peak PL energy over 40Kbar of pressure. The squares designate data taken as the pressure increased; triangles designate data taken as the pressure decreased. There is a large blue-shift with pressure of nearly 170meV with 20Kbars of pressure, followed by a nearly constant peak energy for higher pressure. This leveling off of the PL peak energy at higher pressures is suggestive of the Γ-X band crossing in GaAs at higher pressures. The origin of the apparent hysteresis is not known at this time. The PL intensity is seen to decrease significantly with

Figure 9. Pressure dependence of the energy of the peak PL from porous silicon.

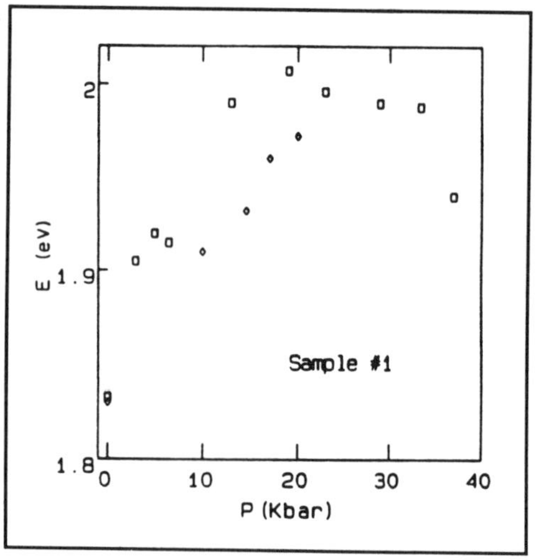

pressure. The large blue-shift is at least an order of magnitude too large to be due solely to the pressure-induced increase in quantum confinement caused by size reduction, as determined from a simple particle-in-a-box calculation using the pressure dependence of the lattice constant. Blue-shifts of the same order of magnitude are seen in samples of porous silicon fabricated by another research group, under different conditions.[26] For porous Si the blue-shift is in the wrong direction to be consistent with the pressure-induced red-shift in the PL from either amorphous Si[27] or the indirect gap in bulk crystalline silicon.[28] Thus, these results provide strong evidence that the origin of the porous Si PL is not a known phase of amorphous Si and that it is not solely due to quantum confinement of the Γ- X bandgap of bulk silicon.

Cathodeluminescence. Under bombardment from high-energy electrons, porous silicon initially emits a bright yellow CL. After a short time, on the order of several seconds, the CL dies. Apparently, the high-energy electrons damage the surface of the microcrystallites. The damage appears to be irreversible. After damage, no PL is emitted either. This effect was confirmed independently by a group at IBM.[29] Although studies by other groups have indicated that hydrogen can be driven off the microcrystallite surfaces and replaced with a corresponding degradation and restoration of the PL, this CL damage appears to be permanent. Once damaged, the porous Si apparently does not respond to the usual methods of restoring the PL.

Summary of the results on porous silicon and device implications. Our results provide strong support for the quantum confinement model of the PL, in which the photon energy of the PL from the silicon microcrystallites is shifted into the visible due to the physical confinement of the quantum mechanical wave functions in the small particles. At the same time our data raise serious questions with the proposals that porous Si PL originates from amorphous Si or from siloxene. However, our results also indicate that other processes involving states with higher degrees of localization than provided by the physical confinement of the microparticle, such as those associated with traps, impurities, or defects, may be equally important in the mechanism for the PL. Once the nature of these more localized states is understood, they may be the key to other novel device applications. Our structural studies do not see quantum wire structures, but rather somewhat irregular microcrystallites. These microparticles appear to be separated from each other by hydrogenated silicon or silicon oxides. Electrical conductivity through a porous silicon film will therefore be via sequential tunneling into and out of the microcrystallites or through defects around them. These studies do not appear to support a high-void fraction in our samples.

Our studies confirm the results reported by others concerning the importance of the passivation and surface chemistry of the microparticles and the variation in time and from sample to sample of the chemistry in the film. Other fabrication techniques for nanoparticles for device application will therefore have to consider surface passivation as an integral part of the fabrication process. The damage caused by high-energy electrons indicates that the usual way of exciting EL films may require reexamination for these materials. Other excitation techniques, such as high-frequency excitation using capacitive contacts, should be considered. Efforts at an LED device design have been initiated at EPSD.[30] Encouraging preliminary results have been achieved elsewhere,[31,32] but significant problems must still be solved. Our experimental results indicate that quantum confinement may significantly affect the usual relationships of energy levels expected from bulk silicon. The binding energies of impurities and defects may be greatly altered, drastically changing the procedure of doping materials for device applications.

Resonant Tunneling Through Microcrystallites

Figure 10 shows conductance-voltage curves for resonant tunneling through a very thin film of silicon microcrystallites.[2] The microcrystallites were fabricated by recrystallizing a very thin amorphous silicon layer in an oxygen ambient. The conductance spikes between -10 and -20V are due to resonant tunneling through the 3-D confinement of the microcrystallite. The steps at higher negative voltages are due to resonant tunneling through 1-D

Figure 10. Conductance-gate voltage curve for resonant tunneling through silicon mocroparticles.

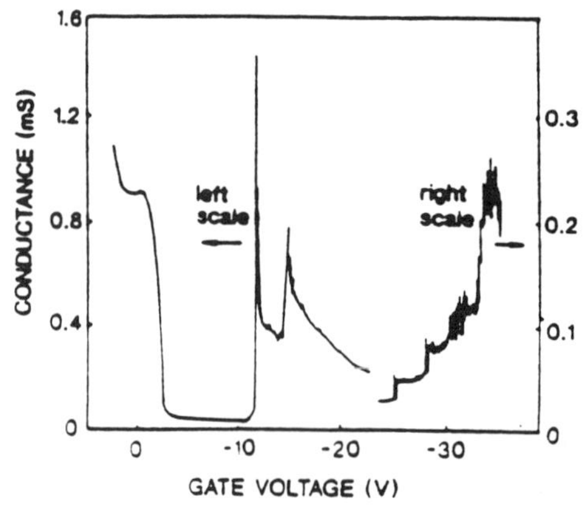

confinement, which occurs when the wavefunctions in neighboring crystallites overlap to form an essentially one dimensional state. The large separation between the spikes below -20V is due to charge accumulation in the crystallites, in a phenomenon similar to "coulomb blockade." Resonant tunneling devices are inherently resistant to noise, and their basic physical mechanism operates well into the millimeter wave spectrum. These demonstrations of resonant tunneling and coulomb blockade-like effects in a film of silicon microcrystallites open the way for new classes of fast tunneling devices compatible with silicon very large-scale integration (VLSI) technology.[33]

Nanoparticle Fabrication

In addition to the two fabrication methods previously discussed for microcluster particles of silicon, two additional novel methods have been under investigation at EPSD. The first method is based on the pulsed power technology that has been developed and perfected here at EPSD. A piece of silicon is subjected to a fast very high current, high voltage pulse. The silicon is vaporized in a manner similar to an "exploding wire" plasma. The silicon atoms recrystallize as they leave the plasma, and the resulting microcrystallites are collected.[34] Figure 11 is a scanning electron microscope (SEM) micrograph of a preliminary shot. In the figure, crystallites smaller than 100nm can be seen on the collection substrate. A similar exploding wire machine using metal wire targets, operating in Russia, has demonstrated that

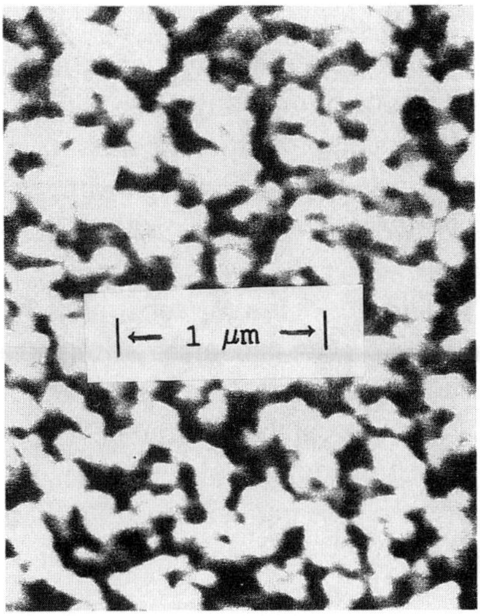

Figure 11. SEM micrograph showing microcrystallites of Si formed in the pulsed power fabrication process.

metal microclusters of 10-nm size can be generated by this method.[35] It is expected that the EPSD technique for silicon can be optimized to produce silicon microparticles of this size. In a preliminary experiment, a second method uses a laser ablation technique to deposit a film of microstructures with diameters ranging down to values less than 100nm.

Nonlinear Optics Effects

Researchers from MICOM have predicted large nonlinear optics parameters for very small clusters of metals, semiconductors, and superconducting material.[36] Active research is being pursued at MICOM to develop such material for optical device applications, such as harmonic generation and optically bistable switches. Experiments on porous silicon and other silicon microcrystallites produced at EPSD are being coordinated with the MICOM research group to measure the nonlinear optic parameters and assess the potential for optical devices.

Summary

It is abundantly clear that these new materials have tantalizing new properties for opto-electronic device applications. The strong PL from porous silicon may well reflect quantum confinement involving localized defects or impurity states. The sharp conductance-voltage features of silicon microparticles are most readily explained by resonant tunneling. Semiconductor microparticles are predicted to have nonlinear properties of interest in optical devices. Semiconductor device physicists typically use concepts of bandgaps, band offsets, impurity doping energy levels, and carrier transport characteristics in visualizing and formulating device concepts. In these new materials consisting of semiconductor microparticles or microstructures at the nanometer scale, the values of these key quantities can become significantly altered from the expected bulk semiconductor values. Our results make a significant contribution to the expanding body of research defining these new properties. Out of this research will come new procedures and techniques for utilizing these novel materials in devices, and the research will open the way for entirely new device concepts and applications.

Acknowledgments

The authors wish to thank M.J. Wade for assistance with fabrication; L. Calderon and M. Taysing-Lara for assistance with data and data analysis; D.W. Eckart, L. Casas, and T. Monahan for assistance with sample analysis; and E.H. Poindexter for assistance with composition. We especially wish to thank M. Weiner, L. Kingsley, and T. Burke for the use of the photo of the preliminary results of their pulsed power fabrication process, and we wish to recognize the novel laser ablation fabrication process being developed by A. Tauber.

While preparing the information in this chapter, authors J.F. Harvey and W. Zhou were under a National Research Council-EPSD associateship, and H. Shen was affiliated with Geo-Centers, Inc., Lake Hopatcong, NJ 07849. Research conducted by R. Tsu was supported in part by the U.S. Army Research Office and the Office of Naval Research.

References

1. L.T. Canham, *Appl. Phys. Lett.* **57**, 1046 (1990).
2. Q. Ye, R. Tsu, and E.H. Nicollian, *Phys. Rev.* **B 44**, 1806 (1991).
3. V. Lehmann and U. Goesele, *Appl. Phys. Lett.* **58**, 856 (1991).
4. A. Halimaoui *et al.*, *Appl. Phys. Lett.* **59**, 304 (1991).
5. K.H. Jung *et al.*, in *Light Emission From Silicon*, S.S. Iyer, L.T. Canham, and R.T. Collins, Eds. (*Mater. Rsch. Soc.* **256**, San Francisco, 1992), p. 31.

6. M.A. Tischler *et al.*, in *Light Emission From Silicon*, S.S. Iyer, L.T. Canham, and R.T. Collins, Eds. (*Mater. Rsch. Soc.* **256**, San Francisco, 1992), p. 189.
7. T. Motohiro *et al.*, in *Light Emission From Silicon*, S.S. Iyer, L.T. Canham, and R.T. Collins, Eds. (*Mater. Rsch. Soc.* **256**, San Francisco, 1992), p. 53.
8. S.M. Prokes *et al.*, in *Light Emission From Silicon*, S.S. Iyer, L.T. Canham, and R.T. Collins, Eds. (*Mater. Rsch. Soc.* **256**, San Francisco, 1992), p. 107.
9. M.W. Cole *et al.*, *Appl. Phys. Lett.* **60**, 2800 (1992).
10. V. Lehmann, H. Cerva, and U. Goesele, in *Light Emission From Silicon*, S.S. Iyer, L.T. Canham, and R.T. Collins, Eds. (*Mater. Rsch. Soc.* **256**, San Francisco, 1992), p. 3.
11. R.T. Lareau, unpublished data.
12. L.T. Canham *et al.*, *J. Appl. Phys.* **70**, 422 (1991).
13. F. Rong and E.H. Poindexter, unpublished data.
14. M.B. Robinson *et al.*, in *Light Emission From Silicon*, S.S. Iyer, L.T. Canham, and R.T. Collins, Eds. (*Mater. Rsch. Soc.* **256**, San Francisco, 1992), p. 17.
15. V. Petrova-Koch *et al.*, in *Light Emission From Silicon*, S.S. Iyer, L.T. Canham, and R.T. Collins, Eds. (*Mater. Rsch. Soc.* **256**, San Francisco, 1992), p. 41.
16. Y.H. Xie *et al.*, *J. Appl. Phys.* **71**, 2403 (1992).
17. R. Tsu, H. Shen, and M. Dutta, *Appl. Phys. Lett.* **60**, 112 (1992).
18. J.F. Harvey *et al.*, *Bull. Am. Phys. Soc.* **37**, 565 (1992).
19. J.F. Harvey *et al.*, in *Quantum Well and Superlattice Physics IV*, G.H. Doehler and E.S. Koteles, Eds. (SPIE Vol. 1675, Bellingham, WA, 1992), p. 134.
20. R. Tsu and L. Ioriatti, *Superlatt. Microstruct.* **1**, 295 (1985).
21. H. Shen *et al.*, *Bull. Am. Phys. Soc.* **37**, 669 (1992).
22. P.L. Bradfield, T.G. Brown, and D.G. Hall, *Phys. Rev.* **B 38**, 3533 (1988).
23. X.L. Zheng *et al.*, *Bull. Am. Phys. Soc.* **37**, 670 (1992).
24. M. Taysing-Lara *et al.*, *Bull. Am. Phys. Soc.* **37**, 670 (1992).
25. D.C. Morton *et al.*, *Bull. Am. Phys. Soc.* **37**, 669 (1992).
26. W. Zhou *et al.*, *Appl. Phys. Lett.* **61**, 1435 (1992).
27. B.A. Weinstein, *Phys. Rev.* **B 23**, 787 (1981).
28. B. Welber *et al.*, *Solid State Commun.* **29**, 1021 (1975).
29. J.L. Batstone *et al.*, *Bull. Am. Phys. Soc.* **37**, 564 (1992).
30. M.F. Tompsett and R. Tsu, patent disclosure, CECOM docket no. 4728.
31. N. Koshida and H. Koyama, *Appl. Phys. Lett.* **60**, 347 (1992).
32. N.M. Kalkoran and F. Namavar, in *Light Emission From Silicon*, S.S. Iyer, L.T. Canham, and R.T. Collins, Eds. (*Mater. Rsch. Soc.* **256**, San Francisco, 1992), p. 89.
33. J.F. Harvey, R.A. Lux, and R. Tsu, patent disclosure, CECOM docket no. 4741.
34. J.F. Harvey *et al.*, patent disclosure, CECOM docket No. 4800.
35. C.G. Thornton, private communication, 1990.
36. Y.Q. Li *et al.*, *J. Opt. Soc. Am.* **B 6**, 814 (1989).

16

Optical Control of Resonant Tunnel Diode Devices

Robert A. Lux
James F. Harvey
Peter G. Newman
Dana J. Sturzebecher
Thomas P. Higgins
Thomas E. Koscica
Arthur C. Paolella
Anthony Deanni
Wayne H. Chang
U.S. Army Research Laboratory
Electronics and Power Sources Directorate
Fort Monmouth, NJ 07703-5601

Direct optical control of resonant tunnel diodes (RTD) by optical fiber promises low-noise operation of military mm-wave systems at frequencies previously not feasible and with remote antenna and phased array applications. RTDs have been operated as oscillators as high as 400GHz and as mixers above 2THz. GaAs RTDs operating at 10GHz have been optimized to produce power densities of 5000W/cm^2, with conversion efficiencies of about 10 percent. Power conversion efficiencies of 30 percent appear feasible using InGaAs RTDs frequency. RTDs can operate at higher frequencies than field effect transistors (FETs), heterogeneous bipolar transistors, and other two terminal devices such as Gunn and impact avalanche and transit time diodes. Lack of isolation between input and output in a two-terminal device and the low impedance of the RTD make their integration into conventional FET-based microwave and millimeter monolithic integrated circuit designs awkward. However, recent results in optical control of the RTD oscillators show that isolation of the RTD and antenna circuit on a separate chip from the rest of the system provides the capability for remote antenna operation and optically synchronized antenna arrays using optical fiber. In effect the optical signal acts as a third terminal, allowing isolation of input and output signals. Direct optical frequency modulation and direct optical injection locking of RTD oscillators have been demonstrated for the first time. The authors propose a physical mechanism for the optical control of RTDs that can impact military mm-wave radar and communication systems.

Introduction

Direct optical control of microwave semiconductor devices has received a large amount of attention because of the potential applications in phased array antennas for radar and communications, for antenna remoting, in interfacing optical fiber systems, and in isolating control and rf signals in microwave systems. A recent review article discusses these topics, including optical frequency tuning of Gunn, impact avalanche and transit time diode (IMPATT), trapped plasma avalanche-triggered transit diode (TRAPATT), field effect transistor (FET), and high electron mobility transistor (HEMT) oscillators.[1]

The direct optical control of resonant tunnel diodes (RTD) via an optical fiber promises low-noise operation of military mm-wave systems at frequencies that were previously infeasible and with the capability of remote antenna and phased array applications. RTDs have been operated as oscillators as high as 400GHz and as mixers above 2THz.[2,3] GaAs RTDs operating at 10GHz have been optimized to produce power densities of 5000W/cm^2 of chip area, with conversion efficiencies of about 10 percent. Power conversion efficiencies of 30 percent appear feasible using InGaAs RTDs.[4] These output power densities are roughly comparable to FET oscillators at this frequency. RTDs can operate at higher frequencies than FETs, heterogeneous bipolar transistors (HBT), and two-terminal devices such as Gunn and IMPATT diodes. In addition, RTDs have very low-noise characteristics. The lack of isolation between input and output in a two-terminal device and the very low impedance of the RTD make the integration of these devices into conventional FET-based microwave and millimeter monolithic integrated circuit (MMIC) designs very awkward. Optical control of the RTD oscillators would make it possible to avoid this problem by optically isolating the RTD and antenna circuit, on a separate chip, from the rest of the system and provide the capability for remote antenna operation and optically synchronized antenna arrays using optical fiber. In effect, the optical signal acts as a third terminal, allowing isolation of input and output signals. Optical control of RTDs as switching devices has been reported. This chapter presents the results of an Electronics and Power Sources Directorate (EPSD) investigation of optical control of these devices as microwave sources.

Device Structure

The material for the RTDs was fabricated at the EPSD Molecular Beam Epitaxy (MBE) facility. The basic structure of the devices is shown in Figure 1. The thickness of the undoped spacer layers was varied from 50nm to

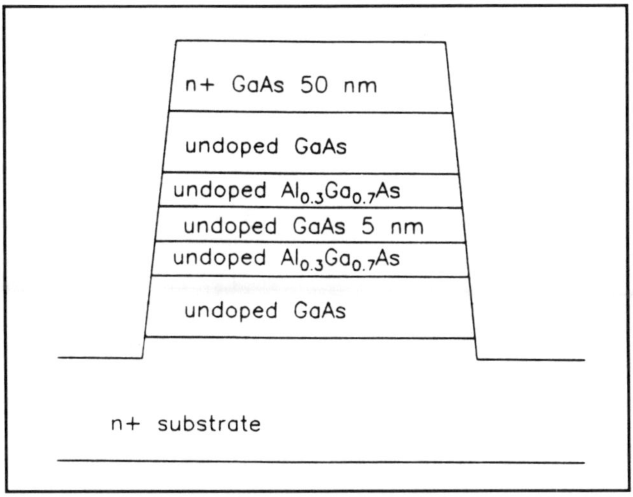

Figure 1. Layer structure of RTD material. Undoped GaAs layers varied from 50nm to 250nm.

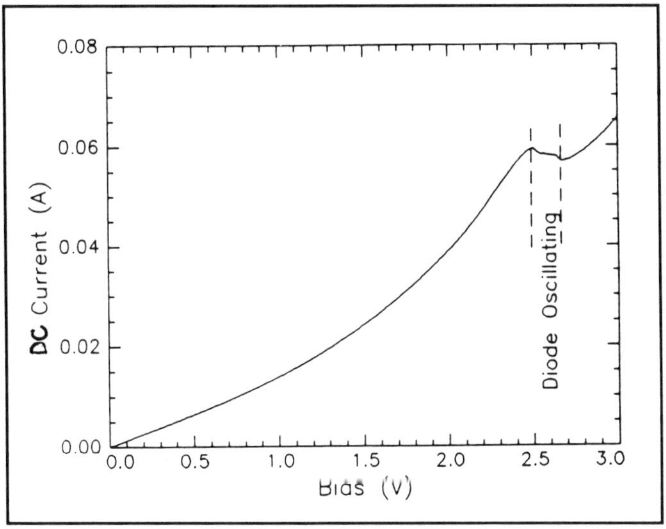

Figure 2. DC current versus bias voltage for RTDs.

250nm. Mesas of diameters from 2μm to 32μm were etched. The current versus bias voltage curves of all devices are similar, with the value of the peak current being approximately proportional to the area of the mesa. Oscillators could be made from all sizes of mesa and all spacer layers. Results reported here will be for a structure with a 50nm top spacer and a 250-nm lower spacer. Either Au-Ge or Pd-Ge contacts were made to the top of the mesa and to the bottom of the n^+ substrate. No significant difference was seen in the current versus voltage curves (I-V) between the two types of contact. However, the Pd-Ge contact has two advantages that make it more useful for future devel-

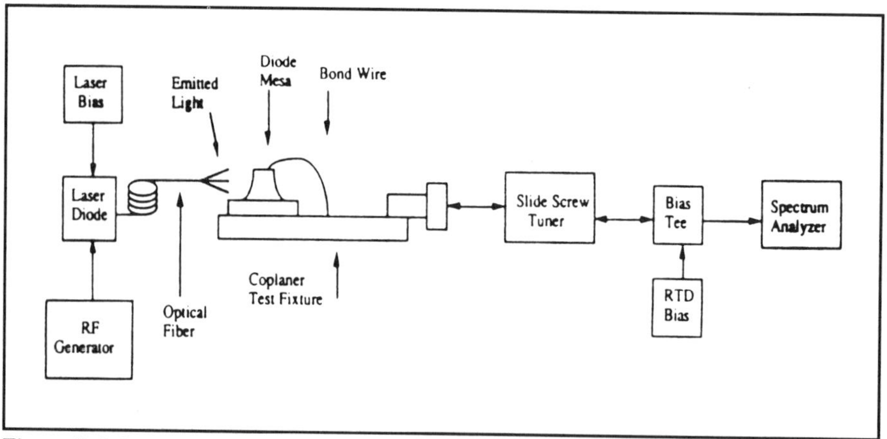

Figure 3. Schematic of RTD test fixture.

opment: diffusion of the metal into the semiconductor is limited, and the contact shows better high temperature stability. Figure 2 shows a room temperature DC I-V curve for a 32-μm diameter diode. The portion of the curve labeled "diode oscillating" does not represent a true DC state but is the average current in a diode oscillating with the parasitic impedance of the curve tracer.

Experimental Setup

The RTD was mounted in a 50-ohm coplanar test fixture with a wire bond (25μm Au wire, 1-meter long) from the top of the mesa (Figure 3). When biased in the negative differential resistance region, the RTD oscillated at a free-running frequency determined by internal and external capacitances and the external inductance, which is primarily due to the bond wire. The stub tuners were adjusted for maximum power output. An Ortel 1210C pigtail laser diode of 840-nm wavelength was used as the source of illumination. Maximum intensity at the exit of the fiber was 4mW, with the illumination covering the edge of the diode mesa and the underlying substrate. Various laser rf generators allowed the laser to be amplitude modulated from DC to 3GHz.

Frequency Modulation

The diode was illuminated with 2mW DC optical power through a variable attenuator. It was found that frequency of oscillation was a function of both the bias voltage and the illumination power, as shown in Figure 4. For

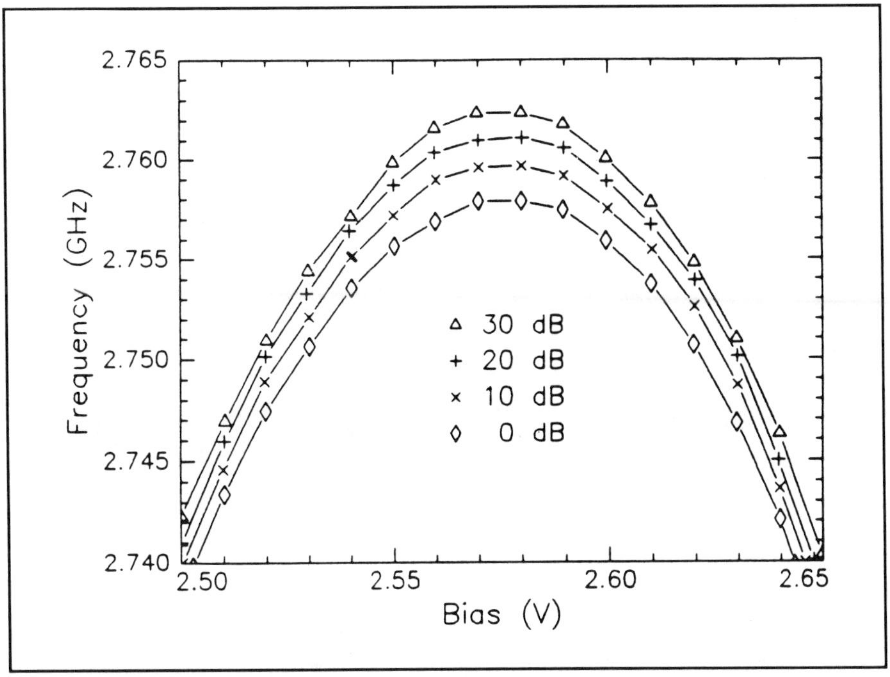

Figure 4. RTD oscillator free-running frequency as a function of bias and DC optical input. The four curves are for different optical attenuation values inserted in the fiber link.

any bias voltage, the frequency shift is a nonlinear function of the optical power, each 10-dB increase producing an approximately equal frequency shift.

The oscillator frequency can follow a high frequency modulation of the laser. Figures 5 and 6 show frequency spectra of an oscillator with 5kHz square wave optical modulation and 75MHz sine wave optical modulation, respectively.

Injection Locking

As with other microwave oscillators, it is possible to electrically injection lock a free-running RTD oscillator to an rf signal. This led us to attempt injection-locking with an optical signal. With a laser source of 4mW at the exit of the coupling fiber, we found that when the frequency of optical modulation approached the free-running RTD oscillator frequency, the oscillator locked to the optical modulation frequency. Figure 7 shows the RTD oscillator spectra when the oscillator is free-running or locked to an optical signal, respectively. On locking there is the typical injection locking reduction of

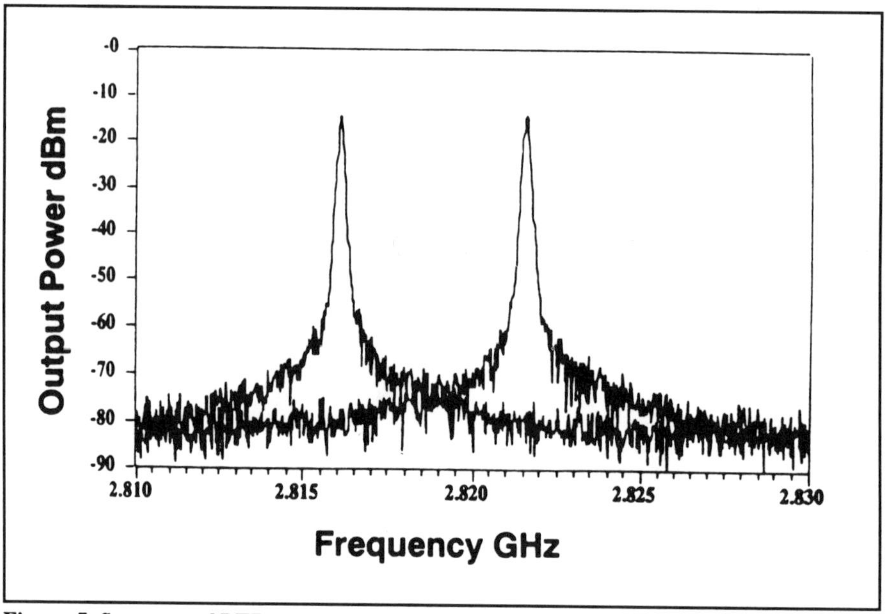

Figure 5. Spectrum of RTD oscillator modulated by 5kHz square wave optical signal.

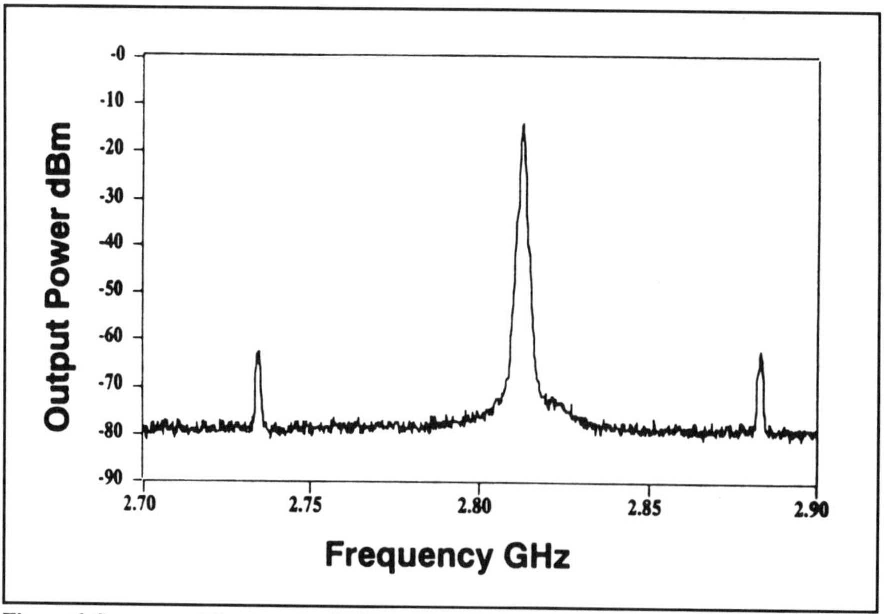

Figure 6. Spectrum of free-running RTD oscillator with 75MHz sine wave optical modulation.

noise sidebands and narrowing of the central peak. The total locking bandwidth was 150kHz.

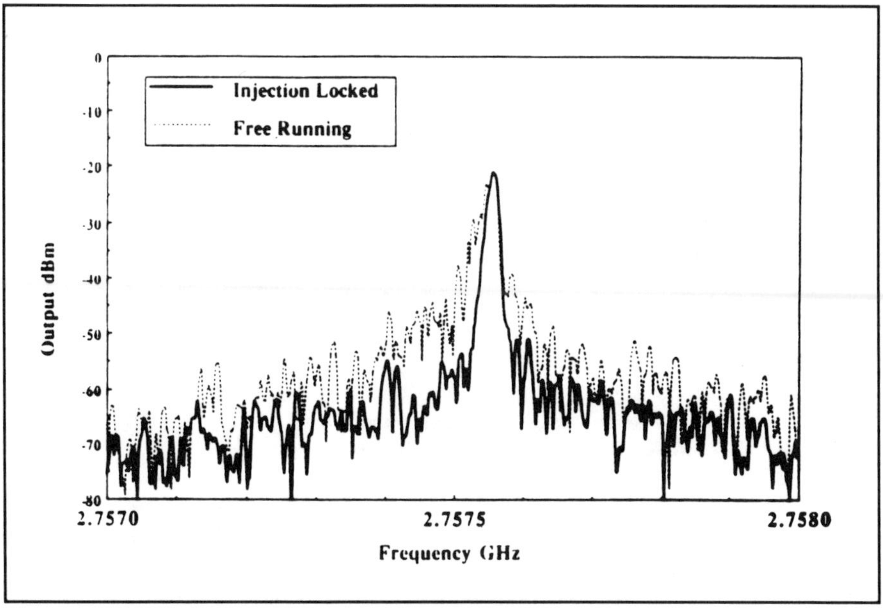

Figure 7. Injection locking of RTD oscillator by a 4mW optical signal.

Mechanism of Optical Control

We propose three mechanism to explain the observed optical effects: (1) photoconductivity reduces the resistance of the diode external to the barrier layers; hence, the reduced IR voltage drop changes the effective diode bias; (2) holes generated below the barrier layer drift into the barrier region and are trapped, resulting in a changed electric field in the device; and (3) DX center traps in the AlGaAs barriers are ionized by the light, resulting in a dipole field within the barrier structure.[5]

Although all three mechanisms may contribute in some cases, we believe the third is not likely in our diodes. All the diodes reported here have a thick metal top on the mesa; hence, the only way the barrier region could be illuminated is through the edge of the mesa. This would be a very inefficient process. Further, there would be a large difference in optical effects, depending on the diameter of the diodes. We see no such effect.

Photoconductivity in the substrate also does not seem to be a satisfactory explanation. As shown in Figure 4, to produce the observed optically induced frequency shifts would require bias shifts of about 50mV. This is greater than the total voltage drop in the substrate when operating in the oscillator region; hence, an increased substrate conductivity would not produce the observed effects.

Conclusions

Our work at EPSD has produced the first demonstration of optical control in microwave RTDs. Although these results are a proof of principle demonstration, the high-frequency, low-noise capability of RTDs makes them a potential candidate for future Army systems.

References

1. A.J. Seeds and A.A.A. de Sales, *IEEE Trans. Microwave Tech.* **38**, 577 (1990).
2. E.R. Brown *et al.*, *Appl. Phys. Lett.* **55**, 1777 (1989).
3. T.C.L.G. Sollner *et al.*, *Appl. Phys. Lett.* **45**, 1319 (1984).
4. V.K. Reddy *et al.*, in *Quantum Well Injection Time (QWIT) Diode Oscillators*, R.J. Temkin, Ed. (Fifteenth International Conference on Infrared and Millimeter Waves, SPIE, 1990), pp 88-90.
5. T.C.L.G. Solner, E.R. Brown, and H.Q. Le, *Lincoln Lab. Jour.* **1**, 89 (1988).

17

High-Sensitivity Infrared Hot-Electron Transistors

K. K. Choi
M. Taysing-Lara
W. Chang
U. S. Army Research Laboratory
Electronics and Power Sources Directorate
Fort Monmouth, NJ 07703-5601

Research on long wavelength infrared detection using intersubband absorption in III-V materials has been progressing rapidly in the last several years. However, the operating temperatures of the detector arrays still need to be kept below 77K due to the large thermally assisted tunneling current at high temperatures. In order to reduce the dark current, we developed a new infrared hot-electron transistor (IHET) structure, in which a thick quantum barrier is used as an electron energy high pass filter to preferentially filter out the dark current. By adopting a thin (300Å) InGaAs base layer to obtain a high photocurrent transfer ratio, we achieved a detectivity of 1.4×10^{10} cm$\sqrt{\text{Hz}}$/W at 77K, with a cutoff wavelength of 9.5μm; this is two times as large as the companion state-of-the-art GaAs quantum well infrared photoconductor (QWIP), and it exceeds the limit imposed previously. Combined with the large reduction in dark current, the noise equivalent temperature difference of the detector arrays can then be reduced by a factor of 36 compared with the QWIP detector arrays. Besides enhancing the sensitivity of the detector, the transistor reduces the system noise by providing a photovoltage gain of 9.7 and a power gain of 3.2. The device can also be used in a quasi-photovoltaic mode to facilitate readout coupling.

Introduction

Recently, thermal imaging has been successfully demonstrated using GaAs/AlGaAs multiple quantum well infrared photodetectors (QWIPs). However, the operating temperature tends to be lower than 77K due to the large dark current existing at this temperature.

We propose to use an additional electron energy filter to reduce the dark current of a QWIP device and hence increase the device's operating temperature. The resultant device constitutes an infrared hot-electron transistor (IHET). When the infrared detection is dark current limited, we will show that the detectivity of the IHET is increased by a factor of $\alpha_p/\sqrt{\alpha_d}$ in comparison with the original photodetector, and the noise equivalent temperature difference is reduced by a factor of α_p/α_d, where α_p and α_d are the photocurrent and the dark current transfer ratios, respectively.

Furthermore, in the voltage coupled readout circuits, we will also show that the IHET structure increases the voltage responsivity of a QWIP by providing an internal photovoltage gain, and the IHET can be DC coupled to a high gain amplifier without a light intensity modulator.

We will discuss the basic operating principles and the state of the art of QWIP and IHET devices.

A basic QWIP device consists of a number of periods of GaAs quantum wells and $Al_xGa_{1-x}As$ barriers. The quantum wells are doped to 10^{18} cm^{-3} and the barriers are undoped. The well width is adjusted so that the spacing of the lowest two subbands matches the incoming photon energy. The barrier width is wide enough to suppress the tunneling current from the ground subband. The Al molar ratio x, on the other hand, determines whether the second subband is bounded or extended. The multiple quantum wells are sandwiched between two n doped layers as electrical contacts. For an IHET structure, an additional $Al_yGa_{1-y}As$ layer is grown on the top of one of the contacts as an electron energy filter, followed by another n-doped layer as the collector.

In the following text, we will describe the electrical and optical properties of a typical IHET device. The device is grown by molecular beam epitaxy on a (100) semi-insulating substrate. The first layer grown is a 6000-Å-thick n^+-GaAs layer as the emitter contact layer. Next, an infrared sensitive QWIP structure is grown, consisting of 50 periods of GaAs wells and $Al_{0.25}Ga_{0.75}As$ barriers. The well and barrier widths are 40Å and 200Å, respectively. On the top of the QWIP structure, a thin (300Å) pseudomorphic $In_{0.15}Ga_{0.85}As$ base layer is grown, followed by a 2000-Å-thick $Al_{0.25}Ga_{0.75}As$ barrier and a 1000-Å-thick GaAs as the collector contact layer. All the layers except the barriers are doped to $1.2x10^{18}cm^{-3}$. The band structure is shown in Figure 1.

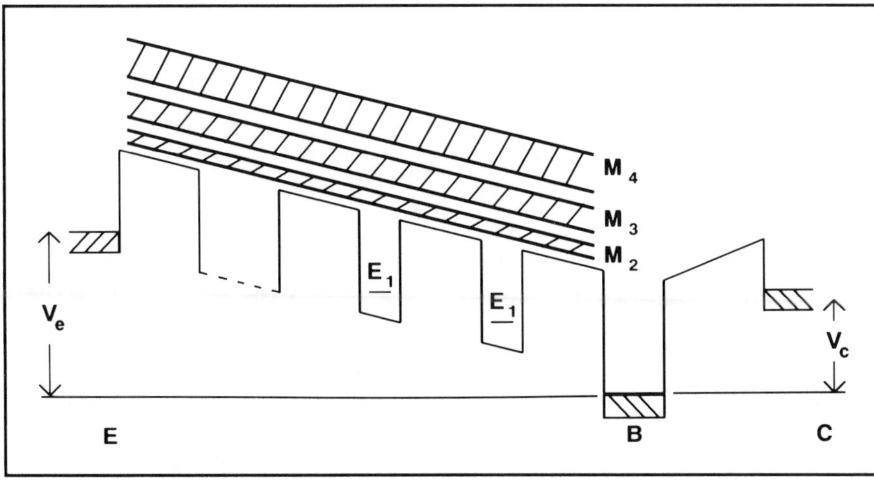

Figure 1. Band structure of detector under an emitter bias V_e and a collector bias V_c. M_n are minibands above the barriers.

Infrared Absorption of QWIP

Infrared absorption by QWIP structures is initiated by electronic intersubband transitions from the ground state E_1 to the first excited state E_2 within the conduction band. Because the dipole moment of the transition is along the superlattice axis, the electric field component of the incident light is necessarily nonzero along this direction. The absorption coefficient α_q can be shown to be:

$$\alpha_q = \frac{\rho_s}{L_p} \frac{e^2}{2\varepsilon_0 m^* c} f \frac{\sin^2\theta}{\eta} \frac{\Gamma/2}{(\omega_q - \omega_0)^2 + (\Gamma/2)^2}, \qquad (1)$$

where ρ_s is the two-dimensional electron density per well, L_p is the length per period, f is the oscillator strength, θ is the angle between the direction of light propagation and the superlattice axis, η is the refractive index, and Γ is the absorption linewidth.

For $\rho_s = 4\text{x}10^{15}$ m^{-2}, $L_p = 500$Å, $f = 0.96$, $\Theta = 45°$, $\hbar\omega_o = 124$meV, $\Delta\lambda/\lambda = 20$ percent, α_{peak} is calculated to be 490cm^{-1}. If light is allowed to make a double pass through the sample, the internal quantum efficiency η for unpolarized light is then equal to 11 percent, estimated from the expression $\eta = 0.5(1-\exp(-2\alpha l))$, where l is the thickness of the QWIP. This estimation is done by assuming the infrared light coupled into the device at a 45° angle. By using proper diffraction gratings for light coupling, η can be made close to unity.[1]

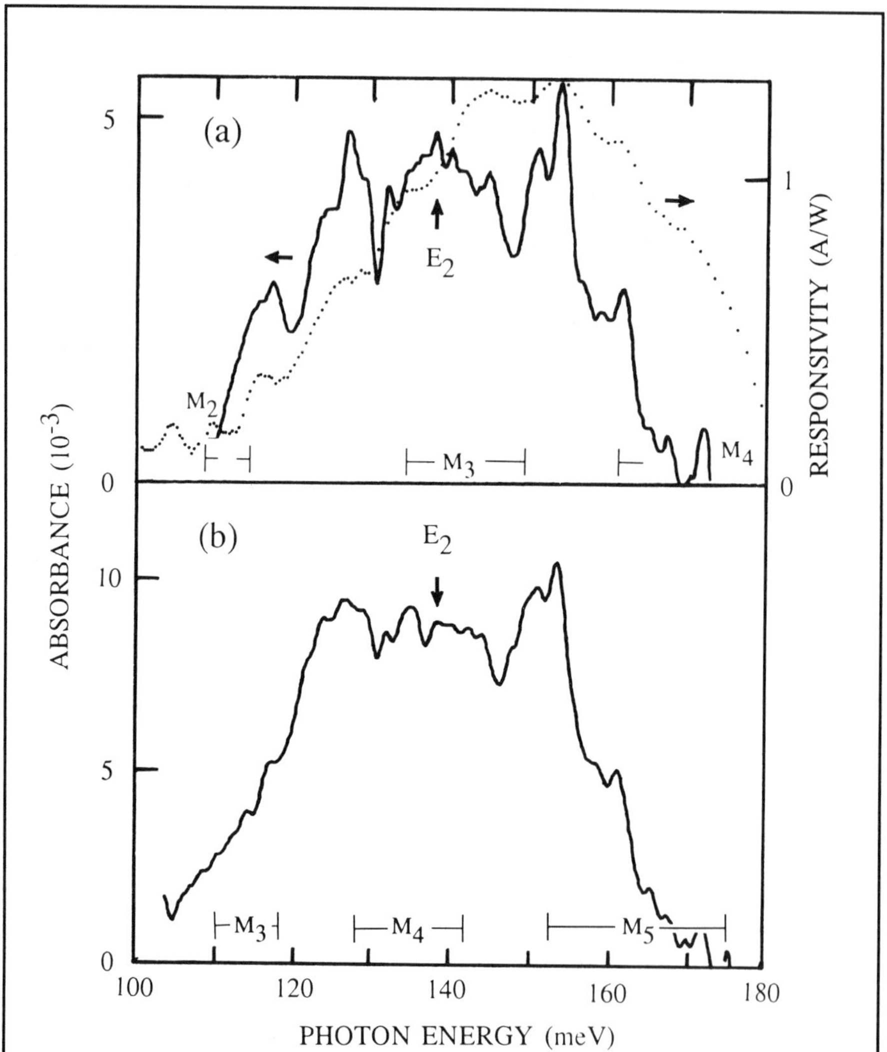

Figure 2. (a) Absorption spectrum at Brewster's angle and photocurrent spectrum of a QWIP with well width 40Å, barrier width 200Å, and Al molar ratio of 0.25. (b) Absorption spectrum a QWIP with well width 40Å, barrier width 300Å, and Al molar ratio of 0.23.

Figure 2(a) shows the absorbance of the mentioned QWIP structures as a function of incident photon energy at Brewster's angle. From the peak absorption, we can estimate η at 45° to be 5 percent, half of the calculated value. On the other hand, the measured η (= 10 percent) of a similar structure is consistent with the calculated value. The second sample has the same nominal well width and doping density but different barrier parameters (width

= 300Å and Al molar ratio = 0.23), the absorption curve is shown in Fig. 2(b). The difference in the value of η of the two samples may be due to the difference in the actual doping density and the uncertainty associated with the background subtraction. Note that the two samples have different miniband structures due to different barrier parameters (locations of the minibands are shown in Figure 2). Despite the differences in the miniband structures, the same peak position at E_2 is observed. This indicates that for single bound state detectors, the optical transition energy is determined by the resonant excited state E_2 of individual wells, which depends only on the well width, but not on the miniband structure. This result shows that the well width is the only adjustable parameter to tune the operating wavelength when the excited state is extended. Also, as shown in Figure 2(b), excessive reduction of the barrier height or increase of the barrier width reduces the detector sensitivity, because the minibands between the top of the barriers and E_2 generate additional thermionic emission current but negligible photocurrent.

On the absorption peak, besides the E_2 peak, there are other modulation structures. These modulation peaks, absent in the normal incidence, match the oscillatory structures in the responsivity curve shown in Figure 2(a) measured at temperature equal to 10K and at 45° light coupling angle. This result indicates that the absorption modulations are due to electron transitions, which can be attributed to the localized states created as a result of coupling between E_1 and the minibands.

Dark Current Characteristics of IHET

The dark current characteristics of the detector at 77K are shown in Figure 3. For this device, the emitter area is $7.92 \times 10^{-4} cm^2$, and the collector area is $2.25 \times 10^{-4} cm^{-2}$. The device configuration is shown in the insert of Figure 3. The emitter current I_e is the current measured at the emitter normalized according to the ratio of the emitter area to the collector area, which also represents the current level of a QWIP device. The electron current injected to the base is separated into two branches according to the energy of the electrons by the energy filter located between the base and the collector. The lower energy electrons are blocked by the barrier and drained through the base; the higher energy electrons overcome the barrier and are collected at the collector and form collector current I_c. The dark current transfer ratio α ($\equiv I_c/I_e$) as a function of emitter voltage V_e at a constant collector voltage V_c (= 0 V) depends on the electron energy distribution ρ. In fact, the conducting electron distribution at a different temperature can be measured by thermally stimulated hot-electron (TSHE) spectroscopy,[2] in which r, defined by $d\alpha/dV_e$,

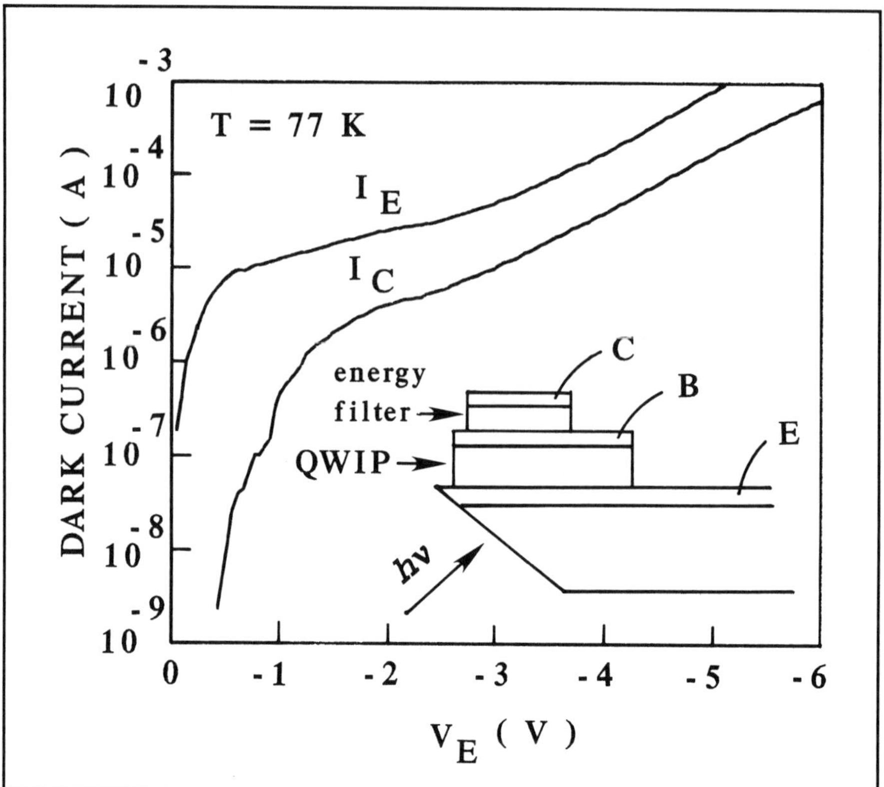

Figure 3. Emitter dark current I_E and collector dark current I_C as a function of emitter bias V_E at 77K. Insert shows schematic device configuration.

is measured as a function of V_e at different temperatures. Figure 4 shows the TSHE spectra at temperatures equal to (a) 100K, (b) 77K, and (c) 50K. r can be shown to be proportional to the hot-electron distribution, with the higher energy electrons appearing at the lower V_e. The data show that the peak of the distribution at 77K is located between E_1 and E_2, and hence indicate the dominant transport mechanism at this temperature is thermally assisted tunneling.[3] At finite temperature, the ground state electrons located at the high energy Fermi distribution tail can be scattered by impurities and tunnel through the quantum barriers in the QWIP structures.

Opto-Electronic Properties of IHET

The photoresponse of the detector can be characterized using either a CO_2 laser or a glowbar-monochromator. Figure 5 shows the current responsivity

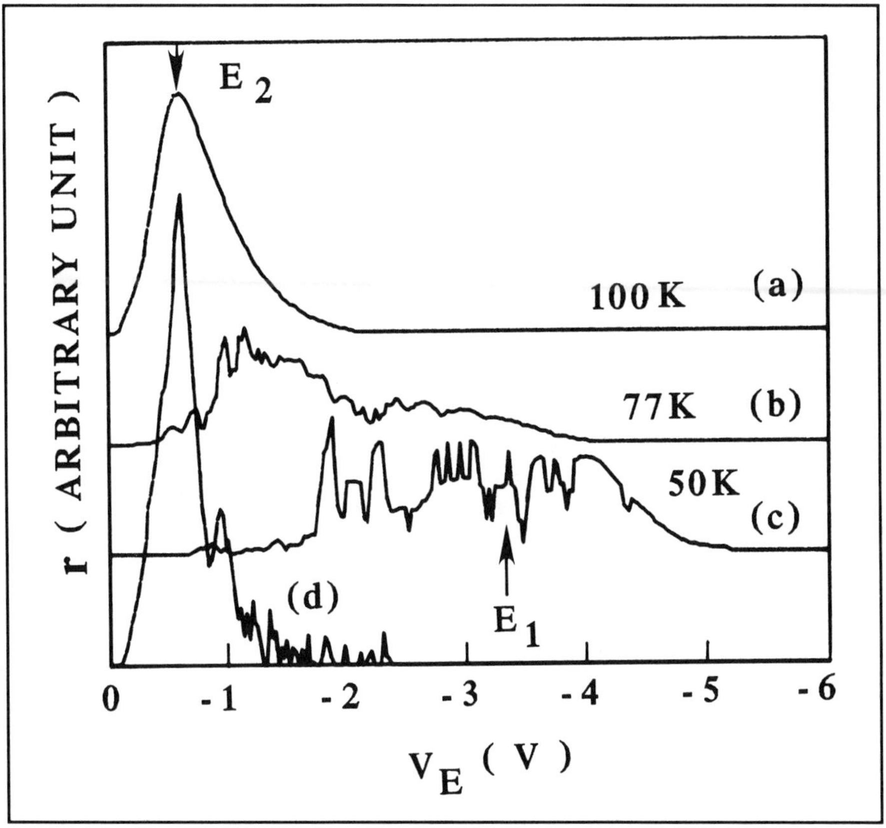

Figure 4. Parameter r (defined in text) as a function of V_E at (a) 100K, (b) 77K, and (c) 50K. Parameter r for photocurrent measured at 90K is indicated by (d). Curves have been shifted vertically for clarity.

R as a function of V_c. At low V_e, R is directly proportional to V_e. At larger V_e, R becomes constant. After $V_e = -4$ V, R increases again. In order to understand the characteristics of R, it is important to know the photoelectron distribution in the detector. Curve (d) of Figure 4 is a plot of r_p defined as $d\alpha_p/dV_e$, where α_p is the ratio of collector photocurrent I_{pc} to the emitter photocurrent I_{pe} normalized to the same area. r_p can also be shown to be directly proportional to the photoelectron distribution. In this measurement, the wavelength of the incident light is 9.25μm. The result shows that the width of the photoelectron distribution is very narrow. The estimated full width at half maximum (Γ_ρ) is 17meV. It is in agreement with the next measurement using different incident photon energy.

In the following experiment, the emitter and the collector voltages are

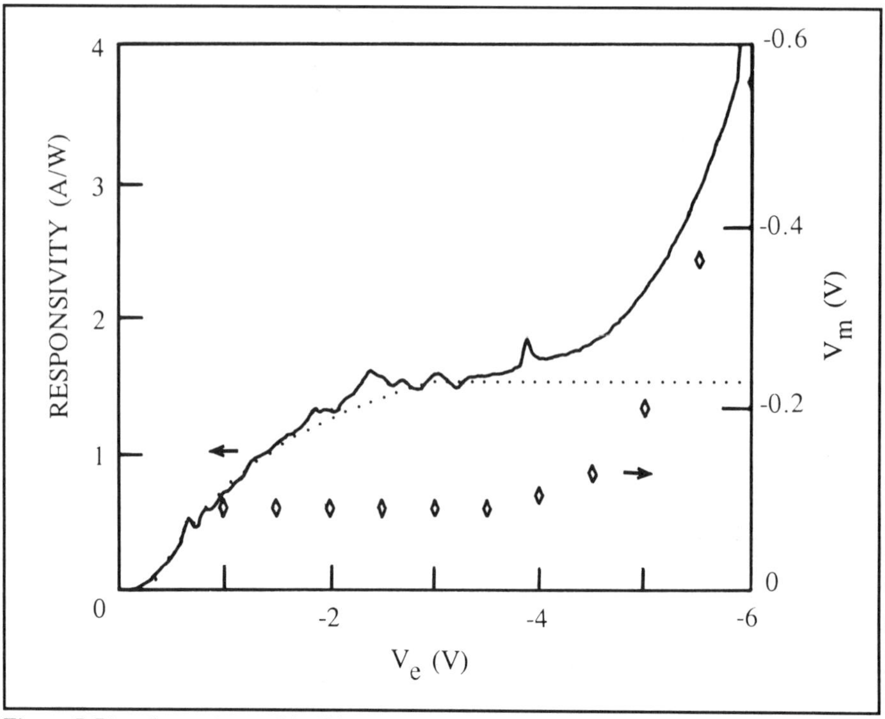

Figure 5. Plot of experimental (solid curve) and theoretic (dotted curve) DC current responsivity R at photon energy equal to 150meV as a function of emitter bias V_e. Diamonds are positions of photoelectron peak eV_m above analyzer barrier shown in Figure 7. R and eV_m show strong correlation.

fixed at -1.2V and 0V respectively. α_p is then measured as a function of incident photon energy. The result is shown in Figure 6. The solid curve is fitted to the experimental data assuming that Γ_ρ is Lorentzian and that photoelectron energy increases with the photon energy. Γ_ρ is then deduced to be 17meV at the front boundary of the analyzer in agreement with the last experiment. When the detector is under 1000K radiation, Γ_ρ is found to be 31meV, consistent with the absorption width.

In order to know the photoelectron distribution at different values of V_e, I_{pc} is now measured as a function of V_c at different at values of V_e. As V_c becomes more negative, the barrier height of the energy filter becomes higher as shown in Figure 1, and less photocurrent will be collected. The rate of change of I_{pc} as a function of V_c, dI_{pc}/dV_c, is then a measure of ρ. The result is shown in Figure 7. Γ_ρ is found to be insensitive to V_e, with the mean value 76meV. However, this apparent Γ_ρ is not the intrinsic width of the photoelectron injection because additional broadening occurs in the thick energy filter.

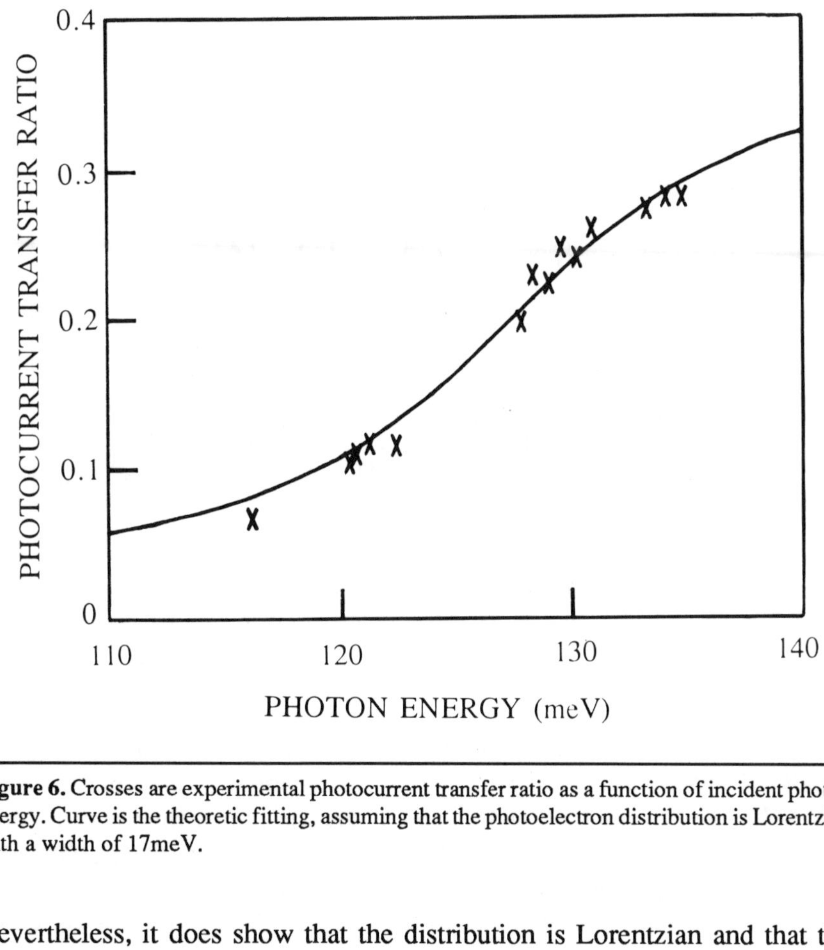

Figure 6. Crosses are experimental photocurrent transfer ratio as a function of incident photon energy. Curve is the theoretic fitting, assuming that the photoelectron distribution is Lorentzian with a width of 17meV.

Nevertheless, it does show that the distribution is Lorentzian and that the energy of the photoelectrons increases with V_e when V_e is large.

These results show that the photoelectron distribution is extremely narrow, indicative of the diffusive transport of the photoelectrons; if the transport were ballistic or quasi-ballistic, the expected energy distribution would be much broader. Hence, the standard theory for photoconductors applies to QWIP devices:

$$R = \frac{e}{h\upsilon}\eta\,(V_e)g(V_e) = \frac{e}{h\upsilon}\,\frac{\eta\,(V_e)}{\dfrac{2v_s}{v_d} + \dfrac{\tau_{tr}(V_e)}{\tau_0}}\,, \tag{2}$$

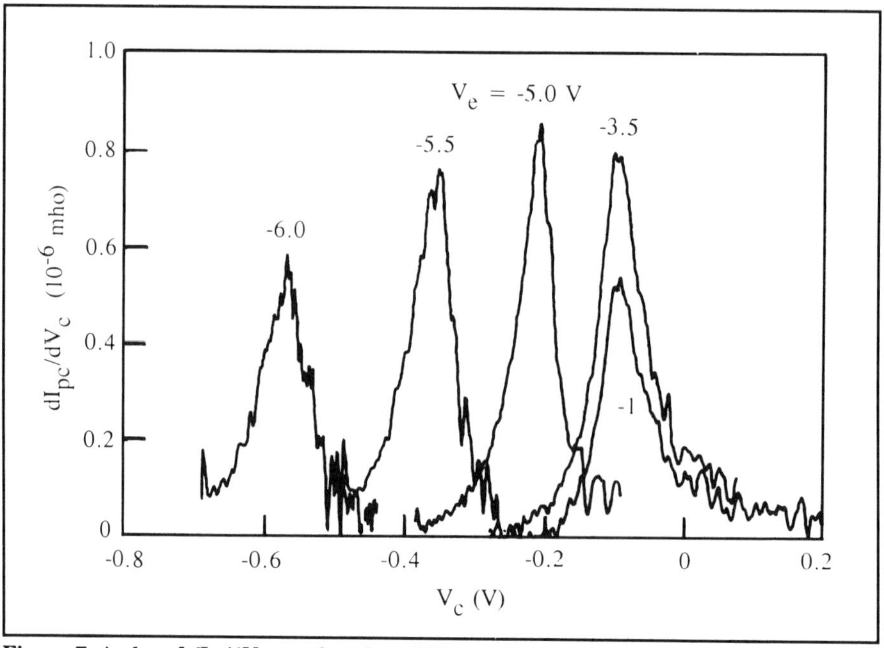

Figure 7. A plot of dI_{pc}/dV_c as a function of V_c at different V_e at temperature equal to 10K.

where R is the current responsivity, η is the quantum efficiency, g is the photoconductive gain, v_s and v_d are the contact recombination velocity and the drift velocity, respectively, and τ_{tr} and τ_o are the transit time and the hot-electron lifetime, respectively. Equation 2 is valid if $v_s << v_d$, that is, if the hot-electron energy relaxation rate is small at the contacts.

In order to apply Equation 2 to the present device, we must know the specific characteristics of the device. For the present device, high field domain (HFD) formation is observed up to 77K. Below a critical emitter voltage (V_k) equal to -3V, an increase in V_e results in an increase in the number of HFDs, but the electric field in the low field regime and the high field regime remains constant.[4] Because the major contribution of photocurrent is from the HFDs, the effective η is then proportional to V_e below V_k and constant fo V_e above V_k. Under a constant electric field below V_k, v_d is also a constant and is actually equal to the saturated drift velocity because the electric field (≅60meV per period) is much larger than the critical field (7meV per period) for maximum drift velocity to occur.[5] Because v_d is constant, τ_{tr}, being the time to travel across the high field regime, is proportional to V_e below V_k and constant above V_k. If we assume v_s and τ_o to be constant, the DC current responsivity can be fitted as shown in Figure 5 for $|V_e| < 4V$ with $v_s/v_d = 0.13$, consistent with the assumption, and $\tau_o/\tau_{tr} = 2.5$ at V_k.

For $|V_e| > 4V$, R is larger than the fitting predicted. In order to understand this discrepancy, it is necessary to examine R at the lower bias. When V_e is small, τ_{tr}/τ_o is less than $2v_s/v_d$; therefore, R is proportional to η and hence to V_e. At larger bias, the reverse is true; R is then proportional to $\eta\tau_o/\tau_{tr}$, which is independent of V_e, as observed. The rising of R after -4V then indicates that τ_o increases at large bias. This conclusion is in fact consistent with the increasing photoelectron peak position at large V_e, as shown in Figure 7. As the average energy of the photoelectrons above the quantum barriers is increased, τ_o is expected to increase. In Figure 5, we plot the peak position eV_m above the analyzer barrier, which shows strong correlation with R. Note that the contact recombination term we included is crucial to the theory since it explains the voltage dependence at low voltages and determines the ultimate gain achievable at high voltages when τ_o becomes very large.

Sensitivity of IHET

In the following, we will estimate the dark current limited detectivity of the detector based on its electrical and optical properties. The detectivity D* is defined as $R\sqrt{A}/i_n$, where A is the area, R is the current responsivity of unpolarized light at a particular wavelength and i_n is the dark current-induced noise current. At 77K and V_e equal to -0.85V, i_n at the emitter is measured to be 2.2pA/$\sqrt{Hz}$ (normalized by the square root of the emitter to collector area ratio). Combined with the measured R shown in Figure 5, D* is evaluated to be $4.0x10^9$cm$\sqrt{Hz}$/W at 8.2μm. Likewise, D* is equal to $5.5x10^9$cm$\sqrt{Hz}$/W and $6.0x10^9$cm$\sqrt{Hz}$/W at V_e = -1.33V and -1.8V, respectively. Note that the value of the gain deduced from the noise current at -0.85V is 0.66, from the relationship $i_n = (4egI_d)^{1/2}$, differing from the value of 3.0 we deduced from Equation 2. The discrepancy is due to the fact that the gain that determines the noise current is the average gain of the entire QWIP, including the high field regime and the low field regime, whereas the gain determined in Equation 2 is the average gain at the high field regime only.

For the D* of an IHET, the current responsivity R is reduced by a factor of α_p compared with a QWIP, but at the same time, the dark current is also reduced by a factor of α_d, where α_p and α_d are the photocurrent transfer ratio and the dark current transfer ratio, respectively. As a result, D* of an IHET is increased by a factor of $\alpha_p/\sqrt{\alpha_d}$ in comparison with the original QWIP. Because the energy of the photocurrent is higher than the energy of the dark current at 77K as indicated in Figure 4, α_p can be much larger than α_d at certain values of V_e. For example, at -0.85V, α_p is 0.33 which is 36 times larger than α_d (=9.1x10^{-3}). Hence, the D* of the IHET is increased by a factor of 3.5 at this bias, and is equal to 1.4x10^{10}cm$\sqrt{}$Hz/W. At lower temperatures, it is 3.0x10^{10}, 1.1x10^{11}, and 7x10^{11}cm$\sqrt{}$Hz/W at 70K, 60K, and 50K, respectively, higher than the corresponding QWIP devices.

Note that the D* of an IHET can exceed the estimate made by Kinch and Yariv[6] for the QWIPs. In their analysis, the thermal electrons include all electrons with energy higher than the barrier height (=190meV in the present case), whereas an IHET is capable of discriminating all electrons up to the energy (=215meV) corresponding to the cutoff wavelength. This difference alone reduces the dark current by a factor of 43, and hence increases the estimated by a factor of 6.6 at 77K.

Besides D*, another figure of merit of a detector array is the noise equivalent temperature difference (NEΔT). When IHETs are used in a detector array with charge coupled device (CCD) readout circuitry, the integration time per frame can be made longer by a factor of $1/\alpha_d$ than in the case with the QWIP, due to the reduced dark current. As a result, NEΔT of the detector array is a factor of α_p/α_d (= 36 in the present case) smaller when the IHET structure is used, due to the increased D* and the longer integration time.

IHET for Photovoltage Amplification

In the last section, we showed the advantages of using IHETs in the charge coupled detector arrays. In the following text, we would like to show that an IHET is also advantageous in voltage-coupled readout circuitry. In scanning detector arrays, high gain operational amplifiers are often used instead of CCDs due to their lower cost. In this case, photovoltaic detectors such as photodiodes are much more favorable than photoconductors, mainly because being under a large DC bias, photoconductors often need AC coupling, and hence require some type of light intensity modulation during the operation. In the present voltage-coupled readout circuit, we will show that the IHET structure can also be used to improve the characteristics of a QWIP in two significant areas, specifically, the IHET provides an internal photovoltage gain that increases the voltage responsivity of a QWIP, and the IHET can be DC-coupled to a high gain amplifier, thereby eliminating the requirement for light modulation.

In order to understand the amplification mechanism of an IHET, let us first examine the photovoltage (V_p^c) measured from a QWIP, given by:

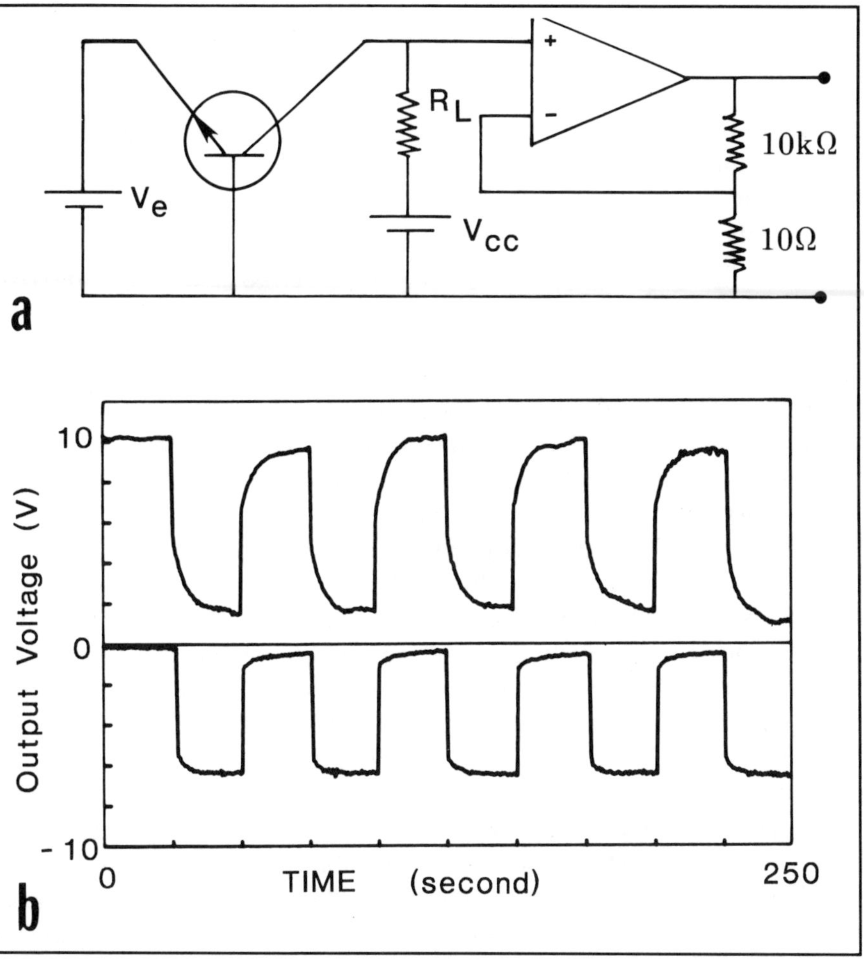

Figure 8. Schematic diagram of an IHET DC coupled to a voltage amplifier, (b) DC output voltage of the amplifier at 11K (lower curve) and at 77K (upper curve). The 77K data have been shifted up by 10V for clarity.

$$V^c_p = i_p \cdot (\frac{1}{r} + \frac{1}{R_L})^{-1}, \tag{3}$$

where i_p is the emitter photocurrent, r is the internal dynamic resistance of the QWIP, and R_L is the external loading resistance. If the same QWIP is incorporated into the emitter of an IHET structure, the photovoltage (V_p^t) measured across R_L, according to the standard transistor theory, is easily shown to be:

$$V^t_p = \alpha_p i_p \cdot \left(\frac{1}{R} + \frac{1}{R_L}\right)^{-1}, \tag{4}$$

where 1/R is the output admittance with input open. In the present case, R = 4.6MΩ is much larger than r = 69kΩ at 77K and V_e = -0.85V. If we assume the loading resistance in both the QWIP and IHET is equal to r and R, respectively, then the photovoltage gain will be equal to α_pR/r which is 22. However, in order to achieve this high gain, the loading resistance in the IHET case is very large. One has to apply a large positive collector bias V_{cc} as shown in Figure 8 to balance out the voltage drop across R_L due to the ambient dark current flow. If we limit V_{cc} to be less than 0.2V, the largest R_L one can adopted is 2MΩ in order to keep the operating point Q above the threshold voltage V_{th} (≅ -0.15 V). The experimental value for the photovoltage gain is 9.7 at V_e = -0.85V when R_L of the IHET is 2MΩ and V_{cc} is 0.2V. The voltage responsivity is measured to be 130kV/W.

For a given R_L, one can adjust V_{cc} such that the operating point Q is on the axis where V_c is zero. Under this condition, V_c is nonzero only when there is a photosignal, similar to the photovoltaic operation. The IHET can then be directly coupled to a high gain voltage amplifier without saturating the amplifier output. Figure 8(a) shows an amplifier circuit with a voltage gain of 1000. Figure 8(b) shows the output voltage of the IHET with R_L = 100 kΩ and V_{cc} = 3mV at T = 10K and 8mV at 77K. These data are obtained by manually opening and closing an optical shutter in a regular time interval, with the initial shutter position closed. Note that the ambient output voltage without light is very close to zero volts, showing the flexibility of the transistor in DC coupling. The measured noise is temperature independent in the present case, which indicates that the light detection process is amplifier noise limited. Finally, it is interesting to note that the schematic diagram in Figure 8(a) may be viewed as a two stage amplifier, because the light signal detected at the emitter is first amplified by the transistor and subsequently by the operational amplifier.

Summary

We have elucidated the basic operating principles of QWIP and IHET devices. In general, the IHET improves the characteristics of a QWIP significantly. The present IHET increases the detectivity of a QWIP by a factor of two, reduces the noise equivalent temperature difference by a factor of 36, increases the photovoltage response by a factor of 9.7, and provides quasi-photovoltaic operation. The detectivity of the present detector is $1.4x10^{10}$cm√Hz/W at 77K. With further optimization of device parameters such as doping density and structure of the QWIP, structure of the energy filter, and scheme of optical coupling, a broadband 10-μm IHET with D* close to 10^{11}cm√Hz/W should be achievable at 77K.

References

1. J. Y. Andersson and L. Lundqvist, *Appl. Phys. Lett.* **59**, 857 (1991).
2. K. K. Choi *et al.*, *Appl. Phys. Lett.* **57**, 76 (1990).
3. E. Pelves *et al.*, *J. Appl. Phys.* **66**, 5656 (1989).
4. K. K. Choi *et al.*, *Phys. Rev.* **B 35**, 4172 (1987).
5. X. L. Lei *et al.*, *Phys. Rev. Lett.* **66**, 3277 (1991).
6. M. A. Kinch and A. Yariv, *Appl. Phys. Lett.* **55**, 2093 (1989).

18

New Optical Switch Stemming From Dipole-Dipole Interactions in Dense Media

Michael E. Crenshaw
Michael Scalora
Charles M. Bowden
U.S. Army Missile Command
Research, Development and
Engineering Center
Redstone Arsenal,
AL 35898-5248

Dense collections of two-level atoms offer the potential of exciting new phenomena, not present in dilute gases, when the density is sufficiently large that induced near dipole-dipole (NDD) interactions become important.

The authors report new results for the response of a dense medium to optical pulses whose temporal duration is much less than a dipole-dephasing time, incident upon a thin film of material composed of atoms or other two-level systems, coupled by NDD interactions. The authors show that the induced inversion that remains after an optical pulse has passed has a nearly step-function response to the peak value of the time-varying optical field. This feature is exceptionally robust with respect to perturbations in parameters such as variations in pulse shape and temporal area. The nearly step function response of the system to the peak field strength of the pulse, independent of pulse shape and area, constitutes a new and unique optical switch, which, unlike optical bistability, does not have hysteresis and is independent of dissipation from incoherent effects. This can lead to unique, ultrafast, ultra-energy-efficient and highly cost-effective optical switches and logic devices for application in all-optical-light-control-by-light optical data processing and optical computing.

Introduction

In dense media, of densities such that there are many atoms (or other two-level systems) within a cubic resonance wavelength, induced near dipole-dipole (NDD) interactions can cause an excitation-dependent renormalization of the resonance frequency.[1-5] Studies of local NDD effects have focused on the steady-state, or stationary, limit in which optical fields vary slowly in an induced-dipole dephasing time or on the time scale of superradiation. In this limit, the field is taken to be constant, and it is possible to derive solutions of the equations of motion that show that NDD interactions may cause bistability that is intrinsic to the material and does not require an external feedback mechanism, such as an optical cavity.[1-3] Optical bistability has important applications for optical switching and digital optical memory elements in optical computing and data processing.[6] This phenomenon forms the basis for devices such as the optical counterpart of the transistor, which constitute the building blocks for all-optical-light-control-by-light digital optical computing.

In this report, and in a recently published summary of our work,[7] we describe novel results obtained for the response of a thin film of material containing or composed of a dense collection of two-level systems (two-level atoms, bound excitons at donor sites) to resonant excitation by an ultrashort pulse (a coherent optical pulse whose temporal duration is much less than an induced-dipole dephasing time). In such a case, the time development of the applied field is important, and the field cannot be taken as constant, as it is in the stationary limit. We find that the induced inversion in the population of the two-level systems that remains in the material after an ultrashort optical pulse has passed has a nearly step-function response to the peak value of the time-varying optical field. The population of the two-level systems, initially in the ground state, always returns to the ground state when the peak field strength is less than the effective dipole moment per unit volume. However, when the peak field strength is in a small region around the effective dipole moment per unit volume, the population is transferred to the excited state and remains in the excited state indefinitely. This result is exceptionally robust with respect to perturbations in parameters such as variations in pulse shape and width. This nearly step-function response of the system to the field strength constitutes a new and unique optical switch, which, unlike optical bistability,[6] does not have hysteresis and is independent of dissipation from incoherent effects, that is, dephasing and spontaneous relaxation. This can lead to unique, ultrafast, and ultra-energy-efficient and highly cost-effective optical switches and logic devices for application in all-optical-light-control-by-light optical data processing and optical computing. All-optical digital systems offer the possibility of very high integrity systems, wide bandwidth, high speed, bulk processing, and elimination of optical-to-electronic interfaces.

In the Model section, the equations of motion for dense collections of two-level systems, resonantly excited by ultrashort coherent optical fields, are presented. These equations, derived by Bowden and coworkers,[1,2,4] incorporate correction terms into the well-known optical Bloch equations.[8] These correction terms, representing contributions to the field experienced by a dipole from nearby dipoles, often cannot be ignored in condensed matter. These correction terms also appear due to the Lorentz local field correction, but for different reasons and conditions.[3-5] The third section presents results of our study of the transient dynamics of dense collections of two-level systems excited by time-dependent coherent optical pulses. No such analysis has previously appeared, except for a short summary of our work.[7] These results, predominately numeric due to the nonlinearity of the equations of motion, indicate a step-function-like response of the system to the peak field strength of an optical pulse. The fourth section discusses possible applications for this response, including optical switching, optical logic, mensuration and material evaluation, and coherent optical pumping. Following the Summary, the Appendix presents a previously unreported analytic solution to the equations of motion in a special case.

Model

The two-level system in quantum optics is a model for two quantum mechanical energy eigenstates connected by a dipole transition. Examples of systems that can be modeled effectively as a two-level system in the optical regime include electronic transitions in atoms, vibrational energy levels in molecules, excitons near the band edge in semiconductors, and laser action.

The optical Bloch equations that follow from the time-dependent Schrödinger equation for a two-level system govern the evolution of the system variables in an ensemble of non-interacting two-level systems.[8-10] Recently, these equations have been extended to include dipole-dipole interactions between two-level systems separated by distances much less than the resonance wavelength.[1-5] For an ensemble of two-level systems interacting via a dipole transition, with an externally applied, classic, time-varying, coherent field, $1/2\ E(t)[\exp(-i\omega t) + \text{c.c.}]$, the optical Bloch equations, in the rotating-wave approximation and extended to include near NDD interactions in dense media, can be written as:[1-5]

$$\frac{\partial u}{\partial t'} = (\Delta - \varepsilon w)\tau_p v, \quad (1)$$

$$\frac{\partial v}{\partial t'} = -(\Delta - \varepsilon w)\tau_p u + \Omega\tau_p w, \quad (2)$$

$$\frac{\partial w}{\partial t'} = - \Omega\tau_p v, \quad (3)$$

where the system variables u, v, and w are ensemble averages. In writing Equations 1 to 3, we have assumed that the pulse is ultrashort, that is, much shorter than the induced-dipole dephasing time T_2, the population-decay time T_1, and the superradiation time τ_R, but slowly varying on the time scale of an optical period. Differentiation is with respect to a dimensionless scaled time $t' = t/\tau_p$, where τ_p is a characteristic time, typically a scaling factor for the temporal width of the pulse. The strength of the NDD interaction is given by $\varepsilon=4\pi\mu^2N/3\hbar$, where μ is the matrix element of the dipole transition, N is the number of two-level systems per unit volume (number density), and $\hbar$ is Plank's constant divided by 2π.[4] A related quantity, the effective dipole moment per unit volume $\mu_V = 4\pi\mu N/3$ is defined for later use. The instantaneous Rabi frequency $\Omega(t) = \mu E(t)/\hbar$ is the field written in units of frequency. The transition frequency ω_0 is the separation of the energy eigenstates divided by $\hbar$ and $\Delta = \omega - \omega_0$ is the detuning of the field carrier frequency ω from the transition frequency ω_0. Throughout this chapter, the field carrier frequency is assumed to be at exact resonance with the transition frequency ($\omega = \omega_0$), except where explicitly stated otherwise.

The extended Bloch equations, Equations 1 to 3, like the ordinary optical Bloch equations, describe the trajectory of a pseudovector whose components are u, v, and w, on a unit sphere, $u^2+v^2+w^2 = 1$. The inversion w represents the difference between the population in the excited state and the population in the ground state. A medium is inverted if there is more population in the excited state than the ground state ($w>0$). If all of the two-level systems are in the excited state, then the medium is in the excited state (fully inverted), which is represented by $u=0$, $v=0$, and $w = +1$. Likewise, the medium is in the ground state ($u=0$, $v=0$, $w = -1$) if all of the two-level systems are in the ground state. The system variables u and$v$ are proportional to the components of the nonlinear polarization in-phase and in-quadrature with the exciting field. The propagation of a laser field through the medium is described by the extended optical Bloch equations and Maxwell's wave equation, coupled via the nonlinear polarization. Although the present work lays the foundation for future studies incorporating propagation effects, we limit the present discussion to local response by assuming that the film thickness in the propagation direction is much smaller than a resonance wavelength.

Results

The inversion is constant, or asymptotically approaches a constant value, after a time-dependent field has passed through the medium ($\Omega(t \rightarrow \infty) = 0$), Equation 3. In the absence of NDD interactions, an ultrashort resonant optical pulse leaves the medium with an inversion [8]

$$w = -\cos\Theta \tag{4}$$

that depends only on the dimensionless temporal pulse area

$$\Theta = \int_{-\infty}^{\infty} \Omega(t)\mathrm{d}t = \frac{\mu}{\hbar}\int_{-\infty}^{\infty} E(t)\mathrm{d}t \tag{5}$$

where we have assumed ground-state initial conditions. Therefore, an ultrashort resonant optical pulse with a pulse area of π (a so-called π-pulse) can be used for coherent optical pumping of a rarefied medium, fully inverting the population. In the process, an amount of energy equal to $n\hbar\omega_0$ is transferred from the field to the medium, where $n=VN$ is the number of two-level systems and V is the irradiated volume. A 2π-pulse returns the population of the two-level systems to the ground state, with no permanent transfer of energy to the medium. A fully inverted medium is often referred to as an amplifying medium because systems in the excited state have stored energy that, under appropriate conditions, can be deposited into an optical field. A medium in the ground state can only extract energy from an optical field and is frequently called an absorbing medium.

To our knowledge, no closed-form solution of the nonlinear differential equations exists for fields of arbitrary temporal profile in the presence of NDD interactions. Therefore, we numerically solved the extended Bloch Equations 1 to 3, using ground-state initial conditions, for the inversion that remains in the dense medium of two-level systems after an ultrashort optical pulse has passed. We find that the asymptotic limit of the inversion (final inversion) has profoundly different characteristics in dense media than the well-known solution, described in the preceding paragraph, has in rarefied media.

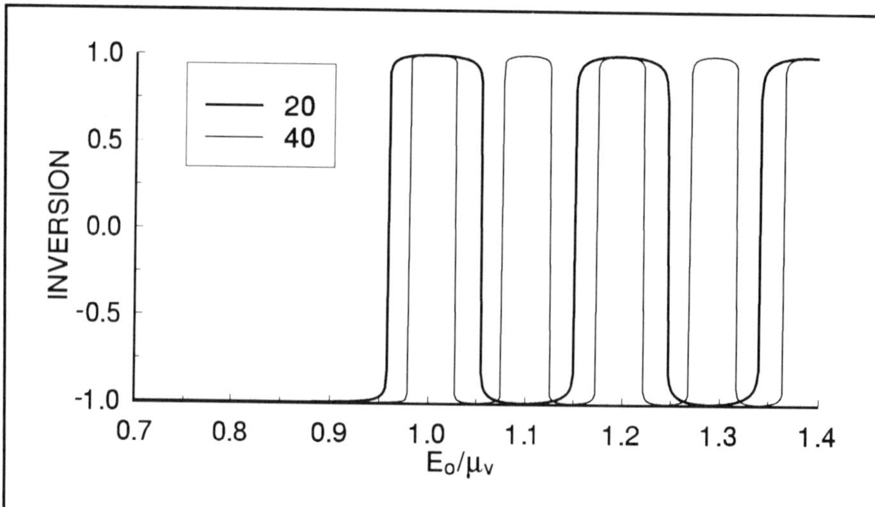

Figure 1. The final inversion, as a function of the normalized peak field strength E_0/μ_V, due to an ultrashort Gaussian pulse for $\varepsilon\tau_p = 20$ and for $\varepsilon\tau_p = 40$.

The final inversion in a dense medium as a function of the peak field strength E_0 (normalized by the effective dipole moment per unit volume μ_V, which is held constant) is shown in Figure 1 for a Gaussian pulse, $E_0\exp(-t^2/\tau_p^2)$, in Figure 2 for a hyperbolic secant pulse, $E_0\text{sech}(t/\tau_p)$, and in Figure 3 for a Lorentzian pulse, $E_0/(1+(t/\tau_p)^2)$. (Note that these curves do not represent a continuous evolution of any system variable, but rather, each point on the curve corresponds to the outcome of exciting the medium with a pulse that has the peak field strength E_0.) In these figures, typical of results obtained over a wide range of parameters, there are steplike transitions in the final inversion as a function of the peak field strength. In each case, the system is left in a fully inverted condition, $w = +1$, when $E_0/\mu_V = 1$. About the point $E_0/\mu_V = 1$ is a region of peak field strength for which the dense medium is left fully, or nearly fully, inverted. Within this region, an amount of energy approximately equal to $n\hbar\omega_0$ is absorbed from the pulse and stored in the medium. The medium is left in an amplifying condition, and this response can be considered analogous to the response of a rarefied medium to the celebrated π-pulse. For peak field strengths on the weak side of this region, the transient field leaves the dense medium in the ground state, in a manner analogous to a 2π-pulse in a rarefied medium. For peak field strengths on the strong side of this region, the final state of the medium alternates between absorbing (ground state) and amplifying (inverted) as a function of peak field strength. Accordingly, we can consider the response of the final inversion to the peak field strength to be approximately a translated square wave, with the first half-cycle approxi-

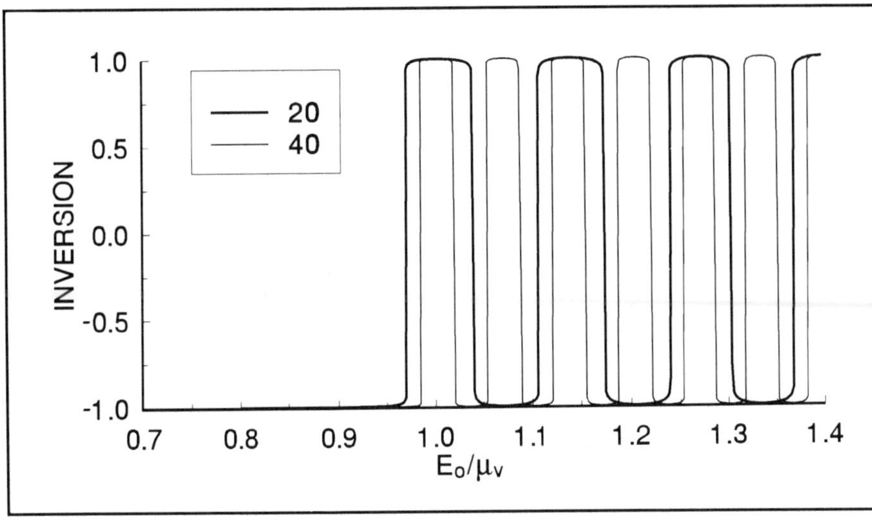

Figure 2. The final inversion, as a function of the normalized peak field strength E_0/μ_V, due to an ultrashort hyperbolic secant pulse for $\varepsilon\tau_p = 20$ and for $\varepsilon\tau_p = 40$.

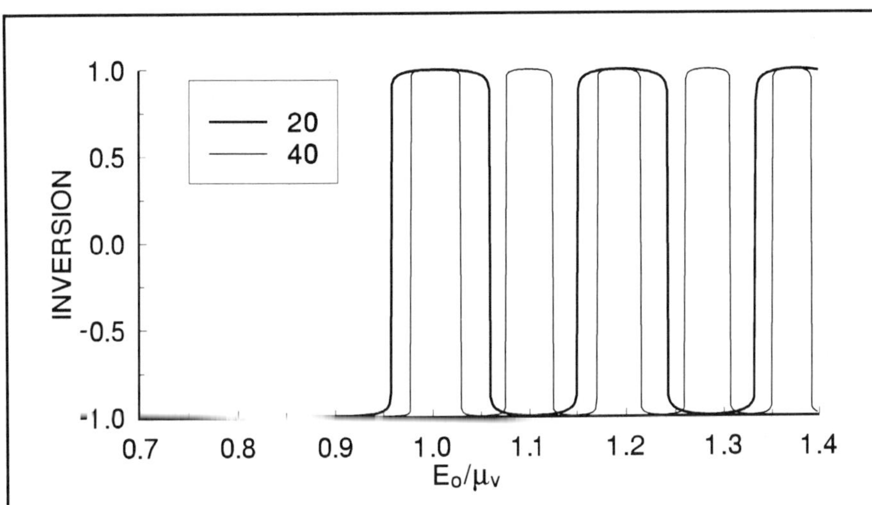

Figure 3. The final inversion, as a function of the normalized peak field strength E_0/μ_V, due to an ultrashort Lorentzian pulse for $\varepsilon\tau_p = 20$ and for $\varepsilon\tau_p = 40$.

mately centered at E_0=μ_V. Although the pulse area is well-defined for any specific pulse shape and dipole moment, it should be noted that the just discussed results in dense media are independent of the pulse area because the temporal width and shape of the pulse, the number density, and the dipole moment are parameters.

We show in the Appendix that the condition $E_0/\mu_V = (1+1/\varepsilon^2\tau_p^2)^{1/2}$

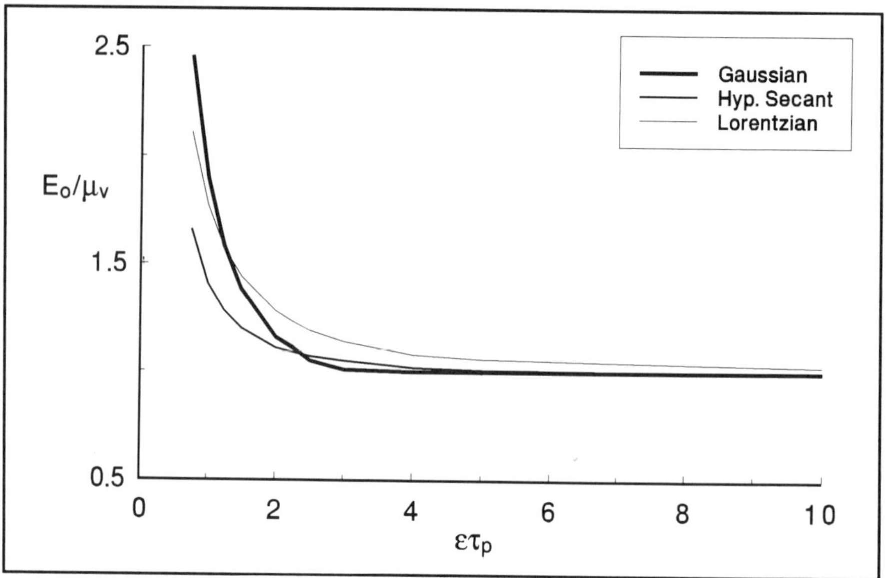

Figure 4. The value of E_0/μ_V, at the first occurrence of complete inversion, as a function of $\varepsilon\tau_p$ for a Gaussian, a hyperbolic secant, and a Lorentzian pulse.

defines a value of the peak field strength for which an optical pulse with a hyperbolic secant shape fully inverts the dense medium. Similar results were obtained computationally for Gaussian and Lorentzian shaped pulses with Figure 4 showing that the value of E_0/μ_V at which the first instance of full inversion occurs can be much greater than unity if $\varepsilon\tau_p$ is small, but rapidly converges to unity for $\varepsilon\tau_p > 1$. In general, we find that with increasing $\varepsilon\tau_p$, the value of the peak field strength at which the first instance of full inversion occurs approaches μ_V; the inversion region becomes better centered about this point, and the transitions between absorbing and amplifying states become sharper and more periodic. Accordingly, the qualitative results are limited to $\varepsilon\tau > 1$.

These results were presented in the context of simple, symmetric pulse shapes at exact resonance in order to maintain clarity in the exposition. Further numeric tests indicate that the important features of our results are remarkably robust. Figures 5 to 7, typical of these results, show that the abrupt transition to a flat inversion region near $E_0=\mu_V$ persists in the presence of pulse asymmetry (providing the applied field is not highly irregular, such as having multiple local maxima) and moderate detuning from resonance.

As long as the applied field is exactly resonant with the transition frequency, the results obtained for the inversion in a medium that is initially in the excited state $w = +1$ are equivalent to the results obtained for a medium that is initially in the ground state $w = -1$ under change of sign in w. Therefore,

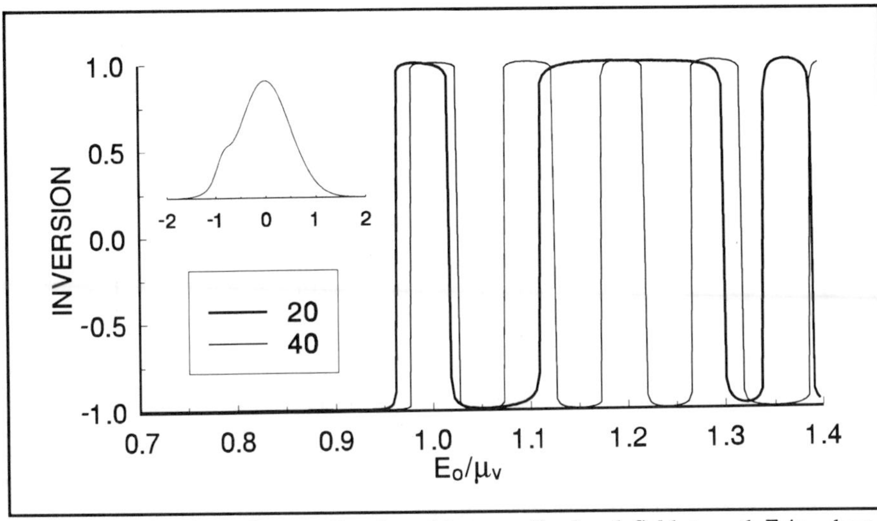

Figure 5. The final inversion, as a function of the normalized peak field strength E_0/μ_V, due to an ultrashort compound Gaussian pulse, $E_0(\exp(-t'^2) + 0.143\exp(-25(t'+0.875)^2))$ for $\varepsilon\tau_p = 20$ and $\varepsilon\tau_p = 40$. The inset shows the temporal profile of the pulse.

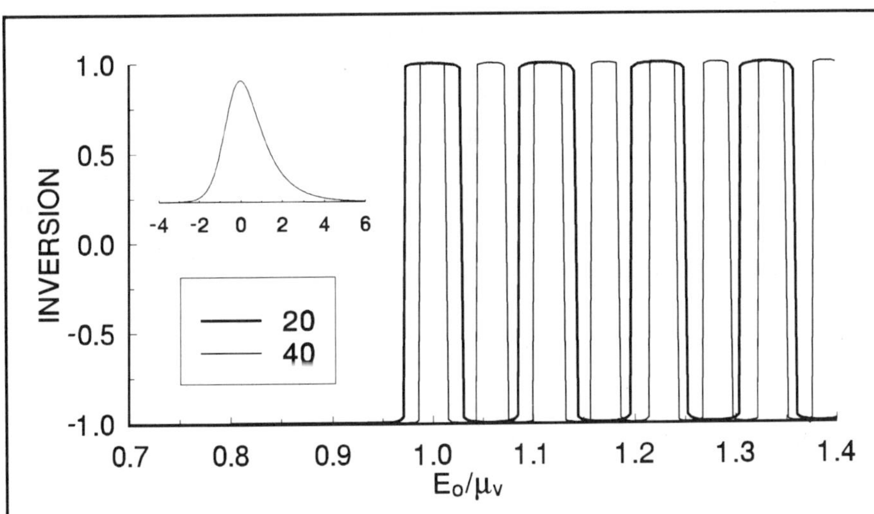

Figure 6. The final inversion, as a function of the normalized peak field strength E_0/μ_V, due to an ultrashort skewed hyperbolic secant pulse, $0.877E_0\exp(0.5t')\mathrm{sech}(t')$ for $\varepsilon\tau_p = 20$ and $\varepsilon\tau_p = 40$. The inset shows the temporal profile of the pulse.

a medium in the excited state can be re-inverted by a pulse with $E_0/\mu_V = 1$, returning the medium to the ground state. Detuning the applied field from resonance breaks this inversion symmetry. The final inversion as a function of E_0/μ_V, where the field carrier frequency is detuned to the high frequency side

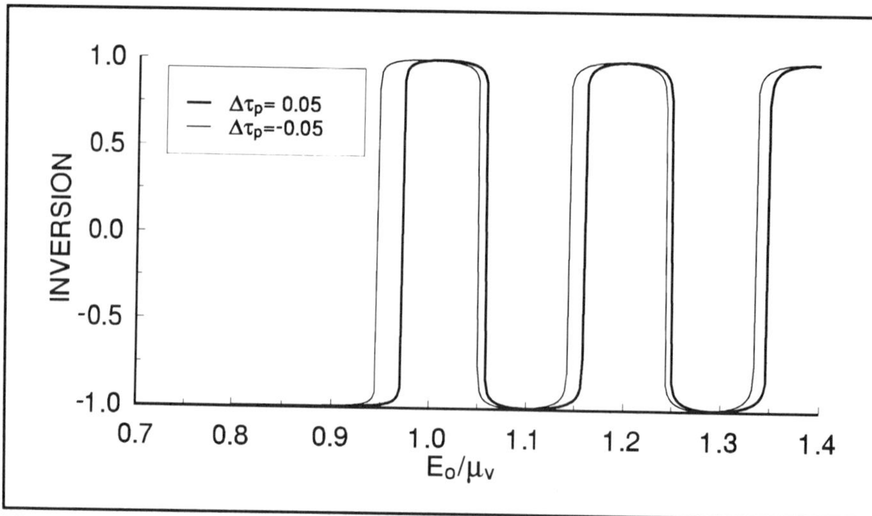

Figure 7. The final inversion, as a function of the normalized peak field strength E_0/μ_V, due to an ultrashort Gaussian pulse ($\varepsilon\tau_p = 20$) detuned from exact resonance by $\Delta\tau_p = +0.05$ and $\Delta\tau_p = -0.05$.

of the transition frequency, is shown in Figure 8 for both ground state and excited state initial conditions. Figure 9 shows the equivalent results with a negative detuning. Figure 8 shows that, by detuning the field carrier frequency from resonance, it is possible to construct an ultrashort coherent optical pulse such that the medium will be left with an inversion near −1 after the pulse has passed, independent of whether the initial condition is the ground state or the excited state. Likewise, Figure 9 shows that, with a negative detuning from resonance, it is possible to select a peak field strength such that the inversion is left near the excited state, again independent of whether the initial condition is the ground state or the excited state.

Applications

Optical switching and optical logic. The steplike transitions in the final inversion as a function of peak field strength suggest that a new kind of optical switch may be devised.[11] The excited state can be interpreted as an "on" state, and the ground state can be regarded as the "off" state. The state of the switch can be probed by an ultrashort pulse, with the probe becoming amplified if the switch is "on" or experiencing absorption if the switch is "off."

The basic switching element comprises a thin film of material, typically condensed matter, containing or composed of a dense collection of two-level

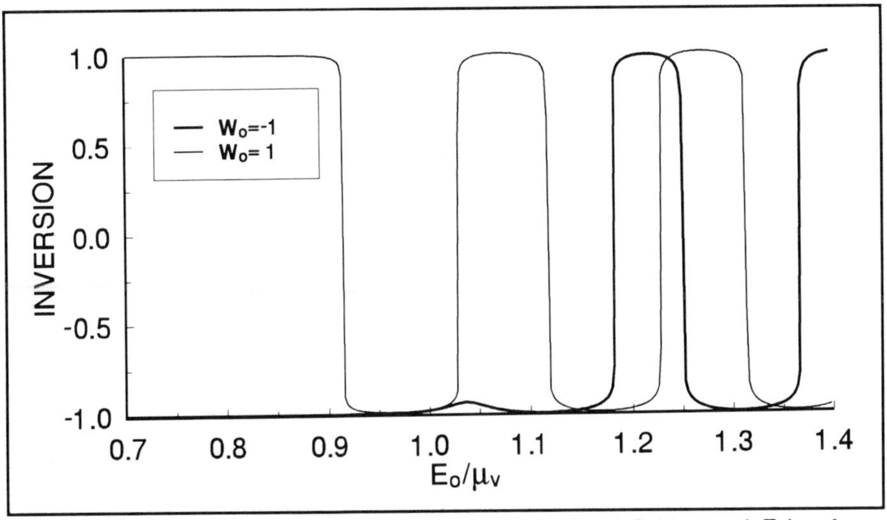

Figure 8. The final inversion, as a function of the normalized peak field strength E_0/μ_V, due to an ultrashort Gaussian pulse ($\varepsilon\tau_p = 20$) detuned from exact resonance by $\Delta\tau_p = +0.2$ for ground state and excited state initial conditions.

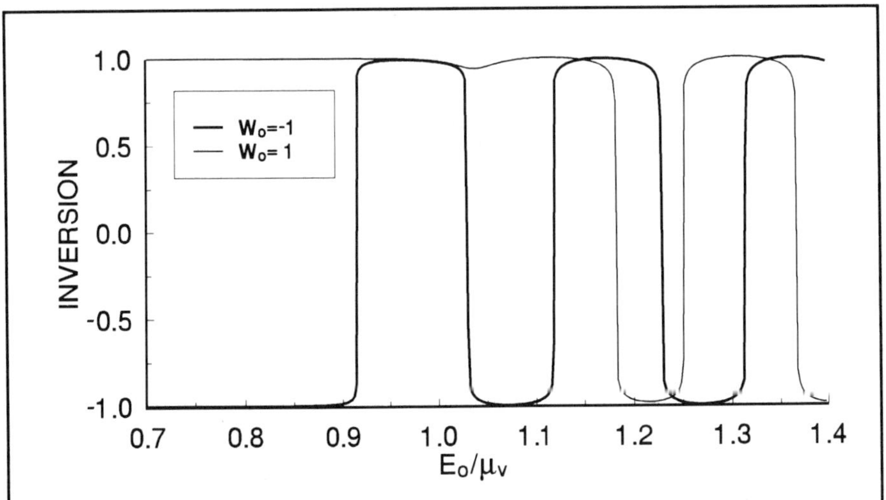

Figure 9. The final inversion, as a function of the normalized peak field strength E_0/μ_V, due to an ultrashort Gaussian pulse ($\varepsilon\tau_p = 20$) detuned from exact resonance by $\Delta\tau_p = -0.2$ for ground state and excited state initial conditions.

systems. The population of the two-level systems is assumed to be originally in the ground state due to decay processes, not included in Equations 1 to 3, that occur on time scales of the order of the induced-dipole decay time T_2. The population can be transferred to the excited state by illuminating the thin film

Table 1
AND Gate (w_0 = -1)

Input A	Input B	A + B	w
0	0	0	-1
$\mu_v/2$	0	$\mu_v/2$	-1
0	$\mu_v/2$	$\mu_v/2$	-1
$\mu_v/2$	$\mu_v/2$	μ_v	1

with an ultrashort resonant coherent optical field with a peak field strength approximately equal to the effective dipole moment per unit volume, $E_0=\mu_V$. In the process, an amount of energy equal to $n\hbar\omega_0$ is absorbed from the field and stored in the medium. A second resonant pulse, again with a peak field strength approximately equal to the effective dipole moment per unit volume, is then used to probe the medium. If the energy of the probe, upon exiting the medium, is greater than the energy input by $n\hbar\omega_0$, then the medium was in the excited state. However, if the energy is reduced by $n\hbar\omega_0$ then the medium was in the ground state. Note that, for exactly resonant pulses, the process of probing the state inverts, or re-inverts, the medium. The basic operation of this switch can be considered to be a digital optical flip-flop. The switch may also be used as a pulse height window discriminator, or comparator, in which pulses with peak field strengths between a lower threshold and an upper threshold change the state of the switch.[12]

There is a clear application of the optical switch just described to a two-input AND gate, with the inputs being identical pulses with peak field strengths of $\mu_V/2$. Assume that the medium is initially in the ground state. The medium remains in the ground state if neither of the pulse inputs pass through the medium. If either of the pulse inputs pass through the medium, then the medium is partially excited, but returns to the ground state. Only if both pulse inputs are present, simultaneously, will the optical field have the necessary field strength, $E_0=\mu_V$, to transport the population of the two-level systems to the excited state. Similarly, both inputs must be present in order to re-invert a medium that is initially in the excited state. The input/output table for the two-input AND gate is shown in Table 1 for ground state initial conditions and in Table 2 for excited state initial conditions. A NOT operation, inverting or re-inverting the medium, is performed by an optical pulse with a field strength of $E_0=\mu_V$. Therefore, the two-input AND operation can be inverted by preceding or following the inputs with an optical pulse such that $E_0=\mu_V$, creating a two-input NAND gate.

Additional logic operations may be performed using a reference-plus-

Table 2
AND Gate (w_0 = +1)

Input A	Input B	A + B	w
0	0	0	1
$\mu_v/2$	0	$\mu_v/2$	1
0	$\mu_v/2$	$\mu_v/2$	1
$\mu_v/2$	$\mu_v/2$	μ_v	-1

Table 3
NOR Gate (w_0 = -1)

Ref.	A	B	A + B + R	w
$\mu_v + \delta/4$	0	0	$\mu_v + \delta/4$	1
$\mu_v + \delta/4$	$\delta/2$	0	$\mu_v + 3\delta/4$	-1
$\mu_v + \delta/4$	0	$\delta/2$	$\mu_v + 3\delta/4$	-1
$\mu_v + \delta/4$	$\delta/2$	$\delta/2$	$\mu_v + 5\delta/4$	-1

inputs scheme. Let δ denote the width, in terms of the variation of peak field strength, of the full inversion region about μ_V, and let the inputs be pulses, each with a peak field strength of $\delta/2$. A NOR operation can be performed by combining the pulse inputs with a reference pulse that has a peak field strength of $\mu_V+\delta/4$. It is assumed that the reference pulse and the pulse inputs have the same temporal development. If neither of the inputs is present, then the reference pulse inverts, or re-inverts, the medium. If either or both of the inputs are incident on the medium simultaneously with the reference pulse, then the peak field strength of the combined field is outside the full inversion region about $E_0=\mu_V$. In this case, the medium is not inverted or re-inverted but is returned to its initial state. The input/output table for the NOR gate is shown in Table 3 for ground state initial conditions and in Table 4 for excited state initial conditions. Similarly, a NAND operation is performed if the reference pulse has a peak field strength of $\mu_V-\delta/4$, as shown in the input/output tables, Tables 5 and 6. It is possible to extend this reference-plus-inputs scheme to OR and AND operations, using reference pulses with peak field strengths of $\mu_V-3\delta/4$ and $\mu_V-5\delta/4$, respectively. However, these latter two operations may not be necessary because preceding or following the NAND or NOR logic operations with the reference pulse inverts the logic operation.

It is generally necessary to be able to assign a specific state to a switch, as well as to be able to change the state. For instance, we may want to load a memory register with a specific bit pattern, or to clear (reset) a register,

Table 4
NOR Gate (w_0 = +1)

Ref.	A	B	A + B + R	w
$\mu_v + \delta/4$	0	0	$\mu_v + \delta/4$	-1
$\mu_v + \delta/4$	$\delta/2$	0	$\mu_v + 3\delta/4$	1
$\mu_v + \delta/4$	0	$\delta/2$	$\mu_v + 3\delta/4$	1
$\mu_v + \delta/4$	$\delta/2$	$\delta/2$	$\mu_v + 5\delta/4$	1

Table 5
NAND Gate (w_0 = -1)

Ref.	A	B	A + B + R	w
$\mu_v - \delta/4$	0	0	$\mu_v - \delta/4$	1
$\mu_v - \delta/4$	$\delta/2$	0	$\mu_v + \delta/4$	-1
$\mu_v - \delta/4$	0	$\delta/2$	$\mu_v + \delta/4$	-1
$\mu_v - \delta/4$	$\delta/2$	$\delta/2$	$\mu_v + 3\delta/4$	-1

Table 6
NAND Gate (w_0 = +1)

Ref.	A	B	A + B + R	w
$\mu_v - \delta/4$	0	0	$\mu_v - \delta/4$	-1
$\mu_v - \delta/4$	$\delta/2$	0	$\mu_v + \delta/4$	-1
$\mu_v - \delta/4$	0	$\delta/2$	$\mu_v + \delta/4$	-1
$\mu_v - \delta/4$	$\delta/2$	$\delta/2$	$\mu_v + 3\delta/4$	1

without knowing the previous contents. We demonstrated in Figure 8 that if the carrier frequency is detuned to the high frequency side of the transition frequency a pulse with a peak field strength just below μ_V can be used to place the medium in the ground state, regardless of whether the initial state is the ground state or the excited state, thereby resetting the switch. Likewise, a pulse can be used to set the switch, placing the medium in the excited state, regardless of whether the initial state is the ground state or the excited state, by detuning to the low frequency side of the transition frequency, as was demonstrated in Figure 9. Although it is not necessary to know the previous state of the switch in order to put it in the desired state, this information can be deduced (query and reset) by determining whether the pulse has gained energy, lost energy, or has no change in energy.

The energy required to excite the medium is clearly $n\hbar\omega_0$. However, the total energy of the switching pulse will generally be greater. The kind of switching times and energies required can be calculated from the condition $\Omega_0=\varepsilon$. Restrictions on the parameters that must be satisfied are $\varepsilon \ll \omega_0$ and $\tau_p \ll T_1$ and T_2. Further, our results indicate that it is reasonable to require $\varepsilon\tau_p > 1$. Assuming a pulse width of one picosecond, a dipole moment of $10ea_0$, and an active device area of $1 \mu m^2$, we determined the total pulse energy to be of the order of a femtojoule at optical frequencies. It is significant that the energy transfer between the medium and the fields is coherent. This allows very high repetition rates without heat damage to the material (heating can be a significant problem in absorptive optical bistability, limiting bit rates).

Mensuration and material evaluation. The steplike transitions between the ground state and the excited state about the point $E_0/\mu_V=\Omega_0/\varepsilon=1$ constitute a unique and unambiguous signature of NDD interaction. Materials in which these effects might be observed include oxygen ions in $KCl{:}O_2^-$ and bound I_2 excitons at donor sites in CdS single crystals.[13,14] Of special significance is the fact that these steplike transitions can be used to develop sensitive techniques for measuring the strength of the NDD coupling parameter ε. One can obtain the value of ε by simply increasing the peak field strength of an ultrashort pulse from shot-to-shot (for example, by removing neutral density filters) and locating the points where the sharp transitions in the state of the medium occur by the change in the transmitted energy. The midpoint between the peak field strengths at these transitions is the effective dipole moment per unit volume μ_V. Likewise, the midpoint between the peak Rabi frequencies $\Omega_0=\mu E_0/\hbar$ at these transitions is the NDD parameter ε. The precision of the measurement depends on τ_p, because this parameter determines the width of the full inversion region. Measurements of the frequency renormalization in dense media have been performed using the reflectivity of an interface between a sapphire window and a dense potassium vapor.[9] A portion of the observed frequency shift was attributed to collisions.

Optical pumping. Coherent optical pumping refers to coherent excitation of a medium of two-level systems that transfers population from the ground state to the excited state. The well-known π-pulse method of coherent optical pumping is not valid in dense media. However, excitation with an ultrashort pulse that has a peak field strength within the full inversion region about the effective dipole moment per unit volume constitutes a method for coherent optical pumping of dense media.[15] Although the width of the full inversion region depends on parameters such as pulse width, pulse shape, dipole moment, and density, this method of optical inversion is less sensitive to perturbations in the parameters than the π-pulse method where the inversion falls off as the cosine of the pulse area. Typical perturbations that might occur are pulses with a spatial profile or that experience mild absorption, or the pulse width, pulse shape, dipole moment, or density may not be known to a high degree of accuracy or may be difficult to control. Applications for coherent optical pumping in dense media include laser gain media, superfluorescent lasing, and coherent spectroscopy. The method may also be used to re-invert media, thereby transferring population from the excited to the ground state.

Summary

We have identified a unique and unambiguous signature of NDD interaction in dense collections of two-level atoms: An ultrashort pulse leaves the population of a dense collection of two-level systems in the ground state or the excited state with a step-function-like dependence on the peak field strength. This result can be used to measure the NDD coupling parameter and for a coherent optical pumping mechanism if propagation effects and medium relaxation times can be neglected, that is for thin films and short optical pulses. A fast, energy-efficient optical switching mechanism has been proposed. This optical switch can form the basis for optical logic devices for application in all-optical-light-control-by-light optical data processing and computing.

Acknowledgments

This work was performed while M. Scalora held a National Research Council-MICOM Research Associateship.

Appendix

Time development of the inversion. Interesting results for dense collections of two-level systems, excited by ultrashort pulses, can be obtained by studying the time development, as well as the asymptotic limit, of the inversion. Time development of the inversion, with E_0 as a parameter, is shown in Figure 10 for an applied field, expressed in units of frequency, given by

$$\Omega(t) = \Omega_0 \operatorname{sech} t/\tau_p = \frac{\mu E_0}{\hbar} \operatorname{sech} t', \tag{6}$$

where the applied field is exactly resonant with the transition frequency. The pulse area, Equation 5, for this field is $\Omega_0 \tau_p \pi$.

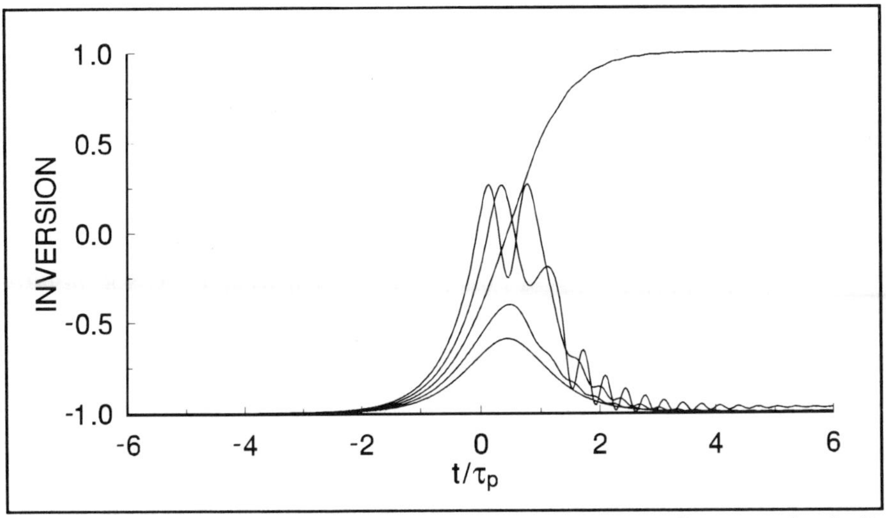

Figure 10. Time development of inversion, due to an ultrashort hyperbolic secant pulse, for E_0 equal to $0.8\mu_V$, $0.9\mu_V$, and $1.1\mu_V$.

The curve in Figure 10 that represents transfer of population from the ground state to the excited state suggests that the extended Bloch Equations 1 to 3 may have a solution in the form

$$u = a\text{sech}t' \tag{7}$$

$$v = b\text{sech}t' \tag{8}$$

$$w = c\tanh t' \tag{9}$$

Substituting the *ansatz* into the extended Bloch Equations 1 to 3 (with Δ=0) yields one set of equations for the derivatives of the system variables, while taking the derivatives of Equations 7 to 9 with respect to t' yields another. Equating these two sets of equations, we obtain

$$a = -\frac{\Omega_0^2\tau_p^2 - 1}{\varepsilon\Omega_0\tau_p^2},\quad b = \pm\sqrt{\frac{\Omega_0^2\tau_p^2 - 1}{\varepsilon^2\Omega_0^2\tau_p^4}},\quad c = \pm\sqrt{\frac{\Omega_0^2}{\varepsilon^2} - \frac{1}{\varepsilon^2\tau_p^2}}, \tag{10}$$

If the medium was in the ground state (u_0=0, v_0=0, w_0= −1) (or the excited state (u_0=0, v_0=0, w_0=+1)) at some time in the indefinite past ($t\rightarrow-\infty$), then Equations 7 to 9 are a solution to the extended optical Bloch Equations 1 to 3 for the field (Equation 6) if, and only if,

$$\frac{\Omega_0}{\varepsilon} = \frac{E_0}{\mu_v} = \sqrt{1 + \frac{1}{\varepsilon^2 \tau_p^2}} \tag{11}$$

at exact resonance. Consistency with the initial conditions clearly requires $\Omega_0 = \varepsilon$ ($E_0 = \mu_V$) in the limit $\varepsilon^2 \tau_p^2 \to \infty$. Further, near-equality of E_0 and μ_V holds for $\varepsilon\tau_p > 1$, in accordance with our numeric results. In the opposite limit of small $\varepsilon\tau_p$, the medium is inverted by a hyperbolic secant pulse when Ω_0 is approximately $1/\tau_p$. In this limit, the pulse area $\Omega_0 \tau_p \pi$ approachs π, *natura non facit saltum.*

The extended Bloch Equations 1 to 3 can be written in the form:

$$\ddot{w} - \frac{\dot{\Omega}}{\Omega}\dot{w} + \Omega^2 \tau_p^2 w + \Omega\varepsilon\tau_p^2 w \left(1 - w^2 - \frac{\dot{w}^2}{\Omega^2 \tau_p^2}\right)^{1/2} = 0 \tag{12}$$

using the identity $u^2+v^2+w^2=1$. Equation 12 contains dynamical frequency and damping terms, leading to the damped oscillations seen in Figure 10 as well as dynamic anharmonicities.

References

1. Y. Ben-Aryeh, C.M. Bowden, and J. C. Englund, *Phys. Rev.* **A34**, 3917 (1986).
2. Y. Ben-Aryeh and C.M. Bowden, *Opt. Comun.* **59**, 224 (1986).
3. M.G. Benedict *et al.*, *Phys. Rev.* **A43**, 3845 (1991).
4. C.M. Bowden, Near Dipole-Dipole Interaction Effects in Quantum and Nonlinear Optics, (International Conference on Quantum Optics, Hyderabad, India, 1991), unpublished.
5. F. Friedberg, S.R. Hartmann, and J.T. Manasseh, *Phys. Rev.* **A40**, 2446 (1989).
6. H.M. Gibbs, *Optical Bistability: Controlling Light with Light* (Academic Press, New York, 1985).
7. M.E. Crenshaw, M.Scalora, and C.M. Bowden, *Phys. Rev. Lett.* **68**, 911 (1992).
8. L. Allen and J.H. Eberly, *Optical Resonance and Two-Level Atoms*, (Wiley, New York, 1975).
9. J.J. Maki *et al.*, *Phys. Rev. Lett.* **67**, 972 (1991).
10. R.P. Feynman, F.L. Vernon, and R.W. Hellwarth, *J. Appl. Phys.* **28**, 49 (1957).
11. M.E. Crenshaw, M. Scalora, and C.M. Bowden, Active Element for Optical Signal Processing and Optical Computing, U.S. Army Missile Command, Invention Disclosure, AMPC4760, 1992.
12. H.V. Malmstadt, C.G. Enke, and S.R. Crouch, *Electronics and Instrumentation for Scientists* (Benjamin/Cummings, Menlo Park, CA, 1981), chapter 4.
13. M. Dagenais and W.F. Sharfin, *Appl. Phys. Lett.* **45**, 210 (1984).
14. M.S. Malcuit *et al.*, *Phys. Rev. Lett.* **59**, 1189 (1987).
15. M.E. Crenshaw, C.M. Bowden, and M. Scalora, Method for Optically Inverting a Dense Collection of Two-level Systems, U.S. Army Missile Command, Invention Disclosure, AMPC4761, 1992.

Advanced Electronic Detonator Modules for Live-Fire Missile Testing

Christopher G. Braun
U.S. Army Laboratory Command
Electronics and Power Sources Directorate
Fort Monmouth, NJ 07703-5601

An advanced electronic detonator module used with high energy Exploding Bridge Wire (EBW) detonators has been developed for the Live-Fire and Lethality Division of the Combat Systems Test Activity (CSTA) and fully proved in live-fire missile testing. For the first time, this module enables the combination of safe, realistic, and lower cost live-fire missile testing because it uses a detonator module that has been sunstantially reduced in size and has an increased reliability and performance over previous methods. This module also represents the first application to come out of several years of basic DoD research and development into the next generation power semiconductor switch.

The heart of this module is a MOS Controlled Thyristor (MCT)—a new kind of high-power electronic switch—that forms the basis for a compact (8x4.5x2-cm board), completely self-contained electronic detonator circuit. Prior to triggering, the MCT is used as an off/opening switch, making the module highly immune to noise or dV/dt. Upon receiving a trigger signal, the MCT switches on, producing a ~1μs FWHM pulse of greater than 300kW at the detonator to ignite the EBW. This detonator module has been used in live-fire test programs conducted by CSTA, including the Improved HELLFIRE missile test and a reactive armor test with an outstanding success rate of 19 of 21 and 30 of 30, respectively.

A tandem detonator module for multiple charge warheads is also under development. Two or more electronic detonator modules are controlled by a sequencer module that provides the safety interlocks and sub-microsecond timing of multiple events.

Prospects for Laser Weapons in Army Combat

Walter R. Sooy
Lawrence Livermore National Laboratory
Livermore, CA 94550

Recent developments in laser technology hold promise for achieving a major reduction in the size and weight of high-power laser weapons. Size and weight have been serious impediments to the use of lasers in battlefield weapon systems roles; hence, these technology advances may at last open the door to the development of practical laser weapons.

Specifically, critical advances have been achieved in the area of solid state laser technology involving major improvements in efficiency and thermal management. The technical components of these improvements are:

(1) The use of diode lasers as pump sources for solid-state lasers. As contrasted with flash lamp pumps historically used with solid-state lasers, diode pumps deposit one tenth as much waste heat in the lasers.

(2) The development of efficient diode arrays, capable of high peak power output.

(3) The development of very effective microchannel coolers that permit the diode arrays to operate at high average power.

(4) The development of manufacturing techniques that have reduced the cost of diode power by a factor of one hundred.

(5) The concept of thermal cyclic operation of the laser. The laser runs uncooled until the temperature builds to some maximum value, then the laser operation ceases while the amplifier is vigorously cooled. Segregating the

functions of lasing and cooling permits each to be separately optimized and higher overall power to be extracted.

In addition, advances from outside the diode pumped solid state laser (DPSSL) area can be applied advantageously to engineering compact weapons: lightweight optics for beam directors, phase conjugation for beam quality control, atmospheric compensation to reduce turbulence-induced beam distortion, compact power generators and energy storage devices, and improved sensors.

DPSSL technology is already making inroads into established low-power laser applications, such as rangefinders, designators, and machining devices, and has become the approach of choice for midpower applications still under development, such as illuminators and antisensor countermeasure devices. High-power laser weapon development has not yet been initiated, but the technology exists to support such a thrust and may be initiated soon.

The major application of this technology will be in air defense, particularly defense against tactical battlefield missiles. It appears feasible to develop highly lethal weapons (radiant intensity of 0.3-1 x 10^{16} watts/steradian) that are compact enough (several tons weight, several cubic meters volume) to be mounted on tracked vehicles, helicopters, or aircraft.

Photonic Integration Technologies for Telecommunications

T.L. Koch
AT&T Bell Laboratories
Holmdel, NJ 07733

Integration is expected to play an important role in future optical transmission, networking, and switching applications. Just as in the case of electronic technology, photonic integration offers the potential for cost reduction, increased reliability, and in some cases increased functionality and performance.

Two fields are emerging within this area. The term opto-electronic integrated circuits (OEICs) has been used to describe devices where transistor circuitry for signal conditioning or amplification has been combined with either an optical detector or source to convert to or from an electrical signal from or to an encoded optical signal. In some instances, arrays of such circuits may be integrated, but there is usually no integrated optical interconnection between optical devices on the same chip. In contrast to this, photonic integrated circuits (PICs) may contain no purely electrical devices at all. Rather, they aim to replace labor-intensive and inefficient single-mode optical connections between different guided-wave semiconductor optical components with lithographically-produced transparent semiconductor waveguides on the same substrate.

Recent research prototype OEICs and PICs have demonstrated system-level performance at AT&T Bell Laboratories. Further research is being conducted on telecommunications circuits and specifically on transmission applications where the system architectures are very clearly defined.

Electric Field Effects on the Optical Properties of Layered Semiconductor Structures

Richard L. Tober
Garnett W. Bryant
U.S. Army Research Laboratory
Signatures, Sensors, Signal and Information Processing Directorate
2800 Powder Mill Road
Adelphi, MD 20783-1197

W. Q. Li
P. K. Bhattacharya
Department of Computer Science and Electrical Engineering
University of Michigan
Ann Arbor, MI 48109-2122

The development of novel opto-electronic devices for military applications depends on the ability to understand and control basic properties of semiconductor structures. The effect that electric fields have on carrier confinement, for example, is important for controlling optical modulators and signal processing components. Therefore, optical modulation techniques have been used to investigate the effects that an applied electric field has on the optical properties of regions near compositional boundaries in artificially layered structures.

Electric field modulated spectra obtained from heterojunction samples have oscillatory structures that depend not only on the magnitude of the applied field but on confinement near the heterojunctions as well. Experiments indicate two field regions, each with distinct spectral lineshapes.

Spectra obtained with relatively small DC electric fields have rapid oscillatory structure whose period does not significantly depend on the magnitude of the applied field. In contrast, the period of the bulk Franz-Keldysh oscillations that emerge with large fields does. Theoretical calculations suggest that a crossover from the low- to high-field behavior occurs when carriers can freely move through the crystal.

Potential profiles determined from effects of doping, compositional variations, and applied fields were used to calculate spatially dependent band-to-band overlap matrix elements. The agreement with experimental results indicates that information regarding barrier width and height can be obtained from optical spectra. This provides valuable feedback for the design and implementation of opto-electronic devices for military needs.

Transient Energy Transfer in Wide Gap Semiconductors

Steven S. Cotariu
John E. LaSala
Photonics Research Center
United States Military Academy
West Point, NY 10996

Wide gap semiconductors are important in the creation of photonic devices that operate in the visible part of the spectrum. A visible light optical limiter, of interest to the Army for detector and eye protection, has been demonstrated in bulk ZnSe, a II-VI semiconductor with a cutoff wavelength in the blue. Our research investigates nonlinear dynamics in ZnSe and other wide gap semiconductors that contribute to optical limiting and other beam interactions of potential use in visible light applications.

An intense pulse of visible light produces high concentrations of free charge carriers through two photon absorption (TPA) in II-VI semiconductors. Free carriers are responsible for the strong optical limiting behavior that has previously been reported. In our work, we have developed another mechanism mediated by free carriers produced through TPA, the transfer of energy from a strong pump pulse to a weak probe pulse in polycrystalline ZnSe and ZnS. Gain as high as 20 has been observed in weak 25ps pulses of 532nm light crossing a high intensity (3 GW/cm^2) pump pulse in ZnSe. Lower gain has been observed in ZnS with comparable pump intensity. The process is mediated by a transient free carrier volume grating generated by TPA of the interfering beams. Gain depends strongly upon polarization, intensity, and relative arrival time. The gain is not highly dependent on crossing angle, and there are no crystal orientational dependencies. A 1D code successfully accounts for the observed gain features.

Advanced Computing

19

Robotics and Artificial Intelligence

R. W. Brockett
Division of Applied Sciences
and The Center for
Intelligent Control
Harvard University
Cambdige, MA 02138

The goal of control/cybernetic engineering is to create systems capable of highly flexible or "intelligent" response to changing circumstances. Until recently, basic electromechanical and computer engineering needed to build these systems has not been been available at reasonable cost. Now, we can implement algorithms based on sound mathematic principles using available computing elements, and implement systems that incorporate vision sensors, real-time distributed control, and statistically sound theories of adaptation and learning. Sensor/computational systems designed to emulate human capabilities in perception (image and speech understanding and handwriting recognition) have proved to be more difficult than initially anticipated. Progress in practical resolutions of some of these problems has resulted from using highly nonlinear signal processing techniques. Examples include variational theories of image segmentation, system identification methods applied to speech models, information theoretic methods applied to data compression, and computational complexity applied to learning. Many problems that arise when these perceptual modules are integrated into real-time systems are unsolved. We need a significant change in how we define and solve programming problems for robotic devices to achieve their potential for reducing human labor and their acceptance in the workplace. An important aspect of this viewpoint will be methods that give sensing and feedback control a more prominent role, going beyond the idea that a robot is just a position controlled device. Only by pursuing this direction can we achieve the degree of code portability and reuse necessary for the development of quality, cost-effective systems.

Introduction

The goal of control/cybernetic engineering, going back to the 1940s and before, has been that of implementing systems capable of exhibiting highly flexible or "intelligent" responses to changing circumstances.[1] Both the basic electromechanical engineering and the computing capabilities required to build practical systems of this type have been out of reach until quite recently, but today there exist rapidly expanding possibilities for making progress toward these long-held goals. It appears that the implementation of mathematically principled algorithms is possible on the computing hardware now, or about to be, available. Important problems under investigation include questions about the use of vision sensors, real-time distributed control, and mathematically tractable theories of adaptation and learning. Successes include demonstrations of vision guided automobiles operating at high speed, highly effective speech recognition systems, and the use of robotic devices in manufacturing.[2]

Experience with sensor systems designed to emulate human capabilities has shown that the basic perceptual problems associated with attempting to understand the meaning of pictures, spoken sounds, and handwriting are much more difficult than many of the early researchers anticipated. With hindsight, one might say that this was to be expected on the basis of the fact that quite a large percentage of the human brain is devoted to these tasks. The significant progress now being made toward the resolution of some of the basic problems results, in large part, from the discovery and aggressive use of highly nonlinear signal processing techniques. Examples include the variational theories that have been brought to bear on the image segmentation problem, the theories of learning based on computational complexity, and information theoretic-based approaches to perceptual problems. Attempts to incorporate perceptual modules into larger systems, however, often raise problems about communication and distributed computation that are not yet solved.[3]

Although the present ideas about how to program robots have led to significant successes in particular domains, it seems likely that a significant change in the way programming problems are framed and solved will be necessary before robotic devices can reach the next level of acceptance. This new viewpoint will need to regard sensors and feedback control in a more integral way, going well beyond the idea that a robot is just a position-controlled device. Only by vigorously pursuing this direction will we be able to achieve the degree of code portability and code reuse necessary for the development of economic, high quality software. Moreover, it seems clear that the computing will need to be distributed. There are many technical problems here, and their solution will require the combined efforts of computer scientists, robotic engineers, experts in perception, and experts in automatic control.

Communication, Computation, and Control

Although for several decades it has been felt that computation, communication, and control are the basic methodologies for the nonphysical aspects of electrical engineering, there have been rather few examples of applications in which the use of these disciplines is truly integrated. Up until now they have coexisted in many systems, but they usually do not interact in any very significant way. This is a result of technologic limitations. The most demanding applications of automatic control have tended to involve feedback control of dynamic systems. In such cases time is critical. Until recently the computational hardware required to achieve real-time control has been expensive and difficult to use. Only a few specialized applications have received the attention required to implement a close coupling between computing and control. Communication theory has been very successful in developing appropriate strategies for dealing with uncertain channels, applying digital methods to improve fidelity, and using coding theory to store and retrieve data with great speed and reliability. Those areas that appear to have the most potential for use in control problems, such as algorithmic approaches to rate distortion trade-offs and coding to a fidelity criterion, have only come in to their own rather recently.[4]

Certain robotic applications, such as those that call for the use of vision systems to guide robots, now require the use of computing, communication, and control in an integrated way. The computing that is to be done must be opportunistic, that is, it must be tailored to fit the needs of the specific situation being encountered. The data compression that is needed to transmit television signals to a computer must be be done with a view toward how the results will be used by the control system. It is both technologically difficult and potentially dangerous to build complex systems that are controlled in a completely centralized way. For this reason we need to decide how to distribute the control function over the communication system. Recent work on the theory of communication protocols has made available better methods for designing efficient distributed algorithms; perhaps this work can be adapted in such a way as to serve the needs of robotic applications.

Recently there have begun to appear commercial devices that provide reasonably wideband wireless local area communication; they have enormous potential for simplifying and enhancing the operation of mobile robots. These systems can be divided into those that are free-space, line-of-sight systems versus free-space, omni-directional systems. The advantage of the omni-directional systems based on radio links is that they do not require pointing control. The advantage of the directed beam systems based on infrared transmission is that they are relatively immune to cross-talk and interference, and they do not flood the area with their signal. Because of this technology one can now build tetherless vision guided robots with remote computational support. In effect, wireless communication allows the experimenter to give brains to mobile robots and legs to computational engines.

Motion Description Languages

At present there is almost no transportability of robotic motion control languages. The idea of vendor-independent languages that apply with no change to a wide range of computing platforms and peripherals has not yet been made to work in the field of robotics. The clear success of such notions when applied to operating systems, languages, networks, disk drives, and printers suggests that it is, however, only a matter of time. The main new ingredient associated with robotic programming is the necessity for designing and building symbol-to-signal transducers. By this we mean systems that accept symbol strings as inputs and produce analogue signals or motions as outputs.[5] Because the motion generated by a robot is usually its reason for being, it is reasonable to think of motion control as being the central problem. This suggests that what is missing is a consensus about how one should structure and standardize a "motion description language." Such a language should, in addition to other things, allow one to implement compliance control in a general and natural way. These ideas suggest several new questions to the control engineer. In terms of the major themes of automatic control, that is, feedback, stability, optimization, and estimation, there are a number of performance issues. For example, at what level of detail in modeling (such as kinematic or dynamic, linear or nonlinear, deterministic or stochastic, and so on) does optimization enter in a meaningful way? Questions of coordination, sensitivity reduction, stability, and so on all arise. One of the more obvious conclusions is that it will be necessary to make readily available to robotic programmers, possibly even in the form of on-screen "help" pages, certain insights about stabilization and control. From another point of view, the main problem in motion control is that of dealing with redundancy. After all, if there is only one way to do a task, then deciding to do it is all that is required; from that point on the whole process can be regarded as simply the unfolding of the inevitable.

The contrast between the sophistication and ease of use of computer programs written for everyday applications such as word processing, graphics editing, and CAD/CAM, and the difficulty of using a computer to help generate the description of a motion that is to be executed by a robot has become so obvious that steps are now being taken to address the problem in earnest. This will not, however, be an easy task. Just as the step from text processing to graphics is made difficult by the two-dimensionality of graphics, as opposed to the one-dimensionality of a string of text, motion generation will be intrinsically more difficult than graphics by virtue of the fact that it is three-dimensional, and it evolves in time. Clues that can be gleaned from the study of motion generation for animation should be studied as we attempt to put in place the data structures, algorithms, and database systems required to support an improved programming environment. In another vein, the use of computers for real-time control is currently driving computer technology as hard or harder than any other application area but it has seemed difficult to formulate a general, theoretically sound approach to this class of problems. It has proved to be difficult to use the inspiration provided by biologic systems to generate designs suitable for implementation with today's computer hardware.

Many of the important algorithmic issues that must be confronted in the implementation of robotic systems involve the change of coordinates necessary to permit an efficient description of relative position. Important questions include:

1. How should a computer carry out coordinate changes?
2. What methods are best for computing inverse kinematics?
3. Is it efficient to implement specific differentiable manifolds as data types?
4. What are economic but adequate sets of coordinate systems for motion control?

Are there sufficiently powerful economic forces at work in this field to provide the incentive to complete the development of such a set of ideas? This is an especially important point because of the size and scope of programming tasks involved. Mathematicians are at home with the idea of writing papers and books meant for human consumption, and while it is required that these be essentially error free, humans are robust enough so that certain types of mistakes and gaps are tolerated. This is not so for computer programs, and the support structure in the form of computer editors and computer graphics people is not readily available. Thus there is, at present, a paucity of mathematic ideas breaking through the computer barrier, polished to the point that they can be useful to those who must put together working systems.

Human Interfaces

The field of robotics, being the most active focal point for the interaction between computer science and control, has become an important test bed for the development of the whole field of automatic control. It seems clear that more extensive use of computer control, be it for factories, automobiles, or homes, will be most effective if it comes with a natural human interface. Having this goal in mind, one should look for interfaces that are not only suitable for the given application but that are sufficiently general so that, with minor modification, they can serve in related applications as well. One may argue that biologically inspired solutions deserve serious consideration in view of their proven robustness and manifest naturality. It may seem ironic that even for the most advanced robots, motion control has not been developed to the same degree as some highly specialized topic such as the control of printers. However, this is consistent with the relative popularity of soft and hard automation and reflects the current unavailability of effective, sensory rich and mechanically elegant, cost-effective robots.

Future Work

We have suggested a number of questions here whose answers could hasten the development of new applications for robotics. Some of these are most properly developed as part of an experimental facility, and some are pure algorithm development. It seems that the future of robotics is bright but that considerable patience and hard work is required.

Acknowledgments

This work was supported in part by the U.S. Army Research Office under Grant DAA03-K-0171, supporting the Brown-Harvard-MIT Center for Intelligent Control.

References

1. N. Weiner, *Cybernetics or Control and Communication in the Animal and the Machine* (MIT Press, Cambridge, MA, 1948).
2. E.D. Dickmanns and B.D. Mysliwetz, *IEEE Trans. PAMI* **14**, 199 (1992).
3. D. Mumford and J. Shah, *Proceedings IEEE Computer Vision and Pattern Recognition Conference*, 1985.
4. A. Gresho and R.M. Gray, *Vector Quantization and Signal Compression* (Kluwer, Boston, MA, 1991).
5. R.W. Brockett, *Proceedings IEEE Conference on Robotics and Automation*, April, 1988.

20

Scene Segmentation and Definition for Autonomous Robotic Navigation Using Structured Light Processing

C. David Brown
U.S. Army Combat
Systems Test Activity
Aberdeen Proving Ground,
MD 21005-5059

Morphologically based processing of images of projected structured light is applied to segment and define a three-dimensional scene. The structured light, viewed by a stereoscopically located, single video camera, results in an image in which the regional three-dimensional characteristics of the scene are represented by corresponding distortion of the structure of the projected light. The segmentation methods are based on this region-specific distortion in the image of the structured light by applying morphologic image filtering techniques to sense and locate these distortions. In addition, an extension of the segmentation process yields a three-dimensional planar fit of the region of interest in the scene. An illustrative example is presented. The scene segmentation and definition capabilities of this method can provide essential input to an autonomous navigation system for an off-road or planetary exploration vehicle.

Introduction

Substantial research has been directed toward the application of machine vision to the problem of the guidance of robots and autonomous vehicles along pathways that are not previously designated by special markings.[1-3] The primary function of the machine vision system is to provide the autonomous robotic navigation system with information describing the location and description of the pathway in front of the vehicle. Color, and its closely associated gray level contrasts, is a commonly employed method of obtaining pathway location; however, during off road navigation, many instances can be imagined where the color of the pathway differs little from that of the background. In these situations, the three-dimensional (3-D) texture, or change in shape, in the scene must be sensed by the visual system, with a region of low texture is designated as pathway, and high texture as background.

Stereo vision employing two video cameras closely models the human vision system for sensing the 3-D characteristics of a viewed scene. However, the processing is complicated by the problem of determining pairs of corresponding points in the two images. Researchers have investigated replacing one of the cameras by a structured light source, such as projected lines or a grid, for autonomous vehicular navigation.[1] This configuration is becoming commonplace in industrial machine vision applications:[4] the correspondence problem is eliminated because each element of the structured light is projected at a known angle and sensed as a definable spot or shape in the camera image, and only the image of a single video camera needs to be processed.

Additional advantages are realized if a laser is used as the projection source. A laser is a highly collimated, high-power source. Therefore, the portions of the scene illuminated by the laser contrast well with portions not illuminated. Because the laser is also monochromatic, narrow band optical filters can be employed to enhance this contrast. Therefore, laser-projected structured light yields an image that is easily thresholded for further processing. The thresholded image then becomes a binary, two-dimensional (2-D) representation of the 3-D texture in the scene, which is then conducive to simplified regional processing to segment the image based on the 3-D texture.

The real advantage of this processing method is that critical processing parameters are defined in terms of 3-D scene measurements, but processing is accomplished almost entirely in 2-D, thus greatly simplifying the processing. For example, acceptable vertical pathway texture should be something less than vehicle ground clearance. Given predefined sensing system parameters such as camera location, lens focal length, laser location, and scan angles, these scene measurements can be projected into the 2-D processing plane and stored in a look-up table to facilitate processing.

When autonomous vehicles are navigating along predefined roadways, it is a reasonable assumption that the grade and cant of the roadway will not exceed the vehicle's capabilities. However, with off-road navigation or planetary exploration, no such assumption can be made about the pathways chosen

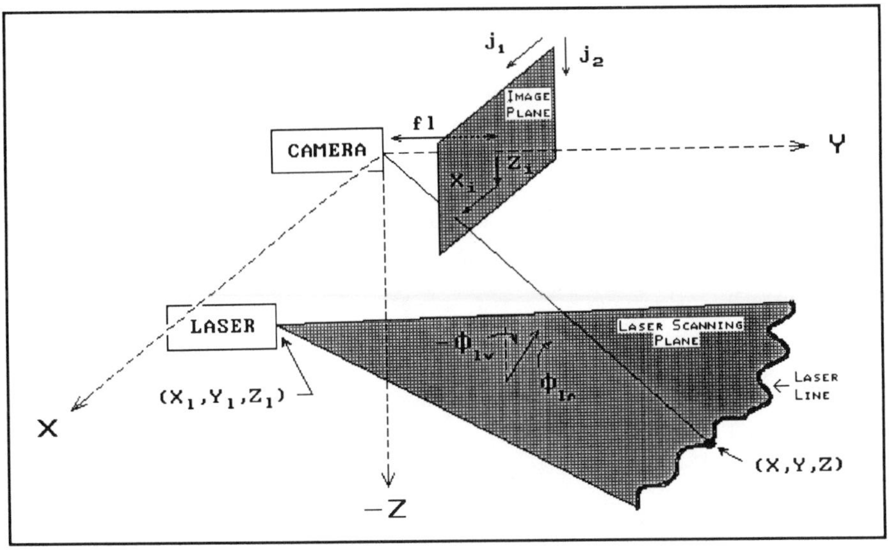

Figure 1. Three-dimensional geometry.

by the autonomous vision system through an unstructured environment. Therefore, the vision system must also have the ability to define the 3-D characteristics of the chosen pathway to determine its acceptability. Further processing of the image of a structured light grid, projected on the scene ahead of the autonomous vehicle, can accomplish the required pathway definition. Thus, this method may provide a valuable augmentation to present autonomous vehicular vision systems to accomplish the difficult task of off-road navigation or planetary exploration.

Background

Figure 1 illustrates the geometric principles of the scene segmentation method. The camera image of a horizontal scan line changes in response to the 3-D texture of the scene. The 3-D texture of the pathway differs greatly from that of the background in the scene, with the pathway texture being considerably more fine and regular. Therefore, the shape of the grid lines varies significantly between the pathway regions and background regions of the image. The application of the above descriptions of image texture and the accompanying segmentation methods provides a first order processing method to identify the pathway boundaries for input to a route planner for an autonomous robot or vehicle.

If adequate geometric parameters are known about the structured light projection and the camera sensing system, then the 3-D characteristics of the scene can be determined.[5] The camera focal point can be chosen as the origin, with the laser scanner focal point at (x_l, y_l, z_l). The laser scan rotation angle (ϕ_{lr}) and vertical projection angle (ϕ_{lv}) are known, and the camera image location (x_i, fl, z_i) is calculated based on the previously described camera calibration. The coordinates (x,y,z) of the laser spot in the scene can then be derived by calculating the point of intersection of the camera ray:

$$\frac{x}{x_i} = \frac{y}{fl} = \frac{z}{z_i} \tag{1}$$

with the laser scanning plane:

$$\frac{x}{x_0} + \frac{y}{y_0} + \frac{z}{z_0} = 1 \tag{2}$$

where the planar intercepts are:

$$x_o = x_l = y_l/\tan(\emptyset_{lv}) \sin(\emptyset_{lr}) + \frac{z_l}{\tan(\emptyset l_r)} \tag{3}$$

$$y_o = \frac{\sqrt{x_l^2 + z_l^2} \sin\left[\phi_{lr} + \tan^{-1}\left(\frac{z_l}{x_l}\right)\right]}{\tan(\emptyset_{lv})} + y_l \tag{4}$$

$$z_o = x_l \tan(\emptyset_{lr}) + y_l \tan(\emptyset_l) \cos(\emptyset_{lr}) + z_l \tag{5}$$

Solving for x, y, and z yields:

$$\begin{bmatrix} \frac{1}{x} \\ \frac{1}{y} \\ \frac{1}{z} \end{bmatrix} = \begin{bmatrix} 1 & \frac{fl}{x_i} & \frac{z_i}{x_i} \\ \frac{x_i}{f_l} & 1 & \frac{z_i}{f_l} \\ \frac{x_i}{z_i} & \frac{fl}{z_i} & 1 \end{bmatrix} \begin{bmatrix} \frac{1}{x_o} \\ \frac{1}{y_o} \\ \frac{1}{z_o} \end{bmatrix} \tag{6}$$

The laser scanner traces an approximately parallel grid on a planar surface normal to the projection axis. This is an essential characteristic of the horizontal scan lines, so that the segmentation procedure can be based on the detection of straight lines in the camera image. Also, the laser horizontal projection angle is not needed to derive the 3-D coordinates of the point in the scene; therefore, vertical scan lines are not required in the structured light grid.

Each scan line is formed as the laser beam sweeps through a plane in space. The vertical projection angles associated with each horizontal line on the structured light grid are those corresponding to the horizontal planes swept by the laser beam as it traces the grid lines. As long as the grid scanning remains unchanged, a look-up table method can be utilized to obtain the laser vertical projection angle and rotation angle corresponding to a particular grid line of interest. In addition, the camera image location as shown in Figure 1 is calculated based on a previously derived camera calibration and the location of the pixel in the camera image.

Utilization of extracted 3-D information from a scene to determine surface contour can be accomplished,[6] and this becomes a valuable consideration in off-road autonomous navigation. Because a path for travel is desired, a significant concern is the planar content of a possible path. As explained previously, it is possible to stereoscopically locate various points throughout a surface of interest. Restricting interest to surfaces of a single planar orientation suggests that a collection of points residing on a surface can indicate the planar nature of that surface. It is likely that stereoscopically extracted points are not coplanar; thus, a description of the surface requires that a plane of best fit be determined. This can be accomplished by implementing a mathematical minimization technique (see Appendix), with the assumption that the surface under analysis consists of a single plane.

Image Analysis

The captured video image of structured light horizontal grid lines represents a two-dimensional grey level image field denoted as: $a(j_1,j_2)$, where j_1 and j_2 are the horizontal and vertical spatial indices of each pixel in the field, and $a(j_1,j_2)$ is the gray level at that spatial image location. Because all desired information in the image is given by the fact that pixels are either illuminated by laser light or not, a binary image is desired to simplify further processing. As explained in the introduction, the image of the structured light is easily thresholded because of the high-power density of laser illumination. The resulting binary image field is denoted as $t(j_1,j_2)$ (see Figure 2).

Extraction of processing subregions. The 3-D texture in the scene is represented by the 2-D shape of the images of the laser lines. Therefore, considerable reduction in processing can be accomplished by only processing subregions of the image containing the laser lines.

The 3-D texture of regions in the scene is proportional to the spatial deviation of the image of the structured light grid lines. Pathway determina-

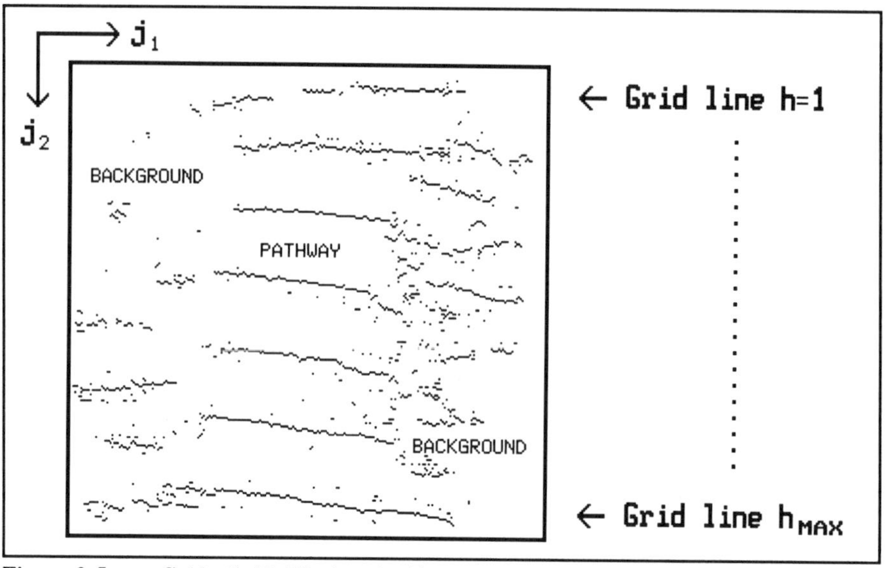

Figure 2. Image field, $t(j_1,j_2)$. Black and white are reversed for clarity.

tion amounts to detection of lines in the thresholded image field. Line detection is accomplished by means of the Hough transform by mapping each image point into a parameter space based on the linear equation $j_2 = aj_1 + b$ where j_2 is the vertical index, j_1 is the horizontal index, a is the slope of the line, and b is the vertical intercept.[7] Alternately, the transformation can be based on the normal form of the linear equation, $j_1\cos\theta + j_2\sin\theta = \rho$, where ρ is the normal distance from the origin to the line, and θ is the angle from the horizontal axis to the normal to the line.[7] Acceptable pathways differ little from level; so the grid line images are near horizontal, and the range of slopes of the lines is small and close to zero. Therefore, the Hough transform based on the linear equation $j_2 = aj_1 + b$ is appropriate and more computationally efficient.

The Hough transform maps every white (binary pixel value of 1) image pixel into a line in the (a,b) parameter space output of the Hough transform using the equation $b_h = j_2 - a_h j_1$. The range of slopes corresponds to acceptable pathway cant, thus excluding pathways with excessive cant. The range of slopes is quantized and sequenced, forming a line in the (a, b) plane corresponding to the (j_1, j_2) image spatial coordinates of each white pixel in the image field. The Hough transform output of a complete image amounts to the addition of the lines associated with each white image pixel. Peaks in the Hough transform parameter space, formed at points where many of these lines intersect, denote strong line tendencies in the image field, $t(j_1, j_2)$. Parameter

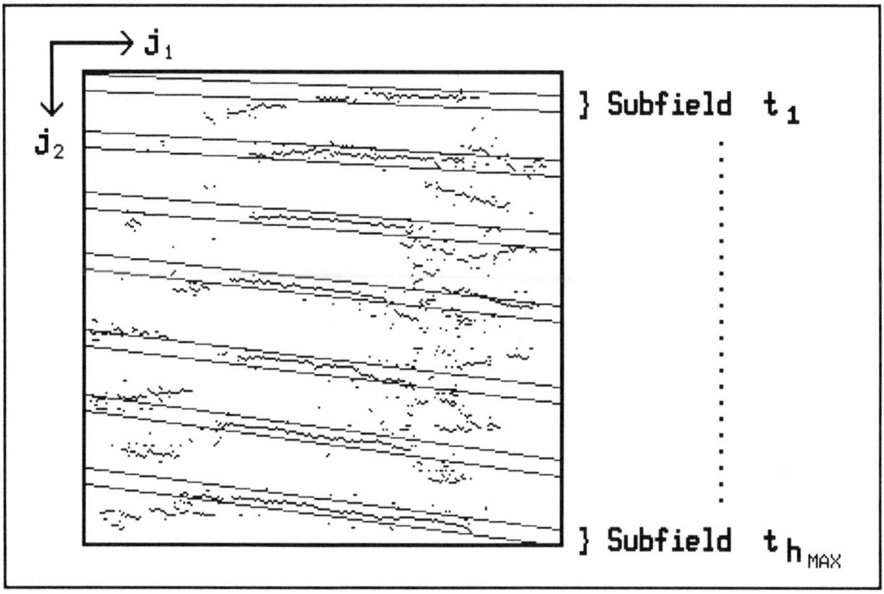

Figure 3. Subregions denoted on image field, $t(j_1 j_2)$.

space coordinates (a_h, b_h) are associated with each peak denoting a line with slope a_h and intercept b_h in the image field.[11]

Image subfields are then extracted from the image field (see Figure 3) where the center line (t_h^0) of each subfield, $t_h(i_1, i_2)$, corresponds to the mapping: $i_1 = j_1$ and $i_2 = a_h j_1 + b_h$, (see Figure 4). Image lines of slope a_h are now horizontal in the subfield, $t_h(i_1, i_2)$. Each subfield is $2p + 1$ image lines high, such that eachline (t^l_h) in a subfield (t_h) corresponds to the mapping:

$$i_1 = j_1 \text{ and } i_2 = a_h j_1 + b_n + l, \ -p \leq l \leq p \qquad (7)$$

Morphologically based image segmentation. The extraction of information from the structured light image involves detection of the image location of a definable shape, such as a contiguous line for sensing texture. Morphology involves the design of an operator to sense this definable shape,[8] and therefore is well suited to this application. Morphology is also very computationally intensive However, because of its wide general applications, hardware implementations of morphological operations are being designed. The increased speed of these implementations makes the use of morphology very attractive for this application.

The autonomous vehicle application requires that only regions of two textures be identified; the low texture pathway region, and high texture

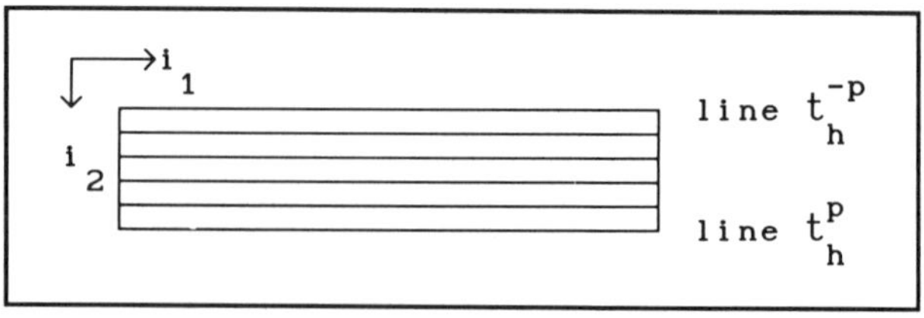

Figure 4. Image subfield, $t_h(i_1,i_2)$.

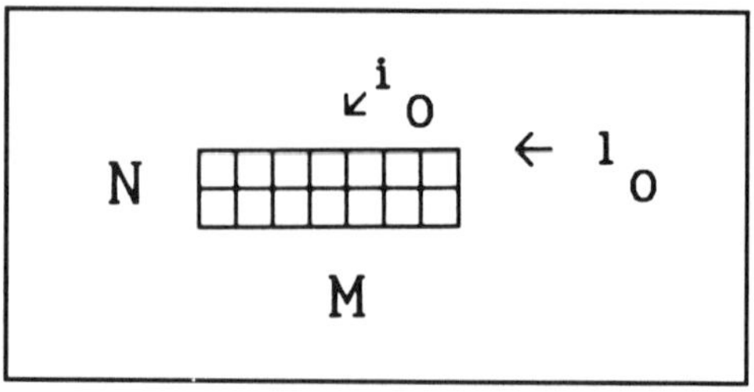

Figure 5. Morphological window operator, w_h.

background. Detection of the pathway within the image subfield can now be viewed as an image filtering problem. The continuous line image across the low texture pathway region can be treated as the object of interest, whereas the broken portions of the line across the background regions appear as noise. Pathway detection amounts to filtering the noise while accurately preserving the edges of the object. A ranked order filter is chosen because it is particularly effective in a situation where the noise consists of strong impulses, such as the broken line clusters encountered in the background regions, and where the image characteristic to be preserved is object edge sharpness.[9,10]

A morphological window operator is defined where M is the width of the window in pixels, and is odd, and N is the height of the window in image subfield lines (Figure 5). These parameters are established in scene dimensions based on vehicle and scene characteristics. For example, the height of the operator projected into the scene should equal the acceptable texture height in the scene, which should be something less than the vehicle ground clearance. The projected operator width should be greater than the widest unacceptable continuous plane in the scene (the width of a large tree or stone).

The relationship of operator parameters, as defined in the scene, to their projection into the image plane is illustrated in Figure 6. The values of these projected parameters in the image plane are functions of a laser plane and location in the image; however, all of these parameters can be calculated in advance and stored in a look-up table. For each laser plane, it is possible to determine in advance the required operator sizes at every pixel in the image. Figure 6 depicts the process used to determine operator size at a particular pixel, for an arbitrary laser plane. The scene coordinates of the point (x, y, z),

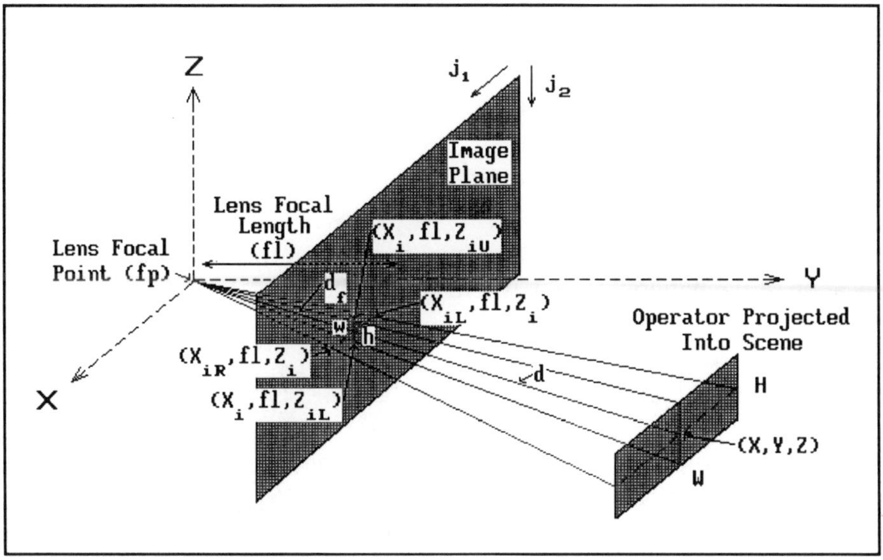

Figure 6. 3-D scene to 2-D Image Projection.

where the ray extending from the focal point through the image plane intersects the laser plane, can be obtained using Equations 1 through 6. The actual coordinates of the pixel of interest, (x_i, fl, z_i), are determined utilizing the results of a camera calibration. Definition of the acceptable texture height, H, and widest unacceptable plane, having a width of W occurs in a plane containing the scene point (x, y, z) that is parallel to the xz-plane. With the operator parameters defined in the scene, calculation of the operator size in the image plane is accomplished using the following equations:

$$w = W\sqrt{\frac{x_i^2 + fl^2 + z_i^2}{x^2 + y^2 + z^2}} \tag{8}$$

$$h = H\sqrt{\frac{x_i^2 + fl^2 + z_i^2}{x^2 + y^2 + z^2}} \tag{9}$$

Subsequently, the height and width of the image plane in actual units, h_i and w_i, and in pixels, J_1 and J_2, are utilized in determining operator size in pixels, as given in the following equations. Note, however, that the resulting operator height and width, N and M respectively, are quantized to whole pixel values; thus, a particular size operator is often applicable to large regions of the image.

$$M = \frac{w}{w_i} J_1 \tag{10}$$

$$N = \frac{h}{h_i} J_2 \tag{11}$$

The window operator is translated throughout the image subfield. The location of the window (i_0, l_0), where i_0 is the horizontal window center and l_0 is the top line, is described by:

$$w_h\,[t_h\,(i_1,i_2);\, i_0;\, l_0] = \{t_h^l(i_1): i_0 - \tfrac{M}{2} \leq i_1 \leq i_0 + \tfrac{M}{2}\,;\, l = l_0,\, l_0 + N - 1\} \tag{12}$$

A ranked order filtering operation is performed at each window location, resulting in a mapping of each image subfield, $t_h(i_1,i_2)$, to a filter output subfield, $x_h(i_1',i_2')$. Each pixel in the filter output subfield corresponds to the filter output:

$$x_h(i_1',i_2') + Rank_T\,[t_h(i_1,i_2) \in W_h(i_0,l_0)] \tag{13}$$

where the $Rank_T$ operation consists of ordering the pixel values within the filter window from 1 to 0 and choosing the pixel value at the rank corresponding to the threshold T. Thus, if T or more pixel values are 1, the filter output is 1. The value of T is related directly to the operator width such that a near continuous line across the operator is detected.

The filter output subfield (Figure 7) is then compressed vertically into a single line $y_h(i_1')$, corresponding to the image field spatial location of the Hough transform extracted line, $t_h(i_1)$, where:

$$y_h(i'_1) = \text{Max}[x_h^0(i_1'),\, x_h^1(i_1'),\, \ldots.x_h^{2p-N+2}(i_1')] \tag{14}$$

Although the ranked filter preserves the pathway edges, or ends of the pathway lines, the filtering operation erodes $M/2$ pixels from each end of each line and both sides of any breaks in the lines. Restoration of eroded pixels is accomplished by morphological dilation, which is a transformation that combines two sets using vector addition of set elements. If A and B are sets in Euclidean N-space (E^N), the dilation of A by B is denoted by $A \oplus B$ and defined by[11]:

$$A \oplus B = [c \in E^N \mid c = a + b \text{ for some } a \in A \text{ and } b \in B] \tag{15}$$

The vertical compression of the filter output subfield into a single line (y_h), results in the dilation needing to be performed only in the single

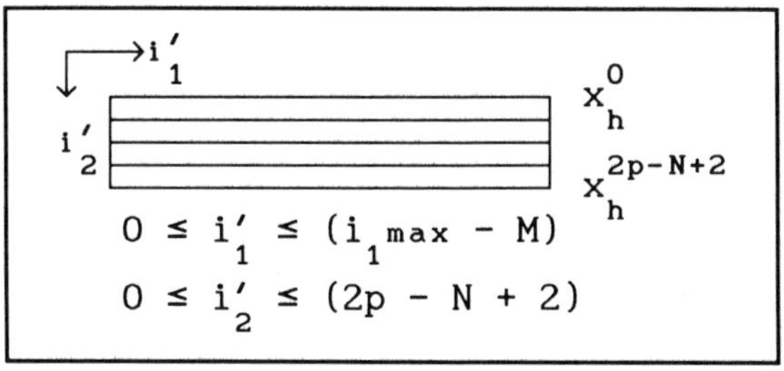

Figure 7. Filter output subfield, $x_h(i'_1,i'_2)$.

horizontal spatial dimension; therefore, $N = 1$ in the previous definition. Each of the resulting filtered, compressed lines is then dilated with an M by 1 structuring element, B:

$$y'_h(i'_1) = y_h(i'_1) \oplus B \tag{16}$$

The dilated line segments $y'_h(i'_1)$, overlay the region of low texture pathway crossed by the structured light horizontal grid line, and the left and right end points $y'_h(i_{1l})$ and $y'_h(i_{1r})$, of each line segment correspond to the left and right boundaries of the pathway in image subfield coordinates.

These end points $y'_h(i_{1l})$ and $y'_h(i_{1r})$ are then mapped to their corresponding image coordinates $a_{hl}(j_{1l}, j_{2l})$ and $a_{hr}(j_{1r}, j_{2r})$ using the transformations:

$$j_1 = i_1 \text{ and } j_2 = a_h i_1 + b_h \tag{17}$$

The points $a_{hl}(j_{1l}, j_{2l})$ and $a_{hr}(j_{1r}, j_{2r})$ constitute a pathway boundary point pair which, when combined with all other such pairs ($h = 1$ to h_{MAX}), defines the pathway location in the image field quantized corresponding to the number of horizontal structured light grid lines.

In practice, pixels are not extracted from the image field and placed in a new location and orientation. Instead, the window operator is structured with slope a_h and translated through the image field $t(j_1, j_2)$ along lines parallel to the line $j_2 = a_h j_1 + b_h$. The output of this process is a boundary point pair detected at each horizontal laser line image location. Connecting all of the left pathway boundary points with straight line segments and all of the right points in the same manner yields an image of the estimate of the pathway boundaries.

3-D point extraction. Image segmentation utilizes only part of the information obtainable from the structured light. The discrete horizontal grid lines correspond to known vertical laser projection angles. As a result of the segmentation processing, the approximate vertical image location of each horizontal laser line corresponds to the vertical position of the center line t_h^0, of the image subfield $t_h(i_1, i_2)$, in which it is contained. Also, the 3-D coordinates of only the points within the pathway boundaries are of interest;

therefore, an arbitrary number of image points can be chosen lying on the image subfield center line between the previously identified left and right boundary points $a_{hl}(i_{1l},i_{2l})$ and $a_{hr}(i_{1r},i_{2r})$ in each subfield. All points in one image subfield will have the same vertical laser projection angle.

As in the segmentation procedure described previously, the chosen subfield points are mapped to their corresponding points in the image field using the transformations shown in Equations 17. By using time sequential images of single laser lines, the grid line that corresponds to each chosen point is identified, such that the vertical laser projection angle of each chosen point can be determined. The image coordinates (j_1,j_2) of each chosen point can then be mapped to camera image plane coordinates using the camera model and calibration previously described. The laser angle and camera image coordinates constitute all the information required to utilize Equation 6 to calculate the 3-D scene coordinates corresponding to each detected point.

Once again, in practice, pixels are not extracted from the image field and placed in a new location and orientation described as a subfield. Instead, point selection is accomplished along each line $j_2 = a_h j_1 + b_h$ from the output of the Hough transform, and need only occur between the previously identified left boundary point $b_{hl}(j_{1l},j_{2l})$ and the identified right boundary point $b_{hr}(j_{1r},j_{2r})$.

Experimental Results

Scene segmentation. The experimental "autonomous vehicle," utilized to test the previously described processing procedures consisted of a tripod mounted video camera and a separately mounted portable holographic laser scanner. Numerous laboratory and outdoor scenes were imaged and processed during design and testing of the described processing. These images and the corresponding results are detailed by Brown.[12] Analysis of one particular full-scale outdoor pathway scene yielded the images shown in Figures 8 to 12. Figure 8 is a captured video image of the uniformly lit scene containing the pathway extending away from the camera in the y direction. The low texture path in the center of the image is nearly the same color as its high texture surroundings and thus, when shown as a 2-D image, is difficult to detect with the human eye or by gray scale methods. Figure 8 also shows the preprocessed, horizontal laser projected light grid overlaid on, and registered with, the scene image.

When the structured light horizontal grid was projected onto the scene, the laser lines became discontinuous in proportion to the 3-D texture in various regions of the scene. As shown in Figure 8, the laser lines became

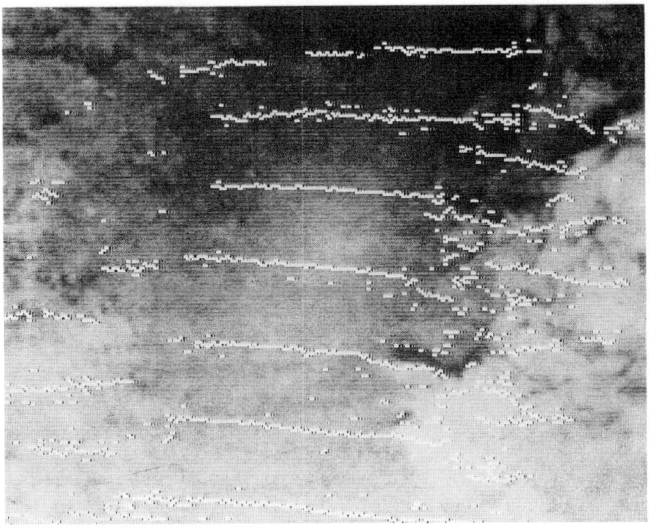

Figure 8. Video image of the scene with overlaid structured light.

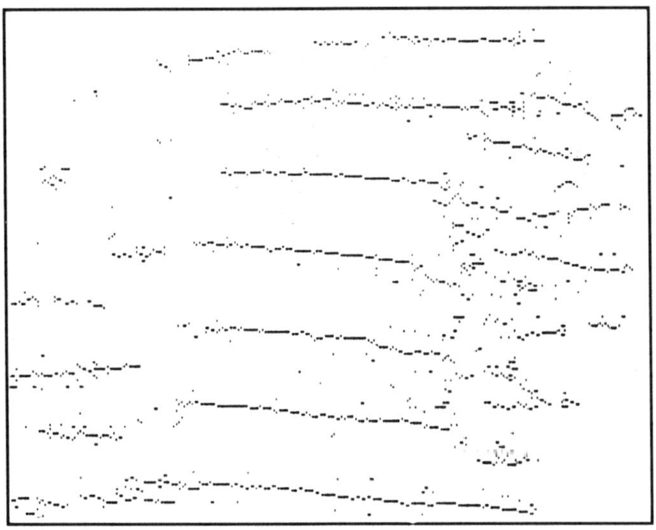

Figure 9. Binary image of structured light.

highly discontinuous over the high texture background in the scene and remained fairly continuous throughout the lower texture pathway in the center of the scene. The globally thresholded binary image field $t(j_1,j_2)$ shown in Figure 9, was utilized for further processing (image polarity has been reversed for clarity).

To locate the laser lines in the image, the Hough transform was performed on the image of Figure 9. Image subfields, shown in Figure 10, were

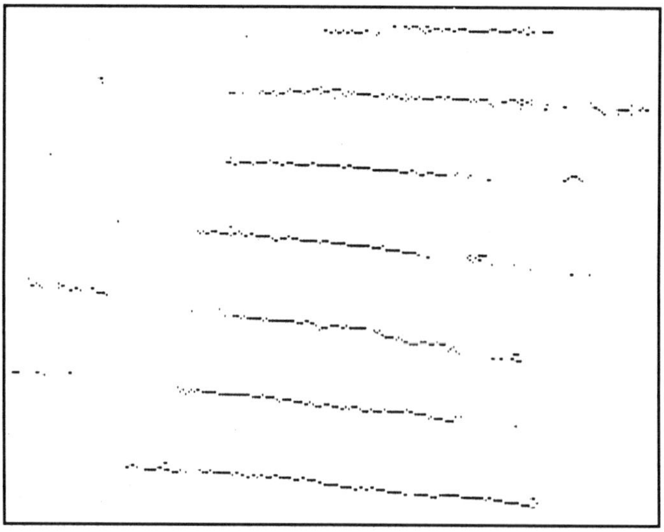

Figure 10. Image subfields extracted with Hough transform.

then extracted from the image field where the center line (t_{h0}) of each subfield $t_h(j_1,j_2)$ corresponded to a line detected by the Hough transform. Each subfield was five ($2p + 1$ with $p = 2$) image lines wide.

The next goal in the processing was to eliminate the apparent image noise associated with the high texture background while preserving the boundary information associated with the low texture pathway. This was accomplished by using a ranked filter with an M by N window operator and subfield that varied in size from image top to image bottom such that their sizes in the scene correspond roughly to sensing 5.6cm wide by 3.1cm high along the path segment. Scaling these dimensions up to a reasonable pathway width would correspond to a 10cm vertical acceptance and a horizontal rejection width of 18cm. Ten centimeters is small but still within a range of acceptable vehicle ground clearance, and 18cm is the reasonably largest probable unacceptable planar surface, such as a flat spot on a tree or a flat rock.

A rank threshold equal to M was chosen because a horizontal contiguous line, with vertical deviations not exceeding N pixels, contained within the window would illuminate at least M pixels. The seven subfields $t_h(j_1,j_2)$ had axes aligned with the subfield's orientation yielding the new coordinates $t_h(i_1,i_2)$ where:

$$i_1 = j_1 \text{ and } i_2 = a_{hj1} + b_h + l,\ -p \le l \le p. \tag{18}$$

In theory, all operations were performed in the (i_1,i_2) plane, however in practice, all steps were performed in the (j_1,j_2) plane. The filter's output,

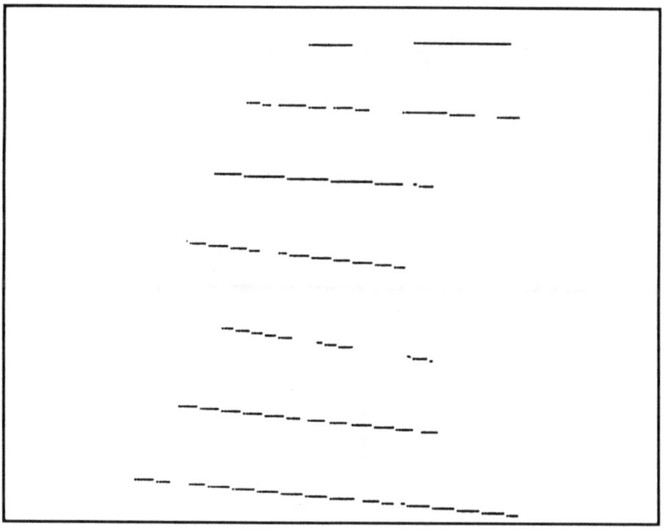

Figure 11. Center line of each subfield following morphological erosion and dilation.

$x(i'_1,i'_2)$, could then be described by Equation 13, and subsequently compressed vertically as described by $y_h(i'_1)$ in Equation 14. The output of such a filter in the (j_1,j_2) plane then corresponded to white pixels located along lines that had position and orientation of the lines previously derived from the Hough transform. Restoration of eroded pixels was achieved by morphological dilation of all $y_h(i'_1)$ with an M by 1 structuring element, B, as defined by Equation 15. The result of this dilation is shown in Figure 11. Using the slope and intercept information obtained from the Hough transform, the left and right boundaries of the path at each laser line were located by searching along each Hough line for the farthest right and left white pixels. All boundary points on the left side of the path were then connected, and all boundary points on the right side of the path were connected, yielding the path outline shown in Figure 12 overlaid on the original scene image.

Pathway definition. Knowledge of the relative positions of all visual system components is a requirement for implementing a stereoscopic visual system. To achieve this, a theodolite was utilized for measuring the horizontal and vertical angles of various points throughout the neighborhood of the experimental "autonomous vehicle." The additional measurement of distances, from theodolite to each point, provided a set of spherical coordinates for each surveyed point. Visual system descriptors included in the survey were the focal points of the laser scanner and camera lens. Eight points along the perimeter of the pathway were also surveyed. The planar fit algorithm described in the Background section and the Appendix was applied to these

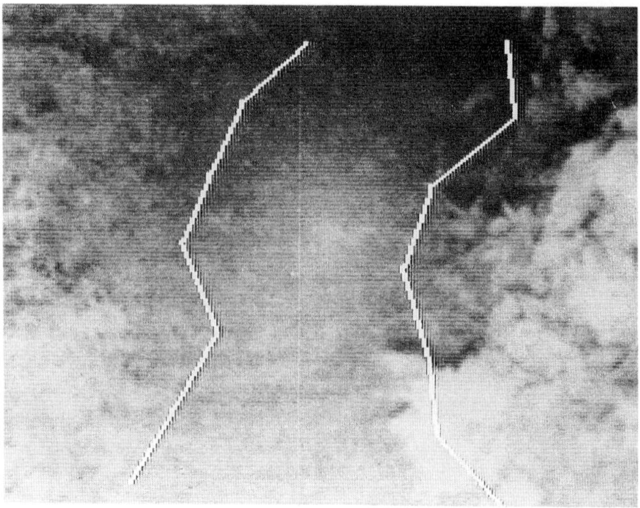

Figure 12. Detected pathway boundaries overlaid on original scene.

points, thus providing a basis of comparison for the resulting 3-D points and planar fit derived by the vision system.

The video camera used in all experiments was calibrated as required. The calibration information was used to calculate the vertical and horizontal image plane displacements corresponding to each chosen point. To derive the laser angles of projection, a look-up table had been constructed as described in the Background section. This table contained the vertical laser angles of projection for every horizontal line in the projected grid. Thus, determination of the particular scan line that contained the point allowed reference to the look-up table to obtain the corresponding angular information.

After the camera image location and vertical laser projection angles of each chosen point had been derived, scene definition could proceed. The method described in the Background section was used to calculate the 3-D coordinates of all the chosen points. Finally, the planar fit technique, described in the Background section and the Appendix, was applied to the calculated points in the sample image, producing the equation of the plane of the sample pathway as derived by the vision system. The derived equation for this particular experimental pathway is given in Equation 19 for comparison with the planar fit of the surveyed data points of the pathway that is described by Equation 20.

$$0.08x + 0.27y - 0.96z - 199.36\text{cm} = 0 \qquad (19)$$

$$0.01x + 0.23y - 0.97z - 197.94\text{cm} = 0 \qquad (20)$$

Discussion

System performance. Figure 12 clearly indicates that the vision system described herein is capable of segmenting a scene in an unstructured environment into regions of background and likely pathway. The detected boundaries are well defined and, as determined by comparison with survey data, do coincide with the actual pathway boundaries. The output of the segmentation procedure consists of a minimal amount of data (seven pairs of points, in this case), which results in a succinct pathway identification that will provide desirable input to a vehicular navigation system.

In addition, the consistency of Equations 19 and 20 demonstrates that this vision system is not only capable of discerning the pathway ahead of the vehicle but also is capable of locating the pathway in three dimensions and defining it as a plane. Although the 4.7° angle between the normals of the two planes appears to be significant to vehicular navigation, it should be realized that the major contributing component is the difference in angles with respect to the x axis. This rotation is caused by the resolution in point location as projected across the width of the pathway. The average width of this particular pathway was 93cm, which is substantially scaled down from an expected width of an acceptable vehicle pathway. With a 93-cm base length, a 4.7° rotation can be caused by a point z component inaccuracy of as little as 6cm. Such a vertical inaccuracy, when measured at a distance from laser to path center of 830cm in the y direction, can result from an angular error on the order of 0.4°. This angular error is beginning to approach laser beam divergences and the projection angle accuracies defined for this experiment. In addition, the 4-cm vertical displacement between the derived and surveyed planes is insignificant in terms of vehicular navigation. In all, the pathway definition is considered correct within the experimental constraints, and it is expected to improve significantly as the pathway size is increased.

Computational simplicity. The described segmentation and definition procedures utilize standard image processing methods, such as the Hough transform and image morphology. Little unique processing is involved, and therefore, implementation of this procedure is well defined and computationally simpler than more elaborate methods. This reduction in computational complexity should result in a substantial increase in speed over current image segmentation methods and should also make it possible to implement the unique algorithms in hardware for still greater processing speed, or on computers that could easily be integrated into an autonomous vehicle system.

Image morphology is beginning to see widespread application, and as a result, considerable effort is being devoted toward hardware implementations that considerably increase the speed of various morphological processes. Hardware implementation efforts are also being directed toward the Hough transform, and this could result in substantial reduction in processing requirements and increased computational speed.

Image definition. The research described herein concentrated on a single planar fit to the region of interest in each scene. This was done for simplicity and because of general applicability to vehicular navigation, where it is likely that any acceptable pathway will be planar within some tolerance. In fact, this tolerance could be used as one condition for pathway acceptability. An extension to the planar fit method, described in the Appendix, can be utilized to determine when the region of interest in the scene contains bifurcations requiring a multiplanar or piecewise planar fit. This extension could merely be a routine that decides when the eigen value associated with the best single planar fit, described in the Appendix, exceeds a predetermined threshold for "planar." An algorithm would then be invoked that would determine the image location of a likely bifurcation and regions of the pathway for two or more separate planar fits. The information could also provide input to a higher level of processing that is capable of deciding the meaning of detected planar orientations and break points, such as a steep slope, cliff, or wall in the pathway. Although the previously described higher level processing could be termed artificial intelligence, and is beyond the scope of this research, it should be noted that the methods described in this paper can provide direct interface with such processing.

Obstacle location and identification. As described in the Background section, the image of structured light responds to the 3-D shape of the scene. The assumption was made during this research that there were no obstacles in the chosen pathway. However, an obstacle in the pathway is simply a distortion in the 3-D shape of the pathway and is therefore sensed by all of the processing methods described. An additional step following the dilation of the filtered image would be required to examine each scan line image for discontinuities that could indicate the presence of an obstacle in the region that is about to be chosen as a pathway. Pathway select versus obstacle alert criteria would be dictated by higher level processing and control systems.

An adaptive method may be applicable whereby initial processing identifies a discontinuity in pathway scan lines that may indicate the presence of an obstacle, and the laser scanner would then be aimed at that area to provide higher resolution scanning for identification. The laser scanner utilized during this research can be constructed with the dynamic capability to direct higher resolution scanning at selected portions of the original larger scan area. The scene definition methods described previously could then be applied to this higher resolution, small scan area to define the obstacle.

Appendix: Planar Fit

Suppose the points (x_k, y_k, z_k) for $k = 1,2,...,n$, on a particular surface are available. Determination of the best fitting plane is achieved by minimizing the sum of the squares of the distances, from each point, to the plane. The plane that is sought is described as $ax + by + cz + d = 0$. The sum minimized is thus

$$\sum_{k=1}^{n} \frac{(ax_k + by_k + cz_k + d)^2}{a^2 + b^2 + c^2} \tag{21}$$

By requiring that the magnitude of the normal to the plane equal unity, $a^2 + b^2 + c^2 = 1$, the process is transformed into a constrained extremum problem. Therefore, the method of Lagrange multipliers can be applied resulting in the equation given below, where λ denotes a Lagrange multiplier.

$$\nabla \sum_{k=1}^{n} [(ax_k + by_k + cz_k + d)^2] = \lambda \nabla (1 - a^2 - b^2 - c^2) \tag{22}$$

Subsequent expansion and manipulation of Equation 22 yields a system of equations that can be expressed in matrix form as illustrated by Equation 23. Note that the terms x_a, y_a, and z_a denote average values of x, y, and z, respectively, that is,

$$x_a = \frac{\Sigma x_k}{n}$$

Also, the multiplier, λ, in the matrix equation is an eigen value.

$$\left[\begin{bmatrix} \Sigma x^2_k & \Sigma x_k y_k & \Sigma x_k z_k \\ \Sigma x_k y_k & \Sigma y^2_k & \Sigma y_k z_k \\ \Sigma x_k z_k & \Sigma y_k z_k & \Sigma z^2_k \end{bmatrix} - n \begin{bmatrix} x^2_a & x_a y_a & x_a z_a \\ x_a y_a & y^2_a & y_a z_a \\ x_a z_a & y_a z_a & z^2_a \end{bmatrix} \right] \begin{bmatrix} a \\ b \\ c \end{bmatrix} = \lambda \begin{bmatrix} a \\ b \\ c \end{bmatrix} \tag{23}$$

Solution to this matrix equation consists of three eigen values along with their corresponding eigen vectors. The normal, $N = \langle a,b,c \rangle$, to the plane of best fit is the eigen vector that corresponds to the smallest eigen value. Then, the parameter, d, is readily calculated by utilizing the resulting d component of the gradient relationship expressed in Equation 22.

Acknowledgments

The author acknowledges the contributions of Richard Marvel, Gonzalo Arce, Charles Ih, and David Fertell throughout various parts of the research and experimentation. Also, Charles Boncelet proposed the method for computation of the planar fit.

References

1. J. LeMoigne and A. M. Waxman, *Computer Vision, Graphics, and Image Processing* **28**, 203-206 (1984).
2. A. M. Waxman *et al.*, *IEEE Journal of Robotics and Automation* **3**(2), 124-141 (1987).
3. R. M. Inigoet *et al.*,*IEEE Transactions on Pattern Analysis and Machine Intelligence* **6**(6), 820-826 (1984).
4. D. Nitzan, *IEEE Transactions on Pattern Analysis and Machine Intelligence* **10**(3), 291-309 (1988).
5. S. J. Gordon and W. P Seering, *IEEE Transactions on Pattern Analysis and Machine Intelligence* **10**(3), 374-386 (1988).
6. P. J. Besl and R. C. Jain, *IEEE Transactions on Pattern Analysis and Machine Intelligence* **10**(2), 167-192 (1988).
7. R. C. Gonzalez and P. Wintz, *Digital Image Processing*, 2nd ed. (Addison-Wesley, Reading, MA, 1987), pp. 130-134.
8. J. Serra, *Image Analysis and Mathematical Morphology* (Academic Press, New York, 1982), pp. 15-20.
9. P. Maragos and R. Schafer, *IEEE Transactions on Acoustics, Speech, and Signal Processing* **35**, 1170-1184 (1987).
10. T. Nodes and N. C. Gallagher, *IEEE Transactions on Acoustics, Speech, and Signal Processing*, 29 (1982).
11. R. M. Haralick, S. R. Sternberg, and X. Zhuang, *IEEE Transactions on Pattern Analysis and Machine Intelligence* **9**(4), 532-550 (1987).
12. C. D. Brown, *Scene Segmentation and Definition for Autonomous Robotic Navigation Using Structured Light Processing*, Doctoral Dissertation, University of Delaware, 1988.

Note. The image segmentation and definition method described in this Chapter is covered under United States Patent No. 5-155-775, issued 13 October 1992.

21

Modeling Human Performance to Predict Unit Effectiveness

Steven R. Hursh
U.S. Army Medical Research and Development Command
Walter Reed Army Institute of Research
Washington, DC 20307-5100

Richard E. McNally
Science Applications International Corporation
626 Town Center Drive
Suite 205
Joppa, MD 21085

Army standard models are used to analyze and predict the impact of new equipment and operational doctrine on the combat effectiveness of military units. Absent from most models is consideration of human performance in modulating the effectiveness of units. The Medical Research and Development Command has initiated a program to incorporate consideration of medical products that improve soldier health and performance. The purpose is to accurately portray the impact of health and performance on unit effectiveness and to document to what extent medical products enhance the ability of military units to carry out their mission.

Medical interventions, both current and future, were modeled for their ability to attenuate expected medical threats and enhance soldier performance and unit effectiveness. First, medical therapies for nerve agent threats greatly improve unit recovery after chemical attack. Soldiers exposed to nerve agents can be partially protected by pretreatments, and casualties can be rapidly returned to duty through the use of nervous system antidotes. Future medical defenses are predicted to restore the effectiveness of military units to 90 to 95 percent of normal within four days of chemical attack. Second, laboratory and field studies of the impact of sleep deprivation on soldier performance have led to a more optimal doctrine for sleep management and fatigue reduction. Unit effectiveness can rapidly deteriorate to less than 50 percent within four days of combat with only four hours of sleep per day. Sleep discipline providing at least seven hours of sleep per day optimizes unit productivity for engagements lasting from four days to three weeks.

Introduction

A variety of Army standard models are used to analyze and predict the impact of new equipment and operational doctrine on the combat effectiveness of military units fighting hypothetical battles. Absent from most models is any consideration of human performance in modulating the effectiveness of units. In 1989, the U.S. Army Medical Research and Development Command (MRDC) initiated a program to incorporate into an Army standard model consideration of medical products, both materiel and concepts, that improve the health and performance of soldiers. The purpose was twofold: first, accurately portray the impact of soldier health and performance on unit effectiveness and, second, document the extent to which medical products enhance the ability of military units to carry out their warfighting mission.

MRDC, through cooperation with the U.S. Army Research Laboratory (ARL), selected the Army Unit Resiliency Analysis (AURA) Model as the first test bed for these concepts. The model was improved to incorporate the latest medical data on soldier health and performance across a range of variables: infectious disease, biologic and chemical weapons effects, heat stress, and sleep deprivation. Medical interventions, both current and future, were modeled for their ability to attenuate these threats and enhance soldier performance and unit effectiveness.

The authors discuss the impact of products that have evolved from the neurosciences. First, modeling has been applied to estimate the battlefield toxicity of potential nerve agent attacks and predict the likelihood and severity of casualties. A variety of medical protective measures have been assessed and shown to improve the resiliency of typical combat units. Second, modeling has been applied to estimate the degradation of individual and unit performance under conditions of continuous operations that permit only limited amounts of sleep. A series of sleep disciplines were modeled, and the net benefits of adequate sleep for overall unit productivity were demonstrated.

Overview of Modeling Strategy

The purpose of medical modeling is to demonstrate the impact of medical research and development products, both concepts and materiel, on the Army's warfighting capabilities using accepted models of battlefield threats and combat effectiveness. There are four goals to this program: (1) demonstrate the impact of medical products for improved unit effectiveness, (2) assist the transfer of medical technology and information for operational planning, (3) generate timely information to researchers and developers as

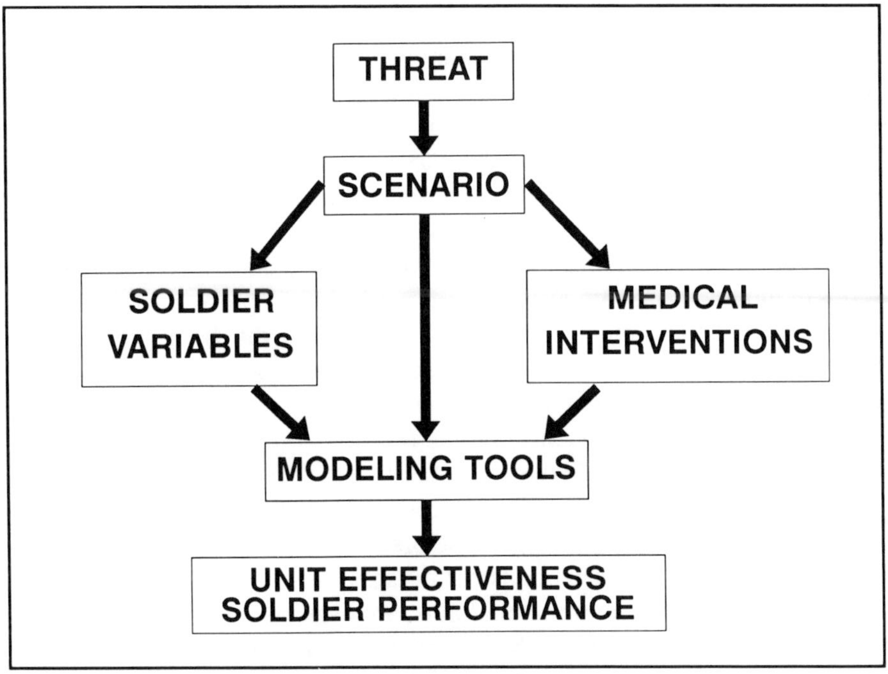

Figure 1. Modeling of medical products. Overview of strategy.

decision aids for planning and milestone decisions, and (4) provide feedback to researchers concerning data gaps and model refinement.

To meet these goals, a common modeling strategy has been followed and is diagramed in Figure 1. First, a specific battlefield threat forms the focus of each modeling task. These threats encompass a broad range of medical hazards to the soldier such as biologic and chemical weapons, infectious disease, stress and fatigue, environmental extremes (heat, dehydration, and so on), and traumatic or accidental injury. The threat is then specified within the context of a battlefield scenario that includes the geographic location, meterologic conditions, weapons characteristics, combat units and tasks, uniform and protective devices, and operational doctrine. For each threat and scenario, appropriate soldier variables must be modeled that represent the likely impact of the threat in that scenario. These human factors include death or incapacitation, rate of return to duty after injury, and performance degradation. Degradation factors may be subdivided further into cognitive factors, physical work capacity, physiologic limits such as water requirements and thermal load, uniform restrictions and encapsulation, and individual variability. The model must be able to simulate the impact of all these factors, individually and in combination. At the same time, medical interventions are specified that can reduce degradation of some or all of these performance factors. The medical interventions may include the following: preventative drugs or vaccines, physical conditioning, therapy drugs, physical barriers, and operational doctrine such as sleep discipline, work schedules, acclimatization, and water discipline. The next step is to select a modeling system that can portray the impact of each medical intervention on these individual soldier variables and then simulate the effects of those soldier variables on unit effectiveness. The modeling system may be a single simulation program or a series of modules that compute different aspects of the problem and feed into a final estimate of unit effectiveness. Ultimately, the final product is a representation of the impact of the medical interventions on casualties, individual performance and unit effectiveness.

To date, a central modeling system for estimating unit effectiveness has been the Army Unit Resiliency Analysis (AURA) developed by the ARL. This model was selected because it contained rudimentary algorithms to represent many soldier performance factors, was an accepted Army modeling system, contained representations for a variety of military units that were coordinated with the appropriate schools, and was open to further improvement by BRL to incorporate the latest medical data and performance representations. The current version of AURA represents the cooperative activities of the proponent laboratory, the medical laboratories of the MRDC, and the primary contractor, Science Applications International Corporation.

Other modeling tools have been used as required. The first case to be discussed in the following text illustrates the use of the Modular Chemical Assessment Structure (MCAS) for estimating chemical and conventional casualties as an input factor to AURA. Other modeling tools have been used to represent biologic weapons effects, heat strain, and non-auditory blast injury.

Modeling of Medical Therapy for Nerve Agent Defense

Introduction. The 1980s were characterized by a reemphasis on the development of measures to increase the ability of U.S. forces to survive and fight on a chemically contaminated battlefield. Ironically, although the Warsaw Pact is now dissolved and the likelihood of facing a Soviet adversary on the battlefield has evaporated, the probability of U.S. forces facing a chemical warfare (CW) threat nonetheless has increased. During the recent Persian Gulf conflict (Operations Desert Shield and Desert Storm), the risk that U.S. forces would be subjected to chemical attacks was greater than at any time since World War I.

The chemical defense research and development programs of the 1970s and 1980s have yielded many improved capabilities for detection of chemical agent and protection against and treatment of injuries caused by CW agents, and improvements are in progress. In addition, the medical community has made substantial contributions to these efforts and is also on the brink of significant improvements. In this section, some of the improved products of medical chemical defense research will be examined. During the past two decades, we acquired the capability to treat nerve agent casualties with oximes, such as pralidoxime chloride (2-PAM). In the 1980s, preexposure drugs for nerve agents were fielded, and several drugs that showed great promise are now nearing transition from the laboratory bench and animal studies to development programs that will place them in the hands of our fighting men and women and their medical providers. Enhanced treatments, including the use of anticonvulsant drugs, reduce damage to the nervous system caused by nerve agents, thereby shortening recovery time. Developments this decade may include improved preexposure compounds, and future improvements might be nerve agent antibodies—products of biotechnology—that will increase the levels of nerve agent that soldiers can tolerate.

Objective. We present an abbreviated discussion of the comparative effectiveness of current, enhanced, near-term, and far-term therapies for nerve agent intoxication. In addition, a "no therapy" approach is presented for comparison. A full report is available.[1]

Approach. The effectiveness of several nerve agent treatment alternatives is addressed. We evaluate the effectiveness of these treatment alternatives with two levels of physical protection from agent effects: no protection (MOPP 0) and full mission-oriented protective posture (MOPP IV).

Figure 2 illustrates the general approach used for these studies. First, CW attack profiles were established (weather, terrain, type of weapon, agent deployed, and so on). Then, the MCAS model was used to develop agent dosage and deposition, chemical casualty data, and return-to-duty distributions. The principal MCAS outputs are numbers of casualties resulting from the chemical attack, given the protection and treatment options applied, and profiles for return of casualties to duty. These data are input into AURA to determine unit effectiveness over time. AURA also accounts for decrements in effectiveness resulting from the burdens of physical protection.

Chemical attack parameters were based on methods developed for the U.S. Army Medical Research Institute of Chemical Defense (USAMRICD) study *Battlefield Levels of Soman 1989* and have been used to predict threat agent dose data for the current study.[2] The former study developed munition expenditure requirements, resulting agent concentration and dose to protected

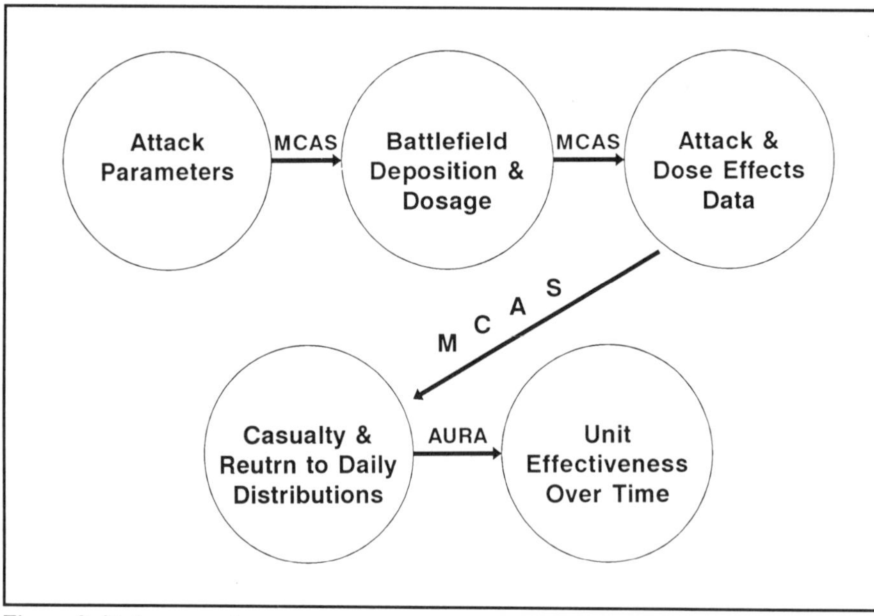

Figure 2. Study approach.

and unprotected personnel, and casualty data for artillery and air-delivered attacks under both summer and winter conditions. The munitions used were 152-mm artillery projectiles with unthickened Soman and bulk fill bombs with thickened Soman.

Two hundred fifty-six rounds of artillery, fired in one volley, were required to achieve the effects studied. Forty bombs, delivered simultaneously from four aircraft, were required. We studied each fire plan in a Central Europe summer day, a Central Europe winter dawn, and a Southwest Asia (SWA) summer day environment, against both an air-delivered bulk chemical bomb and an artillery-delivered chemical shell scenario. The bulk bomb provided multiple line sources over the target. The artillery scenario provided multiple point sources over the target. As an example, the deposition pattern for the artillery attack on a SWA summer day is shown in Figure 3.

Therapies considered. Nerve agents are a group of structurally similar organophosphorus compounds with differing inorganic fragments (cyanide, fluorine, or sulfur) and organic fragments. They share a common phosphorus-oxygen structure. They exhibit the common physiologic effect of interfering with the enzyme acetylcholinesterase (AChE); consequently, the balance of the neurotransmitter acetylcholine (ACh) is drastically altered. ACh is an essential neurotransmitter for the parasympathetic nervous system, involved in breathing, perspiration, and other essential functions, and necessary for the

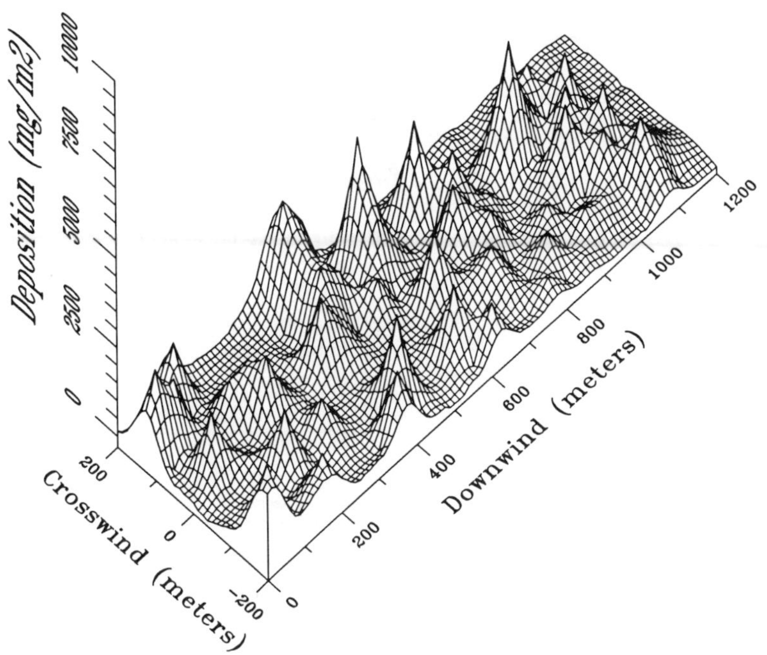

Figure 3. Southwest Asia artillery attack deposition pattern, assuming a Southwest Asia (SWA) summer day environment.

innervation of skeletal muscles, as well as for other central nervous system functions. ACh is released by nerve endings to carry the nerve impulse to the end organs. AChE neutralizes the released ACh after the firing of the nerve impulse. When nerve agents block the action of AChE, the resulting excess of ACh causes an uncontrolled overstimulation of the skeletal muscles, parasympathetic end organs, and other central nervous system end effectors. This, in turn, causes the symptoms that are characteristic of nerve agent poisoning (pinpointing pupils of the eye and resulting dimness in vision, difficulty breathing, drooling, excessive sweating, nausea, vomiting, cramps, involuntary defecation and urination, twitching, jerking, staggering, headache, confusion, drowsiness, coma, convulsions, and death).[3]

The therapies addressed are listed in Table 1. The Mark I Nerve Agent Antidote Kit administers 600mg of the oxime *N*-methylpyridinium-2-aldoxime chloride (2-PAM) and 2mg of atropine sulfate. Every soldier carries three Mark I kits for self- or buddy aid. Atropine is one in a class of drugs that blocks the action of ACh, thereby mitigating the toxic effects at the neurons that result from nerve agent poisoning. 2-PAM is one of several oximes widely

Table 1
Nerve Agent Treatment Alternatives

Option	Drugs Utilized
No therapy	None
Current therapy	Atropine and 2-PAM (Mark I Antidote Kit)
Enhanced current therapy	Pyridostigmine pretreatment, with atropine and 2-PAM
Near-term therapy	Pyridostigmine pretreatment, with atropine, 2-PAM, and anticonvulsant
Far-term therapy	Biotechnology product (nerve agent antibody)

believed to reactivate AChE bound by nerve agents. Pyridostigmine pretreatment is a compound that reacts with AChE, thereby protecting the enzyme from destruction by nerve agents.[4] Anticonvulsant therapy counters convulsions that are a major effect of nerve agent poisoning, which can lead to brain damage and probably contributes to lethality.[5] For purposes of this study, diazepam was the anticonvulsant drug modeled. Nerve agent antibodies are proposed as a future protection. It seems likely that molecular biology will provide an effective nerve agent pretreatment that will reduce agent toxicity in man manyfold—probably by early next century.[6]

Mission-oriented protective posture (MOPP) provides for utilization of the chemical protective ensemble in a manner tailored to the threat of chemical attack. "MOPP 0" is the standard battle dress uniform. MOPP I through IV progressively add or close components until the individual is completely sealed in the full ensemble. MOPP IV protection is at a cost of restrictions to range of motion, tactile sensitivity, and field of vision; also, the additional weight of the ensemble imposes limitations. If the individual has a heavy physical work load, there is yet another burden: increased heat stress as the temperature and relative humidity within the ensemble components increase. In hot climates, this burden is significant for even light-work scenarios. Table 2 describes the protection options considered.

In general, research and development observations have indicated that the alternate treatment regimens described in Table 1, in comparison with the no-treatment or the current-treatment alternatives, will result in improvements to such factors as resistance to injury (threshold, visual incapacitation, incapacitation, death) and time to return to duty. In the current study, we looked

Table 2
Protection Options

Option	Equipment
None	Battle dress uniform (BDU) with normal load-bearing equipment (LBE) and individual weapon
MOPP IV	BDU and all MOPP gear, with LBE and weapon; MOPP donned at time of attack

at the relative effectiveness of the five medical alternatives in terms of unit effectiveness over time for both of the physical protection options.

Modeling Tools

MCAS. MCAS is a chemical agent dose and effects simulation. MCAS develops the vapor and liquid agent challenges on the target, as a function of time, which would result from an attack scenario, given specified meteorologic and terrain conditions. The agent challenges described previously were developed with MCAS. MCAS permits varying the response of the troop population in response to the challenge, based on the protection available and the effectiveness of treatment. Thus, numbers of casualties resulting from the attack scenarios and numbers of soldiers who return to duty after medical intervention can be evaluated with various treatment alternatives. Protective ratios derived for lethal effects under each treatment alternative were developed in coordination with the U.S. Army Medical Research Institute for Chemical Defense and are contained in the full report.[1] Note that anticonvulsant with a treatment regimen, in comparison to the equivalent treatment without anticonvulsant, does not change the protective ratio; rather, it hastens recovery and return-to-duty.

AURA. The principal tool employed to evaluate unit effectiveness with various medical product alternatives is the AURA model.[7] AURA is a one-sided, company/ battalion-level methodology that uses inputs from a variety of accepted models and other sources to assess the effects of many types of attack or conditions on unit performance. AURA describes the unit physically and functionally and may be used to determine the extent to which it can perform its prescribed functions (missions), given the availability of physical assets (persons and equipment). AURA describes units in terms of the combinations of persons and equipment required to perform mission functions.

Table 3
Units Evaluated

Unit	Subunit*
Headquarters unit	Headquarters and headquarters battery, field artillery battalion, heavy division (TOE 06-366J)
Artillery unit	155-mm self-propelled battery, M109 (TOE 06-367J)
Attack helicopter unit	Attack helicopter company, AH-1 Cobra (TOE 17-187J)
Infantry anti-armor	Infantry anti-armor company (TOE Unit 07-248J)
Ammunition supply	Combat service support company (TOE Point 09-64)

*From U.S. Army Table of Organization and Equipment (TOE).

AURA also considers whether alternate persons or equipment would be available to perform mission functions in the event that specific equipment or persons were lost, the effectiveness of those alternatives in relation to the principal means, and similar types of issues. AURA can accept input from other analytic models to establish personnel and equipment losses and, in turn, to determine unit effectiveness. The units examined during this study are listed in Table 3.

AURA was used to develop unit effectiveness for these units, based on their AURA descriptions, degradation due to MOPP, and chemical casualties developed by MCAS. No conventional casualties or other causes of degradation to unit effectiveness were played. MCAS outputs the percentage of the target population in four casualty categories for input to AURA: threshold effects, vision impairment, incapacitation, and death. MCAS also provides the number of days from the attack until incapacitated casualties return to duty. AURA inputs from MCAS reflect the effects of medical protective or therapy drugs that reduce casualties or hasten return to duty. Therapy side effects are also represented, such as the visual effects of atropine. No replacements were played in the AURA simulations. AURA results are sensitive to the ability of a unit to use alternate unit members to accomplish critical tasks. The units studied in these experiments showed a range of sensitivity to personnel losses, with the Ammunition Supply Point showing the least sensitivity and the Headquarters unit showing the most sensitivity to casualties.

Results and analysis. We examined unit effectiveness after Soman attack by two weapon systems, against five types of units, in two levels of protection, applying five nerve agent treatment alternates, under three sets of climatic conditions. Altogether, 300 sets of conditions were represented. For

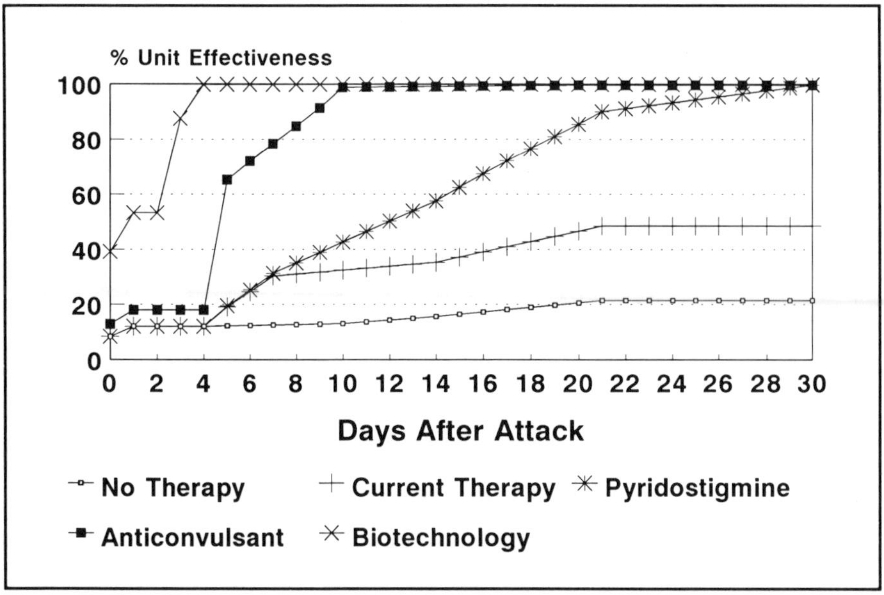

Figure 4. Unit effectiveness after Soman artillery attack, artillery unit, MOPP IV, SWA summer day.

the sake of this abbreviated report, we will examine only results achieved after an artillery attack against the artillery unit and the attack helicopter unit, under Southwest Asia day conditions. We selected these units because, from among units studied, they fall roughly midway in sensitivity to loss of specific skills. Southwest Asia day conditions were chosen because the region is of particular interest at this writing. We used the artillery attack because the results appear to be driven by a uniform distribution of agent over the target area.

We will emphasize a comparison of the current doctrine for treating nerve agent casualties in the field (atropine and 2-PAM), with the enhanced regimen (adding pyridostigmine pretreatment), the near-term improvement (adding pyridostigmine and anticonvulsant therapy), and the far-term treatment (nerve agent antibody). A key point of the analysis is the time to recover combat effectiveness after the attack. For purposes of this analysis, we consider units to be combat-effective if they are ≥80 percent effective.

Figure 4 shows the results when an Artillery unit in MOPP IV suffers the hypothetic artillery attack. The figure plots unit effectiveness as a function of time for all five treatments. With the current therapy, the unit never regains significant capability. Pyridostigmine pretreatment enables the unit to regain significant effectiveness after 17 days. Anticonvulsant therapy has a clearly beneficial effect, restoring the unit to combat effectiveness after seven days. Unit capability improves by more than 50 percent after Day 4, as unit members who do not experience convulsions return to duty. Pretreatment with

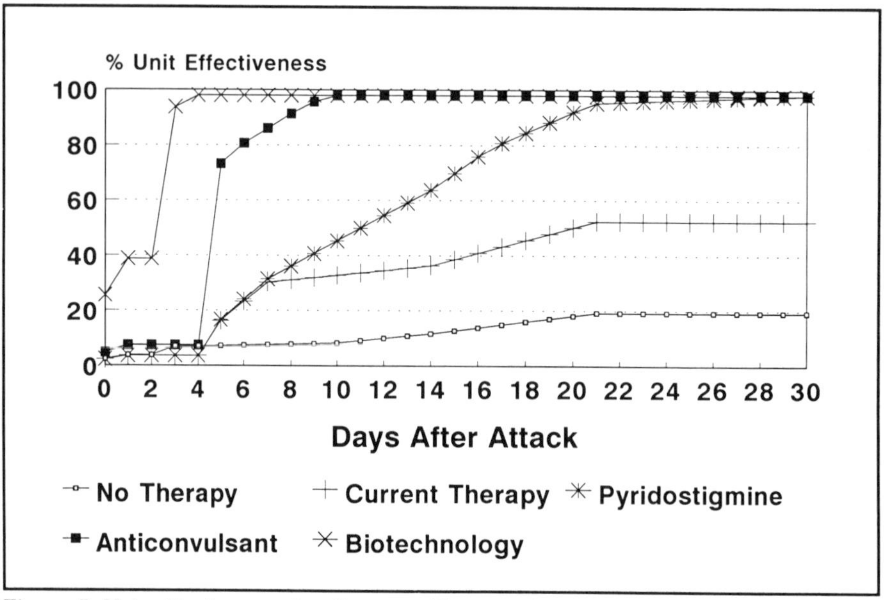

Figure 5. Unit effectiveness after Soman artillery attack, attack helicopter unit, MOPP IV, SWA summer day.

nerve agent antibodies rapidly improves capability after Day 2, as visually impaired personnel recover; at Day 3, the unit reaches 80 percent effectiveness; and after Day 1, at least 50 percent effectiveness is shown.

Figure 5 shows the results of an artillery attack against the attack helicopter unit. The helicopter unit shows about a 10 percent additional reduction in initial postattack effectiveness, which might be due to this unit's higher-than-average sensitivity to vision effects. Current therapy fails to restore the attack helicopter unit. With pyridostigmine pretreatment, the unit is restored to near-full-combat effectiveness. However, the unit remains less than 80 percent effective for 18 days. Anticonvulsive therapy restores the unit significantly by the fifth day. Nerve agent antibodies notably enhance the performance of the unit. Initial loss of effectiveness (25 to 35 percent) is less than with other therapies. The unit is restored to greater than 80 percent effectiveness in three days or less and to more than 90 percent effectiveness in less than four days.

Conclusions. Current therapy is only marginally effective under the attack conditions studied and fails to return to duty the number of casualties required to restore combat effectiveness. Current therapy augmented with pyridostigmine pretreatment will eventually restore combat effectiveness. Without personnel replacements, however, unit recovery will take several weeks. This latter result can probably be accounted for as follows: although atropine plus 2-PAM with pyridostigmine is effective in restoring AChE levels, other effects of nerve agents extend the recovery period of some casualties. Even so, over a 30-day period, this therapy is effective in ultimately restoring combat effectiveness. Current therapy augmented with pyridostigmine pretreatment and anticonvulsive therapy significantly improves time of recovery of units to combat effectiveness. Only one-third to one-half the time required with current therapy enhanced with pyridostigmine pretreatment is required for recovery. We can reasonably expect that use of nerve agent antibodies will further shorten unit recovery times. Other comparisons in the study not detailed here indicated that no therapy option is sufficiently effective to permit troops to operate in a high chemical threat environment without physical protection. No unit studied under MOPP 0 conditions was restored to combat effectiveness during the study period, regardless of medical intervention used.

Evaluation of Sleep Discipline to Sustain Unit Productivity

"Continuous operations," as defined in U.S. Army Field Manual 22-9, is continuous land combat with some opportunity for sleep, although this sleep may be brief or fragmented.[8] The successful development and application of a doctrine for managing sleep and alertness in continuous operations is essential to success on the battlefield. The potential technologic advantages of night vision and electro-optic devices for all weather operations increase the potential for continuous operations. With limited combat personnel available for future contingency operations, the U.S. Army cannot expect to conduct continuous operations by rotating personnel in shifts or by fighting soldiers and units to exhaustion and replacing them with fresh personnel.[9]

To be successful, the combat unit must maintain an efficient balance of sleep and work in order to sustain high levels of productivity in continuous operations. Ideally, every individual would receive adequate sleep to maintain effective performance on a daily basis; this performance must be maintained in all activities during the day. Nominally, the amount of sleep necessary to do this depends on the individual and the tasks to be performed. Continuous operations are routinely conducted by U.S. Navy, U.S. Air Force, and U.S. Army aviation units with six to eight hours of sleep in every 24-hour period. It is difficult and potentially dangerous to conduct unit level exercises to evaluate the long-term impact of the continuous operations with different sleep disciplines. Laboratory measurements of effectiveness on experimental tasks do not readily translate into performance of tasks necessary to maintain operational effectiveness. As a step to examine the implication of different sleep disciplines, an effort was undertaken to define, develop, test, and evaluate a methodology to incorporate sleep into the AURA model. This effort was completed in October 1989.[10] Since that time, other improvements have been incorporated in the sleep modeling process, and results are summarized here.

Objective. The objective of this study was to determine the impact of alternative sleep regimens on two Army units, one performing a mixture of physical and cognitive tasks, the other performing primarily cognitive tasks. The units selected were the M109 artillery battery and the headquaters unit (Table 3). The greatest impact of sleep deprivation is evident in soldiers performing cognitive tasks as opposed to those performing strictly labor-intensive tasks. The results from the headquaters unit, which arranges primarily cognitive tasks, served as a "worst case" view of the effects of sleep deprivation.

Approach. The AURA methodology was used to measure the impact of sleep deprivation on each unit's performance. The AURA model determines, for a given unit under a given set of conditions, unit effectiveness in performing a prespecified mission. This effectiveness value was used as a measure of performance for the current study and a basis of comparison for the cases studied.

Sleep methodology. The sleep methodology within AURA has been documented in an earlier report.[10] A discussion of this methodology and the modifications made to the methodology during the course of this study will be included in this report.

In the sleep methodology used for the previously referenced studies, there were six key variables to identify the sleep cycle within AURA.[11] These variables included the definition of the unit of effective sleep, the maximum allowable accumulation of these effective sleep units, the level of sleep units that are equivalent to 100 percent effective performance, and the use rate of effective sleep units during work activities. The final variable used in the methodology was the size of the reservoir of effective sleep units that can be saved for use at a later time. Three additional variables were developed for use during this analysis. These new variables included a parameter to specify a required work period.This variable allowed us to control sleep periods more easily and directly than did the previous methods of adjusting the minimum sleep units needed to remain awake. Two other variables were added to the AURA model to allow the representation of individual variations in required sleep. These will be discussed in detail in the section on the AURA methodology.

Sleep accumulation function. The basic sleep accumulation function is constructed in four segments (see Figure 6). For the first 10 minutes, no

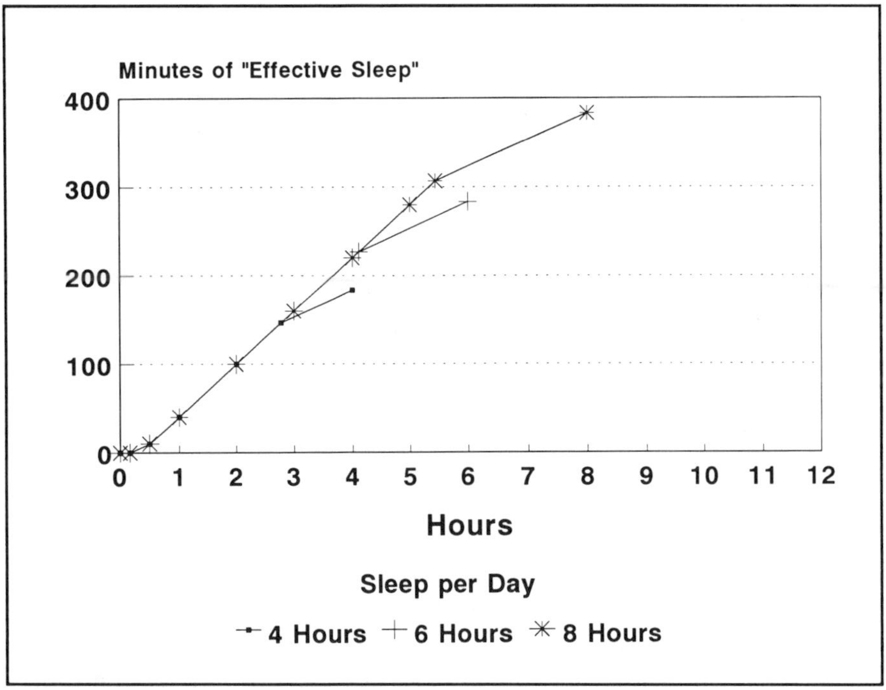

Figure 6. Sleep accumulation function.

"minutes of effective sleep" are generated; for minutes 10 through 30, a person is given half-credit for the amount of time asleep (that is, 10 minutes). So a person at 30 minutes into their sleep period will have accumulated 10 "minutes of effective sleep," reflecting the fact that individuals do not start deriving full benefit from sleep until 30 minutes after they have fallen asleep. From 30 minutes until the point where they have made up 80 percent of the difference between where they started at the beginning of the sleep period and maximum extent of their reservoir (that level defining full restedness), a person is able to accumulate effective sleep on a one-to-one basis one minute of effective sleep for each minute asleep. After replenishing 80 percent of the deficit, a person only accumulates half of a minute of effective sleep for every minute the person sleeps. The benefit to be derived from any period of rest is calculated by solving a set of simultaneous equations described in the main report. For example, if a person is in balance working 18 hours and sleeping six hours, during that six-hour sleep period he would accumulate 283⅓ effective minutes of sleep and be at maximum sleep accumulation at the end of the sleep period. If a person was very tired, having worked more than 18 hours, then a six-hour sleep period could potentially yield 340 "minutes of effective sleep" in a six-hour period because the reservoir remained below the 80 percent threshold of diminishing returns. Finally, once a person sleeps sufficiently long to reach the maximum of their reservoir, further sleep accrues no further benefit to the sleep accumulation function.

Work degradation function. The rate at which sleep units are used is based on the observations of degradations in performance in individuals deprived of sleep over extended periods. The Department of Behavioral Biology of the Walter Reed Army Institute of Research has observed that the maximum time that a person can work without effective rest is four days. Measures of a variety of mental (cognitive) tasks indicate that the rate of correct performance declines an average of 25 percent per day.

It has also been observed that a person in their normal sleep cycle (where they are in balance) will be able to retain 100 percent performance in their every day of work in that environment. A person who has a nominal sleep requirement of six hours of sleep per 18 hours of work will still perform at 100 percent of their capability at the end of the 18-hour work period. With these limits in mind, assuming work is to continue, we modeled a linear drop in the effectiveness at which they were able to perform their job. For this analysis, all jobs in the headquarters unit were considered to be cognitive and, therefore, subject to a performance degradation of 25 percent per day without effective rest. The artillery unit was represented by a range of degradation functions corresponding to the mix of both physical and cognitive tasks; maximum degradation for any task was 25 percent per day.

Sleep requirements distribution. To model individual variations in sleep requirements, the differences in sleep balance points for individuals were assigned according to an approximate normal distribution of requirements. The distribution of sleep requirements was categorized into four classes:

1. Ten percent of the unit was assumed to require 7.5 hours of sleep.
2. Forty percent of the unit required 6.5 hours of sleep.
3. Another 40 percent of the unit required 5.5 hours of sleep.
4. 10 percent of the unit required 4.5 hours of sleep.

These factors resulted in an approximation to the normal distribution, with a mean of six hours and a standard deviation of 45 minutes. The differences in sleep balance point required an adjustment to the size of the reservoir for individuals performing different jobs. The absolute rate at which "minutes of effective sleep" were used was individualized to reflect both the rate of use imposed by the job and the adjusted size of the reservoir for the individual.

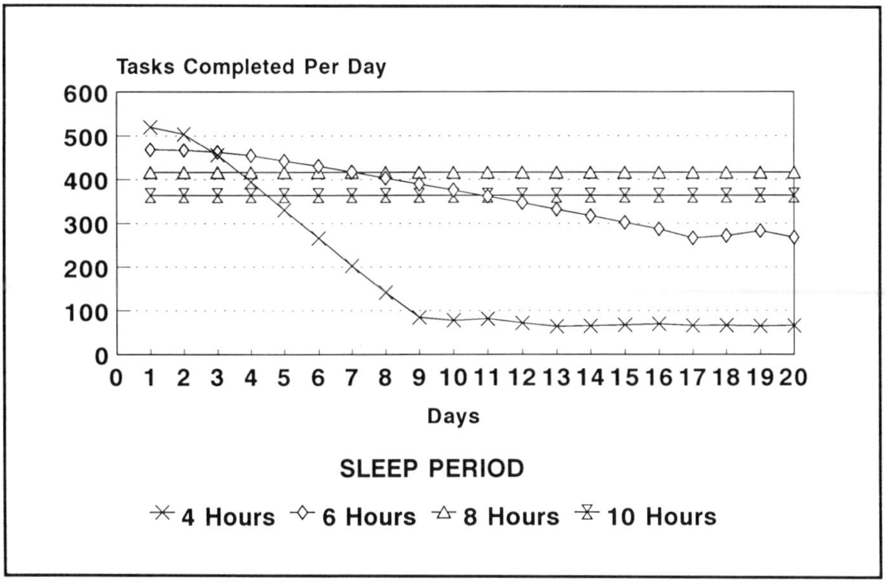

Figure 7. Tasks completed, Headquarters unit.

Sleep disciplines considered. We evaluated a series of sleep regimens or disciplines that represent different doctrinal prescriptions of sleep per day. For the Headquaters unit we examined four, six, eight, and 10 hours of sleep per day; for the artillery unit we tested regimens of four, five, six, seven, eight and nine hours per day.

Unit effectiveness and productivity. The prime output of AURA for this study was unit effectiveness at 4-hour intervals across a 20-day continuous operations scenario. When the unit was at work, AURA estimated the effectiveness of the unit given the current level of fatigue computed, as described previously. When the unit was taking sleep according to the prescribed discipline, AURA portrayed the unit at zero effectiveness. For any given sleep regimen, AURA depicted the effectiveness of the unit when awake and the duration of zero effectiveness when the unit was asleep. Each unit under each regimen showed a daily cycle of work and sleep for 20 days. We calculated for each day an average hourly productivity that reflected both the quality of performance when awake (degraded perhaps by fatigue) and the duration of scheduled sleep. Note that increases in the length of sleep imposed a trade-off; the increased sleep improved performance when awake but lengthened the period of zero productivity when asleep. By comparing average productivity across the range of sleep regimens, AURA provided an estimate of the optimal sleep discipline that maximized productivity.

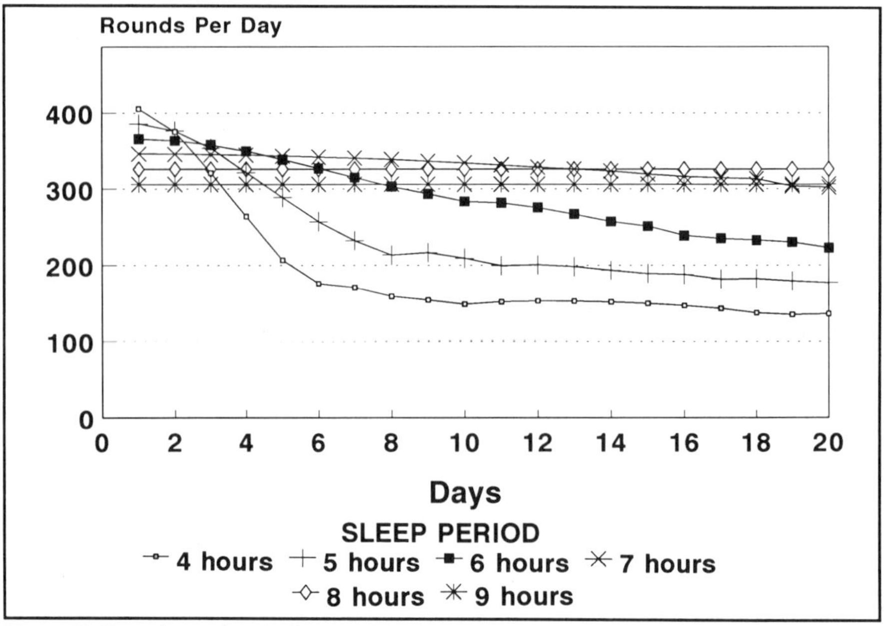

Figure 8. Rounds per day, Artillery unit.

Results. Figures 7 and 8 show the impact of sleep deprivation for a 20-day period. The headquaters unit (Figure 7) is required to perform 26 fire support tasks per hour. Thus, the graph illustrates the number of fire support tasks completed per day for each of the different sleep regimens. In all cases, the effects of sleep deprivation clearly drive the unit's ability to complete its mission over the 20-day period. Of the disciplines studied, eight hours of sleep produced the greatest sustained productivity. Six hours of sleep produced greatest productivity for scenarios lasting less than 10 days. Four hours of sleep sustained performance for only three to four days.

The performance of the artillery unit (Figure 8) was less affected by marginal levels of sleep (for example, four hours) compared with the Headquaters unit, primarily because of the generally less cognitive nature of the tasks and the unit's ability to perform its mission with several alternative work strategies. Nevertheless, performance became unacceptable after four days of continuous operations. Seven hours of sleep produced optimal performance for two weeks, whereas eight hours of sleep sustained the unit indefinitely. For both units studied, sleep regimens greater than eight hours were actually detrimental to overall unit productivity by scheduling more sleep than necessary.

Figure 9 illustrates the impact of various sleep disciplines on the artillery

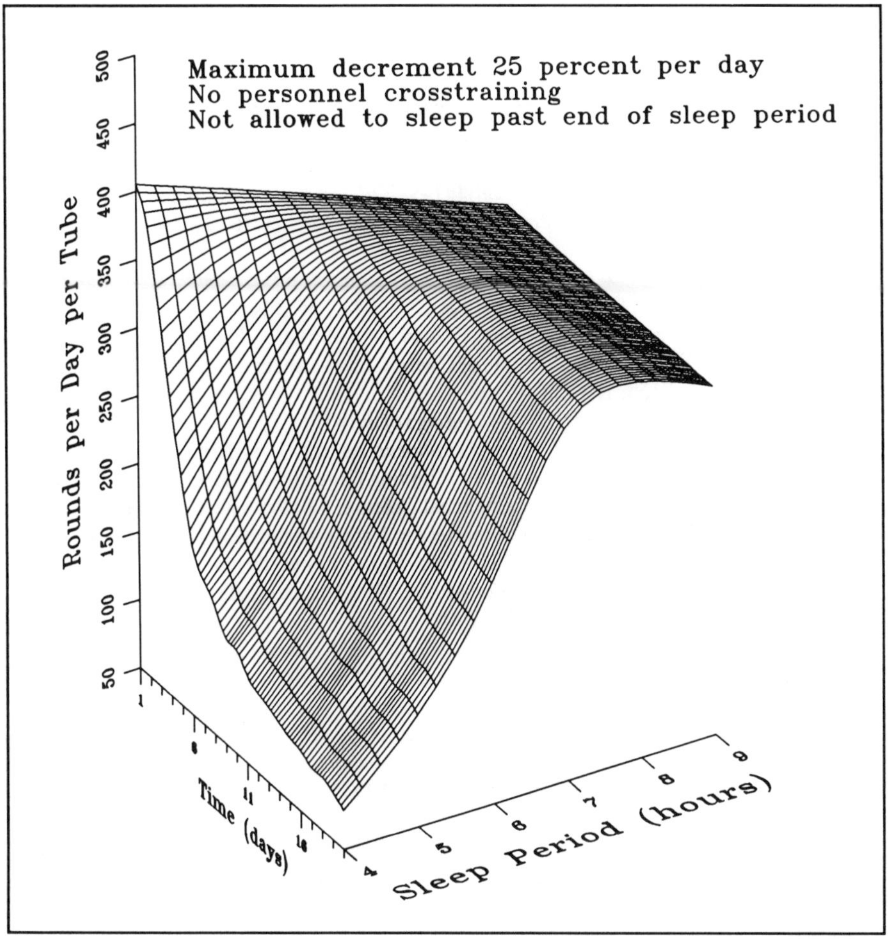

Figure 9. Productivity of artillery unit.

unit. These results show that the four-, five- and six-hour sleep regimens are not sufficient to prevent sleep deprivation effects on unit performance. Although the nine-hour sleep regimen provides optimal performance during work periods, a seven- to eight-hour regimen provides the greatest overall productivity over the 20-day period.

Conclusions. Seven or eight hours of sleep provide sustained performance levels for long-term operations. Six hours of sleep provide high performance for up to 10 days, but this is subject to accumulated degradation in the level of performance over time. Four hours of sleep provide highest levels of performance for short bursts of activity, but after a few days the unit will lose combat effectiveness and require a respite to recover.

References

1. J.P. Bulger *et al., New Medical Therapy Options for Nerve Agent Defense* (U.S. Army Medical Research and Development Command Task Order Report, Contract No. DAAL02-90-C- 0071, T.O.# 0001AE, Science Applications International Corporation, Joppa, MD, 1991).
2. R.E. McNally and D.L. Bareis, *Battlefield Levels of Soman* (U.S. Army Medical Institute of Chemical Defense, Aberdeen Proving Ground, MD, Contract No. DACMD17-88-C-8155, 1989) p. 3.
3. *U.S. Army Field Manual No. FM 3-9 Military Chemistry and Chemical Compounds* (U.S. Government Printing Office, Washington, DC, 1975), pp. 3-9.
4. D.L. Bareis, *Nerve Agent Pretreatment Transition Information Paper: Physostigmine Plus Adjunct* (U.S. Army Medical Institute of Chemical Defense, Ft. Detrick, MD, Contract No. DAMD17-88-C-8155, 1989).
5. G.M. Kiebzak and A.J. Feister, in *Proceedings of the Workshop on Convulsions and Related Brain Damage Induced by Organophosphorus Agents* (U.S. Army Medical Research Institute of Chemical Defense, Aberdeen Proving Ground, MD, 1990), p. 12, 179.
6. Nerve Agent Scavengers: Medical Chemical Defense, *A Bulletin on Research and Development, Volume 3(1)* (U.S. Army Medical Research and Development Command, 1989), p 2.
7. J.T. Klopcic, R.M. Sheroke, and S.K. Price, *Input Manual for the Army Unit Resiliency Analysis (AURA) Computer Simulation Model*, 1990 Update, (Ballistic Research Laboratory Technical Report BRL-TR-3187, 1990).
8. *U.S. Army Field Manual No. FM 22-9, Soldier Performance in Continuous Operations* (U.S. Government Printing Office, Washington, DC, 1983)
9. WRAIR Technical Report No. BB-87-1.
10. R.E. McNally, A.M. Machovec, and D.T. Ellzy, *Evaluation of Sleep Discipline in Sustaining Unit Performance* (U.S. Army Medical Research and Development Command Task Order Report, Contract No. MDA903-88- D-1000, Science Applications International Corporation, Joppa, MD, 1989).
11. J.T. Klopcic, *The AURA Fatigue and Heat Stress Algorithms* (Technical Report BRL-MR-3802, Ballistic Research Laboratory, 1989).

22

Optical Neural Network Device Architecture

N. Gupta
J. Pham
G. J. Simonis
R. P. Leavitt
M. R. Stead
U.S. Army Research Laboratory
Optical Processing Branch
2800 Powder Mill Road
Adelphi, MD 20783-1197

There is a critical need for developing novel processing systems to perform aided target recognition based on the neural processing architecture employed by the human brain. The emerging field of artificial neural networks (NNs) is based on using many relatively simple but highly interconnected nonlinear processing elements that perform summing and thresholding on weighted signals. The problem addressed here is to find a compact monolithic optoelectronic implementation of a feed-forward NN in GaAs.

We have implemented an optical waveguide-based neural architecture using molecular-beam-epitaxy-grown GaAs material. The weighting and thresholding functions are based upon the electroabsorption property of a semiconductor superlattice embedded in the waveguide core. The device emulates the basic neural decision-making process. It accepts multiple inputs, applies different adjustable weights to them, sums these weighted inputs, and uses an optoelectronic thresholding switch to make a binary decision. So far, the architecture has been successfully implemented to design and test a two-input/one-output neuron and is being expanded to include more complex single neurons and an NN. During the course of this implementation, we have designed and tested a very-high-dynamic-range (250dB/cm) optoelectronic modulator used for weighting and a highly nonlinear optoelectronic (25dB) switch. A very high on/off ratio (2300) is obtained for the neuron device.

Introduction

The problem of pattern recognition is of utmost importance to the U.S. Army for aided target recognition (ATR). Aside from military applications, pattern recognition is used for many civilian applications, such as speech recognition, machine recognition of handwriting, robotic vision, communication systems, automated control, design and manufacturing, radar data analysis, and so on.

The human brain is the most outstanding system for performing pattern recognition. It can recognize a pattern even with partial or incomplete information and in relatively ill-defined situations. On the other hand, a digital computer, which is very fast in number manipulation and other serial operations, is very inefficient in pattern-recognition tasks. It would require an enormous memory, a very large knowledge base, and a rule-based expert system to do relatively simple ATR tasks using traditional methods on electronic computers. The emerging field of artificial neural networks (NN) technology is based on the present knowledge of how a biologic brain functions as a parallel processor using a large number of relatively simple but slow (millisecond) processing elements (neurons), working in concert with each other with a massive number of interconnections.[1-3] Such an NN can be used to perform processing at several levels in ATR problems—from input data preprocessing (tasks such as data recovery, noise removal, signal-to-noise-ratio (SNR) enhancement, and signal and/or image segmentation) through final target identification. Much work is going on in the software and hardware implementations of NNs. Both electronic and optical approaches are being used for hardware implementations.

Most of the optical implementations have been based on use of a vector-matrix multiplication concept to simulate neural functioning.[4-6] Most of these use a free-space architecture and require many optical components that cannot be integrated monolithically. We have implemented an optical neural architecture in GaAs that is very compact and monolithic and is inspired by the natural fiberlike structures common in all biologic neurons. It is based upon a newly emerging basic technology of semiconductor quantum wells and superlattices, which are being investigated for developing novel optical modulators and switches.[7-8] The device emulates the basic neural decision-making process. It accepts multiple inputs, applies different adjustable weights to them, sums these weighted inputs, and uses an optoelectronic thresholding switch to make a binary decision. So far, the architecture has been successfully implemented to design and test a two-input/one-output (2-to-1) neuron and is being expanded to include more complex single neurons and an NN. During the course of this implementation, a very-high-dynamic-range (250dB/cm) optoelectronic modulator and a highly nonlinear optoelectronic (25dB) thresholding device have been designed and tested. A very high on/off ratio (2300) that is mostly input-limited is obtained for the neuron device. The research described in this paper was carried out starting from the basic physics research to a device design application in our laboratories.

Requirements of Neural Networks

The success of biologic systems appears to be due to their complex and massively parallel architectures. A three-pound human brain consists of over 10 billion neuron cells communicating among themselves through networks of over 100 trillion synaptic interconnections. The working principles of neural functions are essentially the same from a squid's brain to a human brain. The main difference is in the number of cells and interconnections.

Each neuron has a relatively simple structure and functioning method. A typical neuron consists of a cell body, ranging from about 5 to 100μm in diameter, from which emanates one major fiber, the axon, and a number of fibrous branches, the dendrites. The axon carries the outgoing signal from the neuron, and near its end it usually branches out extensively. The dendrites receive the input signals from other neurons, where they form a contact, called a synapse. A weight is applied to each signal, and the cell body sums these weighted signals and performs a simple thresholding operation based on this sum to send a signal on its axon or not. In general, one neuron receives input from thousands of other neurons through its dendrites and similarly feeds its output to thousands of other neurons through its branching axon. The time scale involved is relatively slow (~milliseconds). Figure 1(a) shows typical biologic neurons, and Figure 1(b) shows schematics of a neuron's operational principle. The basic neural operation may be summarized as follows:

1. A single neuron has multiple inputs, multiple interconnections, and a single output.
2. Each input has an adjustable synaptic weight associated with it that may have either positive or negative value.
3. The neuron must sum weighted inputs and exhibit a nonlinear transfer function in issuing an output signal.

An artificial NN consists of a massively interconnected network of such processing elements. For mathematical simplification, the network is divided

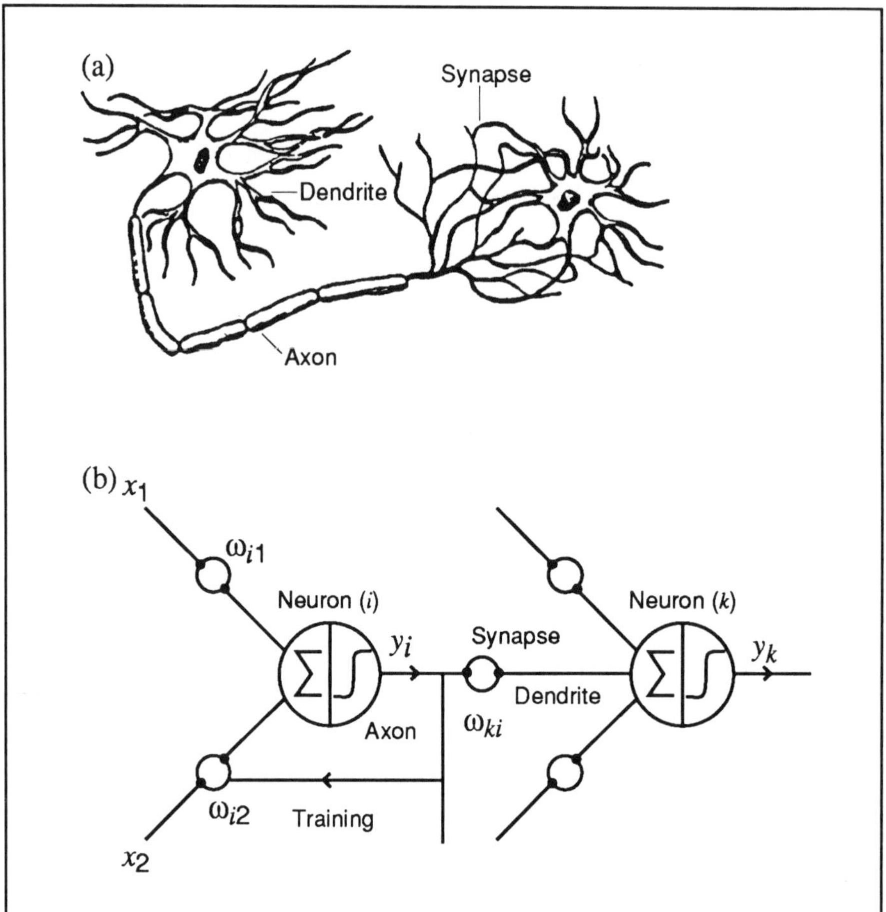

Figure 1. Neuron systems: (a) two biologic neurons, and (b) schematic of neural operational function showing inputs (x_j), weights (ω_{ij}), summing, thresholding, and output (y_i) for neuron (i).

into a number of layers; each layer is made of a number of neurons or nodes. Every neuron in each of these layers is connected to every neuron in the next layer for a network that is fully connected. The number of connections could be less, depending upon the application. Every NN has an input and an output neural layer. There can be some additional layers in between these two, known as the hidden layers.

In general, an NN is very robust: its performance does not degrade even if some of its interconnections are not being used. An NN model is specified by the net topology, node characteristics, and training or learning rules. These rules specify an initial set of weights and indicate how these weights should be adapted during use to improve performance. Such training can be supervised or unsupervised.

The operating principle of an NN can be mathematically represented by a nonlinear function, f, as follows[1]:

$$y_i = f\left(\sum_{j=1}^{N} \omega_{ij} x_j - \alpha_i\right), \quad (1)$$

where x_j are N inputs, y_i is the output of ith neuron, ω_{ij} are the weights applied by the ith neuron to the input from the jth neuron, and α_i is a threshold term. The nonlinear function used in NN modeling is usually a hard limiter such as a sigmoid function.

Much work is going on in the software and hardware implementation of NNs. Both electronic and optical approaches are being used. Optical implementations have a very promising future because of the inherent parallelism of light. Much of this work uses a free-space architecture to implement the vector-matrix multiplication concept embedded in Equation 1.[4-6] Most of these architectures contain at least three planes: the first one, corresponding to the input neurons, contains an array of discrete sources of light [either one- or two-dimensional (1-D or 2-D)], such as spatial light modulators (SLMs), light-emitting diodes (LEDs), and so on; the second contains a 2-D array of interconnection elements representing the weight matrix, that is, an SLM (either fixed or programmable) or a hologram (2-D or volume); and the third, corresponding to the output neurons, contains an array of discrete photodetectors for summing the weighted inputs with a built-in quadratic nonlinearity (sometimes an additional thresholding is provided at this stage using electronics or some other scheme). Some approaches also include hidden layers. Most of these implementations are of the tabletop type where, due to size, mechanical stability is a problem. Recently, this kind of architecture has been implemented in GaAs.[6]

In 1989, the Defense Advanced Research Projects Agency (DARPA) conducted a study on the state of the art in NN research and its future direction.[3] It concluded that a marriage of optics and electronics could be the best way to actually implement the NN architectures and that the most promising technology for the implementation of such optoelectronic NNs is based on GaAs and other III-V semiconductor alloys and superlattices. In this paper we present an optical implementation of a feed-forward NN architecture that is being carried out in a monolithic GaAs/AlGaAs waveguide structure, where both the weighting and thresholding are provided by the room-temperature macroscopic nonlinear behavior of an embedded superlattice by application of small electric voltages.

Optoelectronic Implementations of NNs

In a hardware implementation of an NN, the most important functional components are discrete sources of light; interconnections; and weighting, summing, thresholding, and output signals. GaAs-based integrated optics offers components that can perform these functions: laser diodes, low-loss single-mode rib waveguides, power combiners/splitters, modulators, thresholding devices, detectors, and others. A waveguide structure is similar to an optical fiber. The propagating light mode is confined in a very thin region, the core, which is surrounded by layers of material with lower refractive indices known as cladding.

In our waveguide-based architecture of an NN, we have first implemented a single neuron. The sources of light are discrete laser diodes. The light is coupled into an N-to-1 rib waveguide power combiner structure. The simplest of such structures is a 2-to-1 power combiner, which can be a small-angle Y-junction. The device architecture contains at least three electrically active components (two modulators and a thresholding element), which are electrically isolated. Ideally, the waveguide parameters should be chosen so that the light propagating through the rib waveguide is single-mode in both transverse and lateral directions. The transverse mode confinement is due to the waveguide core and the lateral confinement is provided by the rib walls. Inclusion of a superlattice in the waveguide core gives rise to electroabsorption. As each of the input signals passes through the device, an electric field is applied to change the transmission of the core material, and the light is modulated (a synaptic weight is applied). The output guide of the power combiner sums the weighted signals coming from various input guides. It also contains a nonlinear switch in which the waveguide section acts both as a detector and a modulator.[7,8] The nonlinear switch output depends on the amount of light passing through it; for a higher intensity of light, its transmission is very high, whereas for a low intensity of light, it is very low.

This power combiner is the basic functional building block of our neural architecture. A more complex structure can be made by cascading many such building blocks. This implementation depends upon designing several integrated optical components: low-loss single-mode rib waveguides, Y-junctions, crossing waveguides with various angles of crossing, modulators, and

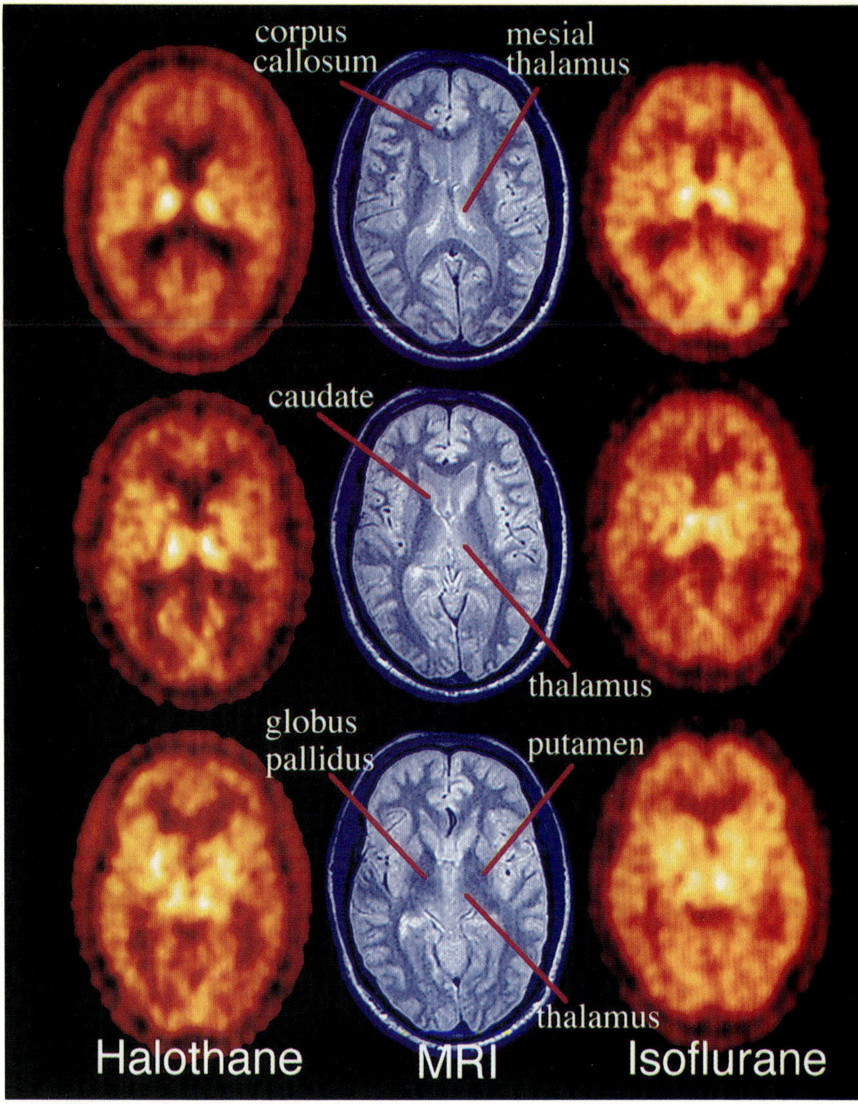

Color Plate 1. Kornguth, Chapter 1, Figure 2, p. 44.

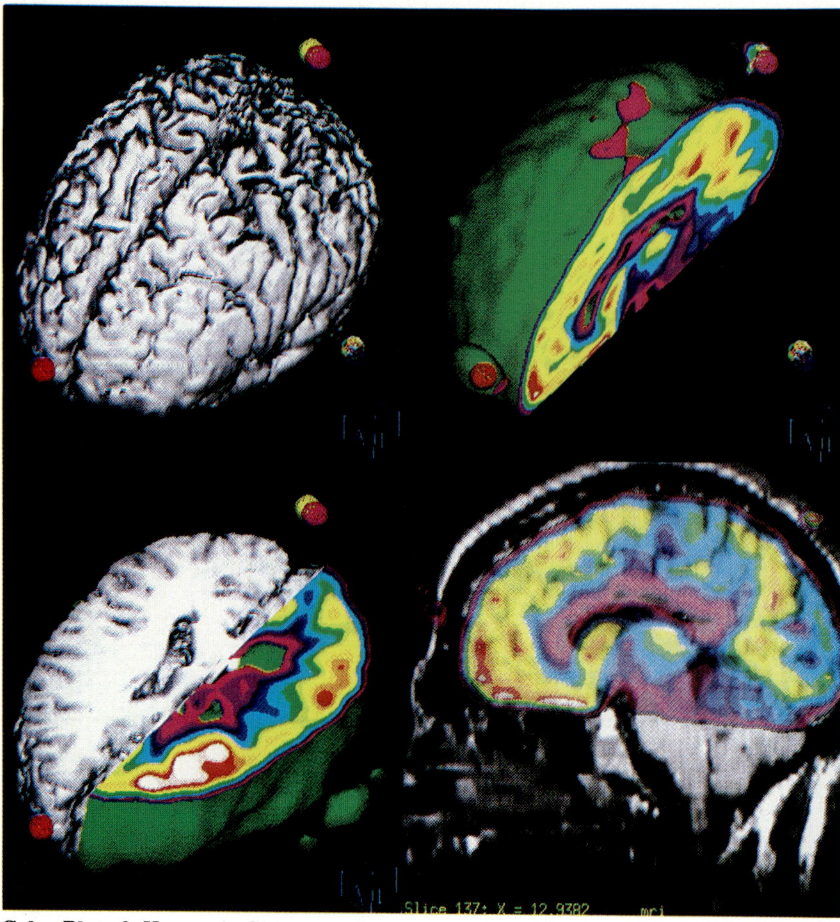

Color Plate 2. Kornguth, Chapter 1, Figure 3, p. 44.

Color Plate 3. Bishop *et al.*, Chapter 40, Figure 1, p. 646.

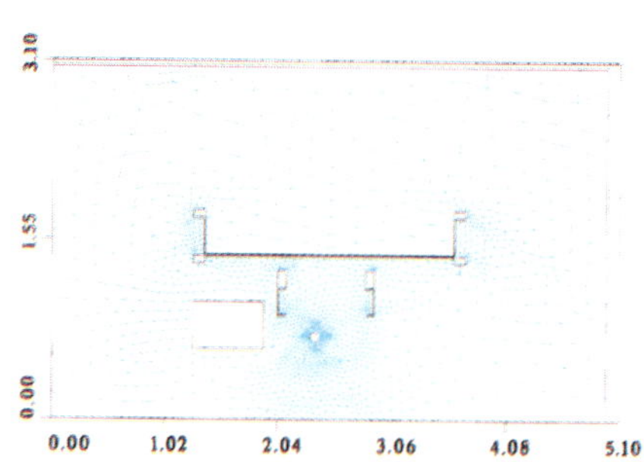

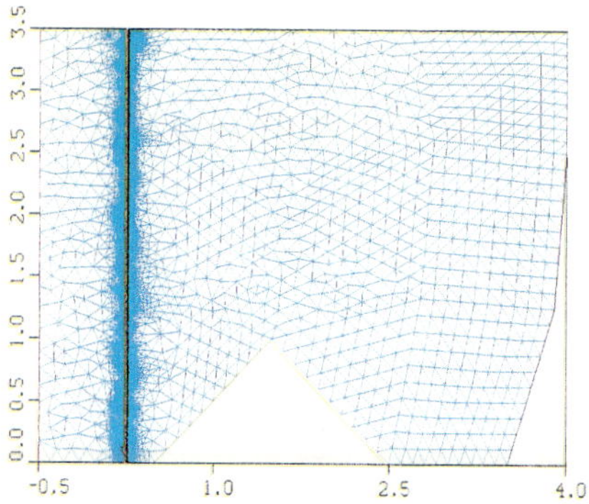

Color Plate 4. Bishop *et al.*, Chapter 40, Figure 3, p. 648.

Range

Pink:	3.8839 - 3.5495
Red:	3.4659 - 3.1316
Yellow:	3.0480 - 2.7136
Green:	2.6300 - 2.2956
Blue:	2.2121 - 1.8777
Purple:	1.7941 - 1.3761
Black:	1.3761 - 1.000

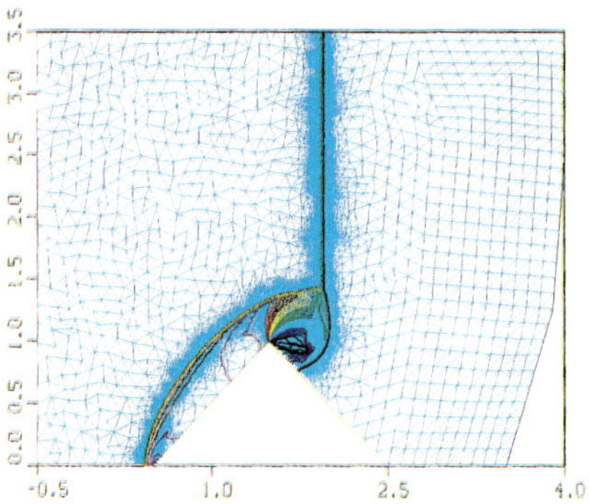

Color Plate 5. Bishop *et al.*, Chapter 40, Figure 4, p. 649.

Range

Pink:	7.4867 - 6.6298
Red:	6.4156 - 5.5586
Yellow:	5.3444 - 4.4875
Green:	4.2733 - 3.4164
Blue:	3.2021 - 2.3452
Purple:	2.1310 - 1.2741
Black:	1.0599 - 0.2030

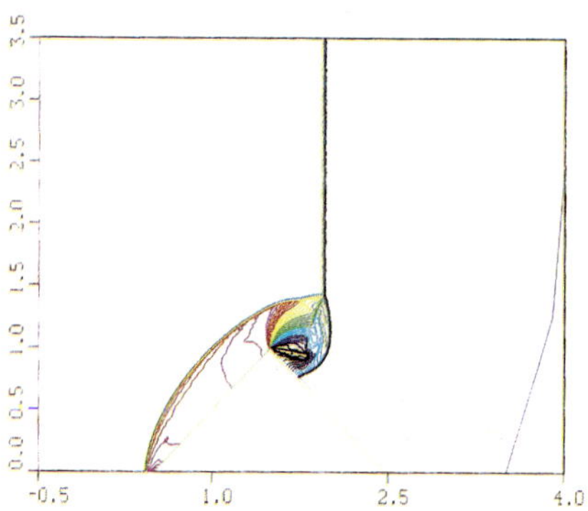

Color Plate 6. Bishop *et al.*, Chapter 40, Figure 5, p. 650.

Range

Pink:	7.4867 - 6.6298
Red:	6.4156 - 5.5586
Yellow:	5.3444 - 4.4875
Green:	4.2733 - 3.4164
Blue:	3.2021 - 2.3452
Purple:	2.1310 - 1.2741
Black:	1.0599 - 1.2030

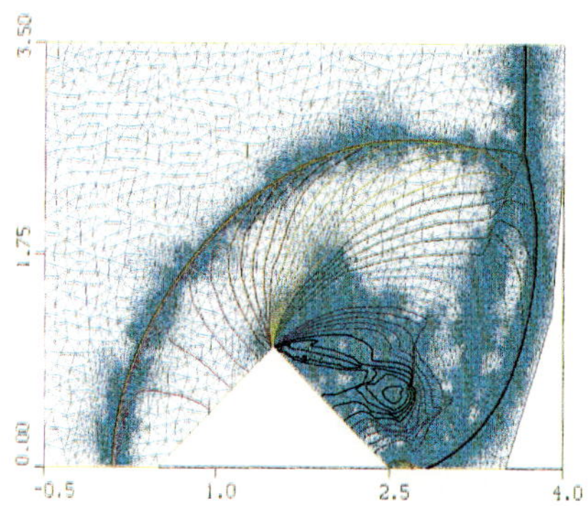

Color Plate 7. Bishop *et al.*, Chapter 40, Figure 6, p. 651.

Range

Pink:	7.5143 - 6.6526
Red:	6.4372 - 5.5755
Yellow:	5.3601 - 3.4994
Green:	4.2829 - 3.4212
Blue:	3.2058 - 2.3441
Purple:	2.1287 - 1.2670
Black:	1.0516 - 0.0822

Color Plate 8. Bishop *et al.*, Chapter 40, Figure 7, p. 652.

Range

Pink:	7.5143 - 6.6526
Red:	6.4372 - 5.5755
Yellow:	5.3601 - 4.4984
Green:	4.2829 - 3.4212
Blue:	3.2058 - 2.3441
Purple:	2.1287 - 1.2670
Black:	1.0516 - 0.1899

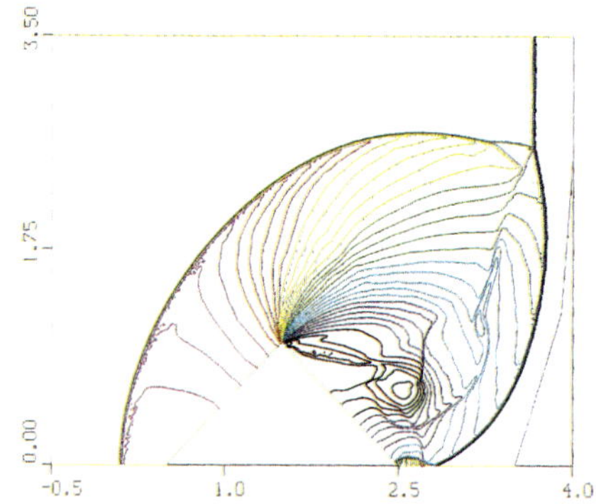

Color Plate 9. Bishop *et al.*, Chapter 40, Figure 8, p. 653.

Range

Pink:	$1.57x10^{8}$ - $1.37x10^{8}$
Red:	$1.34x10^{8}$ - $1.15x10^{8}$
Yellow:	$1.12x10^{8}$ - $9.20x10^{7}$
Green:	$8.91x10^{7}$ - $6.94x10^{7}$
Blue:	$6.65x10^{7}$ - $4.67x10^{7}$
Purple:	$4.39x10^{7}$ - $2.41x10^{7}$
Black:	$2.13x10^{7}$ - $1.51x10^{6}$

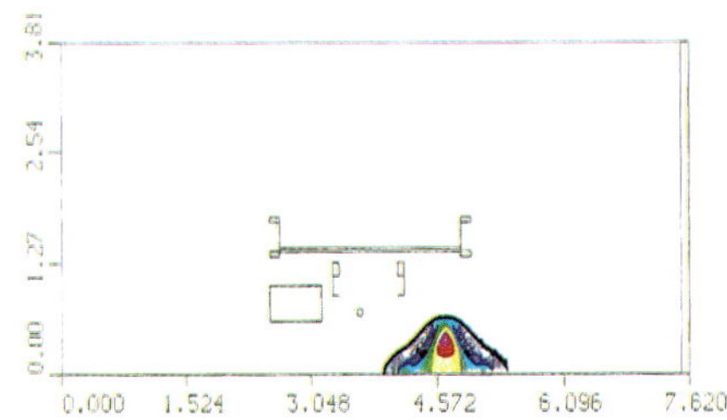

Color Plate 10. Bishop *et al.*, Chapter 40, Figure 9, p. 653.

Range

Pink:	$9.53x10^{8}$ - $4.19x10^{8}$
Red:	$3.72x10^{8}$ - $1.64x10^{8}$
Yellow:	$1.46x10^{8}$ - $6.40x10^{7}$
Green:	$5.69x10^{7}$ - $2.50x10^{7}$
Blue:	$2.22x10^{7}$ - $9.77x10^{6}$
Purple:	$8.69x10^{6}$ - $3.82x10^{6}$
Black:	$3.40x10^{6}$ - $1.49x10^{5}$

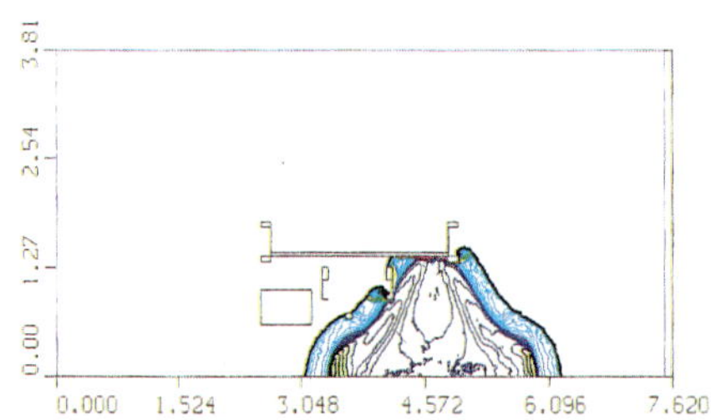

Color Plate 11. Bishop *et al.*, Chapter 40, Figure 10, p. 654.

Range

Pink:	$2.58x10^{7}$ - $1.38x10^{7}$
Red:	$1.26x10^{7}$ - $6.78x10^{6}$
Yellow:	$6.20x10^{6}$ - $3.32x10^{6}$
Green:	$3.04x10^{6}$ - $1.63x10^{6}$
Blue:	$1.49x10^{6}$ - $8.00x10^{5}$
Purple:	$7.32x10^{5}$ - $3.92x10^{5}$
Black:	$3.59x10^{5}$ - $1.92x10^{5}$

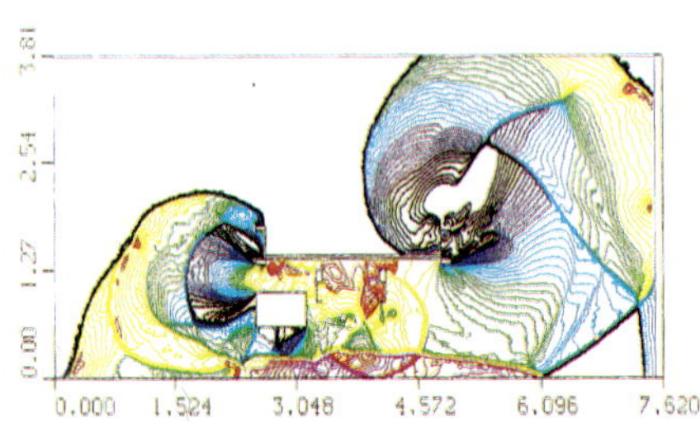

Color Plate 12. Bishop *et al.*, Chapter 40, Figure 11, p. 655.

Range

Pink:	$3.57x10^{6}$ - $2.25x10^{6}$
Red:	$2.10x10^{6}$ - $1.32x10^{6}$
Yellow:	$1.24x10^{6}$ - $7.80x10^{5}$
Green:	$7.30x10^{5}$ - $4.60x10^{5}$
Blue:	$4.31x10^{5}$ - $2.71x10^{5}$
Purple:	$2.54x10^{5}$ - $1.60x10^{5}$
Black:	$1.50x10^{4}$ - $9.42x10^{4}$

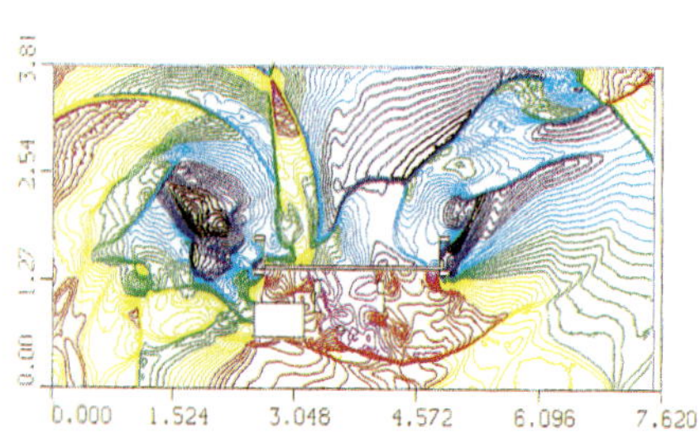

nonlinear switches. During the course of this investigation, such components have been designed, tested, and integrated monolithically to characterize the performance of a single 2-to-1 neuron.

Properties of Superlattices

Advances in III-V optoelectronics research have been particularly rapid in the area of semiconductor laser diodes, detectors, modulators, amplifiers and switches. The ability to grow thin (<10-nm) high-quality layers of compound semiconductors using molecular-beam epitaxy (MBE) has led to the discovery of several new optical and electrical phenomena. Novel opto-electronic devices that depend upon quantum mechanical phenomena have been demonstrated. A quantum well (QW) is a thin heterostructure of a low-bandgap material, such as GaAs, sandwiched between two layers of a higher bandgap material, such as AlGaAs, to constitute an electron well. The thickness of the low-bandgap material is around 30 atomic layers or ~10nm. Large changes in optical absorption due to single and multiple QWs in the presence of an external electrical field have given rise to new optical modulators and switching devices.[7,8] More recently, attention has been focused on a periodic structure of very thin QWs (width ~3nm), which are separated by layers of similar width of higher bandgap material. Such a structure of coupled QWs is called a superlattice (SL).

These SLs show very strong macroscopic nonlinear effects at room temperature due to quantum mechanical effects. In such a structure, the electrons and holes are not localized, that is, the energy band structure is continuous. When an electric field is applied, this continuous band structure breaks up into a series of minibands known as the Stark ladder.[9,10] The separation between these levels depends directly on the applied field, the period of the SL, and the electronic charge. When such a structure is illuminated by light, the absorption of light corresponding to these miniband transitions takes place, as shown in Figure 2. This makes the SL absorb light at wavelengths to which it was transparent in the absence of the applied field. Below-gap voltage-modulated absorption devices can also be made when these transitions are made to coincide with the incident photon energy. A great deal of materials engineering is possible with these structures. For example, one can change the layer thickness or the potential well depths to create devices with very different operating characteristics. Very-high-dynamic range modulators are being designed and tested using such SLs.[11,12]

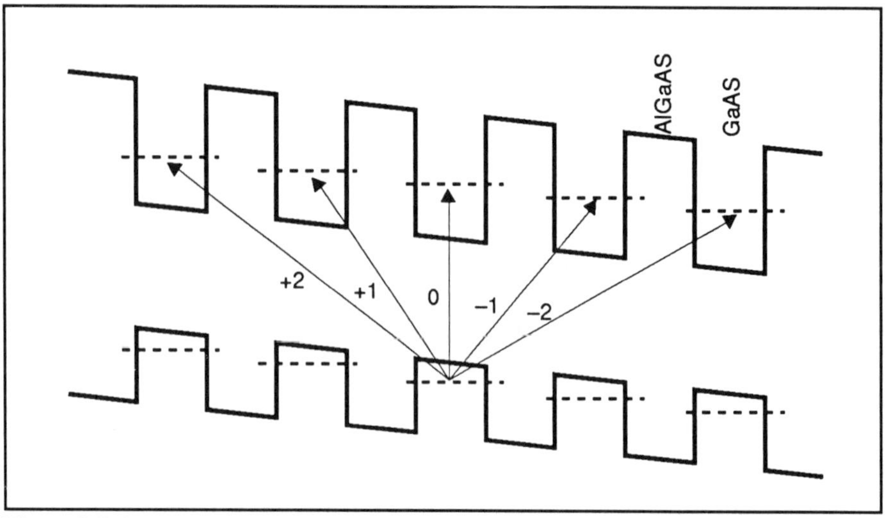

Figure 2. Below-gap transitions for a Stark ladder, giving rise to electroabsorption in presence of an applied field.

Material Growth

The GaAs/AlGaAs structure used in this research was grown by MBE. It is grown in a *p-i-n* heterostructure waveguide configuration, as shown in Figure 3. The waveguide core is undoped; it consists of an SL 52.5-nm thick, with 100nm of $Al_{0.2}Ga_{0.8}As$ (n = 3.53) on either side. The 10-period SL contains alternate layers of 2.5-nm GaAs and 2.5-nm $Al_{0.4}Ga_{0.6}As$. The top cladding region contains 500nm of $Al_{0.4}Ga_{0.6}As$ (n = 3.39), out of which the first 125nm next to the core are undoped and the rest is *p*-doped. The bottom cladding region also consists of 1250nm of $Al_{0.4}Ga_{0.6}As$, including 125nm of intrinsic layer next to the core, and the rest is *n*-doped. The doping levels are around $10^{18}/cm^3$. These layers are grown on n^+-doped GaAs substrate with an *n*-doped 1000-nm GaAs buffer layer. The topmost layer in this structure is the 50-nm p^+ GaAs cap. Incorporation of the *p-i-n* diode structure is very useful in applying an external electric field to the waveguide core. The doping is spaced back from the core region to reduce free-carrier-induced losses for the propagating optical mode. The thickness of the core region is chosen to give a single transverse electric (TE) or transverse magnetic (TM) propagation mode at a 780-nm wavelength in the transverse direction, as shown in Figure 4. From photoluminescence measurements, the bandgap of this material is found to be 1.75eV, or 708nm.

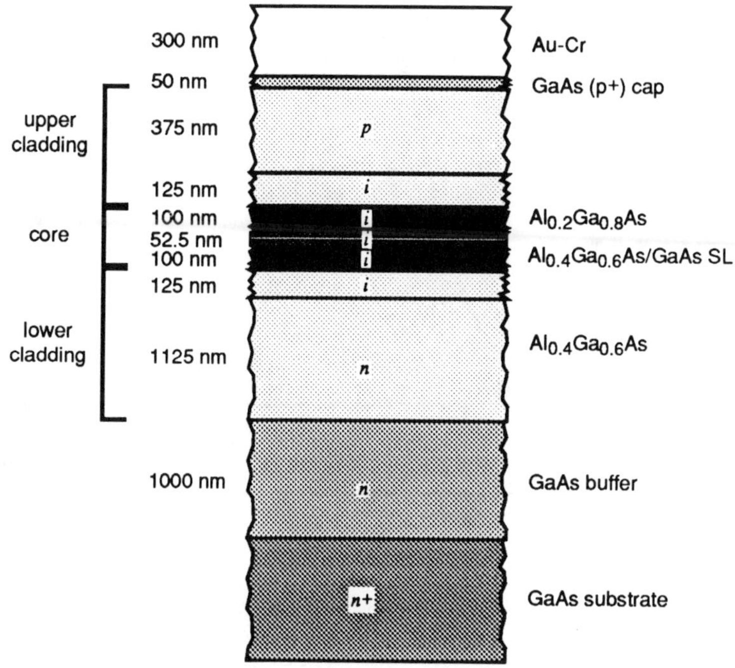

Figure 3. Layer structure of MBE-grown material (not drawn to scale).

Device Fabrication

A self-aligned single-mask lift-off technique was used to fabricate 5-μm rib waveguides in a Y-junction configuration. This technique consists of using photolithography to pattern the resist mask on the MBE-grown material, depositing 300nm of chrome/gold film by *e*-beam evaporation through the developed resist, lifting off the undesired chrome/gold by developing the resist, and then etching the rib pattern by using a standard GaAs chemical etchant, with the metallization providing the electrode and the self-aligned mask. Figure 5(a) shows the device dimensions for a single device; Figure 5(b) shows a 3-D illustration of the Y-junction optical neuron, and Figure 6 shows a microscope photograph (magnification 24X) of the chip containing 10 neuron devices. Each device is 4mm long; the two input guides are separated by 50μm. The height of the rib walls was chosen to be 550nm, which

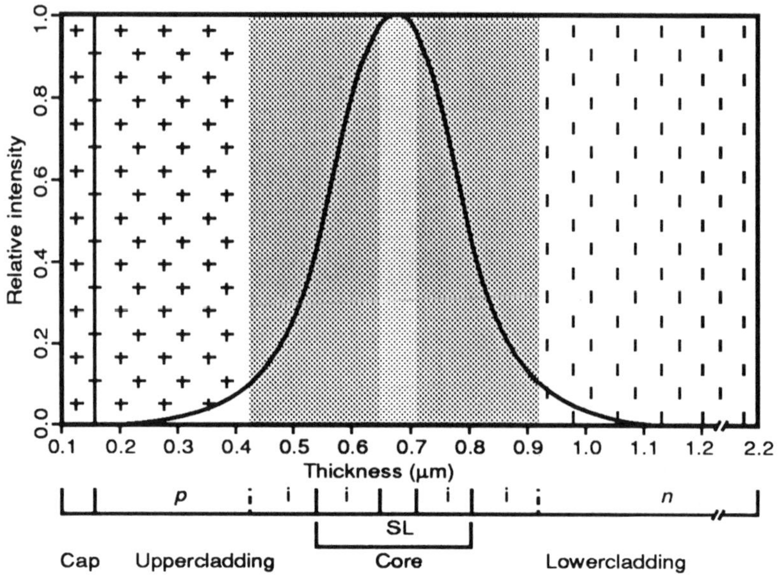

Figure 4. Profile of single transverse mode in waveguide core, showing how intensity of light is distributed as a function of device thickness. Core and cladding regions are shown with their doping profile.

Figure 5. (Opposite page) 2-to-1 waveguide neuron: (a) Geometric layout showing dimensions of various parts (not drawn to scale). (b) Schematic 3-D illustration of functional building block; rib waveguide structure, three electrically isolated active components (M1, M2, and M3), and electrical contact pads are clearly shown.

is enough for electrical isolation of the device but makes the device multimode in the lateral direction, thus degrading the SNR slightly. Two cuts (3 x 5μm), as shown in Figure 5(a), were included in the Y-junction mask. During etching, these resulted in tthe breaks needed to electrically isolate the three active components, labeled M1, M2, and M3, as shown in Figure 5(b). The entire chip is cleaved at both ends to provide optical-quality coupling facets. The chip is mounted on a conducting holder that acts as the bottom *n*-contact, and *p*-contacts are provided by the top gold layer. Three 100 x 200μm bonding pads connected by 10-μm-wide conducting channels were also incorporated in the device. The chip is mounted on a holder with pins. Gold wires 25-μm thick are bonded to the *p*- and *n*-contacts so that these can be connected to pins for easy electrical access.

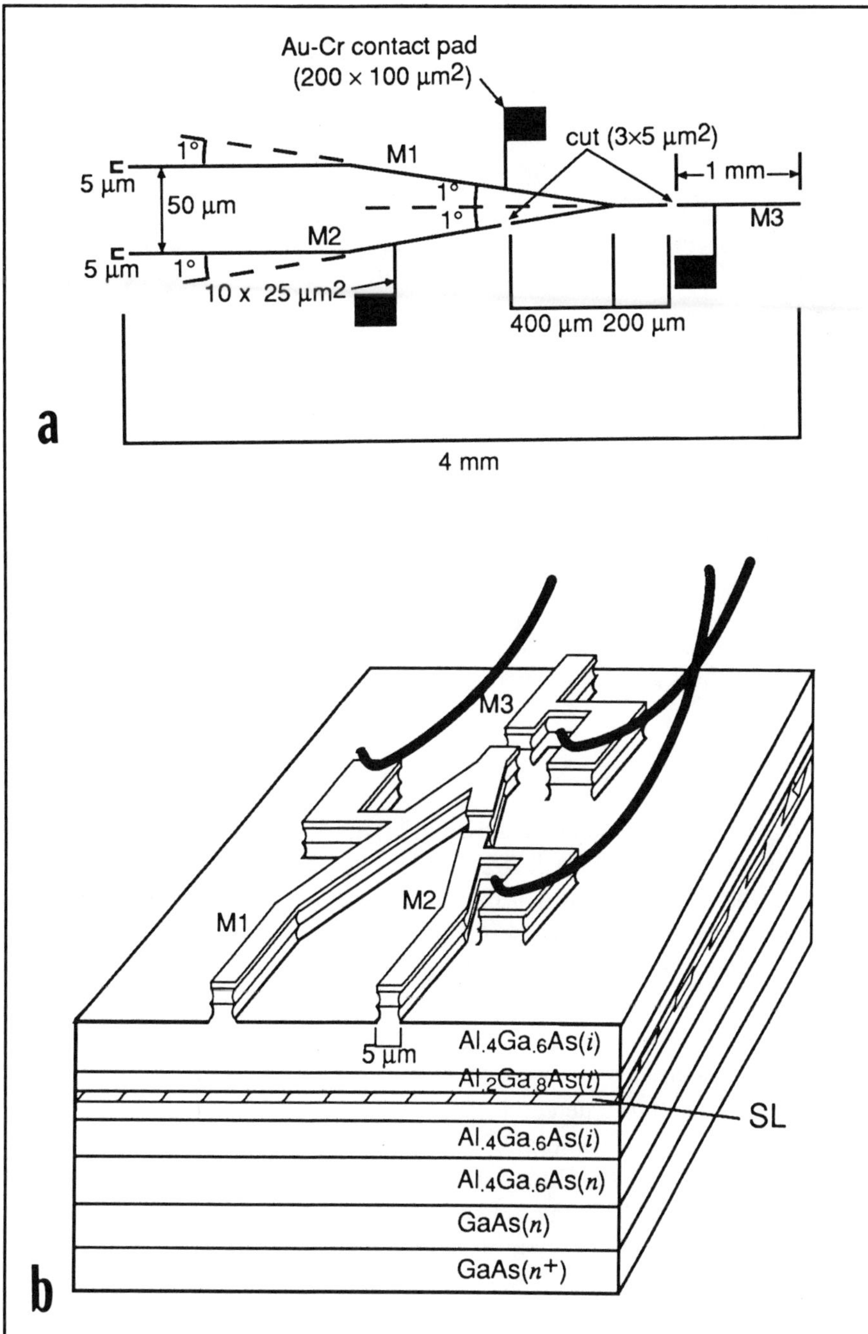
Au-Cr contact pad
(200 × 100 µm2)
cut (3×5 µm2)
1 mm
M1
M2
M3
5 µm
50 µm
1°
10 x 25 µm2
400 µm 200 µm
a
4 mm
M3
M1
M2
5 µm
Al.4Ga.6As(i)
Al.2Ga.8As(i)
SL
Al.4Ga.6As(i)
Al.4Ga.6As(n)
GaAs(n)
GaAs(n+)
b

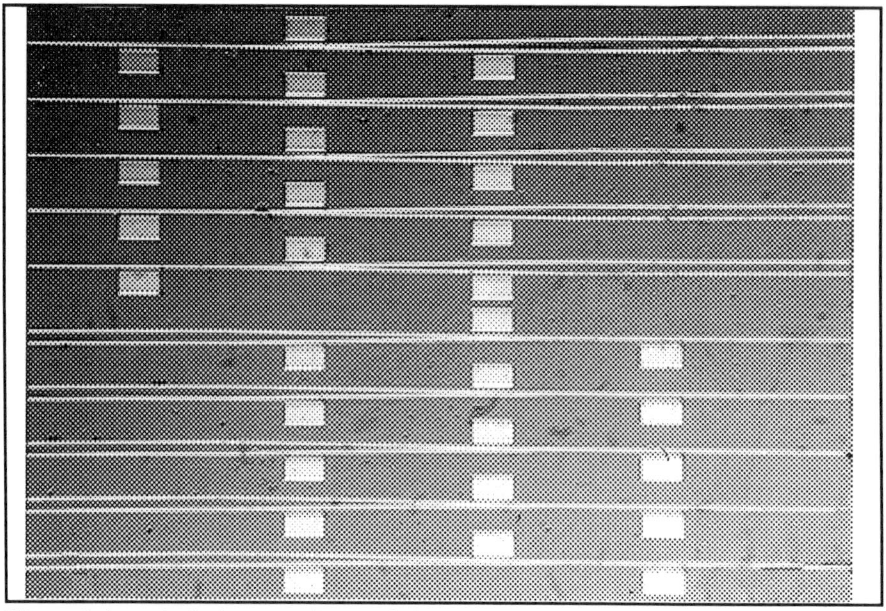

Figure 6. An optical microscope image of 10 waveguide neurons with a magnification of 24X. First five neurons are arranged from right to left; rest are in opposite direction. Each neuron device has dimensions given in figure 5(a).

Experimental Setup

The experimental setup is shown in Figure 7(a) and (b). Light from two 780-nm diode lasers (Sharp LT024) is end-fire coupled into the two input guides of the device using a 40X microscope objective to focus the two incoming beams of light. We separate the two focused spots by 50μm by choosing suitable position alignments for the two lasers. The two laser beams used are totally incoherent with each other. Only TE-polarized light is coupled into the waveguide. The output of the waveguide device is collected by a 10X microscope objective and focussed onto a spatial filter. The light is next focused by a lens onto a detector connected to a power meter and a charge-coupled device (CCD) camera. The image is processed using a PC with a frame-grabber. The spatial filter is used to block out light not coming from the waveguide. Another CCD camera along with a microscope objective is used for the top viewing of the sample and is very useful in positioning and alignment of the laser inputs. The electrical voltages are applied to the device using a four-wire cable, which is connected to the contact pins on the sample holder. The electronic bias circuit is designed to limit the current flowing in each of the three components to 70μA. Because the positioning requirements

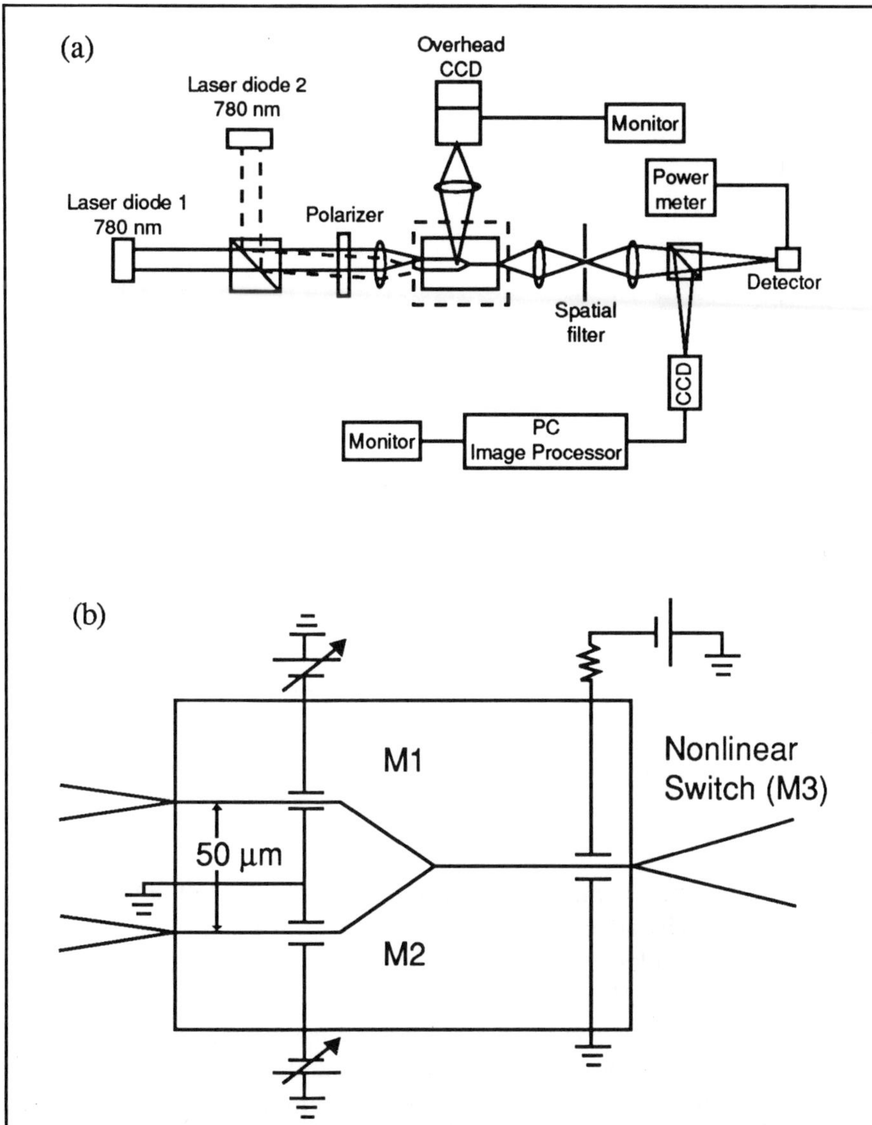

Figure 7. Optical neuron: (a) measurement setup, and (b) expanded view of dotted area in part (a), showing prototype device and electrical circuits for its three electrical components.

for the optical coupling of the laser beams are very demanding (the focused spot has to be coupled into a 5- x 0.25-µm region on the input facet), a piezoelectric transducer (PZT) three-axis controller is used for fine position alignment of the sample.

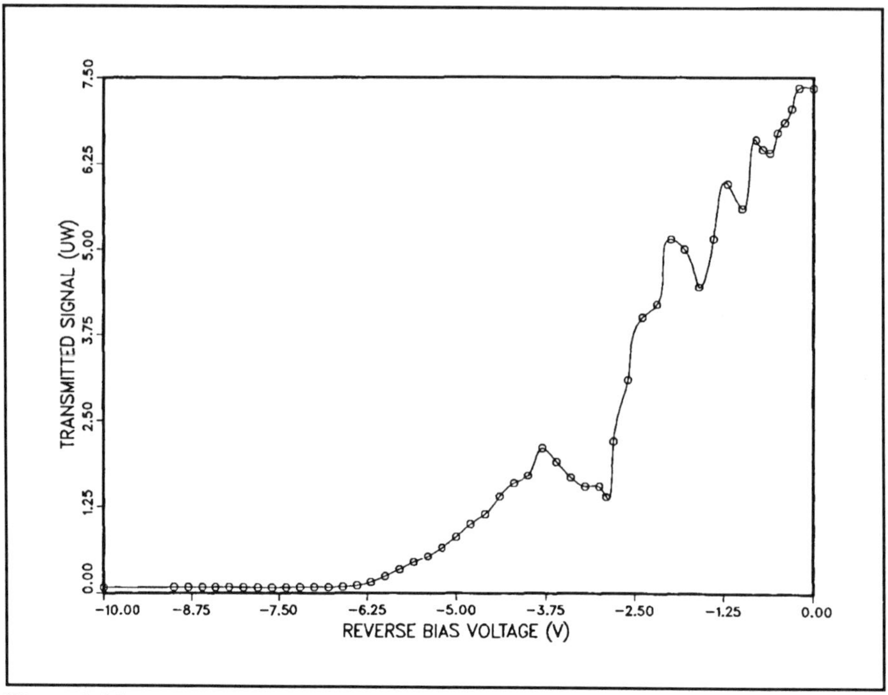

Figure 8. Measurement results showing transmitted output intensity as a function of reverse-bias voltage at 780nm for TE polarization of incident light. Modulator M3 is 1-mm long.

Results

The three active components on the device, M1, M2, and M3 [see Figure 5(b)], were first characterized as modulators. Figure 8 shows the results for M3, which is 1-mm long. The output intensity of the device is measured as a function of the applied bias. The device transmits less light as the reverse-bias voltage is increased, because of the absorptions caused by the miniband transitions. Transmission modulation as large as 250dB/cm has been observed for TE polarized light at 780nm. M1 and M2 perform similarly. Similar results have also been obtained for a range of wavelengths from 777 to 810 nm using a titanium-sapphire laser.[12] The transmitted signal out of a modulator corresponds to weighted input signals, according to Equation 1.

We also tested M3 to perform as an optical thresholding device by using it both as a modulator and a detector.[7,8] This was done by inserting a large series resistor in the modulator reverse-bias circuit at a fixed voltage. In this configuration, when the incident light is absorbed by the device and a current starts to flow in the circuit, there is a voltage drop across the resistor that reduces the device voltage, making it transmit more light. Now, if the incident

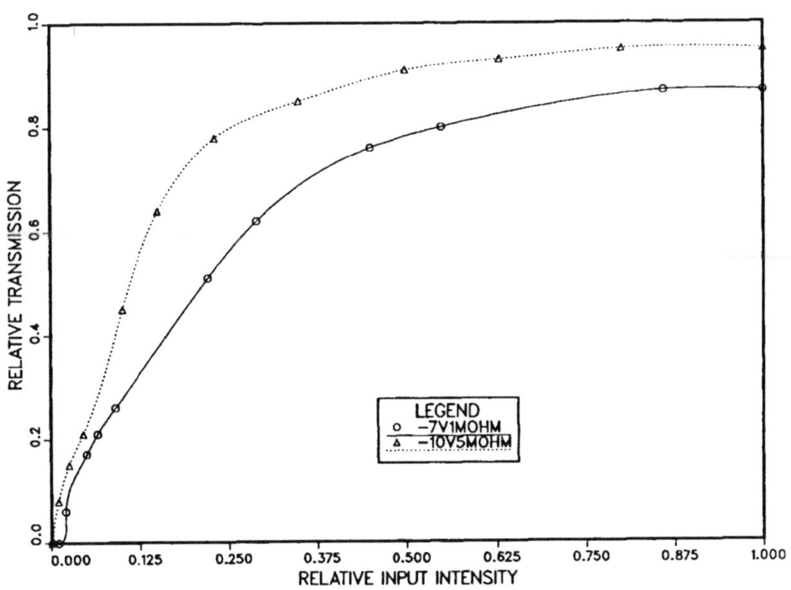

Figure 9. Measurement results showing performance of nonlinear switch. Relative transmission is plotted as a function of relative input intensity for a 1-mm-long component (M3) with -10-V bias and 5-MΩ series resistor, and with -7-V bias and 1-MΩ series resistor.

light on the device is reduced in intensity, it reduces the current flowing through the circuit (one electron-hole pair is created for every photon absorbed), which reduces the voltage drop across the resistor and increases it across the device, making it transmit less light. Measuring the change in relative transmission of the device as a function of the relative input intensity is a very useful measure of its nonlinear behavior. Figure 9 shows the results obtained under two different reverse bias voltage/resistor settings: one with -10V and 5MΩ, and second with -7V and 1MΩ. The relative transmission changed from 0 to 0.95 for the first curve and from 0 to 0.87 for the second one. The modulation depth for this switch is around 25dB. These results clearly show that the nonlinear behavior can be dynamically adjusted based on the system requirements.

Finally, the performance of a 2-to-1 waveguide neuron was evaluated, and the results look very good. The nonlinear switch was run with -10V and 5MΩ. The device was trained to test a logic scheme in which the two sets of weights were selected for different input intensities so that the device is on when the total input intensity of both lasers ($x_1 + x_2$) is over a threshold value; otherwise it was off. The results are listed in Table 1.

Table 1.
Performance of 2-to-1 Optical Neuron

Input set	M1 (V)	M2 (V)	M3	Output Intensity	Neuron Signal (*y*)
1	0	0	0 V	2.4 μW	On
2	0	0	-10V, 5MΩ	2.3 μW	On
3	0	0	-10V, 5MΩ	230.0 nW	On
4	-4.25	0	-10V, 5MΩ	1.0 nW	Off

The device was run with four different sets of operating conditions. In the first set, the output of the device for a particular input intensity was measured to be 2.4μW without any of the active components on. Second, M3 was turned on with -10V and 5MΩ, and the output was 2.3μW for the same conditions. Third, the input was dropped by a factor of seven (the threshold value), and the output was then 230nW. Fourth, M1 was turned on with -4.25V, and the output was 1nW. The neuron on/off ratio of 2300 was obtained by comparing set 4 with set 2, and it is mostly input signal limited.

As discussed earlier, the waveguide implementation of an NN requires designing low-loss single-mode rib waveguides. The propagation losses for the MBE-grown material were measured using the standard cutback method. In this method, several lengths of planar waveguide samples were used to measure the loss. The measured losses are ~43dB/cm at 780nm. The rib waveguide losses would be even higher. Additional losses are contributed by the coupling of input signals into the guide and the bends in the structure.

Conclusions and Discussion

An optoelectronic implementation of an NN in GaAs has been described, and experimental results have been presented. A 2-to-1 neuron device was fabricated and characterized for performance. The results look very good. A 250-dB/cm below-gap optical modulator and a 25-dB optoelectronic nonlinear switch have been tested. These high-dynamic-range modulators have many potential applications. The nonlinear device described in this work can be used as a thresholding detector device, and an array of such devices can be used in other optical implementations of an NN architecture. In the future, laser diodes may be integrated on the chip. The single neuron in this architecture has multiple inputs, unlike many other implementations.

Several problems were encountered in the development. The two most important ones are the high propagation loss of the material, and the packing density of the neurons due to the relatively large area (4mm x 200μm) required in the fabrication of each neuron. Because small-angle Y-junctions are used, the waveguide needs to be relatively long (1.43mm) to provide 50-μm separation between the two input guides (see Figure 5). Each modulator also needs to be relatively long for this reason, as well as for adequate modulation. These problems have an impact on the future design of more complex single neurons and an NN.

The first problem can be addressed by growing the *p*-doped region in the upper cladding much farther back from the core region. We have already accomplished this by growing new MBE material with propagation losses of <10dB/cm. With this material, it will be possible to design a single lateral-mode rib waveguide to give better SNR.

The approach to solving the next problem is twofold: First, material is grown in which the core region is made up of 100-percent SL instead of only 20-percent SL, as previously discussed. Such a material has already been grown. This may increase the modulation and thresholding by as much as a factor of five, which means that the total device length can be reduced correspondingly. The second approach is to design a new compact type of power combiner that uses either S-bends or right-angled bends.[13,14] This work is also proceeding. New approaches for electrical contacts are also being explored.

After a 2-to-1 neuron has been optimized and characterized, more complex neurons (4-to-1) and a small NN will be designed and tested.

Acknowledgments

We thank Fred Towner of Martin Marietta Laboratories, Baltimore, for his assistance in the molecular-beam epitaxy sample growth.

References

1. R. P. Lippmann, *IEEE ASSP Magazine*, 4-22 (April 1987).
2. B. Widrow, R. G. Winter, and R. J. Baxter, *IEEE Trans. Acoust. Speech Signal Process.* **36**(7), 1109 (1988).
3. *DARPA Neural Network Study* (AFCEA International Press, Fairfax, VA, 1988).
4. N. H. Farhat, *IEEE Circuits and Devices Magazine*, 32-41 (September 1989).
5. D. Psaltis *et al.*, in *Optical Processing and Computing*, H.A. Arsenault, Ed. (Academic Press, Inc., San Diego, 1989), chapter 8.

6. Y. Nitta *et al., Photonics Tech. Lett.* **4**(3), 247 (1992).
7. D.A.B. Miller, *Opt. Photonics News,* 7-15 (February 1990).
8. D.A.B. Miller, *Opt. Quantum Electron.* **22**, S61 (1990).
9. R. P. Leavitt and J. W. Little, *Phys. Rev.* **B41**, 5174 (1990).
10. R. P. Leavitt and J. W. Little, *Phys. Rev.* **B42**, 11,784 (1990).
11. E. Bigan *et al., IEEE J. Quantum Electron.* **28**, 214 (1992).
12. M. R. Stead *et al.*, Characterization of Stark-Wannier waveguide modulator, paper to be presented at Conference on Lasers and Electro-Optics (CLEO), 1992, unpublished.
13. C. Rolland, D. M. Adams, and D. Yevick, *IEEE Photonic Tech. Lett.* **2**, 404 (1990).
14. C. T. Sullivan, *III-V Optical Waveguide Technology* (document Air Force WRDC-TR-90-5027, Honeywell Systems and Research Center, Bloomington, MN 1991).

23

Time-Resolved Infrared Diode Laser Spectroscopy:

A State-Specific Probe of Atom-Molecule Reaction Dynamics

Harry E. Cartland
David W. Veney
Photonics Research Center
United States Military Academy
West Point, NY 10996-5000

Scott L. Nickolaisen
Curt Wittig
U.S. Army Research Office
Center for the Study of
Fast Transient Processes
University of
Southern California
Los Angeles, CA 90089-0482

Because of their continuous tunability and narrow (sub-Doppler) linewidth, lead salt diode lasers have proved themselves useful tools in applications where species-specific, and even state-specific, monitoring is required in a spectrally congested environment. We have used time-resolved infrared diode laser spectroscopy to state selectively monitor CO produced in three atom-molecule reactions; H + CO_2, H + OCS, and O + OCS. In the latter two systems, anharmonicity causes significant spectral overlap between CO product vibrational transitions and the C-O stretching fundamental (ν_3) of the abundant OCS bath molecule. Nevertheless, in all cases, nascent CO state distributions could be determined from transient absorption signals through the use of an initial slope technique. Although the systems studied encompass a wide range of collision energy (0.2 to 2.4eV), barrier height (1350 to 9110cm^{-1}), and reaction enthalpy (-17671 to +8960cm^{-1}), the amount of energy partitioned into CO internal excitations is surprisingly constant at ≈15 percent of that available. The nascent CO state distributions are substantially colder than predicted by statistical theories, with preferential channeling of energy into center-of-mass frame translation.

Introduction

Simple atom-molecule reactions such as:

$$H(2.4eV) + CO_2 \rightarrow CO + OH \tag{I}$$

$$H(1.4eV) + OCS \rightarrow CO + SH \tag{II}$$

$$O(0.2eV) + OCS \rightarrow CO + SO \tag{III}$$

are of interest because of the role their surfaces play in combustion and atmospheric chemistry. In addition, their relative simplicity makes them amenable to computational studies, allowing a rigorous comparison between experiment and current theories of dynamics.

An extensive literature already exists on Reaction I, and has recently been reviewed.[1] Most previous studies have focused on laser-induced fluorescence (LIF) detection of the OH fragment and include measurements of product state distributions and a determination of center-of-mass frame (CM) translation using sub-Doppler techniques. However, in one instance, CO(v=0,1) distributions have been measured under laboratory conditions comparable to those employed here using VUV LIF.[2] Also of note is the *ab initio* HOCO potential energy surface described by Schatz and co-workers,[3] which was parametrized for the experimentally determined barrier of 9110cm^{-1}.

In contrast, Reactions II and III have not been the subject of such thorough scrutiny. Earlier studies[4,5] have aimed at measurement of rate constants, yielding activation barriers for these reactions of 1350cm^{-1} and 1540cm^{-1}, respectively.

Through the application of time-resolved diode laser spectroscopy, our goal is to develop a detailed understanding of the factors influencing reactivity and energy disposal in this important class of abstraction Reactions. These factors include collision energy, thermochemistry, and most important, the potential energy surface connecting reactants to products.

Experimental Methods

The UV laser pump/infrared (IR) diode laser probe technique has been described elsewhere in detail.[1] The apparatus used for the study of Reaction I was not identical to that used for Reactions II and III, but the differences were minor. A general description of the technique is provided, and specific details are added as necessary.

Translationally "hot" H atoms were generated in a 2-m pressure stabilized flow cell by UV laser photolysis of an appropriate precursor: H_2S (193nm), HI (278nm), and NO_2 (355nm) for Reactions I, II, and III, respectively. The pulse energy, 10 to 150mJ depending on the experiment, was monitored with a calibrated photodiode to allow normalization of the measured nascent product distributions. The premixed sample consisted of 20 percent of the appropriate atom source in neat reactant and was maintained at 200 to 400mTorr to minimize the collisional relaxation of the nascent product.

A collinear cw IR diode laser then state selectively probed the CO(v,J) Reaction product. The output of the IR diode was collimated and passed through a monochromator to select a single laser mode with a nominal linewidth of ≈ 0.0003 cm^{-1}. Because of anharmonicity, virtually any CO(v,J) level is accessible by temperature and current tuning of the diode output, although more than one diode is usually required to cover the wavelength range of interest. The distributions reported here are based on a sampling of even and odd J states in both the P and R branches.

Prior to entering the sample cell, $\approx$10 percent of the IR beam was split off and passed through a reference flow cell. The reference cell and a Ge etalon permitted precise wavelength calibration. Passing a discharge through the cell generated a reference population in higher vibrational states ($v \leq 9$), allowing excited product states to be probed in the sample cell.

Standard line locking techniques were used to prevent drift of the diode laser. The reference detector signal was input to a lock-in amplifier. By using phase-sensitive detection, a small dither in the diode output produced a first derivative error signal, which was fed back into the diode current controller to drive the laser output to line center.

The sample and reference beams passed through long wavepass filters ($\lambda > 2.5\mu m$) before being focused onto $N_2(l)$ cooled InSb detectors. The signal level was boosted by impedance matched preamplifiers and line drivers. The reference signal was input directly to the lock-in amplifier, while the sample signal underwent another stage of amplification and filtering prior to being input to a transient digitizer. The digitizer recorded the absorption transients at 10^8 samples s^{-1}, and hard wire averaged them in the CAMAC crate controller. Between 200 and 1500 transient signals were averaged at a repetition rate that was empirically adjusted to avoid sample depletion effects. The averaged data was stored, processed, and displayed on a PC AT computer using original software written in ASYST. Using a photodissociation process that produced CO, the detection system risetime was measured at ≈70ns.

For time-resolved CO(v,J) lineshape measurements, a fraction of the IR beam was directed to a vacuum spaced confocal etalon. Instead of dithering the diode laser wavelength, the etalon scanner driver was modulated thereby dithering the fringe positions in time. The line locking technique previously described was used to slave the laser output to an etalon fringe, which was subsequently stepped across the absorption feature of interest. Simultaneous acquisition of a room temperature CO reference lineshape served to calibrate the step size in terms of wavelength.

Results

Some typical v=0 signals from Reaction II, the H+OCS Reaction, are shown in Figure 1, where the raw data have been converted to absorbance units. These transient signals clearly illustrate the reaction and CO relaxation dynamics. The two-component rise for low J is evidence of a small nascent population, increased by rotational relaxation, and to a lesser extent vibrational relaxation, as equilibrium is reached. In contrast, high J levels show substantial initial populations that rapidly decay.

To interpret the data, we developed the following simple model, shown here illustrating Reaction II, H+OCS, but of general applicability:

$$H^* + M \rightarrow H + M \tag{1}$$

$$H^* + OCS \rightarrow CO(v,J) + SH \tag{2}$$

$$H^* + OCS \rightarrow CO(v',J') + SH \tag{2'}$$

$$CO(v,J) + M \rightarrow CO(v',J') + M \tag{3}$$

$$CO(v',J') + M \rightarrow CO(v,J) + M \tag{3'}$$

Reaction 1 is collisional deactivation but effectively includes all processes that remove translationally "hot" atoms, such as diffusion out of the probed volume and various radical reactions. Reactions 2 and 2′ represent reactive scattering into the monitored state CO(v,J) and all other states CO(v′,J′). Modeling the CO manifold as a quasi-two-level system, consisting of the probed state and a "bath" state with which energy is exchanged via

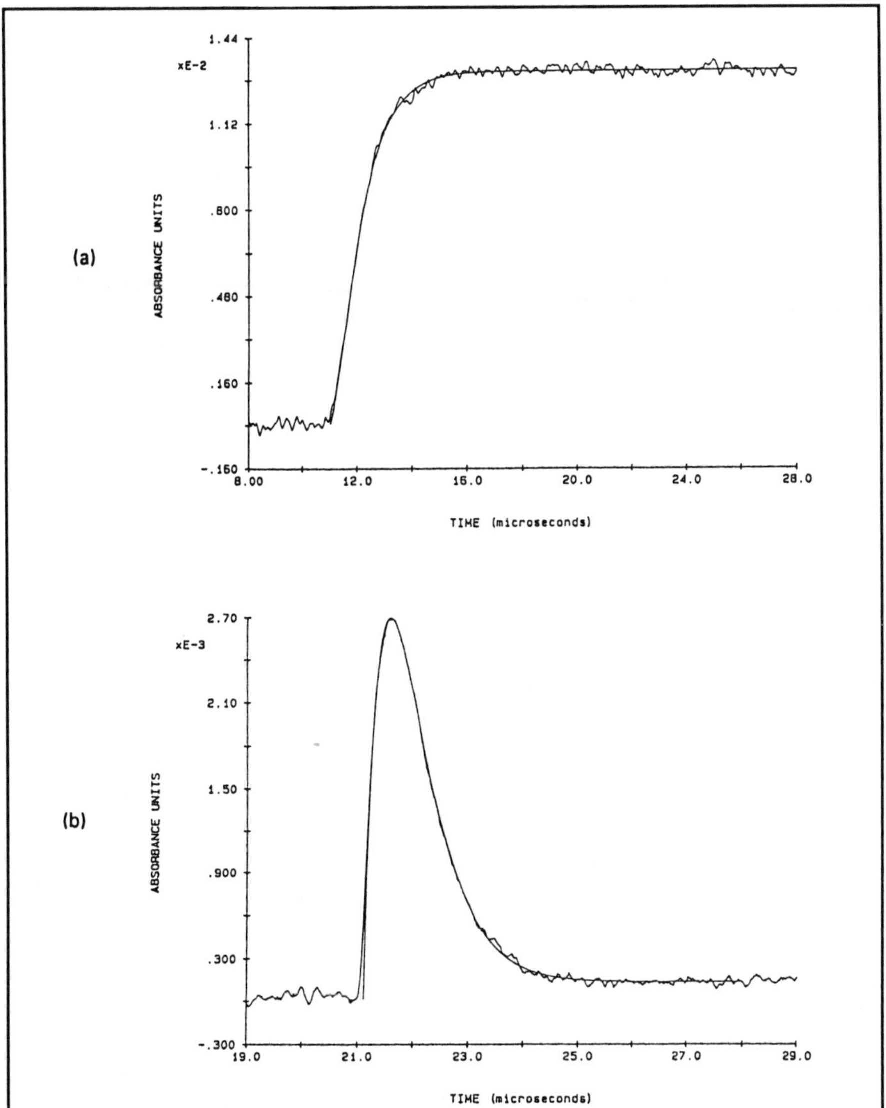

Figure 1. CO (v=0) transient absorption signals from H+OCS → CO+SH; (a) P(10), (b) P(35).

Reactions 3 and 3′, yields an analytic expression for the time-dependent CO(v,J) population [CO(v,J,t)]. As the fits in Figure 1 show, the transient absorption signals are well modeled by this simple scheme.

The derivation of the analytic expression for [CO(v,J,t)] is discussed elsewhere in detail.[1,6] Here we consider only two limiting cases. First, in the long-time limit we find:

$$[CO\ (v,J,t)]_{t\rightarrow\infty} = [H^*]_0\ \frac{(k_{2'}\chi_{ocs} + k_{2'}\chi_{ocs})}{(k_1 + k_2\chi_{ocs} + k_{2'}\chi_{ocs})}\ \frac{k_{3'}}{k_3 + k_{3'}}$$

where $[H^*]_o$ is the initial "hot" atom number density and χ_{OCS} is the mole fraction of the OCS collision partner. This expression has a straightforward interpretation: the concentration of CO(v,J) after attainment of equilibrium is a product of the initial "hot" atom concentration, the branching ratio for reactive scattering, and the equilibrium fraction of molecules in that state. Of more interest, though, is the short time limit:

$$[CO\ (v,J,t)]_{t\rightarrow\infty} = k_2\chi_{ocs}[H^*]_0\ [M]\ t$$

where [M] is the total sample number density. Initially, the CO(v,J) density is linear in time, with a proportionality constant dependent on the state-specific rate constant k_2. Therefore, a plot of the initial slopes of the transient signals gives a relative distribution of state-specific rates, which is equivalent to plotting the actual nascent CO(v,J) number densities.

Confidence in the distributions obtained in this manner requires that the model incorporate all essential features of the kinetics. In Reactions II and III, for example, the unusually high relaxation rate of vibrationally excited CO in OCS ($P\approx0.1$ for CO(v=4)) must be considered. Elsewhere, we explore the effect of processes specific to the individual systems.[1,6] Inclusion of additional kinetics in the model of course improves the fit, but in no case will any of these secondary processes change the limiting expressions.

The time-resolved absorption technique used in the experiments is not as sensitive as multi-photon ionization (MPI) or LIF, but it does present one advantage. A straightforward distinction between gain and absorption clearly indicates the presence of population inversion, which raises an important point. Absorption techniques measure population differences. This means that the initial slope of the absorption signal is proportional to the difference of the upper and lower state-specific rate constants. By probing successively higher vibrational levels until no further absorption is observed, the nascent rotational distribution of the highest populated vibrational level is directly determined. Working backward, the difference measurements yield the distributions in lower vibrational levels. It is important to make these corrections, particularly for Reaction II where excitation of CO is observed up to v=4.

Using the initial slope method, CO rotational distributions were deter-

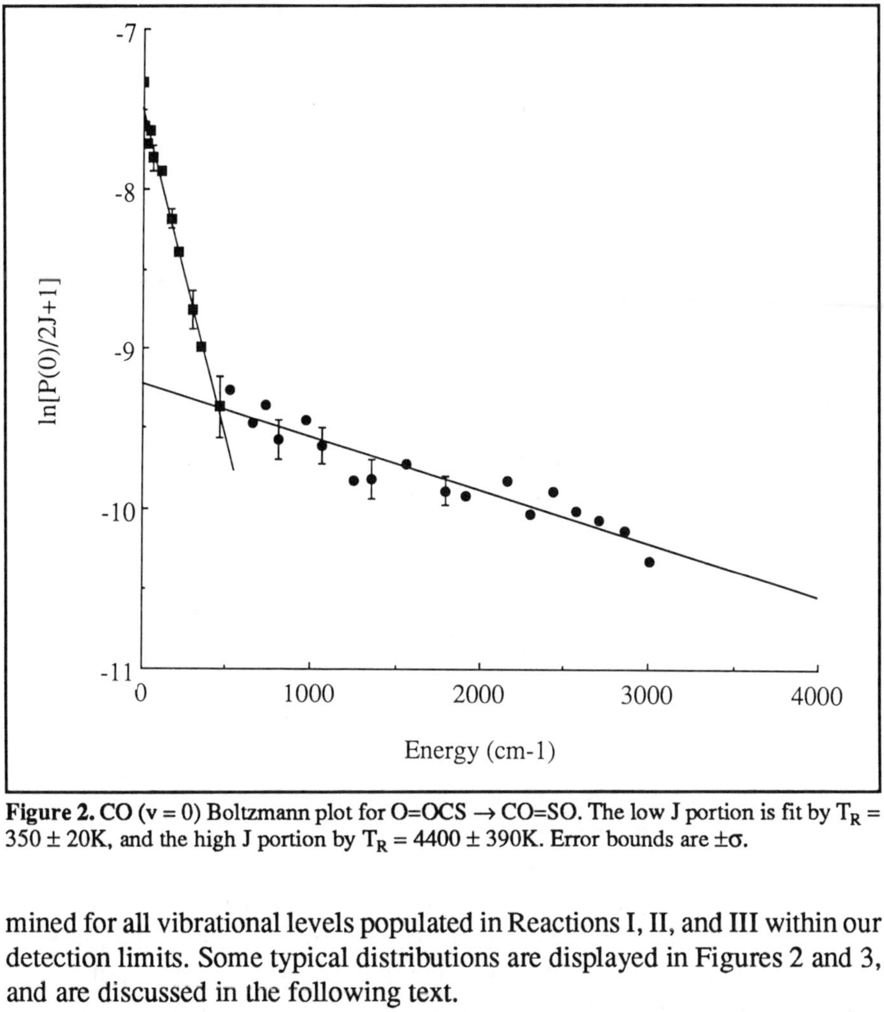

Figure 2. CO (v = 0) Boltzmann plot for O=OCS → CO=SO. The low J portion is fit by T_R = 350 ± 20K, and the high J portion by T_R = 4400 ± 390K. Error bounds are ±σ.

mined for all vibrational levels populated in Reactions I, II, and III within our detection limits. Some typical distributions are displayed in Figures 2 and 3, and are discussed in the following text.

Figure 2 shows a CO(v=0) rotational Boltzmann plot for Reaction III, O+OCS. Strictly speaking, Boltzmann distributions describe systems in equilibrium. This is certainly not the case for the nascent distributions reported here, but Boltzmann plots are often good approximations and can be quite revealing, as the Figure shows. The plot in Figure 2 is clearly bimodal with a low J (energy) temperature T_R=350K, and a high J temperature T_R=4400K. Because the low J portion of the distribution has a temperature near ambient and the rise of the low J signals is delayed from the photolysis pulse, we attribute it to rotationally relaxed CO. Thus, a rotational temperature T_R=4400K is more representative of the nascent CO(v=0) distribution. Relaxed CO absorption was also observed in v=1, yielding a limit on the vibrational

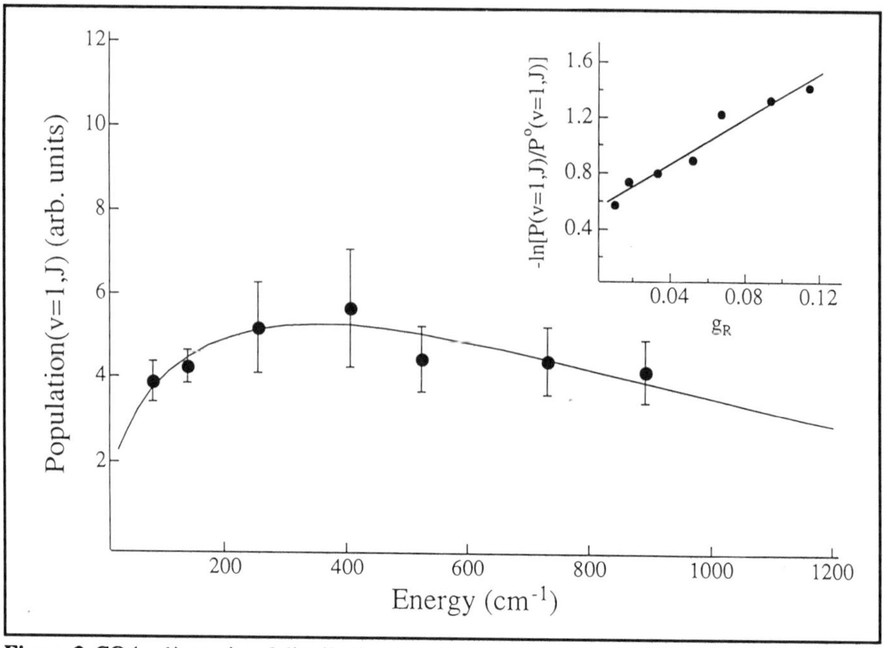

Figure 3. CO (v=1) rotational distribution from $H + CO_2 \rightarrow CO + OH$. The solid line is the back-calculated distribution from the surprisal analysis (inset).

branching ratio [v=1]/[v=0] ≤ 0.05, although poor S/N precluded determination of the nascent rotational distribution.

A better way to represent the data is a surprisal plot, or in terms of a surprisal parameter θ. The surprisal analysis procedure, developed by Levine and Bernstein,[7] can be used to compare an observed rotational distribution P(v,J) to a statistical maximum entropy distribution P°(v,J). Calculation of the "prior" distribution P°(v,J) involves summing over states and is well understood; the details are discussed elsewhere for systems of interest here.[1,6] Surprisal plots are typically linear and, by definition, have positive slopes, or surprisal parameters θ_R, for distributions that are "colder" than statistical. A surprisal plot for CO(v=1) from Reaction I is shown in the inset in Figure 3, where the ordinate g_R is a reduced energy variable. The large positive rotational surprisal parameter θ_R=9 indicates a substantial bias against rotational excitation in the CO product. The solid line overlying the rotational distribution in Figure 3 represents the back-calculated distribution derived from the surprisal analysis and fits the experimental data well. The CO(v=0) rotational distribution was also found to be cold, θ_R=7. Integration of the rotational distributions yields a vibrational branching ratio [v=1]/[v=0] = 0.4. For Reaction I, no signal was detected for 3←2 transitions, indicating that little or no population of v>1 occurs.

Table 1
Summary of Nascent CO Product Distributions

Reaction	CO Vibrational Quantum Number	Relative Vibrational Population	Rotational Boltzmann Temperature T_R (K)	Rotational Surprisal Parameter θ_R
$H + CO_2 \rightarrow CO + OH$	0	1.0	1500 ± 110	7.2 ± 0.7
	1	0.4	1030 ± 50	9.0 ± 0.6
$H + OCS \rightarrow CO + SH^*$	0	1.00	1400 ± 140	12.0 ± 1.4
	1	0.17	780 ± 45	21.3 ± 1.3
	2	0.10	590 ± 100	24.0 ± 3.8
	3	0.07	405 ± 20	51.9 ± 5.6
$O + OCS \rightarrow CO + SO$	0	1.00	4400 ± 390	3.7 ± 0.5
	1	≤ 0.05	—	—

Error limits are $\pm\sigma$.
* $T_v = 3890 \pm 830$ K; $\theta_v = 3.6 \pm 1.1$

Experimentally measured CO distributions for the three reactions studied are summarized in Table 1.

Time-resolved sub-Doppler lineshape measurements on the CO(v=0) product from Reaction II indicate substantial partitioning of energy into CM translation. A total of six transitions from J=7 to J=30 were examined: R(7), P(10), R(16), R(19), R(22), and R(30). A 3-D plot for R(19) is shown in Figure 4(c). Cross sections at 100nsec intervals were fit to Gaussian lineshapes, from which E_{CM} was determined. Because the S/N ratio is initially poor due to the small amount of product present, a preliminary estimation of the nascent E_{CM} was made by fitting a plot of E_{CM} versus time to a single exponential function, and extrapolating back to t=0. Such a plot for R(19) is shown in Figure 4. In all cases, the exponential fit appears to have underestimated the nascent CM kinetic energy. We take the average E_{CM} thus determined, 7180cm^{-1}, as a lower limit for CO(v=0).

Discussion

For Reaction I, H+CO_2, the results are consistent with those of Weston,[8] but disagree with the study of Rice and co-workers[3] who found a [v=1]/[v=0] ratio of 1.2. Using VUV LIF, Rice *et al.*[3] also reported an inverted v=1 rotational distribution peaked at J=23, which should have manifested itself as gain in the experiments discussed here: none was observed. Although the

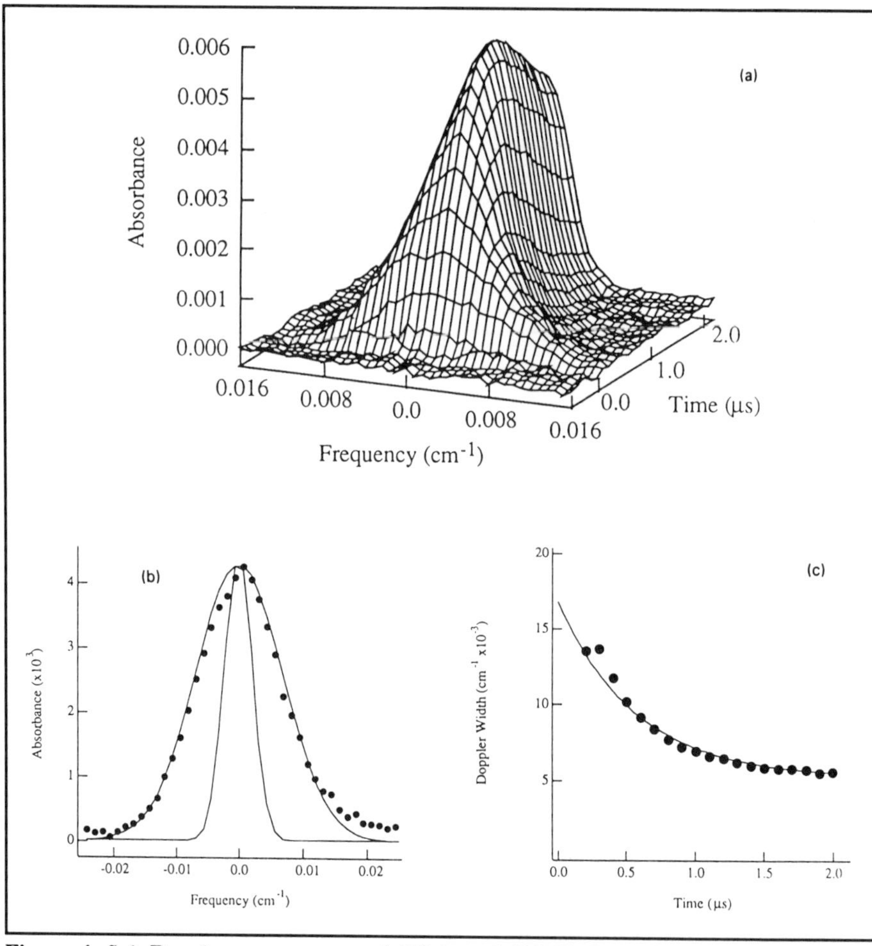

Figure 4. Sub-Doppler spectroscopy of CO (v = 0, R(19)) from H+OCS → CO+SH; (a) absorbance as a function of frequency and time, (b) lineshape, Gaussian fit, and 298K reference lineshape at 500nsec, (c) Doppler width.

observations here are not in quantitative agreement with those of Rice *et al.*,[3] there is agreement that CO internal excitation is less than statistical. Previous studies of nascent OH distributions have shown that they too are colder than statistical,[1] whereas Chen and co-workers[9] were able to demonstrate substantial partitioning into CM translation for OH(v=0; low J) using sub-Doppler techniques. In sum, these results indicate a propensity for CM translation at the expense of product internal excitation, although it is also important to note that both the picosecond spectroscopy studies of Scherer and coworkers[10] and the surface calculations of Schatz and co-workers[3] implicate a HOCO intermediate in this reaction.

Table 2
Potential Surface Features and Partitioning of Available Energy

Reaction	$H + CO_2 \rightarrow CO + OH$	$H + OCS \rightarrow CO + SH$	$O + OCS \rightarrow CO + SO$
CM Collision Energy E_{CM} (cm^{-1})	19170	11070	1650
Reaction Enthalpy ΔH_{298} (cm^{-1})	+8960	-3930	-17670
Activation Barrier E^* (cm^{-1})	9110	1350	1540
Transition State Energy $E_{CM} - E^*$ (cm^{-1})	10060	9720	110
Available Energy $E_{CM} - \Delta H_{298}$ (cm^{-1})	10210	15000	19320
CO Internal Exitation $E_{int, CO}$ (%)	15	13	16
XY Internal Exitation[a] $E_{int, XY}$ (%)	23	<39	—
Translational Energy E_{CM} (%)	62	>48	—

[a] XY = OH, SH, SO as appropriate

The picture of Reaction II, H+OCS, is not so complete. Predissociation of SH precludes studying this fragment with LIF; however, SD distributions in the analogous D atom reaction have been measured and are, like OH, found to be cold and nonstatistical.[11] By conservation of energy, the sub-Doppler lineshape measurements place an upper limit of 5900cm^{-1} for the amount of available energy channeled into the SH fragment. Again, this system displays energy disposal propensity towards CM translation. Unfortunately, there is no HOCS potential surface to aid interpretation of the experiments, but the data are consistent with near collinear collision geometry for the reactive process.[6]

The least well understood is Reaction III, O+OCS. Further experimentation is necessary to elucidate the dynamics of this reaction. SO product distributions must be measured, and a determination of the extent to which energy is partitioned into CM translation must be made. An OSCO potential surface would be useful in addressing the role of reaction intermediates. Despite the relative dearth of information on Reaction III, the data obtained thus far are consistent with the trends set by Reactions I and II; CM frame translation at the expense of product internal excitation.

Factors pertinent to a discussion of the dynamics for these systems are listed in Table 2. With a total of ≈10,000 to 20,000cm^{-1} available to the products, the fraction channeled into CO internal degrees of freedom is surprisingly small, ≈15 percent. Although the data set comprises a wide range of collision energy, activation barrier, and reaction enthalpy, the nascent CO vibrational and rotational distributions are found to be cold when compared with statistical theory. Thus, intramolecular vibrational redistribution (IVR) is not complete along the reaction coordinate; nor, as the data in Table 1 indicate, has equilibration occurred between rotational and vibrational degrees of freedom.

The data suggest the following. First, only a fraction of the total energy "pool" of the system is available for internal excitation of the product fragments, or the phase space for the reactive process is somehow restricted. Second, in that fraction available for product excitations, energy is conserved to the extent that partitioning into vibrational degrees of freedom occurs at the expense of product rotation, or the absence of coupling between rotations and vibrations causes rotational temperature to drop with increasing vibrational quantum number. And third, the majority of the available energy is repulsively released in the exit channel and is directed primarily into relative translational motion of the separating fragments. Further study of the potential energy surfaces of these atom-molecule systems may reveal the dynamic source of the observed nonstatistical behavior.

For Reactions I and II, because of the large difference in mass, it is unlikely that the H atom will simply abstract an O or S atom in passing. Although the required slowing of the H atom to form a chemical bond implies the existence of an intermediate, it does not guarantee statistical distribution of available energy among product degrees of freedom. In the energy regime of these experiments, ≈10^{4}cm^{-1} at the transition state, RRKM estimates of reaction rates are on the order of 10^{13} sec^{-1}.[12] Because IVR may not be complete on this time scale, the applicability of statistical theory to these experiments is questionable, even though a local minimum exists on the surface.

An explanation of nonstatistical behavior in Reaction III presents a challenge of another sort. A lack of data makes interpretation difficult, but a different mechanism is almost certainly operative since the amount of energy over the activation barrier, 110cm^{-1}, falls within the realm where statistical theory could apply. The reaction mechanism for Reaction III must explain a CO product rotational temperature that is significantly higher than its vibrational temperature, unlike Reactions I and II, where the opposite is true. Two mechanisms are consistent with the observed partitioning. In the first, a short-lived OSCO intermediate forms with energy preferentially deposited in a degree of freedom that becomes product rotation (incomplete IVR). In the second, a long-lived OSCO intermediate forms (IVR complete), and rotational excitation is the result of a torque on the CO fragment in the exit channel. It will be interesting to see if this O atom reaction shows the same propensity toward product fragment CM translation that the H atom reactions do: there is good indication that it will.

Conclusion

Using a sensitive and selective IR diode laser probe technique, we have studied reactive scattering of H and O atoms with CO_2 and OCS through state-resolved monitoring of the CO product. The CO rotational and vibrational distributions obtained under arrested relaxation conditions are significantly colder than those predicted by statistical theory. Results for these atom-molecule reactions, coupled with earlier studies, indicate incomplete IVR even if intermediates are present, with preferential partitioning of energy into CM frame translational degrees of freedom.

Acknowledgments

This work was supported by the Army Research Office and the Office of Naval Research.

References

1. S.L. Nickolaisen, H.E. Cartland, and C. Wittig, *J. Chem. Phys.* **96**, 4378 (1992).
2. J.K. Rice, Y.C. Chung, and A.P. Baronavski, *Chem. Phys. Lett.* **167**, 151 (1990); J.K. Rice and A.P. Baronavski, *J. Chem. Phys.* **94**, 1006 (1991).
3. G.C. Schatz, M.S. Fitzcharles, and L.B. Harding, *Faraday Disc. Chem. Soc.* **84**, 359 (1987).
4. S. Tsunashima *et al.*, *J. Phys. Chem.* **79**, 775 (1975); J.H. Lee, L.J. Stief, and R.B. Timmons, *J. Chem, Phys.* **67**, 1705 (1977).
5. R. Atkinson *et al.*, *J. Phys. Chem. Ref. Data* **18**, 881 (1989); W.B. DeMore *et al.*, *Chemical Kinetic and Photochemical Data for Use in Stratospheric Modeling, Evaluation, No. 8*, (JPL Publication 87-4, Jet Propulsion Laboratory, Pasadena, CA, 1987).
6. S.L. Nickolaisen, State Resolved Product Dynamics in Atom-Molecule Interactions, PhD. Dissertation, University of Southern California, 1991.
7. R.D. Levine and R.B. Bernstein, *Acc. Chem. Res.* **7**, 393 (1974).
8. R.E. Weston, private communication.

9. Y. Chen *et al., Chem. Phys. Lett.* **159**, 426 (1989).
10. N.F. Scherer *et al., J. Chem. Phys.* **92**, 5239 (1990).
11. D. Hausler, J. Rice, and C. Wittig, *J. Chem. Phys.* **91**, 5413 (1987).
12. G. Hoffman *et al., Israel J. Chem .* **30**, 115 (1990).

24

Navier-Stokes Predictions of Pitch Damping for Axisymmetric Shell Using Steady Coning Motion

Previous investigations have proposed that the side force and moment acting on a body of revolution in steady coning motion could be related to the pitch-damping force and moment. Current research shows the side force and moment to be directly related to the pitch damping force and moment coefficient for a specific combination of spinning and coning motion. This information improves the methodology because the pitch damping coefficients can now be computed using a single computation. The authors apply this approach to produce predictions of the pitch damping for axisymmetric shell. The flow fields about these projectiles undergoing steady coning motion are successfully computed using a parabolized Navier-Stokes computational approach that makes use of a rotating coordinate frame. The governing equations are modified to include the centrifugal and Coriolis force terms due to the rotating coordinate frame. From the computed flow field, the side moments due to coning motion, spinning motion, and combined spinning and coning motion are used to determine the pitch-damping coefficients. Computations are performed for a generic shell configuration. Predictions are made for various length-to-diameter ratios and supersonic Mach numbers, and comparisons are made with results obtained using a previously published inviscid code. Results are also obtained for a series of ogive-cylinder configurations that have been tested in an aerodynamics range located at the U.S. Army Research Laboratory. The computational predictions show excellent agreement with the range data.

Paul Weinacht
Walter B. Sturek
U.S. Army Research Laboratory
Aberdeen Proving Ground,
MD 21005-5066

Lewis B. Schiff
NASA Ames Research Center
Moffett Field, CA 94035

Introduction

Prediction of the inflight motion of projectiles requires the determination of the aerodynamic forces and moments that act on the body. These aerodynamic forces and moments may be determined by experimental or theoretic means, such as computational fluid dynamics. From a computational standpoint, much of the research effort has focused on determining the static aerodynamics, such as drag and pitching moment. Only a limited number of studies have focused on numeric prediction of dynamic aerodynamic derivatives such as pitch damping force and moment. In the current research effort, a parabolized Navier-Stokes technique has been adapted to predict the pitch damping force and moment using steady coning motion.

By applying linear flight mechanics theory such as that developed by Murphy,[1] it can be shown that aerodynamic side force and moment coefficients acting on a projectile in steady coning motion can be related to the pitch damping force and moment coefficients. Steady coning motion is defined as the motion performed by a missile flying at a constant angle with respect to the freestream velocity vector and undergoing a rotation at a constant angular velocity about a line parallel to the freestream velocity vector and coincident with the projectile center of gravity. This is shown schematically in Figure 1. Coning motion is, in fact, a specific combination of two orthogonal planar pitching motions. The use of steady coning motion to determine the pitch damping aerodynamic coefficients provides an interesting and cost effective approach for determining the aerodynamics that are normally associated with unsteady or time-dependent motions.

Experimental studies of conical bodies undergoing spinning and coning motions were performed by Schiff and Tobak.[2] Their wind tunnel results showed that at low angles of attack, slopes of the side force and moment with angle of attack normalized by the coning rate were in agreement with predictions of the damping-in-pitch force and moment coefficients obtained using linearized theory. They also demonstrated that the Magnus force and moment (variation of side force and moment with spin rate and angle of attack) were negligible; thus, the linear pitch damping coefficients could be determined from the side force and moment due to coning alone for these bodies.

Later, Schiff[3] computed the supersonic inviscid flow about a conical body undergoing coning motion. To compute the flow around the body in coning motion, Schiff made use of a rotating coordinate frame. Within the rotating coordinate frame the flow was steady; thus, the steady Euler equations could be solved. The governing equations were modified to include the

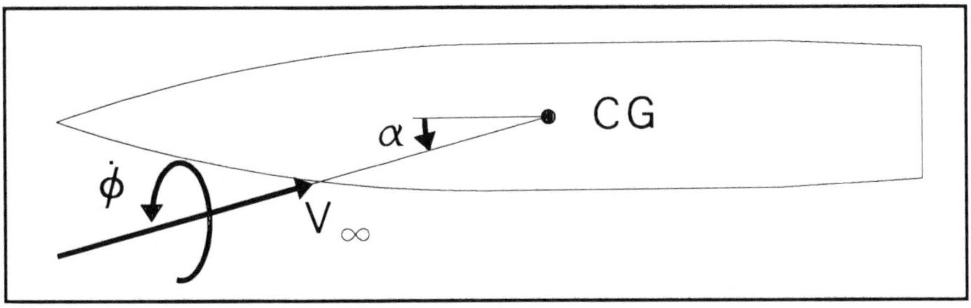

Figure 1. Schematic of coning motion.

centrifugal and Coriolis force terms. His computed results compared well with experimental results and with estimates of pitch-damping coefficients using a linear theory. Later studies by Agarwal and Rakich[4] and Lin[5] also employed rotating coordinate frames to compute the supersonic viscous flow about conical bodies in coning motion. More recently, Weinacht and Sturek[6] performed computations for finned projectiles in coning motion to determine the pitch damping coefficients.

In each of the previous efforts, the pitch damping coefficients were determined from the side moment due to lunar coning motion, which required the Magnus moment to be neglected. The predictions of the pitch damping coefficients presented here were determined from the side moment due to a specific combination of spinning and coning motion that allows the side moment due to this motion to be directly related to the pitch damping force and moment coefficients.

In this chapter, application of this approach has been made to axisymmetric shell. The flow field about these projectiles in steady coning motion has been successfully computed using the parabolized Navier-Stokes (PNS) computational approach of Schiff and Steger.[7] The computations are performed in a rotating coordinate frame similar to that employed originally by Schiff. From the computed flow field, the side moments due to coning motion, spinning motion, and combined spinning and coning motion are used to determine the pitch-damping coefficient. Computations have been performed for a generic shell configuration, a secant-ogive cylinder boattail. Predictions are made for various length to diameter ratios and supersonic Mach numbers.The PNS predictions are compared with predictions made using a Euler code originally reported by Schiff. Results are also presented for a series of ogive-cylinder configurations that have been fired through an aerodynamics range located at the U.S. Army Research Laboratory, Aberdeen Proving Ground, MD. Comparison between PNS results and range data are made.

Theoretic Background

As was discussed previously, steady coning motion is defined as the motion performed by a missile flying at a constant angle with respect to the freestream velocity vector (angle of attack) and undergoing a rotation at a constant angular velocity about a line parallel to the freestream velocity vector and coincident with the projectile center of gravity. This is shown schematically in Figure 1. With respect to a nonrolling coordinate frame, the vertical and horizontal components of the angle of attack, α and β, can be shown to be sinusoidal, constant amplitude pitching motions as the projectile rotates about the freestream velocity vector. Because the components of the angle of attack differ in phase by a quarter cycle, the total angle of attack, $\alpha_t = \sqrt{\alpha^2 + \beta^2}$, is constant. It can be seen then that coning motion is composed of a specific linear combination of two orthogonal planar pitching motions.

Side moment for lunar coning motion. In the current study, there are two particular types of coning motion of interest. The first motion is described as lunar coning motion. In lunar coning motion, the angular velocity of the body results only from the rotation of the body about the freestream velocity vector. By applying linear flight mechanics theory, the side moment produced by lunar coning motion can be written in terms of other aerodynamic coefficients as shown:

$$C_n = \delta(\frac{\dot{\phi} l}{V})\ [\ C_{n_{p\alpha}} + (C_{m_q} + C_{m_{\dot{\alpha}}})] \tag{1}$$

The side moment coefficient, C_n, represents the net moment that causes rotations of the body out of the plane of the angle of attack. For lunar coning motion, the side moment consists of contributions from the pitch damping moment coefficient, $C_{mq} + C_{m\dot{\alpha}}$ which is a moment proportional to the angular rate, and the Magnus moment coefficient, $C_{np\alpha}$ which accounts for a side moment due to flow asymmetries from a combination of spin and angle of attack. This relationship is identical to that presented by Schiff and Tobak[2] for bodies of revolution. Assuming that the side moment due to lunar coning motion and the Magnus moment can be determined, this relationship will allow the pitch damping coefficient to be determined.

Because the projectile body does not rotate with respect to the pitch plane while undergoing lunar coning motion, the flow, when observed in the coning coordinate frame, will be steady for both axisymmetric and non-axisymmetric bodies. If the Magnus moment can be neglected, lunar coning motion can be used to determine the pitch damping coefficients for a wide class of symmetric flight bodies using steady flow approaches. As an example, pitch damping predictions for finned projectiles have been obtained using lunar coning motion.[6]

Side moment for combined spinning and coning motion. A second type of coning motion can be formulated that allows the side force and moment to be directly related to the pitch damping force and moment. In this motion, the angular velocity of the body is described by a rotation of the body about the freestream velocity vector as well as a rotation of the body about its own axis (spinning motion). The rate of spin is equal in magnitude to the coning rate, but opposite in sign.

For this type of coning motion, the side moment can be written as shown:

$$C_n = \delta\left(\frac{\dot{\phi} l}{V}\right)\left(C_{m_q} + C_{m_{\dot{\alpha}}}\right) \tag{2}$$

Here, the side moment is directly proportional to the pitch damping moment coefficient. Despite the simplicity of this expression, the Magnus problem has not been entirely removed from the problem, however. This is because the motion in the coning coordinate frame (computational frame) involves coning and spinning motions. Thus, the computational approach must be capable of capturing the Magnus effect; otherwise, the pitch damping predictions will be in error by the degree to which the Magnus force and moment are improperly determined. Additionally, because of the combined spinning and coning motions, the fluid motion will only be steady for axisymmetric projectile geometries.

Similar expressions relating side force due to coning to the pitch damping force and Magnus force for the cases of lunar coning and combined spinning and coning can be developed using the same approach as previously discussed.

Computational Approach

Computation of the viscous flow field about the axisymmetric shell configurations was accomplished by solving the thin-layer Navier-Stokes equations using a PNS technique. The computations were performed in a coordinate frame that rotates at the coning rate of the projectile. The fluid flow relative to the rotating coordinate frame does not vary with time, allowing the steady (nontime varying) Navier-Stokes equations to be applied. The solution of the steady Navier-Stokes equations can be performed at a reasonable computational cost. In order to implement the rolling coordinate frame, the governing equations were modified to include the effects of centrifugal and Coriolis forces. The steady thin-layer Navier-Stokes equations are shown:

$$\frac{\partial \hat{E}}{\partial \xi} + \frac{\partial \hat{F}}{\partial \eta} + \frac{\partial \hat{G}}{\partial \varsigma} + \hat{H}_c + \hat{H} = \frac{1}{Re} \left(\frac{\partial \hat{S}}{\partial \varsigma} + \hat{S}_c \right) \tag{3}$$

Here, $\hat{E}$, $\hat{F}$, and $\hat{G}$ are the inviscid flux vectors, $\hat{S}$ is the viscous flux vector, $\hat{H}_c$ and $\hat{S}_c$ are inviscid and viscous source terms due to the cylindrical coordinate formulation, and $\hat{H}$ is the source term containing the Coriolis and centrifugal force terms which result from the rotating coordinate frame. Each of these matrices are functions of the dependent variables represented by the vector $q(\rho, \rho u, \rho v, \rho w, e)$, where ρ and e are the density and the total energy per unit volume, and u, v, and w are the velocity components in axial, circumferential, and normal directions. The source term containing the Coriolis and centrifugal force terms is also a function of the angle of attack, coning rate, center of gravity position, and spatial coordinates.

The pressure, p, which appears in the flux terms, can be related to the dependent variables by applying the ideal gas law.

$$p = (\gamma - 1) \left[e - \frac{\rho}{2} (u^2 + v^2 + w^2) \right] \tag{4}$$

The turbulent viscosity, μ_t, which appears in the viscous matrices, was computed using the Baldwin-Lomax turbulence model.[8]

The thin-layer equations are solved using the PNS technique of Schiff and Steger.[7] Following the approach of Schiff and Steger, the governing equations, which have been modified here to include the Coriolis and centrifugal force terms, are solved using a conservative, approximately factored, implicit finite-difference numeric algorithm as formulated by Beam and Warming.[9]

The computations presented here were performed using a shock fitting procedure reported by Rai and Chaussee.[10] This procedure solves the five Rankine-Hugoniot jump conditions, two geometric shock propagation conditions, and one compatibility equation to determine the values of the five dependent variables immediately behind the shock, as well as the position of the shock. By including the implicit part of the source term due to the rotating coordinate frame in the circumferential inversion, the shock fitting procedure of Rai and Chaussee can be used without modification, as long as the free-stream conditions are modified to account for the rotation due to the coning motion.

At the body surface, no-slip, constant wall temperature boundary conditions were applied. For the cases with spin, the tangential velocity, v, was set equal to the local velocity of the body surface due to solid body rotation.

The computational results presented here were obtained using a grid that consisted of 60 points between the body and the shock. In the circumferential direction, gridding was performed over a 360 degree sector due to the lack of symmetry from the combination of angle of attack, spin, and coning motion. Thirty six grid points were used in the circumferential direction. Grid resolution studies showed virtually no difference in the computed aerodynamic coefficients as the number of grid points were increased from 36 to 72 circumferential points using cylindrical coordinates. The computations were performed using a Cray X-MP supercomputer and typically required less than 20 minutes of CPU time for complete calculation over a single configuration.

Results

Computations were performed to determine the aerodynamics of several axisymmetric shell configurations in steady coning motion. The first set of predictions is for a generic shell configuration; a secant ogive cylinder boattail (SOCBT). Calculations were performed over a range of Mach numbers and body lengths. The PNS results were compared with Euler code results. A second set of results was obtained for a series of ogive-cylinder configurations (Army-Navy spinner rocket, ANSR) that were fired through an aerodynamics range located at the Ballistic Research Laboratory. The computational results are compared with the aerodynamics determined from the inflight motion of the projectile.

Results for the SOCBT. A schematic of the SOCBT is shown in Figure 2. The PNS results were compared with pitch damping results obtained using the inviscid code developed by Schiff.[3] Results were obtained for three body lengths (5, 6, and 7 calibers) and a range of supersonic Mach numbers. The center of gravity position used in the calculations was located at sixty percent of the body length from the nose. The PNS (viscous) results were obtained using the combined spinning and coning motion approach, whereas the Euler (inviscid) results were obtained using lunar coning motion. The use of lunar coning motion in the inviscid approach, instead of combined spinning and coning motion, is a consequence of the viscous nature of the spin boundary condition.

Figure 3 shows the variation of the pitch damping moment coefficient as a function of Mach number for the three body lengths. In the supersonic regime, the pitch damping moment shows a decreasing trend with increasing

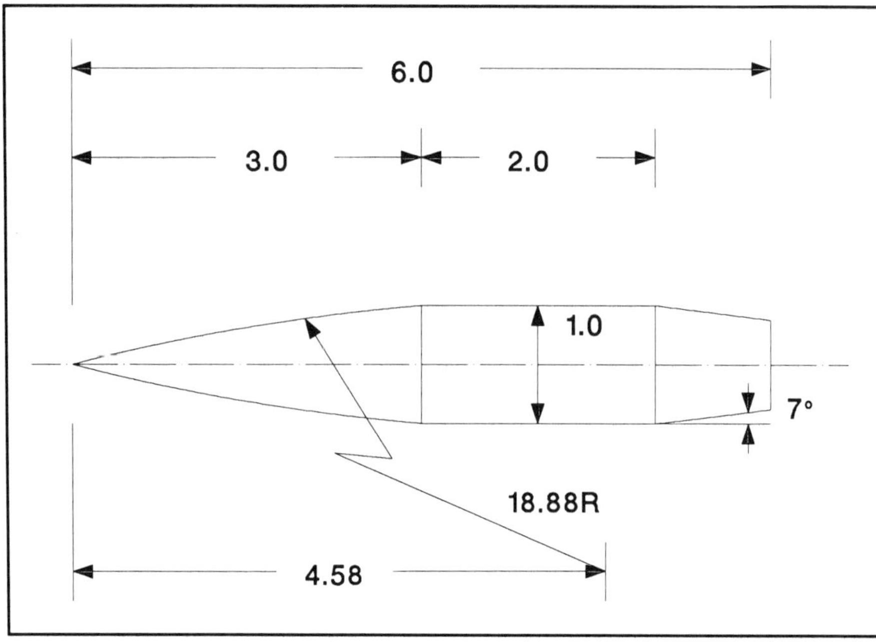

Figure 2. Schematic of SOCBT configuration.

Mach number. The inviscid results at L/D=7 show a maximum near Mach 2. The results also show a significant increase in the pitch damping moment with increasing body length. Although not shown here, the computed results show that the boattail reduces the pitch damping moment by 20 to 30 percent compared with a completely cylindrical afterbody.

The comparison of the pitch damping moment predictions from the PNS approach and the Euler approach shows very good agreement across the range of parameters considered here. This is not surprising because the pitch damping appears to be primarily an inviscid phenomenon. However, the degree of agreement between the PNS and inviscid code is not an absolute indicator of the magnitude of viscous effects, because there does appear to be some code-to-code variation of the predicted pitch damping that is on the order of the viscous effect. The relevance of these differences is probably not significant, especially in light of difficulties in measuring this coefficient experimentally. The predicted variation of the pitch damping force coefficient with Mach number for the SOCBT configuration is shown in Figure 4. These results were obtained from the side force due to combined spinning and coning motion. The results show a strong decrease in the coefficient with increasing Mach number across the range of parameters examined here. The

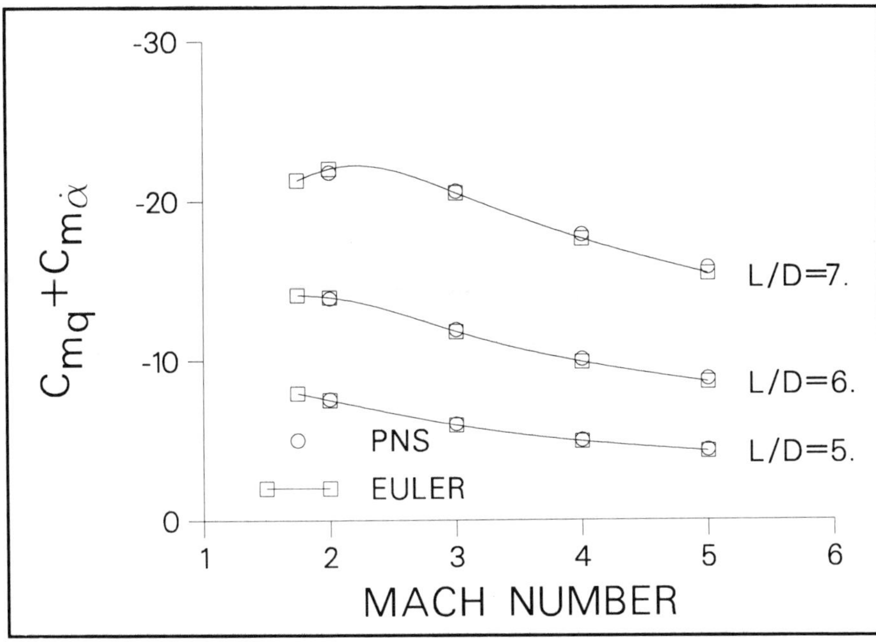

Figure 3. Pitch damping moment coefficient versus Mach number for various body lengths, SOCBT configuration.

results also show an increase in the coefficient with increasing body length. The PNS results are in good agreement with the Euler results.

Results for the Army-Navy spinner rocket. Computations were performed for the ANSR series of projectiles. These projectiles were fired in an aerodynamics range, and the aerodynamics were determined from the projectile motion.[11] The projectiles consisted of a two-caliber ogive nose with several different length cylindrical bodies, as shown in Figure 5. The total body lengths were five, seven, and nine calibers. For each body length, projectiles were fabricated and fired with three different center of gravity (CG) locations. This allowed the aerodynamic forces to be determined from the variation of the aerodynamic moments with CG location.

Figure 6 shows the variation of the pitch damping moment coefficient with CG location for the five, seven, and nine caliber bodies at Mach 1.8. Figure 7 shows a similar plot at Mach 2.5. In both of the figures, the computed results are compared with the experimental measurements. The computational results are in excellent agreement with the experimental data. The computational predictions were made by computing the flow field for each of the CG locations. This was performed because the coning motion produces a rotation

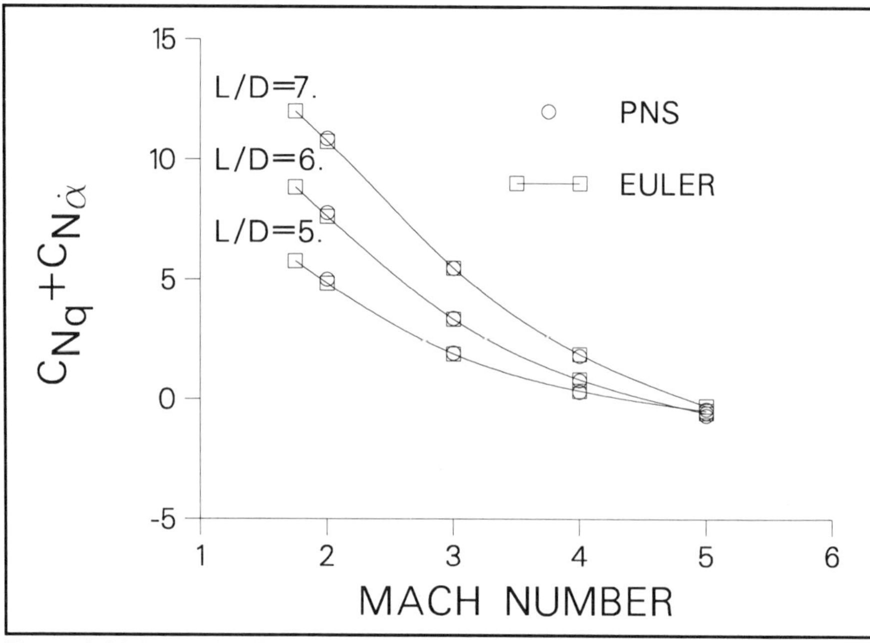

Figure 4. Pitch damping force coefficient versus Mach number for various body lengths, SOCBT configuration.

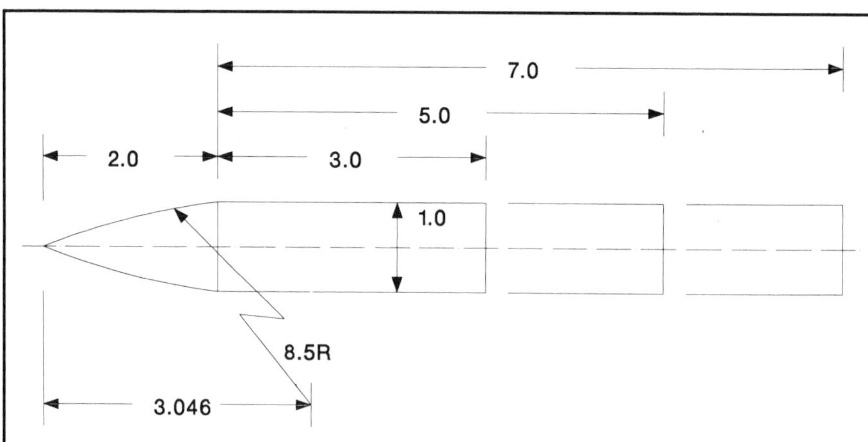

Figure 5. Schematic of the Army-Navy spinner rocket (ANSR).

about the CG, producing a different flow field in each case. These results are shown by the open triangular symbols. However, once the aerodynamics of a given configuration are determined, the CG translation relationships[1] can be applied to predict the aerodynamic coefficients for the same configuration

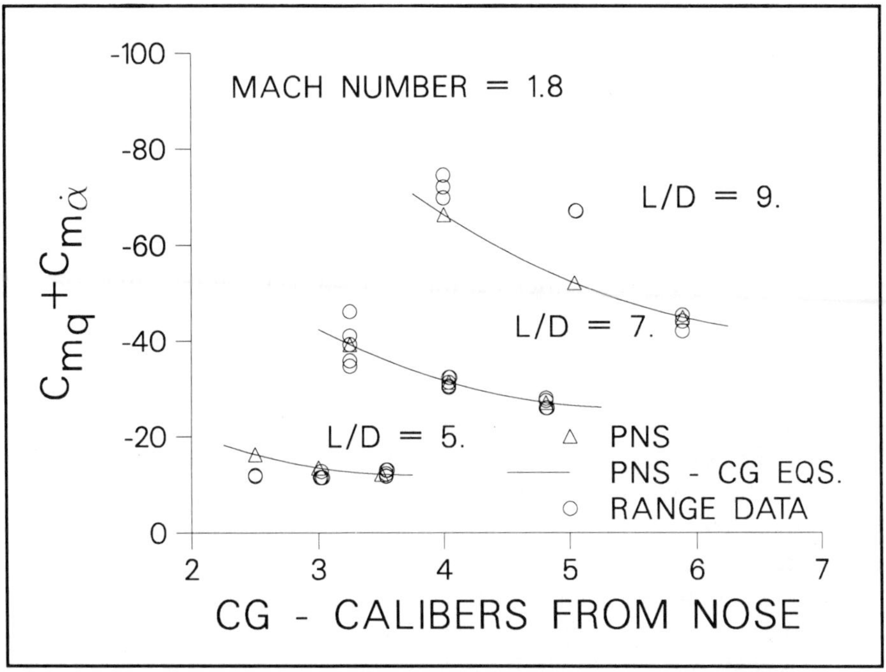

Figure 6. Pitch damping moment coefficient versus CG position, Mach 1.8, ANSR.

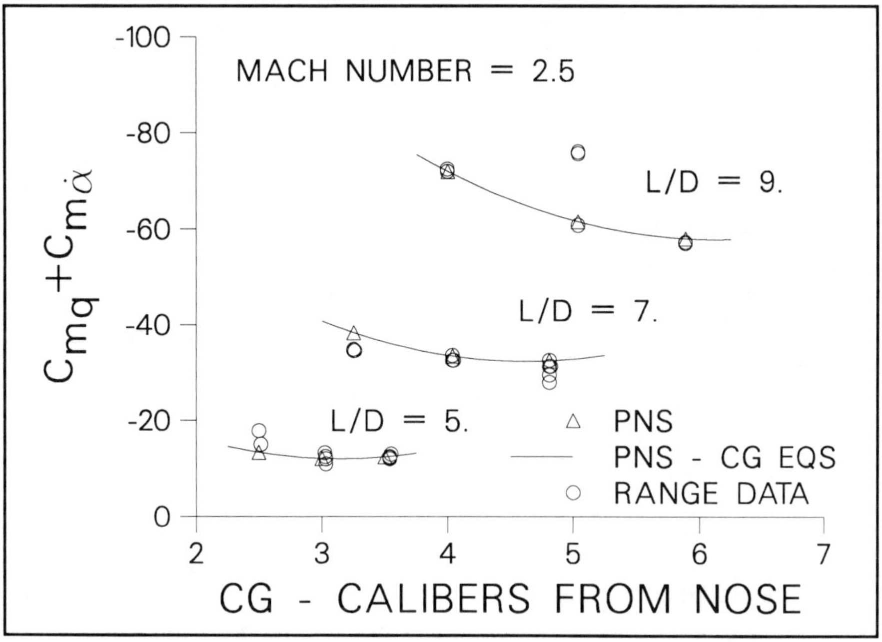

Figure 7. Pitch damping moment coefficient versus CG position, Mach 2.5, ANSR.

with a different CG location. Several of these relationships are shown below:

$$\begin{aligned} \hat{C}_{N\alpha} &= C_{N\alpha} \\ \hat{C}_{m\alpha} &= C_{m\alpha} - s_{cg}C_{N\alpha} \\ \hat{C}_{N_q} + \hat{C}_{N\dot{\alpha}} &= C_{N_q} + C_{N\alpha} \\ \hat{C}_{m_q} + \hat{C}_{m\dot{\alpha}} &= C_{m_q} + C_{m\dot{\alpha}} - s_{cg}\,(C_{N_q} + C_{N\dot{\alpha}}) \\ &\quad + s_{cg}C_{m\alpha} - s_{cg}^2 C_{N\alpha} \end{aligned} \tag{5}$$

The aerodynamic coefficients for the modified configuration are denoted by the "^" whereas the aerodynamic coefficients for the baseline configuration are shown on the right-hand side of the equations. The center of gravity shift, s_{cg} , is in calibers and is positive for a CG shift toward the nose. Using these relationships and the predicted aerodynamic coefficients for the middle CG position, the variation of the pitch damping moment coefficient with CG location was determined. This variation is shown in Figures 6 and 7 by the solid line. The difference between the pitch damping moment coefficients predicted from the CG translation relationships and the pitch damping moment as determined from the direct computations is less than 0.1 percent. This serves as a consistency check for the computational approach.

It is noted that at both Mach numbers, for the middle CG position of nine caliber body, there are several experimental data points that deviate from the trend shown by the predictions and the apparent trend shown by the experimental data. The cause of this is unknown.

As just seen from the center of gravity translation relationships, the pitching moment coefficient, $\hat{C}_{m\alpha}$, varies linearly with the CG shift. The slope of the variation is the normal force coefficient slope, $\hat{C}_{n\alpha}$. By firing projectiles with the same external shape but with different center of gravity positions, the normal force can thus be determined from the variation of the pitching moment with CG location. A similar approach can be used to determine the pitch damping force coefficient from the variation of the pitch damping moment coefficient with CG location. Because the pitch damping moment varies in a nonlinear fashion with the CG shift, a modified damping moment is defined as shown below. When the left hand side of this equation is plotted as a function of CG shift, the results should be a line with slope equal to the pitch damping force coefficient.

$$\hat{C}_{m_q} + \hat{C}_{m\dot{\alpha}} - s_{cg}\hat{C}_{m\alpha} = C_{m_q} + C_{m\dot{\alpha}} - s_{cg}\,(C_{N_q} + C_{N\dot{\alpha}}) \tag{6}$$

It should be noted that for all practical purposes, determination of the pitch damping force from range firings can only be obtained from the variation of the pitch damping moment with CG locations because the pitch damping

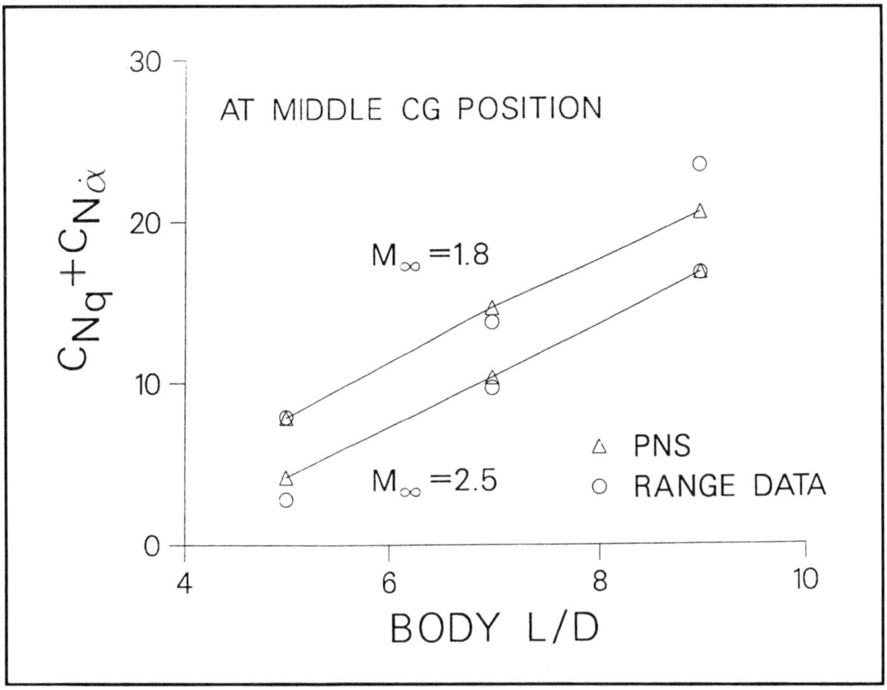

Figure 8. Pitch damping force coefficient versus body length, middle CG position, ANSR.

force coefficient contributes very little to the inflight motion of the projectile. Hence, direct determination is not practical.

From the CG variation of the pitch damping moment, the pitch damping force coefficient was determined from the experimental measurements. Figure 8 shows the variation of the pitch damping force coefficient with body length for the middle center of gravity location. Note that, unlike the normal force coefficient, the pitch damping force varies with CG position. The agreement between the computational predictions and experimental results is within the experimental accuracy and show the correct variation with body length and Mach number.

As mentioned previously, pitch damping predictions were obtained using the combined spinning and coning motion that allows the pitch damping force and moment to be determined directly from the side force and moment. The expected differences between applying combined spinning and coning motions, and lunar coning motion is reflected in the Magnus moment coefficient. In the current effort, the Magnus force and moment have been computed for the ANSR configuration, and comparison has been made with range data obtained from the same series of firings as shown in Figures 9 and 10. The

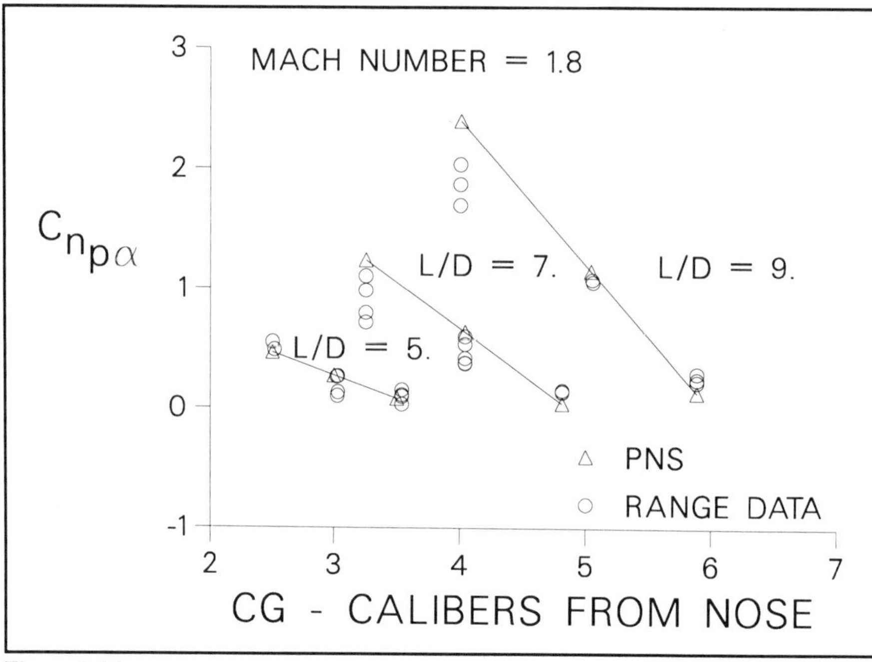

Figure 9. Magnus moment coefficient versus CG position, Mach 1.8, ANSR.

computed results were obtained for a fully turbulent boundary layer. (There is some evidence from the experimental program to indicate laminar flow over a portion of the body, particularly near the nose.) The computational results are in reasonable agreement with the experimental data. The predictions show that determining the pitch damping coefficient from the side moment due to lunar coning motion and completely ignoring the Magnus moment will result in errors of less than five percent for this configuration.

Results were also obtained to determine the pitch damping coefficient from separate calculations involving lunar coning and spin. The results, not shown here, are in excellent agreement with the results obtained using a single calculation utilizing combined spinning and coning motion. The maximum difference between the two approaches was less than 0.1 percent. This demonstrates the lack of coupling between the spinning and coning motions over the range of coning rates, spin rates, and angles of attack considered here. (It interesting to note that at higher angles of attack, a nonlinear variation of Magnus moment with angle of attack was predicted. Even at these angles of attack, no coupling between the coning and spinning motions was observed.)

Predicted in-flight motion of the ANSR. Prediction of individual aerodynamic coefficients, such as pitching moment, lift, or pitch damping moment coefficients, can provide useful information about some of the aerodynamic

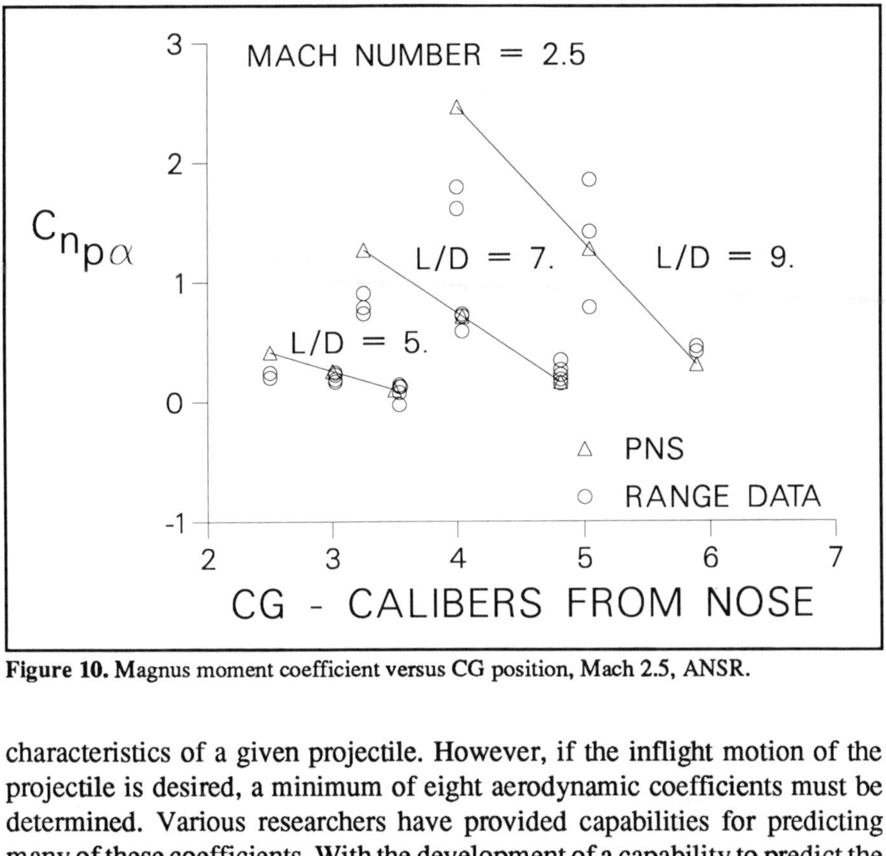

Figure 10. Magnus moment coefficient versus CG position, Mach 2.5, ANSR.

characteristics of a given projectile. However, if the inflight motion of the projectile is desired, a minimum of eight aerodynamic coefficients must be determined. Various researchers have provided capabilities for predicting many of these coefficients. With the development of a capability to predict the pitch damping force and moment coefficients, it is now possible to determine all of the necessary coefficients required to predict the inflight motion of axisymmetric shell using computational fluid dynamics (CFD) techniques. To date, this capability has been demonstrated for supersonic, small angle of attack flight. Extension of the capability to predict pitch damping in the transonic and subsonic flow regimes appears possible and is the subject of ongoing research.

As a demonstration of this capability, the inflight motion of the ANSR configuration was predicted using coefficients determined using the computational approach. The predicted motion was compared with the motion obtained from the range firings. Figure 11 shows the angular motion of the five caliber long projectile launched at Mach 1.8. This three-dimensional plot displays the vertical and horizontal components of angle of attack, α and β, as a function of the distance downrange, Z. The thickened curve displays the three-dimensional motion. Two-dimensional projections of the motion in the vertical and horizontal planes are also shown. It is useful to regard the curves

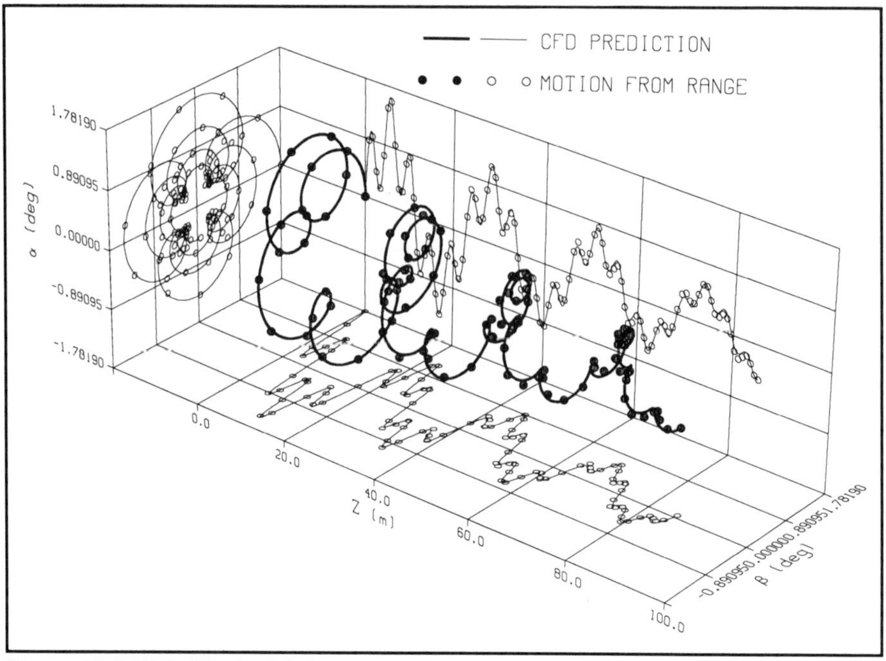

Figure 11. Inflight Yawing Motion for ANSR projectile, L/D=5, Mach 1.8.

shown here as the path traversed by the nose of the projectile as it flies downrange. The motion predicted using the CFD derived aerodynamic coefficients is denoted by the solid lines, whereas the motion obtained from the range firings is shown by the circular symbols. The agreement between the two motions is excellent.

Figure 12 shows a similar result for the inflight motion of a seven caliber long version of the ANSR launched at Mach 2.5. Again, excellent agreement is observed.

Conclusions

A computational approach for predicting the pitch damping coefficients using steady coning motion has been successfully applied to several axisymmetric shell configurations. Through the use of a combined spinning and coning motion, the pitch damping force and moment have been obtained directly from the side force and moment using a single calculation. This approach does not require that the Magnus force or moment be ignored or determined from an auxiliary calculation, as in the case of lunar coning motion.

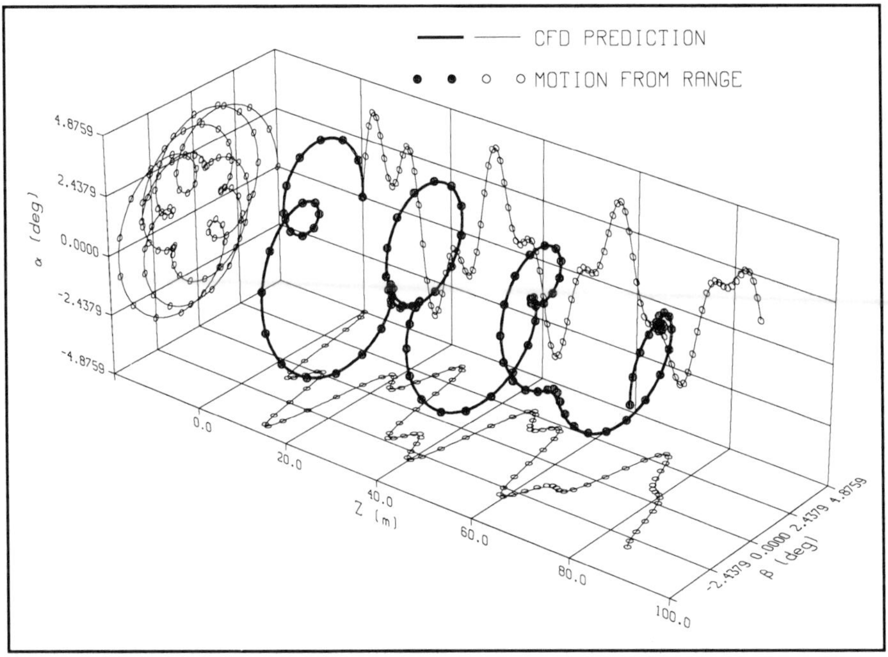

Figure 12. Inflight Yawing Motion for ANSR projectile, L/D=7, Mach 2.5.

The computational predictions for the SOCBT configuration showed good agreement with results obtained with a previously published inviscid code. The results showed an increasing trend in the pitch damping coefficient with increasing length-to-diameter ratio and a decreasing trend in the coefficient with increasing Mach number. The predictions of both the pitch damping force and moment coefficients for the the Army-Navy spinner rocket were seen to be in excellent agreement with the data obtained from aerodynamic range testing. The computational results predicted the correct variation in the pitch damping moment coefficient with changing center of gravity location, body length, and Mach number. For each of the configurations examined here, the effect of viscosity on the pitch damping coefficients was small.

The capability for predicting individual aerodynamic coefficients, such as the pitch damping force and moment coefficients, is significant because it allows particular aspects of the projectile aerodynamic performance to be evaluated. The ability to predict the pitch damping coefficients has additional significance because a complete capability based on numeric solution of the Navier-Stokes equation now exists for determining the basic set of aerodynamic coefficients required to predict the supersonic, small angle of attack, inflight motion of axisymmetric projectiles. For the Army-Navy spinner rocket, the predicted inflight motion obtained from the computed aerodynamic coefficients showed excellent agreement with the inflight motion obtained from range firings.

The use of the combined spinning and coning motion is currently being applied to time marching codes for the prediction of pitch damping at subsonic through low supersonic velocities. In this velocity regime, it is expected that viscous effects will be of greater importance than in the supersonic regime examined here.

References

1. C.H. Murphy, *Free Flight Motion of Symmetric Missiles*, Report No. 1216 (document AD A442757, U.S. Army Ballistic Research Laboratory, Aberdeen Proving Ground, MD, July 1963.)
2. L.B. Schiff and M. Tobak, *AIAA Journal* **8**(11), 1953-1957 (1970)
3. L.B. Schiff, *AIAA Journal* **10**(11), 1517-1522 (1972).
4. R. Agarwal and J.V. Rakich, *Computation of Supersonic Laminar Viscous Flow Past a Pointed Cone at Angle of Attack in Spinning and Coning Motion* (AIAA Paper 78-1211, AIAA 11th Fluid and Plasma Dynamics Conference, Seattle, WA, July 1978).
5. T.C. Lin, *A Numerical Study of the Aerodynamics of a Reentry Vehicle in Steady Coning Motion* (AIAA Paper 78-1358, AIAA Atmospheric Flight Mechanics Conference, Palo Alto, CA, August 1978).
6. P. Weinacht and W.B. Sturek, *Navier-Stokes Predictions of Pitch Damping for Finned Projectiles Using Steady Coning Motion* (AIAA Paper 90-3088, AIAA 8th Applied Aerodynamics Conference, Portland, OR, August 20-22, 1990).
7. L.B. Schiff and J.L. Steger, *AIAA Journal* **18**(12), 1421-1430 (1980).
8. B.S. Baldwin and H. Lomax, *Thin Layer Approximation and Algebraic Model for Separated Turbulent Flows* (AIAA Paper 78-257, 16th Aerospace Sciences Meeting, January 1978).
9. R. Beam and R.F. Warming, *AIAA Journal* **16**(4), 85-129 (1978).
10. M.M. Rai and D.S. Chaussee, *New Implicit Boundary Procedure: Theory and Applications* (AIAA Paper 83-0123, 23rd Aerospace Sciences Meeting, Reno, Nevada, January 1983).
11. C.H. Murphy and L.E. Schmidt, *The Effect of Length on the Aerodynamics Characteristics of Bodies of Revolution in Supersonic Flight* (Report No. 876, U.S. Army Ballistic Research Laboratory, Aberdeen Proving Ground, MD, August 1953).

Operator-in-the Loop Simulation for Acquisition

Edward J. Haug
Carver Distinguished Professor
The University of Iowa
Iowa City, IA 52242

Recent developments in high-speed dynamic simulation of vehicle systems using recursive dynamics and parallel computers are outlined, as they contribute to creation of a real-time operator-in-the-loop simulation at an engineering level of realism. It is shown that the combination of new analytical and parallel computing methods has yielded a speed-up of factor of 1,000 since 1987, which now enables engineering simulations to be run in real-time. Implementation of these methods with simulator motion systems and computer image generation systems developed for aircraft flight simulation are shown to permit engineering design trade-offs to be made, with the operator included in a highly realistic simulation of designs, before they are committed to hardware. Ground-vehicle driving simulators that now exist or are under development support a revolutionary new opportunity in both civilian and military vehicle acquisition. Unique characteristics of ground vehicle environments and the driver's use of visual, motion, and audio cues to effectively control and operate a vehicle are shown to be within the realm of feasible implementation in advanced ground-vehicle driving simulators. This situation is contrasted to high performance aircraft simulation in which motion cues are either impossible to simulate or are not of value. Projects are under way to establish the feasibility of driver-in-the-loop simulation for acquisition, for example, virtual prototyping. Experience gained in existing simulators to support design decision making is also being investigated.

High-Ratio Bandwidth Reduction of Video Imagery for Teleoperation

Dale R. Shires
Frank F. Holly
Phillip G. Harnden
U. S. Army Research Laboratory
Human Research and Engineering Directorate
Aberdeen Proving Ground, MD 21005-5001

To lessen the risks to the soldier and as a force multiplier, some of the Army's future combat vehicles will be teleoperated. Video or forward-looking infrared sensors will be mounted on the vehicle with visual information about the surrounding scene transmitted back to a remote control station. Since visual imagery is a large user of bandwidth, and bandwidth is very limited on the battlefield, there is a need to find a means of reducing this data transmission rate. We have accomplished this by developing a process combining three distinct techniques.

First, the video imagery to be transmitted from the vehicle to the control station is collected from a stereo camera pair, mounted on the vehicle, digitized, and then compressed. Transmission requirements are reduced by using only one field from each camera. These fields are merged to form one video frame for compression. Transmission requirements are reduced by using only one field from each camera. These fields are merged to form one video frame for compression and transmission. The compression technique employed follows the Joint Photographic Experts Group standard. In this process, pixel intensities are transformed into their frequency components through the use of the Discrete Cosine Transform. Many of the erroneous high frequency components are then discarded through a quantization step, and remaining coefficients are run-length encoded using Huffman codes. The overall compression ratio achieved with these steps is 25:1.

Second, passive range finding is performed after the compressed images are received and decompressed at the control station. The frame is broken down into its corresponding left and right image planes, with the goal being to determine the stereo disparity map of the images. These disparity values (the horizontal pixel difference between matching points in the left and right images) are used in conjunction with known parameters from the camera geometry to determine the x (horizontal), y (vertical), and z (distance) world locations of points on the video monitor. The stereo matching technique currently used is a stochastic one, which is based upon simulated annealing and produces a dense disparity map.

Third, a compression ration of 90:1 is achieved by transmitting only one frame every 3 seconds rather than the usual 30 frames per second. By itself, this is unacceptable from a human factors viewpoint because of the large voids in the visual information presented to the remote operator. Therefore, we have developed an image processing algorithm for this step that creates dynamic imagery (synthetic optic flow) during these 3-second intervals. That is, each period consists of one "real" frame and 89 synthetic ones.

In computations vision, optic flow is often used to obtain information about the environment by analyzing the relationships among movement vectors. For example, motion parallax can be used to obtain information about relative distances, and this information in turn can be used to segment a three-dimensional scene into its object components. In this instance, however, optic flow is used in the synthetic rather than the analytic sense; this same relationship between scene information and movement vectors can be inverted to produce optic flow given certain information about the environment. This synthetic optic flow generation on the display is done through the creation of "brightness objects," which, in the initial or "real" frame, are co-extensive with the image pixels. In the synthetic frames, each brightness object retains its initial brightness from the real frame throughout all 89 synthetic frames but is expanded and moved downward and outward in each frame. The amount of movement and expansion is determined by the direction of the vehicle. The results of these steps is smoothly flowing imagery which appears very close to that produced by full bandwidth transmission. However, the bandwidth reduction ratio achieved by this system is 25 x 90 = 2250:1.

An Approach to Mitigating Atmospheric Effects on Image-Based Pattern Recognition by Neural Networks

Donald W. Hoock, Jr.
U.S. Army Research Laboratory
Battlefield Environment Directorate
White Sands Missile Range, NM 88002-5501

Low levels of battlefield haze and smoke can seriously degrade performance of image-based pattern recognition by neural networks. Research at the U.S. Army Atmospheric Sciences Laboratory has led to a novel method to mitigate the impact of atmospheric transmission and contrast losses by directly including the atmosphere in the neural net design and training procedure. An advantage to this approach is that, once trained, the neural net classifier does not need special inputs about the prevailing atmospheric conditions in the field. Equations for the effects of contrast and transmission losses on the outputs from image-based, single and multilayer perceptron neural network classifiers were first derived mathematically. Changes to the neural net configuration that would minimize this impact were then identified. The resulting method was found to require that the atmosphere itself, in the form of images containing no target, be included as a separate recognition class. However, instead of training the network to recognize an image as belonging to this atmospheric class, it is trained to ignore the influence of this class on the other target class outputs. For investigation purposes, single and multilayer neural networks were trained using standard back-propagation methods to recognize objects both with and without using the atmospheric mitigation method. The mitigation method tested successfully both on noisy images degraded by simulated atmospheric losses and on actual images of targets degraded by white fog oil smoke, black carbon smoke, and fine clay dust. Of course, a sufficiently obscured target cannot be classified by any technique. However, for atmospheric transmissions down to 20 percent and for a wide range of path radiance levels, the method has shown a significant performance improvement.

Using Iterative Orthophoto Refinements to Correct Digital Elevation Models

F. Raye Norvelle
U. S. Army Corps of Engineers
Fort Belvoir, VA 22060-5546

Advances over the past 15 years in computer speed, memory size, and disk storage have allowed corresponding advances in the performance and speed of digital stereo correlation methods. Digital correlation methods are used to generate Digital Elevation Models (DEMs) from digitized, stereopairs of aerial photographs at rates exceeding 200 elevations per second. These elevations have to be edited, however, and advances in editing techniques have not kept pace with those of digital correlation. The purpose of this summary is to introduce a new technique for editing DEMs that is commensurate in accuracy and speed with the correlation process that produces the DEMs. The editing technique, "Iterative Orthophoto Refinements (IOR)," was developed by the U.S. Army Topographic Engineering Center (TEC) and is an out-growth of ongoing research efforts in digital stereo correlation.

The IOR process primarily uses software modules that generate orthophotographs and perform digital stereo correlations. The process is based on the condition that an identical pair of orthophotographs can be generated from a stereopair of source images and a DEM. Provided that the orientation data for the original images are correct, any geometric mismatches between the stereopair of orthophotographs will be due to errors in the DEM. The mismatches (x-parallax) can be measured automatically using the digital correlation software, converted to equivalent elevation error, and used to update the original DEM. This procedure is iterated (usually two or three times) until the refinements to the orthophotographs are negligible in size and consequently, the DEM is correct.

Aerodynamic Heating of Unsteady Hypervelocity Projectile

Rao Yalamanchili
U.S. Army Research, Development and Engineering Center
Picatinny Arsenal, NJ 07806-5000

Recent and novel concepts for launching hypervelocity projectiles, such as electromagnetic propulsion, are currently being investigated. It is anticipated that these projectiles will be subjected to severe heat because of aerodynamic behavior of projectile traveling at hypersonic speed through dense air, and also due to thermal effects generated during launch. Defense critical technologies list include, among twenty technologies, the following five that are pertinent to this project: computational fluid dynamics (CFD), hypervelocity projectiles, simulation and modeling, parallel computer architecture, and software producibility. An accurate modeling of thermodynamic phenomena associated with the launch and flight of an unsteady hypervelocity projectile is presented by simultaneous consideration of CFD models, transient three-dimensional heat conduction in projectile, and equation of motion of projectile. In general, CFD model, thermal model, and model of projectile motion are not independent. Each and every model requires initial and boundary conditions to be obtained from the solution of other models. These models are coupled and are to be solved simultaneously by future supercomputers in order to make it realistic. Because its accurate solution is beyond the scope of existing supercomputers and modeling methodologies, salient features and individual effects can be examined while preparing for an accurate debut in the near future.

Toward this goal, the importance of unsteady effect and validity of analogy between momentum and energy transfer are examined for unsteady turbulent flow pertinent to small and large caliber systems. In all cases, the analogy underestimates as much as 100 percent for small times; however, the same analogy overestimates the convective heat transfer coefficients up to 50 percent for large times. Therefore, simulation of unsteady behavior and without any analogy is recommended in any analysis.

Research has shown that it impossible to form numerous implicit, Crank-Nicholson, and explicit finite difference schemes for transient three-dimensional heat transfer of hypervelocity projectiles based on new as well as other Laplacian finite difference approximations. One of the objectives behind the derivation of new finite difference approximation for Laplacian is to compare the most popular numeric methods, for example, finite element and finite difference methods for three-dimensional heat transfer problems. It is important to note that there is a need to find a best scheme in case one tackles the combined transient three-dimensional heat conduction and convection problems simultaneously because of limitation on supercomputers storage and speed in existence today. Even though the classic (h-version) finite element method is found not to be as good as several finite difference schemes, the modern p-version finite element method yields a much higher convergence rate, especially for singular problems, such as melting. Also, simple physics of melting is demonstrated on typical kinetic energy projectiles by transient three-dimensional finite column code developed.

New Directions in Computational Dynamics at TARDEC

Roger A. Wehage
James L. Overholt
U.S. Army Tank-Automotive Command
Research, Development and Engineering Center
Warren, Ml 48397-5000

The U.S. Army Tank-Automotive Command (TARDEC) uses high-resolution, computer-based vehicle models to determine upper bounds on safe operating performance envelopes. Validated model predictions are used to answer numerous technical evaluation questions regarding design and product improvements. More stringent performance specifications are being placed on Army vehicles to meet changing battlefield requirements, and emerging technology is pushing prototype vehicle designs toward increased subsystem complexity to meet these demands. The operator's influence on system performance is more critical than ever because military vehicle design trends are placing major emphasis on systems that adapt to human capabilities. Thus, computer-based vehicle models must be able to accurately account for operator inputs when they are significant. The most feasible solution is to put the human directly into the dynamic vehicle model. This requires real-time simulation and graphic user interface capability, because the operator thinks and reacts in real-time. Real-time simulations are now feasible with the development of a methodology called Symbolically Optimized Vehicle Analysis System (SOVAS), which generates high resolution models. A second methodology called Dynamic Response—Interactive Vehicle Emulator (DRIVE), which interfaces human operators with SOVAS-generated vehicle models, is also being developed. Through research and development, TARDEC is solving the Army's vehicle dynamic modeling and simulation needs as well as investigating its future requirements.

Advanced Materials Science

25

Correlation of Ballistic Performance to Shear Instability Studies in High Strength Steels

John H. Beatty
Morris Azrin
U.S. Army Research Laboratory
Materials Directorate
Watertown, MA 02172-0001

Ballistic failure of thin, high hardness armor plates often occurs through a shearing mechanism called plugging. Unstable, adiabatic shearing creates a cylindrical "plug" of armor that is pushed and ejected through the rear of the armor plate along with the projectile. The concentration of strain within this narrow shear band precludes significant deformation of the material away from the band, so only a small portion of the armor is actually decelerating the projectile. Accordingly, this mode of penetration absorbs very little energy, and the ejected plug and projectile can cause considerable damage to personnel and equipment. The central areas of interest to this study are the localization of the shear deformation and microstructural influences on this localization process. We have examined a high-strength steel, VAR 4340, in which only one microstructural component was purposely varied (the size and distribution of grain refining carbides) while other changes to the material's microscopic substructure were minimized. A variety of laboratory shear tests at various strain rates show how these microstructural changes altered the material's shear behavior which, in turn, explains the changes in ballistic performance.

Introduction

Steel is the most commonly used armor material because it is economical and can be easily fabricated. It often serves a dual role as both a structural member and protective armor in combat vehicles. For these reasons, steel will continue to be an important component of many advanced armor systems for the foreseeable future.

The two properties of steel that most affect its ability to resist penetration by projectiles are the hardness and cleanliness of the steel. For many years, the general relationship between increased resistance to penetration at higher hardnesses has been known,[1,2] with no other single metallurgical factor as important. However, reduced resistance due to shattering of steels harder than 52-54 on the Rockwell C scale (HRC) drastically limits their usage above this hardness range. The cleanliness of the steel also affects the ballistic response;[3] cleaner steels have fewer impurity inclusions. Inclusions degrade both the resistance to penetration and shattering resistance of the armor simultaneously. Until recently, the dominant factors of hardness and cleanliness have obscured the influences of many other metallurgical factors on ballistic performance.

The advent of processing techniques such as Vacuum Arc Remelting (VAR) and Electroslag Remelting (ESR) allow steelmakers to achieve greatly reduced impurity levels and inclusion contents, and this in itself has led to improvements in ballistic properties.[3] These techniques have also allowed for the examination of very "clean" steels to understand the influences of other microstructural components without the overshadowing presence of the inclusions. Recent studies[4] in VAR and ESR 4340 steels have suggested that grain-refining carbides can play an important role during ballistic penetration, and this study is directed towards obtaining a fundamental understanding of how and to what degree these carbides alter the ballistic behavior of high strength steels.

Background

Microstructures of armor steels. Armor steels are used in a wide variety of metallurgical conditions, depending upon the particular application. Rolled Homogeneous Armor (RHA)[5] and High-Hard Armor[6] are the two most common types, each being available in different thicknesses and hardnesses.

To obtain optimum ballistic properties, steel should be processed to produce a fully martensitic microstructure. Martensite is a phase that can form in armor steels upon rapid cooling (quenching). Martensite is extremely hard and strong, which is beneficial to ballistic properties. However, in the as-quenched state, the martensite is usually too brittle, and must be tempered to improve the toughness of the steel. The increase in toughness gained through tempering comes at the expense of a lower hardness. For each application, the trade-off between hardness and toughness should be appropriately balanced to provide sufficient structural integrity and maximum ballistic protection.

The resulting microstructure of steel armor after quenching and tempering is quite complex, and a typical example is shown in Figure 1, along with a schematic. The martensite in these steels is arranged in a lath structure, with individual laths ranging from several to tens of microns in width. The laths are found in several orientations or "packets" within each martensite colony. In addition to the lath structure, other important microstructural components are present. Grain boundaries remain from the austenitic phase that preceded martensite formation, and these are referred to as prior austenite grain boundaries. Along these prior austenite grain boundaries, grain-refining carbides (usually between 0.5 to 0.05μm in diameter) are found. This microstructural component is the central focus of the present study. Grain refining carbides should not be confused with temper carbides that are precipitated during tempering and are much smaller, needle-shaped precipitates. Finally, impurity inclusions can also be found within conventionally produced steels. These particles typically range in size from 1 to 10μm in diameter, and are very detrimental to ballistic properties. As previously mentioned, refinement techniques such as VAR or ESR remove most of these impurities. The steel used in this study, VAR 4340, has little, if any, of these inclusions, so the effects of other microstructural factors on ballistic properties can be examined.

Ballistic failure of thin armor plate. A shearing mechanism called plugging is often the predominant failure mode during ballistic penetration of thin, high-hardness steel armor plate. Figure 2 illustrates this failure mode. The projectile impacts the armor, and the material surrounding the impact zone starts to undergo plastic shear deformation. Kinetic energy from the projectile is transferred to the plate by the plastic deformation, and the armor in this region begins to increase in temperature. Because the deformation is occurring very rapidly, there is insufficient time for the steel to conduct heat away from the deforming zone so the deformation can be considered adiabatic. Two processes are competing with each other at this stage. The deformed substructure within the material causes stiffening (work hardening). Simultaneously, the increase in temperature created by the rapid energy absorption softens the material. At some point in time, the rate of thermal softening surpasses the rate of work hardening, and the material becomes

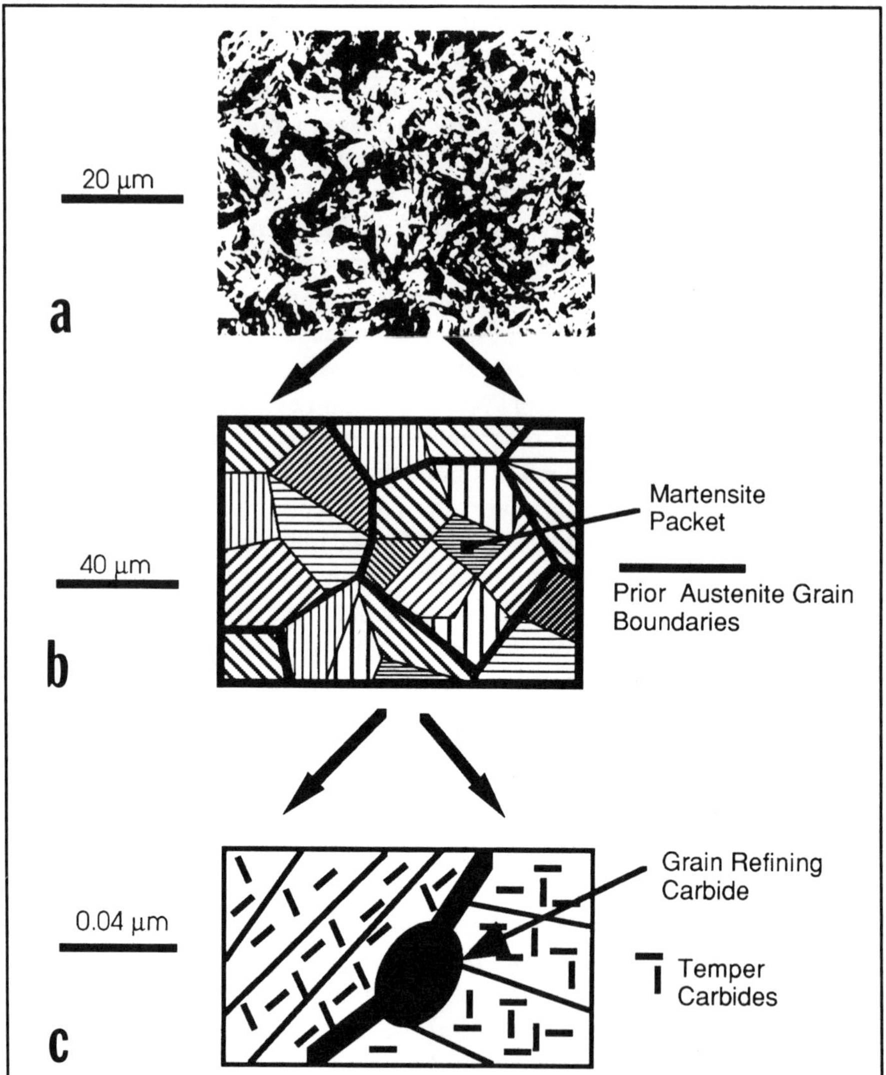

Figure 1. Microstructure of VAR 4340 steel. **a**, optical micrograph showing lath structure; **b**, prior austenite grain boundaries and packets of lath martensite; **c**, grain-refining carbides are found along prior austenite grain boundaries; temper carbides are found within the laths.

unstable. The load bearing capacity in this area is reduced by the thermal softening, which leads to even more severe deformation, which in turn causes even more thermal softening, and so on. This shear instability causes the localization of the shear strain into a narrow region known as an adiabatic shear band.

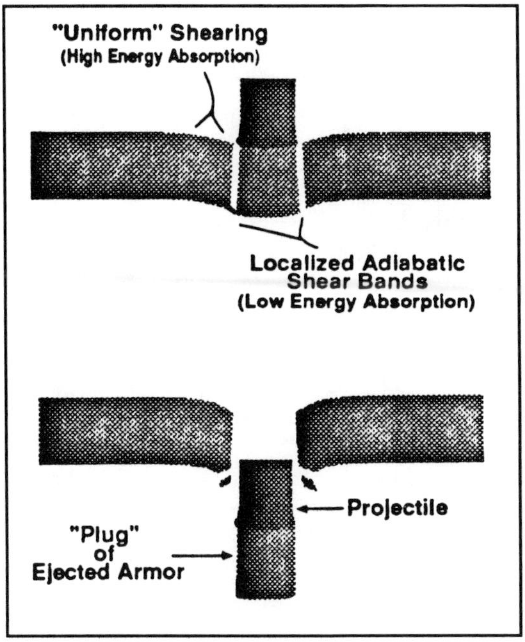

Figure 2. Cross-sectional depiction of failure of armor by plugging. The projectile strikes the armor and creates localized shearing. A plug of armor is ejected through the rear face of the plate.

The adiabatic shear band will grow to surround a cylindrical plug of armor. Because the material within the band continues to thermally soften at an increasing rate, little energy is absorbed by the plate from this point on. The remaining kinetic energy of the projectile is divided between itself and the plug, and both the plug and projectile exit the rear of the armor plate at high velocity, causing damage to any personnel and equipment in their paths. Delaying or preventing the onset of the adiabatic shear banding will allow the armor to absorb more of the projectile's kinetic energy, which in turn can prevent complete projectile penetration of the armor.

General Approach

It is the aim of this study to characterize how microstructural factors affect the formation of adiabatic shear bands. To accomplish this, it is necessary to control the microstructures being evaluated, and to examine their response to shear deformation over a variety of strain rates. An innovative heat treatment has been used to alter the size and distribution of the grain refining carbides, while leaving the rest of the microstructure virtually unchanged. Laboratory shear tests have been performed over a variety of stress states and

Table 1
Alloy Composition, Weight Percent

C	Mn	P	S	Si	Cr	Ni	Mo	Cu
0.42	0.46	0.009	0.001	0.28	0.89	1.74	0.21	0.19

Table 2
Heat Treatment Schedule

1st Austenitization Temperature,°C 1hr	Oil Quench	2nd Austenitization Temperature, °C 15 Min	Oil Quench	Temperature °C 2hrs
845	X	845	X	200
925	X	845	X	200
1010	X	845	X	200
1090	X	845	X	200

strain rates. Finally, the ballistic performance under plugging conditions has also been evaluated. Performing these shear instability tests on several carefully controlled armor microstructures produces a more complete understanding of the plugging process. This should eventually lead to improved armor alloys and armor processing techniques.

Experimental Procedures

The alloy used in the current study is a VAR 4340 steel. Its composition and the four heat treatment schedules utilized are shown in Tables 1 and 2, respectively. Double austenitization treatments were used to produce four different microstructures, all with a hardness of 52HRC. A unique distribution of grain refining carbides is produced during the initial austenitization because the carbide dissolution and coarsening rates vary at each temperature. The second short austenitization is performed to obtain a uniform prior austenite grain size. The time of the second austenitization is not sufficient to significantly change the carbide distribution. Thus, we are able to produce four microstructures with equivalent hardnesses and similar prior austenite grain sizes but with unique distributions of grain refining carbides. These microstructures have been examined in our earlier studies.[7,8] Microstructural characterization included optical microscopy, scanning electron microscopy, and transmission electron microscopy.

Ballistic testing was also performed on each microstructure. A standard

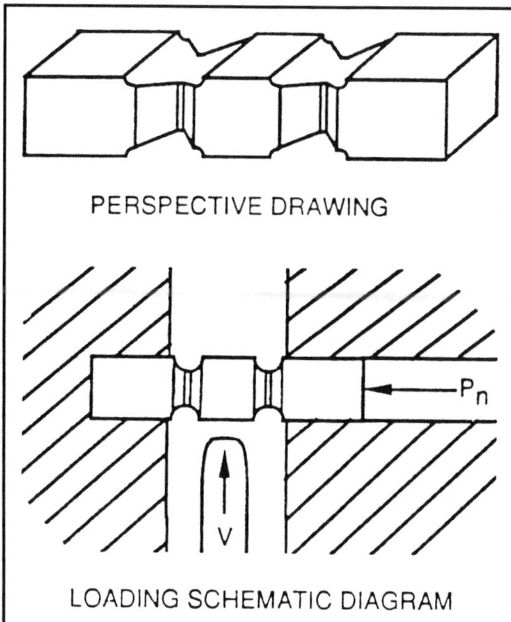

Figure 3. Schematic of the double-linear-shear specimen and its loading configuration.[7,9] For both quasistatic and dynamic tests. An axial load, P_n, can be imposed. "V" indicates the velocity of the plunger or pendulum causing shear. Only quasistatic data are reported in this paper.

V_{50} ballistic performance evaluation was made using two 12"x12"x3/8" plates of each microstructure with 0.50 caliber APM2 projectiles fired at 0° obliquity. The V_{50} is defined as the velocity at which a given projectile has a 50 percent probability of creating a "complete" penetration of the armor plate. A complete penetration is one in which some fragments of either the projectile or the armor plate perforate a thin aluminum witness plate placed behind the target armor. Commonly, as is the case for the current tests, the measured V_{50} is the average velocity of six different shots (three partial and three complete penetrations) that fall within a narrow specified velocity range of 125 feet per second.

The ballistic data will be compared to shear instability tests over a variety of strain rates. Shear instability tests were completed by the authors and other co-workers as part of a comprehensive study and have been reported in detail previously.[7-9] A brief description of each technique follows.

Quasistatic shear tests were performed using thin-wall torsion samples and double-linear-shear[7,9] specimens. The double-linear-shear specimen configuration is shown in Figure 3. For each microstructure, a shear stress-displacement curve was obtained. Additionally, scribe lines through the gage length of each specimen were used to obtain a strain profile for each sample by measuring the angle of deflection of the scribe line. Each test was inter-

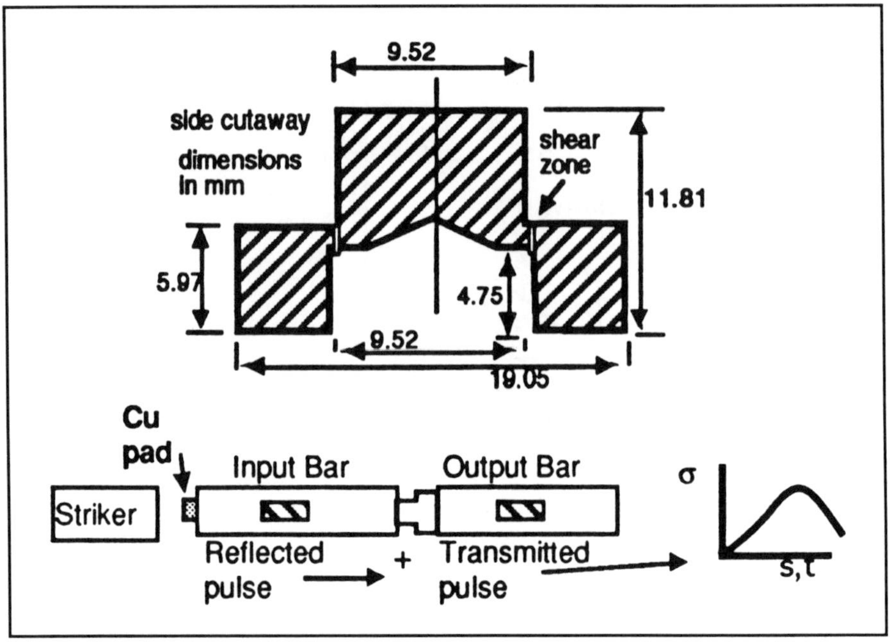

Figure 4. Side cutaway of Hat specimen with split Hopkinson apparatus.

rupted once the load decreased approximately 2 percent from the maximum. The maximum uniform shear strain was found using the strain profiles, and is referred to as the shear instability strain, γ_i.

High strain rate shear behavior was characterized using a hat-shaped specimen (developed by Meyer *et al.*[10,11]) in a split-Hopkinson compression bar (Figure 4). These experiments were part of a joint program with the University of California, San Diego.[8,12] This cylindrical sample creates adiabatic shear bands, while allowing experimental control of the strain-rates and/or displacements induced. The reflected and transmitted elastic waves in the bar allow for the measurement of stress-displacement curves for each microstructure. From this information, the energy required to initiate unstable shearing can be determined.

Results and Discussion

The quasistatic shear behavior data is given in Figure 5. The greatest resistance to unstable shearing is found for the 925°C austenitizing temperature. Cowie[7] showed that the instability strain was proportional to the carbide particle size/average particle separation distance ratio. This behavior is consistent with theories of destabilization by microvoid nucleation softening. The

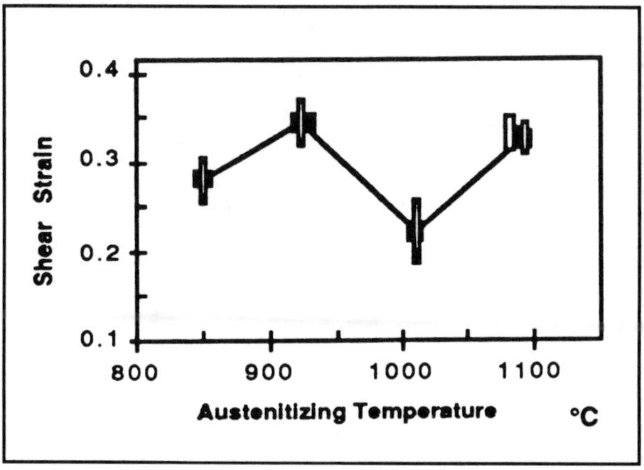

Figure 5. Shear instability strain as a function of the first austenitizing temperature for VAR 4340. The solid squares indicate the instability strain; the vertical bars give the approximate error. A two-step austenitization was used to produce four microstructures, all at 52HRC with a constant prior austenite grain size.

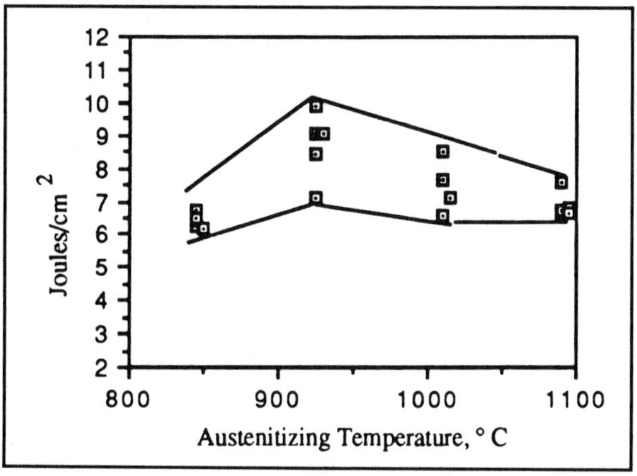

Figure 6. Energy absorbed to the point of instability per unit area sheared versus austenitizing temperature, using the Hat specimen in the split-Hopkinson apparatus.

particles themselves cause local tensile stress fields when the surrounding material undergoes pure shear. These local tensile fields are responsible for nucleating voids. Both the particle size and and the particle separation are related to this instability. This mathematical functionality suggests that the interaction of microvoids that formed on particles of average size creates the initial softening of the material and localization of the shear strain.

This view of microvoid softening in high strength steel is also supported by the stabilizing influence of hydrostatic compression on shear deformation.[9] The application of a compressive stress component reduces the local tensile stresses created at the carbides and delays the nucleation of voids. Therefore, the resistance to unstable shearing is aided by the application of compressive stresses at moderate strain rates.

The high strain rate shear behavior is shown in Figure 6. Because all four microstructures have equivalent shear strengths, the energy absorbed to the

Figure 7. Transmission electron micrograph of material taken from the center of an adiabatic shear band formed in a hat specimen. The structure is microcrystalline, with a crystallite size of 8 to 20nm.

point of instability is directly related to the shear instability strain. Therefore, Figures 5 and 6 can be qualitatively compared. The 925°C austenitization temperature still produced the best results, but the dependence on the carbide size and distribution is unclear. At the present time, there is no explanation for the observed relationship at high strain rates.

Microstructural characterization studies of specimens deformed at high strain rates may aid in the determination of the relationship between carbides and adiabatic shear in high strength steels.[8] Figure 7 shows a transmission electron micrograph of material taken from the center of a deformed Hat specimen. An extremely fine microcrystalline structure is observed, and no carbides can be detected within the shear band. The microstructure is similar in appearance to ball milled metallic powders. It is anticipated that continued analyses of deformed microstructures will give additional insight into the microstructural evolution within the shear band. This insight could lead to the production of improved microstructures for armor applications.

Ballistic V_{50} results for all four microstructures are given in Figure 8. The V_{50} performance closely resembles the results of the Hat test. Again, the 925°C austenitizing temperature produced the best microstructure, and a single peak in resistance to unstable shear was observed. These results and previous work by Meyer *et al.*[10,11] suggest that the Hat test can serve as a controlled laboratory analogue to ballistic testing of thin armor plate if

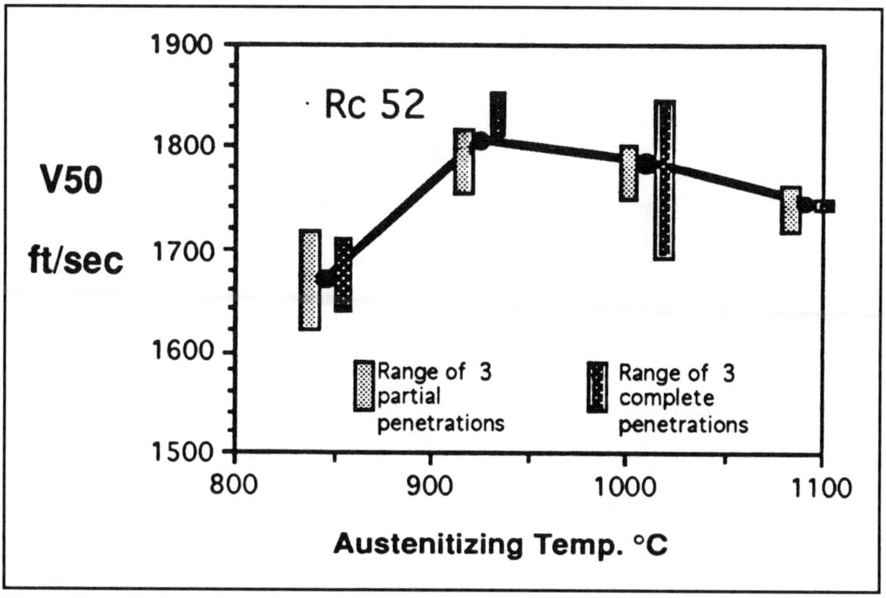

Figure 8. Ballistic results of all four microstructures. The ballistic results are in good agreement with the hat test data given in Figure 6.

plugging failures predominate. This provides us with an excellent experimental method for studying microstructural effects on armor performance. It is also important to recognize that the carbides have significantly affected the ballistic performance. Although the mechanisms remain unknown, these small components of the microstructure demonstrate that proper microstructural control must be maintained in order to optimize ballistic properties. These same results should help to develop new, unique microstructures that can lead to improved armor materials for future Army applications.

Conclusions

- Grain refining carbides significantly affect the shear behavior of high strength VAR 4340 steel.
- At low-to-moderate strain rates, a microvoid softening model can explain the observed experimental results.
- High strain rate Hat tests and ballistic evaluations show good agreement, so the Hat tests may be used to study armor plugging.
- The tailoring of common and unconventional microstructures should be explored to maximize the survivability of future Army systems.

References

1. S.J. Manganello and K.H. Abbott, *J. Materials* **7** (2), 231 (1972).

2. J.F. Mescall and H. Rodgers, in *Proceedings of the 34th Sagamore Army Materials Research Conference* (U.S. Army Materials Technology Laboratory, August 30-September 3, Lake George, NY, 1987), pp. 287-314.

3. C.F. Hickey, T.S. Thomas, and A.A. Anctil, *Comparison of Ballistic Performance of a Split Heat of ESR and VAR 4340 Steel* (document AMMRC TR 85-20, U.S. Army Materials and Mechanics Research Center, Watertown, MA, 1985).

4. E.T. Myskowski and A.A. Anctil, *Effect of Microstructure on Ballistic Performance of ESR 4340 Steel* (MTL TR 89-73, U.S. Army Materials Technology Laboratory, Watertown, MA, 1989).

5. MIL STND 12560G, Plate, Wrought, Homogeneous, 24 January 1985.

6. MIL STND 46100C, Plate, Wrought, High Hardness, 20 October 1986.

7. J.G. Cowie, *The Influence of Second Phase Dispersions on Shear Instability and Fracture Toughness of Ultrahigh Strength 4340 Steel* (document MTL TR 89-20, U.S. Army Materials Technology Laboratory, Watertown, MA, 1989).

8. J.H. Beatty *et al.*, *Formation of Controlled Adiabatic Shear Bands in AISI 4340 High Strength Steel* (document MTL TR 90-54, U.S. Army Materials Technology Laboratory, Watertown, MA, 1990). Also published in *Shock Waves and High Strain-Rate Phenomena in Materials*, M.A. Meyers, L.E. Murr, and K.P. Staudhammer, Eds. (Marcel Dekker, New York), in press.

9. J.G. Cowie, M. Azrin, and G.B. Olson, *Metall. Trans. A.* **20A**, 143 (1989).

10. K-H Hartman, H.D. Kunze, and L.W. Meyer, in *Shock Waves and High-Strain-Rate Phenomena in Metals*, M.A. Meyers and L.E. Murr, Eds. (Plenum Press, New York, 1981).

11. L.W. Meyer and S. Manwaring, in *Metallurgical Applications of Shock-Wave and High-Strain-Rate Phenomena*, L.E. Murr, K.P. Staudhammer, and M.A. Meyers, Eds. (Marcel Dekker, New York, 1986).

12. J.H. Beatty *et al.*, in *Proceedings of the Army Solid Mechanics Conference* (U.S. Army Materials Technology Laboratory, Plymouth, MA, November 1991).

26

Solid-State ^{31}P Magnetic Angle Spinning NMR Studyof G-Agent Simulants Adsorbed on Synthetic Resin Catalysts

George W. Wagner
U.S. Army Edgewood Research, Development and Engineering Center
Aberdeen Proving Ground, MD 21010-5423

William T. Beaudry
J. Richard Ward
U.S. Army Edgewood Research, Development and Engineering Center
Aberdeen Proving Ground, MD 21010-5423

A powder containing sorptive and reactive resins developed for the removal and subsequent catalytic degradation of toxic organophosphorus esters is characterized using dimethyl methylphosphonate (DMMP), p-nitrophenyl diphenyl phosphate (PNDP), and ^{31}P Magic Angle Spinning (MAS) Nuclear Magnetic Resonance (NMR). Two distinct resonances are detected for DMMP adsorbed on the resin mixture, which indicates that the ester partitions between the sorptive and reactive resin components. Preliminary ^{31}P MAS NMR magnetization-transfer experiments reveal that molecular exchange of DMMP between these sites is occurring at a rate of about 0.5 sec^{-1} at room temperature. PNDP hydrolyzes on the resin powder, apparently via its strong base sites, to yield diphenyl phosphate (DPP) and p-nitrophenoxide. The anionic DPP product is tightly bound to the resin, which suggests the formation of a stable intermediate with the reactive quaternary ammonium groups. In agreement with this observation, the reactive sites of the resin did not "turn over" when reacted with excess PNDP. These results demonstrate that the reaction with PNDP is not catalytic.

Introduction

Characterizing chemical reactions on synthetic resins is of current interest in our laboratory, because such resins may be tailored to adsorb and react with toxic organophosphorus esters that inhibit acetylcholinesterase activity.[1] In particular, macroreticular resins possess a porous network that should enhance adsorption and provide sites to attach reactive functional groups.[2,3] The U.S. Army recently adopted a kit for skin decontamination using macroreticular resins[4] in lieu of natural clays such as "Fuller's earth." The resin employed, termed XE-555, is actually a mixture of three components: a high surface area, carbonaceous macroreticular styrene/divinylbenzene resin, and two macroreticular styrene/divinylbenzene ion-exchange resins. One ion-exchange resin is acidic and the other is basic. The strong acid resin contains sulfonic acid groups, whereas the strong base resin has quaternary ammonium groups. In solution, the toxic acetylcholinesterase inhibitors, such as sarin or soman, are rapidly hydrolyzed in strong bases and also exhibit acid-catalyzed hydrolysis.[5]

In order for the XE-555 resin mixture to function catalytically, molecular exchange must exist between the sorptive and reactive resins, and the reactive sites must be able to effect a heterogeneous reaction with the adsorbed liquid. Previous studies in our laboratory with O,O'-diisopropyl phosphorofluoridate (DFP) have suggested partitioning in the resin mixture.[6] Evidence for this partitioning is the apparent low reactivity observed for DFP adsorbed on the resin, despite the apparent high reactivity with the individual reactive resin components. The nature of this partitioning must be understood in order to design effective systems that both adsorb and react with liquid contaminants.

This study employs ^{31}P Magic Angle Spinning (MAS) nuclear magnetic resonance (NMR) to probe the adsorption and reaction of organophosphorus esters at the molecular level. ^{31}P MAS NMR is perhaps uniquely suited for this purpose, because it can distinguish various molecular adsorption sites[7-10] as well as determine the occurrence and rate of molecular exchange between these sites.[11-15]

Experimental Design

Dimethyl methylphosphonate (DMMP), purchased from Aldrich, was used without further purification. Para-nitrophenyl diphenyl phosphate (PNDP), provided by F. R. Longo, U.S. Army Edgewood Research, Development and Engineering Center (ERDEC), was found to be greater than 99 percent pure by ^{31}P NMR. Ambergard^R XE-555 Resin, Amberlite^R IRA-900 Strong Base Resin, Amberlyst^R XN-1010 Strong Acid Resin, and Ambersorb^R XE-348F Carbonaceous Synthetic Adsorbent were obtained from Rohm & Haas Company, Philadelphia, PA. The latter three materials are similar to the components in the XE-555 resin mixture. The IRA-900, XN-1010, and XE-348F resins, manufactured in the form of small beads, were crushed with a ball mill to yield fine powders. The strong base IRA-900 resin, initially in the chloride form, was exchanged with an aqueous 5 percent NaOH solution to yield the hydroxide form. The resin was subsequently washed with deionized water to remove NaCl and excess NaOH and then filtered and dried in air at room temperature.

Solution ^{31}P NMR spectra were obtained at 81MHz on a Varian XL-200 NMR spectrometer. ^{31}P MAS NMR spectra were obtained at 101MHz using high-power proton decoupling on a home-built 6T FT-NMR spectrometer equipped with a Doty Scientific variable temperature 5mm double-tuned MAS probe. ^{31}P spin-lattice relaxation times (T_1) of less than two secondswere observed for adsorbed DMMP and PNDP, which are greatly reduced from their solution values. The short T_1 values alleviate the need for the cross-polarization (CP) technique and therefore the more quantitative conventional FT experiment was used. All chemical shifts were referenced to an external sample of 85 percent H_3PO_4. Positive chemical shifts were at higher frequency (lower field) than the reference. Various reaction products were solvent extracted from the resins, and their structures were determined using multinuclear and two-dimensional ^1H-^1H COSY NMR techniques.

^{31}P MAS NMR magnetization-transfer (MT) experiments were performed at 81MHz on the XL-200 using a Doty Scientific high-speed 7mm double-tuned MAS probe without proton decoupling. The MT spectra were recorded for DMMP vapor loaded on the XE-555 resin using the method described by Ford *et al.*[13] A FORTRAN program, MT-NMR,[13] obtained from H. O. Spivey, Oklahoma State University, was used to fit the MT data to yield the rate constants for DMMP molecular exchange.

Results For Dimethyl Methylphosphonate (DMMP)

The first phase of the study involved identifying the various adsorption sites on the XE-555 resin. DMMP is an ideal candidate for this purpose because it is a small, simple molecule that is stable except at very high or low pH. Therefore, the molecule is not expected to react with the acid or base groups on the resin. This surface probe permits NMR spectra to be obtained that are not complicated by interfering reaction products and allow the straightforward observation of chemically different adsorption sites.

^{31}P NMR spectra of DMMP vapor adsorbed on the XE-555 resin are shown in Figure 1. Spectrum 1(a) was obtained under normal solution NMR conditions and reveals two overlapping peaks at 38 and 29ppm. The relative

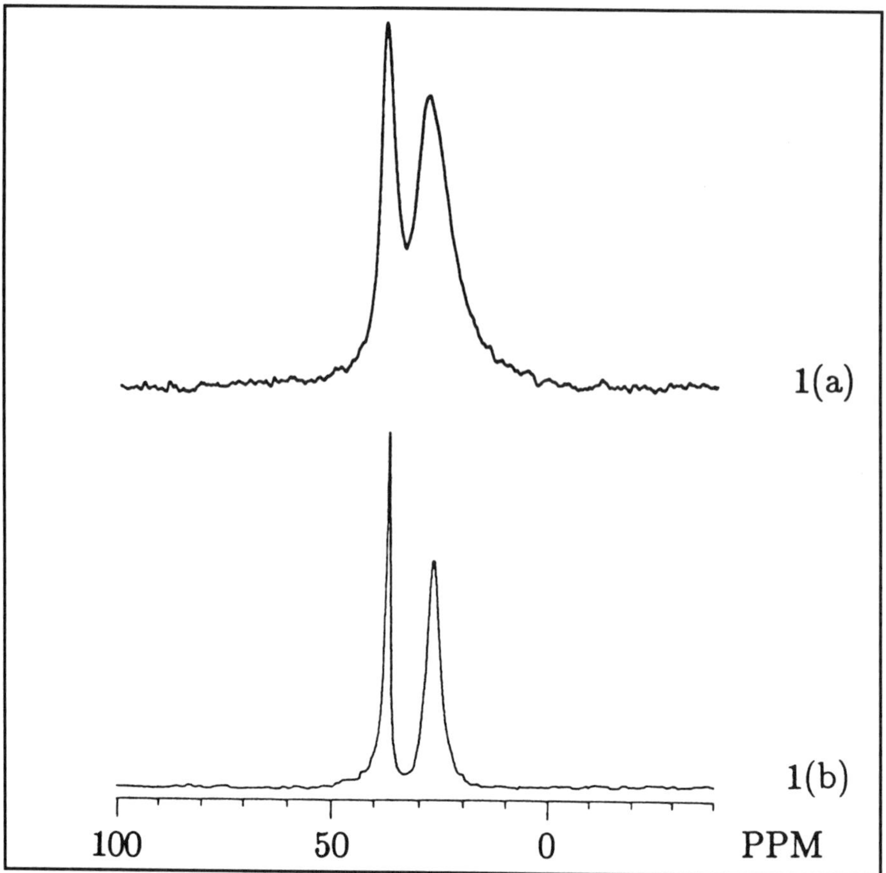

Figure 1. ^{31}P NMR spectra of DMMP vapor loaded on the XE-555 resin obtained using (**a**) normal solution NMR conditions and (**b**) MAS with high-power proton decoupling.

narrowness of these resonances immediately indicates that DMMP is undergoing fast isotropic "liquid-like" motions on the resin, which average the broad chemical shift anisotropy (CSA) powder pattern expected for static organophosphorus esters.[16-19] Spectrum 1(b) was obtained using MAS with high-power proton decoupling. This technique greatly enhances the resolution of the twin peaks as compared with the spectrum obtained without MAS. The narrow downfield resonance occurs at 36.2ppm, which is the same value observed for DMMP in methanol solution. Methanol 2N HCl (80:20 methanol/H_2O) extractions of the DMMP from the resin result in only the presence of DMMP in the filtrate; thus, DMMP does not react with the resin but, in fact, resides in two different adsorption sites. The spectra further show that any

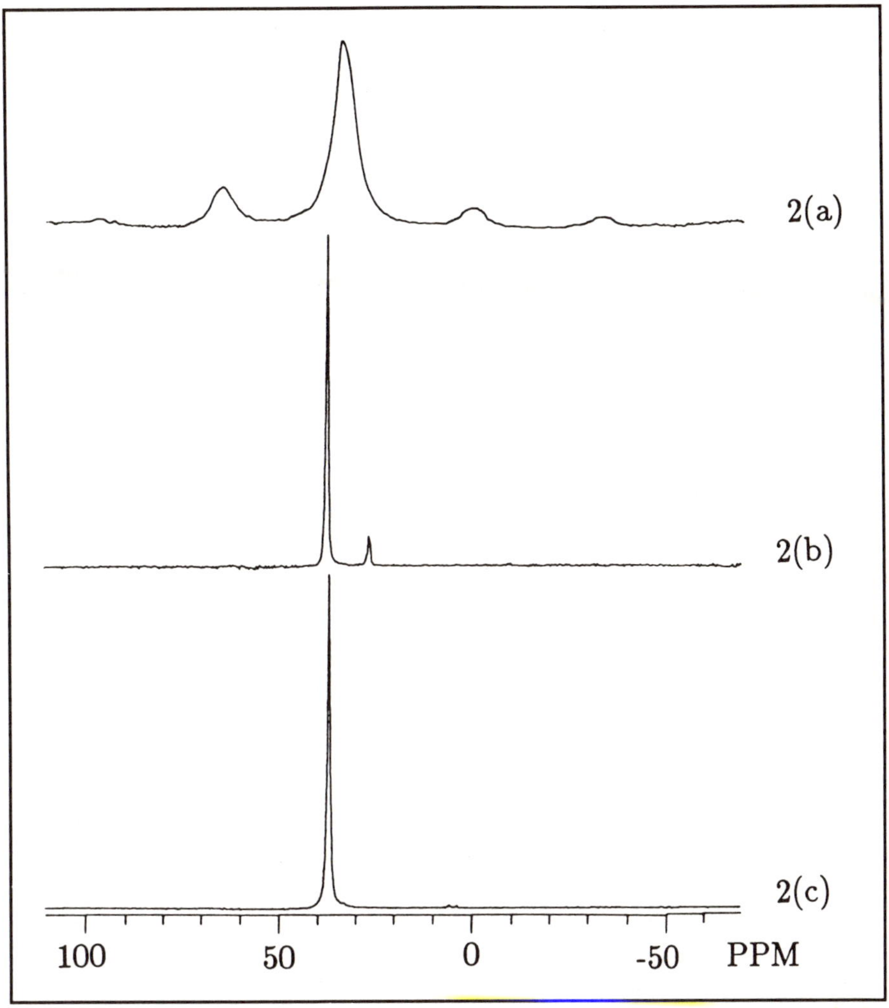

Figure 2. ^{31}P MAS NMR spectra of DMMP vapor loaded on (**a**) carbonaceous macroreticular adsorbent, 15 weight percentage; (**b**) macroreticular strong base resin, approximately 3 weight percentage; and (**c**) macroreticular strong acid resin, 44 weight percentage.

molecular exchange occurring between the two sites is slow on the NMR time scale by virtue of the presence of both individual resonances.[20] The width of the peaks in the lower MAS spectrum (~100Hz) places an upper limit on the exchange rate of 300sec^{-1} at room temperature.

In order to identify the nature of the two adsorption sites, ^{31}P MAS NMR spectra were obtained for DMMP vapor adsorbed on the three individual resins similar to the components in the XE-555 resin. These spectra are shown

Table 1
^{31}P MAS NMR Assignments for Organophosphorus Esters Adsorbed on Various Resins

Ester	XE-555	XE-348F	IRA-900(OH$^-$)	XN-1010
DMMP	36.2* 26.3†	33	36.9 25.5‡	36.3
PNDP	-17.9	-17.8	-17.1	-17.9
DPP	-10.1	not observed	-9.6	not observed
DPPA	not observed	not observed	not observed	-10.7
PNPPA	not observed	not observed	not observed	-11.7

*Ion exchange resin components.
†Carbonaceous adsorbent component.
‡Methyl methylphosphonate hydrolysis product.

in Figure 2. In the case of the carbonaceous macroreticular adsorbent, 2(a), a rather broad resonance is detected at 33ppm (slightly shifted from the solution value of 36ppm) along with CSA spinning sidebands, indicating that the molecules are rather rigidly held. Conversely, DMMP remains highly mobile on the macroreticular strong base [2(b)] and strong acid [2(c)] ion-exchange resins, yielding sharp resonances at 36.9 and 36.3ppm, respectively. In the case of the strong base resin, a second signal occurs at 25.5ppm, which is due to a slow forming hydrolysis product, methyl methylphosphonate. ^{31}P NMR spectra obtained for DMMP adsorbed on various carbonaceous adsorbents yield chemical shifts ranging from circa 25 to 33ppm. Therefore, the second upfield signal observed for DMMP adsorbed on XE-555 is assigned to DMMP adsorbed on the carbonaceous adsorbent component. A complete listing of ^{31}P MAS NMR assignments for DMMP adsorbed on the various resins is given in Table 1.

^{31}P MAS NMR spectra obtained from XE-555 samples with various DMMP weight percentage loadings (spectra not shown) indicate that adsorption on the carbonaceous adsorbent is favored during small weight percentage vapor loadings. Conversely, the ion-exchange resins are allowed to fill at large weight percentage liquid loadings. The relative ratio of the two resonances in these spectra did not change over a several day period, indicating that the DMMP partition achieved during the loading process is stable at room temperature.

A reasonable explanation for the adsorption behavior of DMMP on the XE-555 resin is that the carbonaceous adsorbent possesses a limited number of relatively strong adsorption sites. This is supported by the observation of

CSA spinning sidebands for DMMP adsorbed on XE-348F [Figure 2(a)], which indicates that the molecules possess limited mobility.

The possibility of chemical exchange between the carbonaceous adsorbent and the reactive ion-exchange resins was further investigated by variable temperature ^{31}P MAS NMR. Variable temperature spectra recorded up to 50°C for DMMP adsorbed on the XE-555 resin (not shown) yield no features attributable to molecular exchange. On the contrary, at the higher temperatures the lines actually sharpen due to the increased rate of motion of the molecules within their respective adsorption environments. If molecular exchange were occurring between the sites at a rate approaching the width of the resonances (~100Hz), then a significant line broadening contribution would be evident as the temperature is raised. Therefore, molecular exchange is very slow (less than $300sec^{-1}$) even at 50°C. Site exchange might be observable by NMR above this temperature; however, the quaternary ammonium hydroxide groups begin to decompose at approximately 60°C. The variable temperature spectra were reversible within the temperature range studied.

Preliminary magnetization-transfer experiments[13,14] are consistent with DMMP exchange occurring between the carbonaceous adsorbent and the ion-exchange resins at rates of approximately $0.5sec^{-1}$ at room temperature, although we are experiencing difficulty in obtaining exact fits of experimental data to a simple two-site exchange model. It is perhaps not surprising that our system is more complex given the very heterogeneous nature of the resin. The observed exchange rate increases with increasing DMMP weight percentage loading in the 2 to 14 percent range initially examined. We are continuing these experiments in order to further characterize the exchange process.

Results for P-Nitrophenyl Diphenyl Phosphate (PNPD)

Unlike the DMMP surface probe, PNDP is expected to be more unstable with respect to hydrolysis[21-31] and, thereby, to react with the acid and base groups in the XE-555 resin. In basic solution, the major hydrolysis products are diphenyl phosphate (DPP) and p-nitrophenoxide (a yellow chromophore), although significant amounts of p-nitrophenyl phenyl phosphate (PNPP) and phenoxide also form.[21,24] For acid catalysis, diphenyl phosphoric acid (DPPA) and p-nitrophenol are the sole reported products.[31] PNDP is a white solid at room temperature, which necessitates the use of a solvent in order to load the compound onto the resin.

^{31}P MAS NMR spectra of PNDP adsorbed on the XE-555 resin from a

95 percent (v/v) ethanol/water solution are shown in Figure 3. Two peaks are present at -10.1 and -17.9ppm in the spectrum of a one-day-old sample [3(a)]. The large upfield peak at -17.9ppm, which has the same chemical shift observed for PNDP in solution,[21] is assigned to PNDP adsorbed on the XE-555 resin. The spectrum observed for adsorbed PNDP is different than the spectrum of neat PNDP solid,[32] indicating that the ester does not merely precipitate as the ethanol evaporates but is rather highly dispersed on the resin. The lack of spinning sidebands for the two peaks indicates that both adsorbed PNDP and the other adsorbed species are highly mobile and "liquid-like." The small downfield peak at -10.1ppm is due to the hydrolysis product, DPP, as confirmed by extraction and analysis by solution NMR;[21] however, no PNPP product is detected. Thus, the observed DPP hydrolysis product demonstrates that the reactive sites are accessible to PNDP.

Solvents play a major role in polymer-supported solution-phase catalysis,[2,33] and ordinarily the XE-555 resin would only contain adsorbed water. Therefore, the effect of coadsorbed solvent on the reaction was investigated in the following experiment. A PNDP/ethanol loaded sample was allowed to dry in air just to the point where no visible solvent remained; a portion of the sample was then removed and placed in a capped 5mm NMR tube. The rest of the sample remained exposed to air in order to allow further unrestricted loss of ethanol and equilibration with ambient moisture. The portions were allowed to stand for 23 days before their ^{31}P MAS NMR spectra were recorded. These spectra are also shown in Figure 3. For the capped sample [3(b)], the conversion of PNDP to DPP is nearly complete, with only a small signal remaining for unreacted PNDP. Note that the signal for DPP lacks CSA spinning sidebands, indicating that the product is still very liquid-like. However, the spectrum of the air-exposed portion [3(c)] reveals a reduced conversion along with a DPP product that is being very rigidly held by the resin.

^{1}H MAS NMR indicates that no significant amount of highly mobile ethanol remains at the end of the drying period; however, less mobile ethanol would not be detected as a result of severe linebroadening due to ^{1}H-^{1}H dipolar interactions. The capped sample does contain considerably more mobile water than the air-exposed sample, but water is ruled out as capable of enhancing hydrolysis since partially dried samples rewet with water do not yield increased PNDP conversions. Therefore, these results suggest that coadsorbed ethanol, although of limited mobility, is responsible for promoting PNDP hydrolysis, most likely by promoting PNDP migration rather than allowing release of DPP to regenerate "used" sites. This interpretation is supported by the fact that no turnover is observed when excess PNDP is allowed to react with the XE-555 resin in ethanol solution even though smaller

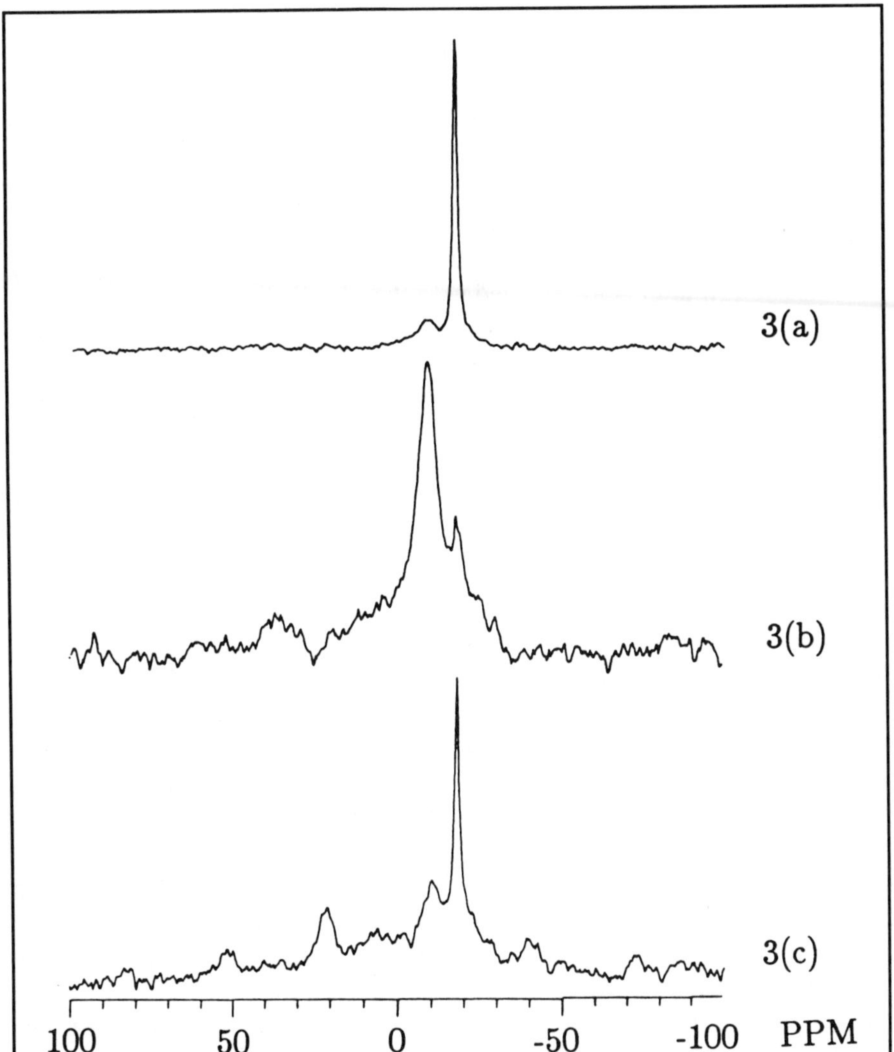

Figure 3. ^{31}P MAS NMR spectra of PNDP adsorbed on the XE-555 resin from ethanol solution. (**a**) one-day-old sample, capped to retain solvent; (**b**) 23-day-old sample, capped to retain solvent; (**c**) same as (**b**) except that the sample remained exposed to ambient humidity.

amounts react completely within four days. Also, DPP product is not extracted by ethanol or water (acidification is necessary in order to dislodge the product from the resin).

To determine the reactivity of PNDP with the various components of the XE-555 resin, ^{31}P MAS spectra were also recorded for PNDP adsorbed on the carbonaceous macroreticular adsorbent and the strong acid and strong base macroreticular ion-exchange resins. These spectra are shown in Figure 4. PNDP adsorbed on the carbonaceous macroreticular adsorbent [4(a)] yields a single peak at -17.8ppm. As expected, no reaction products are observed on this purely sorptive component. PNDP adsorbed on the strong base resin [4(b)] hydrolyzes in the same manner as on the XE-555 resin, as indicated by a peak at -9.6ppm, which is similarly assigned to the hydrolysis product DPP. In support of this assignment, the initially tan strong base resin rapidly becomes yellow after being slurried with the PNDP/ethanol solution, consistent with the release of the p-nitrophenoxide chromophore. Thc strong base resin further parallels the XE-555 resin behavior in that the spectrum of an air-exposed, 15-day-old sample [4(c)] shows a poor conversion along with a large CSA for the immobilized DPP product, whereas the capped control sample [4(d)] yielded a much better conversion along with a highly mobile, liquid-like DPP product. Again, these findings are consistent with the DPP anion remaining bound to the quaternary ammonium ion-exchange sites and enhanced PNDP hydrolysis due to its greater mobility in the presence of coadsorbed ethanol.

Spectra obtained for PNDP adsorbed on the strong acid resin, Figure 5, are quite different than the spectra obtained for both the XE-555 and strong base resins. Again, the peak at -17.9ppm in the spectrum of a one-day-old sample [5(a)] is due to PNDP adsorbed in the macroreticular region. The peaks at -10.7ppm and -11.7ppm are assigned to the acid catalyzed hydrolysis products, diphenyl phosphoric acid (DPPA) and p-nitrophenyl phenyl phosphoric acid (PNPPA), based on NMR analysis of these species in solution. The spectrum taken after the sample had been kept for 40 days in a capped 5mm NMR tube [5(b)] shows further hydrolysis of PNDP with DPPA apparently being the favored product. Solvent extraction of the strong acid resin yields the same relative amounts of PNDP, DPPA, and PNPPA in the filtrate, indicating that the products are easily removed and not tightly bound to the sulfonic acid ion-exchange sites. Spectrum [5(c)] was obtained from the same sample after it had been exposed to air for six days and shows further DPPA and PNPPA formation, along with the appearance of a third peak at -4.4ppm. This signal can be attributed to phenyl phosphoric acid formed from further hydrolysis of the PNPPA product,[21] although insufficient amounts were available for spectroscopic identification. Even after long-term air exposure, the neutral acid products on the strong acid resin remain highly mobile and liquid-like in the absence of coadsorbed ethanol, indicating an increased mobility of the neutral species as compared with the anionic base catalyzed products.

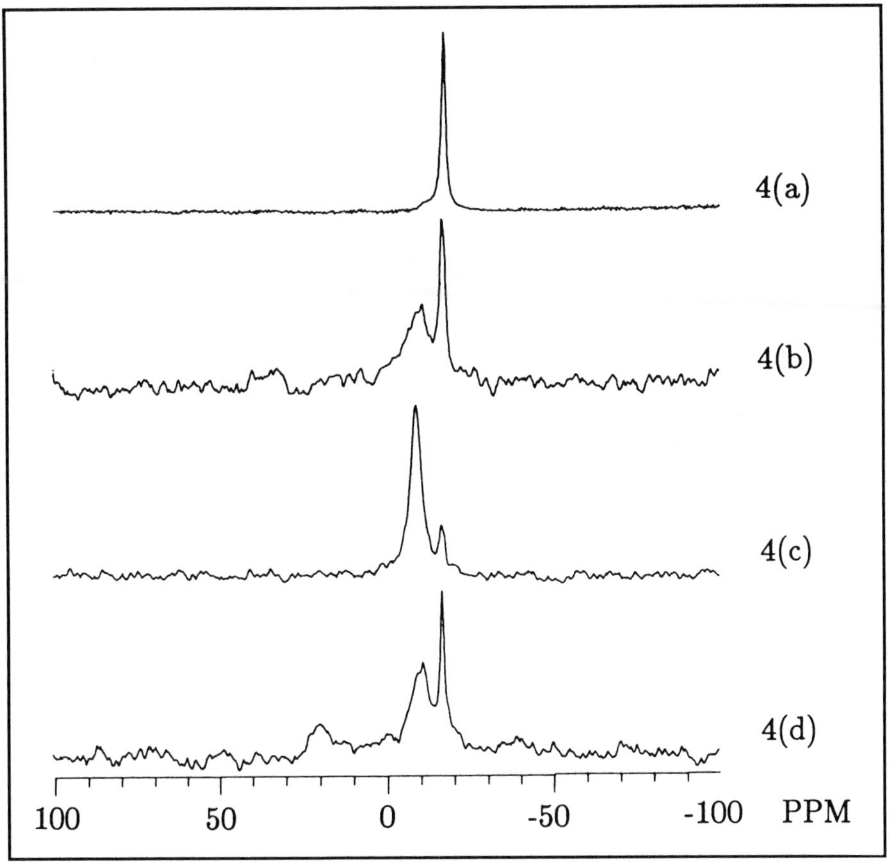

Figure 4. ^{31}P MAS NMR spectra of PNDP adsorbed on (**a**) carbonaceous macroreticular adsorbent, one day, capped to retain solvent; (**b**) macroreticular strong base resin, one day, capped to retain solvent; (**c**) macro-reticular strong base resin, 15 days, capped to retain solvent; and (**d**) macroreticular strong base resin, 15 days, exposed to ambient humidity.

Conclusions

^{31}P MAS NMR spectra of DMMP adsorbed on the XE-555 resin provide direct evidence for the partitioning of adsorbed organophosphorus esters between two different adsorption sites. Comparison of spectra obtained for DMMP adsorbed on strong acid and strong base ion-exchange resins and a carbonaceous macroreticular adsorbent (the three components of the XE-555 resin) allow the assignment of the two adsorption sites to the carbonaceous adsorbent and the ion-exchange resins. Variable temperature spectra obtained for DMMP adsorbed on the XE-555 resin indicate that DMMP exchange

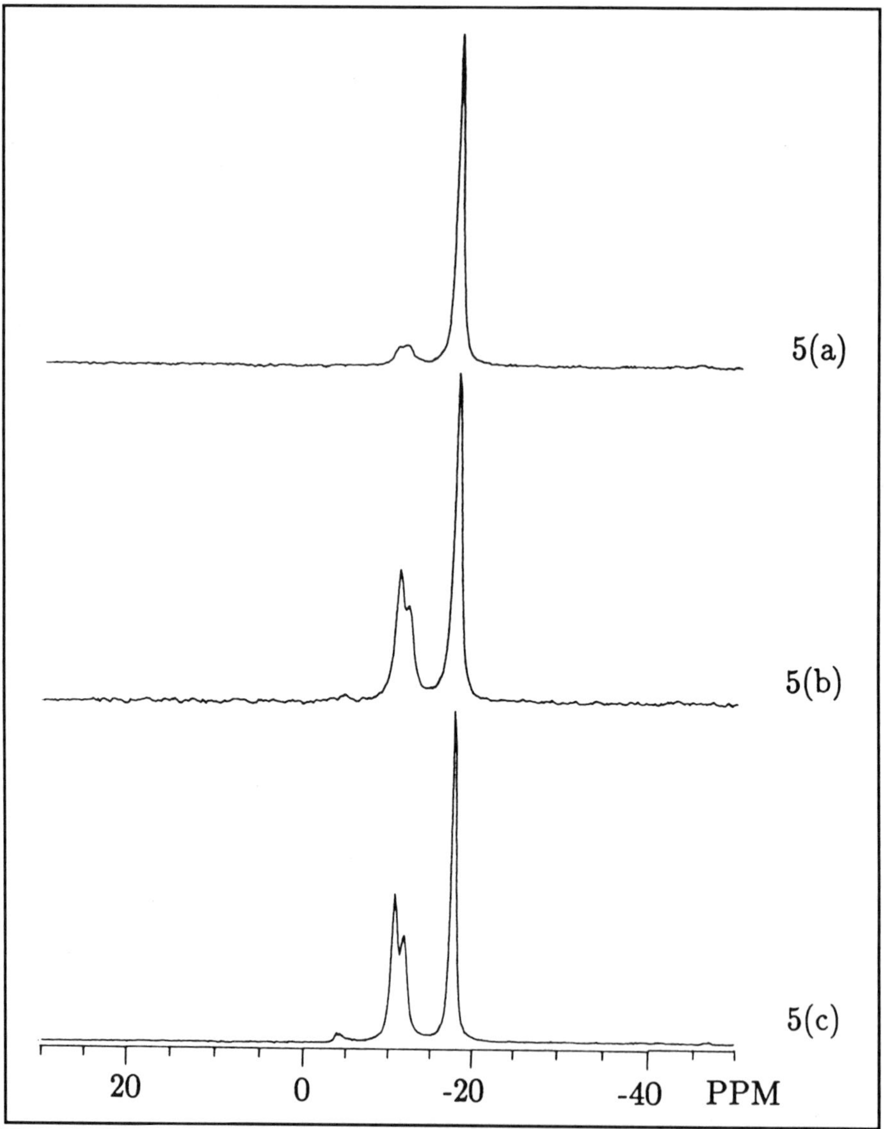

Figure 5. ^{31}P MAS NMR spectra of PNDP adsorbed on the macroreticular strong acid resin. (**a**) one day old, stored in a capped 5mm NMR tube; (**b**) 40 days old, stored in a capped 5mm NMR tube; and (**c**) sample exposed to air for six additional days.

between the carbonaceous adsorbent and the ion-exchange resins cannot be occurring at a rate greater than 300sec^{-1}. In agreement with this limit, initial magnetization-transfer experiments indicate that the exchange rate for DMMP is approximately 0.5sec^{-1} at room temperature. These experiments are continuing in order to develop a better understanding of the exchange process.

The results obtained for the hydrolysis of PNDP adsorbed on the XE-555 resin (from ethanol solution) reveal that the anionic DPP product appears to be tightly bound to the quaternary ammonium ion-exchange sites rendering the reaction stoichiometric. PNDP hydrolysis is enhanced by the presence of coadsorbed ethanol consistent with an increased rate of diffusion to available reactive sites. Product immobilization does not occur for acid catalyzed hydrolysis on the strong acid resin where the neutral DPPA and PNPPA products remain highly mobile, even in the absence of coadsorbed ethanol. However, the analogous acid catalyzed process does not occur on the XE-555 resin. The reason for the apparent lack of reactivity of the XE- 555 sulfonic acid groups remains unclear.

Acknowledgments

We acknowledge D. Ralph Leslie (Defence Science and Technology Organization, Melbourne, Australia) for initiating this project as a visiting scientist in our laboratory. The ^{31}P MAS NMR experiments were conducted at the National Institutes of Health, Bethesda, MD, and we are grateful to D. Torchia for the use of the 6T NMR spectrometer and H. Cole for her helpful assistance with the operation of the instrument. This work was done while one of the authors (G.W.W.) held a National Research Council-CRDEC Research Associateship.

References

1. T. C. Bruice and S. J. Benkovic, *Bioorganic Mechanisms*, Volume II, Chapter 6 (W.A. Benjamin, New York, 1966), pp. 100-159.
2. A. Guyot, *Pure & Appl. Chem.* **60**, 365 (1988).
3. D. C. Sherrington, *Pure & Appl. Chem.* **60**, 401 (1988).
4. Ambergard R XE-555 Resin Material Data Safety Sheet (Rohm and Haas Company, 1989).
5. J. R. Cox, Jr. and O. B. Ramsay, *Chem. Rev.* **64**, 317 (1964).
6. D. R. Leslie, W. T. Beaudry, and L. L. Szafraniec, *Sorption and Reaction of Chemical Agents by a Mixed Sorptive/Reactive Resin* (document CRDEC-TR- 292, U.S. Army Chemical Research, Development and Engineering Center, Aberdeen Proving Ground, MD, October 1991).
7. L. Baltusis, J. S. Frye, and G. E. Maciel, *J. Am. Chem. Soc.* **109**, 40 (1987).
8. L. Baltusis, J. S. Frye, and G. E. Maciel, *J. Am. Chem. Soc.* **108**, 7119 (1986).
9. J. H. Lunsford, W. P. Rothwell, and W. Shen, *J. Am. Chem. Soc.* **107**, 1540 (1985).

10. W. P. Rothwell, W. X. Shen, and J. H. Lunsford, *J. Am. Chem. Soc.* **106**, 2452 (1984).
11. H.-H. Limbach *et al.*, *J. Magn. Reson.* **77**, 84 (1988).
12. A. Schmidt and S. Vega, *J. Chem. Phys.* **87**, 6895 (1987).
13. W. T. Ford *et al.*, *J. Magn. Reson.* **63**, 298 (1985).
14. W. T. Ford, M. Periyasamy, and H. O. Spivey, *Macromolecules* **17**, 2881 (1984).
15. D. Suwelack, W. P. Rothwell, and J. S. Waugh, *J. Chem. Phys.* **73**, 2559 (1980).
16. N. E. Burlinson, B. A. Dunell, and J. A. Ripmeester, *J. Magn. Reson.* **67**, 217 (1986).
17. T. M. Duncan and D. C. Douglass, *Chem. Phys.* **87**, 339 (1984).
18. A.-R. Grimmer and U. Haubenreisser, *Chem. Phys. Lett.* **99**, 487 (1983).
19. J. Herzfeld, R. G. Griffin, and R. A. Haberkorn, *Biochem.* **17**, 2711 (1978).
20. R. K. Harris, *Nuclear Magnetic Resonance Spectroscopy*, Chapter 5 (Pitman, London, 1983), pp.
21. B. A. Burnside *et al.*, *J. Org. Chem.* **53**, 2009 (1988).
22. B. L. Knier *et al.*, *J. Solution Chem.* **17**, 77 (1988).
23. R. A. Mackay *et al.*, *J. Phys. Chem.* **91**, 861 (1987).
24. S. H. Gellman, R. Petter, and R. Breslow, *J. Am. Chem. Soc.* **108**, 2388 (1986).
25. R. A. Moss, K. Y. Kim, and S. Swarup, *J. Am. Chem. Soc.* **108**, 788 (1986).
26. D. A. Jaeger and D. Bolikal, *J. Org. Chem.* **50**, 4635 (1985).
27. D. R. Jones, L. F. Lindoy, and A. M. Sargeson, *J. Am. Chem. Soc.* **106**, 7807 (1984).
28. R. A. Moss, K. W. Alwis, and J.-S. Shin, *J. Am. Chem. Soc.* **106**, 2651 (1984).
29. C. A. Bunton, F. de Buzzaccarini, and F. H. Hamed, *J. Org. Chem.* **48**, 2457 (1983).
30. C. A. Bunton and L. Robinson, *J. Org. Chem.* **34**, 733 (1969).
31. C. A. Bunton, S. J. Farber, and E. J. Fendler, *J. Org. Chem.* **33**, 29 (1968).
32. W.T. Beaudry, G.W. Wagner, and J.R. Ward, *J. Mol. Catal.* **73**, 77(1992).
33. F. Svec, *Pure & Appl. Chem.* **60**, 377 (1988).

27

Magnetron Ion Etching for GaAs Device Processing

G.F. McLane
M.W. Cole
H.S. Lee
A. Lepore
D.W. Eckart
R. T. Lareau
U.S. Army Research Laboratory
Electronics and Power Sources Directorate
Fort Monmouth, NJ 07703-5601

M. Meyyappan
Scientific Research Associates, Inc.
Glastonbury, CT 06033-6058

M. Namaroff
J. Sasserath
Materials Research Corporation
Orangeburg, NY 10962

Magnetron reactive ion etching was performed on gallium arsenide (GaAs) wafers using silicon tetrachloride ($SiCl_4$) and Freon-12 (CCl_2F_2) etch gases and was found to be an attractive processing technique for GaAs device fabrication in Army application areas such as monolithic microwave integrated circuits, improved multiple quantum well infrared wavelength detectors, and smart munitions. $SiCl_4$ produced high etch rates, smooth surfaces, and vertical sidewalls appropriate for fabrication of small dimension GaAs devices. Etched surfaces were essentially residue free, but sidewalls exhibited the presence of etch product residues that may be responsible for the excellent anisotropic etching properties of $SiCl_4$. Transmission electron microscopy (TEM) measurements revealed low concentrations of etch-induced defects, and Schottky diode measurements indicated minimal degradation of surface region electrical properties. CCl_2F_2 produced high etch rates, smooth surfaces, and sloped sidewalls appropriate for via connection processing. Residues from etch gas products were found to be present on CCl_2F_2 etched surfaces. TEM and Schottky diode measurements showed low concentrations of defects and minimal degradation of surface region electrical properties, respectively.

Introduction

For many microelectronic device applications plasma assisted (dry) etching has replaced wet chemical etching as the preferred method for providing fine line submicron patterns and selective etch fabrication steps. For device fabrication it is necessary that the plasma etching process provide smooth etched surfaces with minimal damage and degradation of surface region electrical properties, along with anisotropic etching characteristics to produce the desired etched sidewall angle. Etch-induced damage may be either in the form of etch gas surface residues or subsurface crystal defects.

Reactive ion etching (RIE) is a plasma processing technique that has been used for etching silicon and various III-V compound semiconductors. In this process, reactive species are created in the plasma and combine with the wafer surface atoms to form volatile chemical species that can then be removed. Positive ions also generated in the plasma impinge anisotropically on the negatively biased wafer mounted on the cathode and provide energy to promote gas-surface chemical reactions and help remove surface etch products. This anisotropic characteristic of RIE results in faithful transferral of the mask pattern to the semiconductor wafer, with good fidelity of the pattern being reproduced throughout the etch. RIE eliminates mask undercut, which is a major disadvantage of wet etching. However, ion bombardment of the wafer surface may lead to etch-induced damage that can be detrimental to device operation.

Magnetron reactive ion etching (MIE) is similar to RIE, with the addition of a magnetic field to confine the plasma electrons close to the wafer. This minimizes electron loss to the chamber walls and increases ionization efficiency and reactive species generation. Compared to RIE, the magnetic field of MIE allows a higher density plasma discharge to be created that is able to sustain itself at relatively low cathode bias voltages, resulting in low ion bombardment energies and therefore minimal etch-induced wafer damage. In addition, the relatively high density discharge of MIE allows it to operate at lower etch gas pressure than RIE. This leads to anisotropic etching with smaller feature size capability because lower pressures provide reduced ion scattering and associated lateral etching and allow reactive species and wafer etch products to be transported more easily to and from high aspect ratio, small-dimension etched wells.

MIE is a relatively new technique, and although it is a modification of RIE with a magnetic field added, the knowledge base previously developed for RIE is not directly transferrable because the extent of ionization, electron density, and energy distribution is altered considerably by the magnetic field. Hence, there is a strong incentive to investigate the plasma reaction mechanisms and wafer sample etching characteristics of the MIE technique. Although MIE has been investigated for etching Si and SiO_2, there has been relatively little work done on exploiting this technique for gallium arsenide (GaAs) processing. Because the chlorides of Ga and As are relatively volatile, chlorine (Cl)-based chemistry is the choice for etching GaAs. We report here the results of a study of MIE using silicon tetrachloride ($SiCl_4$) and Freon-12 (CCl_2F_2) etch gases. Although some work has previously been reported for CCl_2F_2,[1] this is the first report of magnetron ion etching of GaAs using $SiCl_4$.

We have investigated MIE to determine its effectivness as a processing technique for the fabrication of GaAs devices. Etched wafers are characterized in terms of surface smoothness, sidewall angle, etch product surface residues, subsurface crystal defects, and surface region electrical properties, because these are all important factors in terms of device performance. Wherever possible, we compare our MIE results with results reported by others for RIE.

Experimental Procedures

The etch experiments were performed in a magnetron reactive ion etch chamber having a powered cathode (13.56MHz) containing two bar magnets, with two additional magnets located outside the chamber at the top of the chamber cover. The radio frequency (rf) power source is capacitively coupled to the cathode, resulting in a negative DC voltage bias being developed on the cathode during etching. The cathode is water cooled and covered with an aluminum pedestal that supports the wafer sample. Wafers were thermally heat sunk to the cathode with high vacuum grease. The magnets provide a uniform magnetic field at the wafer location. The chamber is typically evacuated to base pressures in the range 10^{-6}torr.

Etch depth of patterned wafers was measured using a Dektak profilometer after removal of the mask. Etched surface characterization was performed using scanning electron microscopy (SEM), transmission electron microscopy (TEM), Auger electron spectroscopy (AES), and Schottky barrier diode I-V measurements. GaAs wafer samples with (100) surface orientation and $1 \times 10^{17}/cm^3$ n-type carrier concentrations were used for Schottky barrier diode measurements. Semi-insulating GaAs wafers with the same surface orientation were used for all other measurements.

Results

$SiCl_4$ etch gas. Figure 1 shows the etch rate and cathode self-bias voltage as a function of power density for 2 millitorr pressure and 15 standard cubic centimeters per minute (sccm) $SiCl_4$ flow rate. The etch rate initially

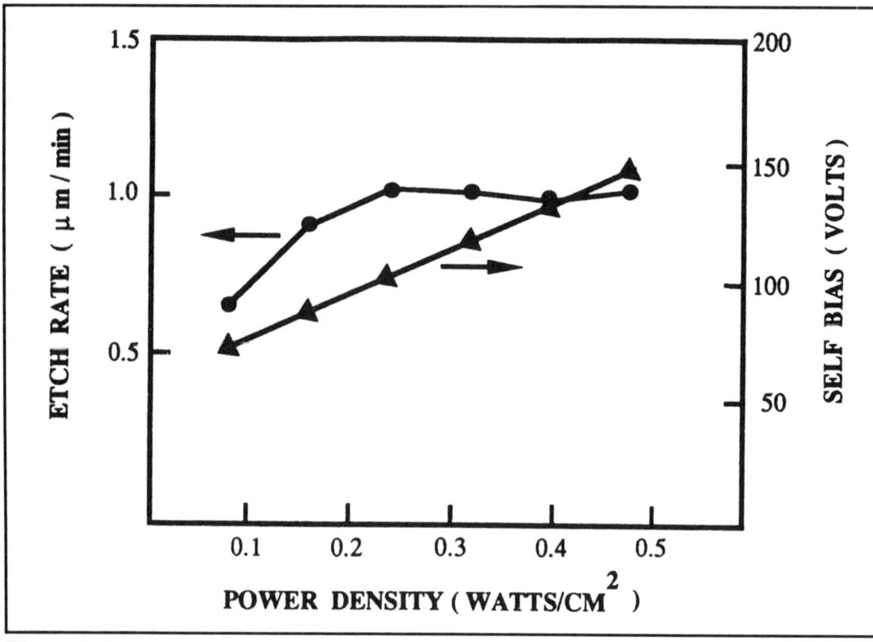

Figure 1. GaAs etch rate and cathode bias voltage as a function of power density for 2 millitorr pressure and 15sccm $SiCl_4$ flow rate.

increases with power and then levels off at higher power densities, eventually reaching an etch rate of approximately one micron per minute. An increase in power density usually results in an increase in electron density and consequently in Cl atom production, which contributes to the observed increase in etch rate at low powers. Additionally, the cathode self-bias voltage increases linearly with power, as shown in Figure 1, resulting in increased ion bombardment energies. Increased ion bombardment energies can result in higher etch rates by increasing sputter removal of etch products or by providing additional activation energy necessary for the formation of volatile chlorides of Ga and As in the first place. At the higher power levels of Figure 1, however, the increased ion bombardment may interfere with Cl atom adsorption on the surface, as pointed out by Pearton *et al.*,[2] and prevent the etch rate from increasing further. Figure 2 shows etch rate as a function of pressure for 0.16W/cm^2 power density and 15sccm $SiCl_4$ flow rate. The observed increase in etch rate with pressure is due to an increase in reactant species production resulting from an increase in density of available etch gas. The etch rates reported here are generally higher by a factor of five or more when compared with results reported on $SiC1_4$, RIE at comparable power densities, and self-bias levels.[3,4] High etch rates allow short processing times in the hostile plasma environment. Because the eventual number of defects in a dry etched

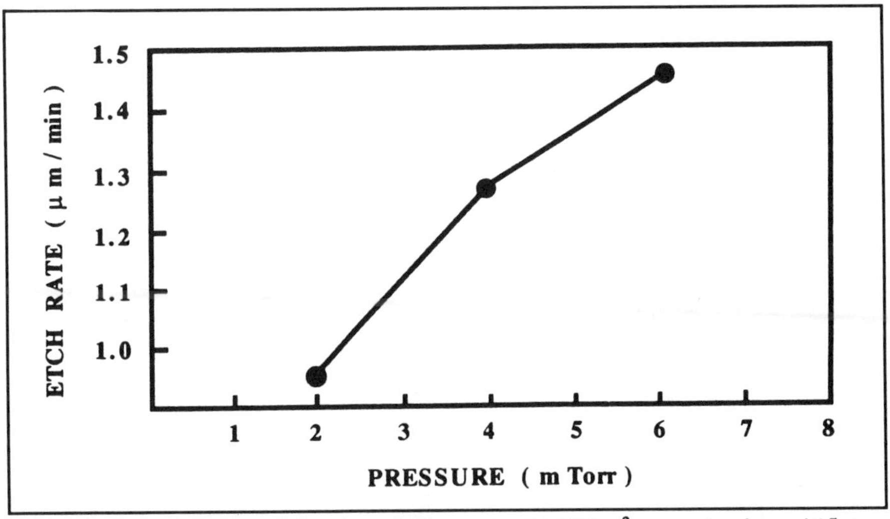

Figure 2. GaAs etch rate as a function of pressure for 0.16W/cm^2 power density and 15sccm $SiCl_4$ flow rate.

Figure 3. Scanning electron micrograph of a patterned GaAs sample etched for two minutes at 2 millitorr pressure and 0.24W/cm^2 power density with 15sccm $SiCl_4$ flow rate.

sample is a balance between defect generation due to ion bombardment and the removal of wafer material and surface defects due to etching, high etch rates may reduce damage accumulation.[5]

Figure 4. High magnification scanning electron micrograph of Figure 3.

Figure 3 shows a scanning electron micrograph of a patterned GaAs sample etched for two minutes at 2 millitorr pressure and 0.24W/cm^2 power density with 15sccm $SiCl_4$ flow rate. The 200-nm wide line masks were fabricated by direct electron beam writing on SAL-601, a novolac-based electron beam resist that was used directly as the etch mask. The etched surface is clean and smooth, and the sidewalls are vertical, with an etch depth of approximately two microns. Striations on the sidewalls are due to non-uniformities of the mask edge being replicated into the sample. Figure 4 shows a scanning electron micrograph of the same sample at higher magnification, showing more clearly the quantum wirelike etched structure. Figure 5 shows another patterned sample etched under the same conditions in $SiCl_4$. These SEM results show that $SiCl_4$ etching of patterned GaAs wafers results in a surface morphology appropriate for fabrication of small-dimension, high-speed devices, in addition to quantum wire and quantum dot structures.

Etched surface profile and subsurface lattice defects were investigated using TEM. Figure 6 shows a TEM micrograph of a sample etched under the same conditions as previously discussed. The surface is extremely smooth, with a peak-to-valley distance along the surface of only 4nm. The defects are in the form of small dislocation loops 0.7 to 2.0nm in diameter, extending a depth of 300nm below the surface, with asurface density of 10^{10}/cm^2. This is an order of magnitude lower than defect densities reported for RIE under similar conditions.[6]

Figure 5. Scanning electron micrograph of a patterned GaAs sample etched for two minutes at 2 millitorr pressure and $0.24W/cm^2$ power density with 15sccm $SiCl_4$ flow rate.

Figure 6. Transmission electron micrograph of GaAs etched for two minutes at 2 millitorr, and $0.24W/cm^2$, 15sccm $SiCl_4$.

To determine the effect of etch-induced damage on the surface region electrical characteristics, Schottky barrier diodes were fabricated on unpatterned etched surfaces, and diode I-V characteristics were compared with those of a Schottky diode formed on a control (unetched) sample. Schottky contacts were formed by electron beam deposition of a titanium/platinum/gold (Ti/Pt/Au) structure. Schottky diode measurements for GaAs samples etched in $SiC1_4$ are shown in Table 1 for several values of pressure and applied power density, along with results for a control sample. Ideality factor (n) and

Table 1
Schottky Diode Measurements for GaAs Samples Etched for Two Minutes at 15sccm $SiCl_4$ Flow Rate

Power Density (W/cm2)	Pressure (millitorr)	Ideality Factor, n	Schottky Barrier height, ϕ_B (eV)
Control	—	1.07	0.786
0.16	2	1.12	0.766
0.16	4	1.13	0.775
0.16	6	1.17	0.764
0.24	2	1.11	0.778
0.40	2	1.16	0.769

Schottky barrier height (ϕ_B) are obtained from the slope and intercept, respectively, of the measured forward bias ln I vs V curve.[7] Ideality factor is normally close to unity, whereas Schottky barrier height depends upon semiconductor material, semiconductor surface states, and metal contact material. These parameters are of particular interest because they are sensitive to Schottky barrier surface region carrier concentration, defect levels, and traps. Etch-induced surface region defects that can affect device performance are thus manifest in changes in the values of n and ϕ_B. Minimal changes in these parameters indicate low etch-induced wafer damage. Table 1 shows that n increases and ϕ_B decreases for the etched samples relative to the control sample, but most importantly the deviations from the control sample values are small. This indicates minimal etch-induced degradation of surface region electrical properties. Schottky diode measurements performed on GaAs samples etched under similar conditions using RIE also showed an increase in n and a decrease in ϕ_B, but the increases in n were much larger than the results reported here.[2,8,9] This indicates that MIE results in less surface degradation than RIE.

Auger electron spectroscopy (AES) measurements were performed on $SiCl_4$ etched samples to determine the possible presence of etch gas residues. Figure 7 shows AES measurements on a horizontal etched surface. The surface is essentially residue free, with only the constituent Ga and As atoms being observed. These residue-free etched surfaces make magnetron reactive ion etching of GaAs with $SiCl_4$ an attractive etching process for device fabrication. In contrast, Stern and Liao,[4] found several monolayers of chlorinated or oxidized Si present on GaAs layers reactively ion etched in $SiCl_4$. Figure 8 shows Auger measurements performed on the vertical sidewalls of the MIE sample of Figure 7. This data indicates the presence of Si, carbon, and

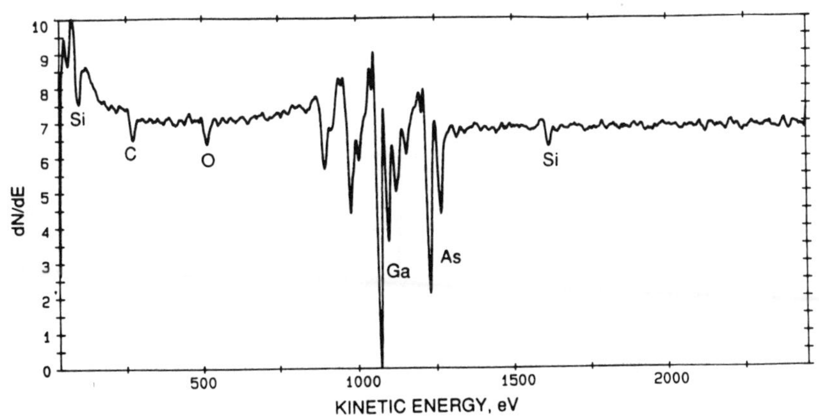

Figure 7. Auger electron spectroscopy measurements on GaAs horizontal surface etched for two minutes at 2 millitorr, 0.24W/cm^2, 15sccm $SiCl_4$.

oxygen on the sidewalls, which could be responsible for the excellent anisotropic etching properties of $SiC1_4$, because this residue may prevent lateral chemical etching of sidewalls.

CCl_2F_2 etch gas. Figure 9 shows the etch rate as a function of power density and pressure for $CC1_2F_2$ with a flow rate of 20sccm. The etch rate increases with an increase in applied power density and a decrease in pressure. An increase in power density results in an increase in Cl atom production similar to $SiC1_4$ etching, leading to increased etch rates. Unlike $SiC1_4$, however, the etch rate for $CC1_2F_2$ increases as the pressure is lowered. In the case of pure chemical etching, a decrease in pressure would result in an etch rate decrease, since the production of active chemical species diminishes due to less etch gas being available.

In Figure 9 at 0.4W/cm^2, cathode bias voltage increased from 95 volts to 105 volts as pressure was lowered from 20 millitorr to 10 millitorr, resulting in an increase in ion bombardment energies. Under these conditions, it appears that as pressure is lowered the increase in the ion bombardment component of etching is a more dominant etch rate mechanism than a possible decrease in the available reactant species. The net result is an increase in GaAs etch rate with CCl_2F_2 as pressure is lowered. We have reported previously[10] that x-ray photoelectron spectroscopy (XPS) measurements reveal the presence of GaF_3 and other non-stoichiometric (Ga_xF_y) gallium fluorides on the surfaces of samples etched with CCl_2F_2. Because gallium fluorides are relatively non-volatile, these etch product species may require enhanced ion bombardment for their removal. Indeed, Klinger and Greene [11] have reported

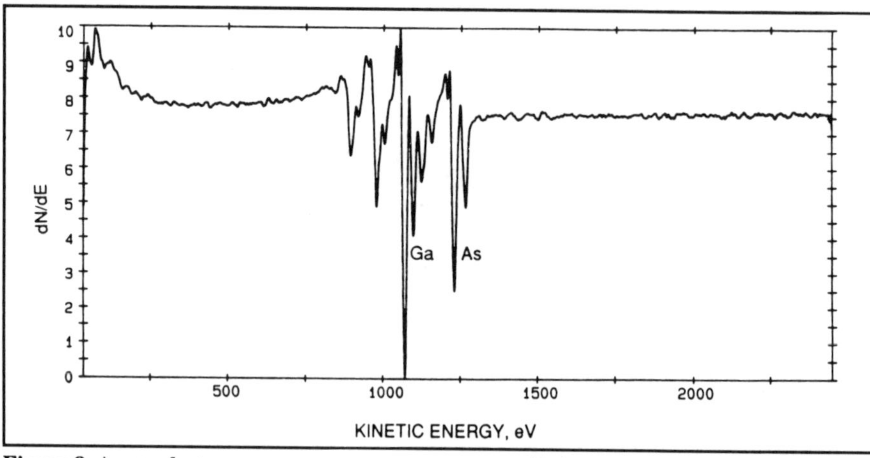

Figure 8. Auger electron spectroscopy measurement on GaAs vertical sidewall etched for two minutes at 2 millitorr, 0.24W/cm^2, 15sccm $SiCl_4$.

Figure 9. GaAs etch rate as a function of power density and pressure for 20sccm CCl_2F_2 flow rate.

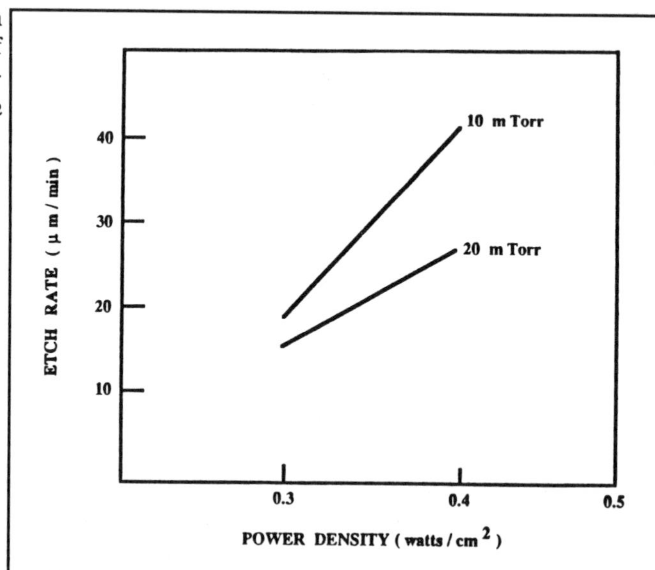

that the removal of Ga in discharges containing fluorine occurs primarily by ion-assisted desorption of GaF_x radicals, where x=1 or 2. The etch rates reported here for MIE are generally much higher than reported for RIE under similar conditions.[12]

Figure 10 shows a scanning electron micrograph of a pattern etched for 30 minutes in 20sccm of CCl_2F_2 at 4 millitorr and a power density of 0.52W/cm^2. The etch mask consisted of an array of nickel/titanium layers in the shape

Figure 10. Scanning electron micrograph of a patterned GaAs sample etched for 30 minutes at 4 millitorr, 0.52W/cm^2, 20sccm CCl_2F_2.

of squares 10 microns on a side. The etch depth is approximately five microns. The etched surface is clean and very smooth. The sidewalls are straight and exhibit negative undercutting, that is, they slope away from the mask edge. This negative undercut profile has been previously observed in RIE of GaAs with CCl_2F_2,[11] and could be due to the redeposition of surface material or formation of polymers on the sidewall as the sample is etched. Negative undercut characteristics are desirable for the fabrication of via hole connections for MIMIC applications, because they facilitate the subsequent sidewall metallization step. For via hole connection applications, etch rates can be increased by using higher power densities and through the addition of Cl_2 to the etch gas.

Figure 11 shows a TEM micrograph for an unpatterned GaAs sample etched for four minutes under the conditions of Figure 10. The surface morphology is extremely good, with peak to valley distances along the surface of only about 9nm. The defects are due to small dislocation loops 2 to 4nm in diameter which extend to a depth of about 70nm below the etched surface. Most of the defects are concentrated in the top 10nm where the density is about 10^{10}/cm^2; total defect count is lower than for RIE under similar etching

Figure 11. Transmission electron micrograph of GaAs sample etched for 4 minutes at 4 millitorr, $0.52 W/cm^2$, 20sccm CCl_2F_2.

Table 2
Schottky Diode Measurements for GaAs Samples Etched for 4 Minutes, $0.31 W/cm^2$, 20sccm CCl_2F_2

Pressure (millitorr)	Ideality Factor, n	Schottky Barrier height, ϕ_B (eV)
Control	1.12	0.747
2	1.12	0.816
4	1.14	0.792
10	1.12	0.772

conditions.[13] Also, GaAs samples etched in CCl_2F_2 using the RIE technique under similar conditions exhibit rough surfaces[13] with peak-to-valley distances along the surface of 170 to 200nm, more than an order of magnitude rougher than for MIE. The defects extend to a distance of 220nm into the sample, much deeper than for MIE. The low damage in MIE is due to the reduced cathode DC bias. For example, the dc bias voltage for MIE under the etching conditions of Figure 10 is 148 volts compared with 380 volts reported for RIE under similar conditions.[13]

Schottky diode measurements for GaAs samples etched in CCl_2F_2 are shown in Table 2 for several values of pressure, along with results for a control sample. Both the ideality factor and Schottky barrier height increased slightly for the etched samples compared to the control sample, indicating minimal etch-induced degradation of surface region electrical properties. For RIE etched samples, n increases and ϕ_B decreases under etching, but the increase in n is somewhat higher than the results reported here for MIE.[14] The increase in ϕ_B for the CCl_2F_2 etched samples reported here is atypical for etched GaAs samples, because ϕ_B normally decreases upon etching. An explanation for this result is obtained from Auger electron spectroscopy measurements performed on a CCl_2F_2 etched surface as shown in Figure 12. Carbon, fluorine, chlorine, and oxygen are seen on the surface in addition to gallium and arsenic. These

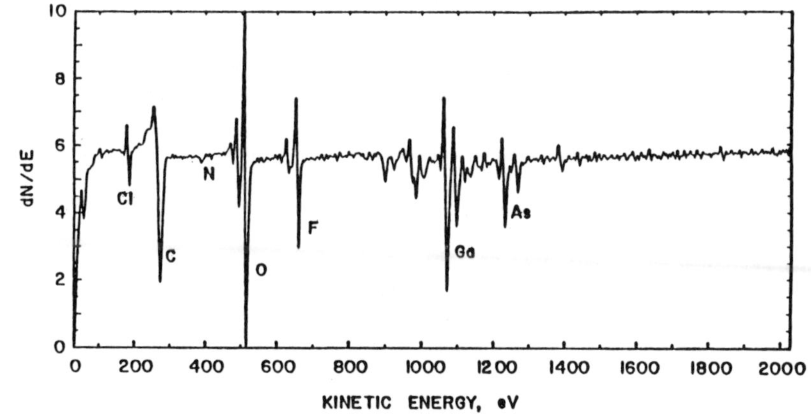

Figure 12. Auger electron spectroscopy measurement on GaAs horizontal surfaces etched for 4 minutes at 4 millitorr, $0.52 W/cm^2$, 20sccm CCl_2F_2.

residual etch gas contaminants formed on the surface could effectively be producing a metal-insulating layer-semiconductor structure instead of the intended metal-semiconductor Schottky barrier diode. This thin insulating residue acts as an additional tunneling barrier, resulting in an apparent increase in ϕ_B. Similar effects have been seen for GaAs reactive ion etched in CF_4 and CHF_3,[15] and also in $CC1_2F_2$:0_2 mixtures.[13] These residues are strictly surface impurities, however, since they were easily removed by a light sputtering step.

Conclusions

Magnetron reactive ion etching has been found to produce higher etch rates, smoother surfaces, and less wafer damage and electrical property degradation than standard (unmagnetized) reactive ion etching. This relatively new plasma etching method has the potential for being an important processing technique for the fabrication of GaAs devices for Army applications.

References

1. EJ. Contolini and L.A. D'Asaro, *J. Vac. Sci. Technol.* **B4**,706 (1986).
2. SJ. Pearton *et al.*, *J. Vac. Sci. Technol.* **B5**, 607 (1990).
3. S. Salimian *et al.*, *J. Vac. Sci. Technol.* **B9**, 114 (1991).

4. M.B. Stern and P.F. Liao, *J. Vac. Sci. Technol.* **Bl**, 1053 (1983).
5. G.F. McLane *et al.*, *J. Appl. Phys.* **69**, 695 (1991).
6. M.W. Cole *et al.*, *The Journal of Scanning Microscopy* **14**, 31 (1992).
7. S.M. Sze, *Physics of Semiconductor Devices* (John Wiley and Sons, Inc., New York, 1969).
8. R. Van Daele, D. Lootens, and P. Demeester, *Vacuum* **41**, 906 (1990).
9. T. Mantei and J.J. Ibara, *J Appl. Phys.* **61**, 4885 (1987).
10. M. Meyyappan *et al.*, *38th Annual AVS Symposium and Topical Conference*, Seattle, WA, Nov. 11-15, 1991.
11. R.E. Klinger and J.E. Greene, *J. Appl. Phys.* **54**, 1595 (1983).
12. S.W. Pang, *J. Electrochem. Soc.* **133**, 784 (1986).
13. S.J. Pearton *et al.*, *J. Appl. Phys.* **65**, 1281 (1989).
14. S.J.Pearton *et al.*, *J. Appl. Phys.* **66**, 3839 (1989).
15. S.W. Pang *et al.*, *J. Vac. Sci. Technol.* **Bl**, 1334 (1983).

28

The Infiltration of Cotton Fibers With Sol-Gel Solutions, Forming Oxide and Non-Oxide Ceramics

Richard J. Ruzga
Vincent F. Hock
U.S. Army Corps of Engineers
Research Laboratory
Champaign, IL 61826-9005

The objective of this research was to employ cotton to fabricate both oxide and non-oxide ceramics from cotton fibers infiltrated with ceramic sol-gel solutions. Cotton fiber contains a high amount of carbon and carbon containing components. The capillarity and absorbency of the fibers make them ideal for infiltration with ceramic materials. The fibers can be infiltrated in either single fiber form or fabricated into complex structures prior to infiltration.

Depending on the atmospheric conditions during heat treatment, the cotton fibers could be oxidized or reduced, thus becoming a source of carbon. Because the fibers were infiltrated with silica in the form of sol-gel, oxidizing atmospheres yielded SiO_2, while reducing atmospheres yielded SiC. The production of SiC by this method is obviously significant because most current procedures involve the pyrolysis of expensive polymer precursors. This procedure has many potential advantages: the cost savings compared with conventional fabrication of SiC would be significant, and the ability to shape the ceramic fiber matrix prior to processing would allow fabrication of vehicle panels, machine parts, ballistic surfaces, radar dishes, helmets, and high-temperature woven and fibrous insulation.

Introduction

Many investigators have used naturally occurring ceramic and organic materials to create various novel processing techniques. The objective of this research was to employ cotton to fabricate both oxide and non-oxide ceramics. Cotton fiber contains a high amount of carbon and carbon-containing components. The capillarity and absorbency of the fibers make them ideal for infiltration with liquid ceramic materials. The method investigated dealt with producing silica sol-gel solutions for infiltrating cotton fiber structures. The fibers themselves could be infiltrated in either single fiber form or fabricated into complex structures prior to the infiltration process. Depending on the atmospheric conditions during subsequent heat treatment, the cotton fibers can be oxidized or reduced, becoming a source of carbon.

Because the fibers were infiltrated with silica in the form of a sol-gel solution, oxidizing atmospheres yielded SiO_2, while reducing atmospheres yielded SiC. The production of SiC by this method is obviously significant because most current procedures that produce SiC involve the pyrolysis of expensive polymeric precursors.

As a result of this unique material processing technology,we believe that the resulting array of carbon filaments provide reinforcement superior to that obtained through synthetic fibers, because the ceramic matrix is extended into the hollow structure of the cotton fiber. Moreover, due to a lack of porosity, a synthetic fiber bonds only with the ceramic at its surface.

The potential uses for this new material include areas currently proposed for ceramic composite materials within the Department of Army (vehicle panels, machine parts, ballistic surfaces, radar dishes, personnel helmets, and high-temperature woven and fibrous insulation). Given that the supply of advanced carbon materials comes from either overseas sources or from petroleum products, this research is significant because the raw materials are domestic, thereby militarily secure, readily available, and beneficial to our country's agricultural base.

Fiber Chemistry and Properties

The cross-section of a cotton fiber consists of a number of concentric layers, which can be broadly classified as belonging to either the outermost region (primary wall) or the inner region (secondary wall). It is the primary wall, in contact with the physical environment, that largely determines the

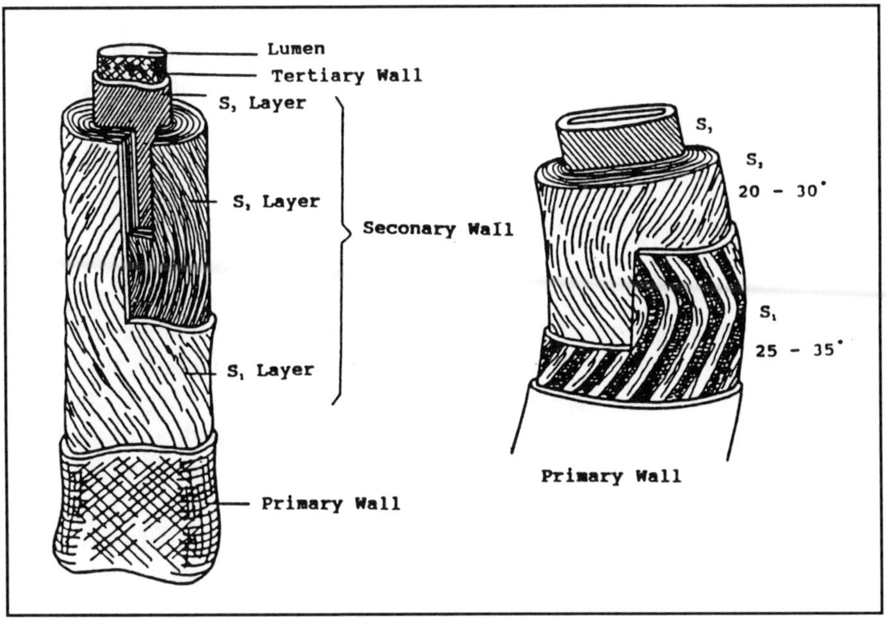

Figure 1. Cotton fiber morphology showing layers.

processing behavior of the fiber. Nevertheless, the body of the fiber has a modifying effect on the periphery of the cotton hair and cannot be dismissed. The fiber morphology can be concisely summarized into six distinct regions, beginning with the outermost layer and continuing in toward the center of the fiber, as illustrated in Figure 1.

Cellulose is the primary chemical component of the cuticle, although, in addition, other noncellulosics in the form of ash, pectic, and proteinaceous materials can be found distributed inhomogeneously throughout the fiber.[1] The interaction of liquids with the solid surfaces of cellulose fibers is facilitated by the substantial accessibilities of internal surfaces. Jeffries *et al.*[2] found the accessibility of liquids to the internal surfaces increases progressively from raw cotton to mercerized cotton. Mercerization is a technique used to increase luster, dye absorption, and strength in cotton materials. Molecules of average dimensions 30Åx8Åx3Å have access through the network system of openings to the extremely large internal surfaces of the fiber.[3] The molecules cannot penetrate within the crystalline regions but are apparently well distributed through the more porous amorphous portions and over the large crystalline surface areas.

The mere immersion of cotton in a sol-gel solution is not sufficient for infiltration. When infiltration does take place, it continues until equilibrium is reached between the cotton and the sol-gel solution under quiescent ambient conditions. Ordinary infiltration depends upon a balance of effects but is controlled mainly by heat and the addition of sodium chloride or sulfate.

Conversion of Cellulose to Carbon

Carbon fibers were first produced by Edison[4] almost a century ago through the carbonization of bamboo and cotton cellulose materials that were formed into the required shape before being placed in the carbonizing chamber. Later, Edison made consistent filaments by dissolving cleaned commercial cotton in a solution of zinc chloride. Superior filaments were made by depositing pyrolytic carbon on to such a base filament by "flashing" the filament in a hydrocarbon atmosphere.

Carbon fibers in the form of felt, wool, or cloth have found many applications in high-temperature furnaces, corrosion resistant materials, packings, electric heating elements, and refractory composite materials for use in the aerospace industry. Cellulose and rayon are the most widely used precursors for making carbon fibers. The chemical decomposition of the fiber precursors is very complex and not fully understood. Cellulose yields 15 to 30 weight percent carbon fiber and does not melt during decomposition; hence, the physical form of the precursor is maintained. However, much shrinkage accompanies the breakdown of the organic structure of the precursor.

The breakdown of cellulose can be modified by varying the heating rate, the furnace atmosphere, and the moisture content. The type of cellulose used also has an influence on the end product obtained. Additions of catalysts can alter the mode of breakdown; those which contain halides allow faster heating rates and can increase the carbon fiber yield up to 40 percent. The temperature range 250 to 300°C is very critical in the process because the bulk of the weight loss takes place within this temperature range. A large shrinkage accompanies the release of volatiles, whose rate of release must be controlled for optimum results.

Ford and Mitchell[5] have described heating conditions suitable for making graphite fibers and textiles, the process being carried out in a protective atmosphere:

1. Temperature rise from 10°C/hr to 50°C/hr in the range 100°-400°C.
2. Temperature rise up to 100°C/hr in the range 400°-900°C.
3. Heating to about 3000°C until graphitization occurs.

The heat treatment atmosphere can play an important part in determining the mode of breakdown of the cellulose and the final structural properties of the graphite fiber. Inert atmospheres such as nitrogen demand slow heating rates if good yields are to be obtained. Shindo, Nakanishi, and Soma[6] showed that nonoxidizing acid vapors such as HCl allow faster heating rates and produce yields up to 40 percent. These halides act as catalysts to assist the conversion to carbon. Metal halides may also be used in the manufacture of active carbon fibers from cellulosic materials.

Sol-Gel Processing Techniques

Sol-gel refers to a dispersion of colloidal particles (10 to 1000Å) in liquid solutions. When the viscosity of the sol-gel increases sufficiently, usually through the partial loss of its liquid phase and polymerization (gelation), the sol-gel becomes rigid. Advantages of sol-gel processing routes for the formation of ceramics include better homogeneity and purity, lower temperature of preparation, and novel procedures. Some disadvantages are large drying and firing shrinkages, lengthy processing times, and expensive starting materials.

Sol-gel processing techniques can lead to the formation of two distinct types of gels: polymeric and colloidal. Polymeric gels are formed from metal alkoxides that have at least three replaceable functional groups. These gels polymerize, without precipitation, into three-dimensional networks. Colloidal gels are formed from sol-gels of stable colloidal particle suspensions. Gelation occurs as the agglomerates grow to consume the host liquor. Various starting compounds have been used to prepare silica gels, with the most frequently used starting materials being silicic acid and alkoxysilanes. The gels prepared from different starting compounds have different volume shrinkages, oxide contents, and physical properties and microstructures after heat treatment.

The formation of a polymeric gel from metal alkoxides in the sol-gel process generally involves the following two steps:

Hydrolysis[1] (1)

$$M(OR)_n + xH_2O \longrightarrow M(OH)_x(OR)_{n-1x} + xROH$$

Polycondensation[2] (2)

$$\text{-M-OH} + \text{H-O-M} \longrightarrow \text{-M-O-M-} + H_2O \quad \text{(dehydration)}$$

$$\text{-M-OH} + \text{R-O-M} \longrightarrow \text{-M-O-M-} + ROH \quad \text{(dealcoholation)}$$

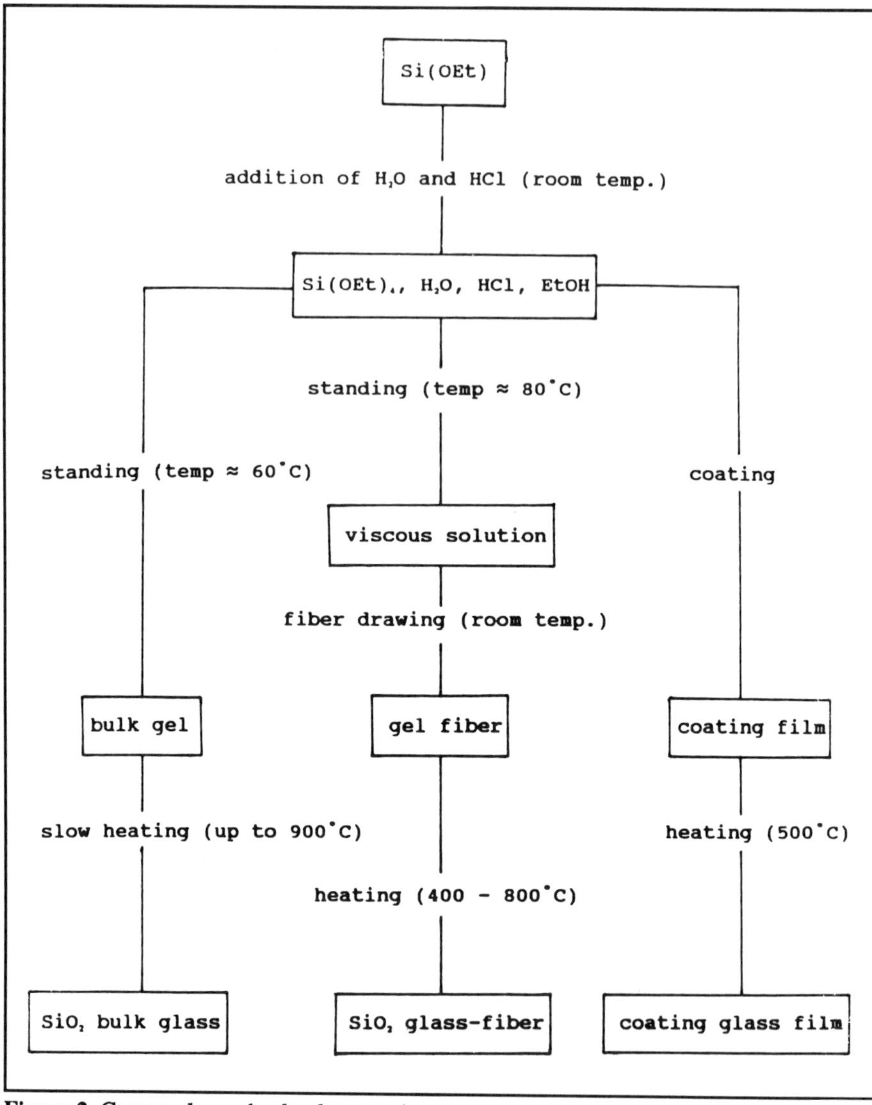

Figure 2. Commonly used sol-gel processing routes.

These reactions tend to occur simultaneously, and their competitive/complementary rates (depending on different factors) often determine the nature of the final product. Gelation can be brought about either by (1) bringing the polymer species into close proximity and increasing the probability of cross-linking between them by removing the solvent, or (2) allowing the system to age, bringing about further dehydration and hydrolysis by the random collision of the polymeric species in solution. Continued cross-linking brings about further network formation, expulsion of organic groups, and finally, gelation. We attempted to produce oxides and non-oxides by combining the processing routes shown in Figure 2.[7]

Experimental Procedures

In our experimental procedure, we used the following molar ratios:

2mol H_2O to 1mol alkoxide
5mol ethanol to 1mol alkoxide

These molar ratios were maintained throughout the processing, and the exact batch mixtures were experimentally determined to work best with the amount of catalyst used. The quantity of the acetic acid catalyst remained constant and was not changed with respect to the size of the sol-gel batch.

The approach involved thoroughly mixing 730ml of ethanol with 90ml of deionized water in a reaction kettle for 10 minutes at room temperature on a magnetic stirring hot plate. Next, approximately 557ml of tetraethylorthosilicate (TEOS) was added to the solution, which was then heated between 50° to 60°C. The solution was allowed to mix within this temperature range for 5.5hr, and then 5 drops of concentrated acetic acid were added. The resulting sol-gel solution was then stirred for 16 hours as it cooled to room temperature.

A sample of the sol-gel solution was extracted, weighed, and then dried to powder form at approximately 100°C in a drying oven. This powder was calcined at 600°C for 24 hours to convert the mixture to pure silica powder. The silica powder was weighed to determine the SiO_2 yield of the sol-gel reaction. The theoretic value for silica yield from a sol-gel reaction is approximately 29 percent. The silica yield from the sol-gel solution prepared by the method just described ranged from 8 to 12 percent, slightly less than half the theoretic value.

Two forms of cotton fibers in mat form were used: raw fibers and bleached mercerized fibers. The mats were sectioned into samples with dimensions of 6"x 8". Care was taken to ensure that the edges of the mats remained clean cut and that there was no delamination between the fiber layers. Various sample thicknesses were created by stacking the individual sheets on top of one another. We use two-sheet thicknesses in some of the experiments performed and used more thicknesses in others. We found that delamination occurs during heat treatments if the sheetss are stacked too thickly. This is due to insufficient infiltration. To prevent delamination, we sandwiched the sample between two perforated sheets of stainless steel cut to the sample dimensions. This assembly was placed in a 6"x8"x3" high stainless steel cage to support the sides of sample during infiltration.

The cage was then placed in a 8.5"x10.5"x6.5" stainless steel tank. A stainless steel weight was placed on top of the assembly to compress the sample slightly. String was attached to the cage so that the sample could be lifted and removed easily from the tank after it had been infiltrated with the solution. Enough silica solution was then added to the tank to completely cover the sample. A rubber gasket was then placed between the tank and the tank lid, and the lid was secured to the tank with C-clamps. The lid was equipped with two standard 24/40 openings, into which a condensing tube and glass stopper were placed. A third opening was used for inserting a thermometer to monitor temperature during infiltration. A magnetic stirring bar was placed under the cage to stir the solution.

The samples were immersed in the solution for 24 hours, removed, pressed, and then placed in a drying oven at 100°C for approximately 24 hours. The cotton samples generally showed a weight gain of approximately 120 percent due to absorption during the infiltration process. The dried samples were then treated according to heat treatment schedules employed by Ford and Mitchell.[5]

Some cotton samples were oxidized in a furnace in ambient air. Other samples were converted to carbon by flowing argon gas into the furnace chamber as a protective atmosphere. The furnace chamber was purged for approximately 1.5 hours prior to heating the sample under the argon atmosphere; then the argon flow rate was controlled using a flowmeter. The argon gas was channeled through the refractory base plates via an alumina tube, with the sample placed between the inlet tube and outlet tubes. An alumina bell-shaped cover was placed over the gas tubes and the sample.

We used an alternative two-step method to investigate the formation of silicon carbide in the case of infiltrated cotton samples that were fired very quickly to temperatures less than 400°C. The resulting cotton mat showed very little shrinkage when the conversion took place in argon atmosphere. The predominantly carbon structure was re-infiltrated with colloidal silica and dried in a drying oven. This composite sample was then heat treated under argon up to 1000°C.

An optimal atmosphere would be one created by a reducing gas such as 10 percent hydrogen in nitrogen. Another problem that possibly could be eliminated with a strongly reducing atmosphere is the action of the oxygen

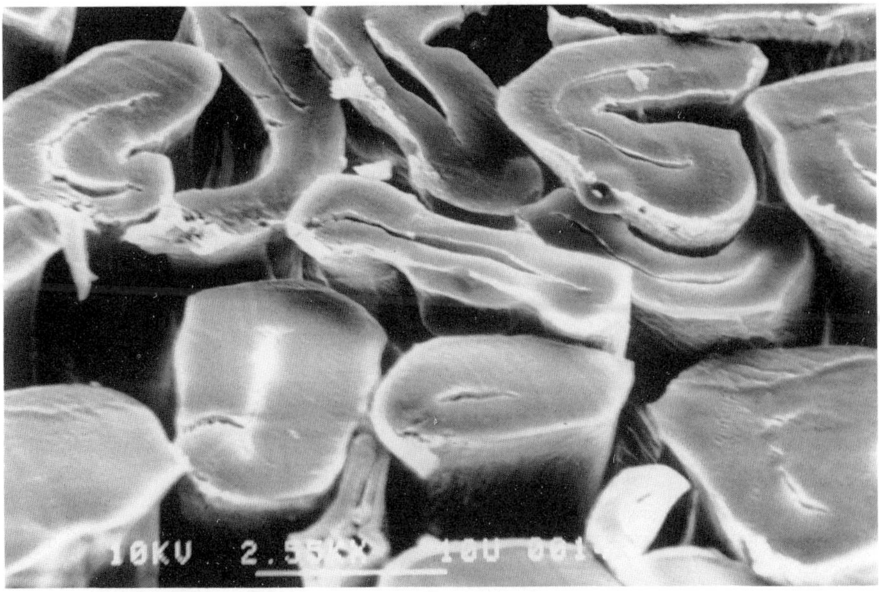

Figure 3. SEM of raw cotton fiber.

Figure 4. SEM of mercerized cotton fiber.

contained within the silica sol-gel solution, which tends to reoxidize the carbon structure in the two-step process. In this case, the sample loses carbon in the form of carbon dioxide and becomes silica.

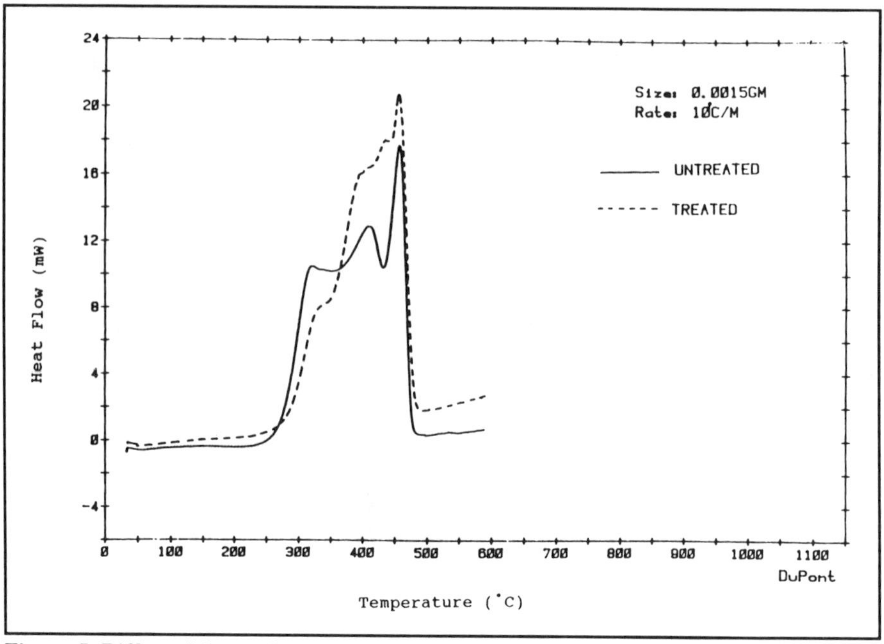

Figure 5. Differential scanning calorimetry plot comparing raw and mercerized fibers.

Results and Discussion

The process of mercerization, previously discussed, alters the physical and chemical makeup of the cotton fiber by increasing the absorbency of the cotton. We therefore believed that this form of cotton fiber would increase the uptake of the sol-gel solution. Figures 3 and 4 are scanning electron micrographs (SEMs) of the two fiber forms, raw and mercerized, respectively. Figure 3 shows that the lumen is only slightly open due to collapse of the fiber upon drying. It appears that the infiltration material must be on the order of a few hundred nanometers to impregnate this central core. Figure 4 reveals that the mercerized fibers are swelled and that the lumen is completely lost. This fiber form is less dense than the raw fiber form and contains more capillaries.

An attempt to examine the effects of mercerization included using a differential scanning calorimeter to determine the energy required to fully combust the fiber. Figure 5 is a differential scanning calorimetry (DSC) plot that compares the power required to pyrolyze an equivalent amount of the two cotton fiber forms. Both forms of the cotton fiber degrade in the temperature range of 250°C to 475°C. The curve corresponding to the raw fiber is broader than that for the treated fiber because raw fiber contains waxes and other impurities that are removed during the process of mercerization. Because

Figure 6. SEM of composite material formed at the bottom region of the sample.

mercerized cotton cellulose has a more structured form than raw cellulose, the corresponding heat flow needed for degradation is greater, thus appearing graphically as a higher maximum point.

Cotton fiber samples infiltrated with the sol-gel solution and subsequently heat treated generally had shrinkages of three times their original dimensions and were very powdery. The samples had very little structural integrity due to incomplete infiltration. An investigation into whether the center of the fiber was actually infiltrated was difficult to conduct. Procedures whereby 5 percent MgO was added to the sol-gel were conducted to increase the strength of the resultant sample. The procedure appeared to increase sample strength when used in conjunction with alumina sol-gel mixtures, but proved to be of little value for silica sol-gels.

Two approaches were investigated for the production of non-oxide fibers. The first consisted of using argon as a protective atmosphere. This atmosphere proved to be inadequate in reducing the oxygen in the fiber and sol-gel mixture, thus leaving silicon carbide. A good candidate reducing gas would be 10 percent hydrogen in nitrogen. Care must be taken to ensure proper ventilation and minimize the possibility of combustion. The furnace used in this investigation had inlet and outlet gas ports on the set plate. Therefore, the gas flowed most heavily near the bottom of the sample.

It is apparent in Figure 6 that a fibrous structure remains and that silica clumps surround these fibers; that is, silicon carbide did not form, rather, a

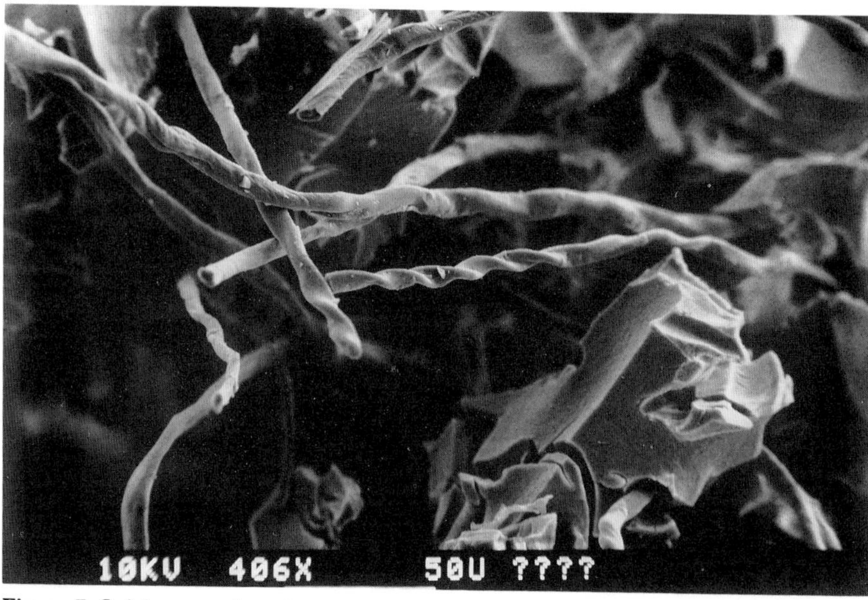

Figure 7. Quick conversion of cotton to carbon.

composite of silica and carbon fibers did. This composite form may prove to be of use in some applications if either the silica existed in a continuous phase with carbon fiber running through it, or if the two phases were to react.

The second approach investigated for the formation of silicon carbide involved pyrolyzing the cotton fiber sample by increasing the temperature to 750°C at a rate of 2°C/min under an argon atmosphere. This sample was then reinfiltrated with the original sol-gel solution and dried at 100°C for 24 hours. This method, as mentioned earlier, was also performed with colloidal silica. The result, Figure 7, is similar to that which was formed from the one-step process. One difference is that the second infiltration appears to greatly increase the amount of the matrix silica phase. Figure 7 shows that many of the original cotton fiber structural characteristics are retained during the quick conversion of the cotton to carbon.

Summary

We investigated the possibility of forming both oxide and non-oxide ceramic fibers by infiltrating cotton fibers with sol-gel ceramic precursors. Silicon-carbide fibers were produced using the non-oxide approach in small quantities, although the majority of the samples retained excess silica.

Heat treatments of the cotton fiber must be done with careful time/temperature control through the temperature range of 250°C to 450°C. Efforts to produce ceramics by this method should incorporate a heating rate similar to that used by Ford and Mitchell.[5] The success of the method is also strongly contingent upon the ability to maintain the necessary atmospheres and temperatures required for reaction throughout the entire sample.

Fibrous silica ceramics can be produced using this method but are so brittle that they crack under the slightest force. It appears that complete infiltration does not occur; rather, the voids between fibers become filled with the sol-gel solution. We initially investigated mercerized cotton fiber samples; however, the bulk of our work was conducted with raw fiber. This introduced the problem of full infiltration of the fibers, as well as coating them. These samples exhibited less shrinkage in comparison to the treated fiber.

We found that in order to fully reduce the sample, the atmosphere must not only be oxygen-free, but must also reduce additional oxygen from the gel. The two-step process in which the cotton is first converted to carbon and then re-infiltrated with either the initial sol-gel or with colloidal silica offered possible solutions to this problem. The best approach would be to perform this process under a highly reducing atmosphere.

In order to fabricate near-net shapes, the processing conditions must be designed to decrease the shrinkage that accompanies the heat treatments. From this investigation it was also concluded that other forms of silica should be used in the infiltration procedure, including colloidal suspensions.

Based upon our initial results, it appears feasible to fabricate an oxide or non-oxide fiber or woven mat via the infiltration of cotton fibers using sol-gel solutions. These fibers or mats can then be fabricated into usable shapes for Army applications.

Conclusions

Fibrous oxide ceramics can be processed, but they are brittle and crack due to incomplete infiltration.

In the one-step process for forming non-oxide materials, it was apparent that a fibrous structure remained and that the composite consisted of a silica matrix with carbon fibers. The two-step non-oxide process resulted in the formation of small amounts of silicon carbide, with the remainder being silica. The major difference between the one-step and two-step processes was that the second infiltration results in an increase in strength due to a higher loading of the silica sol-gel solution.

References

1. K.E. Duckett, *Surface Characteristics of Fibers and Textiles*, Volume 2, (New York, 1975), pp. 67-108.
2. R. Jeffries *et al.*, *Cellulose Chemical Technology* **3**, 255-274 (1969).
3. Mauersberge, *Matthew's Textile Fibers*, 5th edition, (Wiley & Sons, New York).
4. T.A. Edison, U.S. Pat. 470925 (15.3.1892)
5. C.E. Ford and C.V. Mitchell, U.S. Patent 3107152 (15.10.63).
6. A. Shindo, Y. Nakanishi, and I. Soma, *High Temperature Resistant Fibers from Organic Polymers*, (Interscience, New York, 1969), p. 305.
7. S. Sakka, *American Ceramic Society Bulletin* **64**, 1463-1466 (1985).

Advanced Composite Interceptor Kill Vehicle Structures

Larry Atha
Douglas H. Ennis
U. S. Army Strategic Defense Command
Huntsville, AL 35807-3801

The weight reduction, stiffness gains, and primary bending mode frequency increase of structures formed from properly chosen composite materials, relative to metal counterparts, are well known. In 1987, the U.S. Army Strategic Defense Command initiated a development and evaluation program to determine if advanced composites could be used for the main structural components of interceptor kill vehicles. The initial stage of this work focused on fabrication of a demonstration structure for the Ground Based Interceptor (GBI) Exoatmospheric Re-entry Vehicle Interceptor System (ERIS) flight design. The objectives were threefold: selection of the optimum composite materials, advanced bonding and joining technology development, and evaluation of structural performance.

The results exceeded model predictions. The completed resin matrix composite structure mass was 52 percent lower than the baseline design. Modal testing demonstrated a 200 percent increase in stiffness and a 41 percent gain in first-mode bending frequency. The current phase of this effort focuses on cost-effective, bulk-quantity fabrication techniques. Single-step technology has been successfully applied to a thermoset-based structure. Two-step molding and assembly have been demonstrated for a GBI-X class thermoplastic structure. Process reliability evaluation and verification of all test article performance parameters are in progress. Test results and future program directions are available.

Advanced Materials Science:

Evaluation of New Materials

Thomas W. Eagar
Richard P. Simmons
Materials Processing Center
Massachusetts Institute of Technology
Cambridge, MA 02139

The dramatic growth of materials science and engineering over the past two decades has made it possible to engineer new materials with specialized properties in a very short period of time. It has been suggested that we have changed from a philosophy of "designing *with* materials" to "designing *of* materials." Indeed, many new products and systems are conceived before the necessary materials are available; it is assumed that the materials properties that are required can be developed. While it is often correct that materials can be designed to meet these applications, it is frequently found that the cost of these specialized materials is excessive. In addition, as materials become ever more specialized, the market volume decreases, which further increases the costs. As new designs require operation of the materials closer to their limits, the failure rates increase. While the growth of materials science and engineering has brought, and will continue to bring, major advances in our standard of living, it is essential that we use sound judgment in selecting the new technologies that are most promising. Several of the criteria that should be considered in evaluating the technologic and commercial potential of advanced materials are available from our offices.

Ceramic Matrix Fiber Reinforced Composites

Paul Daschka
Richard Ruzga
John Petreanu
Ashok Kumar
U. S. Army Corps of Engineers
Construction Engineering Research Laboratory
P.O. Box 9005
Champaign, IL 61826-9005

This study investigated the use of a fiber-reinforced ceramic matrix composite material (CERHAB) for its ability to repair metallic components damaged by corrosion, cavitation, and erosion. The material's cavitation and erosion resistances were tested using a vibratory cavitation unit and an erosion unit. Scanning electron microscopy (SEM) was used to examine the interface between the repair material and the substrate. Several application methods were used to evaluate CERHAB, including furnace coating, microwave dip coating, and laboratory flame spray. Under laboratory conditions, CERHAB showed a good potential for use as a repair material on metallic components damaged by corrosion. In addition, in-situ vitrification of lead based paint may be accomplished by spraying the silica glassy compositions, followed by fusing in place.

Materials Processing and Manufacturing Technologies

Dick J. Wilkins
University of Delaware
Newark, DE 19717

Process technologies must be given prominent attention in the 21st century. The use of total quality management and concurrent engineering for product, process, and technology development is influencing the emerging role of manufacturing science. Of special importance will be manufacturing science's ability to quantify and simulate the effects of processing decisions, which will lead to intelligent manufacturing and a new way of interaction between government and industry.

Manufacturing technology is the application of the accumulated "know-how" along with the "know-why" derived from manufacturing science. Some trends that are beginning to change manufacturing include concurrent engineering.

Concurrent engineering. As defined by Winner at the Institute for Defense Analysis, concurrent engineering is "a systematic approach to the integrated concurrent design of products and their related processes, including manufacture and support. It causes developers to consider all elements of the product life cycle from conception through disposal, including quality, cost, schedule, and user requirements." This interaction is especially critical in the design of composite components because of the physical coupling among materials, shape, and processing. Certain shapes are much more efficient. In bending, for example, tubes are more efficient than rods. Likewise, certain manufacturing processes are more efficient in producing certain kinds of shapes, while only certain materials are available in the form needed for a given manufacturing process. Interactions among shape, process, and materials are critical in choosing the best solution to a given product development.

Tools of concurrent engineering are just as applicable to process development, but the tools of manufacturing science are much less developed than the analytic methods used to provide structural efficiency. However, computer simulation is changing that trend. Use of such simulations will bring the process specialist to an equal footing with the structural analyst in his ability to show the team the benefits of certain process or tooling decisions.

Manufacturing science. Manufacturing science utilizes the tools of science and engineering in its process of transforming raw materials into useful structures. For example, it typically employs heat and pressure, thus using the disciplines of solid mechanics to define temperature changes, materials science to track changing microstructure, physics to define forces and motions, and chemistry to define products and reactions.

Process models are the communication media of manufacturing science: they incorporate incoming material stream variations with variations in the process settings in order to predict the manufactured result. Models provide inexpensive experimental observations and economize the process by evaluating controls related to cost. They allow a process to be investigated in terms of robustness, or resistance to noises and variations, and can run a process that produces high quality output even when subjected to extreme input variations.

Intelligent manufacturing. Intelligent manufacturing is derived from the use of a process model to anticipate the effect of changes in the process and to select appropriate sensors to control the process while subjecting it to a variety of inputs. The model can initially be a rough functional model that can be refined on an as-needed basis, but the best process models are based on laws of physics and chemistry. Temperature and pressure sensors are commonly needed, and new noninterfering sensors are being developed with ultrasonics and acoustic emission. Using the process models, controls are devised to provide a completely automated response to a given set of variations. Expert systems and neural nets are often used to drive the controls.

Implementation of intelligent manufacturing is a logical outcome of concurrent engineering. But concurrent engineering does not go far enough because it assumes the traditional adversarial procurement process. The rationale that most of the costs are locked in at the beginning of the development process brings forth the new rationale of concurrent requirements.

Concurrent requirements. In military procurement, the buyer works out the requirements in advance of the bidding process, so that detailed bids can be prepared and the contract awarded to the lowest bidder. In the future, adversarial procurement will be bankrupt, and the government will have to work concurrently with the suppliers to set the best mix of requirements. Fortunately, the tools of concurrent engineering, manufacturing science, manufacturing technology, and so forth, analyzed through the aid of model processing, will be developed to a level to permit concurrent requirements to be feasibly implemented.

Battlefield Environment

Summaries

29

Onset and Nature of the Fracture of Uniaxially Compressed Gun Propellants:

A Small-Angle Neutron and X- Ray Scattering Study

Robert J. Lieb
Samuel F. Trevino
U. S. Army Research Laboratory
Weapons Technology Directorate
Aberdeen Proving Ground, MD 21005-5066

U.S. Army Research, Development and Engineering Center
Picatinny Arsenal, NJ 20895-5000

National Institute of Standards and Technology
Gaithersburg, MD 20899-0000

In an effort to understand the mechanism of fracture of gun propellant grains, the responses of two propellant formulations, M30 and M43, have been studied. Specimens were prepared from extruded solid strands and uniaxially deformed at about 100sec^{-1} to 2, 5, 10, and 20 percent strain. Axial and ortho-axial sections of these damaged specimens were investigated with small angle neutron and x-ray scattering. Monochromatic beams irradiated the specimens, and the resulting scattering patterns were recorded as a function of scattering angle with a two-dimensional detector. These measurements were performed using instruments located at the National Institute of Standards and Technology. The scattering from both propellants showed substantial asymmetry and dramatic changes as a function of strain. These results directly reflect the surface area that exists between the regions of different materials and also any new surface area created by fracture. Interpretation of the measurements has for the first time revealed important aspects of the mechanism of fracture failure in propellants.

Introduction

Pressure generation within guns is known to depend on the amount of surface area of the propelling charge. Generation of fracture surface area can occur when mechanical loads are applied to the propellant grains during ignition and combustion. Many studies[1-6] have been undertaken to establish the link between the mechanical response to deformation and the surface area generation that results from mechanical damage. Results of these studies have led to improvements in performance and propellant vulnerability responses but have not revealed the nature of fracture onset and growth within the propellant.

The measurement of surface area created by mechanical damage has been performed in various ways. Picnometery and the closed bomb burning of damaged specimens have been used, and they offer insight into the extent of fracture damage of propellant specimens. However, in order for the results to be optimally used, the experimental results must be related to the influence that the damage has on propellant combustion. Problems occur when the results are interpreted. For example, the surface area measured by the picnometer may not be accessible to the flame, or conversely, the flame may avail itself to much more area than the picnometer can measure. Closed bomb analysis seems to provide the most direct measurement, but even there, dynamic effects, such as the ease of the ignitability of freshly fractured surfaces or the nature of the generation of fracture surface area under the influence of the dynamic environment, raise questions about applicability of the experimental results to the operational environment situation. It is imperative, therefore, that as much as possible be known about the fracture process, including its onset and early development, so that the applicability of test methods and results can be properly made. One method that can be used to investigate the boundary surfaces of a system is by measuring the effects of the scattering of radiation by that system.

Neutron and x-ray scattering by a specimen is dependent upon the scattering cross-section of the materials in the specimen and on the contrast between materials at boundaries. In a homogeneous material, fracture will increase the scattering cross section by the introduction of the fracture surface. In a composite composed of materials of different contrast, scattering will occur at interface boundaries. This scattering will be increased by fracture occurring either by separation of the component materials or by bulk fracture.

These scattering features offer a method by which the onset and nature of the fracture process can be investigated. Because the greatest change in scattering due to fracture would be expected to occur in composite materials, composite propellants M30 and M43 were selected for this investigation. Specimens were prepared and uniaxially compressed at a high rate of deformation to various strain levels that ranged from 0 to 20 percent. The changes in the structure of these materials caused by this treatment were investigated by small angle scattering of both neutrons and x-rays. These measurements revealed the onset and nature of growth of the damage. This information will offer significant insight into the fracture process and will help direct future research efforts aimed at uncovering important processes that occur during propellant combustion.

Neutron and X-Ray Small-Angle Scattering

The theory and practice of scattering of neutrons and x-rays are well established.[7] Improvements in the instrumentation and adaptations of the theory to understand the physics in new materials continues. Here, we will focus on those aspects that address the present work.

Both types of radiation are able to scatter coherently from centers of which the material is composed. For neutrons, these scattering centers are the nuclei, and for x-rays, they are the electrons of the atoms in the material. Because the scattering is coherent, it carries information about the geometric properties of the conglomerate of scattering centers. An essential quantity that describes the radiation is the wave vector Q whose magnitude is given by:

$$Q = \frac{4\pi}{\lambda} \sin\theta, \tag{1}$$

where λ is the wavelength of the radiation and 2θ is the scattering angle. The scale, r_s, of the geometry to which the radiation is sensitive is of order:

$$r_s = Q^{-1} \tag{2}$$

Typical values of the radiation wavelength range from 0.1 to 1nm (1 to 10Å) so that the study of structures with geometric scales of 100nm requires scattering angles of about 1° or less (thus, small angle scattering). The structures of this scale size will reveal themselves if they exist in a matrix that scatters the radiation, that is, has contrast. This matrix may consist of different materials or a material and voids. The only requirement is that the contrast, $\Delta\rho_{ij}$, between the two materials i and j be non-zero. The contrast ρ_i for material i can be calculated from measured properties of the scattering centers of which it is composed, and is given by:

$$\rho_i = \frac{N_A D_i \sum_k n_k b_k}{\sum_k W_k} \tag{3}$$

where N_A is the Avogadro constant, D_i is the density of material i, the sums are over the k atom types in material i, n_k is the number of atoms of type k, W_k is the atomic weight of atom type k, and b_k is the scattering length of atom type k for the radiation of wavelength λ. Note that the dimension of ρ is L^{-2}. For neutrons, the values of b_k for the various materials are available as measured quantities, whereas for x-rays $b_k = Z_k b_e$, where Z_k is the atomic number of atom k and b_e is the scattering length of a single electron for x-rays, namely, the classical electron radius. These quantities are usually available for the material of interest so that the contrast is calculable. Note that the contrast of material i is different for neutrons and x-rays, but more importantly is that the relative contrast $\Delta\rho_{ij}$ of materials i and j will be different (unless one is singularly unfortunate) for neutrons and x-rays. This last observation is particularly useful when the system under study consists of three components (as is the case here).

The dependence of the scattering intensity I(Q) on Q is different for different magnitudes of Q even for the same material. This can be understood from Equation 2. The region of interest in the present work is known as Porod scattering and is valid for the range of Q sensitive to the surface area of the geometric structure. In this region the intensity is given as:

$$I\,(Q) = 2\pi \frac{\sum_{ij} \Delta\rho_{ij}^2 \, S_{ij}}{Q^4} \tag{4}$$

where the sum is over pairs of materials ij between which there exists a boundary of specific surface S_{ij} (surface area per unit volume of sample). In the present work the propellants consist of three materials, the filler, *F*, the binder, *B*, and voids, *V*. There are therefore three specific surface areas to be determined: S_{FB}, S_{FV} and S_{BV}. There are only two measurements that are available, namely, I_n and I_x, the neutron and x-ray intensities. As indicated previously, voids can exist in the binder, filler, or between binder and filler. If these voids had been accessible to the outside, they could have been filled with a contrasting liquid, making available independent data. This was tried without success. A simple model, in which each void presents an equal surface to each of the filler and binder materials, is used here. During manufacture the filler is wet by the binder and extruded. Voids within the binder or filler alone comprise a very small amount of the interface surface. It is thought that any new fracture surface would occur *between* the binder and filler, as is usually seen in these fractured materials. These observations help justify considering this model, in which $S_{FV} = S_{BV} = S_1$. Now, the two measurements of the neutron and x-ray intensities yield the equation:

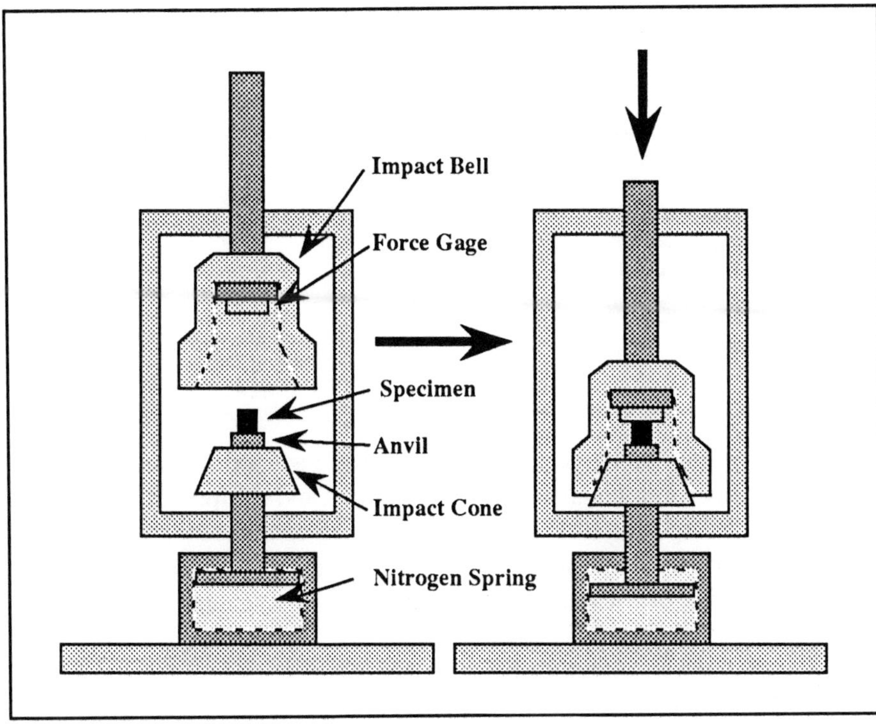

Figure 1. Servohydraulic tester.

$$I_m = 2\pi \frac{\left[(\rho^2_{F_m} + \rho^2_{B_m}) \frac{S_1}{2} + (\rho_{F_m} - \rho_{B_m})^2 S_{FB} \right]}{Q^4}, \quad (5)$$

where *m* represents either neutrons or x-ray radiation. Inversion of these equations yields a direct measure of the two specific surfaces.

Experimental Procedure

Preparation of propellant specimens. The specimens were prepared by cutting (using a diamond saw) extruded solid sticks of propellant into right circular cylinders (diameter = 12.7mm) with a length of 10.0mm. Two sets of five specimens were prepared for each propellant, and each set was uniaxially strained to 0 (undamaged), 2, 5, 10, and 20 percent. The compression was conducted in a servohydraulic test fixture illustrated in Figure 1. The unique feature about this machine is that compression can be arrested at a selected

Table 1
Nominal Percent Composition of Propellants

Fillers	M30	M43
Nitrocellulose	28	4
NC nitration level	12.6	12.6
NItroglycerin (NG)	22	—
Nitroguanidine (NQ)	48	—
Ethyl centralite (EC)	2	—
RDX (ground)	—	76
Cellulose acetate butyrate (CAB)	—	12
Plasticizer	—	8

strain by adjusting the height of the anvil. As the actuator moves down, the force gage comes into contact with the specimen and compression begins. It is halted when the inside surface of the impact bell contacts the mated surface of the impact cone. The nitrogen cylinder then absorbs the residual system energy by moving the piston. Therefore, the level of strain can be selected by adjusting the anvil height. The compressive strain rate was $100 sec^{-1}$.

The nominal formulation for each propellant is given in Table 1. The filler in M30 is nitroguanidine (NQ), which is in the form of long needle-like crystals of diameter from 5 to 10μm. The nitrocellulose (NC) and nitroglycerin (NG) form an intimate mixture and act as a binder for the system. The filler in M43 is RDX and is of significantly different form than the filler in the M30 system. The RDX is ground into small (1-20μm) ellipsoidal crystals that are held together by the intimately mixed cellulose acetate butyrate (CAB)/ NC/plasticizer binder. The adhesive binding strength in this system is much weaker than that of the M30 system. Figure 2 shows scanning electron micrographs SEM) of split propellant specimens that reveal the undamaged, intrinsic, propellant morphology.

After compression, the propellant cylinders were cut into 1-mm thick scattering specimens, as illustrated in Figure 3. Two types of specimens were made. Type A was cut perpendicular to the cylinder axis, and Type B was cut parallel to the axis and through the center of the cylinder. The two types were needed to detect in greater detail the way in which damage occurs. Note that the alignment of the NQ, which occurs during the extrusion process, is along the axis of the grain. This orientation is presented differently to the beams in the Type A and Type B specimens (hereafter referred to as simply A or B), as shown in Figure 4.

Scattering measurements. Both the neutron and x-ray spectrometers are

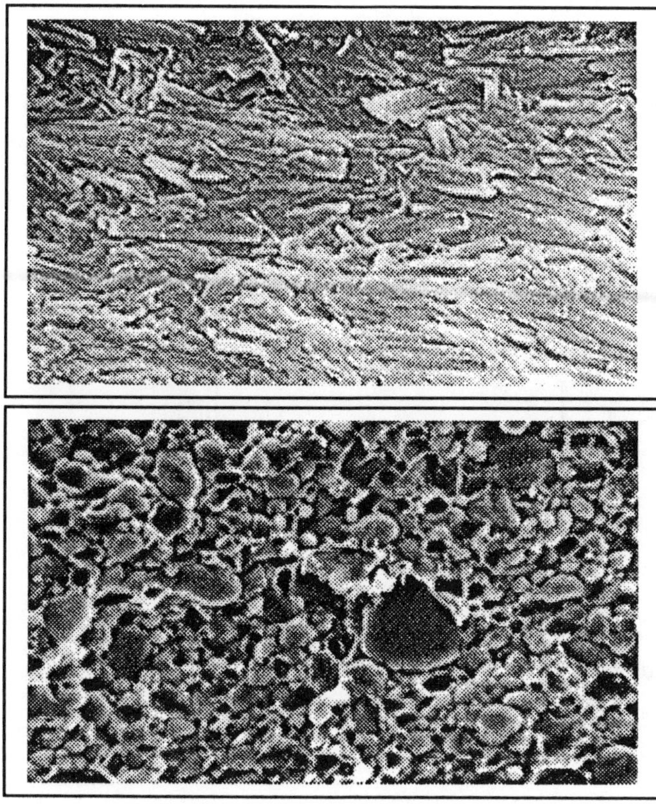

Figure 2. Scanning electron micrographs of the propellant specimens. Top: (**a**) is M30, bottom: (**b**) M43

located at NIST. Figure 4 is a schematic appropriate for both instruments. The method of obtaining the monochromatic beam of radiation differs for each technique, but the subsequent characteristics are essentially the same. The neutron beam wavelength was 8Å and the x-ray was 1.5Å. The direction of the neutron beam is defined by a series of circular slits separated by distances on the order of 1m. The distance between the sample and detector is about 8m. These two geometries determine the angular resolution of the instrument. The detector is a two-dimensional, position sensitive device that collects data simultaneously in a plane perpendicular to the incident beam. The data are stored in digital form for subsequent analysis. This includes the ability to obtain averages for constant scattering angle as functions of the azimuthal angle. If asymmetry exists in the scattering pattern, sector averages for azimuthal angles can be readily performed. The data collection time for the present set of measurements totaled approximately six hours for each scattering run.

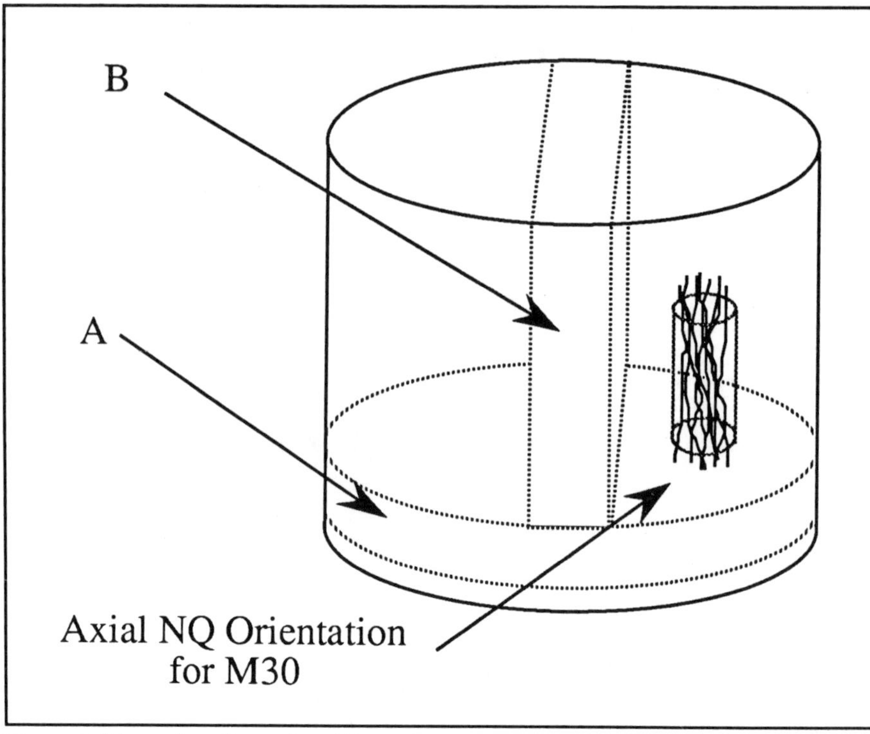

Figure 3. Illustration of type A and B specimens.

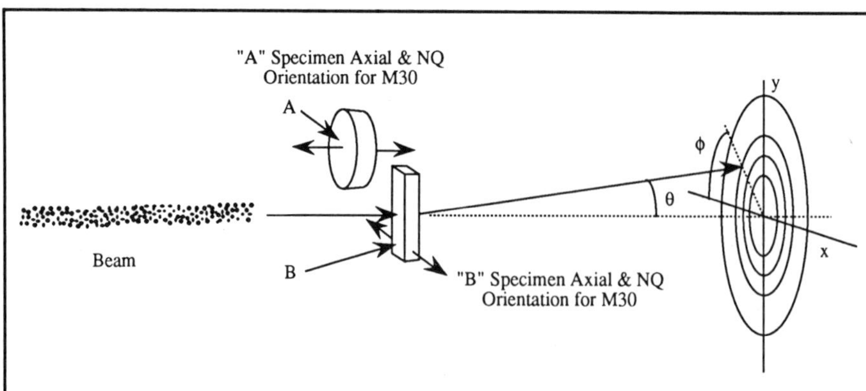

Figure 4. Schematic of scattering experiment.

Results

Both the neutron and x-ray scattering patterns were qualitatively similar. Different amounts of spherical asymmetry were obtained for the different

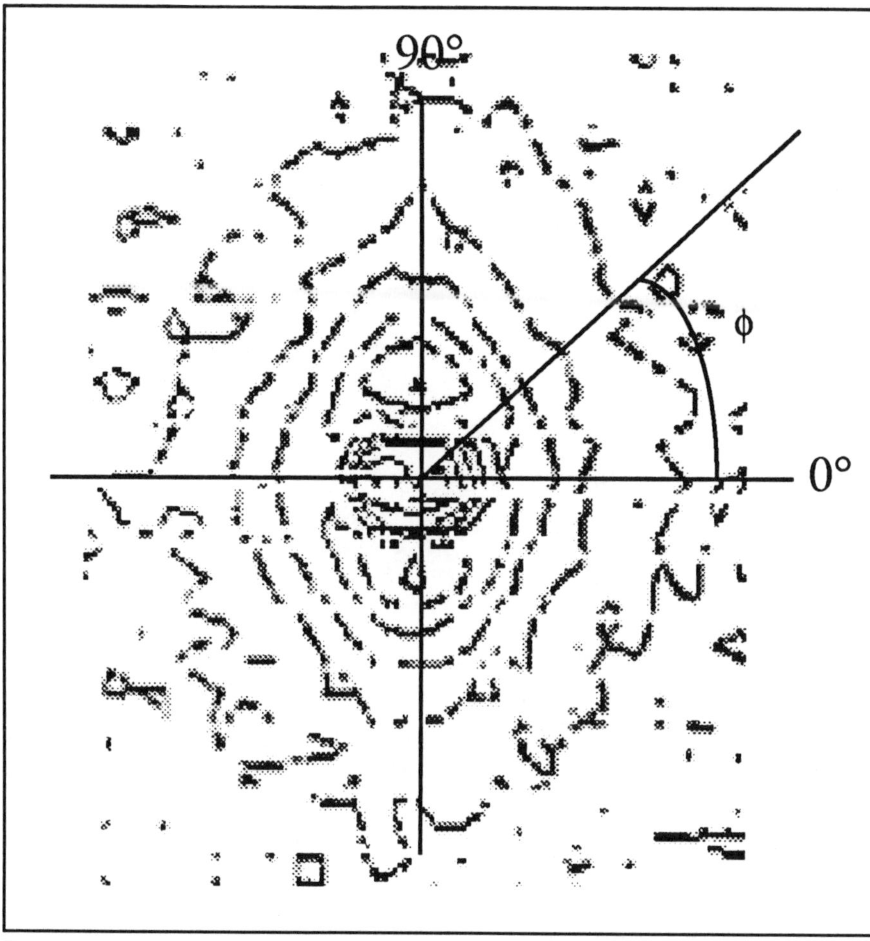

Figure 5. Undamaged B type scattering results for M30 showing NQ alignment.

degrees of strain. The B specimen of unstrained M30 produced the greatest degree of asymmetry. A plot of the raw two-dimensional data is presented in Figure 5. The asymmetry is due to the orientation of nitroguanidine needle-like crystal along the specimen axis. For all the B specimen data, sector averages, over 30° centered at ϕ, with ϕ equal to 0, 30, 60, and 90° were obtained. The data were corrected for detector efficiency and background and normalized to standard samples from which absolute cross-sections were determined. Sector averages were not performed for the A specimens because they showed symmetric scattering. The coefficient of Q^{-4}, which is associated with the contrast and specific surface factor in Equation 5, was obtained by plotting IQ^4 versus Q^4, fitting a straight line, and determining the intercept.

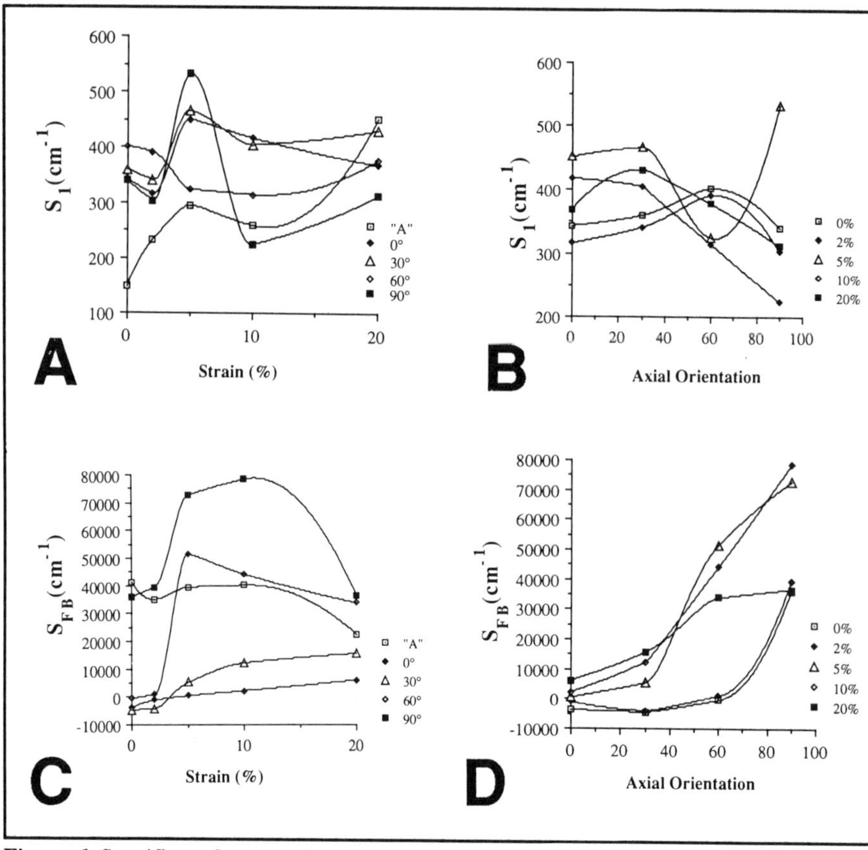

Figure 6. Specific surface area for M30 propellant.

This was done for both neutron and x-ray data. Equation 5 was inverted to obtain S_1 and S_{BF}. Figures 6 and 7 present the values of these specific surface areas for both propellants as functions of strain for the A and B specimens, and for sector averages for B specimens. For the B specimens, ϕ equal to 0° corresponds to the specimen axial direction.

Analysis

M30. Figure 6 shows the specific areas S_1 and S_{FB} plotted against strain and axial orientation. The specific area is the interface area per unit volume of material and is given in units of cm^{-1}. S_1 for the A specimen indicates that there were some voids present in the undamaged material. Because previous M30 propellant micrographs indicated that there were no voids in the binder,

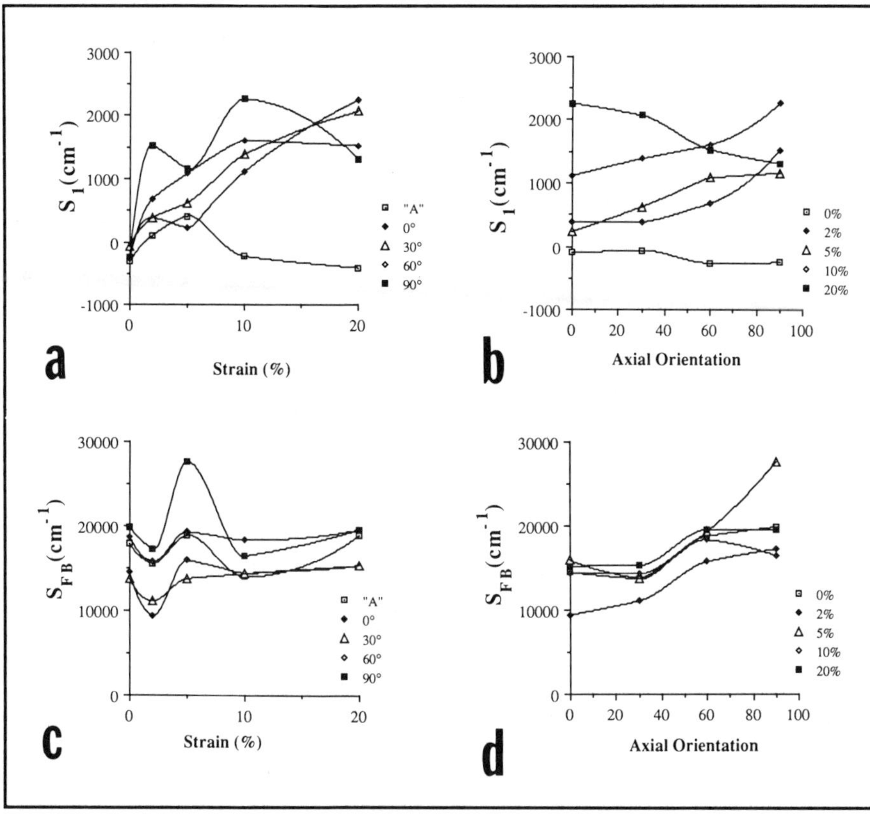

Figure 7. Specific surface area for M43 propellant.

micrographs of crystalline NQ from another investigation were studied. Figure 8 shows voids that are present on the surface of the NQ crystals. This micrograph and others from this series indicate that hollow NQ crystals could result from these elongated voids being internalized. The mismatch in the S_1 values for the A and B specimens at ϕ equal to 90° could be explained from such internal voids. Now, if spherical voids were present, or if the detector had infinite resolution, these values would be the same. However, with elongated voids, as indicated here, larger surfaces are measured when averaging S_1 along the axial instead of the direction perpendicular to it and the area vector.

As strain increased, the A specimen showed an increase in S_1 that leveled off and then increased again at higher strain. This is consistent with separation of binder and filler with the internal voids of the NQ staying unchanged. For the B specimen at ϕ equal to 0°, not much change was observed, possibly because of the insensitivity to separation in this direction. However, in the other three directions for ϕ there was a slight decrease in S_1,

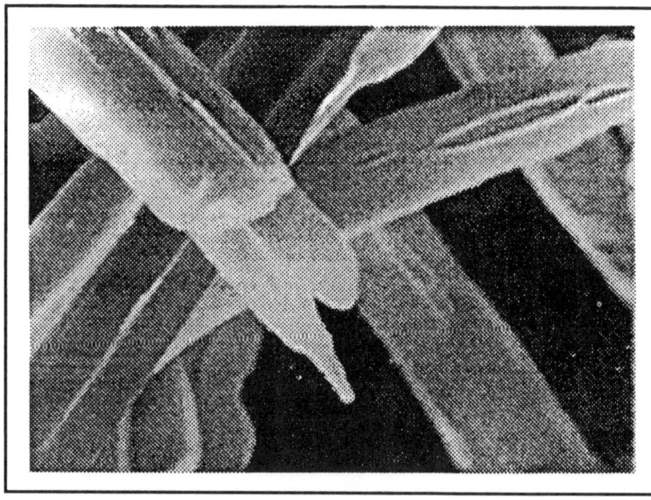

Figure 8. Micrograph of NQ crystals showing surface voids.

with a significant rise and then a decrease or leveling out, and then a final increase at high strain. This can be thought of as an early collapsing of voids or a closing of previously separated surfaces; then, after 2-percent strain, a significant separation and reorientation of the NQ crystals occurred within the solid. The most difficult feature to explain is the rapid drop in S_1 at ϕ equal to 90° at five-percent strain. There is no apparent reason as to why the value is reduced so much.

In Figure 6b, the S_1 values for the B specimen are plotted against axial orientation, with strain level of each curve being held constant. The general trend indicated that initially there was a slight reduction in the S_1 due to void collapse or recombination of surfaces. At five-percent strain, significant separation occurred, which recombined as strain was increased. At 20 percent, S_1 increased again, and possible reorientation of the NQ had begun.

In Figure 6c, the S_{FB} values at zero strain show a good match between the A and B specimens at ϕ equal to 90°. Also, the values were very low along 0° and 30°, as they should be with little surface between binder and filler in this direction. The rise in values with increasing strain may be attributed to the reorientation of NQ crystals. The perplexing feature of this plot is the enormous increase in S_{FB} at 60° and 90° at five-percent strain. A mechanism accounting for this increase could be the precipitation of NG, which is a liquid, from the binder mixture. This would create new binder-filler surface and still allow separation surfaces to be created. The decrease of S_{FB} at higher strains is in line with continued separation, and the magnitudes need not be reflected in corresponding increases of S_1.

The change in values of S_{FB} with axial orientation follow what was expected at each strain level. All values were low in the $\phi = 0°$ axial direction (axial direction) and increase as ϕ increases. This was expected and reflected the alignment of the NQ crystals. However, the unexpected increase in S_{FB} with strain is again clearly displayed.

M43. The specific area values for M43 are displayed in Figure 7. This propellant showed no voids in the undamaged specimens. In the B specimen, S_1 showed increasing values at increasing strains, and generally higher values in damaged specimens at increasing ϕ values. This indicated separation of the binder-filler interface, because the binder deforms plastically at 20°C, and no fractured RDX has been observed under these conditions in micrographs of damaged grains. Two features in the S_1 curves need to be commented upon. The first concerns the divergence of the S_1 values for the A and B specimens for $\phi = 90°$ at increasing strain. These values should track with each other because the surface area created here is due to binder-filler separation, and axial alignment is much reduced from that found in the M30 propellant. These data suggested a significant difference in the degree of damage in the two scattering specimens, which was subsequently shown to be the case. The mechanism of failure is by shear failure along 45° cones, as illustrated in Figure 9. Because the A specimen was taken from the end of the compression specimen, the center region of A had few areas of failure. The B specimen, on the other hand, contains the central portion of the compression specimen, at which the 45° failure cone is directed. So the B specimens would be expected to contain greater amounts of damaged material. The second feature concerns the decreasing S_1 value with axial orientation at 20-percent strain (see Figure 7b). It is thought here that RDX crystals were becoming so loose that they were able to rotate. This would cause what was observed, a more rapid increase in S_1 at lower ϕ values and reduction in S_1 at higher ϕ values, due to the ellipsoidal shape of the RDX.

The S_{FB} curves plotted against strain show similar changes at increasing strain levels, and generally show higher values as ϕ increases. Note that in this case the A and B specimens at $\phi = 90°$ matched well, as expected. The curves show an initial decrease, indicating separation, as did S_1. However, at five-percent strain a dramatic increase in S_{FB} was found that was followed by a decrease and then leveling. A possible mechanism for this observation is illustrated in Figure 10. From SEM micrographs, it is known that many small RDX crystals are present in the matrix. If tightly packed agglomerations of small particles lodge between larger crystals, they could burst during the plastic shear failure and create new S_{FB} surface, as shown in Figure 10. The separation process continues, as indicated in the S_1 values, but this process is in competition with the shear process. The changes in S_{FB} with axial orientation showed these same features and made another observation clear, as well.

Figure 9. Type B specimen at 20 percent strain showing 45° failure surfaces.

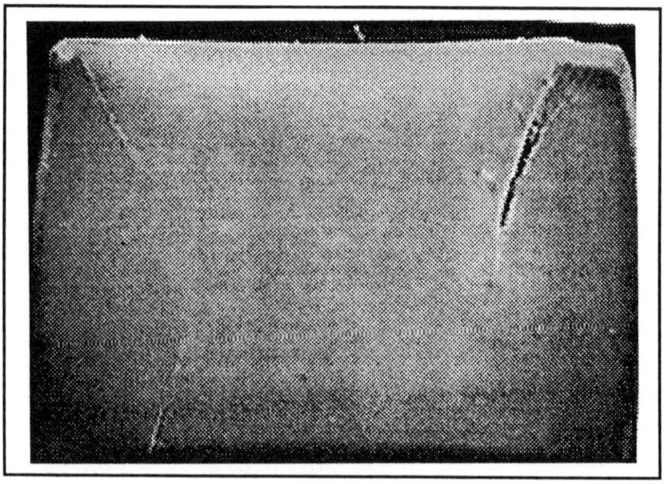

The ellipsoidal shapes of the RDX crystals appeared to align with the axial direction of the specimen. This was indicated by the increasing values of S_{FB} at increasing ϕ for each strain level (see Figure 7d). These observations provided a consistent picture and added significant insight into the failure process for the M43 propellant.

To find what this process requires, assume that one percent of the RDX is agglomerated and the eccentricity of the particles making up the agglomerate is 0.5. This assumption is reasonable because MicroTrac™ data[8] indicates that 3.1 percent of the crystals are smaller than 0.9μm for this type of RDX. This is the lowest size distribution category reported. Now, the average increase in surface area per cm^3 of propellant from two- to five-percent strain is $5.7 \times 10^3 cm^{-1}$. A quick calculation of the particle size in these agglomerations, required to produce the area increase indicated in these calculations, gives the following. Using the surface area and volume formulas for an ellipsoid, the major axis would be need to be about 0.05μm in order to produce enough area. This is quite a small size. However, if precipitation of RDX occurred when the solvent from the extrusion process was evaporated, particles of this size could be deposited.

One additional piece of information that adds credence to the interpretation of these measurements is found in a recent determination of the surface area of the RDX crystals,[9] which was made before mixing the crystals and binder. The value of the surface area density, S_d, was found to be $1.86 \times 10^4 cm^2/g$. This can be converted to specific density for M43 by using

$$S_{FB} = S_d \, \delta_F \, v_{F'} \ , \tag{6}$$

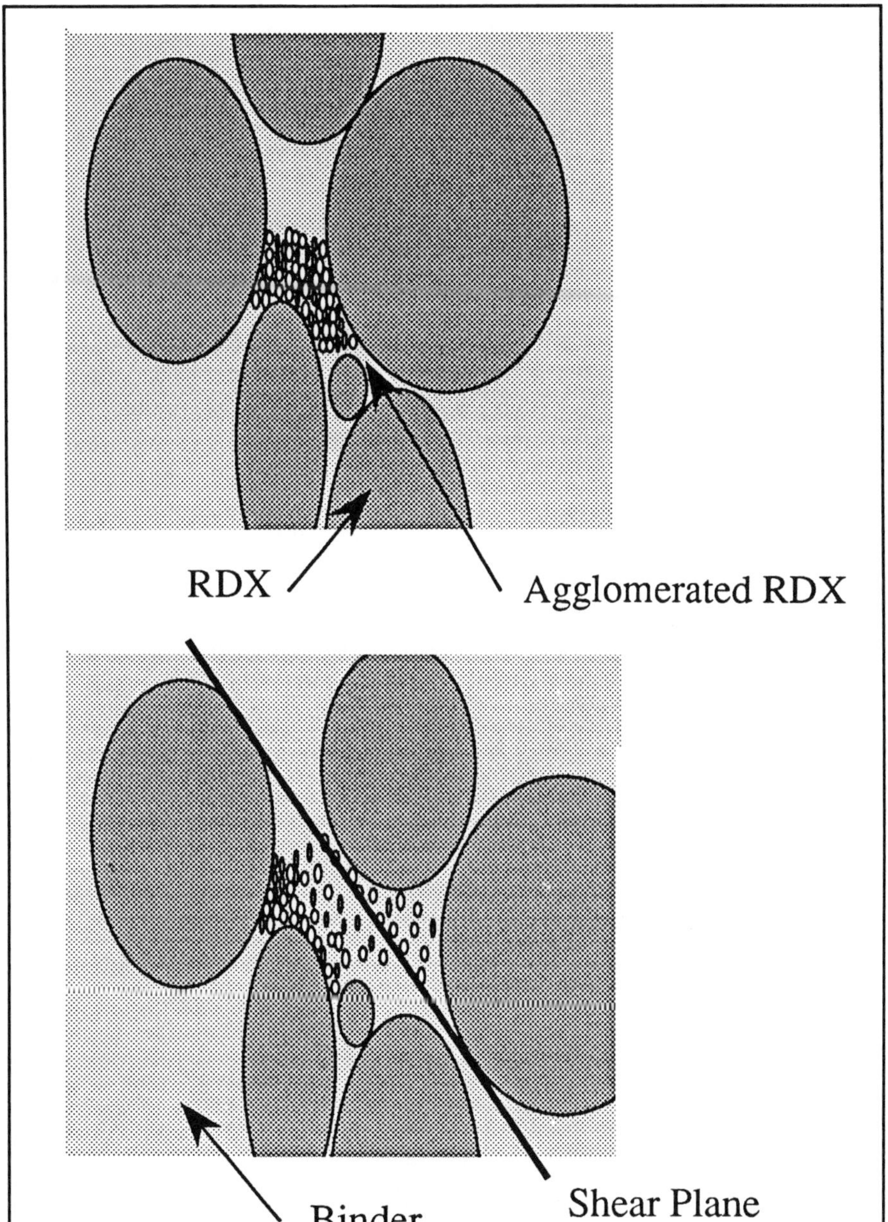

Figure 10. Illustration of a mechanism that would increase S_{FB} with increasing strain.

where δ_F and v_F are the mass density and volume fraction of the filler in the M43, respectively. The density of RDX is $1.82 g/cm^3$, and its volume fraction in M43 propellant is 0.675. When these values are used, the specific surface value received is $2.3 \times 10^4 cm^{-1}$. This value is in excellent agreement with the values measured here for the undamaged specimens (see Figure 7c). The fact that this value is larger than observed in the scattering experiments lends support for the agglomeration model.

Conclusions

A new method has been found for investigating the changes in surface morphology of filled energetic materials. Neutron and x-ray scattering were used to measure the binder-filler and void surfaces in M30 and M43 gun propellant as a function of strain and axial asymmetry of the specimen.

For M30 propellant, scattering measurements showed many features of the propellant system, including: previously noted axial alignment of the NQ crystals; the large binder-filler surface area orthogonal to the crystal axis; the strain level at which fracture begins; and the nature of the failure mechanism (binder-filler separation). Also indicated in the data were internal voids within the NQ crystals, closing of separated portions of damaged propellant after a certain strain level, and the reorientation of NQ crystals at high strain. One result that is not well understood was the indication of dramatic increases in binder-filler surface area. This was unexpected and would require an unusual occurrence, such as the precipitation of nitroglycerin (a liquid) from the binder mixture. Such an occurrence would help resolve the apparent inconsistencies in the data. Efforts to understand these results are continuing.

It seems that for M30, the collapse of preexisting voids, or the mating of preexisting binder-filler separations began immediately upon deformation. Between two- and five-percent strain significant separation damage began, and at the same time some mechanism (such as the precipitation of NG) created more binder-filler surface. As strain continued beyond 10 percent, separation continued, and the trend was for greater void area and decreased binder-filler surface.

The most significant morphologic features observed for M43 propellant were the following: First, no voids were indicated before compression. Second, axial alignment of the major axis of the ellipsoidal RDX particles along the specimen axis was indicated. This alignment was much less than that found for M30, but still observable. Third, damage in the A and B specimens was markedly different due to the shear failure occurring in the specimen and the method of specimen preparation.

Results suggested that as deformation occurred in M43, separation began and generally continued throughout the deformation. Some new binder-filler surface seems to be generated at about five-percent strain. This could have arisen from either a recombination of a separated interface, precipitation of plasticizer on a much smaller scale than for M30, or the breakup of agglomerated RDX, as previously discussed. As compaction continued, the separation grew, and there was an indication that, somewhere between 10- and 20-percent strain, RDX crystals became free to rotate in the matrix.

This information was largely unknown before these experiments were performed. These new insights will help guide the development of propellants and aid in the development of methods for increased performance and reduced vulnerability response.

The scattering techniques that were developed will permit more extensive progress in future scattering efforts. This work should include a study of low temperature deformation to investigate the agglomerated RDX conjecture. If the suggestion is correct, the increase in binder-filler surface noted in this study should not appear in the specimens deformed at lower temperature. The lower temperature should not permit plastic flow of the binder and should thereby eliminate the creation of new surface. Other experiments on single and double-based propellants will offer similar insight into the fracture mechanisms of those systems and provide methods for improvement.

References

1. G.A. Gazonas, A. Juhasz, and J.C. Ford, *Strain Rate Insensitivity of Damaged-Induced Surface Area in M30 and JA2 Gun Propellants* (Technical Report BRL-TR-3251, U.S. Laboratory Command, Ballistic Research Laboratory, Aberdeen Proving Ground, MD, 1991).
2. P. Lu *et al.*, Shaped Charge Jet Impact on Gun Propellants Study I - Temperature and Mechanical Properties Effects, 1991 Propulsion System Hazards Subcommittee Meeting, March 1991.
3. R.J. Lieb and M.G. Leadore, *Mechanical Failure Parameters in Gun Propellants* (Technical Report BRL-TR-3296, U.S. Laboratory Command, Ballistic Research Laboratory, Aberdeen Proving Ground, MD, 1991).
4. G.A. Gazonas, *The Mechanical Response of M30, XM39, and JA2 Propellants at Strain Rates from 10^{-2} to 250 sec^{-1}* (BRL-TR-3181, U.S. Ballistic Research Laboratory, Aberdeen Proving Ground, MD, 1991).
5. R.J. Lieb, in *1987 JANNAF Structures & Mechanical Behavior Subcommittee Meeting* **1** (CPIA Publication 463, Chemical Propulsion Information Agency, March 1987), pp.51-62.
6. A.W. Horst, in *1981 JANNAF Structures and Mechanical Behavior Subcommittee Meeting* **1** (CPIA Publication 351, Chemical Propulsion Information Agency, December 1981).
7. O. Glatter and O. Kratky, Eds., *Small Angle X-ray Scattering* (Academic Press, New York, 1982).
8. R. Simmons, Naval Surface Warfare Center, Indian Head, MD, private communication, 1992.
9. S. Caulder, Naval Surface Warfare Center, Indian Head, MD, private communication, 1992.

30

Thermal Decomposition of RDX and HMX in Condensed Phase: Isotope Scrambling and Deuterium Kinetic Isotope Effect Studies

Richard Behrens, Jr.
Combustion Research Facility
Sandia National Laboratories
Livermore, CA 94551-0969

Suryanarayana Bulusu
Energetic Materials Division
U.S. Army Armament
Research, Development, and
Engineering Center
Picatinny Arsenal,
NJ 07806-5000

Understanding the thermal decomposition kinetics and mechanism of 1,3,5-trinitro-hexahydro-s-triazine (RDX) and octahydro-1,3,5,7-tetranitro-1,3,5,7-tetrazocine (HMX) is essential to improve the combustion and sensitivity properties of formulations that use these nitramines. Following previously published determination of the decomposition products and their rates of formation, attention is now focused on the bond-breaking and bond-forming steps in the decomposition process. To this end, the inter- versus intramolecular origin of the decomposition products of RDX and HMX has been traced by isotopic cross-over experiments using mixtures of differently labeled (^{2}H, ^{13}C, ^{15}N, and ^{18}O) analogues of each compound.

From an analysis of the measured fractions of isotopic scrambling in different products, the extent of deuterium kinetic isotope effect (DKIE) of each product, and the temporal behaviors of the rates of product formation, several primary reaction pathways were identified for decomposition of RDX and HMX.

The authors present detailed data only on RDX and discuss its important pathways of decomposition and the probable rate-controlling step.

Introduction

The cyclic nitramines 1,3,5-trinitro-hexahydro-s-triazine (RDX) (Structure I) and octahydro-1,3,5,7-tetranitro-1,3,5,7-tetrazocine (HMX) (Structure II) are energetic ingredients that are used in various propellants and explosives.

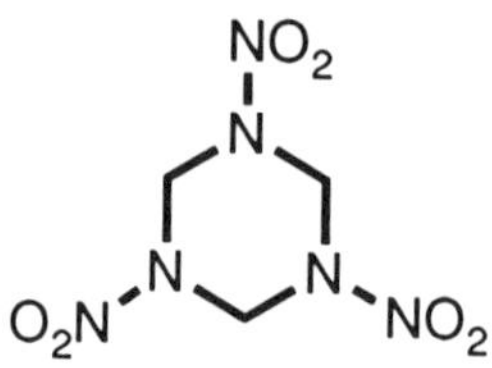

(I)

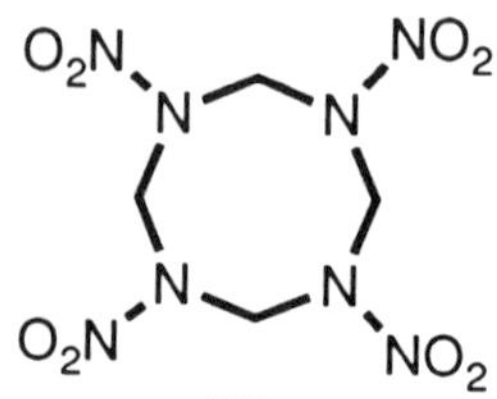

(II)

Understanding the complex physicochemical processes that underlie the combustion of these materials can lead to methods for modifying the propellant and explosive formulations in order to obtain better ignition, combustion, or sensitivity properties. Our work strives to provide a link between the physical properties and molecular structures of different nitramines and their combustive behavior. This requires the understanding of the reaction kinetics and transport processes in both the gas and condensed phases. The thrust of our work is to obtain an understanding of the physical processes and reaction mechanisms that occur in the condensed phase so that the identity and rate of release of the pyrolysis products from the condensed phase can be predicted, as a function of pressure and heating rate, based on the physical properties and molecular conformation of the material.

Reviews[1-3] of the literature on RDX and HMX have discussed the roles of unimolecular decomposition and autocatalysis on the thermal decomposition of these compounds. Previous work[4-7] has shown that HMX decomposes in the condensed phase and that the identity and rates of release of the pyrolysis products are determined by complex physical processes coupled with several different chemical decomposition mechanisms.[6,7] The physical processes include the formation and containment of the gaseous pyrolysis products in bubbles within the HMX particles and the subsequent release of the gases as escape paths form within the HMX particles. The long containment time of decomposition products within the HMX particles allows secondary reactions between the decomposition products and the HMX.

In contrast to HMX, RDX has a lower melting point (202°C for RDX compared with ≈280°C for HMX), so that decomposition occurs primarily in the liquid phase, with a faster release of the gaseous products from the sample than for HMX. Although both RDX and HMX have been found[6,8,9] to form

H_2O, HCN, CO, CH_2O, NO, N_2O, amides (such as methylformamide), a nonvolatile residue (NVR), and the mononitroso analogues of the parent compound [1-nitroso-3,5,7-trinitro-1,3,5,7-tetrazocine (ONTNTA) from HMX and 1-nitroso-3,5-dinitro-hexahydro-s-triazine (ONDNTA) from RDX], only in experiments with RDX[8-11] have NO_2 and oxy-s-triazine (OST) been observed. The precise structure of OST has not been determined, but it is likely to be one of the following:

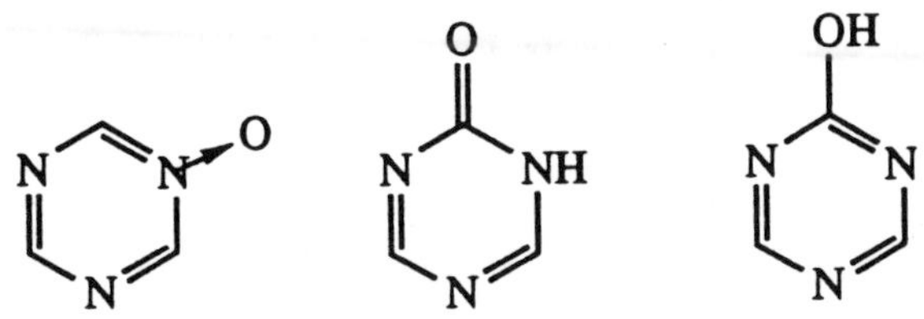

(III)

We present the results of measurements of isotopic scrambling of products from mixtures of isotopic analogues of RDX and of the observed deuterium kinetic isotope effects (DKIE). Together, these two types of results are used to probe the decomposition mechanisms of RDX.

The basic mechanisms that have been considered in the past to describe the decomposition of RDX fall broadly into three categories shown in the following reactions. These include the least energy pathway[11] involving the N-N bond cleavage,

$$RDX \rightarrow NO_2 + H_2CN + 2H_2C = N\text{-}NO_2, \qquad (R1)$$

the concerted triple scission pathway,

$$RDX \rightarrow 3H_2C = N\text{-}NO_2 \qquad (R2)$$

found in infrared multiphoton dissociation (IRMPD) experiments with RDX,[12] and the HONO elimination pathway[13]

$$RDX \rightarrow 3HONO + 3HCN. \qquad (R3)$$

Secondary reactions of methylenenitramine, R4 and R5, have been postulated

$$H_2C = N\text{-}NO_2 \rightarrow HONO + HCN \qquad (R4)$$

$$H_2C = N\text{-}NO_2 \rightarrow CH_2O + N_2O \quad \text{(C-N bond scission)} \qquad (R5)$$

for years to explain the formation of HCN, CH_2O, and N_2O and were shown to occur in the IRMPD experiments[12] with RDX.

Recent thermal decomposition studies of RDX[9,10] have shown that it is also important to account for the formation of OST and ONDNTA in the decomposition mechanism because they may be intermediates from which the final products are formed. To address these issues, mixtures of isotopically labeled compounds are used to track the breaking and formation of chemical bonds that lead to the generation of the thermal decomposition products of RDX. We also use DKIE to probe the role of carbon-hydrogen bonds in the formation of the different decomposition products. We present results from the previously described simultaneous thermogravimetric modulated beam mass spectrometry (STMBMS) measurements between RDX, RDX-d_6, and mixtures of unlabeled RDX (RDX-ul) and the isotopically labeled analogues, RDX-d_6, RDX-^{13}C, and RDX-$^{15}N_6$, and those between RDX-^{13}C and RDX-^{18}O. The same measurements are used to determine the DKIE on the pyrolysis products during the course of the decomposition and to determine the amount of isotopic scrambling from different isotopic mixtures of RDX. We report the calculated fraction of isotopic scrambling observed for most of the products formed in experiments with different combinations of the labeled RDX compounds.

Experimental

Instrument description. The STMBMS apparatus and basic data analysis procedure have been summarized in a previous work[9] and are described more extensively elsewhere.[14-16]

Sample preparation. Samples of RDX with different isotopic labels (see Table 1) were synthesized by the general method previously reported.[17] The isotopic enrichment of each of the isotopomers and the amounts of each one used in a given experiment are given in Table 1.

Samples for experiments with mixtures of isotopic analogues of RDX were prepared by mixing equal amounts in acetone solution, followed by evaporation of the solvent.

For each mixed-isotope experiment listed in Table 1, experiments were run on each of the isotopic analogues separately. The results from these experiments are used to check for the presence of ion signals at mass-to-charge ratio (m/z) values that should originate only from the mixing of isotopes in the scrambling experiments. If there is a measurable amount of ion signal present in the experiments with the separate isotopic analogues, then the results of the separate experiments are used to correct the data from the isotope scrambling experiments.

Table 1
Experimental Parameters*

Compound	Enrichment (%)	Experiment reactant quantity (mg) 1	2	3	4
RDX-ul†		2.04	2.68	3.35	
RDX-d_6	97.3	1.98	2.6		
RDX-$^{15}N_6$	98.5			3.53	
RDX-^{13}C	98.1				2.26
RDX-^{18}O	3.7				2.26
Electron energy (eV)	-	18	20	20	20
Orifice dia. (cm)		0.00254	0.00254	0.00254	0.00254
Temp. range‡		198 - 207	198-207	198 -208	198-207

* All mixtures of isotopic analogues of RDX are recrystallized from an acetone solution.
† ul = unlabeled.
‡ Temperature range over which the decomposition of RDX occurs in °C. Heating rate for all experiments is 0.5°C/min in the temperature range of the decomposition.

Analysis

The data analysis procedures used to identify the pyrolysis products and determine their rates of formation have been described previously.[14-16,20] Because this chapter examines the extent of isotopic scrambling and the DKIE on the relative rates of formation of the various products, it is sufficient to compare the ratios of the ion signals measured at different m/z values for the various products. The structures of the more complex decomposition products discussed here are reasonably assumed structures based on the mass spectra of different isotopic analogues of RDX.[9] OST has been studied previously by tandem mass spectrometry,[10] but a definitive structure could not be assigned.

The relative size of the ion signals at the different m/z values associated with an individual decomposition product is dependent on the amount of each isotopically labeled analogue used in the experiment, the fraction of each analogue that participates in forming isotopically scrambled products, the cracking factors in the ion fragmentation process, and the fraction of isotopic enrichment in each isotopically labeled analogue.

Briefly, the amount of each product isotopomer formed from the decomposition of a mixture of r moles of isotopomer 1 and s moles of isotopomer 2 of RDX is given by:

$$rRDX_1 + sRDX_2 \rightarrow x\kappa r\,(1 - f)\,A_{11} + \kappa s\,(1 - f)\,A_{22} + \kappa f\,(xr + s)\,A_{12} \quad (R6)$$

A_{11} and A_{22} are pyrolysis products formed from RDX_1 and RDX_2,

Table 2
H_2O Isotopic Scrambling Results

Isotopes	Experiment Number	m/z	Relative Intensities	Temporal Correlation
H/D	2*	18	64±11	Y
		19	100	Y
		20	243	Y

* Results for isotopic scrambling of H_2O from Experiment 1 are not included because of interference from extraneous signal at m/z = 18.

respectively; A_{12} is the set of mixed isotope products; κ is moles of product A formed per mole of RDX_2, x is the ratio of moles of product A formed from RDX_1 to the moles of product A formed from RDX_2 (for deuterium-labeled RDX, this corresponds to the DKIE factor); and f is the fraction of the product molecules produced from either RDX_1 or RDX_2 that participate in isotope scrambling.

Reaction 6 is then combined with the following expression that accounts for the relative number of ions formed in the mass spectrometer at different m/z values from the product isotopomer:

$$S_k^+ = \sum_{i=1}^{\#\,\text{isotopomers}} F_{i,k}\, A_i^+ \tag{1}$$

where S_k^+ is the ion signal at m/z=k; A_i^+ is the total number of ions formed from the isotopomer A_i with mass m_i; and $F_{i,k}$ is the fraction of the total number of ions formed from A_i that appear at m/z=k.

From reaction R6 and Equation 1 expressions for the amount of ion signal at each different m/z value associated with a decomposition product are formulated and used to construct equations based on the ratios of the ion signals at different m/z values for a given product. These equations are then solved for the values of f and x.

Results

The isotopic scrambling results for the products H_2O, CH_2O, N_2O, OST, and ONDNTA, based on experiments outlined in Table 1, are summarized in Tables 2 to 6, whereas the mixing fractions (crossover), *f*, and the observed DKIEs, *x*, for the deuterium labeled products are presented in Table 7.

Table 3
CH_2O Isotopic Scrambling Results

Isotopes	Experiment Number	m/z	Relative Intensities	Temporal Correlation
H/D	1	29	49±3	Y
		31	15±3*	Y
		32	100±3	Y
	2	29	61±5	Y
		31	16±5	Y
		32	100±5	Y
$^{13}C/^{18}O$	4	31	100	Y
		32	5.4±0.6	Y
		33	1.4±0.1†	Y

* Corrected for 31^+ contribution from pure RDX-ul.
† Corrected for the 33^+ contribution observed in a separate experiment with RDX-^{13}C. The corrections are made using the CH_2O (^{13}C m/z=31) signal for normalization.

Table 4
N_2O Isotopic Scrambling Results

Isotopes	Experiment Number	m/z	Relative Intensities*	Temporal Correlation
$^{14}N/^{15}N$	3	44	100±10	Y
		45	-4.1±2	na
		46	93.9±11	Y

* Ion signal at each m/z value is corrected for overlapping ion signals that are observed in experiments with individual RDX-ul and RDX-$^{15}N_6$ samples.
na - not applicable.

Figures 1 through 6 show the relative intensities as a function of temperature and the temporal correlations of groups of ion signals at various m/z-values that are used to determine the extent of the DKIE and isotopic scrambling for the decomposition products formed from RDX. Because the heating rates were linear with time, the curves also reflect the temporal correlations between the ions signals. All of the data presented in Figures 1 through 6 clearly show that the decomposition products are first observed when the RDX starts to melt at 198°C. As shown in a previous work,[9] decomposition of RDX is significantly inhibited in the solid phase.

H_2O. Scrambling of the hydrogen isotopes is observed in the formation of water from a mixture of RDX-ul and RDX-d_6. Analysis of the data in Table 2 shows that the fraction of water that participates in mixing is 1.1±0.1, which represents complete mixing, and that the DKIE is 1.5±0.1. Because low

Table 5
Oxy-s-triazine (OST) ($C_3H_3N_3O$) Isotopic Scrambling Results

Isotopes	Experiment Number	m/z	Relative Intensities*	Temporal Correlation
H/D	1	97	100±3	N
		98	4±3	na
		99	4±3	na
		100	45 to 71±3	N
	2	97	100±2	Y
		98	3.1±2.1	na
		99	2.8±0.4	na
		100	38 to 59±2	Y
$^{14}N/^{15}N$	3	97	100±2	Y
		98	1±2	na
		99	1±2	na
		100	100±2	Y
$^{13}C/^{18}O$	4	97	85.4±4	Y
		99	5.4±0.5	Y
		100	100.0	Y
		102	0.8±0.8	N

* Ion signals at m/z-values of 98 and 102 have been corrected for ion signal due to fragmentation of RDX in ionizer of mass spectrometer.
na - not applicable.

electron energies of 18 and 20eV are used in the mass spectrometer, contributions from OD^+ are ignored. The results are calculated assuming only the molecular ions, H_2O^+, HDO^+, and D_2O^+ contribute to the measured signal. Because H and D exchange is rapid in water and is likely to occur before detection of the product, the observation of complete H/D scrambling does not necessarily imply that water is produced in a bimolecular reaction between either RDX itself or RDX and a decomposition product.

CH_2O. As indicated in Table 7, the fraction of RDX that forms formaldehyde with hydrogen and deuterium exchange is about 7 percent for the experiments. It can be seen from Figure 1 that there is a high degree of temporal correlation between the ion signals from the isotopic analogues of formaldehyde. The observed DKIE for formaldehyde formation is 1.05±0.1, which implies no deuterium kinetic isotope effect. The ion signals used in the analysis are for m/z values of CH_2O^+, CHO^+, and their isotopic analogues. Measured ion fragmentation factors for CH_2O and CD_2O were used to calculate the ratio of CHO^+/CH_2O^+ and CDO^+/CD_2O^+ formed in the ionization process.

Experiment 4, with a mixture of RDX-^{13}C and RDX-^{18}O, shows that 45±7 percent (Table 7) of the formaldehyde formed is from analogues.

Table 6
ONDNTA [$C_3H_6N_3(NO_2)_2NO$] Isotopic Scrambling Results

Isotopes	Experiment Number	m/z*	Relative Intensities†	Temporal Correlation‡
H/D	1	132	100±14	Y
		133	7±5	Y
		134	0§	na
		135	5±2	Y
		136	90±5	Y
	2	132	97±19	Y
		133	7±4	Y
		134	0§	na
		135	8±1	Y
		136	86±3	Y
$^{14}N/^{15}N$	3	132	90 to 98±17	N
		133	95 to 70±5	Y-135
		134	0d	na
		135	86 to 69±5	Y-133
		136	85 to 94±6	N

* m/z values are for an ion formula of $C_2H_4N_2(NO)(NO_2)$ that is formed by fragmentation of ONDNTA in ionizer of mass spectrometer.

† Ion signals at all of m/z values have been corrected for contributions from fragmentation of RDX in the ionizer of mass spectrometer. In addition, ion signals at m/z-values of 133 and 135 have been corrected for contributions from isotopomers that arise directly from individual isotopic analogues of ONDNTA in absence of scrambling.

‡ Y indicates that ion signals at these m/z values are temporally correlated with all other m/z values listed in experiment. N indicates that ion signals are not temporally correlated with other ion signals in group. Y-nos. indicate that ion signal is correlated with another m/z-value in group indicated by number.

§ There is no evidence of transfer of two atoms between product molecules.

na - not applicable.

However, scrambling of the ^{13}C and ^{18}O in formaldehyde may be the result of polymerization and depolymerization of formaldehyde after its initial formation and prior to its detection. The ion signals of $^{13}CH_2O$ (31), $CH_2{}^{18}O$ (32), and $^{13}CH_2{}^{18}O$ (33) were used to calculate the value for *f*.

N_2O. Experiment 3 (Table 4), with a mixture of RDX-ul and RDX-$^{15}N_6$, shows that none of the N_2O is formed from mixing between the two differently labeled RDX analogues. The ion signals of $^{14}N_2O^+$, $^{14}N^{15}NO^+$, and $^{15}N_2O^+$ (Figure 2) are used to calculate the mixing fraction.

OST. Three different kinds of isotopic scrambling experiments were used to derive mixing information for OST. The results are tabulated in Table 5 and shown in Figures 3 to 5. Analysis of the data presented in Table 5 shows no mixing in any of these experiments.

The experiments with RDX-ul and RDX-d_6 (Figure 3, Table 7) show a strong DKIE. The relative size of the DKIE ranges from 2.2 during the initial stage of formation of OST down to 1.3 at the final stage of formation of OST.

Table 7
Mixing Fractions and DKIE Results from RDX Isotopic Scrambling Experiments

Decomposition Products	Scrambled Isotopes	Mixing fraction *f*	DKIE x^*
H_2O	H/D	1.1±0.1	1.5±0.1
CH_2O	H/D	0.075±.02	1.05±0.1
	$^{13}C/^{18}O$	0.45±.07	
N_2O	$^{15}N_6/^{14}N_6$	-0.04±0.03	
OST	H/D	0±.05	2.2 to 1.3†
	$^{15}N/^{14}N$	0±.05	
	$^{13}C/^{18}O$	0±.05	
ONDNTA	H/D	0.250±.05	1.05±0.2
	$^{15}N_6/^{14}N_6$	1.0 to 0.61‡	

* Values for x are calculated only for the H/D exchange experiments.
† DKIE for OST decreases as sample undergoes decomposition due to increase in amount of RDX-d_6 relative to the amount of RDX-ul.
‡ Fraction of NO that exchanges during the formation of ONDNTA decreases as sample size decreases.

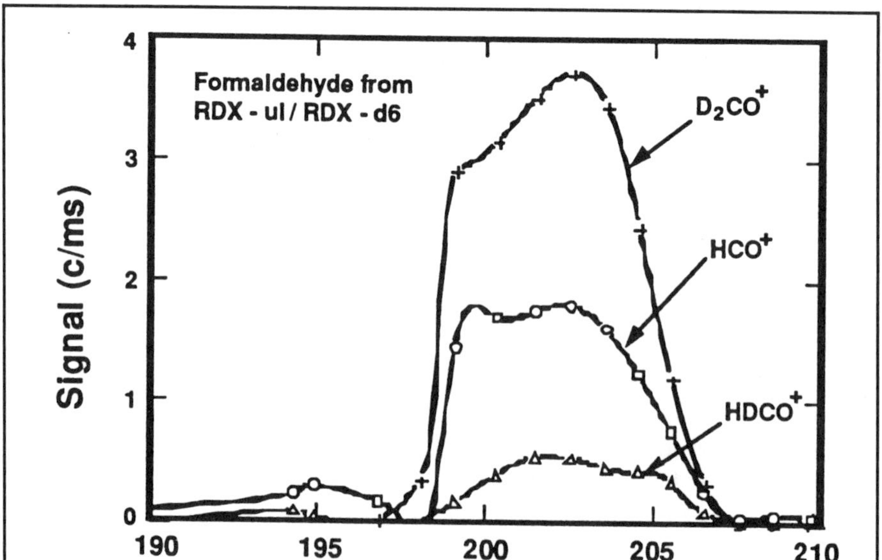

Figure 1. Ion signals from CH_2O and its isotopic analogues formed during decomposition of a mixture of RDX-ul and RDX-d_6 (Experiment 1 in Table 1). Between 193 and 210°C, heating rate of sample is 0.5°C/min. Rapid increase in rate of evolution of CH_2O at 198°C occurs as the sample melts.

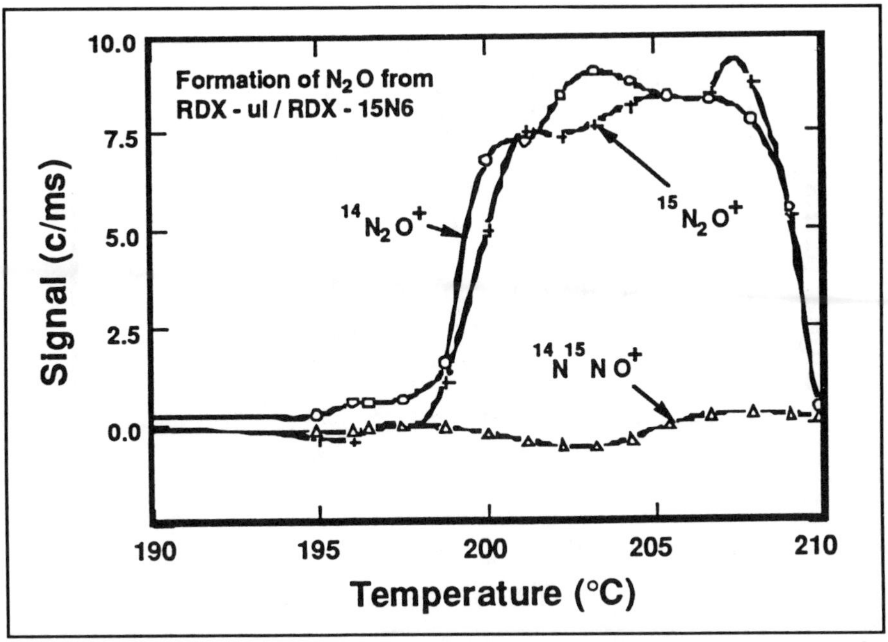

Figure 2. Ion signals from N_2O and its isotopic analogues formed during decomposition of a mixture of RDX-ul and RDX-$^{15}N_6$ (Experiment 1 in Table 1). Between 195 and 210°C, heating rate of sample is 0.5°C/min. Rapid increase in rate of evolution of N_2O at 199°C occurs as the sample melts.

This decrease in the DKIE is probably due to the increase in the relative amount of RDX-d_6 compared with RDX-ul as the sample decomposes.

ONDNTA. ONDNTA represents a product (mol.wt. 206) that is formed by the net loss of a single atom (oxygen) from RDX. The results from two different kinds of isotopic scrambling experiments for ONDNTA are tabulated in Table 6, and the results of the $^{14}N/^{15}N$ isotopic exchange measurements from ONDNTA are shown in Figure 6. The isotopic analysis of this product from the mixing experiments was based on its major fragment ion, m/z = 132 ($C_2H_4N_4O_3$) and its isotopic variants.

The data indicate that 25±5 percent of the RDX participates in scrambling of H and D in the formation of ONDNTA, despite the fact that the product results from the loss of a single oxygen atom. But because there is no DKIE observed in the formation of this product, hydrogen is not involved in the rate-determining step for this product.

The $^{14}N/^{15}N$ mixing in the ONDNTA product can be readily seen in Figure 6. During the initial stages of the decomposition, the ion signals at m/z-values of 132, 133, 135, and 136 are approximately equal, thus indicating

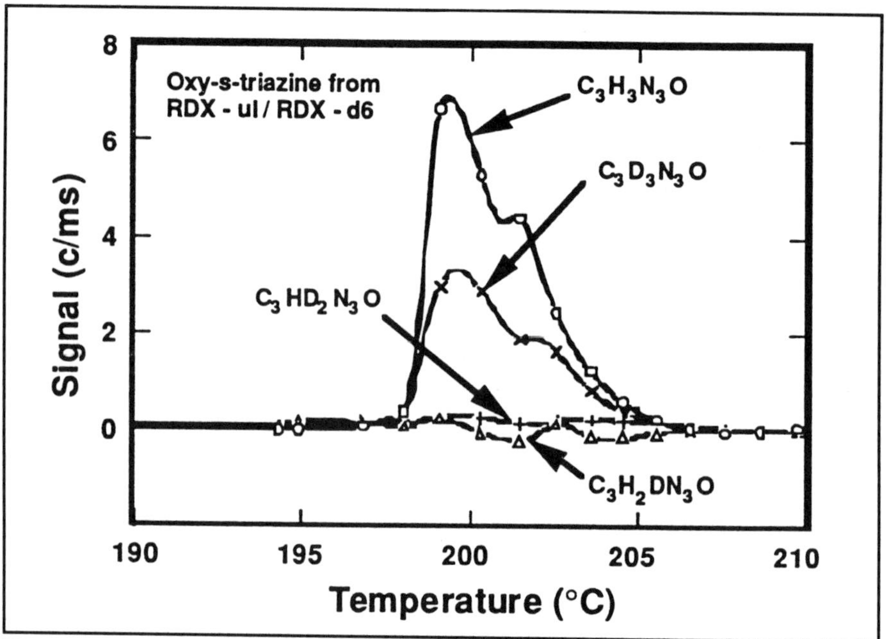

Figure 3. Ion signals observed from oxy-s-triazine and its isotopic analogues formed in decomposition of a mixture of RDX-ul and RDX-d_6 (Experiment 1, Table 1). Heating rate of sample is 0.5°C/min. Rapid increase in rate of OST evolution at 198°C occurs as sample melts.

that the mixing is complete with respect to the N-NO bond. However, during the latter stages of the decomposition, the relative sizes of the ion signals at m/z-values of 132 and 136 are larger compared with the ion signals at m/z-values of 133 and 135, thus indicating that the extent of RDX that participates in mixing with respect to the N-NO bond falls to 60 percent.

Other products. Other products formed during the decomposition of RDX do not involve scrambling of bonds and are not presented here.

Discussion

As indicated in the introduction, this paper focuses attention on the deuterium kinetic isotope effects and isotopic scrambling results of the decomposition products from mixtures of isotopically labeled RDX samples (see Table 1). However, it is helpful and necessary to interpret these results in a broader perspective, taking into account the relative amounts of all the products (see an earlier paper[9]), the role of physical factors, and a comparison with similar data obtained from HMX. Although the use of isotopes has been helpful in identifying several product species in these decomposition studies

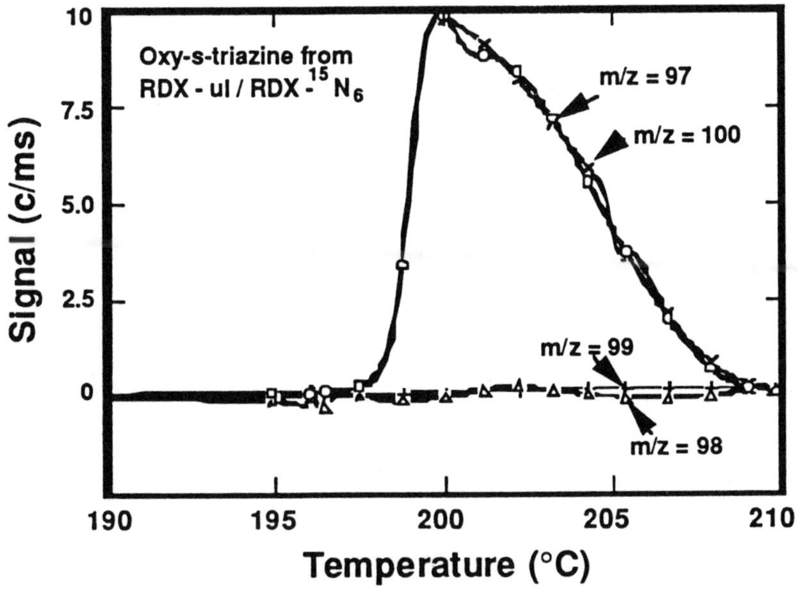

Figure 4. Ion signals from oxy-s-triazine and its isotopic analogues formed during decomposition of a mixture of RDX-ul and RDX-$^{15}N_6$ (Experiment 3, Table 1). Between 195 and 210°C, heating rate of sample is 0.5°C/min. Rapid increase in rate of OST evolution at 199°C occurs as sample melts.

and in bond-making and bond-breaking processes, they also uncover limitations in their use that preclude drawing any conclusions about reaction mechanisms.

It has already been stressed in the preceding chapters in this series that several decomposition channels can be identified in both HMX and RDX decompositions. In view of this, it is pertinent to consider the relative importance of these channels of decomposition and understand the similarities and differences between HMX and RDX imposed by the respective structural similarities and differences in physical properties (see introduction).

N_2O and CH_2O. In RDX decomposition, only a small amount of isotopic scrambling was observed for the C-H bonds of formaldehyde (Table 7 and Figure 1) and the N-N bonds in N_2O (Table 7 and Figure 2), which are two major products. This is consistent with previous work[21] on RDX-$(^{15}NO_2)_3$ in which it was found that N_2O consisted mainly of a ^{14}N-^{15}N-O configuration. If the N-N bond is broken in the reaction path that leads to the formation of N_2O from RDX, then the results with RDX-$^{15}N_6$/RDX-ul should have shown mixed isotopes. That this was not the case. Table 6 shows clearly that as in the case of HMX, N_2O formed in the decomposition originates both directly from

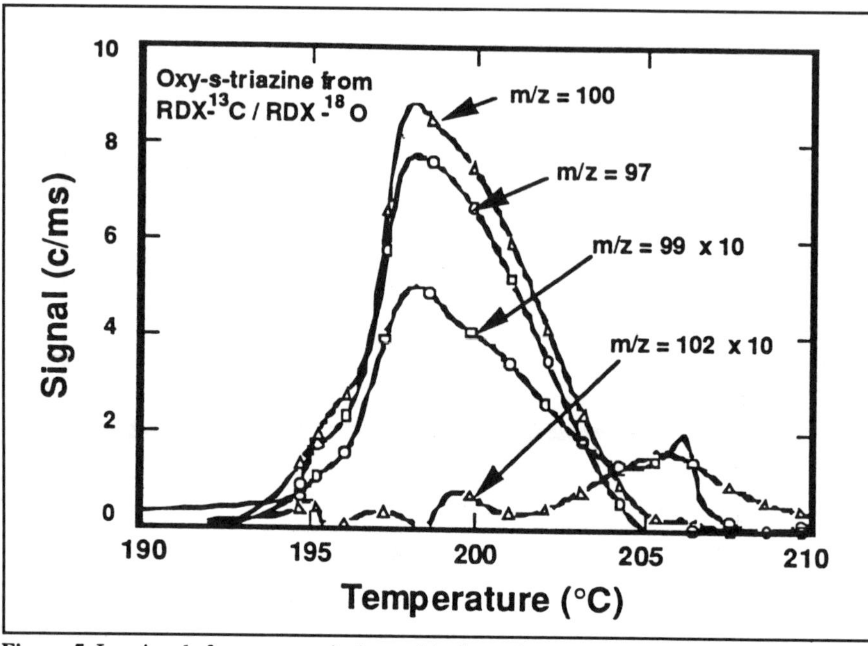

Figure 5. Ion signals from oxy-s-triazine and its isotopic analogues formed during decomposition of a mixture of RDX-^{13}C and RDX-^{18}O (Experiment 4, Table 1). Between 195 and 210°C, heating rate of sample is 0.5°C/min. Rapid increase in rate of OST evolution at 196°C occurs as sample melts.

the parent molecule and indirectly through an intermediate (ONDNTA) without rupture of the N-N bond.

Although the C-H bonds in CH_2O exhibit only a minor degree of mixing, the $^{13}C/^{18}O$ atoms showed 45 percent scrambling. This means that 45 percent of all the C-O bonds in the CH_2O product molecules are formed from atoms drawn from different isotopic molecules. However, this scrambling can readily occur in a CH_2O polymerization-depolymerization process anytime prior to exiting from the reaction cell. In the case of HMX, the degree of mixing is even higher[7] (80 percent), presumably because CH_2O is held for longer periods within the solid particles compared with the decomposition of RDX in a melt. The minor amount (< 7 percent) of H-D scrambling likewise could have occurred after the initial formation of CH_2O due to the labile nature of the hydrogens. The lack of a DKIE for CH_2O shows that C-H bond cleavage is not involved in the rate-limiting step of the reaction pathway leading to its formation, thus providing further evidence of several parallel reaction pathways controlling the decomposition of RDX.

The isotopic scrambling results of N_2O and CH_2O, taken together, point in the direction of a methylenenitramine decomposition to N_2O and CH_2O

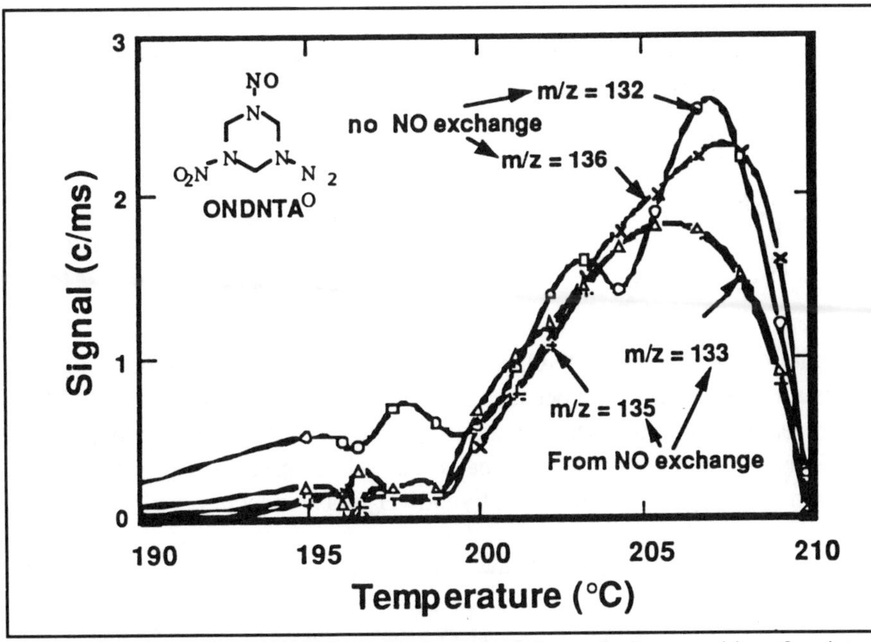

Figure 6. Ion signals from ONDNTA and its isotopic analogues in decomposition of a mixture of RDX-ul and RDX-$^{15}N_6$ (Experiment 3, Table 1). Heating rate of sample is 0.5°C/min. Larger variation in ion signal at m/z = 132 is due to large correction made to signal for contribution of $C_3H_4N_3(NO_2)$ ion dissociation fragment formed from evaporating RDX-^{15}N.

(Reactions R1, R2, and R5), as in the case of IRMPD of RDX.[12] However, in the IRMPD gas-phase study,

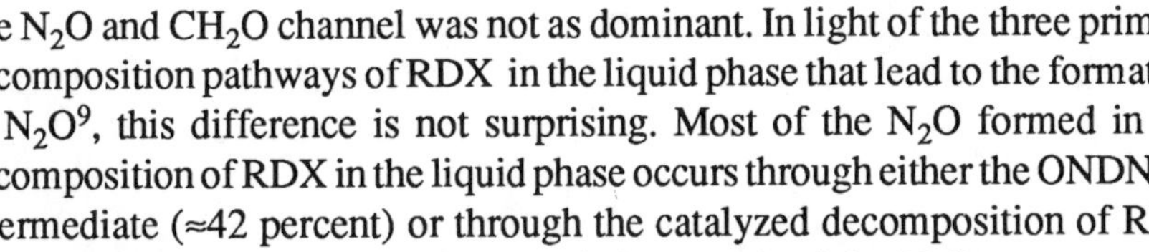

the N_2O and CH_2O channel was not as dominant. In light of the three primary decomposition pathways of RDX in the liquid phase that lead to the formation of N_2O[9], this difference is not surprising. Most of the N_2O formed in the decomposition of RDX in the liquid phase occurs through either the ONDNTA intermediate (≈42 percent) or through the catalyzed decomposition of RDX (≈42 percent). Only a small fraction (≈5 percent) of the N_2O appears to be formed directly from RDX. Whether the N_2O formed via either the ONDNTA pathway or the catalytic pathway passes through a methylenenitramine intermediate is uncertain at this point.

H_2O formation. The H_2O formation rate displays a strong kinetic isotope

effect (Table 7), suggesting that it is formed in a reaction pathway, such as Reaction 6 in which hydrogen is involved in the rate-limiting step.

In the following scheme,

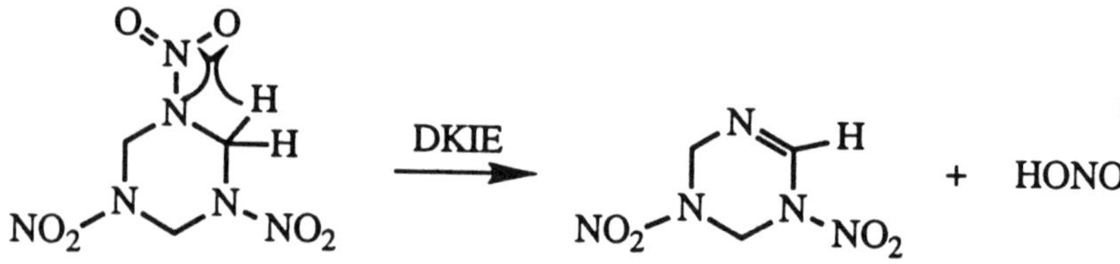

(R6)

the reaction generates HONO, accompanied by a primary kinetic isotope effect with deuterium label because a C-H bond scission is necessary. HONO, under reaction conditions, can undergo one or both of the two possible reactions, R7 and R8. In the former case, a hydroxyl radical is generated, which on hydrogen abstraction R9 yields H_2O as a product.

$$\text{HONO} \longrightarrow \text{OH} + \text{NO} \qquad \text{(R7)}$$

$$2\ \text{HONO} \longrightarrow \text{H}_2\text{O} + \text{NO} + \text{NO}_2 \qquad \text{(R8)}$$

$$\text{OH} \xrightarrow{\text{H - abstraction}} \text{H}_2\text{O} \qquad \text{(R9)}$$

Alternatively, the decomposition of HONO, R8, can directly yield H_2O and two oxides of nitrogen, all of which have been detected among the products. The latter may be expected to seek other radical species to react with, such as the radical left behind on loss of an NO_2 group from RDX. In this case, NO_2 will reform the RDX molecule while NO will lead to the formation of ONDNTA detected in the products.

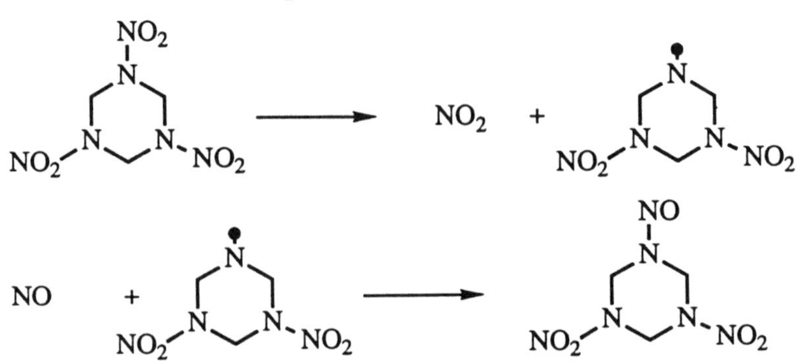

ONDNTA. It is interesting that ONDNTA formation as a product (Tables 6 and 7) shows a substantial mixing of the nitrogen isotopes making up the N-NO bond and 25 percent mixing of H/D isotopes. The former is expected from R11, but it is not immediately apparent why H/D mixing is observed. The explanation for this may be that the hydrogens next to a nitroso group become labile (acidic) and exchangeable. In fact, this exchangeable nature of the hydrogens is taken advantage of in the synthesis of the nitroso precursor to RDX-d_6.

There is also nuclear magnetic resonance evidence showing a substantial contribution of the resonance structure with charge localization on the ring nitrogen atom that would make the adjacent hydrogen more acidic and susceptible to hydrogen abstraction; this promotes decomposition of ONDNTA in the earlier stages of the reaction[9] due to reaction with other decomposition products.

The mixing of the nitrogen isotopes in the N-NO bond of course suggests the breaking of the N-NO_2 bond and reforming it with a nitroso group. This mechanism is quite analogous to the one proposed for the formation of dimethylnitrosamine.[22,23] A radical addition, R •, to one of the oxygens of the nitro-group, followed by loss of the RO group to give nitroso-amine, as proposed by Melius and Binkley[24] is not valid here because it would not predict mixing of isotopes in the N-N bond.

Oxy-s-triazine (OST). Oxy-s-triazine has proved to be a most intriguing product from RDX to rationalize:

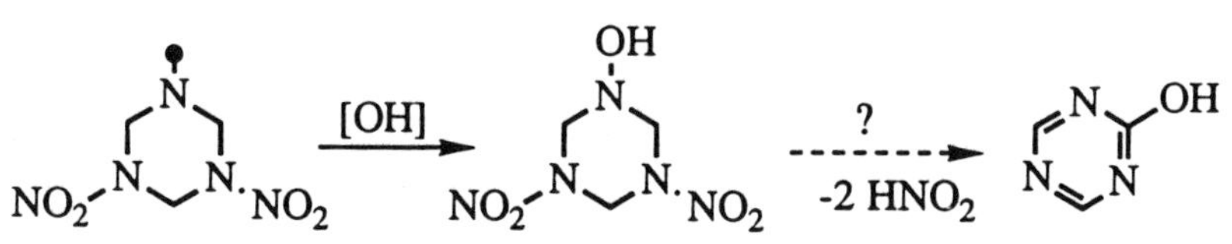

It would be reasonable to expect the radical produced in R10 to pick up a hydroxyl radical and eventually lead to OST by loss of two HNO_2 molecules and internal bond rearrangement. However, the first objection to this would be the excess hydrogen atom on the ring. But, more important, OST is unique in that it fails to show mixing of any of the isotopes used (H/D, $^{14}N/^{15}N$, or $^{13}C/^{18}O$). This requires that it be formed in a strictly intramolecular process. It appears, therefore, that OST represents an additional decomposition pathway that is most likely a unimolecular one. It can be seen in Figures 3 to 5 that the signal from OST abruptly increases at the melting point of RDX and then progressively decays as the sample is consumed. In order to take place, this reaction has to result in a concerted loss of two HONO molecules and an HNO radical by the RDX molecule.This reaction route to OST accounts for the total lack of mixing of the isotopes, the consequent intramolecular nature of the product, the unimolecular kinetics apparent from Figures 3 to 5 and the observed DKIE. It is interesting to note that OST is the only decomposition product from RDX that solely exhibits unimolecular decomposition kinetics (other products are produced by both the unimolecular and other decomposition pathways).

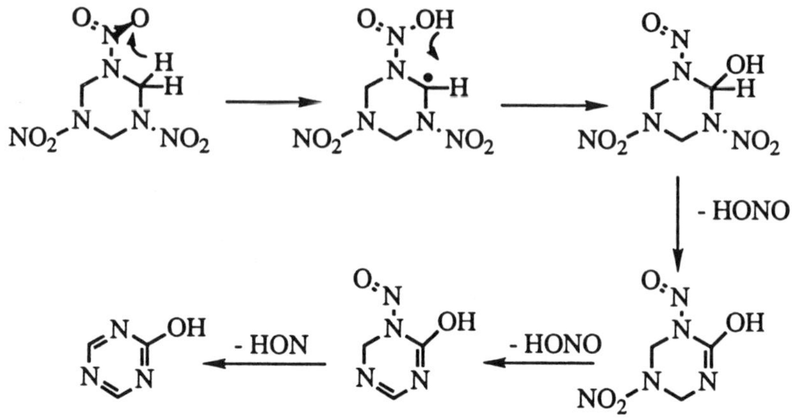

Summary

The results from isotope crossover experiments with RDX and the observed deuterium kinetic isotope effects (DKIE) lend further support for the four primary decomposition pathways described in a previous paper[9] and may be summarized as shown in the following scheme:

	Products	Path	Fraction	DKIE	$^{14}N/^{15}N$ Exchange
RDX →	OST + H_2O + NO + NO_2	1	30%	Yes	No
RDX →	NO_2 + H_2CN + 2 N_2O + 2 CH_2O	2	10%	No	No
RDX —[NO]→	ONDNTA → N_2O + CH_2O + other	3	35%	No/Yes	Yes
RDX —Catalyst→	N_2O + CH_2O + NO_2 + NH_2CHO	4	25%	No	No

Two pathways are first-order reactions solely in RDX. Pathway 1 produces predominantly OST, NO, and H_2O and accounts for approximately 30 percent of the decomposed RDX, and Pathway 2 produces predominantly N_2O and CH_2O, with smaller amounts of NO_2, CO, and NH_2CHO and accounts for 10 percent of the decomposed RDX. Pathway 3 consists of formation of ONDNTA by reaction between NO and RDX, followed by the decomposition of ONDNTA to predominantly CH_2O and N_2O. Pathway 4 consists of decomposition of RDX through reaction with a catalyst that is formed from the decomposition products of previously decomposed RDX. Reaction Pathways 3 and 4 each account for approximately 30 percent of the decomposed RDX.

The lack of isotopic scrambling and the presence of a primary DKIE in OST support a unimolecular mechanism in Pathway 1. The primary DKIE in the formation of OST shows the importance of intramolecular hydrogen transfer in the rate-limiting step that controls Pathway 1.

Reaction Pathway 3 is supported by the scrambling of the $^{14}N/^{15}N$ in the nitroso group of the ONDNTA formed during the decomposition of RDX. In addition, the lack of a DKIE in the formation of ONDNTA also supports this pathway. Furthermore, the presence of H/D scrambling is consistent with the presence of more labile hydrogen on the the neighboring carbon atom of the N-NO group in ONDNTA.

The lack of nitrogen scrambling in N_2O is consistent with the conclusion that methylene nitramine collapses to N_2O and formaldehyde without cleavage of the N-N bond. However, the lack of direct evidence of methylenenitramine makes its formation through this intermediate uncertain.

The lack of a DKIE in the formation of CH_2O supports the conclusion that hydrogen is not involved in the rate-limiting steps of Pathways 2, 3, and 4 that lead to its formation. Unlike the case of HMX decomposition, which is catalyzed by H_2O, there seems to be no evidence of autocatalysis by H_2O in the RDX case. H_2O formation is not the sole contributor to the observed DKIE as in the case of HMX. It is, therefore, apparent that the decomposition details of RDX and HMX are significantly different due to the fact that the former occurs in the liquid phase and the latter in the solid phase. Furthermore, RDX undergoes isotopic scrambling of nitrogen in the formation of ONDNTA, whereas for the analogous compound in HMX, ONTNTA, only 25 percent of the molecules are formed by mixing of the nitrogen isotopes. This implies that the formation of ONTNTA in HMX is influenced by the "cage effect" in its solid state decomposition, whereas the formation of ONDNTA in RDX occurs in the liquid phase, and the cage effect does not play a role. This also predicts that if HMX is decomposed above its melting point, it also would exhibit more scrambling of the nitrogen isotopes.

References

1. T.L. Boggs, in *Fundamentals of Solid-Propellant Combustion*, K.K. Kuo and M. Summerfield, Eds. (Progress in Astronautics and Aeronautics, Vol. 90) (AIAA Inc., New York, 1984), p.121
2. M.A. Schroeder, *Critical Analysis of Nitramine Decomposition Data: Product Distributions from HMX and RDX* (document BRL-TR-2659, Ballistics Research Laboratory, 1985).
3. M.A. Schroeder, *Critical Analysis of Nitramine Decomposition Data: Activation Energies and Frequency Factors for HMX and RDX Decomposition* (document BRL-TR-2673, Ballistics Research Laboratory, Aberdeen Proving Ground, MD, 1985).
4. S. Bulusu and R.J. Graybush, *C. R. Ind. Chim. Belg.* **32**, 647 (1967).
5. R.J. Karpowicz and T.B. Brill, *AIAA Journal* **20**, 1586 (1982).
6. R. Behrens, Jr., *J. Phys. Chem.* **94**, 6706 (1990).
7. R. Behrens, Jr. and S. Bulusu, *J. Phys. Chem.* **95**, 5838 (1991).
8. R. Behrens, Jr., *CPIA Publ.* **457**, 231-240 (1986)
9. R. Behrens, Jr. and S. Bulusu, *J. Phys. Chem.* **96**, 8877 (1992).
10. A.P. Snyder *et al.*, *Mass Spec.* **24**, 15 (1989).
11. C. F. Melius, *J. de Physique, Coloque C4* **48**, 341 (1987).
12. X. Zhao, E.J. Hintsa, and Y.T. Lee, *J. Chem. Phys.* **88**, 801 (1988).
13. R. Shaw and F.E. Walker, *J. Phys. Chem.* **81**, 2572 (1977).
14. R. Behrens, Jr., *Rev. Sci. Instrum.* **58**, 451 (1986).
15. R. Behrens, Jr., in *Chemistry and Physics of Energetic Materials*, S.N. Bulusu, Ed. (Proceedings of the NATO Advanced Study Institute, Vol. 309) (Kluwer Academic Publishers, Netherlands, 1990), p. 327.
16. R. Behrens, Jr., *Int. J. Chem. Kinetics* **22**, 135 (1990).
17. S. Bulusu, J.R. Autera, and T. Axenrod, *J. Labeled Compounds Radiopharma.* **17**, 707 (1980).
18. J.H. Boyer and G. Kumar, *J. Labeled Compds. Radiopharm.* **22**, 1 (1985).
19. F.J. Brockman, D. C. Downing, and G.F. Wright, *Can. J. Res.* **27B**, 469 (1949).
20. R. Behrens, Jr., *Int. J. Chem. Kinetics* **22**, 159 (1990).
21. S. Bulusu, J.R. Autera, and R.J. Graybush, Proceedings of the 1968 Army Science Conference, Volume 2 (Office of the Chief of Research and Development, West Point, NY, 1968), p. 423.
22. S.E. Nigenda, D.F. McMillen, and D.M. Golden, *J. Phys. Chem.* **93**, 1124 (1989).
23. J.M. Fluornoy, *J. Chem. Phys.* **36**, 1106 (1962).
24. C.F. Melius and J.S. Binkley, in *Proceedings of the 21st Symposium (International) on Combustion, 1987* (The Combustion Institute, Pittsburgh, PA, 1988), p. 1953.

31

Hypervelocity Microelectronics: The Keys to the Information and Force Projection Revolutions

Harry D. Fair
Institute for
Advanced Technology
The University of Texas
at Austin
4030-2 West Braker Lane
Suite 200
Austin, TX 78759

Unprecedented advances in microelectronics have enabled an information revolution providing vital position location, communications, and signal processing in even the most remote regions of the world. In parallel, equivalent advances have occurred in propulsion technologies by using electrical rather that chemical energies to electrically launch materials/projectiles to formerly unachievable velocities. By exploiting these advances in microelectronics and electric launch technologies, new opportunities arise for the development of a new class of long-range or short time-of-flight precision weapons systems. The author reviews the extraordinary developments in solid-state electronics and electric launch technology to explore potential new capabilities that arise from the synergistic coupling of long-range sensors, position location, communications, and miniaturized guidance and control with hypervelocity electric propulsion systems.

The Microelectronic Revolution: Discoveries and Enabling Technologies

Unprecedented advances in microelectronics have enabled a revolution in the availability of vital information such as position location, communications, and signal processing in even the most remote regions of the world. If coupled appropriately to exploit equivalent advances in propulsion technologies, new opportunities arise for the development of a new class of long-range or short time-of-flight precision weapons systems. To set the stage for this discussion, we will review the extraordinary developments in solid-state electronics not only to provide a model and critical insights for parallel advances in propulsion technology, but to explore potential new capabilities that arise from the synergistic coupling of long-range sensors, position location, communications, and miniaturized guidance and control with hypervelocity propulsion systems.

During the past four decades, we havc witnessed a revolution in electronics unparalleled in science and technology (Figure 1). The germanium point-contact transistor was invented on December 23, 1947, by Walter Brattain and John Bardeen under the direction of William Shockley of the Bell Telephone Laboratories. A low-powered signal, like that produced by a radio tube, was supplied to one end of a small piece of germanium, and a high-powered signal was obtained at the other end. It was that simple and that revolutionary.

Despite of their magic, point-contact transistors had significant drawbacks. Each transistor possessed slightly different electrical characteristics, clearly limiting the commercial value. By 1950, Shockley invented the junction transistor, a cylindrical plug of germanium doped with chemical impurities. The junction transistor was immediately a commercial success; it was more stable and reliable, easier and cheaper to make. Bell licensed the device to other manufacturers and by 1952 the Army was buying 5000 transistors a week.

At first, germanium transistors sold for $6 each, at a time when vacuum tubes cost $1 each. But the transistors were so superior, consuming less energy and generating less heat, that vacuum tubes were doomed. By the mid-1950s the price dropped to $2.50; the first transistor radios appeared, and the beaches and cities haven't been quiet since!

Transistors, resistors, and capacitors were manufactured separately and were wired and soldered together on masonite-like circuit boards. Although the reduction in size from vacuum tubes was incredible, the fabrication of circuits from discrete elements was expensive and cumbersome. In 1958, a simple circuit was created from a single thin wafer of germanium containing a transistor, a capacitor, and several resistors—the first integrated circuit—all within the same slice of germanium, and the microelectronic revolution had begun in earnest. More recently, there has been a great deal of interest in gallium arsenide as a chip material due to its higher electron mobility and hardness to radiation. Teraflop computers (10^{12} operations) at terahertz speeds (10^{12}Hz) are now envisioned.

Top left: The birth of modular electronics, 1947: The point contact transistor. *Courtesy of AT&T Archives.*

Top Right: Solid-state electronics become commercial, 1950: The junction transistor. *Courtesy of Texas Instruments.*

Left: The microelectronic revolution begins, 1958: The first integrated circuit. *Courtesy of Texas Instruments.*

Bottom: The revolution flourishes, 1990s: Teraflop computers (10^{12} operations per second) at terahertz speeds (10^{12}Hz). *Courtesy of Rockwell International.*

Figure 1. The microelectonics revolution. Advances in electronics over the past four decades have enabled a revolution in information.

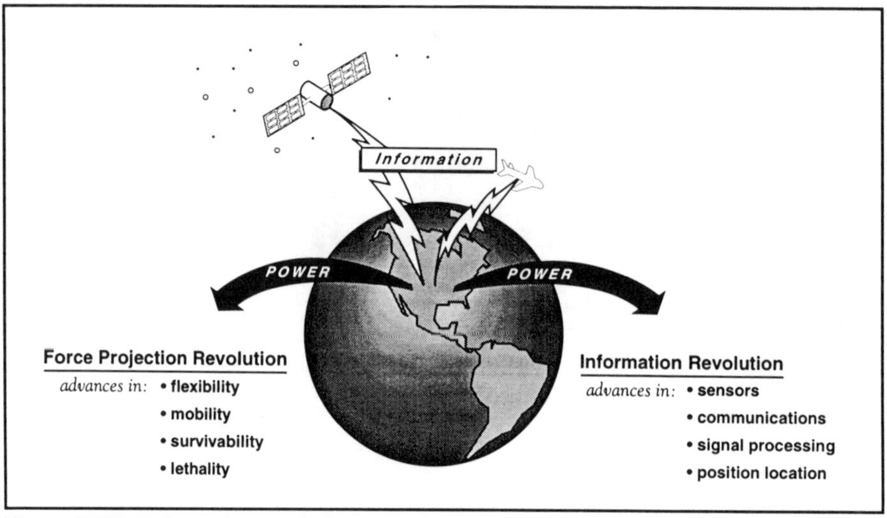

Figure 2. The information revolution must be balanced with the force projection revolution.

The Information and Force Projection Revolutions

These remarkable advances in solid-state electronics have in turn led to an information revolution. Highly integrated, netted, distributed ground, air, and satellite systems will provide immediate intelligence, communication, and targeting information throughout the world. And the revolution is not over. Even further advances are expected in sensors, communications, signal processing, and position location.

This information revolution must be balanced with an equivalent revolution in "force projection" (Figure 2). Capturing and exploiting the awesome commercial advances in microelectronics is a daunting task for the Army. But where is the commercial marketplace for "force projection" technologies? The Army must ensure that there are equivalent advances in force projection technologies not only to exploit new opportunities, but to meet its unique national security needs. Many technologies such as mobility, armor, lethality, and propulsion technologies are clearly outside the focus of the commercial marketplace. With its limited resources, the Army must focus on those key emerging technologies capable of providing the Army the significant advances in combat power and force projection expected of it by the nation.

The Army research development and acquisition community under the leadership of Stephen Conver and George Singley has articulated the Army's requirements for key emerging technologies in the Army Technology Base Master Plan.[1] Additionally, the Board on Army Science and Technology (BAST) of the National Research Council has developed[2] an equivalent set of

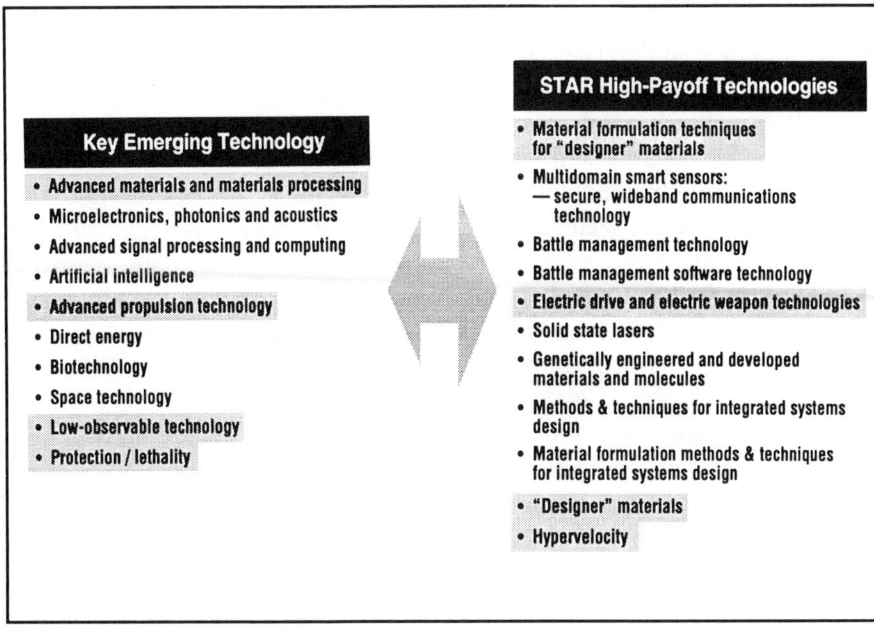

Figure 3. Key emerging technologies that are critical for providing advances in combat power.

high-payoff strategic technologies for the Army (Figure 3). In this chapter, I highlight several of these technologies, which, with sufficient focus, will provide new operational opportunities and the potential for the required revolution in military combat power and force projection. As we shall see, the keystone to these new opportunities is based on the development of advanced means for propelling materials to higher and higher velocities.

The Combat Power Revolution

Even though chemical explosives and propellants have been used in guns and missiles for well over 500 years, there are still improvements that may be achieved in both chemical guns and missiles. Exploiting the advances in microelectronics, particularly sensors and acceleration-hard microcomputers, targeting and precise delivery of lethal quantities of energy at improved range and precision are now in the development stage. The Army is exploring the development of a larger tank gun and longer range artillery equipped with these "smart" munitions propelled by higher energy, lower vulnerability propellants. They are also evaluating guided high-velocity kinetic energy rod warheads propelled by missiles. Additionally, higher-impulse missile propellants can provide the fast burn rate required to extend the keep-out range and sustain high velocities for fast response divert and miss distance corrections for air and missile defense. Gun launched missiles, either for top attack or destruction of advanced frontal armors, integrate compact missile guidance and control with the high launch speed and near-zero dead zone of high-velocity guns. These evolutionary advances provide important near-term opportunities to enhance compact power. But exciting new opportunities arise by integrating miniaturized guidance and control mechanisms with new forms of propulsion that have previously not been practical on the battlefield. The following sections of this chapter will explore the potential for propulsion technologies which, in concert with advanced targeting and guidance and control mechanisms, will provide a revolutionary advance in combat power projection.

To provide a framework for this evaluation, even an extremely simplified treatment of the energetics for conventional guns provides insight to the critical physics and engineering limits and opportunities for improved performance. As shown in Figure 4, the kinetic energy ($\frac{1}{2}mv^2$) imparted to a projectile is obtained by applying a force over a given distance (in this case the length of the gun, *L*). We can quickly see some of the practical engineering limits of this technology. The maximum operating pressure is limited by the strength of the gun tube; the maximum tube length is limited by transportability considerations, and the maximum acceleration is usually limited by the strength of materials in the projectile. To achieve high pressures and maintain them at the base of the projectile, propellants, or more correctly, their gaseous decomposition products, should have high isochoric flame temperatures and low molecular weights. The flame temperature of a typical gun propellant is 3000°K and is limited in large part by the severe gun barrel erosion that occurs due to the melting point of steel at about this temperature. Typical nitrocellulose-based propellants produce gaseous hydrocarbons and oxides with molecular weights on the order of 30 to 40. Thus, the challenge for the gun designer is to seek the means to maximize the pressure, gun tube length, and flame temperature with the lowest molecular weight, consistent with engineering and material limitations. The high performance 120mm tank gun, for example, is capable of accelerating a projectile to an energy of about 9MJ at a velocity of 1.7km/sec.

Development of Light Gas Guns

With this framework, we can understand why, in the early 1960s, two stage light gas guns using hydrogen or helium as the propellant were pio-

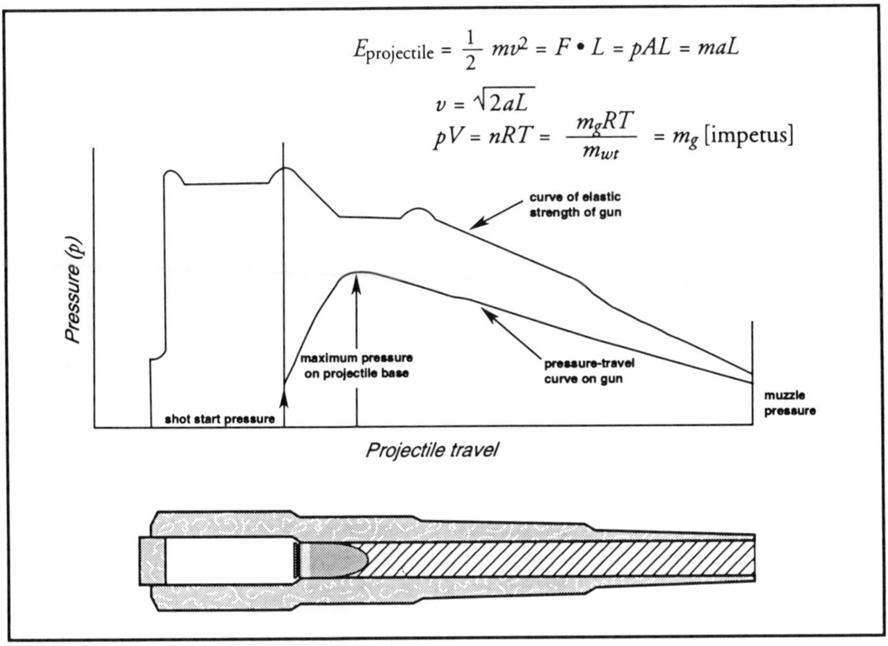

Figure 4. Simplified dynamics of launch (using gas pressure).

neered by Alex Charters and others to develop launch capabilities of small projectiles to velocities > 8km/sec. In a typical light gas gun,[3] a propellant powder chamber is connected to a "pump tube," which is filled to approximately 150psi with a light gas (hydrogen or helium). The propellant accelerates a heavy piston (150 to 200kg for large guns) to a velocity of 500 to 1000m/sec, compressing the gas ahead of the piston. The pump tube terminates in a tapered section which, in turn. is connected to a smaller diameter barrel or launch tube (usually 4 to 5 times less than that of the pump tube, see Figure 5). The gas pressure in the pump tube continues to rise as the piston approaches the barrel until it ruptures a thin steel disk, permitting high pressure gas (typically, 30 to 60,000psi) to reach the base of the projectile.

The compression-heated light gas maintains high pressure at the base of the projectile throughout most of the projectile's travel down the barrel, resulting in significantly higher velocities than can be obtained from conventional (high molecular weight) propellants.

Weaponization of light gas guns has been attempted over the past three decades, but to date, a viable concept has yet to emerge. The overall length is overwhelming (45m for an 80mm gun), and the gun must be separated between pump and launch tubes after firing to remove the piston and install a

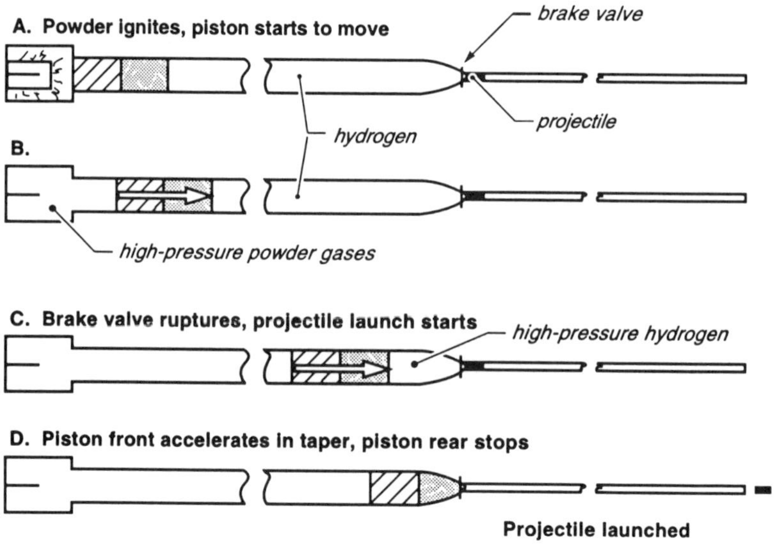

- Heating of hydrogen to high temperatures combined with low molecular weight results in high value of IMPETUS.
- Velocities of ≈ 7 km/s typical.
- Two-stage light gas guns are "common" in the research community.
- Size & weight prevent a practically fieldable system.

Figure 5. Two-stage light gas gun.

new rupture disk and projectile. Handling hydrogen in the field may be manageable but it is at least a serious concern.

A recent innovative new concept called the Electrothermal Light Gas Gun by GT Devices, Inc.[4] replaces the large pump tube with a much shorter high pressure chamber, into which electrical energy is introduced (Figure 6). The high pressure chamber is initially filled to 20,000psi hydrogen. A high-power electric pulse ohmically heats a fuse wire, depositing energy throughout the chamber volume and raising the temperature and pressure of the hydrogen or other light gas. Similarly to "conventional" light gas guns, a rupture disk opens under the increased pressure, permitting the gas to exert pressure on the projectile and accelerating it to hypervelocities. In recent experiments, small masses have been accelerated to about 7km/sec, and the gun appears to be scalable to larger bore sizes.

This new concept may lend itself to rapid fire, because cartridges containing the high pressure chamber, rupture disk, and projectile could be rotated, Gatling gun style, into place behind the barrel.

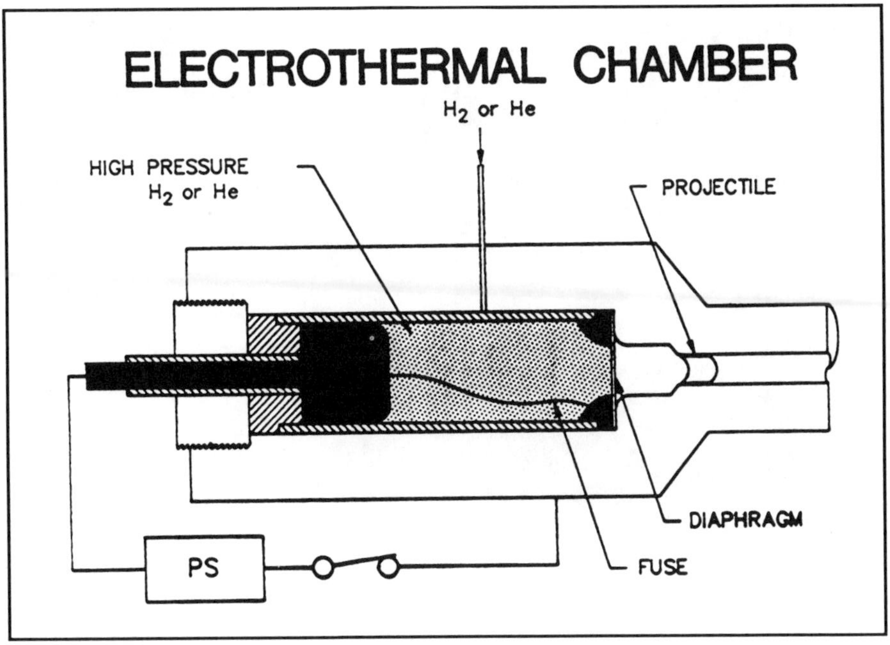

Figure 6. Electrothermal chamber (GT devices).

Although eliminating one of the crucial liabilities of light gas guns (the length and mass of the pump tube), the electrothermal light gas gun still has several major obstacles to overcome before consideration as a serious weapon candidate. Because all of the energy is provided electrically, as is common with all electrically powered guns, a substantial pulsed electric power supply is required, and there is severe concern about handling 20,000psi hydrogen in the field.

Development of Electric Energy Guns and Launches

In recognition of these physics and engineering limits, in the late 1970s the Army initiated a feasibility program[5] to use electrical rather than chemical energy to launch materials to high velocities. Although the fundamental concept was not new, the following developments led the Office of the Secretary of Defense to commission the Army to explore the feasibility of the technology for a variety of military gun and launch applications: the construction of the mass driver (Figure 7) by Kolm and colleagues at MIT's National Magnet Laboratory,[6] the record-braking achievement by Rashleigh and Marshall *et al.*[7] at the Australian National Laboratory in which they successfully accelerated a 10g projectile to over 6km/sec with a homopolar generator-

Figure 7. MIT coil gun. *Photograph courtesy of MIT.*

driven railgun, and a feasibility analysis[8] by an Army Laboratory team at the then Army Research and Development Center at Picatinny.

Following rapidly successful feasibility demonstration experiments, DARPA and the Army developed a jointly funded program to advance the technologies and evaluate potential applications. The initial focus was on determining the physics and engineering hurdles associated with potential tactical electric guns. Railguns, coilguns, and plasma pinch accelerators powered either by capacitors or homopolar generators were constructed and evaluated. Due to aerothermal heating of projectiles as they travelled through the atmosphere, efforts were primarily focused on obtaining velocities on the order of 3km/sec.

Following the establishment of the Strategic Defense Initiative Office (SDIO) in the mid-1980s, efforts were supported to launch guided hypervelocity projectiles to much higher velocities. Ground-based system concepts were evaluated for launching at 4km/sec and space-based systems were designed to launch projectiles to 10 to 15km/sec.

As a result of a Defense Science Board study in 1985, DARPA and the Army initiated an electric gun demonstration program to develop the technology for an antiarmor weapon system. Other defense agencies and military services have joined in efforts to advance the critical component technologies. The Defense Nuclear Agency (DNA) has supported the development of higher energy density capacitors that may, as a class, be important components in providing compact power for these devices.

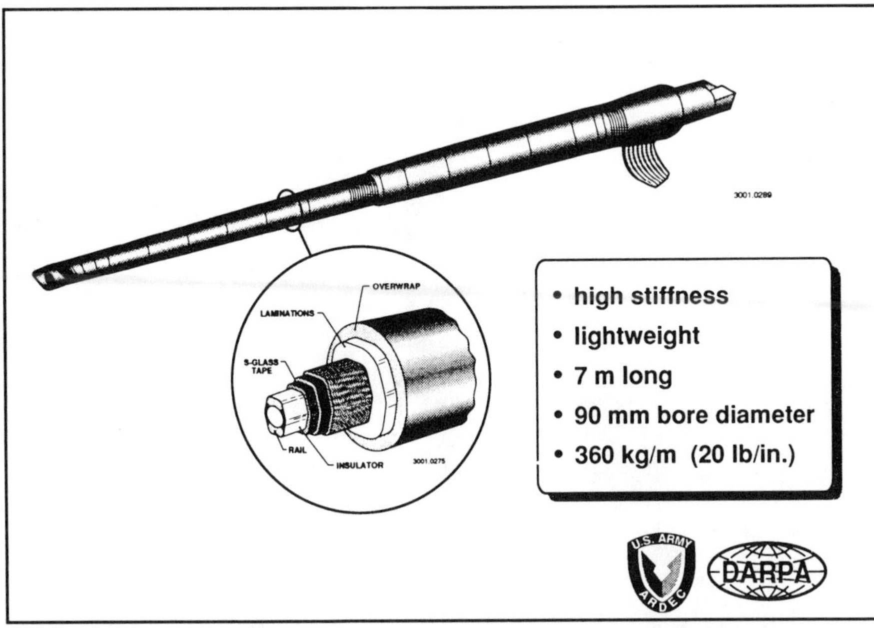

Figure 8. Advanced 12MJ railgun barrel. *Photograph courtesy of UT-CEM.*

Progress has been rapid and spectacular. In less than a decade, launchers have advanced from a series of coils crudely wound by physics graduate students to highly sophisticated high-stiffness, lightweight advanced composite barrels (Figure 8) that exceed the performance of the most advanced conventional high-performance tank guns. With strong industrial participation, railgun barrels are now commercially available that are capable of accelerating sophisticated projectiles (long rod discarding sabots) to energies in excess of those obtained by the M1 Tank Gun (Figure 9) and at velocities of 2 to 4km/sec. Additionally, a high velocity test facility has been sponsored by SDIO/DNA with the goal of accelerating a large projectile on the order of 10km/sec. Although difficult engineering challenges remain (including thermal management, barrel life projectile/launcher interfaces, and others), the feasibility of successfully launching useful projectiles electrically to 3km/sec. has been demonstrated. But the most difficult technical hurdle critical for enabling the development of electrically powered launchers for military applications is the size and weight of the electric power supply. The current commercially available power sources are simply too large and too heavy to power electric guns for tactical applications.

An interesting concept emerged in the mid-1980s that potentially alleviates this problem. Electrothermal and electrothermal-chemical (ETC) guns

Figure 9. High-energy, lightweight Army EML exceeds M1A1 tank gun performance. *Photograph courtesy of SPARTA.*

were designed and constructed in which part of the propulsive energy is added electrically, and part is provided by the energy of chemical propellants. As illustrated in Figure 10, an electrical power pulse is delivered to a chamber, generating a high temperature plasma. The plasma vaporizes the surface of a low molecular weight insulating material on the interior of the chamber, providing a very high temperature, low molecular weight plasma. The high temperature plasma interacts with a working fluid located between the plasma chamber and the rear of the projectile. Working fluids may be inert (such as water), in which all of the energy is then provided electrically, as they may be mildly exothermic so that some energy is provided by chemical reactions in the fluid; but in principle electrical energy is required to sustain the reaction; or they may be highly exothermic in which most of the energy is provided by the chemical propellant (Figure 11). Depending upon the type of working fluid, the electrical power requirements can be 10 to 20 percent of that required for a purely electric system. As a consequence, these devices may be available for tactical applications more rapidly than purely electric launchers. Serious issues concerning the interior ballistics of these guns are yet to be resolved, but intensive modeling and experiments supported by the DNA and

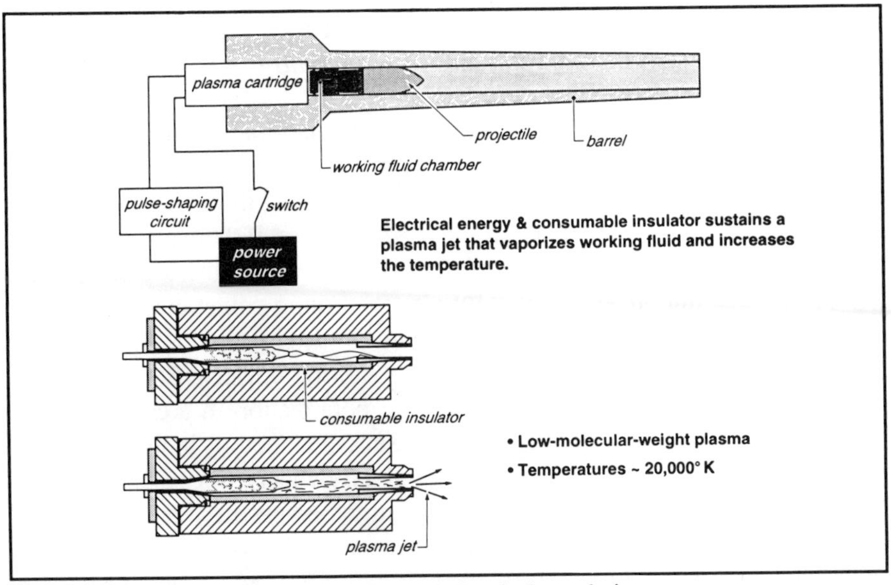

Figure 10. Electrothermal and electrothermal-chemical gun designs.

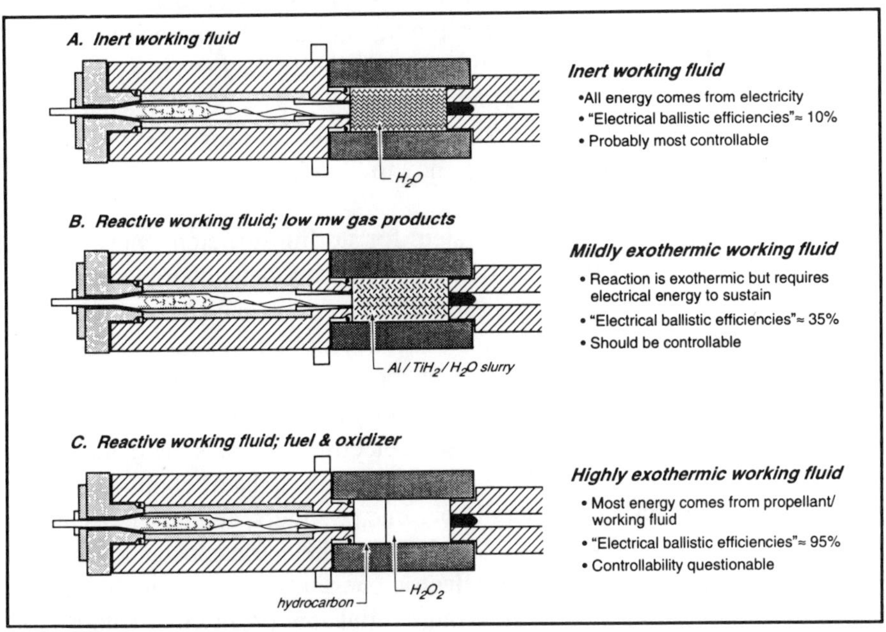

Figure 11. Three types of electrothermal-chemical guns.

the Army are showing excellent progress. A novel variant of the ETC approach is being pursued by Israel,[10] in which the working fluid is replaced with conventional solid propellants. This approach has the advantage of better understood and controlled interior ballistics, but ballistics with much higher molecular weight propellants.

The overarching question is, however, whether the revolution that has occurred over the past four decades in microelectronics can be extended to power electronics; or, are there physics or engineering limits which prohibit a revolution in pulsed-power electronics? And further, if there are no physics or engineering limits prohibiting further advances, are there concepts or approaches that may pave the way for a revolution in power electronics?

Consider the various means for storing energy and evaluating the physics or engineering limits for various energy storing devices. Energy can be stored in electric fields (as in the dielectric polarization in a capacitor), in magnetic fields (as in a conventional or superconducting inductor), or in kinetic energy (as in the rotational energy of a flywheel). Because the energy must be delivered in fractions of a second, high powers are required; thus at this time, electrochemical storage is not considered a viable pulsed power candidate, although batteries could be an important supplemental element in an overall electric power system. The important point is that for ETC or railguns, energy must be delivered in the transit time of a projectile through the gun (several milliseconds) and must be supplied in the form of a controlled high voltage (5 to 20KV for ETC and railguns). Because power is energy per time, it requires an instantaneous power requirement of over 10GW to launch a 20MJ projectile at 50 percent efficiency to 3km/sec.

Thus, the critical questions are:

1. Can we find a medium or system for storing sufficient energy; and
2. Can we transfer this electric, magnetic, or kinetic energy efficiently with the proper controlled voltage and duration in a volume and mass compatible with the requirements for a tactical mobile weapons system?

If we evaluate the current state of the technology for several energy storage concepts and determine the opportunity for improvement, we find that commercially available capacitors use dielectric materials with dielectric constants on the order of 2 and operate at field strengths of about 4MV/cm. The dielectric materials are generally organic polymers, and programs are in progress to determine new polymers with higher dielectric constants. Another class of materials (ceramics) have dielectric constants many orders of magnitude higher (K = several thousand), but exhibit low breakdown voltages or field strengths. Clearly, one approach is to pursue an understanding of dielectric breakdown or failure that could enable this new class of materials to be used for advanced capacitors providing orders-of-magnitude improvement in energy density.

In addition to improved materials, the major revolution in the use of

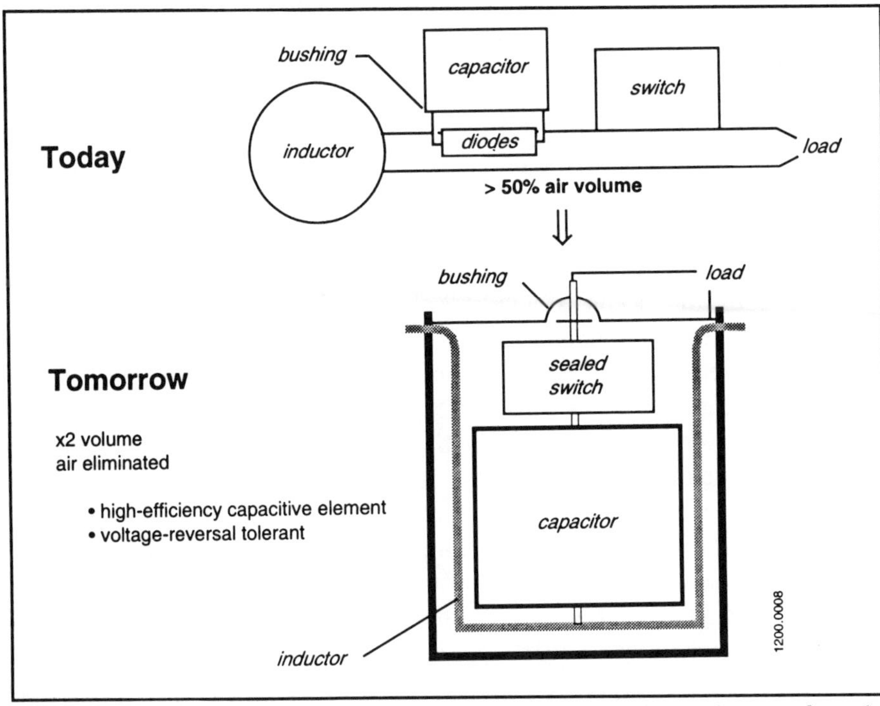

Figure 12. Integrated pulse-forming networks. System power density requirements force the integration of power conditioning elements.

electric fields in power electronics may derive from clever packaging, just as has occurred in the microelectronics revolution. Recall that the same (or very similar) materials have been used throughout these order of magnitude improvements (germanium, silicon, and possibly gallium arsenide). The major advance came from integrating all of the discrete elements into a single chip—the integrated circuit. Simple rearrangement of the components (Figure 12) into modular elements should provide a significant improvement in energy density. But the solid-state power electronic revolution will occur when all of the elements (capacitor, inductor, and possibly batteries, diodes, and switches) are integrated into a single monolithic element. These elements could be designed with specific power characteristics and requirements, much as integrated chips are designed today.

In large continuous-duty power plants, electric generators operate at high efficiency, but there has been no major commercial interest in making these rotating machines smaller or lighter or in optimizing them for pulsed-duty operation. The critical parameters to optimize is the magnetic force density, $v \cdot (J \times B)$, where v = rotor tip speed, J = machine current density, and B = machine magnetic flux density.

Generators employing iron to help concentrate the magnetic field saturate at about 1.5 to 2T. Replacing the iron by lighter weight carbon composite materials enables the field to be increased to about 3T. The current density is limited by the properties of the insulation surrounding the current carrying wires. Even at high efficiency, ohmic heating eventually provides the current density limit due to thermal limitations of the insulating material. Clearly, advances in insulating materials or high temperature superconductors will raise the current density limits. Since the 1930s, turbogenerators have operated with rotor tip speeds as high as 300m/sec. The limiting speed has been associated with the strength of materials used in the rotors (iron and copper coils). Because the energy density increases with the square of the tip speed ($1/2\rho v^2$), an increase in this parameter would be extremely advantageous. During the past decade, Oak Ridge National Laboratory and several industrial laboratories have developed energy storage flywheels from high-strength composite materials and have tested them at tip speeds as high as 1400m/sec. Many of these were large-scale machines storing 6 to 8MJ of energy. All of these high-speed machines were carefully designed to optimize strength and mass distribution for energy storage. The critical issue is whether this composite flywheel technology can be applied to the next generation of rotating electrical machines. A generator employing an advanced composite rotor has been constructed by The University of Texas at Austin[10] and is undergoing full power tests, and a larger machine is under construction. These machines are designed for tip speeds on the order of 400m/sec. An advance even beyond these machines leads to an "inside/out" topology in which the stator is placed on the inside, and the rotating mass is a shell on the exterior of the machine (Figure 13).

A new concept called Star Mirror most clearly attempts to exploit composite flywheel technology. This is a new induction generator design[12] incorporating an extremely complex stator at the interior of the machine and a conducting cylindrical shell rotor at the exterior of the machine (Figure 13B). All of the complex windings are in the stator, and the rotor is a simple cylindrical shell with *no* electrical contacts so there are no brushes or current collectors on any moving parts. This radical design is in its concept development phase but offers the potential for significantly higher energy densities. As with other rotating machine concepts, advances in insulating materials or high-temperature superconductors will even further enhance the performance of this revolutionary machine.

But what are the consequences of these new pulsed power sources and, particularly, how do they influence combat power? A major impact will be on logistics and strategic mobility because hazardous and bulky explosives and chemical propellants are replaced by inert rotating machinery or capacitor

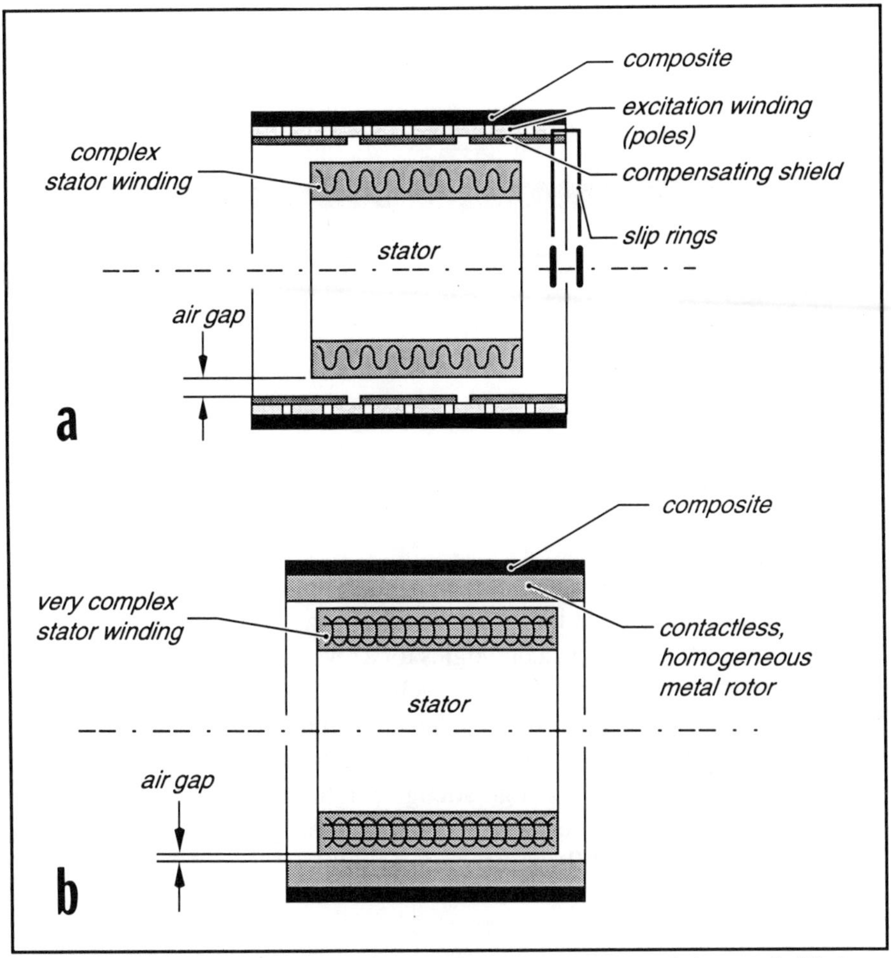

Figure 13. (**A**) Synchronous exterior rotor configuration (compensated alternator), (**B**) Asynchronous exterior rotor configuration (Star Mirror).

pulse forming networks that are unchanged in storage or shipping. But the real revolution will arise from the synergistic combination of distributed information systems providing long-range communications and targeting to this new class of hypervelocity long-range electric weapons providing capabilities not achievable by conventional means. Consider for example (Figure 14) the current long-range indirect fire weapons employed by the Army and Navy. Both have muzzle velocities of 0.8km/sec and a range of 30 to 40km. An electric gun capable of accelerating 100kg to 2.3km/sec[13] would have a range of 500km! Further, a recent study[14] performed for the Strategic Defense Initiative Organization illustrated the role hypervelocity guns can have in an

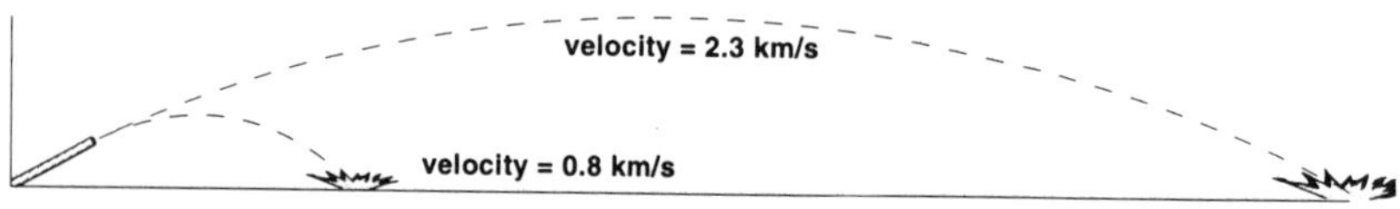

FIRE SUPPORT

		diameter (mm)	mass (kg)	velocity (km/s)	range (km)	kinetic energy (MJ)
Conventional gun system	Army	155	50	0.8	30	15
	Navy	406	800	0.8	40	300
Electric-energy gun system		200	100	2.3	500	260

Data obtained from DARPA Electric Energy Gun System Study • David Hardison, Director • 1989

Figure 14. Lomg-range precision deep-strike artillery.

underlay for theater missile defense due to their high velocity and short time-of-flight. Perhaps the most important advance can arise from the integration of a number of key enabling technologies for a revolutionary new type of high-mobility hypervelocity ground combat vehicle. Technologies including electric drive, electric weapons, active defense, appliqué armors, low signature, indirect vision, direct energy antisensor weapons, and low ground-pressure traction, all integrated into a high-strength, lightweight chassis are rapidly developing and can produce a new type of mobile, lethal, survivable combat vehicle with unsurpassed flexibility and capability (Figure 15).

Summary

Advances in solid-state electronics have created an information revolution, providing significantly improved capabilities in command, control communications, reconnaissance, intelligence, surveillance, and target acquisition. Advances in propulsion can exploit this information revolution to enable a revolution in combat power and force projection. Recently, electrothermo-chemical and electromagnetic launchers have demonstrated revolutionary performance levels as practical weapon system components. To exploit these hypervelocity electric gun systems, technology advances are required in compact pulsed electrical power sources. No physics or engineering limits have been identified that prohibit the development of these power sources, and the Army has recently initiated a Focused Technology Development Program

Technologies

- High-strength, lightweight chassis
- Electric drive
- Electric weapons
- Applique armors
- Pulsed electric power
- Active defense
- Low-signature vehicle and weapons
- Indirect vision systems
- DEW antisensor
- Low-ground-pressure traction

Payoffs

- Long-range transportability
- High mobility
- Improved lethality
- Survivability
- Flexibility

Figure 15. High-mobility ground combat vehicle.

to provide the pulsed power that will enable a revolutionary new class of long-range precision weapon systems.

Acknowledgments

I am grateful to the following persons: Don Berry of the Institute of Advanced Technology (IAT) who provided artwork and helpful discussions on conventional gun systems; William Isbell, Institute Fellow of IAT, who provided important information and insights on light gas guns; Dan Gleesing, IAT Art Director, and Janet Monaco, who provided most of the graphics and word processing; James Asay of SANDIA National Laboratory, who supplied pictures of their light gas gun; Derek Tidman, who supplied diagrams of the GT Devices electrothermal light gas gun; and the National Research Council, which permitted representation of figures from the STAR study.

References

1. Department of the Army, *Army Technology Base Master Plan,* 1992.
2. National Research Council, *STAR Twenty-one: Strategic Technologies for the Army of the Twenty-first Century* (National Academy Press, Washington, DC, 1992).
3. W. Isbell, private communications.

4. D.A. Tidman and D.W. Massey, *IEEE Trans. Magn.* **29**, 621 (1993).
5. H. Fair, *IEEE Trans. Magn.* **18**, 4 (1982).
6. W. Arnold *et al.*, in *Space Resources and Space Settlements*, (NASA SP-428, U.S. Government Printing Office, 1979)
7. S.C. Rashleigh and R.A. Marshall, *J. Appl. Phys.*, **49**, 2540 (1978).
8. H. D. Fair *et al.*, Proposal for the Development of Electromagnetic Guns and Launchers, Propulsion Technology Laboratory Report 07-105-78, submitted to Undersecretary of Defense, Dr. R. Davis, July 1978.
9. H.D. Fair, P.J. Kemmey, and T.F. Gora, Propulsion for Future Munitions Systems, paper presented at the Technology Trends Colloquium, Annapolis, MD, March 29-April 4, 1978.
10. A. Juhasz *et al.*, in *Army Science: The New Frontiers*, D. Kamely, K.A. Bannister, and R.M. Sasmor, Eds. (Borg Biomedical Books, The Woodlands, TX, 1993), p. 515-527.
11. J.R. Kitzmiller *et al.*, *IEEE Trans. Magn.* **29**, 441 (1993).
12. M.D. Driga, *IEEE Trans. Magn.* **29**, 397 (1993).
13. Electrical Energy Gun Systems Study, directed by David C. Hardison for DARPA, 1988.
14. Institute for Advanced Technology Panel, Hypervelocity Gun Technology for Theater Missile Defense, annotated briefing to M.G. O'Neill, Deputy Director, Strategic Defense Initiative Office, August 1, 1991.

32

Solid Propellant Electrothermal Gun Propulsion

Arpad Juhasz
U.S. Army
Ballistic Research Laboratory
Aberdeen Proving Ground,
MD 21005

Stan Smith
U.S. Army
Strategic Defense Command
Huntsville, AL 35807

Zwi Kaplan
D. Melnick
D. Saphier
Soreq Nuclear Research Center
Yavne 70600, Israel

Maxi Blum
Ministry of Defense
State of Israel
Tel-Aviv 61909, Israel

The authors describe the development of a variant of electrothermal-chemical (ETC) propulsion, namely, solid propellant ETC, or SPETC. The technique involves controlling the interior ballistic cycle of granular solid propellants by means of plasma injected into the gun chamber. Plasma power and energy levels are shown to affect the ignition delay times of the propellant, rates of pressurization, and the maximum pressure attained in the gun in a controlled manner. An SPETC interior ballistic (IB) model based on these observations was used to develop plasma injection strategies for tailoring the IB pressurization process and enhancing ballistic performance. Experiments carried out in a 60-mm gun have demonstrated the feasibility of generating desired changes in the interior ballistic pressure-time profiles and enhancements in gun performance using plasma injection techniques. Thus far, experiments have yielded 6.8 percent greater muzzle energy from a 60-mm gun than could have been possible using optimized conventional interior ballistic techniques. Ultimately, the technique promises a 25 to 35 percent enhancement in muzzle energy using existing gun tubes.

Introduction

Hypervelocity gun launcher (HVL) technology requirements have been identified both for future ground-based terminal defense (TD) as well as tactical kinetic energy (KE) applications.[1,2] The common element of an HVL launch for both types of systems is a high projectile velocity leading to short times of flight and increased kinetic energy on target. In the TD application, short time of flight offers quick reaction intercept capabilities against leakers, targets that have penetrated longer range defense architecture. HVL could permit several guided projectiles to be fired from a single launcher or multiple launchers during a single engagement. Compared with conventional guided missile systems, the absence of rocket motors and warheads would make projectiles less expensive and offer the potential of reducing the cost of an engagement, even when several projectiles are launched to increase the probability of success. In the tactical area, increased velocity is important in defeating state-of-the-art armor protections at extended ranges and increasing hit-and-kill probabilities against maneuvering targets. Both electromagnetic (EM) and electrothermal-chemical (ETC) devices have been under study as candidates for HVL. Although ultimately limited in performance as compared with EM, ETC propulsion is generally accepted as the near-term electric solution to the HVL problem.[2]

Most of the developments in ETC propulsion have taken place since 1985. Initial efforts focused on plasma controlled, bulk loaded liquid or gelled propellant systems. Recent work has yielded interesting results with a solid propellant variant of ETC (SPETC). We describe SPETC results that may form the technical basis for a future point-defense element in a terminal defense architecture for theater or national missile defense. Although the technology described is focused on the mission area defined by the Corps Surface to Air Missile (CORPSAM) operational requirement document, it is equally applicable to the antiarmor (A^2) and air defense roles.

One attractive feature of ETC propulsion is the possibility of using existing gun tube and mount technology to attain greater muzzle energies than obtainable from conventional guns. A schematic description of a solid propellant ETC gun system is given in Figure 1.

In addition to the gun tube and mount, an SPETC gun includes an electric pulsed power source, usually in the form of a pulse forming network (PFN), and a plasma injector linked appropriately to the ammunition cartridge to assure ignition of the propellant charge and control of the interior ballistic (IB) process.

The rationale of the ETC concept is that electric control of the IB process can lead to optimum ballistic progressivity, usefully burning greater amounts of propellant by plasma injection at key points in the IB process.[3] Calculations at the Weapons Technology Directorate of the Army Research Laboratory indicate a potential muzzle energy increase of 25 to 45 percent for SPETC versus optimized conventional solid propellant guns. The data presented next

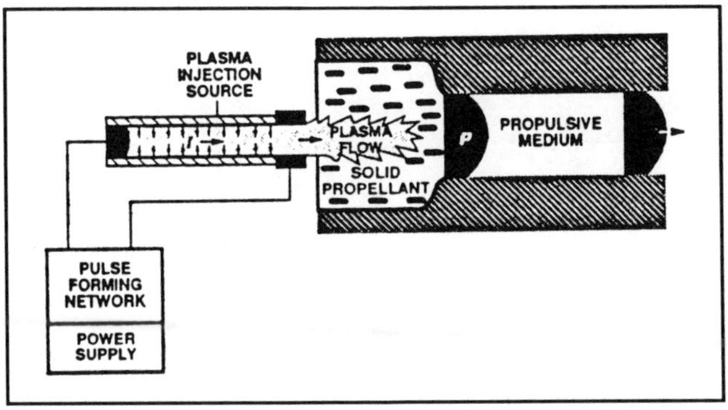

Figure 1. Schematic description of a solid propellant ETC gun.

were obtained as part of a joint U.S.-Israeli research and development effort on SPETC guns. The following main issues will be addressed:

1. The 60-mm experimental gun.
2. Plasma jet-solid propellant interaction.
3. Methodology for improving IB acceleration processes.
4. Experimental results showing improved SPETC guns performance.

Electrical Plasma Source

It is known that an electrical discharge in a dielectric tube can operate as a plasma jet generator in a predetermined, controlled manner.[4] The electrical energy is converted into thermal energy by establishing an electrical discharge in a dielectric tube. This discharge develops a high-pressure, high-temperature plasma that emerges in the form of a plasma jet through a hollow electrode into the gun chamber. A schematic is shown in Figure 2.

A simple model that describes the discharge phenomenon and defines basic analytic relationships between its main parameters has been developed.[5] It permits estimation of plasma temperature, pressure, density, and discharge resistance as well as their dependence on parameters such as tube radius, length, and current density. These relationships are used as practical design tools for plasma jet formation systems matched to the driving power source.

Experimental Setup

The experimental data presented next were obtained using the 60-mm SPETC gun. The main parameters of this gun are presented in Table 1. Figure 3 shows the basic construction of the device.

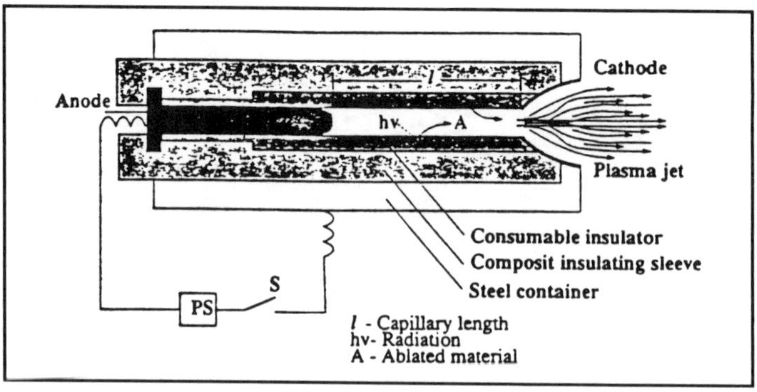

Figure 2. Schematic view of device for high-pressure injector and plasma discharge.

Table 1.
Specifications of 60-mm SPETC Gun

Breech volume	2,250cm^3
Projectile travel	3.60m
Nominal projectile weight	1.35kg
Nominal breech pressure	479MPa (4,700 Bar)

The gun is machined to incorporate an ammunition cartridge. It is mounted on a cradle containing a recoil mechanism. A threaded breech housing allows the insertion of a breech block. The electrothermal (ET) plasma injector is contained in the breech block. The plasma penetrator is located at the base of the ammunition cartridge, permitting axial plasma injection. A flexible electrical connection links the plasma injector and power supply. The power supply itself is comprised of several PFNs. Each PFN can be discharged into the load independently as long as impedance matching rules are used to guarantee a high transfer efficiency of electrical energy to the plasma injector load, permitting a maximum pulse length of 4msec.

Design rules have been developed to generate various pulse shapes as needed for different experiments. One common pulse shape begins with a low power, 100 to 300MW, ignition pulse lasting for 2.5msec, followed by an energy boost at a 1 to 2GW power. An example of this "staircase" pulsed power shape is shown in Figure 4.

In order to provide an efficient energy transfer, adequate impedance matching between the plasma injection unit and the network is required. Impedance matching is done based on the plasma model[5] using network simulation programs and appropriate discharge parameters. When properly done, 95 percent of the stored electrical energy can be transferred to the discharge load.

The following diagnostics are used to characterize SPETC experiments.

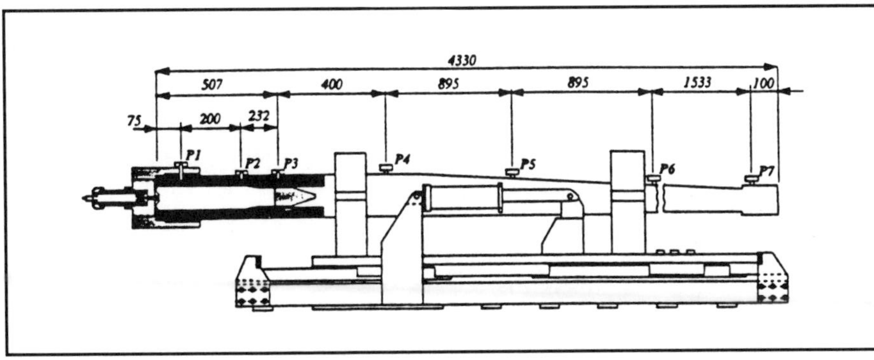

Figure 3. The 60-mm SPETC gun.

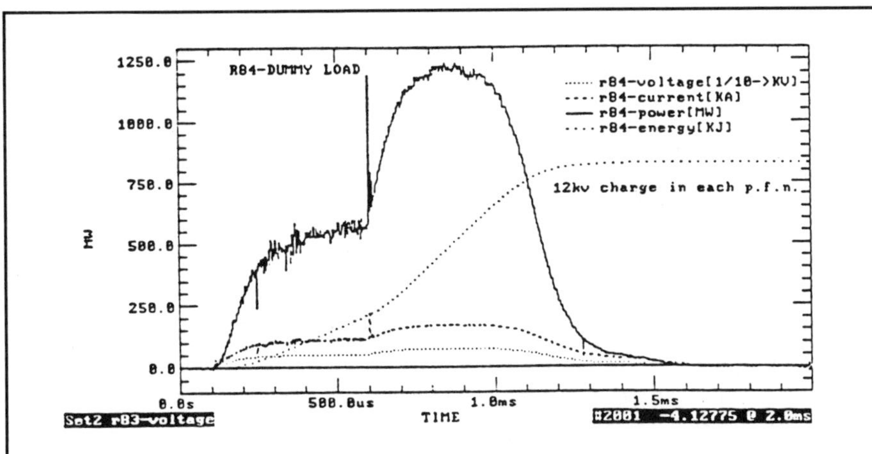

Figure 4. Discharge of two PFNs to generate a staircase pulse shape into a restive load.

The electrical system is characterized for each PFN by monitoring the current, charging voltage, and voltage across the load. Internal ballistics and projectile kinematics are mainly characterized by piezoelectric pressure gauges mounted along the chamber and barrel. These sensors provide the history of the pressure behavior at each location and the arrival times of the projectile at each pressure sensor in the barrel. External ballistic measurements are made with a Doppler microwave system.

Interaction Between Plasma Jets and Solid Propellants

A variety of tests involving propellant type, mass and geometry, projectile mass, free volume in the chamber, and injected plasma characteristics were performed. Initial efforts concentrated on the development of controlled

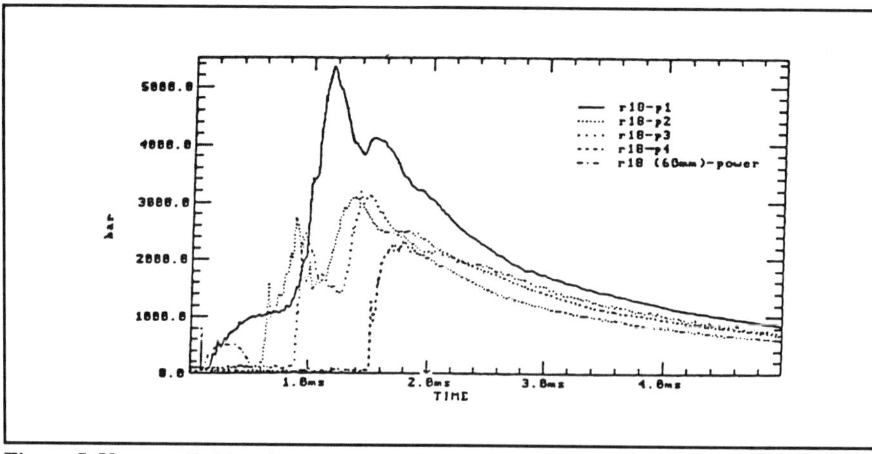

Figure 5. Uncontrolled burning occurred when plasma was injected with no passage for flow.

ignition and smooth ballistic pressure-time histories under plasma injection conditions. Subsequent efforts involved a study of the effects of plasma injection on the interior ballistic process. The data obtained have permitted a tailoring of conditions to obtain enhanced plasma-augmented gun performance.

Early experiments with plasma injected into a cartridge containing a granular propellant charge with no prearranged passage for plasma flow through the bed resulted in uncontrolled burning and resultant pressure waves. Figure 5 shows the results of such an experiment. A maximum breech pressure of 500MPa versus the 200MPa expected was observed. This undesirable behavior is similar to conventional ballistic situations involving localized charge ignition.

To avoid the uncontrolled burning phenomenon, plasma injectors were designed to imitate conventional center core igniter geometry, which yielded far smoother, more controlled ballistic processes. Ultimately, a low mass cellulose acetate (CA) tube, axially located in the cartridge to maintain space for the plasma to spread homogeneously through the charge, was adopted. This configuration yielded the smoothest, most reproducible IB pressure-time behavior. The importance of these experiments is the proof that essential design concepts used in conventional solid propellant charges, that is, avoiding localized ignition and providing for distributed energy transfer in the charge, are applicable to the plasma-ignited situation. These designs greatly facilitate control of the SPETC process.

Once these charge-configuration effects on the plasma augmented ballistic process were understood, a series of experiments were conducted to

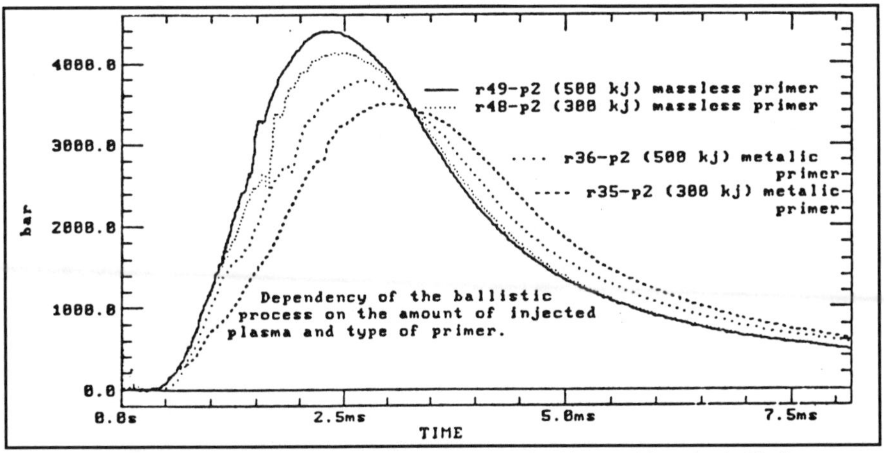

Figure 6. Effects of primer tube and plasma energy on ignition and interior ballistic process.

learn about the influence of the amount of energy as well as the power of the injected plasma on the ballistic process. In line with intuitive expectations, both pressurization rate and maximum pressure attained were found to be functions of the power, duration, and total plasma energy injected into the charge. Moreover, the process proved to be well behaved, in fact quite large amounts of plasma energy, on the order of 1.6MJ, have been safely injected into the 60-mm cartridges filled with solid propellant.

Effects of plasma energy on the IB process obtained using two types of igniter tubes are illustrated in Figure 6. The breech pressure curves of four different firings using the same amount of propellant are presented. Plasma injections of 300 and 500kJ, respectively, were done using both metal polycarbonate and low-mass cellulose acetate igniter tubes. An increase in pressure and rate of pressurization is observed as the amount of injected plasma is increased, indicating enhanced propellant burning as a function of increased plasma energy. The CA tube also led to faster pressurization. The effects of plasma power level on the interior ballistic process were found to be similar. That is to say, the rate of pressure rise as well as the maximum pressure attained in a given experiment significantly increased as the power level of the injected plasma increased.

An interesting series of observations were made concerning ignition delay times, the time difference between the initiating pulse and the onset of pressurization of the plasma augmented ballistic processes. Under plasma injection conditions, ignition delays were found to be significantly shorter and more repeatable than under conventional black powder ignition conditions. In addition, as the amount of injected plasma increased, the delay between the trigger pulse and pressure onset decreased. In otherwise identical experiments involving injection of 34 and 150kJ plasma energy, respectively, ignition delay times of 2 and 0.5msec were noted. These delay times are in sharp contrast to the 10msec ignition delay time obtained using a black powder igniter.

In summary, the effects of plasma on the ignition and combustion properties of solid propellants seem to be well behaved. Plasma ignition appears to offer shorter, more precise ignition delay times than conventional ignition. In addition, increased power and energy levels of the injected plasma result in increases of both the pressurization rate and the maximum pressure attained in the gun. The significance of these findings is in the experimental demonstrations of a new quantifiable mechanism for accelerating the IB process. In conventional IB this would only have been done using either smaller webs or a faster burning propellant composition.

Methodologies for Improved Launch Using SPETC Technique

Insights gained from the experimental data discussed previously were used to develop the strategy for obtaining improved SPETC gun performance. The approach involved the delivery of plasma energy to the propellant charge at various stages of the interior ballistic cycle. Two factors of plasma jet technology, the high-energy density of the plasma and the possibility of adding this energy in a precisely timed and controlled manner, provide the opportunity of influencing the IB process. In principle, ET energy can be added at the ignition phase, before maximum breech pressure is reached, synchronously with maximum pressure or after maximum pressure is past. The basic idea governing the addition of plasma energy is to "widen" the interior ballistic pressure-time profile as close to the maximum pressure as possible, maximizing the area under the curve, and hence the work that can be done on the projectile.

Several strategies could be used for injecting a boost of ET energy to extend the maximum value of the pressure-time curve during the IB process. This injection mode may last as long as the projectile senses the resultant pressure rise. Most simply, the injection may be used to add energy to the gas in the chamber to maintain the maximum pressure value for a longer time, thereby enhancing its ability to do work on the projectile. Energy injected in this way would indeed improve performance, although somewhat less efficiently than the chemical energy of the propellant. Due to the size of PFNs, however, such an energy injection strategy may not be ultimately desirable to the user community. If the plasma energy is used to increase chamber pressure in later portions of the IB cycle to help burn propellant that would otherwise not have been consumed, the plasma energy could be used more efficiently to do work on the projectile. Such would be the case in using plasma augmentation to burn high density charges with an "oversize" web propellant in the gun. Similar arguments may be made for energy addition according to a variety of injection profiles at other stages of the IB process.

Given the complexities of the interior ballistic cycle, mathematical tools simulating the SPETC process are necessary to work out the potential quantitative advantages to be gained from the various plasma injection strategies discussed. Accordingly, the experimental efforts were conducted in cooperation with a theoretic study modeling the SPETC process. The modeling is based on a lumped parameter approach as described by Corner.[6] The basic IB modeling addresses the mass and energy balance of the propellant and the gun; the propellant form function behavior, its surface area history and its relation to the rate of gas production; and the projectile dynamics and pressure gradient along the gun. In addition, other factors such as heat loss, friction, and chambrage effects are addressed. As a first approximation, it has been assumed here that the hot plasma jet injected into the propellant bed interacts with the burning propellant indirectly by affecting the overall energetics of the system. Thus, in this model, plasma injection appears as an additional term in the gas energy balance equation. Computations performed on this basis were sufficient to predict the basic behavior of the SPETC ballistics process.

Experimental Results and Discussion

With the empirically based insights into the effects of plasma on the ignition and burning of propellant in hand and the theoretic IB model mentioned previously, it became possible to address the issue of tailoring the plasma injection to enhance performance from a solid propellant ETC gun. Before discussing the data obtained, however, we defined a meaningful baseline for purposes of comparison. We began with the premise that SPETC needs to provide tangibly improved performance over current conventional capabilities. Given the interest in using existing gun tubes for enhanced performance systems, the benchmark adopted was the greatest muzzle energy that could be delivered by the same gun, using the same projectile, at the same maximum pressure, using conventional ignition and solid propellant charges.

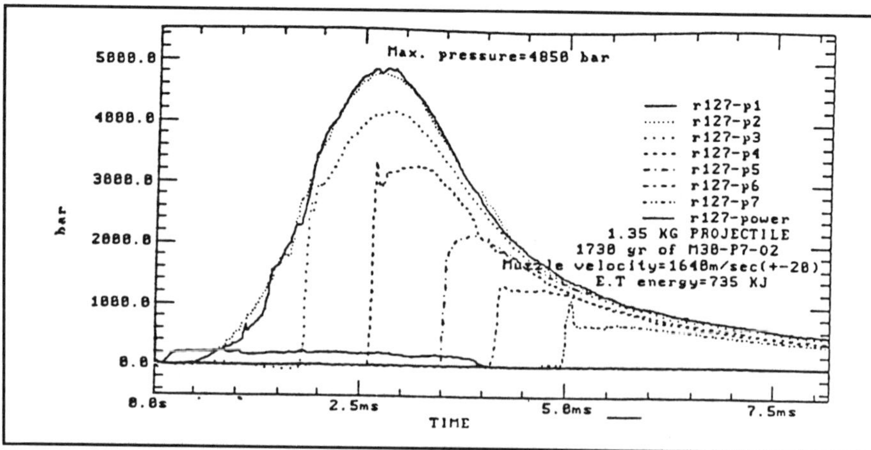

Figure 7. Pressure and power versus time curves for electrical reference shot.

This benchmark is referred to as optimized conventional performance. Using these parameters, a muzzle velocity of 1,620m/sec was established as the baseline for the 60-mm gun described in Table 1. The theoretical value was computed and verified against experimental firings.

The following three experiments present experimental proof for the feasibility of controlling the IB pressurization process and enhancing muzzle energy by injecting a high-power ET pulse at time of maximum pressure.

Figure 7 is considered an electrical reference shot, where a rectangular pulse of plasma at a power of 200MW was injected to cause ignition. The pulse width is about 4msec, that is, about 1msec longer than the time needed to maximum pressure. This large pulse width was chosen specifically to keep the capillary discharge "alive" during the IB process and allow for additional ET energy to be injected in subsequent experiments. The same propellant was used as in the firing in Figure 8. However, due to the influence of the ET pulse on the combustion rate and maximum pressure, the propellant charge mass was reduced by 130g in this firing.

Figure 8 shows the results of a firing in which a second boost of plasma energy was injected into the gun during the interval between 2.5 and 3.5msec. The same propellant shown in Figure 7 was used. The muzzle energy for this particular shot was about four percent higher than for the reference shot. The influence of the ET boosted pulse on the breech pressure is shown in Figure 9. The broadening of the pressure-time curve after maximum pressure is seen to match, with a slight time offset, the stepwise injected plasma boost.

Figure 10 demonstrates the best results achieved so far using the stepwise

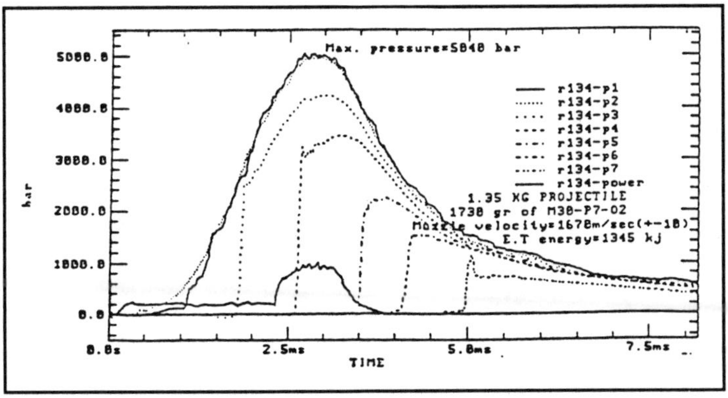

Figure 8. Pressure and power versus time profiles for an ET boosted shot.

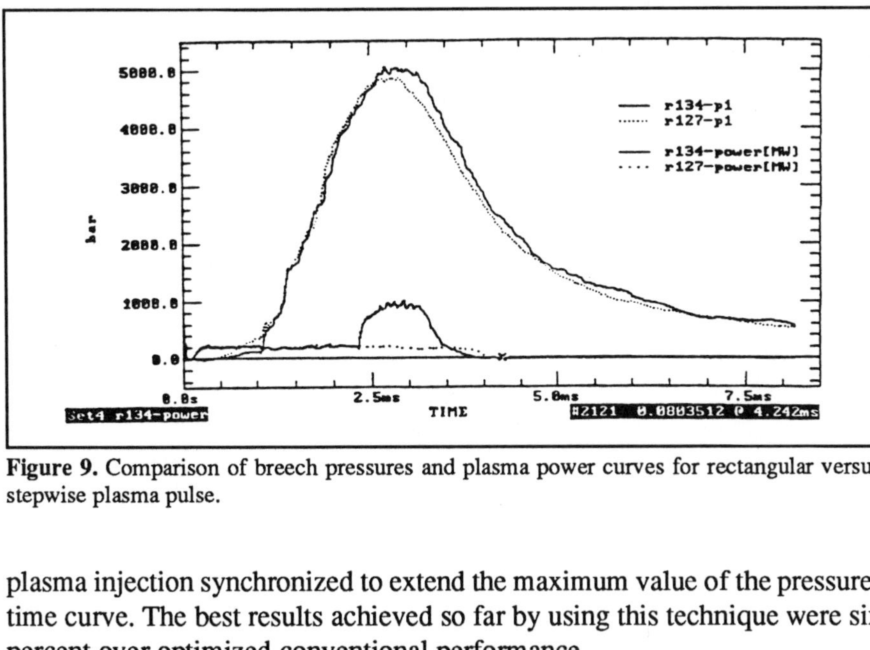

Figure 9. Comparison of breech pressures and plasma power curves for rectangular versus stepwise plasma pulse.

plasma injection synchronized to extend the maximum value of the pressure-time curve. The best results achieved so far by using this technique were six percent over optimized conventional performance.

We investigated another feasible methodology for achieving higher muzzle energy which was based on using different propellant geometries of granulations and webs. Generally, larger web propellants would allow for ET plasma to be injected at both earlier and later stages of the ballistic cycle. It may also permit the use of larger charge masses without exceeding permissible pressures in the gun.

To evaluate firing data in which the maximum pressure did not exactly match the baseline value discussed earlier, conventional IB computations

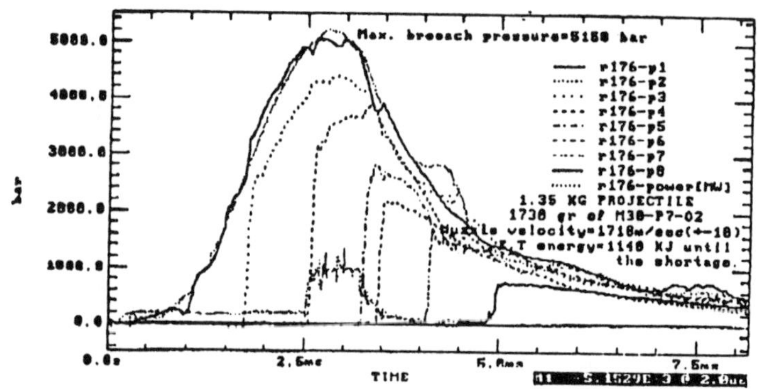

Figure 10. Pressure and power versus time for shot number 176.

Figure 11. Muzzle energy versus maximum pressure, comparing theoretical optimum baseline curve with experimental results.

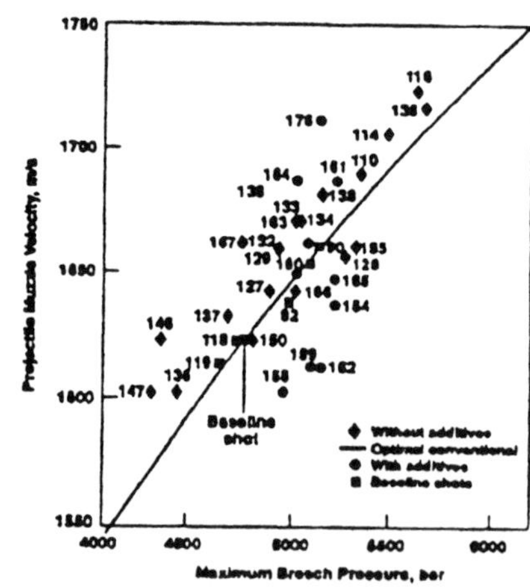

were used to calculate a muzzle energy versus maximum pressure envelope for optimum conventional performance from the 60-mm gun. This tool, the solid curve in Figure 11, could then be used to determine whether the performance from any individual SPETC firing exceeded the performance potential of conventional solid propellant technology, independently of the precise pressure attained in a given test. The results of a number of SPETC firings are plotted with the optimum baseline curve in Figure 11. As indicated in Figure 11, a number of shots fall clearly above the line, indicating that superior performance was achieved. At present, the best unequivocal performance increase documented is 6.8 percent.

Summary and Conclusions

The data presented here are highly encouraging. It is also gratifying that the key features of SPETC, its ability to safely inject large amounts of plasma energy into a solid propellant bed; the controlled positive effects of plasma power and injected energy on the pressurization process; and the effects of the plasma on reducing the ignition delay time of ballistic firings have been independently confirmed in the United States.[7] The controlled results from both the present as well as the U.S. study strongly indicate that solid propellants can retain their physical integrity, hence their form function behavior, while burning under the influence of plasma injection. Results unique to the present study indicate that increased plasma energy injected during the latter portion of the ballistic cycle using a stepwise profile can cause desired changes in the shape of the IB pressure-time curves and attendant enhancements in ballistic performance. Moreover, theoretic results closely match experimental data. Current efforts are focused on increasing the amount of plasma energy injected into the gun and evaluating a wider range of propellant grain types and webs. Given success with these efforts, a near-term muzzle kinetic energy enhancement of 25 to 35 percent over the optimized conventional baseline may be possible.

References

1. R. Hammond, U.S. Army Strategic Defense Command Electrothermal (ET) Program, paper presented at the IAT Workshop on Electromagnetic Power Sources and Launchers, Austin, TX, January 1991.
2. R. Lundberg, ARDEC/FSAC Electric Armaments Program Information Briefing, paper presented at the IAT Workshop on Electromagnetic Power Sources and Launchers, Austin, TX, January, 1991.
3. A. Juhasz, J. Knapton, and K. White, *Process Control Issues in ETC Propulsion*, (CPIA Publication 557, Johns Hopkins University, G.W.C. Whiting School of Engineering, Columbia MD, November 1990.)
4. S. A. Goldsteinet *et al.*, Cartridge Containing Plasma Sources for Accelerating a Projectile, U.S. Patent 4, 715, 261, December 29, 1987.
5. A. Loeb and Z. Kaplan, *IEEE Trans. Magnetics* **25**, 342 (1989).
6. J. Corner, *Theory of Interior Ballistics of Guns* (John Wiley and Sons, New York, 1950).
7. H. McElroy, J. Greig, and A. Juhasz, in *Proceedings of the 5th International Propellants and Propulsion Conference* (Picatinny Arsenal, Dover, NJ, November, 1991).

33

Small-Scale Experiments and Modeling of Laser Ignition of Gun Propellants

Richard A. Beyer
Arthur Cohen
U.S. Army Research Laboratory
Weapons Technology Directorate
Aberdeen Proving Ground, MD 21005-5066

As a part of the Laser Ignition in Guns, Howitzers and Tanks (LIGHT) Program, the ignition of a series of gun propellants has been explored in a series of small-scale experiments. In addition, a model that utilizes heat release kinetics has been used to explore the relative importance of the various physical and chemical effects present during laser ignition of a propellant. The first set of experiments provides data for direct comparison with the model. In that part of the study, it is shown that this model describes the ignition of double-base nitrocellulose propellants adequately. It is also shown that initiation of reaction is independent of the gases present before the event, but that flame spreading can be strongly influenced by the presence of oxygen.

The second series of experiments explores the direct ignition of a series of solid and liquid propellants on a shorter time scale. These measurements demonstrate the applicability of laser ignition to direct propellant ignition (no pyrotechnics or sensitive materials) and have aided in the design of experiments to challenge the next generation of ignition models.

Introduction

The objective of the Laser Ignition in Guns, Howitzers and Tanks (LIGHT) program is to determine the feasibility of replacing or augmenting liquid and solid propellant igniter systems with lasers. Laser ignition of condensed propellants depends on the optical properties of the propellants (absorption and reflection coefficients) and the beam characteristics (wavelength, flux, pulse width and shape). To accomplish the LIGHT objective, it is necessary to identify the beam characteristics for a particular propellant (or energetic material) that produce the required ignition times. It would also be helpful to have a predictive model for the laser-propellant ignition process. We present the initial results of two ongoing investigations of laser ignition of solid and liquid propellants. Most of the experiments were done in closed bombs initially at atmospheric conditions. The first part of this study to be discussed is primarily concerned with evaluating ignition models. A low-power carbon dioxide laser was used, and loading densities were low (~0.0001g/cm^3), leading to relatively long reaction times. The effects of the laser beam power and pulse length and ambient gas composition were investigated. Correlations of measurements with a condensed phase ignition model are discussed. The second part of this study is concerned with identifying the laser beam characteristics appropriate for gun ignition systems. A high-power neodymium laser was used and loading densities were~1g/cm^3 (similar to high pressure bombs), leading to short ignition delays. The pressure records during ignition of two solid and one liquid propellant will be presented.

Evaluation of Ignition Models

Radiant ignition models[1] have previously been developed for solid propellants. It is generally difficult to evaluate the usefulness of ignition models for two main reasons. The first problem is in relating ignition criteria to experimental observations. The second difficulty is that for any particular propellant, neither the values of physical properties nor heat release rates that are required for the calculations are well known. Laser ignition models may have an additional problem, that of describing the details of the laser energy interactions (thermal and nonthermal) with the propellant (and with pyrolysis products). The adiabatic "approximate" thermal radiative ignition model described in Reference 1 has been adapted to compare with experimental data. The model considers condensed phase reactions and in-depth absorption but neglects energy losses (those losses due to surface heat transfer and emission), propellant consumption, and the effect of laser interruption (deradiation). The inability of this model to predict behavior of many solid propellants that are of interest to the LIGHT program (M30, JA2, and LOVA type) suggests that additional complexities must be included in the further development of a more complete model.

Experimental Approach. A continuous wave CO_2 (10.6μm) laser (Synrad model no. 57) with variable power (nominal 10 to 100W) was used as the ignition source for the solids. The response of several propellants at different power (flux) levels and in different gaseous environments has been briefly examined. Data have been obtained for the following propellants: M9 (homogeneous double-base), JA2 (homogeneous triple-base), M30 (composite triple-base), M39 (RDX composite), and HMX2 (experimental HMX composite). The samples were small disks approximately 3/16in wide. The M30 and M39 disks were cut from grains (0.25-in diameter, perforated, with graphite surfaces), and the others were cut from 0.25-in diameter strands. The beam, diameter about 3mm, was directed along the propellant axis and covered approximately 25 percent of the sample surface. Pulse lengths were varied from 0.1 to 15sec. Measurements of light emission and pressure were made using a photomultiplier (300-800nm response) and a piezoelectric transducer, respectively. Measurements were made in open air and in a closed one-liter bomb. The vessel has optical access from four sides, with one containing a zinc selenide (ZnSe) window for transmitting the 10.6μm radiation. The bomb was initially filled with air, nitrogen, or argon. Most of the experiments were done at an initial pressure of one atmosphere. An infrared (IR) microscope-spectrometer (Spectra-Tech Plan 40/Matson Polaris) was used to measure propellant reflectivities and transmittivities of JA2 and M9 at 10.6μm. Propellant reaction products were analyzed by gas chromatography and Fourier-transform infrared (FTIR) spectrometry.

Ignition model. The "approximate" theory for thermal radiative ignition of solids[1] has been used to predict the dependence of propellant "ignition" delays on radiative flux. Reasonable success was obtained for secondary explosives (HMX, RDX) using experimentally determined data for the thermophysical properties and heat release rates.[1] The relevant energy equation and the initial and boundary conditions used for (transparent) solids are:

$$c\frac{\partial T}{\partial t} = ac\frac{\partial^2 T}{\partial X^2} - \frac{\partial q}{\rho \partial X} + \mathrm{Qzexp}\,(-E/RT) \qquad (1)$$

$$q = q_0(1\text{-}f)\exp(-nX) \quad T(X,0) = T_0 \quad \frac{\partial T}{\partial X}(0,t) = 0 \quad T(\infty,t) = \mathrm{T}_0 \qquad (2)$$

where T = temperature, T_0 = ambient temperature, q_0 = radiant flux, n = absorbtion coefficient, f = reflectivity, a = thermal diffusivity, ρ = density, and c = specific heat. The last term is the Arrhenius global heat release rate during ignition (E = activation energy, Qz = zeroth order pre-exponential factor).

The model assumes a condensed phase surface reaction. The ignition delay is calculated as the sum of (inert) "heat up" and (chemical) induction times. The "heat up" time and the corresponding surface temperature are calculated by matching (at the surface) temperatures and time derivatives obtained from analytic solutions of the energy equations corresponding to inert heating and adiabatic reaction. The chemical induction time is obtained from the adiabatic reaction solution. Approximations to these analytic solutions (given in Reference 1) are used in this report to calculate ignition delays as a function of radiant flux. They are valid under conditions where activation energies are much greater than thermal energies, and characteristic lengths for absorption are much less than those for heat conduction during heat up. In all calculations T_0(K)=298, and for all propellants the following values have been used for the thermophysical properties: a(cm^2/sec) = 1E(-3), c(cal/g-K) = .35, ρ(g/cm^3) = 1.6.

Experimental Results

Optical properties. The results of the initial reflectivity measurements indicate that at 10.6μm all propellants reflect less than 10 percent of the incident energy. For the calculations, a reflectivity value of 0.08 has been used. At present, only the M9 and JA2 absorption coefficients (n) have been determined using Beer's Law (scattering neglected) and are equal to 250cm^{-1} and 536cm^{-1}, respectively.

Laser ignition. The laser characteristics corresponding to the emission and pressure traces shown in the following figures, and various delay measurements obtained from these traces are listed in Table 1. *P* is the laser power. The time (relative to start of the laser irradiation) at deradiation (laser off) is denoted by t_l;t_e is the time for start of emission or emission delay. The delay to flame spreading, which is determined from a slope change in the emission-time signal, is denoted by t_f. The exact significance of t_f needs to be determined. t_p is the time for the start of pressurization, and t_d is the final emission decay time. In the figures *e*, *f*, and *d* correspond to t_e, t_f, and t_d, respectively.

The effect of flux level on photomultiplier signals during ignition and flame spreading is shown in Figure 1. In this example, the bomb was filled with nitrogen to one atmosphere; the propellant is M9. Both t_e and t_f delays decrease for M9 as power increases, as expected. The M9 emission decay, t_d, is due primarily to consumption of propellant. The effect of air may be important for direct laser ignition of propellant beds because there is always some ullage in gun chambers. To assess its importance requires experiments

Table 1
Laser Operating Conditions and Delay Measurements for Figures 1, 5, 7 and 9

Figure	Signal	P (W)	t_l (s)	t_e (s)	t_p (s)	t_f (s)	t_d (s)
1	1	2.5	5.80	.19	—	.30	1.65
	2	8.6	1.10	.03	—	.09	1.40
2	1	3.2	1.08	.12	.18	.20	2.18
	2	3.2	1.08	.15	.22	.25	2.22
	3	3.2	1.08	.17	.25	.28	2.46
5	1	1.7	4.34	1.93	—	4.85	4.30
	2	8.6	1.67	.05	—	(.12)	1.70
7		4.2	2.72	.04	—	1.37	3.60

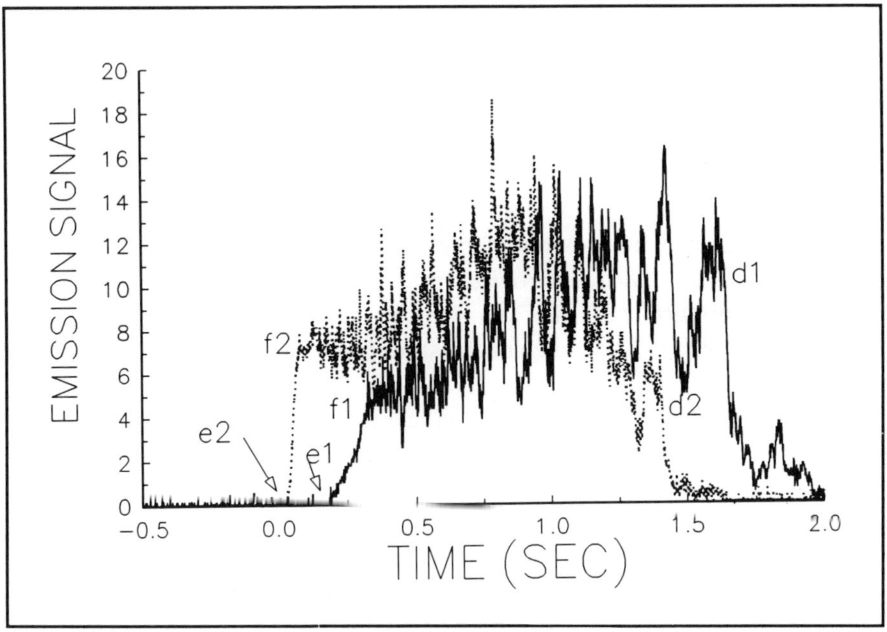

Figure 1. Photomultiplier record of emission for M9 in N_2.

that mimic the restriction of the air supply by neighboring grains or powder. An attempt was made by surrounding the propellant sample with a cylindrical shroud (plastic, 0.75-in ID) with axis coincident with the laser beam, which restricted the path of the incoming air. Figure 2 shows the effect of replacing nitrogen with air and the effect of the shroud on the pressure signals. Signals

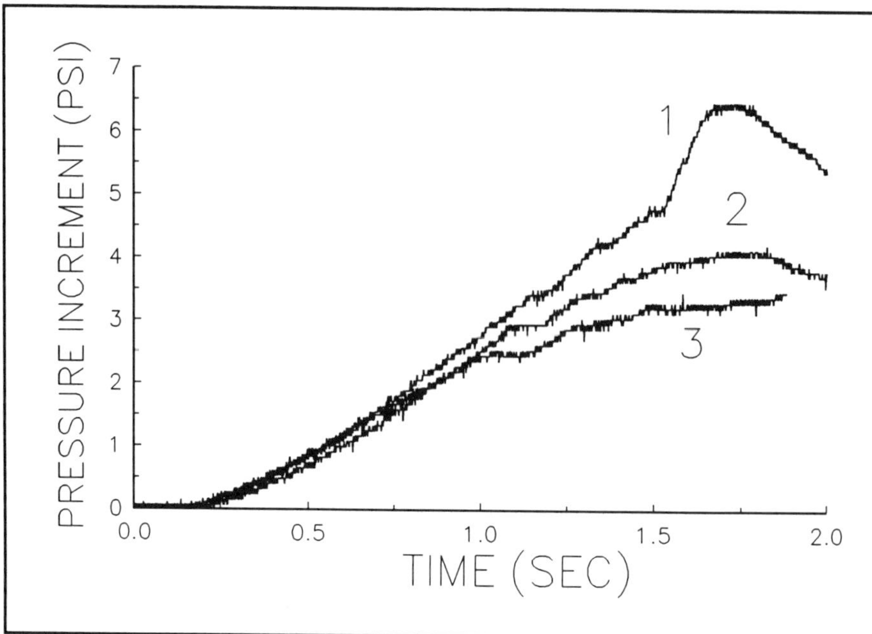

Figure 2. Effect of oxygen on pressure for M9.

1, 2, and 3 correspond to experiments in air, N_2 and air with the shroud in place, respectively. The most obvious effect of the presence of oxygen is the increase in pressure level after deradiation. The temporal behavior is relatively unchanged. The presence of the shroud seems to eliminate the oxygen effect.

The lines in Figure 3 are the calculated ignition delays as a function of flux levels for different heat release rates. *N* refers to a rate determined from conductive (radiant) ignition experiments with the Russian double base propellant N (0.58 NC, 0.28 NG, 0.12 DNT),[1] and "M9" refers to a rate determined from shock tube experiments with M9 (0.58 NC, 0.40 NG) powders in nitrogen (T = 600-700K and pressure = 0.9-3.5atm).[2] For N and M9, values of Qz (cal/g-sec) are 2.5E16 and 2.6E11, respectively. The corresponding values for E (kcal/mol) are 35 and 22. The rate for N propellant was based on surface thermocouple measurements of delays to a jump in temperature that presumably involved flame spreading, whereas the rate for M9 was determined from emission (delays) from small (<10μm), uniformly heated particles that react (explode) without flame spreading. Deriving global Arrhenius parameters from these different "ignition" delay data involve different assumptions or models of the heat-up and heat release processes. The

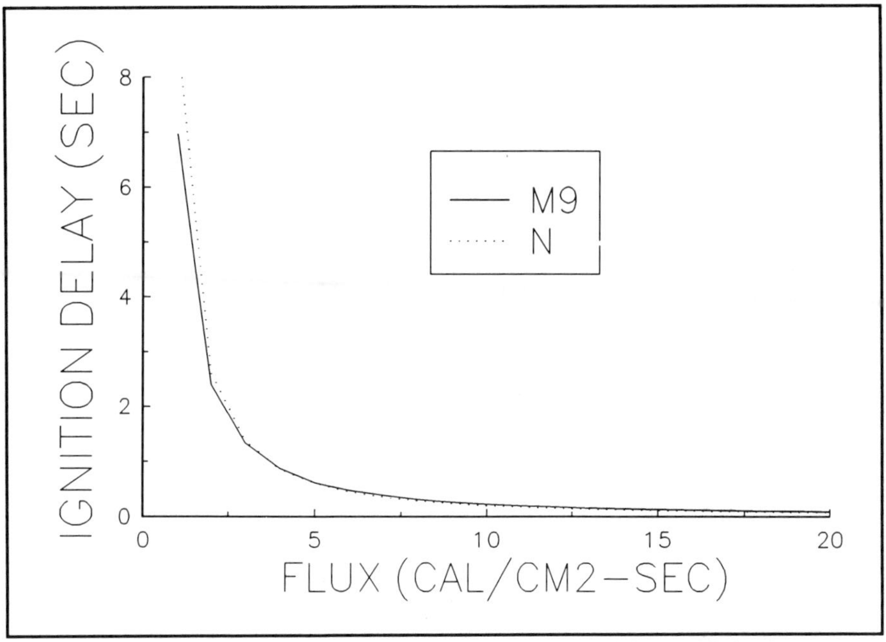

Figure 3. Comparison of approximate model ignition delays for M9 and N kinetics.

rate constants are expected to be valid for thermochemical states obtained under the experimental conditions. The use of M9 and N rates in the calculations are expected to produce ignition delays that correspond to measurements of emission and flame spreading delays, respectively. Although the Arrhenius parameter for the two rate constants differs markedly, the calculated ignition delays are almost identical for flux values greater than 2cal/cm^2-sec.

Figure 4 shows the t_e, t_p and t_f data obtained from experiments with M9 in the bomb under different ambient conditions. The limited amount of data at flux = 10.8cal/cm^2-sec (listed in Table 1 for p = 3.2W) suggest that replacement of N2 by air (with and without a shroud) does not have a great effect on the M9 ignition process. The lines in Figure 4 are the calculated heat-up and ignition delays from the "approximate" model using the M9 kinetics and show that the difference between heat-up and ignition decreases with increasing flux levels. If it is assumed that t_e corresponds to the end of heat-up and that t_p and t_f correspond to ignition, then, considering measurement errors and the uncertainties in thermophysical properties, the predictions of the "approximate" model seem reasonable.

Figure 5 shows the emission signals from experiments with JA2 in argon, initially at ambient conditions, at two flux levels that show the possible

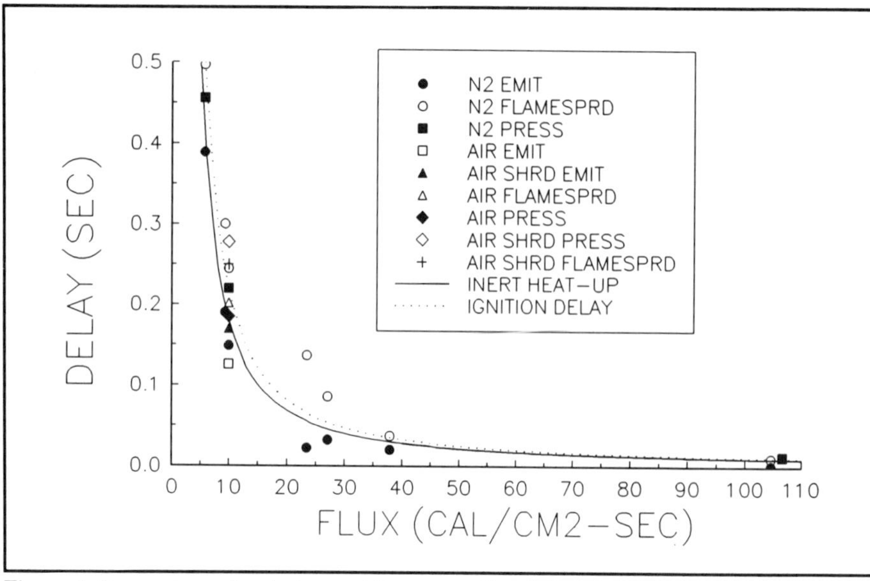

Figure 4. Comparison of emission and flame spreading delays for M9 with calculations using experimental heat release retes.

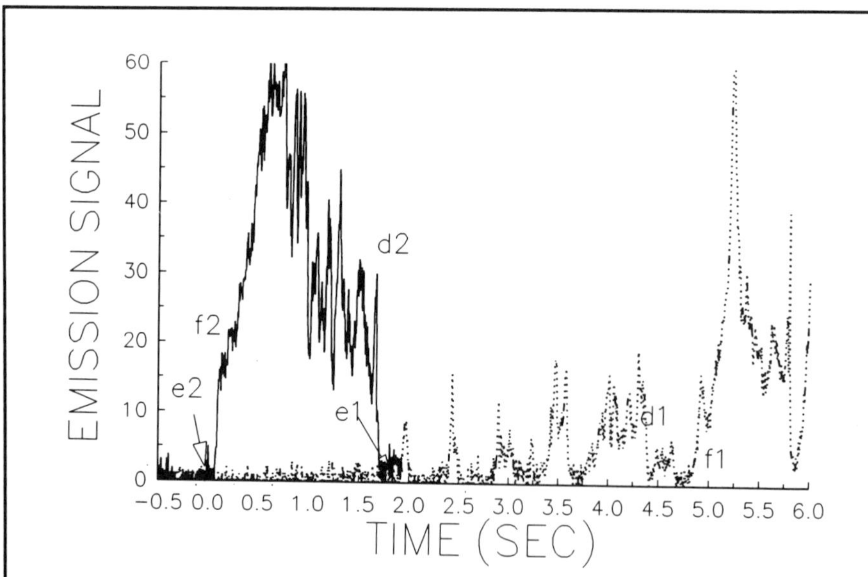

Figure 5. Emission signals showing different effects of deradiation on reaction of JA2 in Argon.

effect of deradiation on propellant reaction. The emission (t_e) and flame spreading delays (t_f) decrease with increasing power, similar to the M9 results,

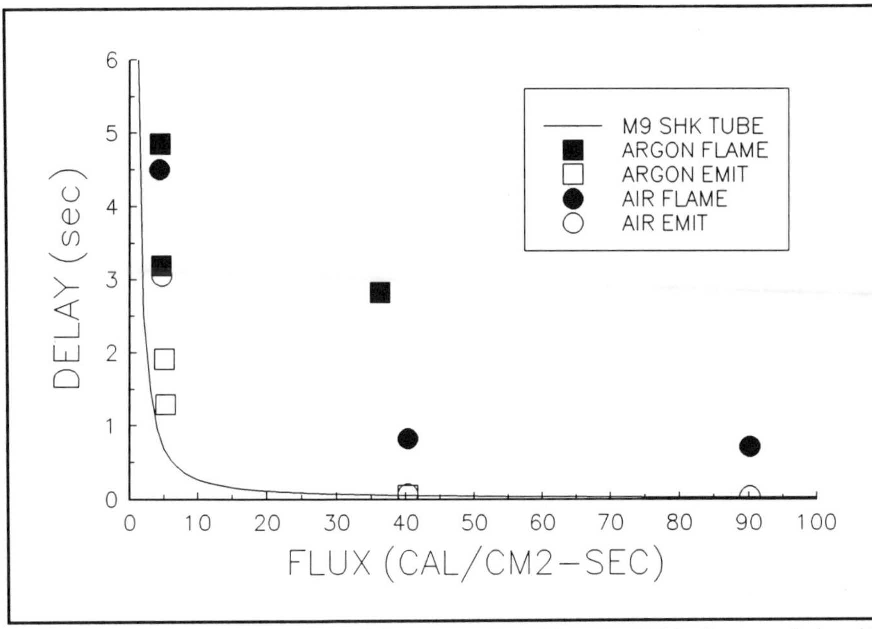

Figure 6. Comparison of emission and flame spreading delays for JA2 in Argon and air, with calculations.

but at 8.6W the emission decay (t_d) corresponds to laser deradiation (t_l), and propellant consumption is not completed. Emission is due to local combustion in the irradiated region, resulting in formation of a crater in the unconsumed propellant. At lower power and longer pulse lengths, reaction decelerates at t_l (corresponding to d1 in the figure) but recovers. Self-acceleration leads to the subsequent sharp increase in emission at (f), which is assumed to be the start of flame spreading. The subsequent decay (not shown) in emission is the result of propellant consumption.

The data in Figure 6 are the emission and flame spreading delays for JA2 in air and Ar initially at ambient conditions. The data at 90cal/cm^2-sec are from an experiment in room air. The others were taken in the bomb. The Arrhenius parameters have not yet been determined for JA2, and the line in Figure 6 is the calculated M9 ignition delays shown previously in Figure 4. The data, near 40cal/cm2-sec suggest that substitution of air for argon shortens flame spreading delays but does not affect emission delays. The differences between calculated delays and measured emission delays are probably within estimated measurement/thermophysical properties errors; except perhaps, at low flux levels, flame spreading delays are much greater than calculations.

Figure 7 shows the emission signal from ignition of M30 in room air in

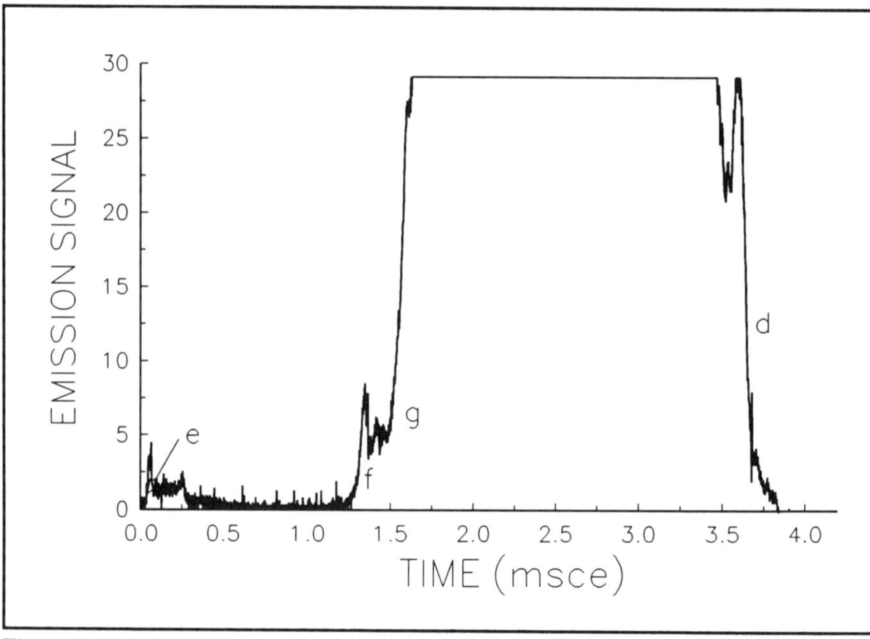

Figure 7. Emission signal showing transition from fizz to flame burning for M30 in air.

which a transition from fizz (f) to flame (g) burning occurs. This transition has only been observed during irradiation and differs from the transition to flame spreading shown in Figure 5 (f1) that can occur after deradiation. The data in Figure 8 are the emission and flamespreading delays for M30 in room air and in the bomb with N_2 at ambient conditions. The line is the calculated M9 ignition delay. The results suggest that emission delays are not greatly different in air or N_2, and except at low flux, values are relatively constant, which is in reasonable agreement with calculations. The flame spreading delays are less in air than in nitrogen (near 35cal/cm^2-sec) and are much greater than calculated values using rates for M9 (and N) propellant.

Discussion and Conclusions

1. Due to space limitations, the results for the nitramine propellants M39 and HMX2[3] have been omitted. For M39, using rate constants derived from differential scanning calorimetry (DSC) experiments, predictions of the "approximate" model were in agreement with t_e but not t_f data. For HMX2, using rate constants derived from either DSC or shock tube experiments, the model was unable to predict either t_e or t_f data.

2. Observations of deradiation and ambient gas effects similar to present

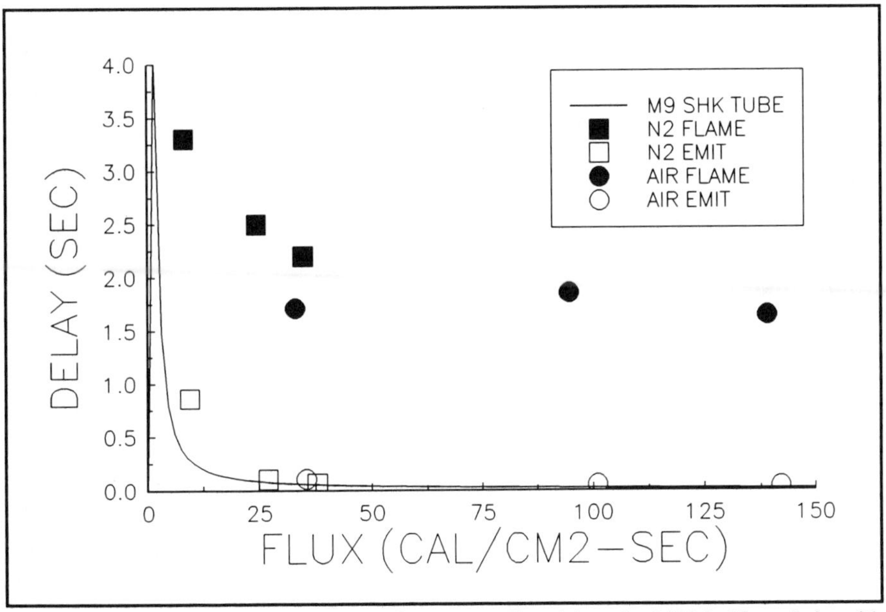

Figure 8. Comparison of emisssion and flame spreading delays for M30 in N_2 and air, with calculations.

observations but at higher pressures (5-20atm) were reported in an early investigation of radiative ignition of double-base propellants.[4,5] The present "approximate" model cannot be used to predict such effects.

3. The substitution of air for inert gases leaves t_e values unchanged but shortens (except for M9) t_f values. This suggests that reaction of gases (O_2) affects flame spreading but not initiation.

4. The good agreement between emission delays (t_e) for radiatively heated M9 disks and calculated values using heat release rates determined from emission delays for shock heated M9 powder strongly suggests that the inverse process may be possible: using the "approximate" model to obtain "useful" rate information for propellants with ignition delays not greatly affected by ambient gases.

5. The decrease in the flux dependence of the calculated ignition delay (for all input rates) as flux levels increase suggests that precise knowledge of initiation rates are not critical inputs into models to predict flame spreading at high flux levels.

6. The laser assisted transition from fizz to flame burning with M30 in air (Figure 7) suggests the possibility that laser absorption by reaction products (CO_2) aids in the transition.

7. Data at higher power are necessary to determine usefulness of ignition models for the LIGHT Program.

These comments emphasize the shortcomings of the "approximate" model. The inability of this model to predict behavior of many solid propellants that are of interest to the LIGHT program (M30, JA2, and LOVA type) implies that the next level of laser ignition models (for many propellants) needs to include explicitly the effect of gas phase reactions on the physics of flame spreading.

Small Chamber Direct Ignition

In addition to the studies described previously to characterize the laser-propellant interaction, an effort has been made to study the direct laser ignition of propellants. The ready ignition of several propellants in the open air has been earlier demonstrated in our laboratory and elsewhere. In the present work, experiments have been started to study ignition in confined vessels that is a step closer to ignition in a gun chamber. This section of the paper discusses primarily the ignition of small samples of solid and liquid propellant that are confined by a chamber with limited ullage. The ignition characteristics are observed and guidelines are set for future work.

Experimental. The chamber used in most of these studies is shown in cross-section in Figure 9. It was machined out of 316 stainless steel with a chamber capacity of $0.1cm^3$. The use of the mylar as a combination window and blowout disk has proved to be advantageous for bench top survey studies. The back wall is formed by the pressure gauge, as shown. No protective grease was used on the gauge because of the very limited volume and concerns that the grease would introduce significant nonreproducible effects on the volume and heat transfer to the walls. Thus, some pressure traces have the characteristic negative shift due to heating of the pressure transducer; this shift is only cosmetic for these studies.

The laser used in these studies was a neodymium glass laser operating at a $1.054\mu m$ wavelength, with pulse length variable from $100\mu sec$ to 10msec and variable energy. For these studies, the light was focused to a 0.125-in diameter spot, with a 2-in focal length lens. Except as noted, the mylar was 0.020-in thick. For some of the studies, a second expansion chamber was created by adding a spacer after the diaphragm with a second diaphragm. The volume of this second chamber was about $0.05cm^3$ or five times the first chamber volume. The point of rupture of the first layer diaphragm is obvious from the sharp pressure drop in each figure discussed next. This dual-chamber arrangement allowed the study of how completely the propellant was ignited and whether sudden, but incomplete, depressurization would extinguish it. Results obtained from this configuration will also be important in the study of

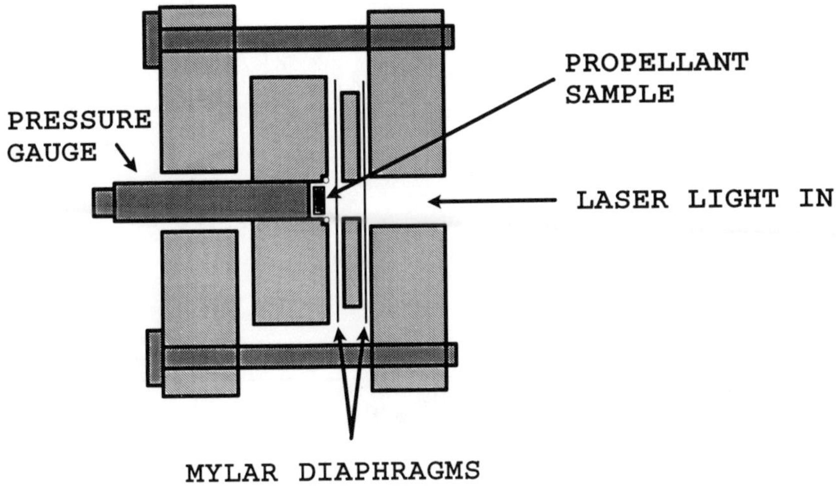

Figure 9. Schematic diagram of the 0.1cm^{-1} test chamber.

laser ignition where irradiated material could be placed in a confined volume to enhance ignition; after ignition the confinement would rupture, and the following processes take place in a depressurized environment.

In contrast to the 10.6-μm wavelength CO_2 laser, most of the propellants are quite transparent near 1μm. Thus, in most of the observations described here, a graphite coating was required on the propellant surface in order to have sufficient energy absorbed to ignite the sample.

Observations

M30 Propellant. The energy to the surface for the M30 propellant ignition event shown in Figure 10 was less than 3J before the apparent ignition time. As can be seen, the second diaphragm broke at lower pressure than the first. The pressure behavior before the burst of the first diaphragm is quite similar to that recorded in the single chamber ignition. In the case of Figure 10, the sample lost 12.1mg, or 9.5 percent of the total mass. The pressure record again shows good sustained ignition. The residual propellant sample from this and similar tests shows the smooth glassy surface and enlarged perforations that are indicative of flame spread over the entire surface.

M43 Propellant. Several LOVA propellants were studied briefly. The

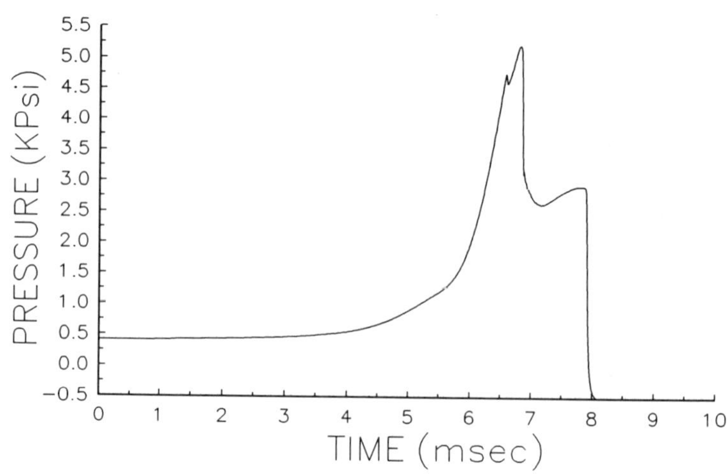

Figure 10. M30 laser ignition in dual chamber fixture.

response of one of these, M43, is shown in Figure 11. Approximately 8.5J of laser energy were incident on the solid surface during 10msec. As can be seen in the plot, the laser raised the pressure sufficiently that full ignition of the propellant took place shortly thereafter. The pressure rise was much slower in comparison with the M30 propellant shown in the previous figure. This result was expected because of the difficulty of igniting the LOVA propellants.

In this case, the original sample was 90.0mg before irradiation and 86.6mg after the event shown. The mass lost by a similar piece of material under the same laser conditions but without confinement showed a mass loss of approximately 0.1mg, which is at our level of measurement uncertainty. The amount of mass lost when an M43 propellant sample was ignited for these brief times varied from 2 to 9mg and appeared to be independent of the original mass. These values of mass loss, as well as the appearance of the samples after the tests, show that vigorous, complete ignition of the samples has taken place.The variability in sample consumption may be related to either relative flame spread rates or differences in diaphragm rupture behavior.

Liquid propellant 1846. Although the liquid propellants are also quite transparent at 1.054μm, most experiments were done without adding graphite. The mechanism of the energy coupling into the liquid is the subject of continuing investigation. This propellant also ignited vigorously and promptly with the laser source. As can be seen in Figure 12, the ignition delay was about

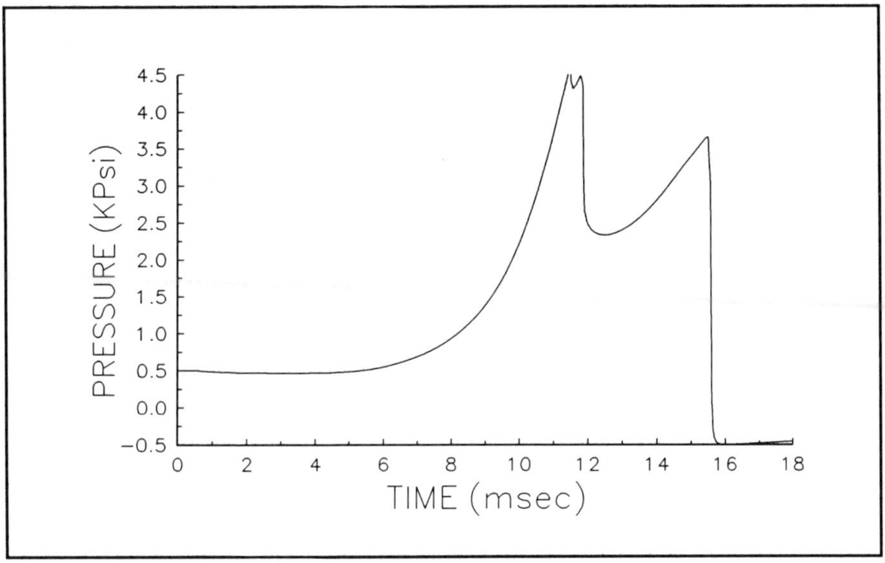

Figure 11. Igniton of M43 propellant in dual chamber device.

5msec for this typical example. The liquid propellant is most obviously different from the solids in that it blew itself out of the chamber almost completely. In the single diaphragm experiments, it behaved much like the M43 in the sense of needing good confinement and pressurization during the laser pulse in order to achieve ignition.

The behavior of LP1846 in the double chamber studies was significantly different from the solid propellants. The main difference is possibly related to the near extinguishment of the LP1846, followed by rapid re-ignition. As can be seen in Figure 12, the second peak is very sharply rising after almost complete depressurization. The experiment shown in Figure 12 incorporated a small (circa 100-μm diameter) vent hole that allowed sustained combustion without exceeding the pressure capacity of the test chamber.

Summary

In the first part of this study, a comparison of experimental and calculated values for ignition parameters has been presented that shows limited agreement and points the direction for future improvements in this research.

In the related study, a variety of propellants have been shown to be readily ignitable by laser light at a wavelength that is producible in compact, efficient package size. Ignition times vary substantially around the typical

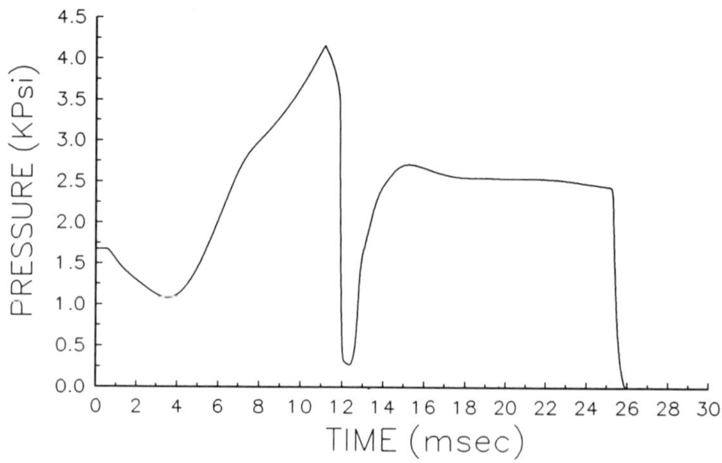

Figure 12. Ignition of LP1846 with vented second chamber.

values shown here; the major uncontrolled variable is presently thought to be the coupling of the energy into the propellant.

Acknowledgments

The authors thank J. Deas, C.S. Miser, K.L. McNesby, and R.A. Pesce-Rodriguez of the Ballistic Research Laboratory (BRL), for help in acquiring experimental data and A.J. Kotlar, also of the BRL, for help with the calculations.

References

1. V.H. Vilyunov and V.E. Zarko, *Ignition of Solids* (Elsevier, New York, 1989).
2. A. Cohen and H.E. Holmes, Convective Ignition of Double-Base Propellants, paper presented at the 19th Symposium (International) on Combustion, The Combustion Institute, 1982, p. 691
3. L. DeLuca *et al.*, *AIAA Journal* **14**(7), 940-946, (1976).
4. L. DeLuca *et al.*, *AIAA Journal* **14**(8), 1111-1117 (1976).
5. A. Cohen and R.A. Beyer (CPIA Publ. 573, Chemical Propulsion Information Agency, Columbia, MD, 1991), pp.435-440.

34

Laser-Based Sensitive Detection of Trace Atmospheric Vapors of Military Interest

Rosario C. Sausa
George W. Lemire
Josef B. Simeonsson
Andrzej W. Miziolek
U.S. Army Research Laboratory
Aberdeen Proving Ground, MD 21005-5066

The sensitive detection and monitoring of trace atmospheric vapors of energetic materials and chemical warfare agents presents an important technical challenge for the defense of our troops, as well as for the security against anti-terrorist attacks and demilitarization safety applications. The authors present the development of a novel, laser-based, ultrasensitive detector of atmospheric vapors of NO_2 containing energetic materials such as propellants and explosives. One laser operating at 226nm is employed to both photofragment the target molecule and detect the characteristic NO fragment by resonance-enhanced multiphoton ionization via its $A^2\Sigma^+$ - $X^2\Pi$ transition. Using a molecular beam, time-of-flight mass spectrometer, the analytic utility of this laser-based sensor is demonstrated on a number of compounds, including nitrogen dioxide, dimethylnitramine, nitromethane, nitrobenzene, TNT, and RDX. A detection limit of 8 and 24 parts-per-billion is demonstrated for RDX and TNT, respectively, using a laser energy of only 100μJ/pulse. An increase in sensitivity is projected by employing higher laser energies and an improved system design. In addition, the authors present and discuss results for the 193nm excitation of trace atmospheric vapors of diethyl methylphosphonothioate, a nerve agent simulant. These techniques should also prove useful for future environmental work in pollution prevention, environmental cleanup, and compliance with government regulations.

Introduction

The detection and monitoring of trace atmospheric vapors of military interest is a major technologic task of growing importance. Examples include propellant and explosives vapor detection for mine countermeasures and anti-terrorist applications, and the detection of chemical warfare (CW) agents. The increasing interest in the environmental aspects of military operations such as pollution prevention, environmental cleanup, and compliance with government regulations require the development of high-performance sensors for various toxic and nontoxic pollutant gases. Laser-based vapor sensors are attractive candidates for these applications because they are highly selective, and they are capable of rapid, real-time monitoring. The ability to manufacture these sensors into rugged and power- efficient devices also makes them attractive candidates.

Many molecules, including CW agents and various energetic materials, are difficult to detect spectroscopically by ultraviolet-visible (UV-VIS) absorption or emission due to the lack of distinguishing structure or absence of any features in the UV-VIS spectral region. As a result, an approach based on the use of UV laser radiation to photodissociate the parent molecules into characteristic fragments is being pursued (see references 1 to 8 for a partial list of representative studies). The photofragments typically include atomic, di- and triatomic species, which generally have structured, readily identifiable transitions in the UV-VIS spectral region. These fragments may in turn be detected either by their prompt emission, if electronically excited during the photolysis pulse by laser-induced fluorescence (LIF), or by resonance-enhanced multiphoton ionization (REMPI). This laser photodissociation/fragment detection approach has been demonstrated previously in our laboratory in a two-laser experiment using organophosphonates that are nerve agent simulants containing organic esters of substituted phosphoric acids.[1, 8-9] Selected organophosphonates were dissociated in a simple flow cell using the focused output radiation of an excimer laser at 193 or 248nm and the characteristic PO moiety, present in all nerve agents, was detected via its $A^2\Sigma^+ - X^2\Pi_r$ transition near 247nm.

Propellants and explosives also contain molecules with characteristic functional groups, namely the NO_2 moiety bonded to either a carbon or nitrogen atom. This functional group is weakly bound to the main skeletal portion of the molecule by approximately 40 to 50kcal/mol, depending on the molecule, and is responsible for the weak and structureless absorption feature observed in the UV near 230nm.[10,11] A perusal of the literature reveals that the UV laser photodissociation of energetic molecules,[12] such as RDX, and model energetic compounds[3,12,13] under collision-free conditions results in the production of NO_2 in the initial step of the photolysis. The NO_2 fragment is usually detected by its prompt emission or by LIF.[3,12-13] However, both of these methods are very inefficient because the absorbed radiation is radiated over a large spectral region, visible to near infrared. The radiative lifetime for these transitions is also very large, typically ~50 to 120μsec,[14] and is indicative of a weak oscillator strength. Moreover, predissociation to NO + O predominates over fluorescence at wavelengths less than 400nm. The NO fragment is better suited for detection because it is readily formed from the predissociation[15,16] of NO_2, and its radiative lifetime[17] (~200nsec) is significantly shorter than the NO_2 states excited in the visible region. The NO fragment can be detected with a high degree of sensitivity and selectivity by (1+1) REMPI and/or LIF via its $A^2\Sigma^+ - X^2\Pi$ transition near 226nm. Thus, one laser tuned to 226nm can be used both for parent photofragmentation and fragment detection.

The laser photodissociation/fragment detection technique is most effective when coupled with pulsed molecular beam (MB) sampling and time-of-flight (TOF) mass spectrometric analysis. A TOF mass spectrometer of modest design can detect a packet of ions of up to 500 atomic mass unit (amu) in less than 100μsec with resolution ($m/\Delta m$) greater than 500, whereas pulsed molecular beam sampling greatly improves the sensitivity and selectivity over ambient samples. The molecules in the beam are colder than those at ambient conditions because they are subject to a supersonic expansion into vacuum through a pulsed, small-orifice nozzle. Thus, they are produced in a few distinct ro-vibrational states, and as a result, their spectral absorption spectrum is considerably less congested and simpler to analyze. Interference from other molecular systems is also minimized, if not eliminated. The pulsed nozzle also greatly reduces the total gas throughput entering the analysis chamber while still maintaining a high density of the sample. Thus, the pumping requirements are reduced, which allows for a more compact design.

In this chapter, we report on the sensitive detection of atmospheric vapor of NO_2 containing compounds based on the 226nm photolysis of the target molecule, with subsequent detection of the NO fragment by REMPI and/or LIF via its $A^2\Sigma^+ - X^2\Pi$ transition. The analytic utility of this novel detector is demonstrated using trace energetic materials and model compounds in an MB/TOF apparatus. Limits of detection are reported for NO_2, dimethylnitramine (DMNA), nitromethane, nitrobenzene, 2,4,6-trinitrotoluene (TNT), and 1,3,5-trinitrohexahydro 1,3,5-triazine (RDX). The 193-nm excitation of diethyl methylphosphonothioate (DEMPS), a simulant for CW agents, was also studied, and the results are presented.

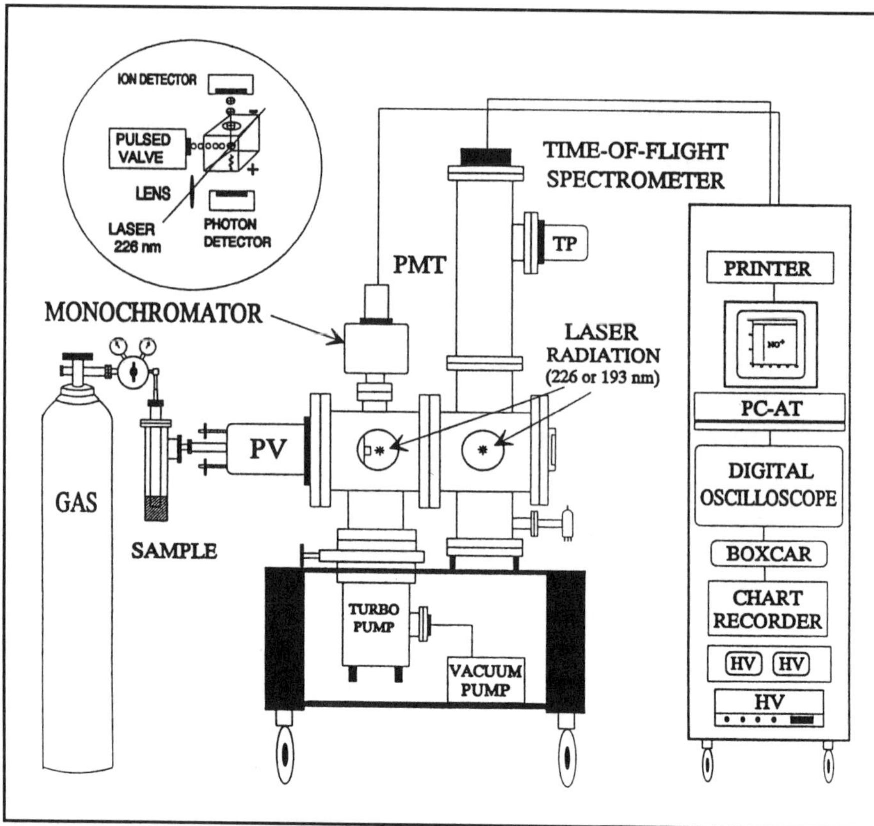

Figure 1. Schematic of experimental apparatus.

Experimental

A schematic diagram of the experimental apparatus used for these experiments is depicted in Figure 1. The apparatus is a general purpose molecular beam apparatus that has been constructed for both laser-induced fluorescence and resonance-enhanced multiphoton ionization studies. There are two main chamber bodies to this apparatus. The first chamber consists of an eight-inch tee with smaller ports added for laser excitation, fluorescence detection, and pressure recording. This chamber is pumped out through the bottom by a 1000-liter per second turbo molecular pump (TurboVac 1000, Leybold-Heraeus, Brewster, NY). As shown in Figure 1, to the left of the tee is a pulsed supersonic valve (R.M. Jordan Co., Grass Valley, CA) that can be translated the length of the chamber and positioned for either LIF or REMPI experiments. To the right of the tee an eight-inch four-way cross serving as the second chamber is mounted. These two chambers are differentially pumped and separated by a skimmer with a 3mm orifice. Mounted on top of the cross chamber is a one-meter commercial time-of-flight mass spectrometer (R.M. Jordan Co., Grass Valley, CA) with reflectron.

The TOF mass spectrometer distinguishes ions of different masses by their different arrival times at the microchannel plate detector located at the end of the flight tube. The kinetic energy of the ions after acceleration is equal and given by the equation:

$$1/2\ mv^2 = Ve \qquad (1)$$

where m is the mass of the ion, v is the velocity, V is the acceleration voltage, and e is the ion charge. Thus, every ion has its own unique velocity, inversely proportional to the square of the mass. The time, t, necessary for an ion to reach the end of a flight tube of length, L, is given by the equation:

$$t = L/v \qquad (2)$$

the difference in time, Δt, which separates ion 1 from another ion 2 is:

$$\Delta t = L/(2Ve)^{1/2}\ (m_1^{1/2} - m_2^{1/2}) \qquad (3)$$

As shown in Equation 3, the difference in times of the ions depends on the difference in the square roots of the masses. In this study, NO was used for the calibration of the mass spectra. For the apparatus employed in this study, a complete mass spectrum of a packet of ions of which the largest is 500amu (1amu = 1.67×10^{-24}g) can be recorded in approximately 72μsec. This calculation is based on a flight tube of length 200cm and an acceleration voltage of 2000V (1eV = 1.60×10^{-12} g-cm^2/sec^2).

For safety precautions, samples of TNT and RDX were handled in aliquots of less than 10mg in a static and shock-free environment. All were used without further purification. They were introduced into the analysis chamber either in an effusive flow, so that the gas in the sampled volume was replenished at a rate sufficient to avoid accumulation of photolysis products, or in a pulsed molecular beam. For the latter, vapors of the analytes were seeded in an atmospheric pressure of air, nitrogen, or argon gas (Matheson Gas Co, East Rutherford, NJ), and the mixture was expanded into the analysis chamber using a pulsed valve with a 0.5-mm diameter nozzle. The valve could be heated up to 150°C and was operated at 10Hz. In the case of RDX and TNT, the sample was heated to approximately 100°C before analysis. A temperature difference of ±10°C results in an uncertainty in the vapor pressure determination which would effect the sensitivity measurement by a factor of two.

The pulse pressure in the analysis chamber was calculated employing the following equation adapted from Reference 18:

$$P_{pulse} = P(r/R)^2(1\text{-}\cos^3\theta) \qquad (4)$$

where P is the backing pressure, r is the radius of the nozzle, R is the radius of the skimmer orifice, and θ is the skimmer transmission angle. This angle θ is equal to the arctangent of the ratio of the radius of the skimmer orifice (R) to the distance from the skimmer to the nozzle (D). A pulse pressure of 180mTorr was obtained when using a backing pressure of 770Torr, a nozzle diameter of 0.5mm, a skimmer diameter of 3.0mm, and a skimmer-to-nozzle distance of 2.0cm.

The samples were photolyzed or ionized using unfocused or focused (f=250mm or 500mm lenses) UV radiation. A broadband excimer laser (Model EMG 150MSC, Lamda Physik, Acton, MA) was employed to generate 193nm radiation, while an excimer pumped dye laser with second harmonic generator (Hyper EX-400/Hyper DYE-3000/Trak-1000, Lumonics Inc., Ottawa, Ontario, Canada) was used to generate the tunable UV radiation near 226nm. Both lasers were operated at 10Hz. Pulsed energies of the order of 5 to 10mJ and 100 to 150μJ for the 193nm and 226nm radiation, respectively, with duration of 15 to 20nsec were employed for excitation. The linewidth of the tunable UV radiation was approximately $0.16cm^{-1}$(fwhm).

The fluorescence signal, whether prompt or induced, was collected at 90° to the photolysis beam and focused onto the entrance slit of a 0.25m monochromator (Model 218, McPherson, Acton, MA) acting as a broadband filter (~8nm fwhm). The monochromator was tuned to approximately 237nm and 500nm for the detection of NO and NO_2, respectively. Schott filters UG-11 and GG-435 were used in place of the monochromator. Signals were detected with a photomultiplier tube (Model 9789QA, EMI, Plainview, NY) and directed into a gated integrator (Stanford Research Systems, Palo Alto, CA). Both fluorescence and ion signals were displayed and monitored in real-time on a 125MHz digital oscilloscope (Model 9400, LeCroy Inc., Chestnut Ridge, NY). A PC-AT computer was used for data acquisition and analysis.

Discussion of Results

Energetic materials. Presented in Figure 2 are the names and structural formulas of molecules commonly used in explosives and propellants, as well as in CW agents. The molecules that were used in this study are also listed. A

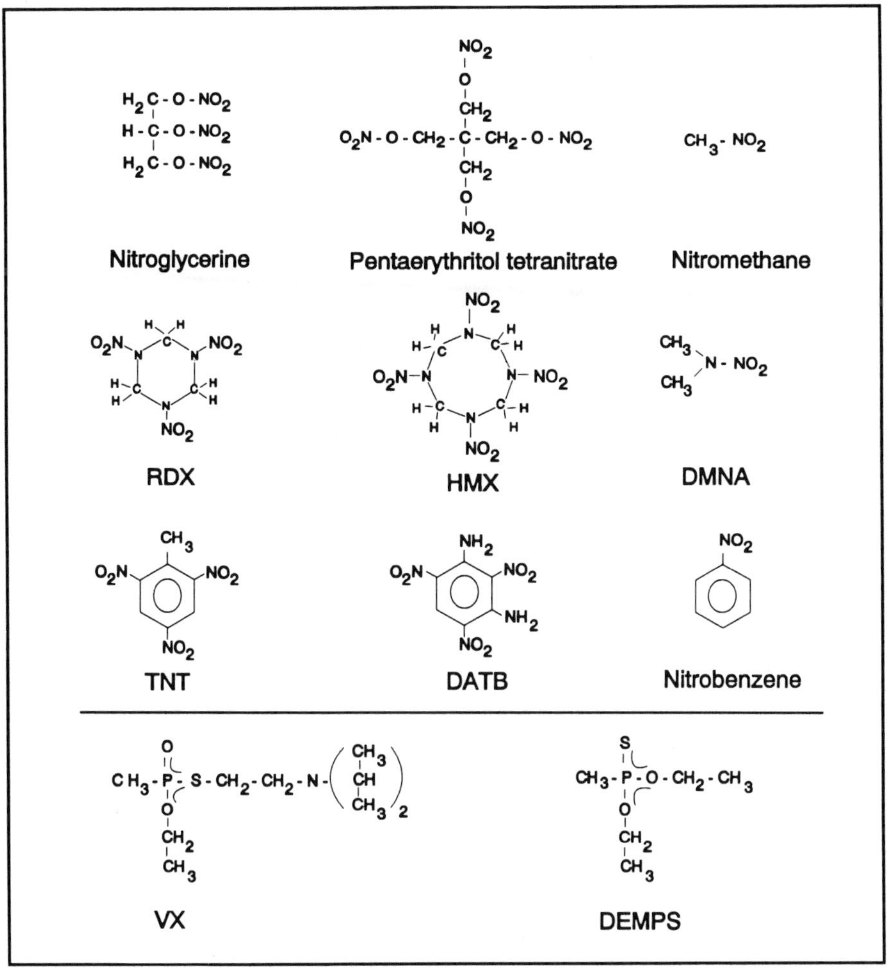

Figure 2. Structural formulas and names of selected compounds of military interest.

cursory inspection of the figure reveals that the energetic materials and model compounds all contain one or more weakly bound -NO_2 functional group. The physical processes underlying our approach to the detection of these molecules may be understood by referring to Figure 3 which shows potential level diagrams of NO_2 and NO. In our approach, the target molecule (R-NO_2) is first photolyzed to NO_2, along with its corresponding companion radical (R), via the process:

$$R\text{-}NO_2 \xrightarrow[226\text{nm}]{h\nu} R + NO_2 \quad (5)$$

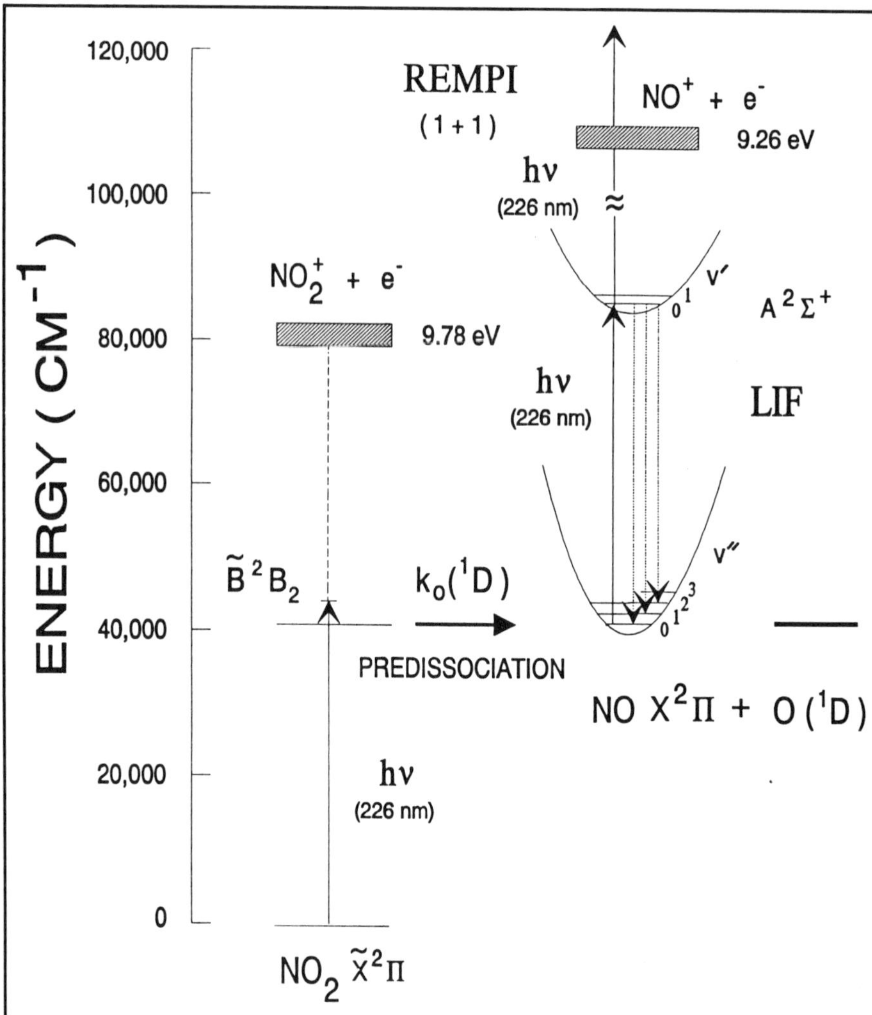

Figure 3. Potential energy diagram of NO_2 and NO showing the multiphoton scheme employed for REMPI and LIF detection of NO photofragment at 226nm.

The NO_2 fragment is then detected by monitoring its predissociative product NO with (1+1) REMPI or LIF and employing its $A^2\Sigma^+$ - $X^2\Pi$ (0,0) band at 226nm.

Presented in Figure 4 is a typical mass spectrum obtained when photolyzing target molecules in a molecular beam using 226nm radiation. The molecular beam was obtained by flowing argon gas at approximately one atmosphere over the vapor of the target compound and expanding the resulting gas mixture into the analysis chamber. The spectrum shown in Figure 4

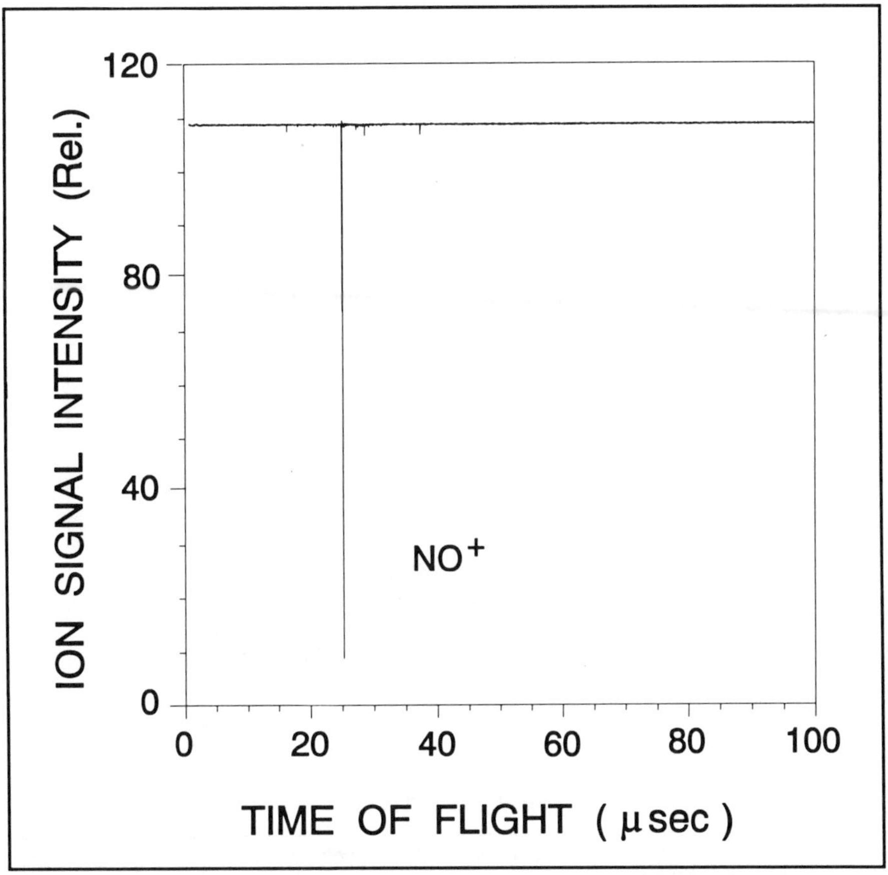

Figure 4. Time-of-flight REMPI spectrum of NO generated from DMNA. Signal generated from 226nm excitation of DMNA (~100μJ).

reveals a prominent peak whose arrival time to the detector is measured as 23.2μsec. This time corresponds to a mass-to-charge ratio (m/e) of 30 and is attributed to the NO^+ ion. Similar results were obtained when using air or nitrogen as the carrier gas. The observed peak is void of any interference and is clearly a result of a REMPI process. No such peak was observed, however, when the laser was tuned to an off-resonance transition.

To verify the mass spectral assignment, an excitation spectrum was obtained by scanning the tunable UV radiation while monitoring the m/c = 30 peak. This spectrum is shown in Figure 5. It reveals numerous rotational lines that are attributed unequivocally to NO $A^2\Sigma^+$ - $X^2\Pi$ transitions.

The detection limits of the various compounds studied employing the fragmentation/REMPI technique that uses 226nm radiation are presented in

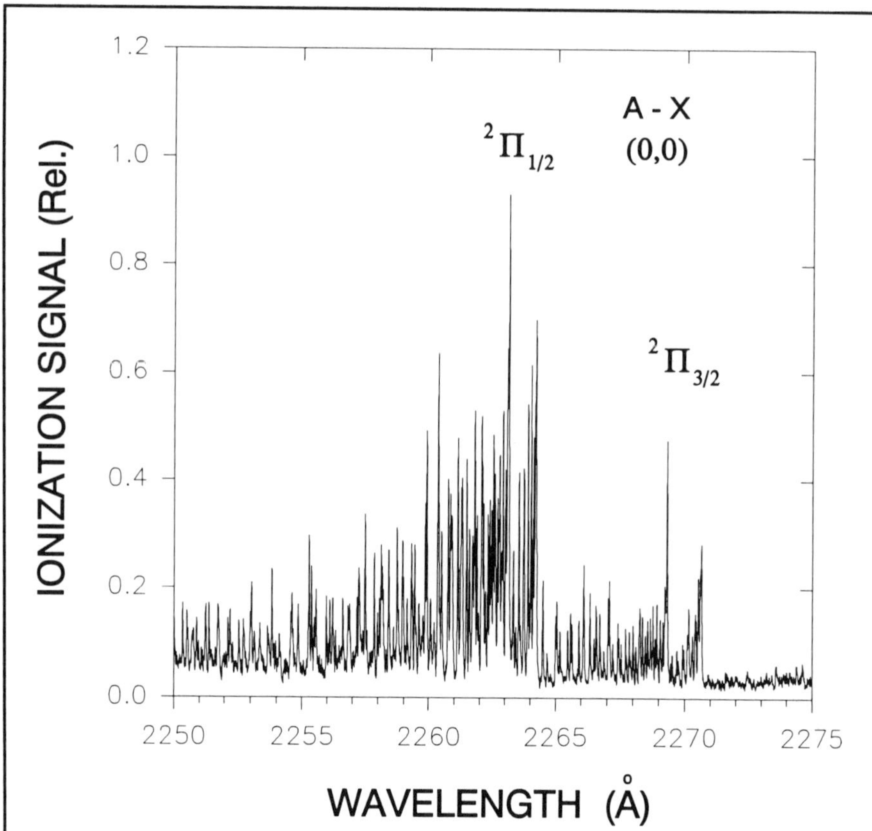

Figure 5. REMPI excitation spectrum of NO in the region of 226nm. NO was generated from a molecular beam of DMNA.

Table 1. REMPI signal production was measured using our present experimental setup, which is far from being optimized, under an identical set of conditions. The limits of detection are defined by the equation:

$$C_L = 3\ \sigma/S \tag{6}$$

where C_L is the concentration (ppb) that produces a signal S (mV/ppb) equal to three times the standard deviation σ(mV) of the background when no analyte is introduced through the nozzle. The limits of detection calculated refer to the gas concentration of the gas prior to introduction into the analysis chamber. Vapor pressure of the compounds were obtained from References 19 and 20 and used to estimate the initial concentration. For one atmosphere

Table 1
Limits of Detection for Various Compounds Studied Using the 226nm Laser Photofragmentation/REMPI Technique with NO Detection

Compound	Limit of Detection (ppb)
NO	8
NO_2	240
CH_3NO_2	1000
DMNA	450
	(2000)*
RDX	8
TNT	24
Nitrobenzene	2400

*Limit of detection obtained by laser-induced fluorescence of NO via its $A^2\Sigma^+$ - $X^2\Pi$ transition.

of backing pressure, the calculated gas volume of the supersonic nozzle (per pulse) is 6.8Torr-cm^3. This corresponds to absolute detection limits of 1 and 3pg for RDX and TNT, respectively, and compares favorably with absolute detection limits reported for TNT that are on order of 200pg.[21] It should be noted that the estimates of the detection limit for the precursors depend on a number of factors, including laser energy, sample introduction, voltage settings of the ion optics and detector, and photodissociation efficiency, all of which affect the rate of signal production. As seen from Table 1, detection limits in the ppb range can be obtained with a relatively small amount of laser energy, approximately 100μJ. The difference in vapor pressure of the various precursors was taken into account when determining these limits. Although the experimental system used for these studies was optimized with respect to these factors in order to obtain the maximum sensitivity possible, the current design of the experimental system could be improved for performing trace analysis. Significant improvements in the sensitivity can be expected with higher laser energies and a supersonic valve/sampling device whose design increases the sample throughput into the photolysis region. The ability to operate the valve at a higher repetition rate (> 50Hz) and elevated temperatures, in order to eliminate condensation of the percursor molecules would also be desirable. A supersonic nozzle meeting these criteria has been designed and demonstrated by Imasaka and co-workers[22] as part of an ionization detector apparatus for gas chromatography. Thus, any difference shown in Table 1 results from the different absorption cross-sections that the target

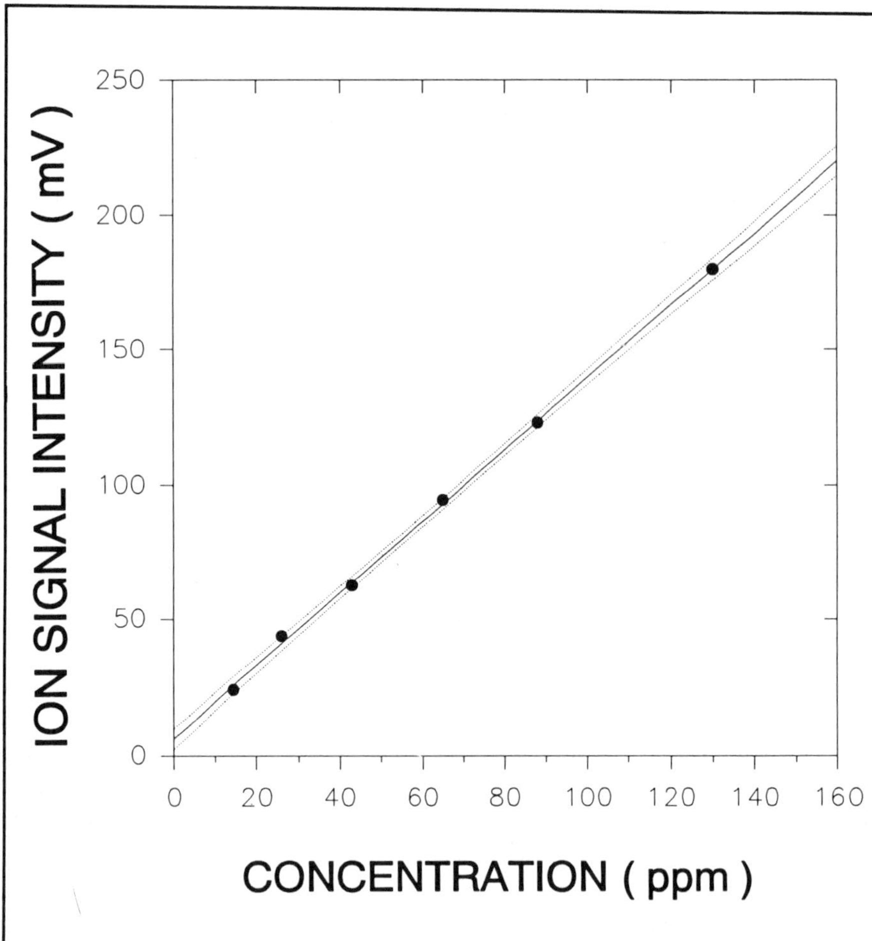

Figure 6. Concentration dependence of NO^+ ion signal. Solid line is a least squares fit of data (•) with 95 percent confidence (...). Its slope yields a sensitivity value of 1.34±0.04mV/ppm (2SD).

molecules have at 226nm and the efficiency in the photodissociative pathways leading to the formation of NO in its $X^2\Pi$ (v'=0) vibronic state. It is clear that RDX, TNT, and NO_2 are particularly efficient in generating REMPI signals. In the case of NO_2, this was not too surprising because Morrison and Grant[15,16] observed that by far the most prominent ion fragment generated by the multiphoton excitation of NO_2 at wavelengths in the region from 425 to 455nm was the NO^+. These researchers' data suggest that the course of excitation in this system is dominated by predissociation into $NO(^2\Pi) + O(^1D)$ at the level of the second photon within the $B(^2B_2)$ state. They contend that this high energy dissociation pathway becomes important soon after the channel becomes energetically accessible.

The selectivity of the laser photodissociation/REMPI technique for the NO^+ in the presence of other species that compete either by forming ions during the REMPI excitation process or from isobaric interferences is given by the following equation:

$$S_T = (R_{EXC}/\Delta r_{EXC})(R_{TOF}/\Delta r_{TOF}) \quad (7)$$

where S_T is the total selectivity, R_{EXC} is the range of UV-VIS spectrum, Δr_{EXC} is the laser bandwidth used in the excitation process, R_{TOF} the range of all possible masses, and Δr_{TOF} is the resolution of the NO^+ ion signal at mass-to-charge (m/e) = 30. A total selectivity value of 5×10^8 is obtained when using values of 500nm and 0.001nm for R_{EXC} and Δr_{EXC}, respectively, and 72μsec (masses 1 to 500amu) and 0.07μsec for R_{TOF} and Δr_{TOF}, respectively.

DMNA was chosen for further study, particularly to investigate the dependence of the concentration and laser energy on signal intensity. Figure 6 shows a plot of concentration versus ion signal intensity of NO^+ from DMNA. The plot reveals a linear relationship, with the slope having a sensitivity value of 1.34 ± 0.04mV/ppm (2SD). This linearity implies that the NO is formed in a collision-free environment and that this technique is suitable for quantification purposes. A detection limit of 450ppb is obtained when using a background value, σ, of 200μV.

The dependence of ion signal intensity on laser energy was also determined using the equation:

$$S_I = C(I_{226})^n \quad (8)$$

where S_I is the ion signal intensity, I_{226} is the laser beam fluence, C is a system-dependent constant, and n the number of photons required to produce the NO^+ signal. The value of n was determined from the slope of a log-log plot of the variables in Equation 8. *A priori*, one might expect the value of n to be as high as 4: two photons required to generate NO in its $X^2\Pi$ state and two photons to ionize it (see Figure 3). However, the slope of the plot yielded a value of $n = 1.62 \pm .01$ (2SD), indicating that the formation of the ion requires at least two photons. This near quadratic dependence suggests that one or more of the intermediate photochemical steps necessary to generate NO^+ from $R\text{-}NO_2$ is saturated. It also suggests that a significant enhancement in the signal can be achieved at higher laser energies.

For comparison, NO was also monitored by LIF via its $A^2\Sigma^+ - X^2\Pi$ (0,0) band following the photolysis of DMNA with 226nm radiation. The limit of detection was estimated as 2ppm, a factor of approximately four worse than that obtained with the REMPI technique. This difference in sensitivity results

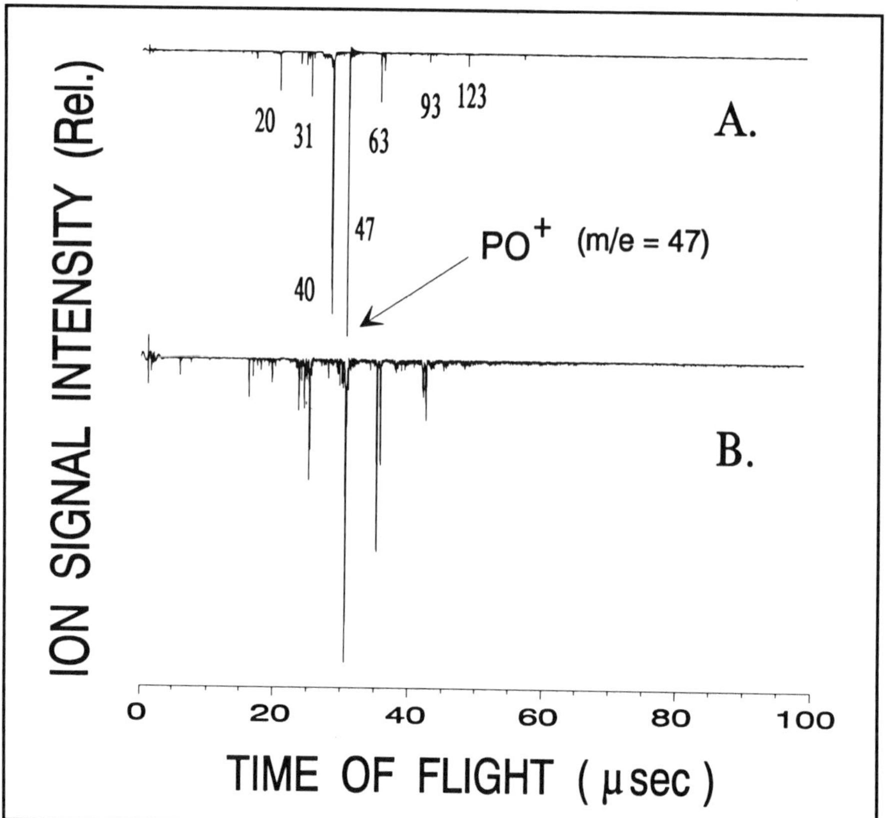

Figure 7. Time-of-flight spectra of DEMPS using 193nm radiation: (A) molecular beam, (B) effusive flow.

mainly from the higher background noise inherent in the LIF measurement. The selectivity of the LIF measurement was also found to be lower than the REMPI measurement. This is attributed to the relatively large bandwidth of the monochromator and interference used, and is not necessarily characteristic of the LIF technique. The large bandwidth was required in order to maintain a sufficient level of sensitivity.

Chemical agent simulant. The CW agent simulant DEMPS was also studied using the MB/MS apparatus. DEMPS was excited using UV radiation resonant with the PO $A^2\Sigma^+ - X^2\Pi$ transition near 247nm. Under our experimental conditions, the PO radical could not be detected. Focusing the laser beam was to no avail. This is not too surprising because the absorption cross-section of DEMPS is relatively small at this particular wavelength, and the laser energy used, approximately 100μJ, is a factor of 10 to 100 less than that typically required for the photofragmentation of other organophosphonates.[1,8] The PO radical was detected, however, when using 193nm radiation with a pulse energy of ~10mJ. The spectra obtained under molecular beam and effusive flow conditions are presented in Figures 7A and 7B, respectively. At this wavelength, the ionization process competes well with photodissociation and, as a result, fragment ions are produced. In both spectra the peak located at 29.0μsec, corresponding to a m/e of 47, is attributed to the PO^+. In Figure 7A, the prominent peaks at 23.6, 33.6, and 47.0μsec correspond to m/e values of 31, 63, and 123, respectively. These values are due to the fragmentation of the parent ion and are assigned to P^+ (m/e = 31), PS^+ and/or PO_2^+ (m/e = 63), and $[CH_3\text{-}(PS)\text{-}O\text{-}C_2H_5]^+$ (m/e = 123). The two peaks at 18.9 and 26.8μsec correspond to m/e values of 40 and 20, respectively. They are attributed to Ar^+ and Ar^{2+}, respectively, generated by the ionization of the carrier gas. The limit of detection for DEMPS employing the PO^+ was calculated as 6ppm.

Conclusion

A novel ultrasensitive detector for trace NO_2 containing compounds has been developed at the U.S. Army Research Laboratory for the purpose of monitoring energetic materials such as propellants and explosives. It employs one laser operating at 226nm and is based on the photofragmentation of the target molecule, with subsequent detection of NO by REMPI and/or LIF via its $A^2\Sigma^+$ - $X^2\Pi$ (0,0) band origin. Using an MB/TOF apparatus, the analytic utility has been demonstrated on a number of compounds, including NO_2, DMNA, TNT, and RDX. Detection limits in the ppb range were shown for RDX and TNT with extremely low laser energies, ~100 μJ/pulse. The ultimate sensitivity is no doubt much better than that shown here. Sample introduction and background noise could also be improved, thus yielding a lower limit of detection.

Ionization using 193nm radiation coupled with MB/TOF spectrometric detection was employed in the detection of trace atmospheric vapors of DEMPS by monitoring the characteristic PO^+ fragment. A detection limit in the ppm range was demonstrated. Ultimate sensitivity is projected in the sub-ppb with the application of higher laser energy. It is difficult, however, to predict the results for CW agents because their photoionization/fragmentation patterns may be quite different. Thus, direct examination will ultimately be required.

Acknowledgments

G.W. Lemire and J.B. Simeonsson gratefully acknowledge support from the U.S. Army Research Laboratory Postdoctoral Research Associateship

Program administered by the National Research Council, and R.C. Sausa acknowledges support from the PIF/OSD Capital Investment Program and Combustion Mission Research Program. We thank S. Bulusu (U.S. Army Research, Development and Engineering Center) and R.S. Long (U.S. Army Chemical Research and Development Center) for providing us with some of the sample compounds, and B.E. Forch (U.S. Army Research Laboratory) for helpful discussions.

References

1. R.C. Sausa, A.W. Miziolek, and S.R. Long, *J. Phys. Chem.* **90**, 3994 (1986).
2. R.C. Sausa, A.J. Alfano, and A.W. Miziolek, *Applied Optics* **26**(17), 3588 (1987).
3. M.J. McQuaid *et al., J. Phys. Chem.* **95**, 2713 (1991).
4. P.J. Dagdigian *et al., J. Phys. Chem.* **93**, 6059 (1989).
5. J. Schendel, R. Hohmann, and E.L. Wehry, *Applied Spectroscopy* **41**(4), 640 (1987).
6. E. Wehry *et al., Applied Optics* **26**(17), 3559 (1987).
7. D.B. Moss, K.A. Trentelman, and P.L. Houston, *J. Chem. Phys.* **96**(1), 237 (1992), and references therein.
8. S.R. Long, R.C. Sausa, and A.W. Miziolek, *Chem. Phys. Lett.* **117**, 505 (1985).
9. A.W. Miziolek *et al., US Army Ballistic Research Laboratory Technical Report*, (document BRL-TR-2792, 1987).
10. M.J. McQuaid and R.C. Sausa, *Applied Spectroscopy* **45**(5), 916 (1991).
11. K.J. Smit, *Journal of Energetic Materials* **9**, 81 (1991).
12. C. Capellos, P. Papagiannakopoulos, and Y. Liang, *Chem. Phys. Lett.* **164**, 533 (1989).
13. J.-C. Mialocq and J.C. Stephenson, *Chem. Phys. Lett.* **123**, 390 (1986).
14. V.M. Donnelly and F. Kaufman, *J. Chem. Phys.* **69**(4), 1456 (1978).
15. R.J.S. Morrison and E.R. Grant, *J. Chem. Phys.* **77**(12), 5994 (1982).
16. R.J.S. Morrison, B.H. Rockney, and E.R. Grant, *J. Chem. Phys.* **75**, 2643 (1981).
17. C.B. McKendrick, C. Fotakis, and R.J. Donovan, *J. Photochemistry* **20**, 175 (1982).
18. J.A. Syage, J.E. Pollard, and R.B. Cohen, Space Division Air Force Systems Command, Report-SD-TR-88-13, February 1988.
19. *Handbook of Chemistry and Physics*, 66th ed. (CRC Press, Inc, Boca Raton, FL, 1986), pp. D196-D212.
20. B.M. Dobratz, *Properties of Chemicals and Explosives Handbook*, LLNL Report-UCRL-52997, March 1981.
21. S.D. Huang, L. Kolaitis, and D.M. Lubman, *Applied Spectroscopy* **41**(8), 1371 (1987).
22. T. Imasaka, T. Okamura, and N. Ishibashi, *Anal. Chem.* **58**, 2152 (1986).

35

High Bandwidth, Wide Field of View Lasercom Demonstration

John C. Johnson
U.S. Army Space and
Strategic Command
Huntsville, AL 35807-3801

Laser satellite networking is a key element of effective communications operations to support both strategic and tactical missions. Lasercom offers a number of important advantages over conventional radiofrequency (RF) satellite communications. The shorter wavelength available using lasers provides higher data rates at less power and smaller apertures, both resulting in lower weight requirements. On the other hand, lasercom entails more difficulty in acquisition and tracking because of the narrow beams used. Technology problems to overcome before intersatellite laser communications can reliably outperform RF communications include acquisition in the presence of significant background light from the earth, tracking to resolutions of a few microradians, high-speed modulation of semiconductor lasers with close to one watt of power in a diffraction limited beam, high bandwidth, low noise detector response, and demonstrated long-term performance. We have developed critical technologies to solve some of these problems and demonstrated them in a laboratory testbed that also supports development and testing of network protocols and algorithms. Our hardware provides new capability in background light rejection by using innovative atomic line filter technology, improved tracking accuracy by using innovative zero backlash Roto-Lok drive gimbal telescopes, and increased communications bandwidth by incorporating multi-link networking protocols.

Introduction

For communications between satellites, higher data rates (> 1 gigabits/sec) can be achieved using optical laser communications than is possible using conventional radio frequency (RF) transceivers (approximately 20 megabits/sec). In addition, these data rates can be achieved by using less power and smaller mirrors (and thus at less weight) by using optical frequencies instead of RF carriers for the signal. The reason for these higher data rates is because of the divergence of the beams, which is proportional to the wavelength in the diffraction limit, and can be made much smaller for the laser transmitter. The laser has a wavelength on the order of one ten thousandth that of the RF signal. The disadvantage of using narrow beams is that acquisition and tracking become more difficult because only very small pointing errors can be tolerated.

Small laser divergence results in smaller size, weight, and power as well as a lower probability of intercept and a high resistance to jamming.

A typical RF system requires a 3-m dish antenna with 150W of transmit power. The overall RF system would weigh approximately 340lb and require on the order of 350W. A typical laser system would require a relatively small telescope (0.15m) and a relatively low transmit power (1W), with a data rate capability of 1000 megabits/sec. The overall laser system would weigh approximately 150lb and require on the order of 100W.

The overall program goals and achievements are summarized in Figure 1. The primary objectives were to demonstrate several critical technologies that would enhance the feasibility of fielding a laser satellite communications network. The first innovation is acquisition in background light using an Atomic Line Filter. The second innovation is the use of high data rate simultaneous video and digital data. The third innovation is the use of a lightweight, low-cost Roto-Lok drive gimbal. The fourth innovation is the development of a high bandwidth, high accuracy tracking system.

Following the demonstration of these critical technologies, an additional goal was to integrate all the critical technology components into a laboratory testbed for the purpose of demonstrating the simultaneous operation of all necessary components. This testbed was designed and laid out such that upgrades to the system are easily accommodated. The testbed was also designed to accommodate networking protocols and algorithms. The testbed, as shown in Figure 2, consisted of three nodes with laser transceivers and two additional nodes that were hard-wired to simulate laser transceivers for the purpose of evaluating network protocols and algorithms.

The optical path of the laboratory laser transceiver is shown in Figure 3.

GOAL	PARAMETER	SOW SPECS	PROGRAM ACHIEVEMENT
• Demonstrate Critical Technologies	Field of View	10 mrad	14 x 19 mrad
– Acquisition in background light using Atomic Line Filter	$\Delta\lambda$ Bandwidth	0.002 nm	0.0024 nm
– High data rate optical communications (diode laser, APD) (U of D scheme for simultaneous video and digital)	Efficiency	---	0.25
	Video Carrier Frequency	1.2 GHz	1.206, 1.212 GHz
	Simultaneous Data	50 MBPS	50 MBPS
	Separate Digital Data	---	1 GBPS
– Lightweight, low cost Roto-Lok drive gimbal			
– High bandwidth, high accuracy tracking	Open Loop Pointing	100 µrad	80 µrad
	Closed Loop Pointing	10 µrad	7 µrad
	Bandwidth	---	> 6 Hz
• Integrate Laboratory Test Bed			
– Demonstrate simultaneous operation of necessary components	# of Transceivers	3	3
	# of Nodes	5	5
– Provide facility to incorporate upgrades	# of Links	10	10
• Support networking protocols and algorithms	U of F Tested Algorithms	---	3

Figure 1. Program goals and achievements.

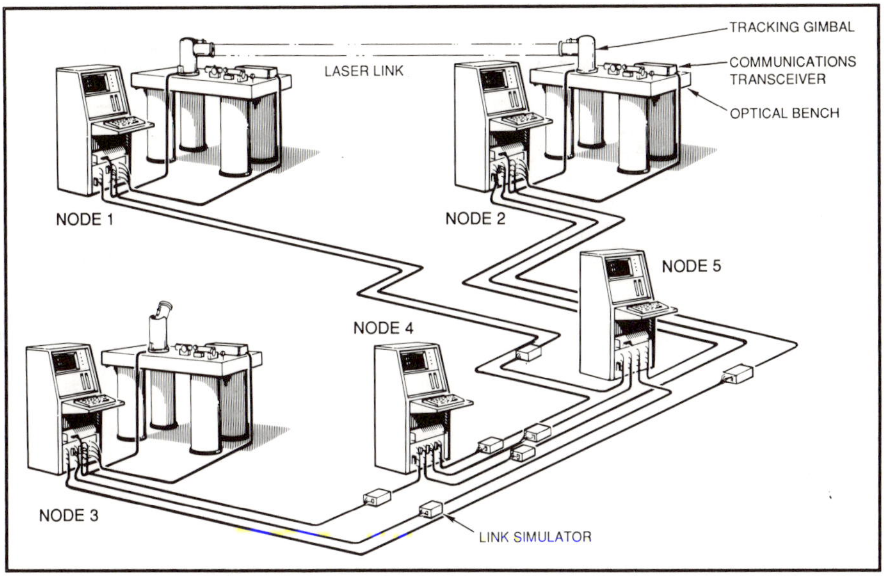

Figure 2. Interactive laboratory demonstration testbed.

A beacon for acquisition and tracking diverges from a small aperture and has a wavelength of 852nm matching an atomic line filter. The tracking signal is received through a 4-in aperture and locked onto with wide and narrow field of view cameras. The 4-in aperture is shared with the transmitted and received communications beams at 780nm, which are separated from the 852nm tracking signal using a dichroic mirror. The communications rate versus range for the laboratory transceiver is shown in Figure 4. Increased data rates available are shown in Figure 4, dashed line, if a higher power laser is used or

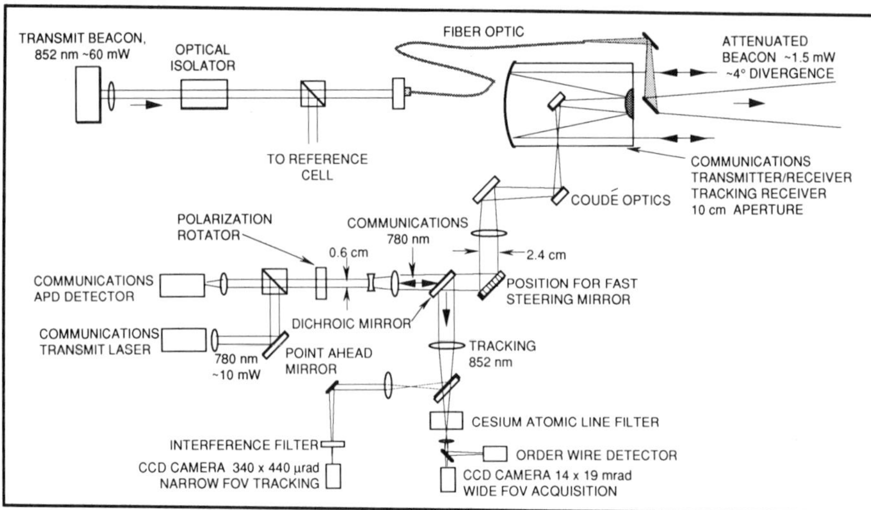

Figure 3. Laboratiory transceiver optical path.

ASSUMPTIONS:

10 cm telescope, area A = 0.006 m^2, diffraction limited divergence θ = 14 µrad

Transmitter peak power P_o = 200 mW [currently realistic, 850 nm]

Transmit efficiency η_o = 0.5, receive efficiency η_r = 0.5

Mispointing efficiency η_p = 0.5 (5 µrad mispoint)

Avalanche photodiode detector with 60 photons/bit for 10^{-8} BER [currently realistic at 10-50 MBPS]

EQUATION:

$$\text{Data rate} = \frac{P_o \eta_o \eta_r \eta_p A}{\pi(\theta/2)^2 R^2} \times \frac{4.2 \times 10^{18}\ \text{photon}}{\text{W} \cdot \text{sec}} \times \frac{1\ \text{MBit}}{60 \times 10^6\ \text{photons}} = \frac{6.82 \times 10^{10}\ \text{MBPS km}^2}{R^2}$$

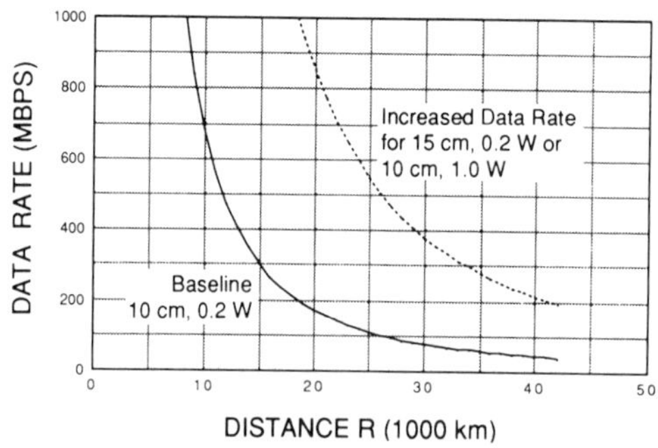

Figure 4. Communications rate versus range for laboratory transceiver (solid line) upgraded to currently available 200mW peak power transmit diode laser. Increased data rates (dashed line) are possible with a higher power laser of larger aperture.

a larger aperture is used. Data rates of one gigabits/sec are possible at ranges of greater than 10,000km.

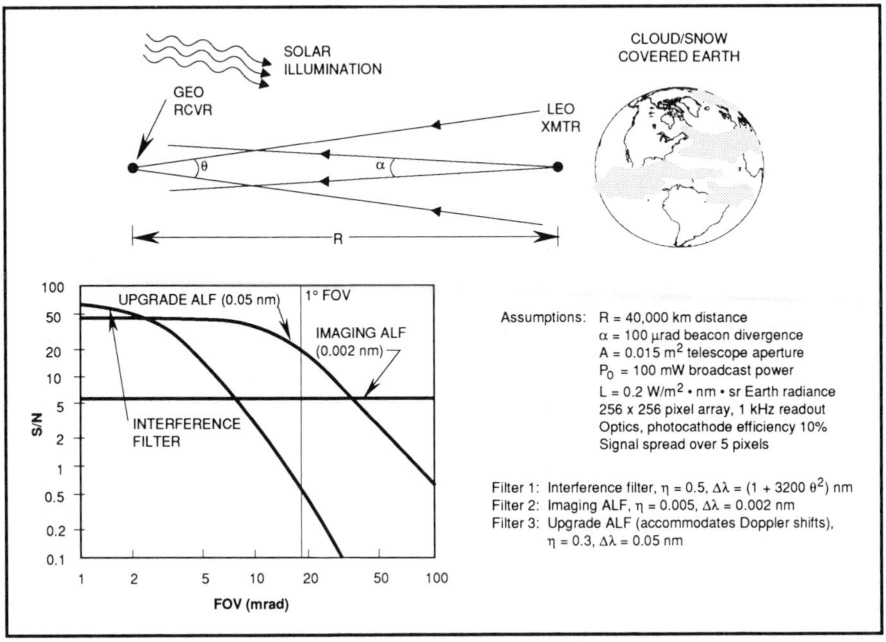

Figure 5. Increased acquisition field of view with atomic line filter. Background light from the sunlit earth is blocked 100 times better than with an interference filter, allowing a 10 times wider instantaneous FOV.

Technical Approach

The three major technical requirements for a free space laser communications system are the ability to do fast acquisition, high accuracy tracking, and high data rate, high reliability communications. Fast acquisition is necessary in a network because links must be broken and reformed as satellites travel in different orbits around the earth, and acquisition should occupy a small fraction of the total link time. Our approach to acquisition is to open up the receiver field of view to about 1° so it is large enough to view the acquisition beacon from the other satellite based on initial ephemeris information without excessive scanning. To have such a wide field of view with the earth in the background requires the use of an extremely narrow bandwidth filter, such as the atomic line filter demonstrated in this program. For high data rate, high reliability communications, we choose direct detection of a modulated laser diode using a low noise avalanche photodiode. Finally, for high accuracy pointing and tracking, we have incorporated the zero backlash Roto-Lok drive.

The background light problem to be overcome in the acquisition stage is illustrated in Figure 5. One satellite (for instance, in geosynchronous orbit) is trying to acquire the tracking beacon of another satellite (in low earth orbit) with the sunlit earth in the background. The instantaneous field of view (FOV) that can be stared at with the acquisition charge coupled device (CCD) camera depends on the amount of light from the beacon, compared with spatial variations in the amount of light from the earth, reflected off of bright objects such as clouds or dark objects such as the ocean. By minimizing the amount of this background reflected light getting to the receiver, the field of view can be maximized. Signal to noise ratios for a given camera pixel are compared for three different light rejection filters. A background earth radiance (clouds) of $0.2 W/m^2$ nm sr is assumed.[1] For subpixel accuracy pointing, a signal to noise ratio (s/n) of better than seven is desired. For the baseline parameters given in the figure, the interference filter (bandwidths ~2nm) gives an s/n of 0.5 at the desired 1°FOV. Atomic line filters[2] can reduce the light transmission bandwidth by a factor of 1000 to 0.002nm. One image preserving line filter developed under this program achieved this bandwidth.[3] Disadvantages of this filter include the necessity for an extra pump laser and a low efficiency of 0.5 percent. Ideally, the filter would have a high throughput efficiency, be simple, and have a bandwidth close to 0.05nm so that the transmit beacon wavelength would not have to be adjusted for relative Doppler shifts between satellites. We have identified a filter that meets these goals and gives an s/n of seven out to a FOV of 2°. It is called a Faraday atomic line filter.[4]

Operational principles of the Faraday filter can be understood by reference to Figure 6. Basically, it consists of a sandwich comprised of two crossed polarizers on the outside and a cesium vapor cell in a magnetic field on the inside. Except at a specific wavelength (852nm) corresponding to an absorption in the cesium vapor, background light is unaffected by the vapor cell and effectively blocked by the crossed polarizers. For wavelengths very near the 852-nm absorption, however, the polarization of the light can be rotated in the vapor cell, thus allowing it to pass through the second polarizer. The path of the light is unaffected, so spatial and temporal information is maintained. The basis for the polarization rotation is the resonant Faraday effect. Because of the applied magnetic field, the index of refraction (and thus propagation velocity) is different for right and left circularly polarized light. A phase delay of 180° is equivalent to a linear polarization rotation of 90°.

Measured and calculated transmission spectra for a Faraday filter are shown in Figure 7. We measured a transmission efficiency of 48 percent for polarized light. (Because of Fresnel reflections, for which AR coatings could be, but were not applied, the optics themselves only had a throughput efficiency of 54 percent.) The total integrated bandwidth for background light was 0.007nm for temperature T=80°C, magnetic field B=100G, and path

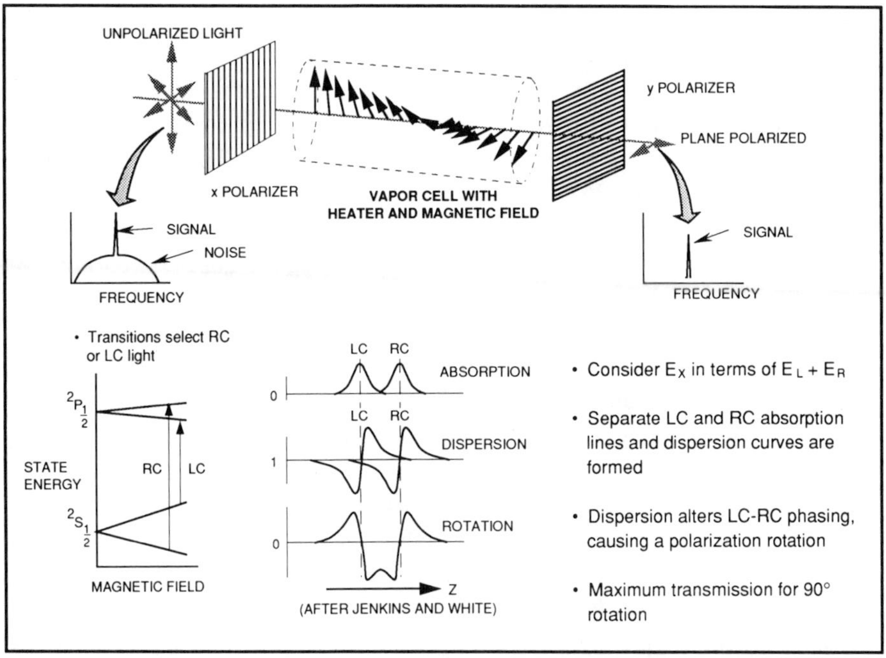

Figure 6. Principles of Faraday atomic line filter operation. Crossed polarizers block background light, while the polarization of the signal wavelength is rotated, allowing it to pass through the filter.

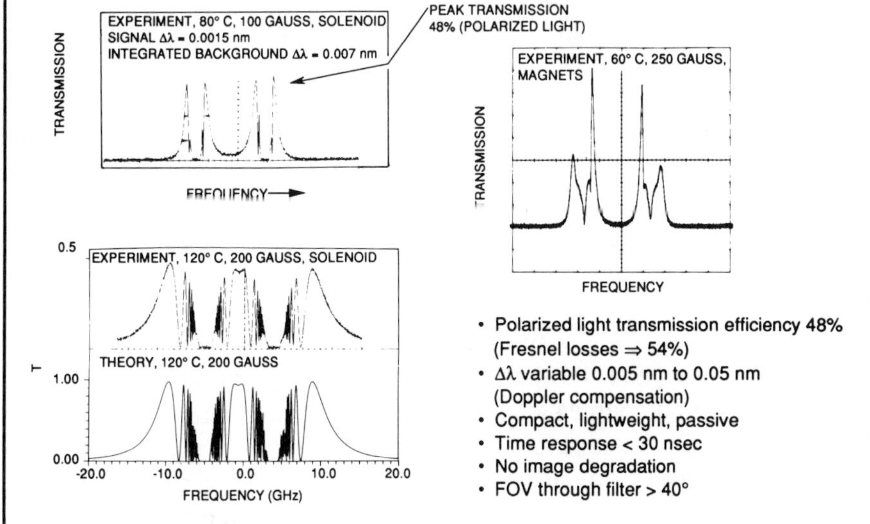

Figure 7. Performance results of Faraday filter. The transmission spectrum is a function of cesium temperature and magnetic field. Our experimental and theoretic spectra agree in detail.

length L=1.2cm. As shown for 120°C and 200G, the transmission spectrum is affected by the temperature and magnetic field. Our experimental results and theoretic calculations agree in detail. Our calculations indicate that a central flat passband about 6GHz wide can be made under appropriate conditions. Intersatellite Doppler compensation requires about ±6GHz, so the transmit beacon wavelength would only need to toggle between a "moving away" and "coming toward" position. For space applications, permanent magnets would be used, as for the 60°C, 250G spectrum in the upper right hand corner.

For efficient high data rate communications (minimizing transmitter power requirements), we plan to use digital modulation of the transmit laser, and direct detection using an avalanche photodiode. An atomic line filter is unnecessary for the communications channel because the transmitter is more collimated (than for the acquisition beacon), and the receiver field of view is small. We were able to reliably transmit and receive a repeating digital waveform at one gigabit per second in the laboratory to demonstrate laser modulation and detection at these speeds. For demonstration purposes, we simultaneously transmitted digital communications at 10 megabits per second (used to test networking algorithms) and video signals upconverted to a 1.2GHz carrier frequency, as shown in Figure 8. By monitoring the video channel, we could see pointing, acquisition, tracking, and establishment of high data rate communications in a time scale of about two seconds.

For very high data rates over long ranges, the transmitted communications laser divergence must be minimized, and the transceiver telescope must be pointed accurately within a few microradians. For accurate closed loop tracking in the presence of satellite jitter, a very low backlash, very high stiffness steering mechanism must be used. Possible choices include a direct drive motor, an offset gear drive, and the Roto-Lok cable drive, which we have chosen. A direct drive requires very high torque (and therefore weight and power) and imparts considerable angular momentum to the host structure. A gear drive is very difficult to make backlash free and will impart jitter to the platform. The Roto-Lok drive overcomes the problems inherent in the other drive systems.

Figure 9 shows a cutaway view of the Roto-Lok drive gimbal developed under this program and shows the principles of its operation. A drive motor shaft (or capstan) is coupled to a large moving circular platform (drum) with many cables (eight for our transceivers). The cables are attached to the drum at their ends, wrapped around the drum, and once around the capstan. Friction keeps the cable from slipping on the drum or capstan, so that as the capstan turns, the drum turns with it. Because there are many cables, any imperfections on individual cables are averaged out, minimizing backlash well below

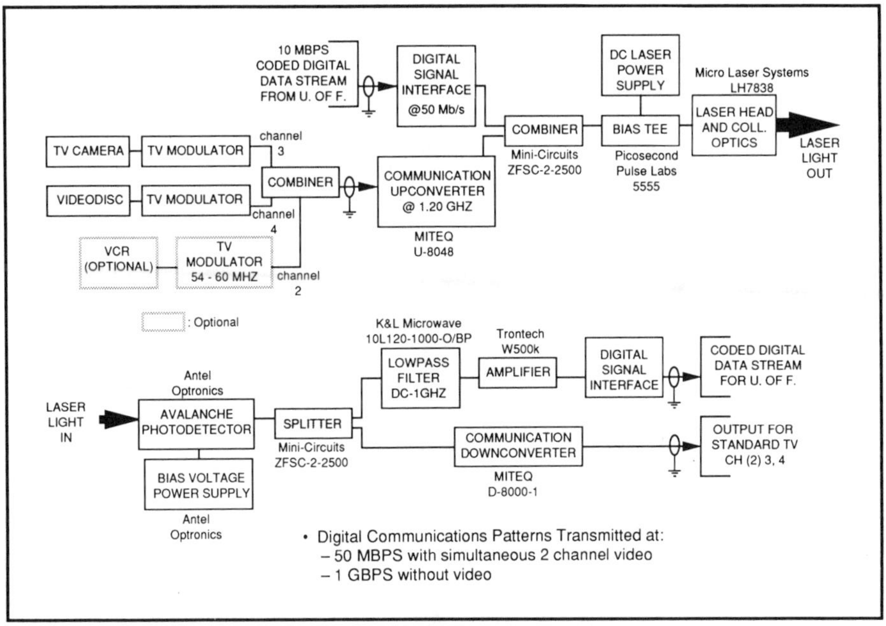

Figure 8. Communications system block diagram. Digital data and upconverted video signals are transmitted simultaneously.

that of gear driven systems with no weight penalty. Cable stretch is minimized because the free span between friction points is very small. Finally, the capstan and drum turn in opposite directions at different speeds determined by the turndown ratio of their relative diameters. By designing the relative moment of inertia to be in this same ratio, a reactionless drive system imparting no angular momentum to the host satellite platform can be built.

The tracking control system used in the laboratory testbed is described in Figure 10. Initial pointing was done with reference to a shaft position encoder on the drive motor, with an angular position resolution of 65μrad (later upgraded to 1.25μrad). After acquiring the beacon laser (in the presence of simulated background light), closed loop tracking was accomplished by centroiding on a CCD camera with a 1° FOV behind an atomic line filter. Fine tracking was done with a second camera with a 400μrad FOV. Tracking performance is shown in Figure 11. With the high resolution shaft position encoder, a tracking accuracy of 7μrad at a bandwidth of about 6Hz was achieved, compensating for atmospheric fluctuations in the room. The bandwidth was probably limited by the 60Hz camera readout rather than the Roto-Lok drive. In further tests incorporating a fast steering mirror and a quadrature avalanche photodiode detector, tracking accuracies of about 1μrad were

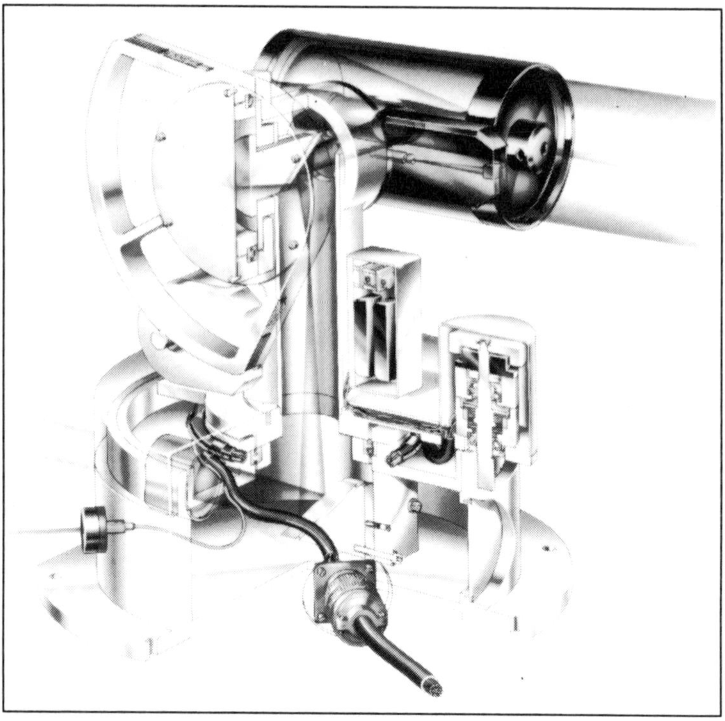

- Low backlash, high stiffness
- The capstan is placed near the drum to increase stiffness by reducing the span of cable

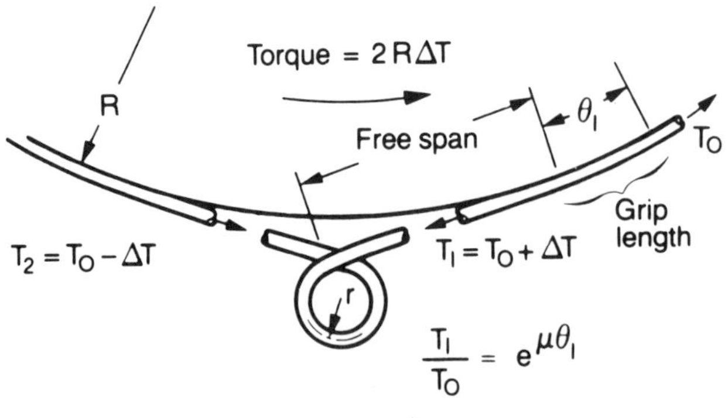

- Beyond the angle θ_1, the cable is unaffected by the load T_1

Figure 9. Roto-Lok drive gimbal telescope principles. Multiple cables wound around the motor shaft (radius r) and the drum (radius R) couple them together in a near zero backlash configuration.

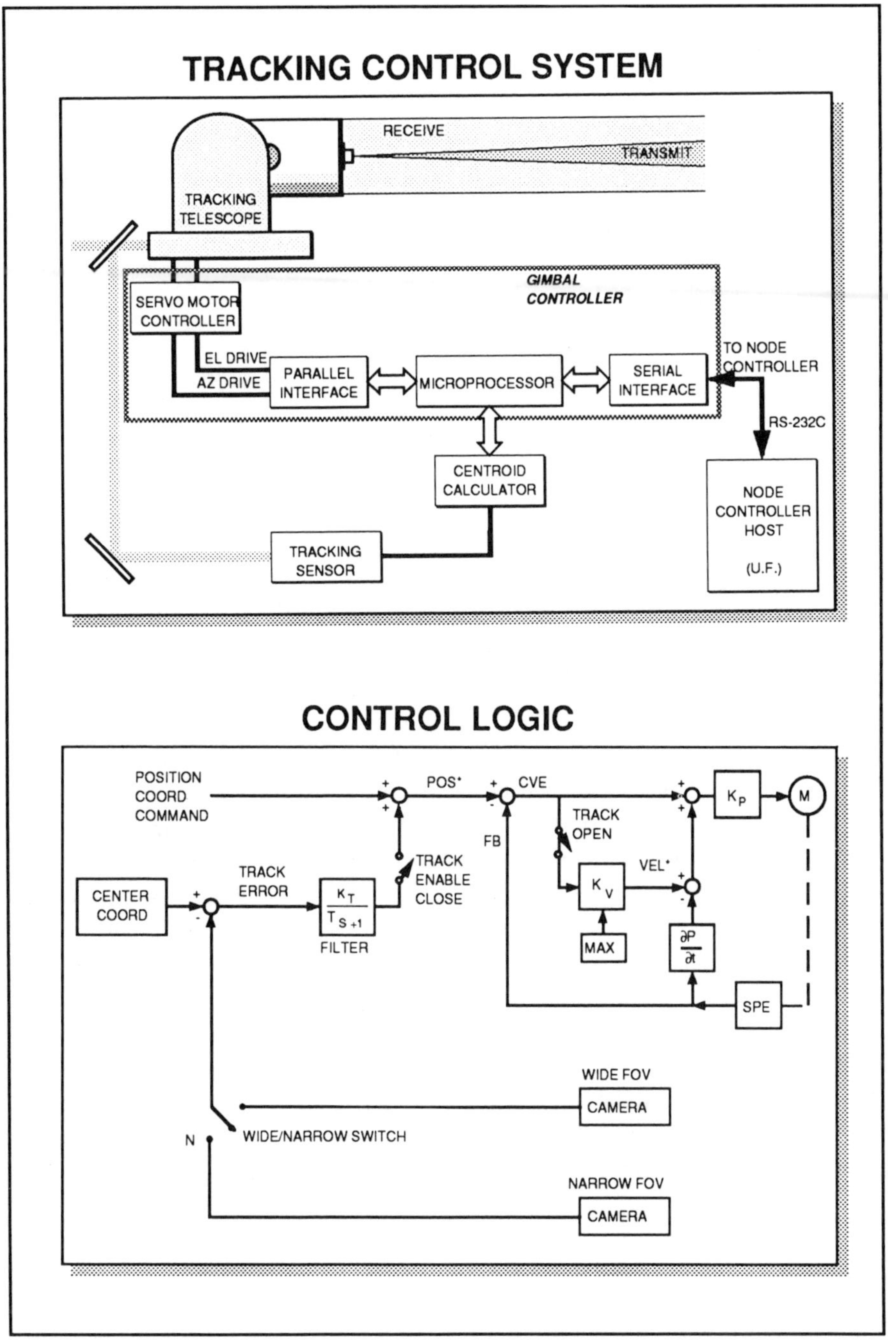

Figure 10. Tracking control system. Control loop uses centroiding offsets from the wide or narrow FOV cameras to null the gimbal to the appropriate shaft position encoder count.

PARAMETER	S.O.W.	TRANSCEIVER 1 &2	HIGH RESOLUTION TRANSCEIVER 3
Wide Acquisition Field Of View	10 mrad	14 x 19 mrad	14 x 19 mrad
Narrow Tracking Field Of View	–	400 µrad	900 µrad
Gimbal Pointing Accuracy (Open Loop)	100 µrad	~400 µrad	~80 µrad
Gimbal Pointing Jitter (Open Loop)	–	± 65 µrad	± 1.25 µrad
Tracking Bandwidth	–	~6 Hz	~6 Hz UPGRADE TO 100 Hz
Tracking System Accuracy (Closed Loop)	10 µrad	~65 µrad	~7 µrad

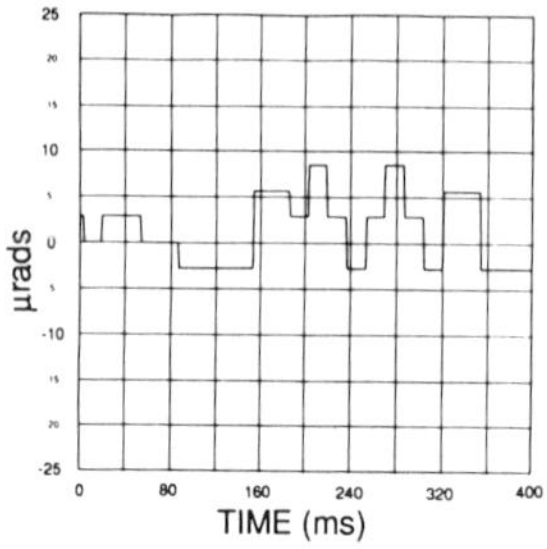

OPEN LOOP ATMOSPHERIC JITTER

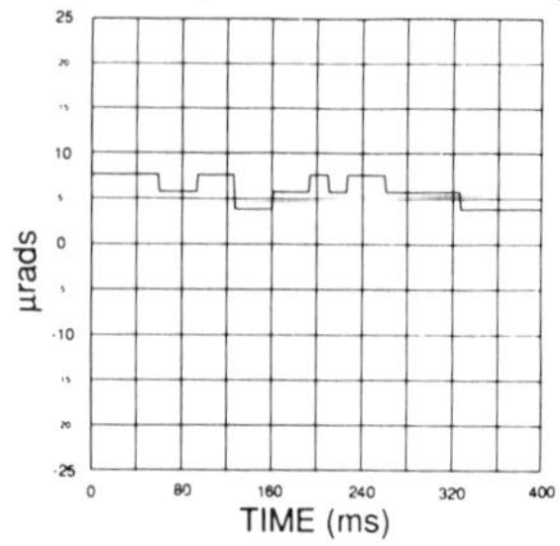

CLOSED LOOP TRACKING

Figure 11. Tracking system performance. With the narrow field of view camera and 1.25µrad resolution shaft position encoder, better than 7µrad pointing accuracy was achieved at 6Hz.

achieved at bandwidths of about 100Hz. Such a system may be needed for 1GBPS communications at 42,000km (geosynchronous) ranges. However, as described later, useful data rates and ranges should be achievable with much less complexity and at lower weight using the Roto-Lok drive gimbal alone.

A final technology issue that needed to be solved for our atomic line filter based system was stabilization of the diode laser acquisition beacon to the wavelength passed by the filter. This can be accomplished by tuning the laser by changing its temperature and current, and closing an electronic control loop based on transmission through a reference filter. However, the temperature and current resolutions necessary are much less severe if optical feedback, as shown in Figure 12, is used. If some of the laser light is diverted through a Faraday filter (matching the receiver of interest) and reflected back into the diode laser, it can automatically stabilize the laser at the wavelength of interest. The figure shows results of a heterodyne experiment showing such locking taking place on one of the Faraday filter transmission peaks.

Conclusions

Laser communications shows great promise for achieving high data rate, low probability of intercept links between satellites. A network of

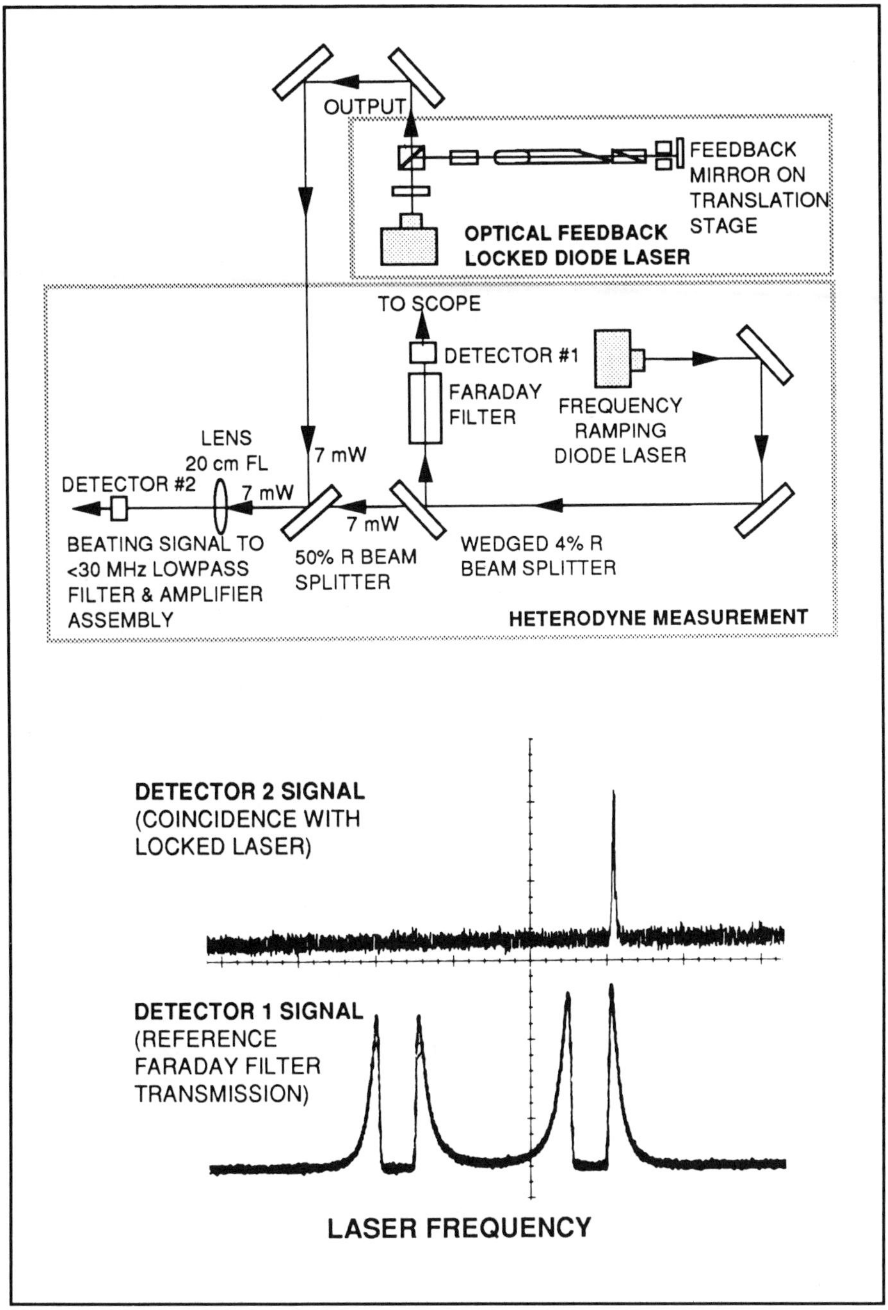

Figure 12. Diode laser wavelength stabilization with optical feedback. Laser wavelength is matched to the atomic line filter passband using optical feedback through a second filter that is part of the laser cavity.

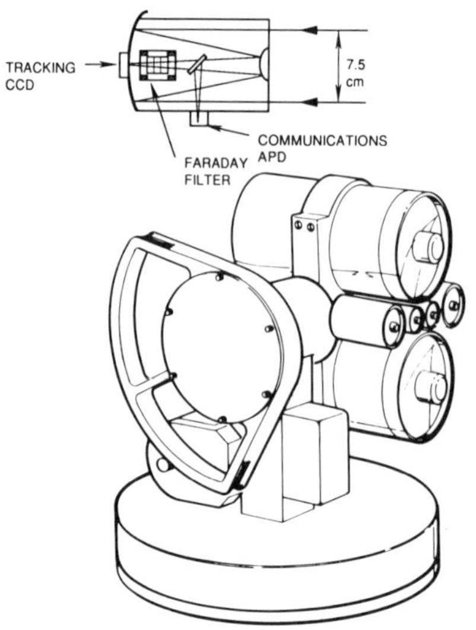

- Wide field of regard gimbal
- Doubly redundant
- No shared optics between transmit and receive, no Coude optics
- Easy to integrate
- 1.2 cm tracking beacon, 50 mW, 1 mrad divergence
- 2.5 cm communications laser, 100 mW, 100 μrad divergence
- 7.5 cm receiver
  - Tracking through 0.05 nm ALF
    256 x 256 pixel, 200 Hz CCD
    20 mrad IFOV, 40 μrad tracking
  - Communications through APD
    200 μrad FOV
  - Point ahead done with CCD centroiding
- Data rate 20 MBPS at 5000 km
- Weight ~ 40 lbs
  Power ~ 40 W

Figure 13. High reliability, lightweight transceiver prototype. Doubly redundant system for reliable lasercom demonstration, with greatly simplified optical path (no shared transmit/receive optics).

surveillance and communications satellites incorporating lasercom could reliably and rapidly transmit critical information around the world, as well as to and from high flying aircraft with short range radio links to the ground. Lasercom promises data rates above 100MBPS at significantly lower power and weight (and therefore launch cost) than is possible with radio frequencies. In this program we have demonstrated innovative technologies that address some of the technical issues associated with intersatellite lasercom, including rapid acquisition, even in the presence of background light, high accuracy pointing and tracking, and high data rate communications. Because of these accomplishments, free space laser communications is ready to come out of the laboratory for implementation, first in aircraft to aircraft links where high data rates are needed, and then in space.

To demonstrate the feasibility of lasercom in comparison with radio frequencies, reliability will be of critical importance. Therefore, a prototype deployable system should be simple and redundant, demonstrating significant data rates (20MBPS) at significant range (5000km) in a compact, lightweight package. Figure 13 shows a functional design for such a system. First, optics are not shared between transmitter and receiver, eliminating complicated systems to avoid cross-talk. Second, the doubly redundant communications and acquisition transmitters have increased divergences (100μrad and 1mrad,

ASSUMPTIONS:
Double redundancy, calculate for single functioning components
7.5 cm receive telescope, area A = 0.004 m^2
Transmit Divergence θ = 100 μrad
Transmitter peak power P_o = 100 mW [currently realistic, 850 nm]
Transmit efficiency η_o = 0.5, receive efficiency η_r = 0.5
Mispointing efficiency η_p = 0.5 (42 μrad mispoint)
Avalanche photodiode detector with 60 photons/bit for 10^{-8} BER [currently realistic at 10-50 MBPS]

EQUATION:

$$\text{Data rate} = \frac{P_o \eta_o \eta_r \eta_p A}{\pi(\theta/2)^2 R^2} \times \frac{4.2 \times 10^{18}\ \text{photon}}{\text{W} \cdot \text{sec}} \times \frac{1\ \text{MBit}}{60 \times 10^6\ \text{photons}} = \frac{4.46 \times 10^8\ \text{MBPS km}^2}{R^2}$$

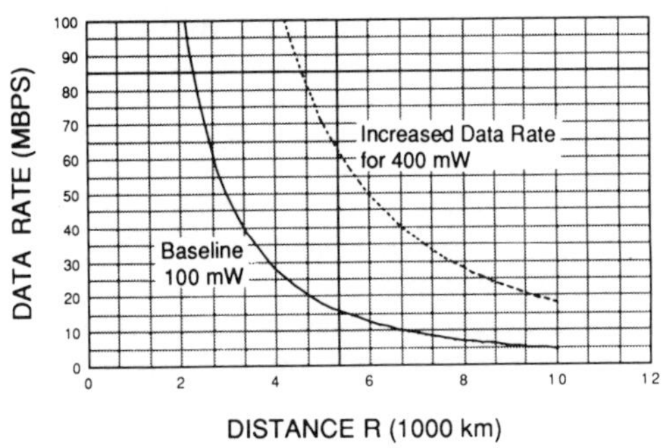

Figure 14. Data rate versus range for lightweight transceiver prototype.

respectively) to reduce pointing accuracy requirements to 40μrad. This can be accomplished with a 1° FOV, 256 x 256 pixel, 200Hz readout CCD coupled to the Roto-Lok drive, eliminating the need for a complicated tracking system. (Subpixel pointing accuracy is easily achieved.) Because of increased laser divergences, small transmitter optics can be used. All lasers and detectors can be mounted directly on the gimbal telescope, increasing efficiency, and reducing interface complexity. Heat can be pumped to the cold side of the gimbal using heat pipe technology. Lightweight permanent magnet Faraday atomic line filters will be used for background light rejection for acquisition and tracking. Finally, the gimbal telescope system will be made reactionless using the Roto-Lok drive technology, so that minimum angular momentum jitter is transferred to the host satellite. Data rate versus range for the transceiver prototype described in Figure 13 is shown in Figure 14. All components are currently available commercially, with the rated noise figures. Although the proposed system is doubly redundant, the data rate is based on only one set of components being functional and would be higher if all components were active.

Acknowledgments

Funding for this program was provided by the Strategic Defense Initiative Organization, Innovative Science and Technology Directorate. We would like to thank Kepi Wu for his encouragement and support. The atomic line filter development and testbed integration was done by Thermo Electron Technologies Corporation, San Diego, under the direction of C.S. Liu and Eric Korevaar. The high data rate communications scheme was provided by Charles Ih of the University of Delaware. The Roto-Lok drive gimbal was developed by Don Carson of Sagebrush Technology, Albuquerque, NM. The networking algorithms used on the testbed were developed at the University of Florida under the direction of Randy Chow and Richard Newman-Wolfe.

References

1. W.L. Wolfe and G.J. Zissis, *The Infrared Handbook,* (Environmental Research Institute, Ann Arbor, Michigan, 1978), pp. 3-46.
2. J.A. Gelbwachs, *IEEE J. Quantum Electronics* **24**, 1266 (1988).
3. E. Korevaar, M. Rivers, and C.S. Liu, *SPIE, Vol. 1059, Space Sensing, Communications, and Networking,* **111** (1989).
4. J. Menders *et al.*, *Optics Letters* **16**, 846 (1991).

36

High Resolution Image Processing of Geophysical Data With Diffraction Tomography

Alan J. Witten
Energy Division
Oak Ridge National Laboratory
Oak Ridge, TN 37831-6200

W. Chris King
Department of Geography and
Environmental Engineering
United States Military Academy
West Point, NY 10996-1695

James R. Ursic
U.S. EPA—Region 5
77 West Jackson Boulevard
Mail Stop HSRLT-5J
Chicago, IL 60604-3590

The need to detect buried objects, locate subsurface voids, or obtain detailed information concerning subsurface structure spans many technical disciplines, including environmental and geotechnical engineering as well as military programs such as tunnel detection and unexploded ordnance disposal. Success has been achieved in the development of technology that employs wave probing techniques to sense subsurface features. Ground penetrating radar (GPR), seismic refraction, and electromagnetic conductivity have demonstrated success; however, technical limitations, primarily in information processing, have limited the general applicability of each of these methods. The authors discuss an enhanced signal processing procedure that can be applied to existing remote sensing technology to greatly expand the detection and imaging capabilities. They present a derivation of the generalized projection slice theorem (GPST), which is the basis for geophysical diffraction tomography, a high resolution subsurface imaging procedure. The technique is then applied to other field surveying methods, such as GPR. The authors conclude by combining the theory of the GPST with a statistically based search algorithm to demonstrate that detection of known features can be accomplished at preselected high confidence levels using the same field systems, but with considerably less data collection.

Introduction

Interest in geophysical remote sensing spans a wide spectrum of military programs, but all have the common thread of needing ways to develop more and better information concerning the subsurface environment. The environmental cleanup and protection program needs information concerning the location of buried wastes that could contaminate groundwater or through other pathways endanger the environment. A special subset of this problem is the location of unexploded ordnance that is obviously not well suited to intrusive investigative techniques. Detection and location of tunnels has had military significance in several conflicts and continues to be important in our deterrence mission in the Republic of Korea. There is reason to expect that rapid and accurate tunnel detection will be important in the small-scale conflicts that are the most likely threats to peace we will face in the immediate future. This list is certainly not a complete set of all possible uses for geophysics, but these examples are sufficient to demonstrate the value of enhanced remote sensing capabilities resulting from the application of advanced signal processing to existing data gathering instrumentation.

In this chapter, we present a method of detection and imaging that has as its input the data collected from the most commonly employed geophysical tools, and applies the *generalized projection slice theorem* (GPST) in the analysis of these data to yield results ranging from greatly improved detection to high resolution images of subsurface features and buried objects. Detection capability can be greatly enhanced by incorporating this more rigorous signal processing procedure into the statistically based maximum likelihood estimator. We also include a theoretic development of the new signal processing procedures and results from the implementation of the theory with a number of commonly used geophysical tools, including a presentation of the improved capability for tunnel detection and location estimation and detection of buried ordnance. High resolution imaging of a variety of buried wastes has already been achieved through the application of geophysical diffraction tomography[1-4], a method that fully exploits the principles of the GPST.

Geophysical Methods

Hazardous waste site characterization is the problem to which geophysics is most widely applied and the purpose to which most developmental research is devoted. The most commonly applied methods are seimic reflection and refraction, ground penetrating radar, electromagnetic induction, resistivity, and magnetometry. Each is discussed in the following text only to the extent needed to understand their basic concepts and uses, and only within the context of the applications previously discussed.

Seismic reflection and refraction. In these methods a seismic pulse is generated at the ground surface, and the reflected or refracted waves are collected over an array of geophones also on the ground surface. Subsurface features are inferred from changes in slope of the first arriving signal as a function of geophone position. Refraction studies are typically applied to characterize site stratigraphy.

Ground penetrating radar (GPR). GPR, also a wave-based method, employs collocated electromagnetic transmitting and receiving antennas that are moved along a line on the ground surface. Data acquired for each measurement line are displayed as a series of parallel plots of amplitude versus time. This manner of presentation is a crude form of imaging a two-dimensional vertical cross-section.

Electromagnetic induction (EM). EM systems use two coils: one is a primary coil to impart a magnetic field into the ground, and a second coil, used to detect the secondary field produced by the currents induced in subsurface conductors by the primary field. A surface map is constructed of the conductivity contours over the surveyed area, and locations of significant change are indicative of subsurface features.

Magnetometry. Ferrous objects produce a localized perturbation in the Earth's magnetic field that can be detected by magnetometers. Magnetometry data is displayed in the same manner as are EM data.

Resistivity. Like magnetometry, resistivity uses a static field; however, here this field is created by electrodes. Arrays, similar in configuration to those used in seismic methods, of both monitoring and injection electrodes are used. Because these electrodes must be driven into the ground, resistivity is relatively cumbersome in field applications.

Enhanced Signal Processing

The focus of this section is the discussion of enhanced signal processing methods that can be applied to data acquired by the conventional geophysical tools described in the previous section. In particular, the goal of these signal processing algorithms is to render these tools more quantitative, yielding accurate size, shape, and location information. Specifically, this section deals with one approach to signal processing based on an analytic relationship between acquired data and an integral transform of the spatial variations in some subsurface quantity. This relationship is known as the GPST[5] because it represents a generalization to arbitrary wavelength of the projection slice theorem that served as the early basis for imaging algorithms used in computed tomography (CT) scanners of diagnostic medicine and nondestructive evaluation. Imaging within the GPST has become known as diffraction tomography (DT), and this form of imaging as well as a companion signal processing algorithm for target detection and location estimation, also based on the GPST, will be discussed later in this section.

Generalized Projection Slice Theorem

Presented here is a derivation of the GPST for the case of propagating scalar waves. The purpose of this derivation is to illustrate the underlying physical concepts and simplify mathematical assumptions associated with techniques based on this theorem. Subsequently, parallel results for other forms of probing energy employed by the geophysical tools described in the previous section are presented.

In general, signal processing algorithms based on the GPST are implemented in the frequency (rather than the time) domain. For this reason, the derivation of this particular GPST begins with the reduced wave equation

$$\nabla^2 u + k^2 u = k^2 u O(\mathbf{x}), \tag{1}$$

where $u = u(\mathbf{r})$ is the wavefield measured over some contour $\mathbf{r}$, $k = \omega/c_o$ is the wavenumber at frequency ω and background wave speed c_o, and $O(\mathbf{x}) = 1 - c_0^2/c^2(\mathbf{x})$ is known as the object profile and is related, in this case, to spatial variations in refractive index. It should be noted that, along with other factors, the distinction among the GPST's for specific geophysical tools lies in the wavenumber k which distinguishes the form of probing energy and the object profile $O(\mathbf{x})$ which characterizes the property of the subsurface inhomogeneities that a particular tool will respond to. It should additionally be noted that implicit in Equation 1 is the assumption of a harmonic time dependence $\exp(i\omega t)$ or, alternatively, a mapping from the time to frequency domain by means of a temporal Fourier transform.

The goal of signal processing algorithms based on the GPST is the exploitation of an analytic relationship between the the measured data, or some associated reduced or processed data, to an integral transform of the object profile. This requires an inversion of Equation 1 in order to express the unknown quantity $O(\mathbf{x})$ as some function of the measured quantity $u(\mathbf{r})$.

Unfortunately, the inversion problem for Equation 1 is ill-posed and extremely sensitive to small errors in the measured values of u. As a consequence, this equation is linearized by invoking a weak scatter approximation. There are several weak scatter approximations that can be used,[6] for illustration purposes, the example derivation considered here is the Born which represents the measured wave field by

$$u(\mathbf{r}) = u_0(\mathbf{r}) + \varepsilon u_s(\mathbf{r}), \tag{2}$$

where u_o is the "illuminating" wavefield, u_s is referred to as the scattered wave field, and ε is a small parameter included to explicitly indicate the assumption that the deviations of the total wavefield u from u_o are small. It is further assumed that the spatial variations in wave speed $c(\mathbf{x})$ do not deviate significantly from the background, c_o, making the object profile, $O(\mathbf{x})$ of order ε. This weak scatter approximation yields the linearized version of Equation 1:

$$\nabla^2 u_s + k^2 u_s = k^2 u_0 O(\mathbf{x}), \tag{3}$$

that can also be expressed in the integral equation form

$$u_s(\mathbf{r}) = -k^2 \int d\xi \; G(r - \xi) u_o(\xi) O(\xi) \tag{4}$$

where G is the Green's function for the scalar Helmholtz operator.

It readily seen from Equation 4 that u_s is derived from measurement and the functions u_o and G can be computed making the only unknown quantity the object profile $O(\mathbf{x})$. In the form given by this equation, the desired inversion can be accomplished be deconvolution of the integral. This deconvolution can be achieved by representing the Green's function by its plane wave expansion,[7] (For simplicity, the remainder of the derivation is given in two dimensions. A similar plane wave expansion can be used in the three-dimensional case, with a derivation of the corresponding GPST that parallels the two-dimensional one given here.)

$$G(\mathbf{r}) = \frac{i}{4\pi} \int \frac{d\alpha}{\sqrt{k^2 - \alpha^2}} e^{i(\sqrt{k^2 - \alpha^2}\,\hat{\mathbf{x}} + \alpha \hat{\mathbf{y}}) \cdot \mathbf{r}} \tag{5}$$

where $\hat{\mathbf{x}}$ and $\hat{\mathbf{y}}$ are unit vectors in the x and y directions, respectively, and measurements of u (u_s) are made along a line parallel to the y-axis. Taking this measurement line to be $\mathbf{r} = x_R \hat{\mathbf{x}} + l\hat{\mathbf{y}}$, (Figure 1) assuming the incident field is a plane wave propagating in the s_o direction, (Again for simplicity, plane wave

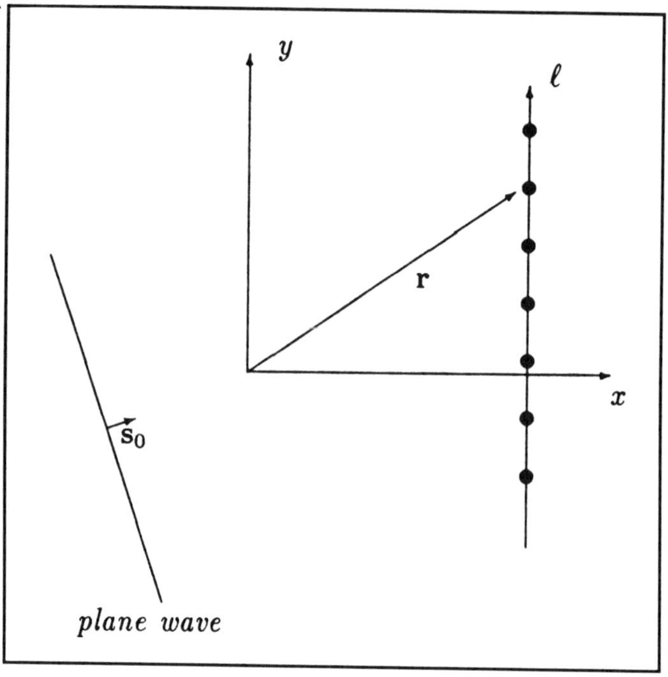

Figure 1. Illustration of geometry and notation used in deriving the GPST. Measurement locations along the line $\mathbf{r} = \mathbf{x}_R + l\hat{\mathbf{y}}$ are indicated by •'s.

illumination is assumed. It can, however, be shown that, for an arbitrary source beam pattern, plane wave illumination can be synthesized by coherently summing the contributions from a multiplicity of distinct source locations.[8,9]) $u_o(\mathbf{x}) = \exp(ik\mathbf{s}_o \cdot \mathbf{x})$, and using the expression for the Green's function given by Equation 5, (Figure 1) Equation 4 becomes:

$$u_s(\mathbf{r}) = \frac{ik^2}{4\pi} \int \frac{d\alpha}{\sqrt{k^2 - \alpha^2}} e^{i(\sqrt{k^2 - \alpha^2} x_R + \alpha l)} \int d\xi \, O(\xi) \, e^{-i(\sqrt{k^2 - \alpha^2}\hat{\mathbf{x}} + \alpha\hat{\mathbf{y}} - k s_0) \cdot \xi}. \tag{6}$$

The deconvolution of Equation 6 can be accomplished by defining the Fourier transform of the data u_s as:

$$\tilde{u}_s(\kappa) = \int dl \, e^{-i\kappa l} \, u_s(l), \tag{7}$$

applying this integral transform to Equation 6, and recognizing that

$$\int dl \, e^{-i(\kappa - \alpha)l} = 2\pi \, \delta(\kappa - \alpha). \tag{8}$$

This procedure yields the final form of the GPST:

$$\tilde{u}_s(\kappa) = -\frac{ik^2}{2\sqrt{k^2-\kappa^2}} e^{i\sqrt{k^2-\kappa^2}x_R}\,\tilde{O}(\mathbf{K}), \tag{9}$$

where:

$$\tilde{O}(\mathbf{K}) = \int d\xi\, O(\xi)\, e^{-i\mathbf{K}\cdot\xi} = \int d\xi\, O(\xi)\, e^{-i(\sqrt{k^2-\kappa^2}\hat{\mathbf{x}} + \kappa\hat{\mathbf{y}} - ks_0)\cdot\xi} \tag{10}$$

The desired result of an analytic expression between measurement and subsurface properties is achieved in Equation 9, which relates the one-dimensional Fourier transform of the scattered field (reduced measured data) to the two-dimensional Fourier transform of the object profile. The key to this result is the plane wave expansion, Equation 5, used in the deconvolution. This step explicitly introduces the concept of coherence by representing the scattered wave as a superposition of plane waves. This aspect of coherence is carried into the GPST where, recalling that $u_o(\mathbf{r}) = \exp(iks_o \cdot \mathbf{r})$, the wave vector $\mathbf{K}$ associated with the integral tranform can be expressed as $\mathbf{K} = k(\mathbf{s} - \mathbf{s}_o)$ which is the difference between a scattered plane wave propagating in the direction $(1/k)(\sqrt{k^2-\kappa^2}\hat{\mathbf{x}} + \kappa\hat{\mathbf{y}})$ and the incident plane wave propagating in the direction $\mathbf{s}_o$. The term multiplying O in Equation 9, $(-ik^2/2\sqrt{k^2-\kappa^2})$ exp $(i\sqrt{k^2-\kappa^2}x_R)$, is the equation for a holographic lens and the imaging procedure discussed in the following subsection is entirely analogous to imaging by optical holography.

Having illustrated the concept and mathematical formulation of the GPST, some general remarks can be made about the GPSTs for the suite of geophysical tools discussed in the previous section. First, GPSTs are specific to both the form of probing energy and the measurement geometry. While, for most cases, the GPST will involve a Fourier transform of the data, $\tilde{u}_s(\kappa)$, it will not necessarily relate this quantity to a two-dimensional spatial Fourier transform of the object profile. The actual form of the integral transform $\tilde{O}$ will depend on the definition of wavenumber k. Only for homogeneous propagating waves will k be real and the associated integral transform be Fourier. Table 1 summarizes the nature of the GPSTs for the five geophysical tools described in the previous section.

It is important to recognize that Table 1 identifies the nature of inhomogeneities that can be resolved by a particular geophysical tool as well as the character of the wavevector $\mathbf{K} = (K_x, K_y)$ which has serious implications regarding the computational ease with which particular types of data can be

Table 1
Summary of GPST Relationships

Method	k	(K_x, K_y)	Transform	O
Seismics	ω/c_0	(Real, real)	Fourier-Fourier	Refractive index
GPR	ω/c_0	(Real, real)	Fourier-Fourier	Refractive index
EM	$\sqrt{-i\omega/\sigma_0}$	(Complex, real)	Generalized	Conductivity
Magnetometry	0	(Imaginary, real)	Laplace-Fourier	Permeability
Resistivity	0	(Imaginary, real)	Laplace-Fourier	Conductivity

ω is frequency
c_0 is a background wave speed
σ_0 is a background conductivity

processed. This topic will be discussed in greater detail in the next two subsections.

Imaging

For wave-based tools (seismics and radar), imaging can be accomplished by numerical inversion of the integral transform indicated in Equation 9 or

$$O(\mathbf{x}) = \frac{1}{(2\pi)^2} \int d\mathbf{K}\ \tilde{O}(\mathbf{K})\ e^{i\mathbf{K}\cdot\mathbf{x}}$$

$$= \frac{2i}{(2\pi k)^2} \int \frac{d\mathbf{K}}{\sqrt{k^2 - \kappa^2}}\ e^{-i\sqrt{k^2-\kappa^2}x_R}\, \tilde{u}_s(\kappa)\ e^{-i(\sqrt{k^2-\kappa^2}\,\hat{\mathbf{x}} + \kappa\hat{\mathbf{y}} - ks_0)\cdot\mathbf{x}} \tag{11}$$

The integration over **K** can be expressed as a sum over view angles s_o and an integration over κ to yield the more computational suitable form

$$O(\mathbf{x}) = \frac{2i}{(2\pi)^2 k} \sum_{s_0} e^{-iks_0\cdot\mathbf{x}} \int d\kappa\, H(k,\kappa,s_0)\ \tilde{u}_s(\kappa)\ e^{i(\sqrt{k^2-\kappa^2}x + \kappa y)}, \tag{12}$$

where the function H contains the mapping from **K** into s_o and κ along with a filter to remove points of double coverage associated with this mapping. The imaging formula defined by Equation 12 is computationally efficient because the integration over κ can be accomplished by fast Fourier transform numerical procedures.

Imaging by means of GPST and the transform inversion given by Equation 12 has become known as *diffraction tomography* (DT) and it is analogous to holographic imaging using a variety of laser beam illumination directions s_o. The GPST, Equation 9, provides a knowledge of $\tilde{O}$ over a wavevector space $\mathbf{K} = \mathbf{K}(s_o, \kappa)$ The extent of this coverage and the resulting image quality will be a function of the number and range of illuminating view direction s_o. For any particular view direction, **K**-space coverage will be a portion of a circular arc. As the view direction changes a region of **K**-space is "swept" out. For any practical geophysical measurement geometry, there will necessarily be gaps in **K**-space that will introduce artifacts into the image. The most significant manifestation is an elongation of imaged features along the direction of the gaps. This has the simple physical interpretation that resolution is always better in the direction normal to that of wave propagation than along the direction of propagation. The previously mentioned gaps occur along the predominant direction of propagation and, consequently, optimal image quality can be achieved by utilizing the broadest possible range of view directions.

The previous discussion is appropriate for wave-based methods such as seismics and cross-borehole radar, which typically employ independently positioned sources and receivers. One important geophysical tool, GPR, uses a collocated source/receiver pair that is moved in unison. For this measurement geometry, individual incident and scattered wave directions cannot be independently controlled and, as a consequence, only a single arc in **K**-space can be realized for a particluar frequency ω. Fortunately, conventional GPR systems are pulsed, offering a reasonable source bandwidth. This bandwidth can be exploited by representing the integration over **K** in Equation 11 as a summation over frequencies ω and an integration over the spatial Fourier transform variable κ. By this means, **K**-space coverage is a series of concentric arcs sufficient to yield good tomographic images.

Although the wave-based imaging procedure described previously is easily implemented, imaging with measurements from other geophysical tools is more difficult, less computationally efficient, and, perhaps, impossible. In the case of resistivity, the integral transform to be inverted (Table 1) is a Laplace-Fourier which, again, can be implemented with standard numeric techniques; however, these techniques do not nearly offer the computational efficiency of the the multidimensional Fourier transform methods associated with wave-based measurements. Because resistivity is a "dc" method, it operates at zero frequency and, for this reason, there is no opportunity to exploit bandwidth. Resisivity studies, however, commonly employ a variety of injection electrodes that serve the same purpose as variable view angles in wave-based methods and allow sufficient **K**-space coverage to yield good image quality. Magnetometry is also a "dc" measurement; however, this measurement is passive, relying on the Earth's magnetic field as a source. As a result, view directions cannot be varied, as is the case in resistivity, and **K**-space coverage is insufficient for imaging.

Most EM tools have a fixed coil separation, making it impossible to independently vary incident and scattered wave directions. Some EM tools, however, can operate at different frequencies, allowing sufficient **K**-space coverage for imaging in a manner similar to that of GPR. The difficulty associated with EM imaging is that the integral transform that must be inverted (Table 1) contains a wavevector component that is complex and, as a result, will require the development of new numeric procedures.

Detection

As a complementary procedure to imaging, the GPST can serve as the basis for target detection and location estimation which, as will be shown later, avoids some of the difficulties and limitations of imaging. The general problem addressed by detection is the quantification of the presence or absence of a known target and, if present, an estimate of its location. The basic concept of the method described here is known as maximum likelihood estimation (MLE), which effects detection and location by correlating measurements with the known target response for a variety of possible locations $\mathbf{x}_0$. Taking the same measurement line that was used in the derivation of the GPST, $\mathbf{r} = x_R\hat{\mathbf{x}} + l\hat{\mathbf{y}}$, defining the measured data to be $u_s(l)$ and the data that would be received if the known target were located at $\mathbf{x}_0$ as $u'_s(l;\mathbf{x}_0)$, a measure of the target's presence and location would be the value $\mathbf{x}_0$ that minimized the square of the difference between u_s and u'_s integrated over all measurement points l. This can be represented as the $\mathbf{x}_0$ that minimizes the function F where

$$F(\mathbf{x}_0) = \int dl \, |u_s(l) - \tilde{u}'_s(l; \mathbf{x}_0)|^2. \tag{13}$$

This procedure for target detection can be immediately implemented; however, it requires the numeric simulation and storage of a massive number of target data "templates" $u'_s(l;\mathbf{x}_0)$. This approach can be prohibitive, particularly if the target is strongly asymmetric and can assume many different orientations.

One means to make this problem more efficient is to use the GPST. The first step in accomplishing this is using Parseval's theorem to express the integral over l as integral over the transform variable κ. This yields

$$F(x_0) = \int d\kappa \, | \tilde{u}_s(\kappa) - \tilde{u}_s'(\kappa; x_0)|^2. \\ = \int d\kappa \, (\tilde{u}_s\tilde{u}_s^* - 2\tilde{u}_s\tilde{u}_s'^* + \tilde{u}_s'\tilde{u}_s'^*), \tag{14}$$

where $\tilde{u}_s'(\tilde{u}_s')$ is defined in Equation 7 and the * denotes complex conjugation. The $\tilde{u}_s\tilde{u}_s$ term is independent of possible target location $\mathbf{x}_o$ and produces only a constant bias in F. Therefore, this term can be dropped and a new function F_L, known as the log likelihood function defined as

$$F_L(\mathbf{x}_0) = \int d\kappa \, (2\tilde{u}_s\tilde{u}_s'^* - \tilde{u}_s'\tilde{u}_s'^*), \tag{15}$$

and the most probable target location $\mathbf{x}_o$ is the one that maximizes F_L.

The GPST is introduced into the detection signal processing algorithm by using it to represent $\tilde{u}_s'(\kappa)$ in terms of the spatial Fourier transform of the object profile $\tilde{O}(\mathbf{K};\mathbf{x}_o)$ for a target centered at a position $\mathbf{x}_o$. For the specific GPST derived earlier, it is recognized that, for propagating waves, $\tilde{u}_s' \; \tilde{u}_s'^*$ is a constant and can be neglected. With this and the shift rule for Fourier transforms which can be used to relate the $\tilde{O}(\mathbf{K};\mathbf{x}_o)$ to the Fourier transform of the object profile at some reference location, here taken to be the origin and denoted by $\tilde{O}_o$,

$$\tilde{O}(\mathbf{K}; \mathbf{x}_0) = \int \; d\mathbf{K} \, O(\mathbf{x}; \mathbf{x}_0) \, e^{-i\mathbf{K}\cdot\mathbf{x}} \\ = e^{-i\mathbf{K}\cdot\mathbf{x}_0} \, \tilde{O}_0(\mathbf{K}), \tag{16}$$

the final form of the log likelihood function becomes

$$F_L(\mathbf{x}_0) = ik^2 e^{-iks_0 \cdot x_0} \int \frac{d\kappa}{\sqrt{k^2 - \kappa^2}} \, e^{-i\sqrt{k^2 - \kappa^2} x_R} \, \tilde{u}_s(\kappa) \; \tilde{O}_0^*(\mathbf{K}) \; e^{i(\sqrt{k^2 - \kappa^2}\, x_0 + \kappa y_0)} \tag{17}$$

There is a striking resemblance between this result and the imaging formula given by Equation 12. Equation 17 is, in fact, an imaging formula; however, rather than providing an image of of a physical quantity $O(\mathbf{x})$, it yields an image of a statistical measure of target location, namely the log likelihood function. As a result, imaging of F_L has been referred to as the *diffraction tomography maximum likelihood estimator* (DTMLE). The holographic lens alluded to in the discussion of imaging is replaced here by a matched filter holographic lens, where features in the data not correlated to the target characteristics as specified by $\tilde{O}_o$ are preferentially rejected. The major difference between the DTMLE and the more simple MLE approach, Equation 13, is that the GPST has been used to eliminate the need to generate a large number of data templates $u'_s(l;\mathbf{x}_o)$ replacing these instead with a single target response $\tilde{O}_0^*(\mathbf{K})$ (or, at worst, one response for each possible target orientation), making DTMLE far more computationally efficient.

The most important difference between target detection with DTMLE and DT imaging (another important difference between DTMLE and DT is that a form of the GPST can be derived for DTMLE without the need for a weak scatter approximation; this makes the formulation of DTMLE exact[10]) is that the DTMLE algorithm uses the GPST directly rather than its definition of an integral transform that must be inverted to reconstruct an image. The implication of this fact on geophysical remote sensing is that target detection and location estimation can be implemented using precomputed filters and computationally efficient fast Fourier transforms for any geophysical tool. Furthermore, the detection problem uses *a priori* information, the target characteristics, and, therefore, does not require as much data as imaging. As a result, DTMLE can be used with only a single view or frequency; however, summing Equation 17 over all available views and/or frequencies will improve the results. This point is particularly important for magnetometry because it allows its use for high resolution target detection where imaging is impossible.

Results

The GPST has served as the basis for a number of geophysical imaging and detection algorithms. These are method-specific, and the current level of technologic maturity depends on both the geophysical tool and measurement geometry.

The most advanced signal processing algorithms are those associated with wave-based measurements. Among these, acoustic geophysical diffraction tomography (GDT) applied in an offset vertical profiling (VSP) geometry [source(s) deployed on the ground surface and receivers emplaced in a vertical borehole] is ready for commercialization. This method has been extensively field tested and an integrated hardware/software system is in use by the U.S. Environmental Protection Agency at Superfund sites. A second such system has been built for use by the U.S. Army Toxic and Hazardous Materials Agency. GDT imaging and DTMLE detection algorithms have been developed and field tested for seismic reflection and refraction, cross-borehole (sources in one borehole and receivers in a parallel borehole) radar and

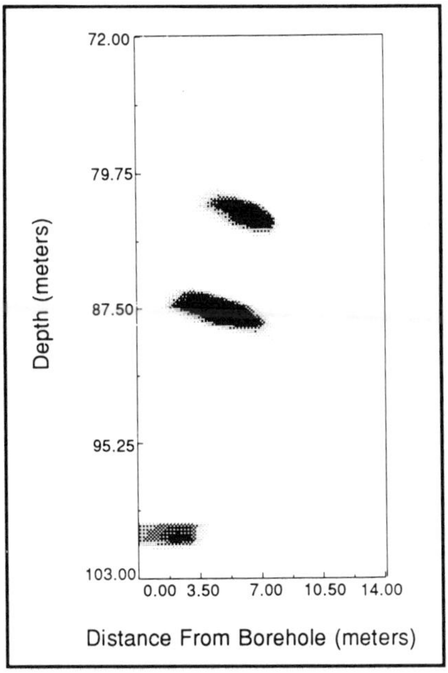

Figure 2. Image of tunnel in the Korean DMZ.[11]

seismics, and GPR. More recently, imaging and detection algorithms for resistivity and detection algorithms for magnetometry have been developed and tested on numerically synthesized data, with field testing anticipated in the near future. Currently, efforts are being directed toward imaging and detection algorithms for EM measurements, with the specific application being the detection of buried unexploded ordnance.

This section presents results, based on both simulated data and actual field measurements, of the applications of enhanced signal processing algorthms. These do not represent the entire extent of experience with GPST signal processing but are selected to be representative of potential applications within the scope of this paper.

Figure 2 is a gray scale GDT image of a vertical cross-section within the Korean Demilitarized Zone (DMZ) constructed by the analysis of cross-borehole radar measurements. In this image, variations in electromagnetic wavespeed are represented as gray levels, with darker gray levels associated with greater wave speed. The black area appearing at a depth of 87.5m and 3.5m from the source borehole is the image of a known tunnel. The dark gray area approximately 7m above this tunnel exhibits nearly the size and contrast of the known tunnel. This feature introduces the concern that a second unknown tunnel exists within this cross-section. It was hypothesized that this feature was not a tunnel but rather, a fracture zone of the type known to occur

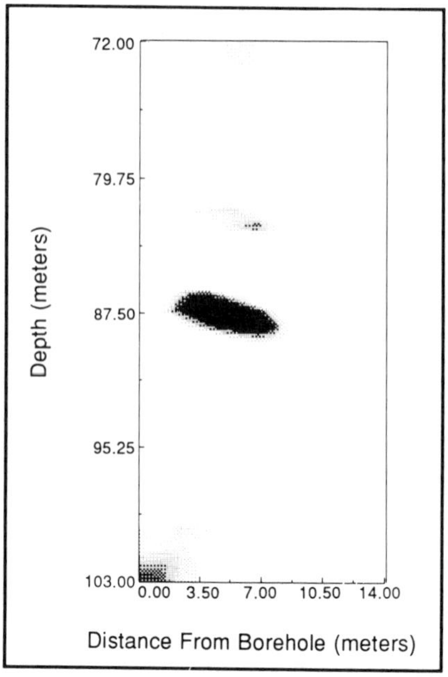

Figure 3. Image of location probability of a tunnel in the DMZ.[11]

in the granite formations within the DMZ. Confirmation of this hypothesis came with the application of the DTMLE algorithm to the same data set. Figure 3 is the corresponding image of the log likelihood function, also displayed as gray levels, with darker gray scales associated with increased probability of the target location. Notice in this figure that the unknown feature of concern above the known tunnel is almost gone because it does not match the characteristics of the target, a circular void 2m in diameter, input to the DTMLE algorithm. The example illustrates the capacity of this algorithm to distinguish a target of the same apparent size but with different characteristics.

The next result presented serves to demonstrate the capability of the DTMLE approach, in this case used with magnetometer measurements to differentiate targets of differing size and depth. Such differentiation is difficult with magnetometry because the diffusion process associated with the detection of features that perturbed the Earth's magnetic field lines yields considerable ambiguity between target size and depth. This is particularly problematic for the use of magnetometry at buried unexploded ordnance sites, because such sites are also expected to be rich in fragments of exploded ordnance that will produce a magnetometer response. These fragments will, in general, be smaller and shallower than the targets of interest, and the problem of concern is the differentiation of unexploded ordnance from fragments.

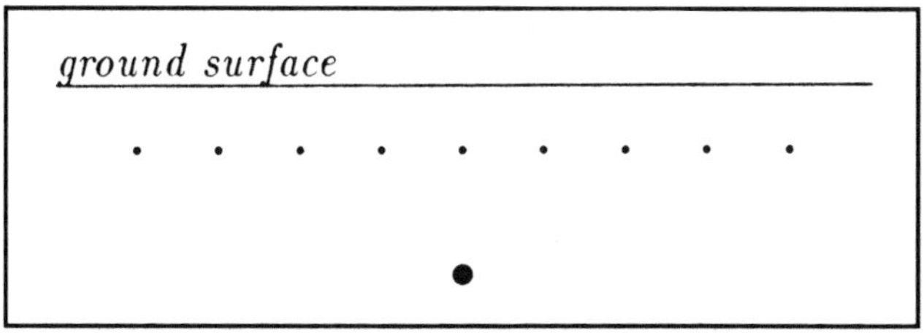

Figure 4. Illustration of target configuration used for simulaiton of magnetometry data.

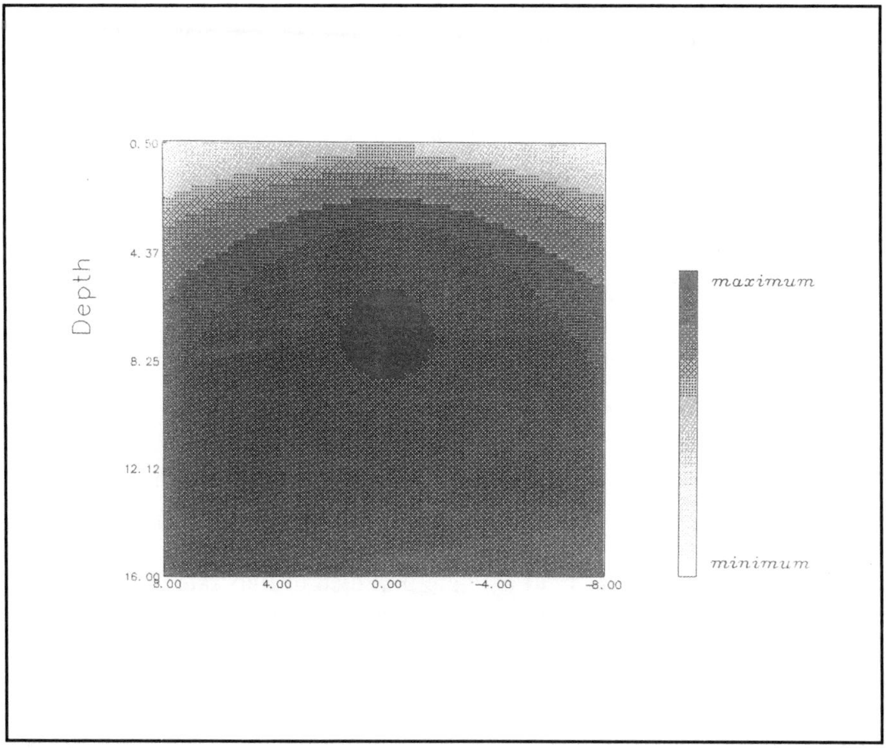

Figure 5. Image of location probability for a large circular disk in the presence of many smaller disks, based on a simulated magnetometer data.

Figure 4 is an illustration of the target configuration used for simulation of magnetometry data. Numerous small, shallow circular targets are present along with a single larger circular target at a greater depth. The DTMLE magnetometer algorithm was applied to this data set, with the target characteristic being the diameter of the larger target. Figure 5 is the image of the target location probability displayed as gray scales. The darkest region of the figure corresponds to the large target location, with no evidence of the presence of the smaller targets. Here the DTMLE matched filter has effected complete rejection of these irrelevant features.

Table 2
Summary of Applicability and Resolution for Enhanced Signal Processing

Method	Buried waste	Tunnels	Ordnance Detection	Resolution
Offset VSP seismic	√	√†		Wavelength
Seismic reflection		√†		Wavelength
Cross-hole seismic		√‡		Wavelength
Cross-hole radar		√‡		Wavelength
Ground Penetrating Radar	√	√	√	Wavelength
Magnetometry	√	√†	√	Measurement spacing
Electromagnetic Induction	√	√†	√	Frequency
Resistivity		√†		Measurement spacing

*VSP = vertical seismic profiling; GPR = ground penetrating radar; EM = electromagnetic induction.
† Shallow tunnels.
‡ Deep tunnels.

Conclusions

The GPST serves as the basis for signal processing algorithms, either to image (GDT) subsurface features or for the detection and location estimation (DTMLE) of known underground targets. A derivation of a particular GPST is given and the relevance of this relationship to various geophysical tools is explored. In addition, several results of the application of these algorithms to both simulated and actual field data are provided to illustrate the resolution that can be achieved and the potential applications for enhanced signal processing using GPSTs.

To conclude, it is worthwhile to return to the intended scope of work presented here and endeavor to remark on the most practical geophysical tools and measurement geometries employing enhanced signal processing on the basis of the three applications considered. These findings are summarized in Table 2 and elaborated in the following text.

A first point that can be made about this table is that various permutations of geophysical tools and measurement geometries employed in a given application with enhanced signal processing are essentially the same as those currently used in geophysical field studies. The use of enhanced signal processing does not affect the suitability of a particular tool to a given application but does increase the resolution that can be achieved, provides more quantitative results, and can significantly improve the prognosis for success in any field study. One major exception to this is the use of seismic imaging at waste sites with an offset VSP geometry. Although seismic reflection and refraction are sometimes used at buried waste sites, they are not employed to locate isolated inclusions such as buried waste. Instead, they are used for stratigraphic applications in characterizing the host site. The application of offset VSP GDT offers a seismic method that has proved quite valuable for imaging buried waste where no seismic-based method has been used before.

The resolution indicated in Table 2 differentiates wave-based methods from other methods. For the wave-based methods, the resolution is wavelength dependent, and subsurface features with length scales on the order of one-quarter of a wavelength can faithfully be detected and imaged. For the "dc" tools, the wavelength is infinite; however, here the resolution is proportional to the spacing between adjacent measurement points. Because these methods employ exponentially damped fields, signal-to-noise ratio considerations limit the actual resolution that can be achieved. EM tools use extremely low frequencies and deal with evanescent waves. For this reason, there is the same signal-to-noise ratio concern as the "dc" tools; however, resolution improves with increasing frequency.

Although it is possible to envision new geophysical tools and measurement geometries for use with enhanced signal processing algorithms, the focus of this study is a more practical one. Here, signal processing algorithms are developed that are driven by particular applications and compatible with existing geophysical methods. These algorithms serve to more fully exploit the available data without alteration of field procedures or implementation efficiency. Furthermore, most signal processing algorithms considered here can be executed in real time and in the field with the current state of digital signal processing technology.

References

1. A.J. Devaney, *Trans. Geosci. Remote Sensing* **GE-22**, 3 (1984).
2. W.C. King, A.J. Witten, and G.D. Reed, *J. Env. Eng.* **115**, 166 (1990).
3. A.J. Witten and W.C. King, *J. Env. Eng.* **116**, 166 (1990).
4. A.J. Witten and W.C. King, *Civil Eng.* **60**, 62 (1990).
5. A.J. Devaney, *Ultrasonic Imaging* **4**, 336 (1982).
6. A.J. Witten and J.E. Molyneux, in *Geophysical Inversion*, J.B. Bednar *et al.*, Eds. (SIAM, Philadelphia, 1992).
7. P.M. Morse and H. Feshbach, *Methods of Theoretical Physics* (McGraw-Hill, New York, 1953).
8. A.J. Witten and J.E. Molyneux, *IEEE Trans. Geosci. Remote Sensing* **26**, 409 (1988).
9. A.J. Devaney and G. Beylkin, *Ultrasonic Imaging* **6**, 181 (1986).
10. G.A. Tsihrintsis and A.J. Devaney, *IEEE Trans. Signal Processing* **39**, 1466 (1991)
11. A.J. Witten, *Inverse Problems* **7**, L49 (1991).

37

Characterizing the Response of Solid Rocket Motors to Fragment Impact

Jamie M. Fisher
U.S. Army Missile Command
Research, Development and Engineering Center
Redstone Arsenal,
AL 35898-5249

As part of a survivability enhancement program for an Army missile system, a methodology for characterizing the response of a solid propellant rocket motor to fragment impact was demonstrated. This methodology combines experimental techniques developed for high explosives with analytic tools derived from basic continuum mechanics.

Wedge tests were conducted on the solid propellant to determine the shock sensitivity and derive its equation of state. These parameters were input into a hydrocode to conduct numeric simulations of fragment impacts. The analyses were used to predict the impact velocity required for a fragment to initiate a detonation in the rocket motor. These predicted values allowed for a tailored test plan to experimentally characterize the detonation threshold for the motor and significantly reduced the number of tests and assets required.

Application of this methodology offers many advantages. This methodology is an alternative to the time-consuming and costly experimental programs normally used to characterize a solid rocket motor's response to projectile impact. The combination of testing and modeling utilized here provides designers with a more accurate data base for investigating mitigative techniques to enhance system survivability.

Background

A survivability enhancement program for an Army air defense missile system was the origination of the work discussed here. In a battlefield environment survivability test series, the missile system reacted violently to a threat stimulus. The system sustained sufficient damage to represent a loss of both system and crew. Following this unacceptable performance, a survivability working group (SWG) was established to develop, investigate, and demonstrate potential survivability enhancements. The solid propellant rocket motor and high explosive warhead were the missile components that posed the greatest risk to the system and crew survivability. The rocket motor, by virtue of its larger size, correspondingly larger volume of energetic material, and the more sensitive nature of its composition, was considered the most vulnerable missile component.

The rocket motor consists of a filament wound, S-glass composite case with a nominal wall thickness of 0.089in. The case is lined and insulated prior to propellant casting. The liner is a two-part system of neoprene and polyvinyl. The insulation, a styrene-butadiene rubber has a thickness of 0.015in. The propellant is a hazards class 1.1, minimum smoke, cross-linked, double-base formulation containing cyclotrimethylene trinitramine (RDX). The rocket motor contains approximately 40lbs of this detonable material.

State-of-the-art technology did not offer an alternative to the detonable material that was capable of providing the required missile performance. Therefore, it became the focus of the working group to develop mitigative techniques involving external barriers to protect the energetic missile components from external threat stimuli. It was agreed that certain threats such as a shaped charge could not be defeated or mitigated significantly without completely redesigning the missile system. As redesign was an unacceptable solution, artillery was selected as a representative, yet defeatable threat stimuli.

In order to develop survivability improvements, it was necessary to define the level of protection required to prevent detonation of the missile's energetic components. By characterizing the sensitivity of the rocket motor and warhead, performance requirements for armoring concepts could be established. By armoring to realistic requirements, an optimized system could be achieved that provided maximum protection with minimal adverse impact on the system as it is currently designed. In the past, overly conservative and unrealistic protection requirements have hampered the development and implementation of armor and other survivability enhancement concepts.

As a member of the working group, the author was tasked to provide a characterization of the rocket motor's sensitivity to fragment impact for use in evaluating potential armor concepts. This characterization had to be accurate, timely, and cost-effective.

Previously proposed test plans involved full-scale motors subjected to fragment spray from detonating artillery shells. These tests are costly, time-consuming, and often yield ambiguous results. In this instance, the cost of a full-scale motor, off the production line, was $14,000. Artillery shells do not fragment in perfectly repeatable patterns; therefore every test is basically unique. Ten tests were suggested to obtain a good statistical data base. This would require a minimum of $140,000 in rocket motor test assets alone.

Testing of this nature requires high-speed instrumentation as well as manual data acquisition. Both increase test time and cost. High-speed cameras are required to record the impact and any resulting reaction. The films must then be reviewed to ascertain the number of fragment impacts, location of impacts, fragment velocity at impact, fragment shape and orientation at impact as well as the time to reaction and nature of reaction. This can be a tedious, time-consuming, subjective task. If the test article does not react, the impact sites must be located, counted, and mapped, and the depth of penetration must be measured; if fragments are recovered, their geometry and mass must be recorded. From this collection of data, generalized statements are made relating the reaction of the test article to impact by various size/mass distributions of fragments from specific artillery shells.

Characterization Procedure

An innovative, more efficient alternative characterization procedure was proposed. This proposed procedure involved the application of experimental and analytic techniques developed for explosives. Although these tools have been widely used in the explosives industry, they have not been applied to Army solid propellant rocket motors. The proposed method was a four-step process. The first step was to obtain equation-of-state data on the solid propellant involved.

In the second step, this data is used to conduct hydrocode simulations of fragment impact on the rocket motor. The simulation results can then be used to design a test plan tailored to provide a characterization of the detonation threshold for several fragment masses/geometries. Finally, in the fourth step, fragment impact tests are conducted to verify simulation results, and further quantify critical mass/velocity combinations, both of which result in motor detonations and provide a required level of protection to evaluate armor

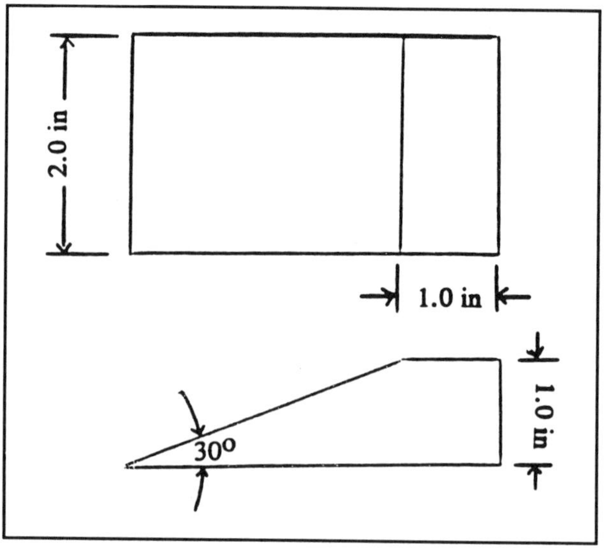

Figure 1. Wedge specimen configuration.

candidates against. This characterization procedure was completed within four months at an estimated total cost of less than $150,000.

Step one: propellant sensitivity testing. The first step was to experimentally characterize the shock sensitivity of the solid propellant so that it could be numerically modeled. This was done by conducting a series of wedge tests. The specimen configuration, shown in Figure 1, is milled from propellant slabs, as illustrated in Figure 2. The propellant wedge is then subjected to a planar shock wave by a donor charge detonation. A streak camera is used to monitor the progression of this wave up the angled face of the wedge. Donor charges are designed to introduce shock pressures of specific magnitudes. If the pressure is sufficient, the shock will transition to detonation. The streak film records both the distance the shock wave traveled before transitioning and the speed of the wave. A series of tests was conducted over a range of input pressures. The input pressures are plotted against the associated run distances to yield a Pop plot. The shock Hugoniot, shock velocity versus particle velocity, is also obtained from wedge test data. The results for the solid propellant tested in this program are shown in Figures 3 and 4. These tests were conducted by Frederick Sandstrom at the Center for Explosives Technology Research of the New Mexico Institute of Mining and Technology in Socorro, New Mexico.

Step two: hydrocode simulations. Experimentally determined Pop plot and Hugoniot relationships serve as inputs into a hydrocode simulation of impact scenarios. The hydrocode used in this exercise, 2DE, was developed at Los Alamos National Laboratory.[1] It is a Eulerian hydrocode that has been

Figure 2. Wedge specimen preparation.

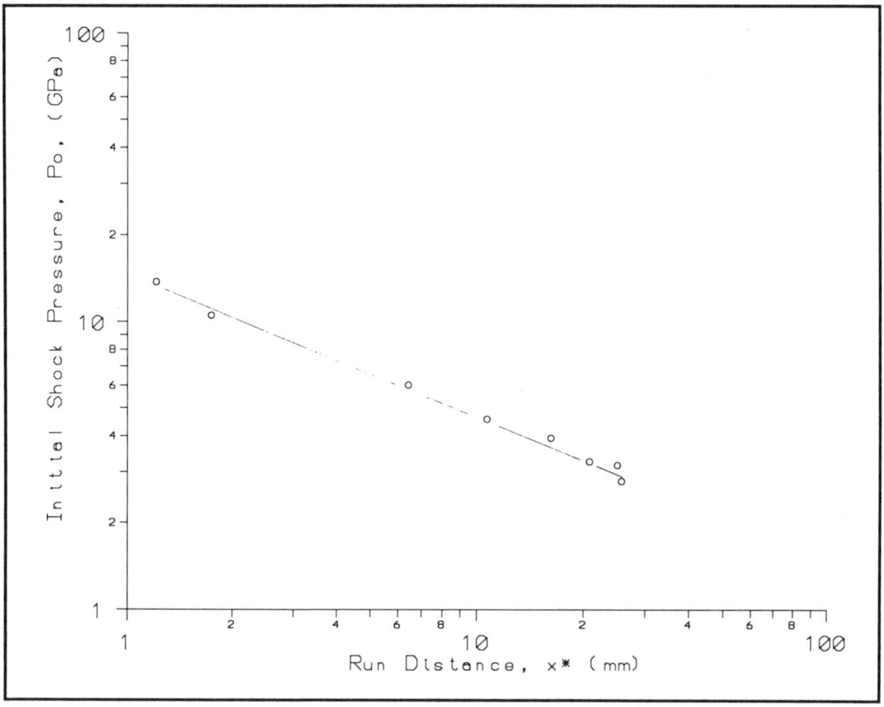

Figure 3. Propellant pop plot: pressure versus run distance.

interfaced with an initiation code known as Forest Fire. Early attempts to model the propagation of shock waves treated all materials as fluids by not

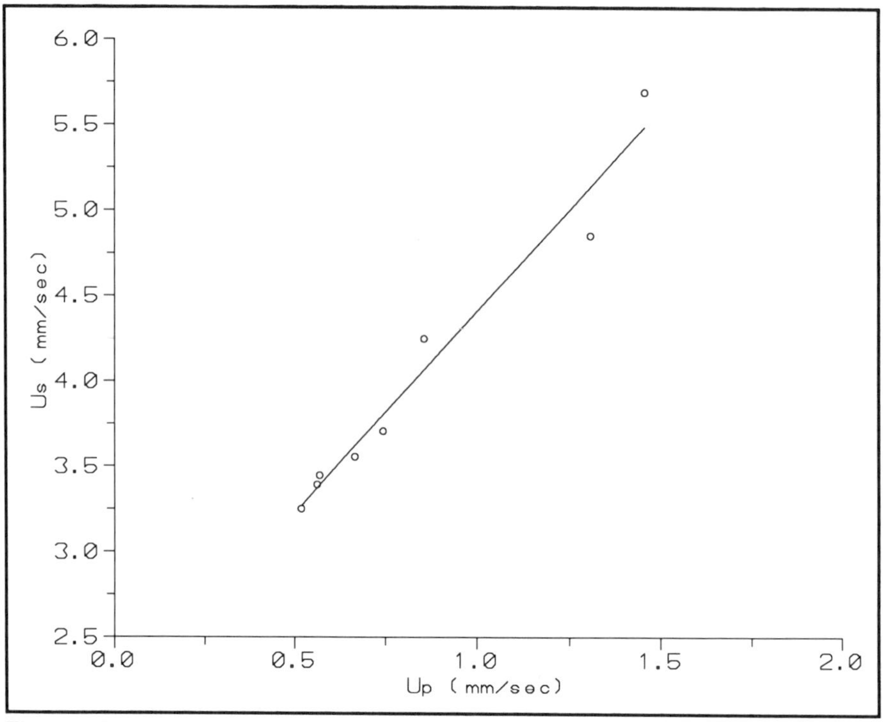

Figure 4. Propellant shock hugoniot: shock velocity versus particle velocity.

including strength effects. The term hydrocodes is a carryover from these early computer codes.

2DE, as are all hydrocodes, is based on classical continuum mechanics, that is, conservation of mass, momentum, and energy. It is a two-dimensional, finite element (FE) code that uses a spatial representation of the problem. As such, the FE grid remains fixed throughout the calculations while material, momentum, and energy are allowed to flow across cell boundaries. As time progresses, the code computes velocity, pressure, mass, density, and so on, by applying the Eularian form of the conservation equations as quantities of energy, mass, and/or momentum flow in and out of the cells.[2]

In applying hydrocodes to explosives, additional input data are required to relate the explosive's material properties, density, and internal energy to pressure. Therefore, an equation of state for both unreacted and reacted material is required. The Forest Fire model utilizes the equations of state and the Pop plot from wedge test data to determine the reaction of an impacted material and the rate of explosive decomposition as a function of impact pressure. Forest Fire was written to predict the initiation and propagation of

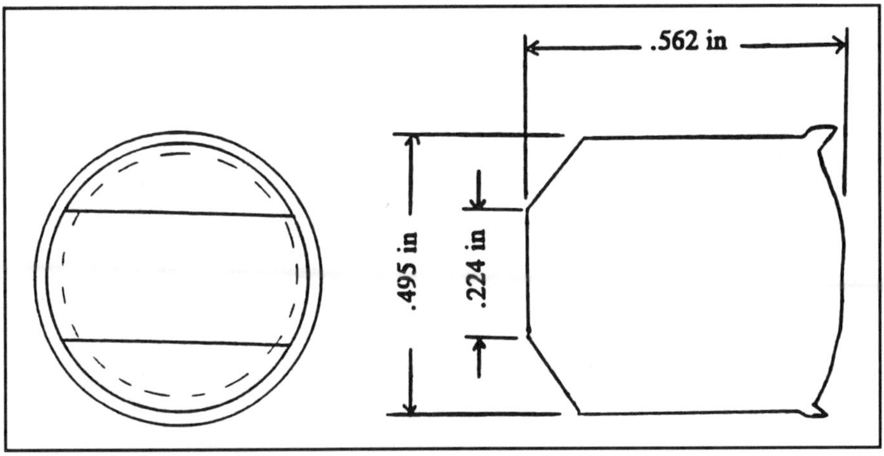

Figure 5. Fragment simulating projectile configuration.

detonations in heterogeneous explosives. When an explosive is impacted with sufficient velocity, the induced shock pressure results in the decomposition of the explosive. This decomposition adds energy to the shock, and detonation occurs.[3]

The numeric simulations conducted for this program modeled the impact of a fragment simulating projectile (FSP) on the rocket motor. The FSP was chosen over a 1/2-in cube fragment because of the standard use of FSPs in armor evaluation. Candidate armors were to be tested against fragment mass/velocity combinations determined from the rocket motor testing. Consistency of the fragment geometry between the two experimental programs was therefore desirable. The FSP is also designed to be shot with standard powder guns and barrels. This simplifies the launching procedure for the experimental programs. The fragment geometry, material, and manufacturing are specified in MIL-P-46593A (Ord).

Three FSP sizes were chosen to obtain detonation threshold velocities for the rocket motor across a band of fragment masses. The three FSPs used were a 44 grain(or caliber 30), a 207-grain (or caliber 50), and a 830 grain (or 20mm). Figure 5 illustrates the geometry of the 207-grain FSP, which is then scaled for the other fragment masses. All three FSPs are shown in Figure 6. The 2DE simulations were conducted by Athena Engineering Company, and yielded the following detonation threshold velocities for the rocket motor: the 44-grain fragment required a velocity of 7743ft/sec to detonate the motor; the 207-grain fragment required 4255ft/sec; and the 830-grain fragment required 2502 ft/sec.

Step three-test plan development. From previous experience, it was

Figure 6. FSPs used in motor impact testing.

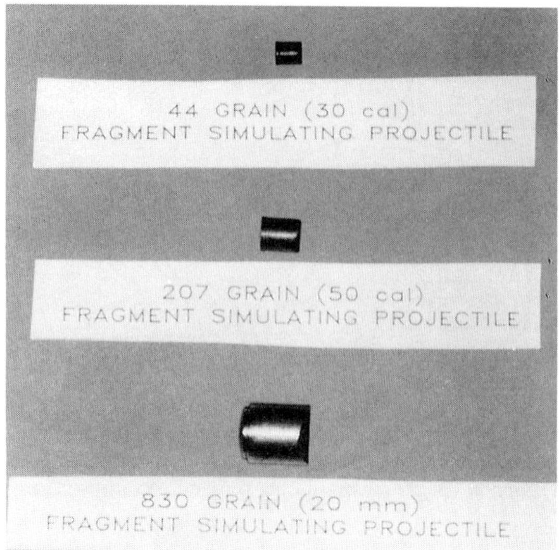

known that a test article with an L/D of one or greater was sufficient to yield the same type of response to impact as a full-scale motor. It was therefore recommended to dissect full-scale motors into thirds. This reduced test costs and provided more test specimens. Nondeliverable rocket motors with discrepancies in the position of the wiring harness were available from the motor manufacturer. These were purchased at a reduced price. The discrepancy in the wiring harness would not affect the motor response to fragment impact. Therefore, these motors were acceptable for this test program. The purchase of these motors resulted in additional program cost savings.

The motors were dissected using a remotely operated, high-pressure grit blast technique. Each motor section was then fitted with two, 1-in thick, aluminum end plates held by four steel tie rods. The sections were potted in place. One plate contained a 1-in diameter hole to serve as a nozzle, providing venting while maintaining internal pressures sufficient to sustain a reaction. The test vessel configuration is shown in Figure 7.

Vulnerability assessments conducted for this missile system were based on the assumption that any fragment with a velocity of 1250ft/sec or higher would cause the rocket motor to detonate. Based on experience, previous tests with this system, and the hydrocode results, this assumption was believed to be overly conservative. Therefore, it was not used as the starting velocity for the experimental motor impact program. Rather, the test plan to experimentally characterize the rocket motor's impact sensitivity was designed to start with the 207-grain fragment at a velocity of 4000 ±20ft/sec. The following

Figure 7. Rocket motor section test article.

shots were planned at 300ft/sec increments until the response went from no detonation to detonation, or vice versa. Once the response change was achieved, the next shots would be at a velocity halfway between those resulting in the two different reactions. This "splitting the difference" proceeds until the detonation threshold velocity is determined to within a maximum of 50ft/sec. This technique is the quickest way to zero in on the desired value, saving both time and test assets. The same procedure would be repeated for the 830-grain FSP. The starting velocity, based on numeric simulation results, would be 2500 ±20ft/sec.

Predicted detonation velocity for the 44-grain FSP was above the capability of the caliber 30 powder gun. Therefore, it was recommended that the starting velocity be the maximum velocity achievable with the standard caliber 30 gun hardware. Further testing would be based on the velocity achieved and the motor response at this velocity. It was expected that the predicted velocity for the 44-grain FSP could not be verified experimentally without switching to a larger barrel and saboting the fragment. The utility of such an exercise would be determined following the completion of all other FSP shots.

Step four:rocket motor impact testing. The rocket motor fragment impact tests were conducted at the New Mexico Institute of Mining Technology. The test set up is shown in Figure 8. Two blast gauges were placed 10ft in front of the test article, one on each side, along 45° angles. Two high-speed cameras were used to record impact and motor response. In addition, two real time video recordings were taken during the entire test procedure. Fragment velocities were determined with Doppler radar as well as multiple velocity screens. The break screens used to determine velocity also provided informa-

Figure 8. Fragment impact test setup.

Table 1
Test Results for 207-Grain FSP*

Impact Velocity	Reaction
3944	Burn
4016	No reaction
4050	Burn
4083	Burn
4283	Detonation
4300	Burn
4326	Detonation
4547	Detonation

* FSP = Fragment simulating projectile.

tion on the orientation of the fragment by the petalling pattern left around the hole as the fragment penetrated the screen. Table 1 contains a summary of the fragment impact velocities and associated motor response for the 207-grain FSP. Figures 9 through 11 show representative examples of the motor responses obtained during this fragment testing. Figure 9 shows a motor that was penetrated by the fragment, ignited, and as a result burned quiescently. Figure 10 shows a motor section that was penetrated by a fragment with no subsequent reaction. Figure 11 shows the aftermath of a motor section that detonated upon impact. As expected, there were no recoverable materials from the motor section, witness plate, or test stand following a detonation.

The detonation threshold velocity for the 207-grain FSP was determined to be 4313ft/sec. The detonation response at the lower velocity, 4283ft/sec, was due to fragment tumbling at impact. Difficulties in reaching the desired

Figure 9. Rocket motor burn response to fragment impact.

Figure 10. Rocket motor with no response to fragment impact.

velocities were encountered with the FSP. The required shot velocities were well above the normal muzzle velocity of a caliber 50 gun, which the FSPs were designed for. The additional pressure required to achieve the higher velocities exceeded the shear strength of the fragment obturator. When the obturator failed, pressure was allowed to leak past the fragment, thus lowering the shot velocity. This problem was overcome by modifying the fragments. The obturator was removed, and a polypropylene pusher plug was attached to the rear of the fragment. This modification facilitated more predictable and repeatable shot velocities, with only a slight drop in the total fragment mass. The modified fragment configuration is shown in Figure 12.

Table 2 contains a summary of the impact velocities and associated

Figure 11. Rocket motor detonation response to fragment impact.

Figure 12. FSP modified configuration.

motor response for the 830-grain FSP. Seven shots were required to determine the detonation threshold velocity of 1900ft/sec. With this large fragment, the motor response was limited to "no response" or "detonation". The fragment impact did not induce burning at velocities below the threshold value, as was seen with the 207-grain FSP. Examining the motor sections that were impacted with no response showed no indication at the entry site or along the bore surface of localized ignition or burning.

Only two shots were conducted with the 44-grain FSP. As discussed previously, the predicted detonation threshold velocity was too high to verify experimentally with a standard setup. The maximum explosive powder load for the caliber 30 gun was used, which yielded impact velocities of 4858ft/sec and 4965ft/sec. In both instances, the fragment failed to exit the motor section. The hot fragment embedded in the propellant web, causing local ignition that spread relatively slowly to the bore surface, igniting the rest of the motor. This is evidenced by the video that shows chuffing at the entry hole for several

Table 2
Test Results for 830-Grain FSP*

Impact Velocity	Reaction
1767	No reaction
1840	No reaction
1853	No reaction
1872	No reaction
1928	Detonation
2036	Detonation
2700	Detonation

* FSP = Fragment simulating projectile.

seconds, then flames appearing in the entry hole, followed shortly by smoke and flames exiting the simulated nozzle. Both motor sections burned for approximately six minutes, which was three times the length of burn experienced with the 207-grain fragment. Due to the benign reaction of these two shots and the good correlation between predicted and experimental results for the 207- and 830-grain FSPs, it was decided to suspend further testing with the 44-grain fragment. It was also noted that the predicted detonation threshold velocity, 7743ft/sec, was well above the fragment velocity from artillery threats of interest to this program.

Application of Characterization Results

The rocket motor impact test program served three major purposes. First, it provided a timely, accurate, and cost-effective characterization of the motor's sensitivity to fragment impact. These data were used to correct and update vulnerability analyses for the missile system. The previously discussed detonation velocity of 1250ft/sec was proved to be overly conservative and invalid for survivability enhancement studies.

Second, the experimental program validated the results of the numeric simulations. Figure 13 shows the predicted detonation threshold velocities plotted as a function of fragment mass. Also shown are the experimentally determined values. For the 207-grain FSP, the predicted value was 1.3 percent lower than the experimental result. For the 830-grain FSP, the predicted value was 24 percent higher than the experimentally determined velocity. These are good correlations between analytic and experimental results of this nature. The results from the hydrocode calculations were certainly more accurate than the detonation threshold velocity value used prior to this test program.

The third and primary goal of the rocket motor impact test series was to

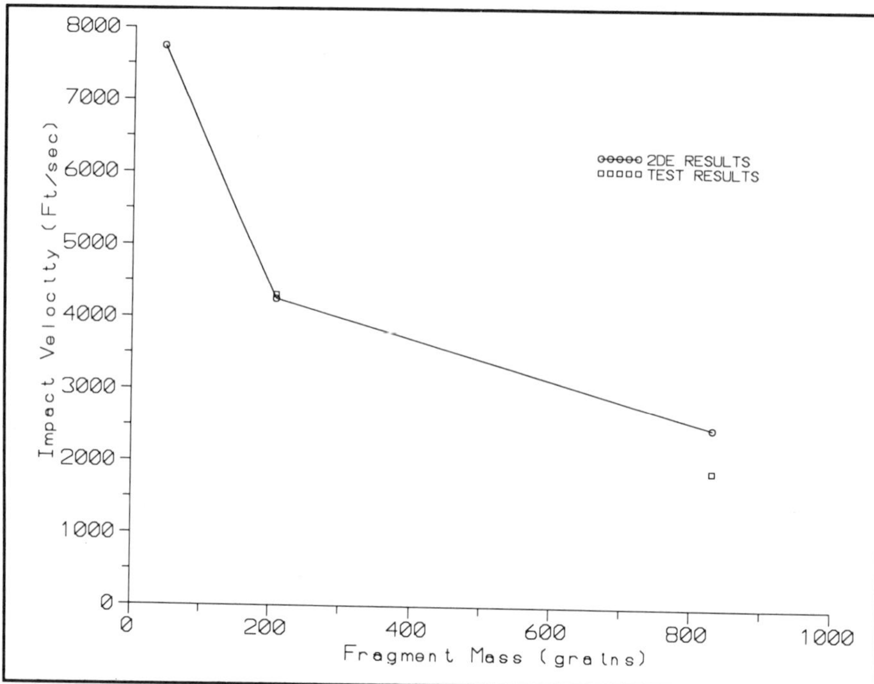

Figure 13. Rocket motor detonation threshold velocity: critical velocity versus fragment mass.

provide a criterion against which to evaluate potential armor solutions. The test program was successful in meeting this goal. The candidate armors were tested with the 207-grain FSP at a velocity of 5000ft/sec. This fragment mass/velocity combination was above the threshold value for the rocket motor and provided a measure of the effectiveness of the armors at reducing the mass/velocity to below the critical value. Armors that were successful at reducing the mass/velocity to values below the detonation threshold offered potential survivability enhancement for the system.

Conclusions

As part of the survivability enhancement program for an Army missile system, an innovative methodology for characterizing the response of a solid propellant rocket motor to fragment impact was demonstrated. This methodology combined experimental techniques developed for high explosives with an analytic tool derived from basic continuum mechanics.

Wedge tests were conducted on the solid propellant to evaluate the shock sensitivity of the material and to derive its equation of state. The experimentally determined parameters were then input into 2DE, a two-dimensional hydrocode, to conduct numeric simulations of impacts against the rocket motor. The hydrocode analyses were used to predict the impact velocity required for a specific projectile configuration to initiate a detonation in the rocket motor. With these predicted velocities, a test plan was tailored to experimentally characterize the detonation threshold for the rocket motor and significantly reduce the number of tests and assets required. In this program, the experimentally determined detonation velocities correlated well with the predicted values, thereby offering validity to the application of these procedures.

Application of this methodology offers numerous advantages. As demonstrated in this program, this procedure is a viable alternative to the time-consuming and costly experimental programs normally used to characterize the response of a solid rocket motor to fragment impact. It also provides a procedure to screen developmental propellants in terms of their shock sensitivity prior to committing them to significant production scaleup. The combination of testing and modeling utilized here also provides a more accurate data base for designers to use in investigating mitigative techniques such as shields or armors to enhance system survivability. In addition, the procedure demonstrated here offers a tool for the evaluation of fragment and bullet impact response of solid rocket motors for Insensitive Munitions (IM) considerations. This procedure offers to IM investigators, as it did to the SWG, a more cost-effective and timely means of characterizing the response of a solid propellant rocket motor.

Future Plans

The success of this innovative methodology that was implemented as part of a survivability enhancement program has sparked interest in its use on other Army systems currently involved in similar activities and as a tool for IM modeling. To further investigate, validate, and improve the effectiveness of this characterization procedure, several follow-up efforts are being planned. The first will include additional modeling and testing of the currently characterized solid propellant against different fragment geometries. Characterization of the shock sensitivity, through wedge testing, of additional Army propellants as well as developmental IM type propellants is underway. Other hydrocodes, both one and two-dimensional, will be evaluated as to their modeling and predictive capabilities. The most effective and useful codes will then be brought in-house to provide researchers with an accurate modeling tool for impact scenarios.

References

1. J. D. Kershner and C.L. Mader, *2DE: A Two-Dimensional, Continuous, Eulerian Hydrodynamic Code for Computing Multicomponent Reactive Hydrodynamic Problems* (document LA-4846, Los Alamos Scientific Laboratory, NM, March 1972.
2. J. Wilbeck *et al.*, *A Short Course on Penetration Mechanics* (Southwest Research Institute, San Antonio, TX, undated).
3. C.L. Mader and C.A. Forest, *Two-Dimension Homogeneous and Heterogeneous Detonation Wave Propagation* (document LA-6259, Los Alamos Scientific Laboratory, NM, June 1976).

38

Advanced Ground Propulsion System Phenomena

Walter Bryzik
Ernest Schwarz
Eugene Danielson
Michael Reid
U.S. Army Tank Automotive Command
Research, Development and Engineering Center
Warren, MI 48397-5000

Unique technology driven engines are essential for developing compact, high-power density ground propulsion systems needed to fulfill future Army ground propulsion requirements.

Dramatic and achievable payoffs are possible in ground propulsion. These payoffs require significant advances in optimized combustion, heat release phasing, and fluid flow/fuel spray interaction. The authors focus on low heat rejection (LHR) "adiabatic" engines. Of particular note, complex interacting mechanisms within the LHR class of engine prove to be extremely sensitive to heat release phasing and its breakdown between premixed and diffusion combustion. Small parameter changes significantly enhance or degrade engine performance, depending on engine geometry, heat transfer, and fluid flow/chemical kinetic phasing interactions. Sensitivities to such small parameter changes are not nearly as great in conventional water cooled engines.

Studies to date reveal that while chemical kinetics play a dominant role in the performance of a conventionally cooled engine, an LHR engine is dominated by mixing controlled phenomena. The combination of shortened ignition delay and diffusion mixing limitations is responsible for producing longer heat release periods in nonoptimized LHR engines. The authors present results of experiments that were designed to develop a better understanding of this heat release behavior.

Introduction

Advanced low heat rejection (LHR) in diesel engines offers dramatic opportunities for improvements in future Army ground vehicle: witness the upgraded M109 Howitzer and Paladin, which use LHR technology.

The LHR diesel combustion chamber is insulated by using high temperature materials on the pistons, cylinder head, valves, cylinder liner, and exhaust ports.[1,2] Greater power and efficiency are derived because thermal energy normally lost to the cooling water and exhaust gas is converted to useful work. By reducing lost energy and essentially eliminating the need for a conventional cooling system, this engine concept will dramatically improve performance and could result in a 50 percent reduction in weight and volume for the entire propulsion system. The engine would have great positive impact, upon military tactical and combat vehicles. Elimination of the water cooling system would greatly improve reliability and maintainability. The engine would not be subject to conventional cooling damage oe sensitive to extreme environmental conditions. Weight reductions allow improved vehicle response, while volume reductions allow reduced armor and improved survivability characteristics. Higher temperature engine operation, quieter combustion, and improved multifuel characteristics are also possible.

With the LHR concept, combat vehicle design becomes far less restrictive because concerns about locating cooling grilles, passages, and associated equipment are eliminated. The cost would be equal to or less than the water-cooled counterpart because water pumps, seals, hoses, and costly water jackets would be eliminated.

To capitalize on LHR technology, two technical challenges must be met: (1) high temperature combustion optimization and (2) structural integrity of high temperature and complex engine components.

Objectives and Methodology

This research is directed at identifying combustion factors responsible for performance differences between conventional water-cooled and LHR engines. LHR combustion has not been adequately studied, and past studies reveal conflicting results.[3-5] Past works have revealed that LHR combustion phenomena are complex and subject to extreme parameter sensitivities. Specifically, the LHR was found to be extremely sensitive to heat release phasing and the contributions to it from premixed and diffusion combustion. For example, small changes in surface thermal conductivity will either significantly enhance or degrade engine performance, depending on engine geometry, heat transfer, and fluid flow/chemical kinetic phasing interactions. The relationship between premixed and diffusion burning in an LHR engine is critical.

High speed in-cylinder pressure data were gathered for an LHR-VT-903, Cummins research engine. This engine is a single cylinder version of the V-8 engine currently powering the Bradley Fighting Vehicle and is representative of high-output military diesel engines. This engine can run as a fully cooled or limited cooled (head cooling only) configuration. In-cylinder pressure data were obtained for varied coolant, intake manifold, and oil temperatures, and for insulated and noninsulated pistons. For the insulated pistons, porous and seal-coated zirconia coatings were examined to determine effects on combustion and performance. Heat release rates were calculated from the pressure data and evaluated for a range of engine operating parameters and design configurations.

It was the intent also to use the combustion and performance data generated to improve in-house diesel cycle simulation codes. The current approach is to use experimental correlations. Predictions with those codes of internal operating conditions and performance of water-cooled diesel engines are very good, but questionable for LHR engines. LHR heat release data will be used to improve the burning rate models in the codes to more accurately predict LHR engine performance.

Test Engine

The test engine was made by cutting the block of a standard V-8 and leaving only two cylinders at the flywheel end. Only one cylinder is active, while the other has a slipper piston for balancing. The front end of the engine, crank, cam, cylinder head, and fuel injection system were modified to match the single cylinder configuration. Engine specifications are listed in Table 1. This engine was configured for cooling of the head alone, thereby simulating LHR or minimum-cooled operation.

A ductile iron piston was used for the standard cooled engine configuration, while two different pistons were fabricated for the LHR tests. The first of the latter was a 0.040-in thick plasma-sprayed zirconia-coated ductile iron piston. The second was similar except the surface was impregnated with chrome oxide to act as a seal coat. The insulated test pistons had the same compression ratio (13:9) as the baseline case.

A synthetic polyester lubricant, Stauffer SDL-1, was employed. SDL-1

Table 1
VT-903 Single Cylinder Research Engine Specifications

Type	4 cycle DI diesel
Bore	5.5in
Stroke	4.75in
Displacement	112.88cu in
Rated speed	2800rpm
Compression ratio	13.9
Rated BMEP	170psi
No. of valves	4
Injectors	Cummins PT

has been extensively used in Army and other LHR engine developments and is adequate for the temperature ranges encountered during these tests. Standard diesel fuel was used for all tests (specification VVF-800, DF-2).

Test Installation

The test research engine was installed in a test cell and was coupled to a Pohl eddy current dynamometer with a 125-hp absorption capability. This dynamometer had an induction motoring section (30hp) for engine starting and friction tests. A Digalog Corp. system was used for dynamometer speed and load control, and a load cell was used for torque measurement.

Test cell support systems for air, fuel, cooling, and oil systems permitted flexible flow and temperature control. Engine operating conditions could be controlled to closely simulate those of a turbocharged, aftercooled engine in a combat vehicle environment. The test cell setup provided the capability to perform parametric studies to determine operating effects on performance and heat release. External air was used to simulate turbocharged conditions. Maximum capability for charge air pressure is 90psi. Intake air was heated with a 120KBTU/hr triple-pass heater system so that intake temperatures could be equilibrated to aftercooled air temperatures. Intake pressure was controlled with a regulator, and exhaust back pressure was maintained with an electrically controlled butterfly valve, permitting an intake-to-exhaust manifold pressure ratio of one to be maintained, which is representative of turbocharged diesel engine applications. Cooling and lubrication systems were installed with pneumatic controllers to allow accurate control of temperatures. Flow rates were controlled with pump bypass circuits. Coolant and lubricant circuits used shell and tube heat exchangers, and steam heating was used to bring these fluids up to temperature without relying on engine heat. A Cummins PT injector was used. Because the engine had no drive for an engine-mounted PT pump, a variable-speed fuel pump was employed. The fueling circuit was equipped with a small needle valve bypassing the pump output to allow fine pressure adjustment. Fuel rate is controlled in a PT system by rail pressure control. Coarse fueling was accomplished by adjusting the pump speed.

Low speed data were acquired with a Fluke system. The engine was fully instrumented for temperature, pressure, speed, load, and flow measurements. Fifty channels were employed for low speed data acquisition. Air, coolant, and lubricant flows were measured with turbine flowmeters. An Elwood system was used for fuel consumption measurement. High speed combustion data were acquired using PEI Inc. PC-based data acquisition hardware and analyzed using PEI Inc. ECA 892 matched software. Heat release analyses conducted offline with a modified Krieger Borman model were developed at the University of Wisconsin. The high speed system was used to record crank-angle (CA) resolved data, including cylinder pressure and fuel injector link load for the injection timing event. This system has four analogue input channels, with a sampling rate of 125,000 samples per second. All high speed data were taken over 100 cycles for each steady-state operation point and were ensemble averaged to provide statistically meaningful results. A Kistler 7061 piezo-electric transducer was employed for cylinder pressure measurement, and a BEI precision optical encoder with .2-CA degree resolution was used for marking and timing.

Test Operating Conditions

The engine was run at 1600, 2100, and 2600rpm at 50 percent and 100 percent load levels for each speed. Torque levels for the baseline correspond to those of the V-8 engine at a 500-hp rating. Coolant, lubricant, and intake manifold temperatures were varied to determine their effect on engine overall performance and heat release.

Engine airflow and air-to-fuel ratio operating conditions are shown in Figure 1 for full load water-cooled baseline tests. For all LHR tests, the fuel rate was held constant, and the airflow rate was adjusted via intake manifold pressure to match the baseline case air-to-fuel ratios. Thus, torque was made the dependent variable. This method was used for the part load cases, although higher air-to-fuel ratios were used, which is representative of standard direct injected diesel engines. Where a compression of 1.0 could not be achieved, baseline and LHR case ratios were maintained at the same level.

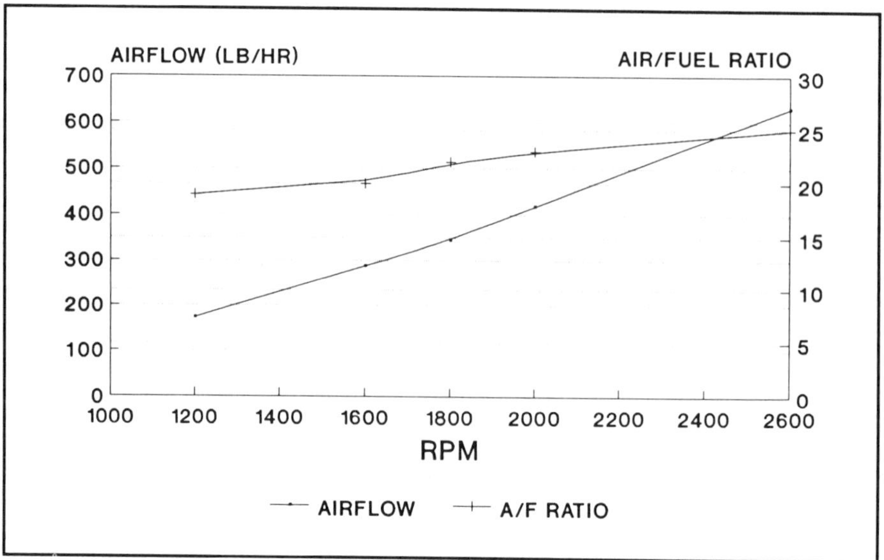

Figure 1. Airflow and A/F versus rpm at full load.

Thermal State of the Test Engine

A thermal finite element analysis (FEA) was conducted to predict temperature across the section of the head. Figure 2 shows temperature variation between valve bridge regions within the firedeck.

The FEA predicted fully cooled peak temperatures agree well with those measured previously.[6] It is concluded that the LHR isotherms are quite representative of actual temperatures, meaning that the LHR head is operating at temperatures 150°F to 200°F hotter than in the fully cooled engine. At these elevated temperatures, material selection could be a critical issue, because the yield strength of gray iron rapidly deteriorates above approximately 600°F.

The liner also sees elevated temperatures because it has no jacket cooling. Figure 3 compares temperatures in the cylinder liner at the top ring reversal location from prior engine testing.[6] The LHR liner is operating at temperatures 200°F to 250°F higher (depending on engine speed) than the liner in the fully cooled engine. Once again, there should be concern about material selection.

Operating conditions in the current test program were similar to those reported,[6] that is, a maximum piston cap temperature of 1394°F, a substrate metal temperature of 750°F, top ring groove temperature of 620°F, and maximum pin bore temperature of 320°F. Piston temperatures would be approximately 200°F cooler for the fully cooled configuration. This assump-

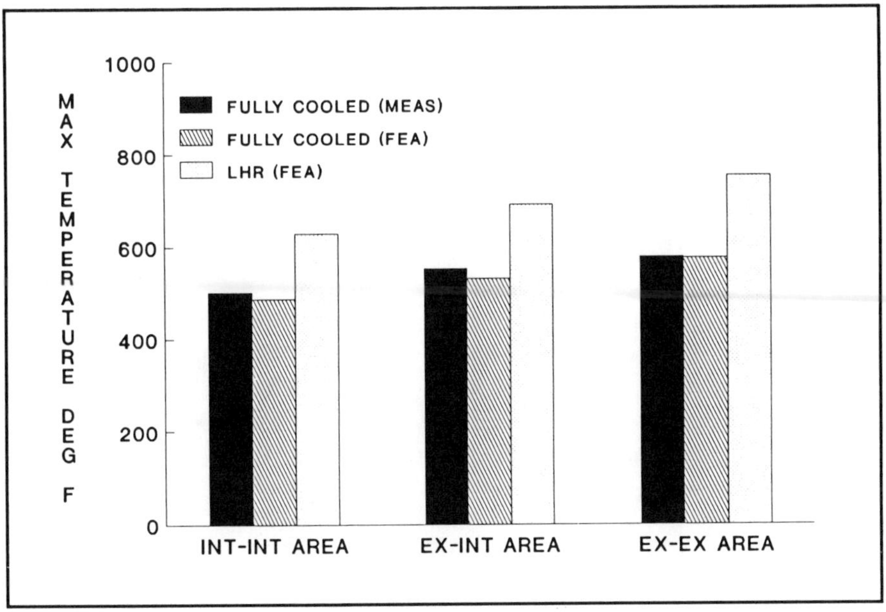

Figure 2. Maximum temperature on firedeck (bridge areas).

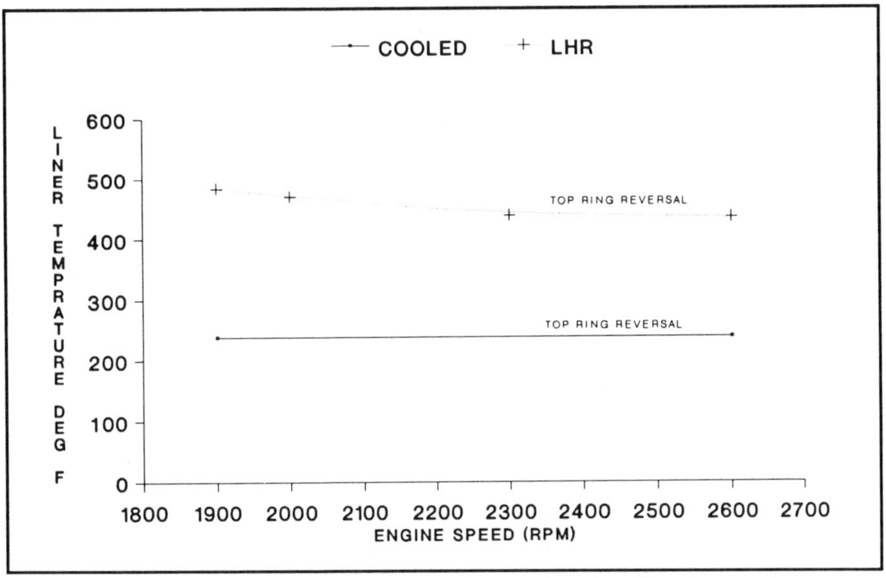

Figure 3. Fully cooled and LHR liner temperatures as a function of engine speed.

tion is based on the general range of temperature differentials already established for the firedeck and liner. Again at these temperatures, piston and coating materials issues need to be carefully addressed.

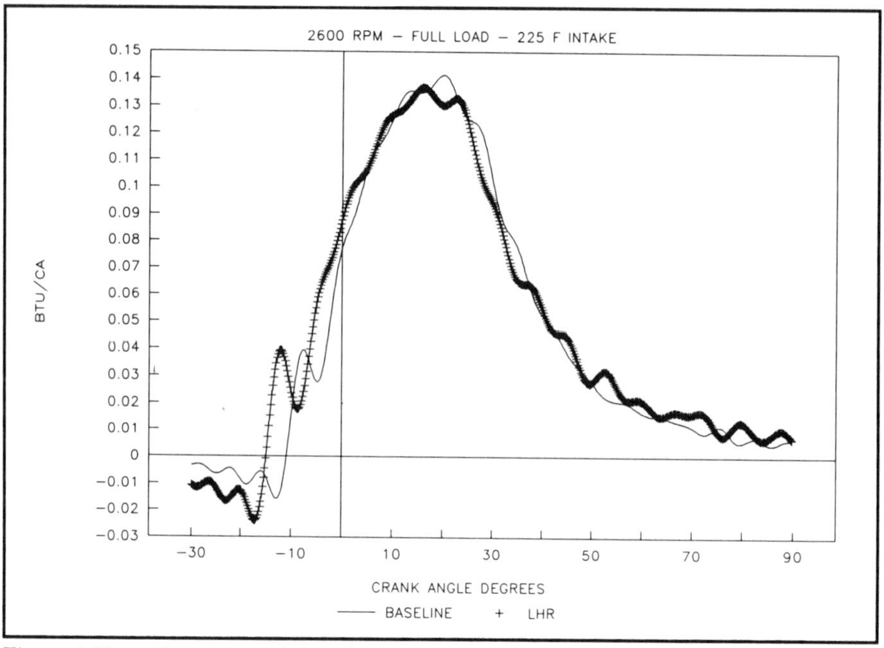

Figure 4. Heat release comparison at 2600rpm under a full load and with an intake temperature of 225°F.

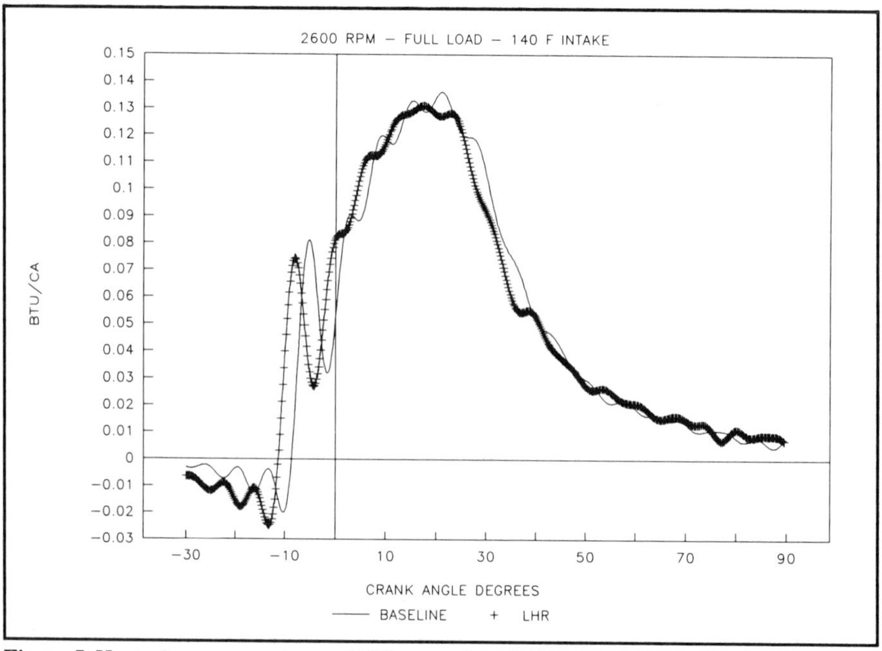

Figure 5. Heat release comparison at 2600rpm under a full load and with an intake temperature of 140°F.

Finally, differences in specific heat rejection to coolant should be mentioned. The fully cooled engine in a multicylinder configuration has a specific heat rejection of 24.5BTU/hp-min, whereas the LHR engine had measured values[6] in the range of 17.0 to 19.0BTU/hp-min, depending upon oil sump, coolant, and charge air temperatures. This is a substantial decrease in heat rejection, thus allowing for smaller cooling systems and less cooling fan parasitic losses.

Combustion Results

Pressure time histories and injection timing events were recorded for all cases. From these data, heat release rates were calculated. Pressure and heat release information is used to analyze the combustion process, which provides insight into the overall performance of the engine.

Heat release diagrams, for the 2600rpm case are provided in Figures 4 and 5. Baseline and LHR cases are compared in each figure. All data for the LHR is at full load, 300°F coolant, and 300°F oil temperatures.

These heat release profiles are typical of quiescent or limited swirl direct injected diesel engines. As crank angle increases, one first encounters the premixed burning peak and then the diffusion burning peak. The diffusion burn portion for all these curves is the predominant mode of combustion. These diagrams reveal the following trends: (1) the premixed burning peak decreases with increasing engine speed for constant intake manifold temperature, (2) the premixed burning peak and burning portion is always greater at a given speed when intake manifold temperature is lower, and (3) for the ceramic minimum cooled case, the premixed burning peak is always less than for the conventional case, with the difference being more at the low speed and very small at the high speed.

Figures 4 and 5 indicate a difference in beginning of ignition between the baseline and LHR cases, with LHR ignition beginning one to three degrees earlier than for the conventional engine. The static injection timing for the Cummins PT injection system was fixed for all test runs. Initiation of injection with this system is a function of static timing and amount of fuel in the injector. The same amount of fuel was injected in the comparative cases. Previous LHR studies have shown that the engine's overhead may become so hot that the injector link expands, causing an advance in injection timing. The link was strain gauged to determine approximate initiation of injection. No apparent change in start of injection was found; therefore, the ignition phase shift can be attributed to a change in ignition delay rather than any dynamic

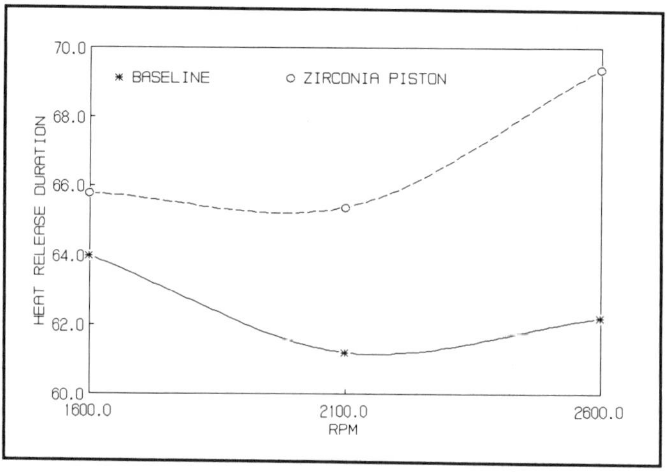

Figure 6. Heat release duration with a 300°F oil temperature and a 300°F coolant temperature.

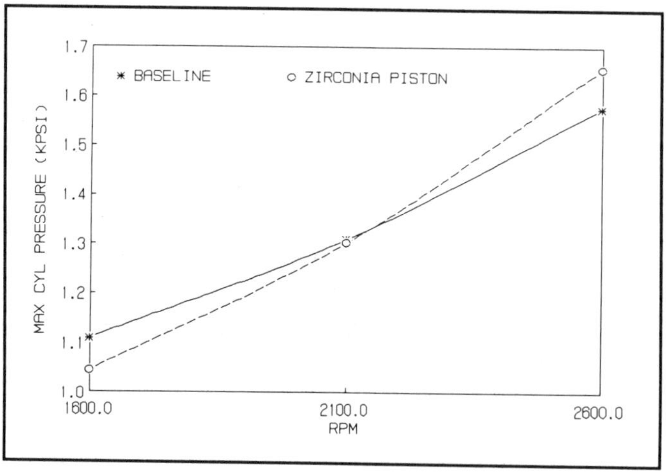

Figure 7. Cylinder pressure comparison with a 300°F oil temperature and a 300°F coolant temperature.

injection change. This makes sense because the higher temperatures at injection would lead to a shorter ignition delay, leading to a decreased premixed fraction, as indicated in the heat release diagrams for full load conditions.

Half-load heat release comparisons were also made. Coolant temperature was 250°F, oil temperature was 250°F, and a charge air temperature 225°F was maintained for the LHR. Trends are not as dramatic as in the full load cases; however, premixed differences are more apparent at the lower speeds, with less premixed burning for the LHR case. Burning duration was longer for the LHR at all speeds.

Heat release summary data for full load conditions are given in Table 2. The heat release duration in crank angle degrees is calculated as the portion of the cycle where 0.5 percent to 0.95 percent of the cumulative net heat release has occurred. The data provide insight into how engine performance is

Table 2
Heat Release Comparisons—Full Load
(300°F Coolant, 300°F Oil)

Case	Speed (rpm)	Intake Manifold Temp (°F)	Peak Cylinder Pressure (psi)	Peak Premixed Net HR Rate (BTU/CA)	Peak Diffusion Net HR Rate (BTU/CA)	Heat Release Duration (CA)
Ceramic	2600	225	1657	0.040	0.137	69.4
Baseline	2600	225	1577	0.040	0.141	62.2
Ceramic	2600	140	1441	0.073	0.131	70.4
Baseline	2600	140	1457	0.081	0.136	66.6
Ceramic	2100	225	1303	0.034	0.135	65.4
Baseline	2100	225	1310	0.043	0.135	61.2
Ceramic	2100	140	1241	0.061	0.131	67.2
Baseline	2100	140	1269	0.087	0.137	64.0
Ceramic	1600	225	1045	0.032	0.115	65.8
Baseline	1600	225	1110	0.053	0.125	64.0
Ceramic	1600	140	1075	0.072	0.119	68.2
Baseline	1600	140	1094	0.106	0.126	63.2

affected by operating parameters. As discussed previously, the premixed burning fraction was generally less for the LHR cases, except at 2600rpm full and half load (not shown) for the hotter 225°F intake manifold temperature conditions, where they were equal. In all cases the heat release duration was greater for the LHR. This is clearly illustrated in Figure 6. Heat release duration increased with speed for these runs. These findings agree with results of experiments by Alkidas,[7] Gatowski,[8] Dickey,[9] Morel *et al.*,[10] and Assanis *et al.*[3]

Of extreme interest is peak cylinder pressure data, because these play a very important role in fuel consumption data presented later. A full load peak cylinder pressure comparison is displayed in Figure 7 for constant 225°F manifold temperature. From Figure 7 and Table 2, cylinder pressure data indicate: (1) at 1600rpm, LHR pressures are lower than the baseline, (2) at 2100rpm, LHR pressures are close to the baseline, and (3) at 2600rpm, LHR pressures are approximately equal at 140°F intake manifold temperatures but greater than the baseline at the 225°F intake.

Performance Results

LHR full load engine performance data are presented in Table 3. Performance parameters include net indicated mean effective pressures (NIMEP)

Table 3
Performance Comparisons—Full Load
(300°F Coolant, 300°F Oil)

Case	Speed (rpm)	Intake Manifold Temp (°F)	Intake Manifold Pressure (in Hg)	Exhaust Manifold Temp (°F)	NIMEP (psi)	NISFC (Lb/NIHP -HR)
Ceramic	2600	225	66.56	1377	208.2	0.343
Baseline	2600	225	64.93	1264	202.5	0.353
Ceramic	2600	140	55.97	1311	195.3	0.343
Baseline	2600	140	55.36	1261	196.9	0.341
Ceramic	2100	225	54.34	1312	188.8	0.327
Baseline	2100	225	52.92	1237	188.2	0.329
Ceramic	2100	140	48.24	1290	182.5	0.341
Baseline	2100	140	46.61	1224	194.3	0.321
Ceramic	1600	225	43.76	1312	162.0	0.353
Baseline	1600	225	42.74	1255	171.7	0.334
Ceramic	1600	140	39.69	1336	169.3	0.351
Baseline	1600	140	38.88	1234	175.9	0.338

and net indicated fuel consumption (NISFC). These quantities indicate the working and pumping loop mean effective pressures and fuel consumptions as calculated from the cycle pressure volume diagrams. Other relevant performance data include intake pressures from which volumetric efficiencies and the engine exhaust temperatures can be calculated.

The LHR engine suffered a loss in volumetric efficiency as boost was added to maintain the constant air fuel ratio of the baseline. This difference was less than 2in Hg in all cases. Other researchers experienced more severe losses for engines of this type with uncooled heads,[11] but volumetric efficiency (based on intake manifold temperatures) can be regained[10] as intake air temperatures are increased. This engine did not experience much loss in volumetric efficiency while the heads were still being cooled. All exhaust temperatures were higher for the LHR case, with the greatest difference of 113°F being experienced at the 2600rpm, 225°F intake manifold point. Again, other researchers[8] have seen higher exhaust temperature differences where less cooling and more in-cylinder surface insulation were employed.

Fuel consumption comparison data are more clearly depicted in Figures 8 to 10. Figure 8 shows that the LHR specific fuel consumption was poorer at 1600rpm, about equal to the baseline at 2100rpm, and better at 2600rpm. A rerun of the 2600rpm case was performed with no change results. Better fuel

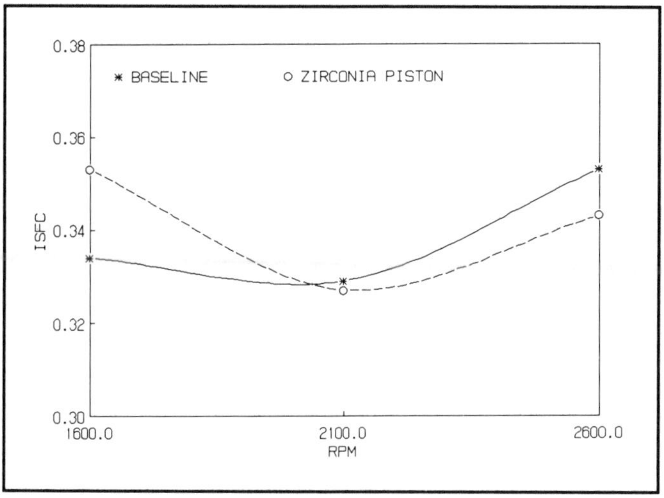

Figure 8. Comparison at full load with intake air at 225°F.

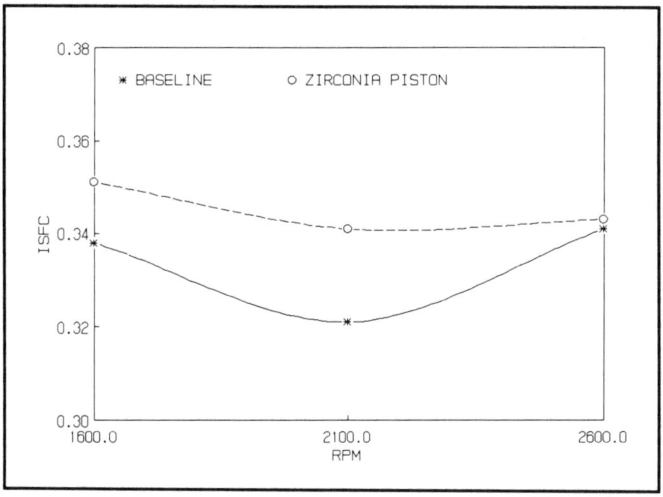

Figure 9. Comparison at full load with intake air at 140°F.

consumption with higher in-cylinder temperatures and higher speeds is confirmed by Figures 9 and 10 at the 2600rpm points. Higher peak cylinder pressures encountered at the higher engine speeds and less variation in the premixed burn for the LHR case would seem to substantiate why fuel consumption is better at the 2600rpm rated speed condition. This was borne out in Figure 10, where at 2600rpm, the indicated specific fuel consumption improved with increased head coolant temperature. Data presented by Ricardo, ITI, and Purdue University[5] on an engine with a similar combustion chamber showed poorer NISFC at low speeds, but better at high speeds and high loads.

The effect of blowby on heat release and performance has been dis-

Figure 10. Coolant comparison at full load with intake air at 225°F.

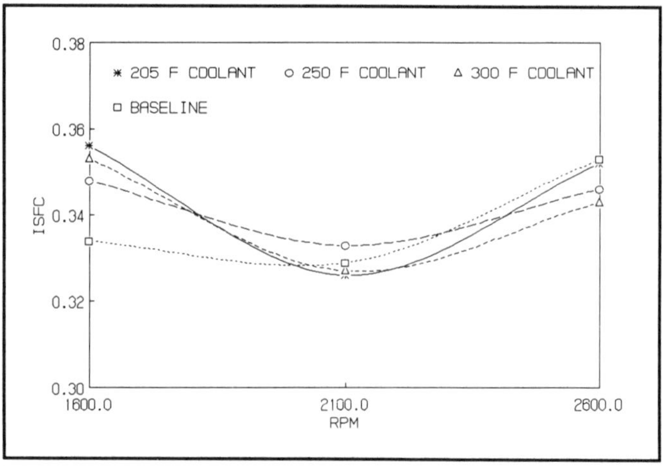

cussed by Alkidas.[7] Crankcase pressure was monitored throughout all the testing conducted here. Crankcase pressures were lower for all LHR test cases. Because this engine had no liner cooling, piston/liner clearances were likely to have shrunk, with resultant lower blowby, particularly at the higher engine speeds. This, along with the ignition advance experienced, could explain why LHR cylinder pressures were higher at 2600rpm.

Ceramic Coating Durability

After each day's test runs, the piston coating and bore were examined to determine their condition because of concerns regarding the harsh operating environments. The zirconia coating lasted approximately 60 hours before it began to spall from the ductile iron substrate. At that point, 50 percent of the coating was missing. Such runs are not reported here.

When the zirconia coated piston failed, a chrome oxide seal coated zirconia piston was installed to determine whether piston surface conditions affected combustion due to fuel absorption or evaporation and burning of fuel from coating. It is plausible this may be causing extended LHR release durations, as has been presented by several experimenters. Unfortunately, this coating failed after a little more than one hour of test time, with no data collected. Upon closer inspection, 30 percent of the coating had delaminated from the center of the piston dome. Further investigation is planned for the piston coating surface effects (porous versus nonporous coatings) on LHR combustion.

Summary

The following comments can be made on the engine tests reported here:

(1) Combustion diagram analysis revealed that ignition delay was shorter, premixed fraction was less, and heat release duration greater for the LHR engine. The volumetric efficiency was less for the LHR engine; however, differences where not substantial (3 percent or less). Exhaust temperatures were greater in all LHR cases.

(2) LHR engine performance results were mixed. Generally, at full load conditions the indicated specific fuel consumption was better (3 percent) at the high speed conditions and worse (5 percent) at the low speed conditions for the highest coolant, oil, and intake manifold temperature test conditions. This corresponds directly with the higher instantaneous peak combustion pressures measured at the high speed conditions. Also, as the head coolant temperatures were increased at the rated 2600rpm condition, the trend indicated better fuel consumption.

(3) Finite element analysis predicted a 200°F higher peak temperature across the LHR head firedeck. The peak temperature was 813°F between the exhaust valves. Previous test work showed an increase of 250°F for the liner top ring reversal temperature and 200°F for the piston dome in an LHR engine.

Conclusions

1. Improved mixing of fuel and air is required to permit optimization of low heat rejection combustion. Specifically, improved time phasing of premixed and diffusion burning are needed. If a desirable fundamental cycle heat release is to be derived. Continued combustion optimization must be pursued to realize the advantages of low heat rejection diesel engines.

2. Heat release in LHR engines exhibits a far greater sensitivity to small changes than in normally cooled engines.

3. While experimentation verification is required, seal coats on porous ceramic coatings may reduce cyclic thermal inertia.

4. The thermal state of the test engine in-cylinder components, even at modest LHR conditions, must be carefully analyzed at the higher temperature conditions in order to obtain acceptable performance and structural and tribologic integrity.

5. Fundamental combustion data are indispensable for improving state-of-the-art burning laws within diesel cycle simulation programs.

References

1. W. Bryzik and R. Kamo, SAE Paper 830314, *SAE Trans.* **92**, 1983.
2. W. Bryzik and R. Kamo, *TACOM/Cummins Adiabatic Engine Program* (SAE Paper 840428, SAE, February 1984).
3. D. Assanis *et al.*, *The Effects of Ceramic Coatings on Diesel Engine Performance and Exhaust Emissions* (SAE Paper 910460, SAE, February 1991).
4. C.S. Reddy, N. Domingo, and R.L. Graves, *Low Heat Rejection Engine Research Status: Where Do We Go from Here?* (SAE Paper 900620, SAE, February 1990).
5. G.L. Borman *et al.*, *Development of a Modular Transient Cycle Analysis Program for the Adiabatic Diesel and Other Compound Diesel Engines* (Technical Report No. 13558, University of Wisconsin and TACOM, June 1991), pp. 57-71.
6. M. Brands, J. Hoenhe, and C. Moore, TACOM Adiabatic Engine Program Review, TACOM, Warren, MI, 1984.
7. A.C. Alkidas, *Performance and Emission Achievements with an Uncooled, Heavy Duty Single Cylinder Diesel Engine* (SAE 890144, SAE, February 1989).
8. J.A. Gatowski, *Evaluation of a Selectively Cooled, Single Cylinder, 0.5 L Diesel Engine* (SAE 900693, SAE, February 1990).
9. D.W. Dickey, *The Effect of Insulated Combustion Chamber Surfaces on Direct Injected Diesel Engine Performance, Emissions and Combustion* (SAE 890292, SAE, February 1989).
10. T. Morel, S. Wahiduzzaman, and E.F. Fort, *Heat Transfer Experiments in an Insulated Diesel* (SAE 880186, SAE, February 1988).
11. M. Woods, E. Schwarz, and W. Bryzik, *Heat Rejection from High Output Adiabatic Diesel Engine* (SAE 920541, SAE, February 1992).

Chromium-Free Carbon:

A New Adsorbent for Chemical Warfare Agent Filters

Robert W. Morrison
U.S. Army
Edgewood Research,
Development and
Engineering Center
Aberdeen Proving Ground,
MD 21010-5423

David T. Doughty
Calgon Carbon Corporation
Pittsburgh, PA 15230-0717

A new, highly effective, environmentally compatible adsorbent has been developed to eliminate hexavalent chromium, a known carcinogen, used as a reactant in all chemical warfare agent filters. Because hexavalent chromium is leachable from ASC carbon, proper disposal of spent filters results in high costs to the Government. The role of hexavalent chromium was to provide reactivity toward the cyanide containing blood agents cyanogen chloride and hydrogen cyanide. Thus, the Chromium-Free Carbon Development Program centered on finding a reactant that could be impregnated on activated carbon and provide protection against cyanide compounds. The other ASC carbon impregnants, copper and a trace amount of silver, were retained in the new formulation to provide reactivity against other agents. Over 100 impregnants were screened to identify five candidate formulations that were then extensively characterized. From this study, the formulation providing the best sorption against a complete spectrum of standard chemical warfare agents was identified. The agent removal role of each impregnant of the formulation has been well defined. Extensive temperature-humidity testing of chromium-free carbon has demonstrated stable performance on exposure to varied field conditions. Improved activated carbon impregnation process technology has eliminated the irritating levels of ammonia that off-gas from military filters. Protective masks and filters were fabricated using chromium-free carbon and these have passed their specification requirements. Implementing chromium-free carbon into the technical data packages of military filters is now in process.

Introduction

Ever since the first use of toxic chemicals as a weapon during World War I, activated carbon impregnated with reactive chemicals has been used to provide respiratory protection for military personnel.[1] One of these early impregnated carbons used copper oxide as the reactant to serve as an acid gas sink. As new chemical threats were identified, additional reactants were added to the carbon to detoxify them. During World War II, hexavalent chromium was added as a reactant to provide protection against hydrogen cyanide and cyanogen chloride. Since that time, some forms of hexavalent chromium have been determined to be human carcinogens, and as a result materials that contain this chemical fall under stringent environmental regulation. Because this hexavalent chromium impregnated adsorbent (ASC carbon) is used in all U.S. military chemical warfare agent filters, the Department of Defense spends over $2,000,000 per year to dispose of spent filters in accordance with environmental regulations. The authors describe the development of an environmentally acceptable adsorbent for use in military chemical warfare agent filters.

Background

In the early 1980s, with the increasing attention to environmental matters, the presence of hexavalent chromium on ASC carbon became an important issue of concern. ASC carbon is activated carbon that has been impregnated with copper, silver, and chromium compounds to provide high levels of filtration performance against all chemical warfare agents. The activated carbon is effective in removing by physical adsorption low vapor pressure chemical warfare agents such as nerve and mustard agents. However, the high vapor pressure agents such as the blood agents cyanogen chloride and hydrogen cyanide are not retained well on activated carbon by physical adsorption and thus, chemical reactants (usually called impregnants) are applied to the surface of the larger pores of the activated carbon. These chemical reactants convert the toxic chemical agents to reaction products (such as copper chloride) that are retained by the carbon and to nontoxic volatile chemicals (such as carbon dioxide) that pass out of the filter in the purified air.

Each year as much as a million pounds of ASC carbon are produced for the manufacture of U.S. military filters used to replace spent filters requiring change-out. These filters are located worldwide and must be disposed of in accordance with environmental regulations. Because of the hexavalent chromium, ASC carbon has been shown to be a hazardous waste under Resource Conservation and Recovery Act (RCRA) regulations.[2] As an RCRA hazardous waste, ASC carbon filled filters are subject to extremely expensive disposal procedures that include not only the costs associated with the actual disposal but also those required to retrieve the filters from their use location.

In addition to the high disposal costs associated with filters containing ASC carbon, the carcinogenic chromium also poses potential physiologic risks to personnel who could potentially come in contact with the carbon.[3] These personnel include those involved in the manufacture of the carbon and filters that contain the carbon, in the event that appropriate protective measures are not used to keep ASC carbon dust out of the breathing air. Even the soldiers, sailors, and airmen who rely on the filters for chemical protection could be at slight risk of physiologic harm in the event that damage to the filter occurs during use, and dust from the chromium impregnated carbon exits the canister into the breathing air.

Beginning of the Chromium-Free Carbon Development Program

To address these concerns resulting from the hexavalent chromium, the U.S. Army Edgewood Research, Development and Engineering Center (ERDEC) began to develop a nonhazardous adsorbent for the filtration of chemical warfare agent vapors. Beginning in 1982, a preliminary screening of options for a chromium-free impregnant formulation was conducted under contract with Westvaco, Inc., North Charleston, SC.[4] Work under this contract demonstrated that alternative nonhazardous impregnant formulations could provide sufficiently high levels of filtration performance to meet filter requirements. Under this preliminary contract, little effort was undertaken to find the optimum impregnant formulation.

In 1986 the Chromium-Free Carbon Development Program intensified with the letting of a contract to Calgon Carbon Corporation, Pittsburgh, PA, with the primary objective of developing a new chemical warfare agent adsorbent that did not contain any hazardous materials yet maintained the current high level of agent filtration performance. The program had the secondary objective of reducing the concentration of ammonia that off-gases from filters when first placed in service. Ammonia, a residual product of the production process, can be evolved at levels that are irritating to the user and can result in reduced confidence in the effectiveness of the protection being provided. The contractual effort consisted of the following phases: impregnant screening, performance optimization, process development, and performance qualification of the carbon.

Screening of Chromium Replacement Options

During Phase I of the Chromium-Free Carbon Development Program, Impregnant Screening, an assessment of the total role of the hexavalent chromium in removing chemical warfare agents was undertaken. This assessment determined that the chromium was only important in providing protection against cyanide-containing blood agents such as hydrogen cyanide and cyanogen chloride.[5] The role of the hexavalent chromium in the filtration of hydrogen cyanide by ASC carbon is to react with the toxic product cyanogen produced by the reaction of hydrogen cyanide with the cupric oxide impregnant. Evidence indicates that the hexavalent chromium plays a role in converting the cyanogen to oxamide, which is retained by the activated carbon by physical adsorption.[6] The role of the hexavalent chromium in cyanogen chloride filtration is not as well understood. Pytleweski[7] has reported that hexavalent chromium complexed with divalent copper results in the decomposition of cyanogen chloride by ASC carbon. For no other chemical warfare agent was a significant role identified for hexavalent chromium in the removal of that agent by ASC carbon. Based on this determination, a literature search was conducted to identify reactants for hydrogen cyanide and cyanogen chloride. The field of potential candidates was initially reduced based on toxicologic considerations. Those candidate impregnants for which carcinogenicity, mutagenicity, or high acute toxicity had been reported were eliminated from consideration. These criteria, of course, were essential, but they did greatly add to the difficulty of the task, because those chemicals that do exhibit high levels of reactivity tend also to be toxic. Several very promising candidates (for example, nickel and cobalt), for which cyanide reactivity had been reported in the literature, were excluded based on toxicity considerations.[8,9] Over 100 different impregnated carbons of sufficiently low toxicity were prepared and tested for cyanide filtration performance. These carbons varied in the actual chemical impregnants used, the loading levels of the impregnants, and specific treatments for the carbon surface. The impregnants consisted of those which stoichiometrically reacted with the agents and those that acted catalytically to decompose the agents.

As a result of the Phase I work, the five most promising candidate impregnant formulations were identified. All of these formulations were prepared on the identical activated carbon substrate as is used for ASC carbon.

By so doing and by limiting the total loading of metal impregnants to 12 percent, the very high physical adsorption capacity that exists for ASC carbon for nerve and mustard agents would not be compromised by the chromium-free formulation. All of these candidate formulations included copper and silver to retain established reactivity toward numerous reactively removed agents (for example, phosgene and arsine). The potential chromium replacements in the five formulations were molybdenum, vanadium, zinc, manganese, and tungsten. All five of these formulations were prepared with and without triethylenediamine as a potential aid to cyanogen chloride removal.

Determination of the Optimum Formulation

Under Phase II of the program, Performance Optimization, the five candidate impregnant formulations were optimized for overall filtration performance against all agents of concern.[5] The most effective loading level for each impregnant in each of the five formulations was determined by statistical contour mapping of agent removal performance as a function of impregnant loading level. As a result of this formulation optimization, an impregnant composition of copper oxide, zinc oxide, silver, and triethylenediamine was initially identified as the best replacement for ASC carbon. This chromium-free carbon formulation was named ASZ-TEDA carbon. During subsequent testing of this ASZ-TEDA carbon to determine the effect of long-term exposure to a variety of temperature and humidity conditions on agent filtration performance, the hydrogen cyanide filtration performance of the carbon was found to fall to unacceptably low levels after exposure to high humidity environments. This poor performance was manifested by premature breakthrough of cyanogen from beds of the carbon when challenged with hydrogen cyanide. The cupric oxide on the ASZ-TEDA carbon was converting the hydrogen cyanide to cyanogen, without an effective reaction path for the decomposition of the cyanogen being provided by the sorbent. The reason for the degradation of the initially high level of hydrogen cyanide performance for fresh carbon to very poor performance after exposure to humidity was not immediately apparent. Spectroscopic analyses (x-ray photoelectron spectroscopy and energy dispersive x-ray spectroscopy) along with scanning electron microscopy were performed to identify differences in the chemical state and distribution of the impregnants before and after humidity exposure.[10] These analyses determined that while the chemical state of the impregnants did not change, the location of the impregnants was altered by the moisture. For the fresh carbon, zinc was heavily concentrated on the external surface of

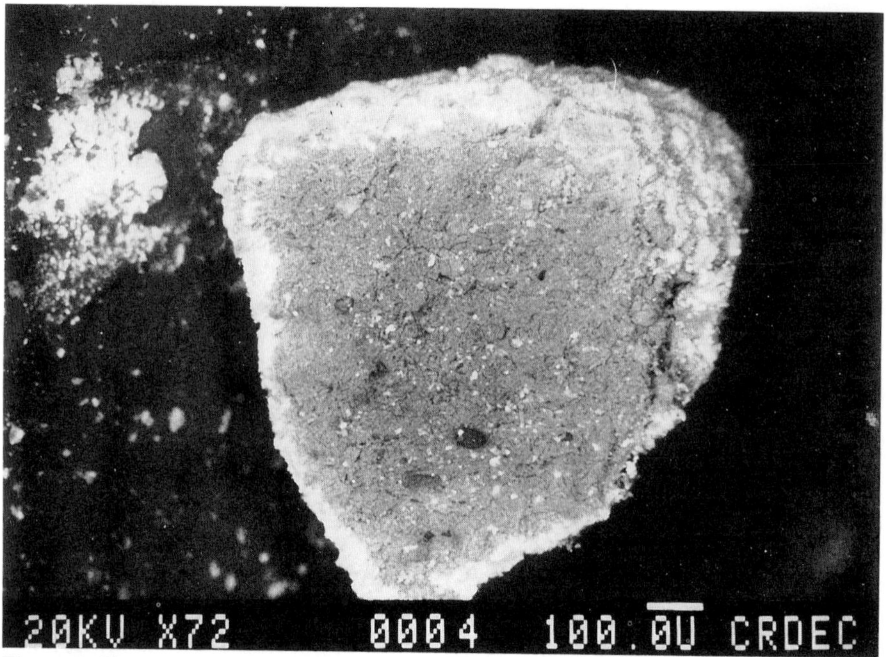

Figure 1. Zinc impregnant distribution (light areas) on a granule of ASZ-TEDA carbon.

the granule, whereas the cupric oxide was dispersed throughout the granule. Figure 1 is a scanning electron microscopy image of the cross-section of an unaged ASZ-TEDA carbon granule. The figure shows a high concentration of impregnants on the exterior surface. Energy dispersive x-ray spectroscopy determined this heavy impregnant concentration to be primarily zinc. Thus, the hydrogen cyanide challenge to the carbon finds the zinc to be more accessible than copper, and reaction to zinc cyanide is favored. No cyanogen is produced from this reaction.[11] As the carbon is exposed to moisture, the cupric oxide was found to migrate to the external surface of the granule. With this increased accessibility of the cupric oxide, the hydrogen cyanide reacts with it to form cuprous cyanide and cyanogen, according to reaction (R1).[12]

$$4\,HCN + 2\,CuO = 2\,CuCN + (CN)_2 + 2\,H_2O \qquad (R1)$$

To address this problem of premature cyanogen breakthrough, a modification to the ASZ-TEDA impregnant formulation was sought. The impregnants considered earlier in the program were rescreened for hydrogen cyanide reactivity. As a result, the addition of molybdenum, in the form of

Table 1
Impregnant Loadings on ASZ-TEDA Carbons

Impregnant	— Loading in Percent — ASZ-TEDA	ASZM-TEDA
Copper as cupric oxide	6	5
Silver in elemental form	0.05	0.05
Zinc as zinc oxide	6	5
Molybdenum (ammonium dimolybdate)	0	2
Triethylenediamine	3	3

ammonium dimolybdate, to the ASZ-TEDA formulation was found to stabilize the hydrogen cyanide filtration performance. The impregnant loadings on this new chromium-free carbon, named ASZM-TEDA carbon, are listed in Table 1 along with those for the precursor ASZ-TEDA carbon.

The ASZM-TEDA carbon was evaluated spectrometrically to acquire an understanding of the role that the molybdenum played in the removal of hydrogen cyanide.[13] This analysis determined that oxidation state of the molybdenum was not changed by exposure to hydrogen cyanide, indicating that the ammonium dimolybdate was apparently not involved in an oxidation-reduction reaction. However, the dispersion of the copper was noted to have been stabilized, thus maintaining the zinc as the most accessible reactant. Figure 2 shows the effect of humid exposure on the copper-to-carbon ratio on the surface of an ASZ-TEDA Carbon granule as compared with that for a ASZM-TEDA carbon granule, as determined by x-ray photoelectron spectroscopy. An attempt was made to identify the nature of any complex of copper with molybdenum that may form on the carbon surface to account for this stabilization of copper dispersion, but this attempt was not successful. Table 2 provides a description of the specific mechanisms by which the ASZM-TEDA carbon removes a number of key chemical warfare agents. From the table, the importance of physical adsorption performance and the wide diversity of chemical reactions provided by the adsorbent is apparent.

Once the final chromium-free carbon formulation had been established, the Industrial Hygiene Office of ERDEC obtained approval from the medical community for the use of the ASZM-TEDA carbon in military filters. The adsorbent was approved for use by the Department of Army Office of the Surgeon General in August 1990. In addition, an ERDEC environmental assessment determined the chromium-free carbon, when discarded, would not be classified as a hazardous waste material under EPA regulations.

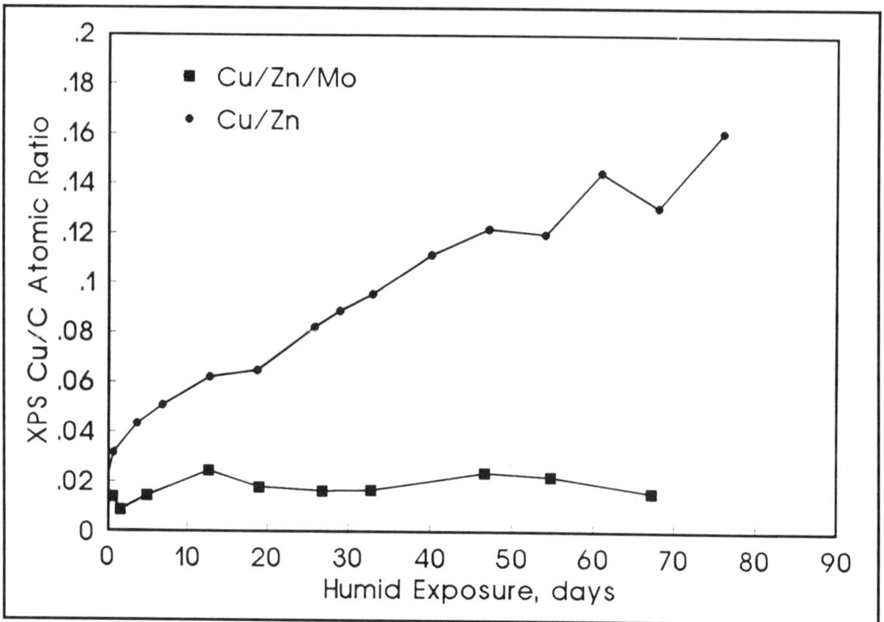

Figure 2. Effect of humid air on the copper-to-carbon ratio on the surface of ASZ-TEDA and ASZM-TEDA carbon granules.

Development of the Best Processes for Applying the Impregnants

The next phase of the Chromium-Free Carbon Development Program was the determination of the optimum processes for applying the impregnants to the pore structure of the carbon. These processes are critical to achieve the desired impregnant loading level on the carbon and chemical state of the impregnants. The processes were optimized such that maximum reactivity was achieved from the impregnant formulation. The carbon impregnation process studies included determination of the optimum impregnant solution concentration, contact time between the impregnant solution and the carbon, drying time and temperature, and heat treatment conditions. The optimum impregnation processes were developed, first through laboratory experimentation, and then through pilot plant studies. The loading of the impregnants on the carbon is controlled primarily through the concentration of the impregnants in the solution into which the activated carbon is dipped and the contact time between the carbon and the impregnant solution. The chemical state of the impregnants is controlled by the specific form of the impregnants dissolved in the solution and the temperature profile used to dry and heat-treat the carbon. Temperature for these steps is ramped up from relatively modest values under which the water is slowly evaporated to much higher values, under which the

Table 2
Specific Mechanisms of Agent Removal Effected by ASZM-TEDA Carbon

Agent	Mechanism of Removal
Nerve	Strong physical adsorption followed by hydrolysis
Blister	Strong physical adsorption followed by hydrolysis
Cyanogen chloride	Weak physical adsorption followed by reaction with copper, zinc. and TEDA impregnants
Hydrogen cyanide	Weak physical adsorption followed by reaction with cooper and zinc impregnants. Reaction with copper forms cyanogen; molybdenum minimizes production of cyanogen.
Arsine	Weak physical adsorption followed by oxidation by copper at low relative humidity and/or catalytic oxidation by silver at high humidity.
Posgene	Weak physical adsorption followed by hydrolysis to HCl and CO_2. HCl then reacts with copper and zinc impregnants.

impregnants convert to their final chemical state. As a result of improvements in drying and heat-treating the carbon, the amount of ammonia that remains on the final product is greatly reduced compared with that of ASC carbon.

A pilot plant scale lot consisting of 3,000 pounds of ASZM-TEDA carbon was produced using the optimum process conditions. The carbon was produced over a one-week period by a continuous operation. The carbon was tested for physical properties and chemical agent filtration performance and was determined to comply with specification requirements. Table 3 compares the properties and performance of the ASZM-TEDA carbon with the specification requirements for the carbon. The specification requirements had been set at values to ensure that filters containing the carbon would meet their performance requirements.

Evaluation of the Filtration Performance of ASZM-TEDA Carbon

Performance evaluation testing of the carbon was concentrated on establishing the level of hydrogen cyanide and cyanogen chloride filtration performance, because as previously described, the filtration of these agents would be the primary impact from the replacement of the chromium impregnant. Samples of the pilot plant production lot of ASZM-TEDA carbon were subjected to long-term exposure in test chambers under various temperature and humidity conditions representing field use. These exposure environments are shown in Table 4.

Table 3
Properties and Performance of ASZM-TEDA Carbon as Compared with Specification Requirements

Property Specified	Specification Value Minimum	Specification Value Maximum	ASZM-TEDA Performance
Cyanogen clhoride life (minutes)	55	NA	67
Cyanogen chloride life after aging (minutes)	45	NA	58
Hydrogen cyanide life (minutes)	28	NA	35
Phosgene life (minutes)	25	NA	38
DMMP life (minutes) (nerve agent simulant)	175	NA	210
Copper content (percent)	NA	6.0	5.0
Zinc content (percent)	NA	6.0	5.1
Silver content (percent)	0.03	0.1	0.048
Molybdenum content (percent)	NA	2.5	1.8
TEDA content (percent)	NA	3.5	3.1
Ammonia off-gassing (milligrams per liter)	NA	0.0010	0.0006
Moisture content (percent)	NA	2.5	1.7
Density (g/cm^3)	NA	0.68	0.61
Hardness (relative value)	85	NA	95
Particle size (millimeter)	0.6	1.7	0.6-1.7

The environmentally exposed carbon was then agent tested at a wide range of bed depths and airflow velocities so that the performance of the carbon in all military filter designs could be assessed. Figure 3 shows a comparison of the effect of tropic exposure on the cyanogen chloride filtration performance of the ASZM-TEDA carbon with that of ASC carbon. The ASZM-TEDA carbon is noted to be much less affected by exposure to this tropic condition. Under the other temperature/humidity conditions (desert, temperate, and arctic), the cyanogen chloride performance of both ASZM-TEDA and ASC carbon is relatively stable. Hydrogen cyanide challenge testing of the exposed carbons showed relatively stable performance under all

Table 4
Exposure Environments for ASZM-TEDA Carbon

Environmental Condition	Temperature (° F)	Relative Humidity (Percent)
Desert	120	10
Tropic	86	88
Temperate	60	70
Arctic	-25	90

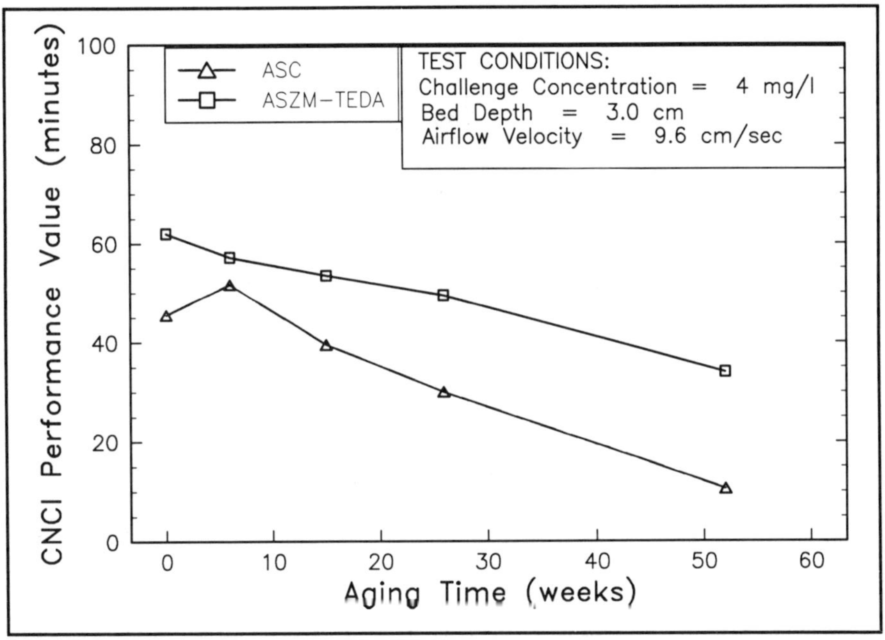

Figure 3. Effect of tropic aging of ASZM-TEDA and ASC carbons on cyanogen chloride filtration performance.

environments of the study. The level of performance against this agent for the ASZM-TEDA carbon was about three quarters of that of ASC carbon but is still at an acceptable level. Figure 4 shows an example of the plot obtained from cyanogen chloride testing conducted at various bed depths and airflow velocities. From such a plot estimates of the performance of specific filters can be obtained from the airflow velocity and carbon bed depth for that filter. The data from this testing indicated that the agent filtration performance of all military filters filled with ASZM-TEDA carbon would meet their perfor-

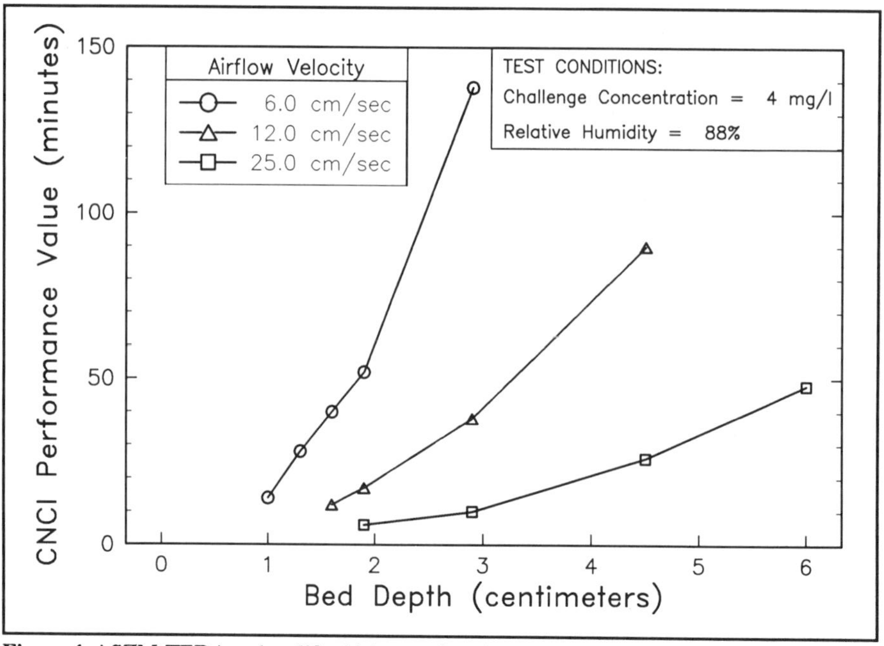

Figure 4. ASZM-TEDA carbon life-thickness plot of cyanogen chloride filtration performance.

mance requirements. Under a follow-on Engineering Study Project, testing of full-scale filters substantiated this conclusion. Currently, the ASZM-TEDA carbon is being implemented into the Technical Data Packages of several of the chemical warfare agent filters. The implementation process will continue until all military filter systems are converted to chromium-free carbon.

Results of the Development Effort

All goals of the Chromium-Free Carbon Development Program have been met. The environmental and physiologic hazards associated with the current ASC carbon have been eliminated. The ASZM-TEDA carbon is not an RCRA hazardous material. The U.S. Army Office of the Surgeon General has approved the material for use in all military filters. Because of its environmentally compatible characteristics, over $2,000,000 per year in military filter disposal costs will be saved. In addition, potential health risks associated with the chromium have been eliminated for personnel involved in the manufacture of the carbon and military filters as well as users of military filters in the event that the filter is damaged.

Extensive agent challenge testing has been conducted to ensure that the performance of filter systems containing this adsorbent is adequate to meet their requirements. As described previously, this testing was concentrated on those agents that for ASC carbon had required hexavalent chromium for their removal. The filtration performance of ASZM-TEDA carbon was determined to be at a level generally comparable to that of ASC carbon under the wide variety of environmental conditions investigated. Under some conditions the performance of the ASZM-TEDA carbon is somewhat better than ASC carbon (cyanogen chloride under tropic conditions), and in some cases the new carbon is somewhat worse (cyanogen chloride under desert conditions), but overall the filtration performance of the two carbons is comparable. Because of the use of the identical activated carbon substrate, the physical adsorption of nerve and mustard agents by the ASZM- TEDA carbon is nearly the same as by ASC carbon. The filtration performance of the new carbon against these agents is slightly less (about 10 percent) because the impregnant TEDA occupies adsorption space in the micropores. This reduction in performance is not a concern because all military filters have excess performance against nerve and mustard agents. The filtration performance of the ASZM-TEDA carbon against the acid gas producers such as phosgene is somewhat greater than that of ASC carbon because of the higher metal oxide impregnant loading for the new carbon.

The objective of reducing the level of ammonia concentration that off-gases from military filters has also been met. Although much effort was spent in striving to completely eliminate the need for the ammonia in the impregnation solution, such an approach was not identified. However, the reduction achieved is very large (about 70 percent), such that the highest levels of ammonia off-gassing from the ASZM-TEDA carbon are less than five parts per million. At this level, the ammonia that off-gases would not be offensive to the filter user.

An additional benefit achieved from development of the new carbon is improved stability from exposure to heat and humidity over that of ASC carbon. This benefit offers the opportunity to reduce filter change-out intervals, which will further reduce production and logistic costs associated with providing chemical warfare protection to military personnel and equipment.

Conclusion

This development effort has resulted in the most significant improvement in the adsorbent used in military chemical warfare agent filters since World War II. The elimination of potential environmental and health hazards associated with military filters will have a tremendous benefit to the environment, to the well-being of personnel potentially exposed to the filter adsorbent, and to the life cycle costs associated with the filters. In addition, the ASZM-TEDA carbon will provide a high level of stable chemical warfare agent protection for military personnel and equipment in a chemical environment.

Acknowledgments

The authors acknowledge the important contributions of Joseph Rossin of Geo-Centers, Inc. for the spectrometric analyses of the chemical state and dispersion of the impregnants on the carbon surface and for his insight into the significance of the results of these analyses.

References

1. W.C. Pierce, in *Summary Technical Report of Division 10, NDRC, Military Problems with Aerosols and Nonpersistent Gases* (National Defense Research Committee, Washington, D.C., 1946), chapter 1.
2. J.R. Conlisk, R.J. Puhala, and L.M. Sturdivan, *Determination of ASC Carbon as a Resource Conservation and Recovery Act (RCRA) Hazardous Waste* (U.S. Army CRDEC Report No. CRDEC-TR-109, 1989).
3. G.A. Lutz, *Physiological and Environmental Effects Relating to the Preparation of Impregnated Charcoal* (U.S. Army CRDEC Contractor Report No. ARCSL-CR-80072, Battelle Columbus Laboratories,1981).
4. E.D. Tolles, *Improved Charcoal Impregnation Technology* (U.S. Army CRDEC Contractor Report No. CRDEC-CR-86066, Westvaco Corporation, North Charleston, SC, 1986).
5. D.T. Doughty, *Development of a Chromium-free Impregnated Carbon for the Adsorption of Toxic Agents* (U.S. Army CRDEC Contractor Report No. CRDEC-CR-118, Calgon Carbon Corporation, Pittsburgh, PA 1991).
6. B.R. Alves and A.J. Clark, *Carbon* **24**, 287-294 (1986).
7. L.L. Pytleweski, *Studies of ASC Whetlerite Reactivity* (U.S. Army CRDEC Contractor Report No. ARCSL-CR-79008, Drexel University, Philadelphia, PA, 1979).
8. J.F. Alder, P.R. Fielden, and S.J. Smith, Carbon, 26, 701-711 (1988).
9. R.J. Grabenstetter and F.E. Blacet, in *Summary Technical Report of Division 10, NDRC, Military Problems with Aerosols and Nonpersistent Gases* (National Defense Research Committee, Washington, D.C., 1946), chapter 4.
10. J.A. Rossin and R.W. Morrison, *Carbon* **29**, 887-892 (1991).
11. M. Smisek and S. Cerny, *Active Carbon* (Elsevier, New York, 1970), p. 188.
12. P.N. Brown *et al.*, *Carbon* **27**, 829 (1989).
13. J.A. Rossin and R.W. Morrison, paper presented at the Annual Meeting of the American Institute of Chemical Engineers, Los Angeles, CA, November 1991, unpublished data.

40

Development of the Mine Resistant Vehicle Using Advanced Computational Methods

Steven S. Bishop
Sergio A. Sergi
U.S. Army Belvoir, Research, Development and Engineering Center
Fort Belvoir, VA 22060-5606

Shmuel Eidelman
Isaac Lottati
Science Applications International Corporation
McLean, VA 22102

Development of the Mine Resistant Vehicle is a joint Army-Marine program to develop a vehicle hardened against antitank (AT) blast land mines. The vehicle will be based on the M925 5-ton cargo truck. Numerical simulations are used to determine the dynamic load produced by the AT mine detonation on the cargo bed of the truck. The simulation of blast structure interaction is done using an adaptive unstructured triangular grid code based on the Second Order Godunov method. Use of the unstructured grid allows detailed descriptions of the vehicle geometry and accurate simulations of the developed shock load. A simulation result for the case of the original flat cargo bed cross sections is shown in the form of pressure contours overlayed on the dynamically evolving unstructured grid. Highly detailed wave interaction patterns were resolved by our code. At the same time, computational resources are allocated only where needed to resolve flow discontinuities or strong rarefaction waves. Computed results point to very high pressures that will develop in the area of the unprotected cargo bed. This is a first step in a long range program that will include analysis of the structural response and crew survivability as a result of blast load and comprehensive analysis of various blast hardening methods.

Introduction

Today, logistic and combat service support vehicles must operate in a variety of environments. Low intensity conflicts are a continuing threat in many regions around the world, and protecting military vehicles from the effects of land mine detonation is of considerable interest. Our study uses advanced computational methods to study the interaction of the five-ton M925Al cargo truck with a land mine. The truck's flat cargo bed measures 4.2m long and 2.4m wide and rests on a chassis positioned about 1.4m above the road. The forward cab rests 1.4m above the road. The considerable standoff distance of the chassis and the cab offers a natural protection from the mine blast effects. In this paper, we will demonstrate the use of Computational Fluid Dynamics (CFD) methods for a detailed study of a land mine blast and M925Al truck cross-section interaction.

Battle Damage Data for Land Mine Incidents

Data for land mine incidents are limited because recording mine encounters is not a high priority during wartime. The primary sources of information to date have been the Vietnam Wart and the Rhodesian Bush War. A second limiting factor in data acquisition is that a battle damage assessment reporting team usually is not operational for an entire conflict or war. The Survivability & Vulnerability Information and Analysis Center (SURVIAC), Wright-Patterson Air Force Base, OH, has compiled a database based on Battle Damage Assessment Reporting Team findings. The available information from Southeast Asia dating from April 1969 to June 1970 is summarized for mine incidents. In this database, there are 547 incidents recorded with various vehicle types, including 40 incidents of wheeled vehicle mine encounters. The vehicle types range from a .25 to a 5-ton capacity wrecker. Fifteen of the recorded incidents involve a 5-ton truck. The data also report that 39 incidents, an overwhelming majority, were from a standard or conventional land mine. One incident resulted from an Improvised mine. Of the 39 vehicles, 34 were rendered inoperable; five vehicles were scrapped.

From a structural standpoint, the location of "hit" or the position of the mine's detonation as it strikes the vehicle is significant. Table 1 shows that the most common location of the land mine incidents for a wheeled vehicle is with a tire.

In order to improve the land mine resistance of the M925Al truck, first we need to understand the details of the blast wave structure interaction. By

Table 1
Damage Type From Wheeled Vehicle/Land Mine Incidents

Number of Incidents	Hit Location
4	Body
1	Body/Wheel
2	Track
32	Wheel
1	Wheel/Fuel Tank

doing so, we can estimate the extent of required changes to the truck configuration for various levels of protection. The primary goal is to assure sufficient protection for the driver and passengers to survive a single incident of mine detonation, with minimal modification to the truck geometry.

To study the details of mine blast effect on the truck, we use advanced methods of CFD represented in the Adaptive Unstructured Grid Upwind Second Order on Triangles (AUGUST) code developed by Science Applications International Corporation (SAIC). This code allows detailed description of the truck geometry and high resolution simulation of shock-wave structure interaction. Using this code, we can accurately predict the pressure and impulse that will affect various areas of the cargo bed or cabin as a result of mine detonation and assess survivability for the personnel. We also can use the pressure/history data from the blast/structure interaction as an input to the available structural analysis program (ANSYS or DYNA) and assess the damage to the trucks inflicted by the mine detonation.

The accuracy of the data provided by the CFD simulation is very important to calculations of the structural response. However, these two problems can be decoupled in time, and the CFD simulation can proceed on the assumption that the structure is rigid. This is possible because the duration of the gas dynamic phase of the blast structure interaction is about 2msec, and the structural response phase starts 10 to 20msec after the detonation.

Unstructured Triangular Grid

Practical numeric simulations of fluid dynamic problems call for modeling flows over complicated shapes. In addition, important flow features such as shed vertices, shock waves, slip lines, and boundary layers usually have widely varied lengths and time scale and need to be resolved. Accurate solution of these problems requires computational grids dynamically adapted to the evolving flow feature, and with full control over solution accuracy in the key regions of the computational domain. It is commonly accepted that only unstructured grids can provide full flexibility in obtaining the local grid resolution sufficient to accurately resolve subscale flow features. The five years since CFD research introduced these grids and methods have produced landmark simulations clearly demonstrating their advantages.[1-3]

The Second Order Godunov method was implemented on rectangular grids[4] and in a few cases on general quadrilateral grids.[5] This approach has limited application, because structured quadrilateral grids have great difficulty describing a complicated computational domain with multiple bodies of different geometries and scales. Recently, we implemented the Second Order Godunov method for unstructured triangular grids.[6] This enables us to combine the robust and accurate numeric algorithm with a gridding technique, which allows us to describe very complex domains with ease and efficiency. In addition, we developed a novel dynamic grid adaptation methodology that only allocates a dense computational grid to regions where enhanced resolution is needed to resolve strong gradients in flow parameters. As demonstrated herein, this enables an extremely economic allocation of computational resources and accurate simulation of a complicated phenomenon that includes multiple reflection and interaction of the shock waves.

Mathematic Model and Integration Algorithm

We consider a system of two-dimensional Euler equations written in conservation law form as:

$$\frac{\partial U}{\partial t} + \frac{\partial F}{\partial x} + \frac{\partial G}{\partial y} = 0 \tag{1}$$

where:

$$U = \begin{vmatrix} \rho \\ \rho\, u \\ \rho\, v \\ e \end{vmatrix}, \quad F = \begin{vmatrix} \rho\, u \\ \rho\, u^2 + p \\ \rho\, uv \\ u(e+p) \end{vmatrix}, \quad G = \begin{vmatrix} \rho\, v \\ \rho\, uv \\ \rho\, v^2 + p \\ v(e+p) \end{vmatrix}$$

Here u, v are the x, y velocity vector components, p is the pressure, ρ is the density, and e is total energy of the fluid. We assume that the fluid is an ideal gas and the pressure is given by the equation-of-state

$$p = (\gamma - 1)\,[e - \frac{\rho}{2}\,(u^2 + v^2)] \tag{2}$$

where γ is the ratio of specific heats. It is assumed that an initial distribution of the fluid parameters is given at $t = 0$, and the boundary conditions defining a unique solution are specified for the computational domain.

The system of governing equations in Equation 1 can be written as:

$$\frac{\partial U}{\partial t} + \nabla\, Q = 0 \tag{3}$$

where Q represents the convective flux vector. By integrating Equation 3 over space and using Gauss' theorem, the following expression is obtained

$$\frac{\partial}{\partial t}\int_{\Omega} uda + \oint_{\partial\Omega} qdL = 0 \tag{4}$$

where $dl = ndL$, n is the unit normal vector in the outward direction, and dL is a unit length on the boundary of the domain. The variable Θ is the domain of computation, and $\partial\Theta$ is the circumference boundary of this domain.

Equation 4 can be discretized for each element (cell) in the domain

$$\frac{(U_i^{n+1} - U_i^n)}{\Delta t}\, A_i = \sum_{j=1}^{M} Q_j^n\, n_j\, \Delta L_j \tag{5}$$

where A_i is the area of the cell; Δt is the marching time step; U_i^{n+1} and U_i^n are the primitive variables at the center of the cell at time n and at the update $n+1$ time step; Q_j is the value of the fluxes across the M boundaries on the circumference of the cell, where n_j is the unit normal vector to the boundary edge j, and ΔL_{jj} is the length of the boundary edge j. The fluxes Q_j^n are computed applying the Second Order Godunov algorithm, and Equation 5 is used to update the physical primitive variables U_i according to computed fluxes for each marching time step $\Delta t'$. The marching time step is subjected to the Courant-Fredrichs-Lewy constraint.

We seek a solution to the system of Equation 1 in the computational domain, which is decomposed into triangles with arbitrary connectivity. An unstructured triangular grid is shown in Figure 1. The grid is used to describe the computational domain around a cross-section of an M925 cargo truck and allows detailed description of a complex computational domain.

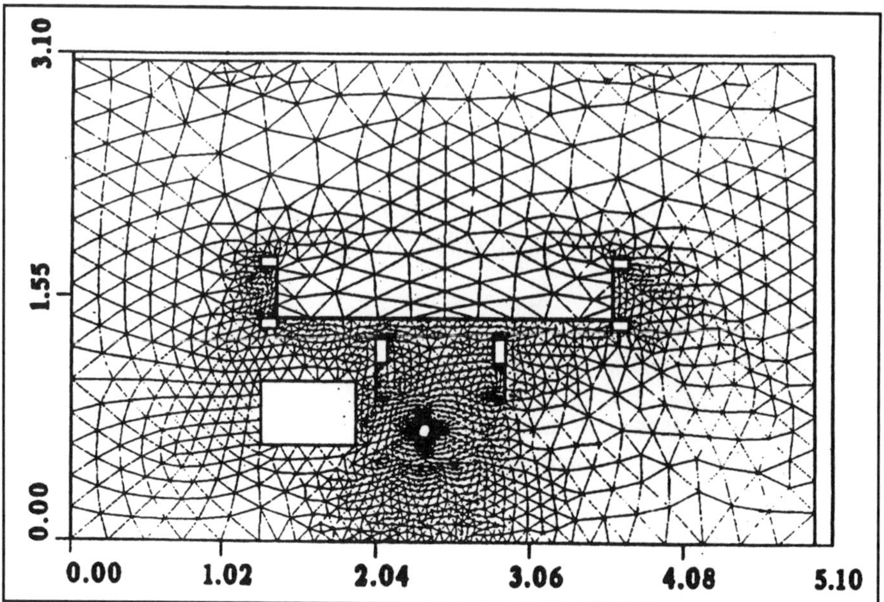

Figure 1. (Color Plate 3) Domain decomposition using unstructured triangular grid. Grid over cross-section of M925Al cargo truck.

The numeric technique for solving Equation 1 on an unstructured grid is described in References 6 and 7. This numeric technique applies to some of the ideas that were introduced in References 5 and 8. The unstructured codes apply the center-based formulation, that is, the primitive variables are defined in the center of the cell, which makes the cell the integration volume, while the fluxes are computed across the edges of the cell. The basic algorithmic steps of the Second Order Godunov method that is used in our code can be defined as follows:

1. Find the value of the gradient at the baricenter of the cell for each gas dynamic parameter U.
2. Find the interpolated values of U at the edges of the cell using the gradient values.
3. Limit these interpolated values based on the monotonicity condition.[7]
4. Subject the projected values to the characteristic's constraints.[9]
5. Solve the Riemann problem, applying the projected values at the two sides of the edges.
6. Update the gas dynamic parameter U according to the conservation Equation 1 applying to the fluxes computed and the current time step.

Figure 2. Interferogram of M = 2.85 shock interaction with a 45° double ramp.[9]

Code Validation Problem

To validate the AUGUST code, we simulated the problem of interaction of Mach 2.85 planar shock wave propagating in a channel with a 45° symmetric double ramp. The simulation results will be compared with experimental and computational results given in Reference 9. Figure 2 shows the experimental interferogram of the problem to be simulated. We will present a number of density contour plots showing flow evolutions in the course of this transient simulation. The ambient pressure and density were set to a unit value, and the Mach number is 2.85. Flow in the channel is from left to right.

Figure 2 is the experimental interferogram of the stage of shock/ramp interaction when the shock wave has just passed the ramp. In Figure 2, we observe a complex flow pattern created by the shock ramp interaction. The flow field contains reflected shocks, recompression shocks, four triple points, strong rarefaction fan, slip lines, and vortices. To simulate accurately all the details of the shock/ramp interaction, the computer code should have high resolution for modeling flow regions with sharp gradients.

Figure 3 shows the computational domain for the shock ramp interaction

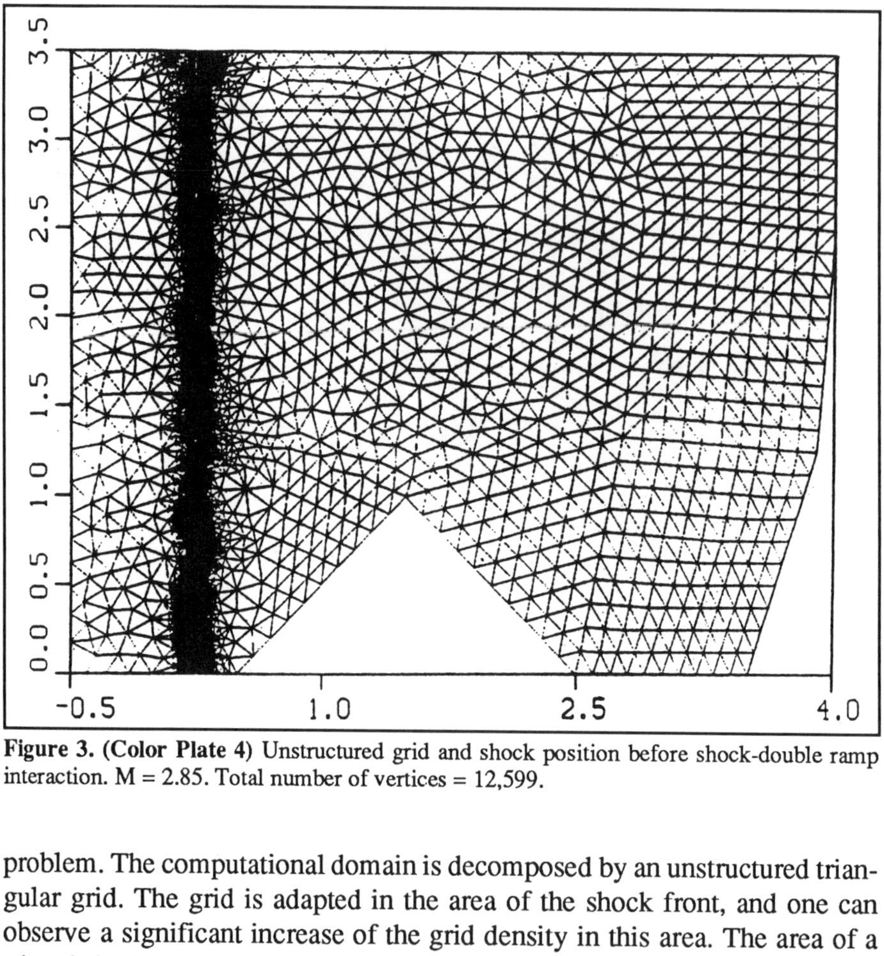

Figure 3. (Color Plate 4) Unstructured grid and shock position before shock-double ramp interaction. M = 2.85. Total number of vertices = 12,599.

problem. The computational domain is decomposed by an unstructured triangular grid. The grid is adapted in the area of the shock front, and one can observe a significant increase of the grid density in this area. The area of a triangle in the dense region of the grid is about 100 times smaller than the area of a triangle in the initial grid. In Figure 3, the solution in the form of density contours is shown at the time just prior to the shock impinging on the ramp. Here it is possible to observe that the grid adaptation algorithm adds cells in front of the shock discontinuity and removes them behind the discontinuity; thus, only the area that requires high resolution is covered by the dense grid.

This methodology minimizes the use of computer resources for a required resolution. The adaptive unstructured grid methodology allows high resolution representation of the flow gradients, and in Figure 3 we can observe a very sharp planar shock.

In Figures 4 and 5, we show the computational grid and the density contours for the solution of the stage at which the main shock front just passes over the apex of the ramp. The density contours overlay the grid in Figure 4 to illustrate the grid adaptivity. For clarity, contour plots are displayed separately from the grid in Figure 5. Figure 4 shows that the grid adaptivity

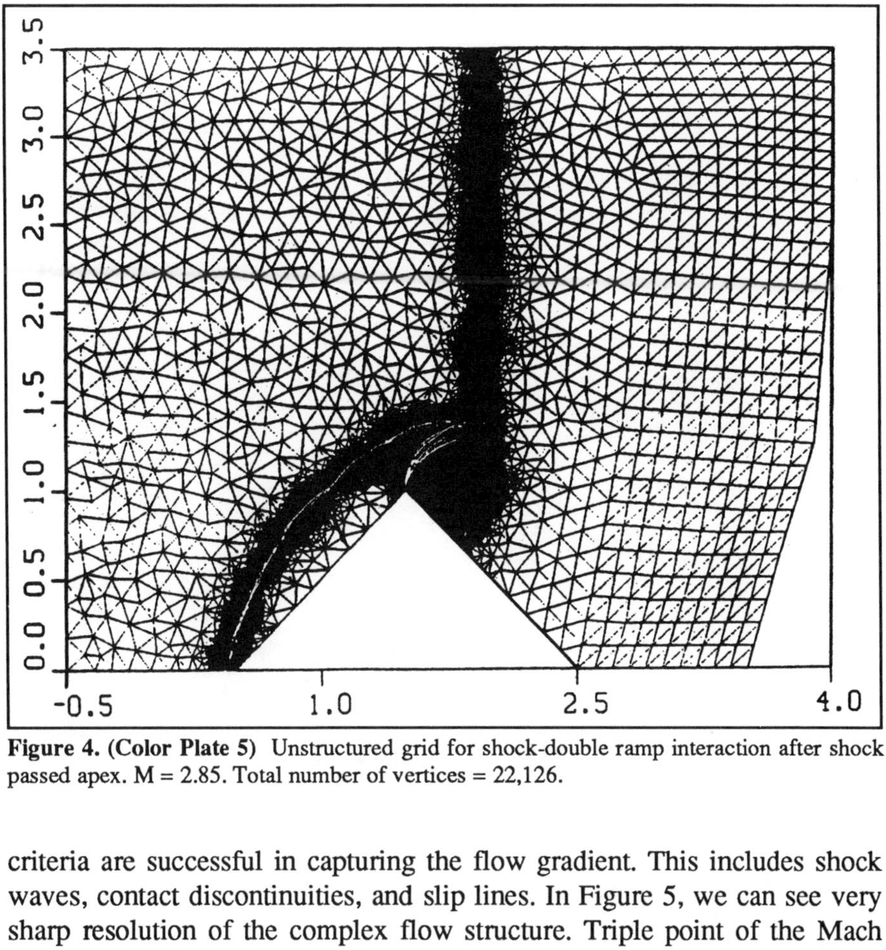

Figure 4. (Color Plate 5) Unstructured grid for shock-double ramp interaction after shock passed apex. M = 2.85. Total number of vertices = 22,126.

criteria are successful in capturing the flow gradient. This includes shock waves, contact discontinuities, and slip lines. In Figure 5, we can see very sharp resolution of the complex flow structure. Triple point of the Mach reflection, slip line, and strong vortex forming as a result of strong expansion around the apex are clearly visible.

In Figures 6 and 7, the simulation results are shown at the stage that corresponds to the time of the experimental results shown in Figure 2. We can observe that the grid is adapting to a changing flow pattern. All the elements of the complex flow are captured by the dynamically adapting grid. The overexpansion around the ramp apex led to formation of the recompression shock, which follows the strong vortex originated at the apex. We also can observe a formation of the triangle of shock waves that originates a vertical slip line that follows the main shock, which is typical for this interaction. All the flow features of the experimental interferogram can be identified in our simulation result shown in Figure 7. Further analysis of the values of the contour levels and the density levels shown in the interferogram validate a numeric result accuracy of more than 4 percent, which is well within the accuracy of the experimental result.

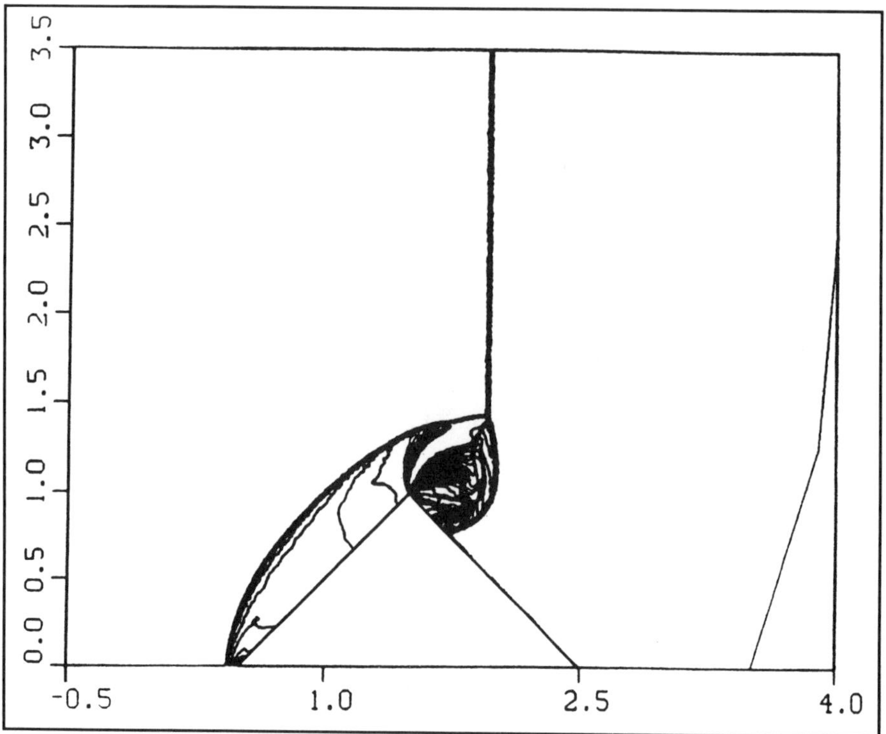

Figure 5. (Color Plate 6) Density contours for stage shown in Figure 4. ρ max = 7.59; ρ min = 0.096; $\Delta\rho$ = 0.214.

In addition to the results shown in Figures 3 through 7, we have validated the AUGUST code for a range of supersonic, transonic, and subsonic transient and stationary flow problems. The results presented in our previous publications[5-7] have consistently demonstrated the high resolution and accuracy capabilities of the code and the ability of the adaptive grid technique to capture the details of complex flow regimes.

Truck-Land Mine Blast Interaction Simulation

We present here the results for the base case of the mine resistant vehicle study: detonation of the mine triggered by the wheels of the unprotected truck. The truck cross-section geometry (see Figure 1) includes cargo bed, support beams, fuel tank, and the drive shaft. We can observe that the unstructured grid allows accurate description of the complex geometry, having elements of different length scale. We have selected this case as a reference case because it will allow us to test our simulation methodology and give data of pressures

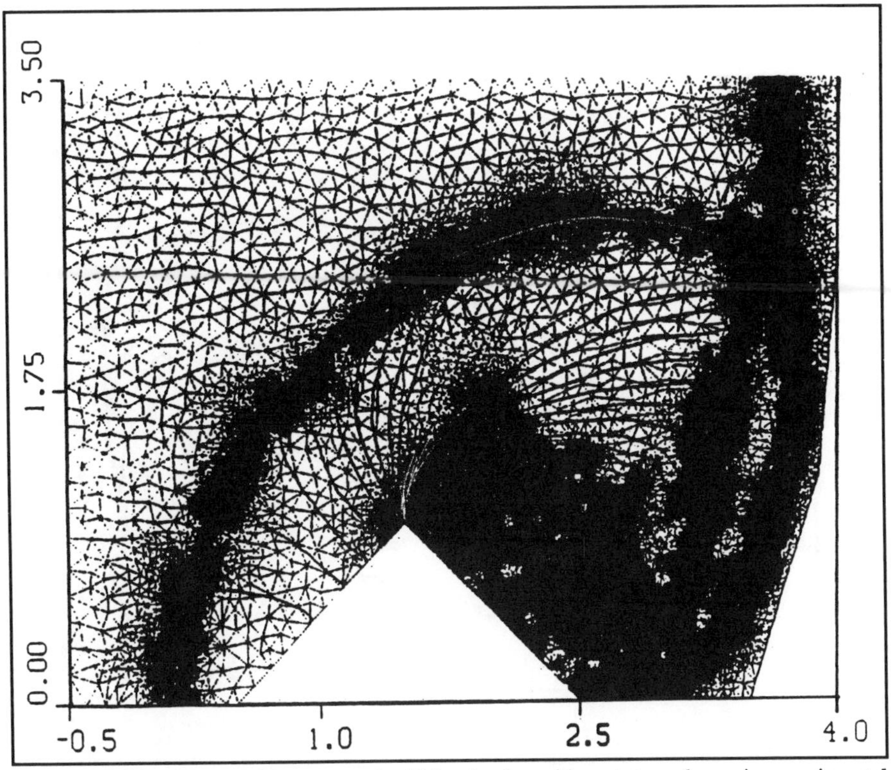

Figure 6. (Color Plate 7) Unstructured grid for shock position at stage shown in experimental interferogram. M = 2.85. Total number of vertices = 79,035.

and impulses that are typical for the unprotected M925A1 truck. We simulate interactions of the cross-section shown in Figure 1 with the blast generated by the detonation of a single M-15 antitank mine (about 20 pounds of TNT).

The planar two-dimensional presentation of the truck cross-section will represent the worst case scenario, because of lack of expansion into the third dimension, and will be typical for the area in the middle of the cargo bed. To take into account three-dimensional expansion of the detonation product from the land mine detonation, we have simulated the initial stages of the blast propagation using two-dimensional axisymmetric formulation. This simulation was done by a structured code that also takes into account the equation-of-state for the detonation products. This code, SOG2DSE (for Second Order Godunov 2 Dimensional for Solid Explosives), was developed by SAIC and is currently used by the U.S. Army Belvoir Research Development and Engineering Center (BRDEC) for predicting the effects of mine detonations. Two-dimensional axisymmetric simulation will give an accurate account of the three-dimensional effects for a symmetric blast propagation. The

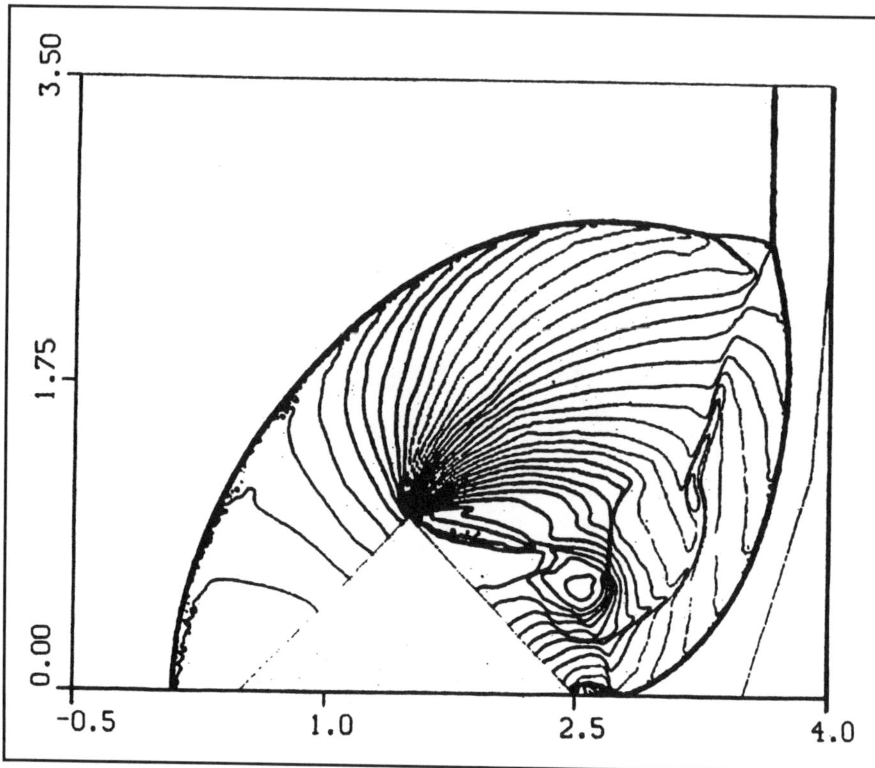

Figure 7. (Color Plate 8) Density contours for stage shown in Figure 6. ρ max = 7.622; ρ min = 0.082; Δρ = 0.215.

axisymmetric simulation was carried out up to the point at which the incident shock encounters the elements of the truck geometry. At this point, the result of the axisymmetric simulation on structured grid is mapped up to unstructured triangular grids. Thus, the simulation on unstructured grid starts at the stage shown in Figure 8, where the results are represented by pressure contours in pascals. Here, the maximal pressure is observed in the region of detonation products and is ≈160MPa. To accommodate the high gradients of pressure, grid density is significantly increased in the area of the detonation products. The initial grid shown in Figure 1 had 2,005 vertices; the grid at the stage shown in Figure 8 has 7,873 vertices. The 5,800 additional vertices were used to cover the propagating blast wave region to assure high resolution.

In Figure 9, the simulation results in the form of pressure contours are shown at the stage when the leading shock wave is impinging on the right corner of the cargo bed. The highest pressure is observed at the point of shock wave reflection and is about 1000MPa at this stage. Also, very strong reflections are observed at the drive shaft and at the right supporting beam. In

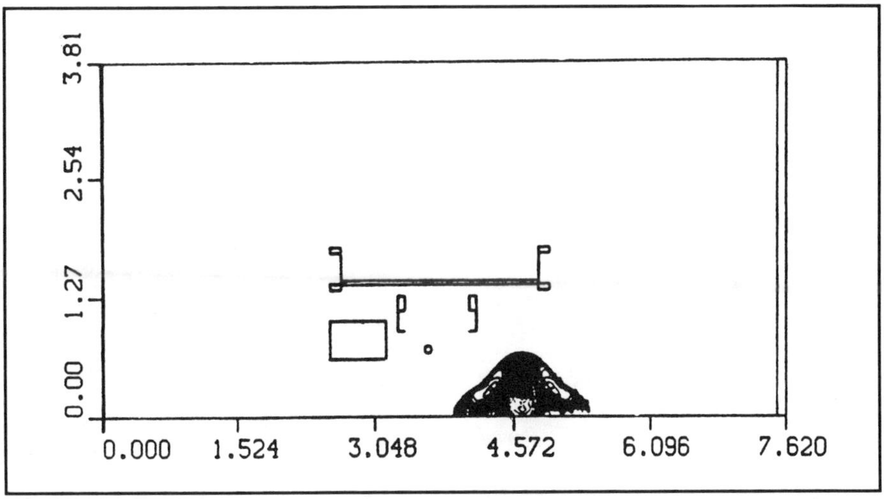

Figure 8. (Color Plate 9) Pressure contours for mine/blast truck cross-section interaction. t = 0.0msec; Pmax= 158MPa; Pmin = 0.099MPa.

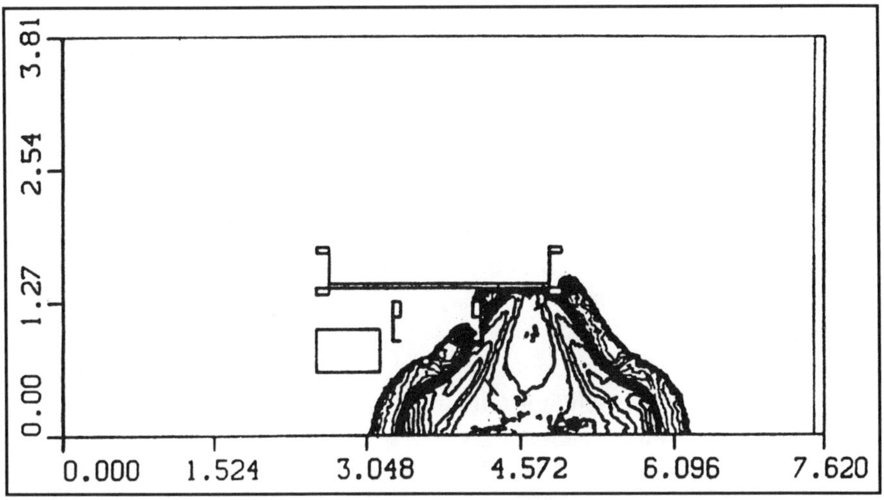

Figure 9. (Color Plate 10) Pressure contours for mine blast/truck cross section interaction. t = 0.147msec; Pmax = 953MPa; Pmin = 0.7MPa.

Figure 9, very high resolutions of the flow gradients are observable. To assure this resolution, the number of grid points used here was 15,604, or about 6,800 more than in Figure 8. This trend will continue throughout the simulation. Because the flow pattern gets more complex, new shocks and discontinuities appear, requiring more and more computing resources. However, the advantage of the adaptive grid is evident because the new grid points are added only when needed in simulation.

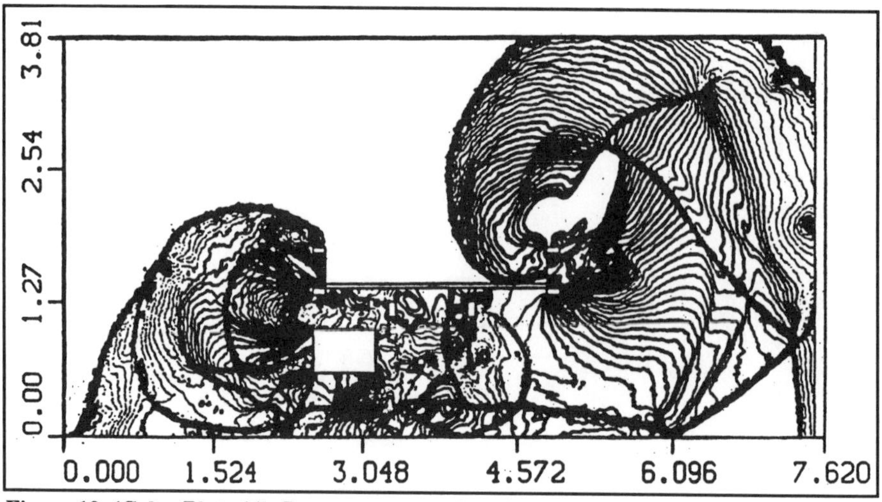

Figure 10. (Color Plate 11) Pressure contours for mine blast/truck cross section interaction. t = 1msec; Pmax = 25.8MPa; Pmin = 0.025MPa.

In Figure 10, the simulation results are shown at t ≈ 1msec. The leading front of the blast wave propagating from the right has reached the cargo bed inner area. At this stage of the simulation, the pressure in the inner volume of the cargo bed is ≈0.4MPa. Highest pressures reaching 25MPa are observed under the truck in the closed areas that impair venting. The strong shocks at the upper side of the cargo bed are followed by areas of strong rarefactions that lower the pressure to the level of 0.025MPa. This combination can produce a severe jolt to the personnel at levels predicted by our simulation, and most probably will be fatal.

Figure 11 shows the simulation results at t ≈ 2.3msec. Here, extremely complex flow patterns, including tens of shock fronts and other discontinuities, are resolved well by the AUGUST code. The collision of the shock fronts from the left and right side of the cargo bed has produced a complex flow that has pressure values up to 0.78MPa. However, the dominant pressure is about 0.4MPa. The higher pressures of about 3.6MPa are observed under the truck in enclosed areas. This high pressure can lead to the truck overturning.

Conclusions

Increasing the mine protection of the M925A1 cargo truck is an important goal of our program. The geometry modifications that can lead to significant increases in personnel survivability should be minimal and cost

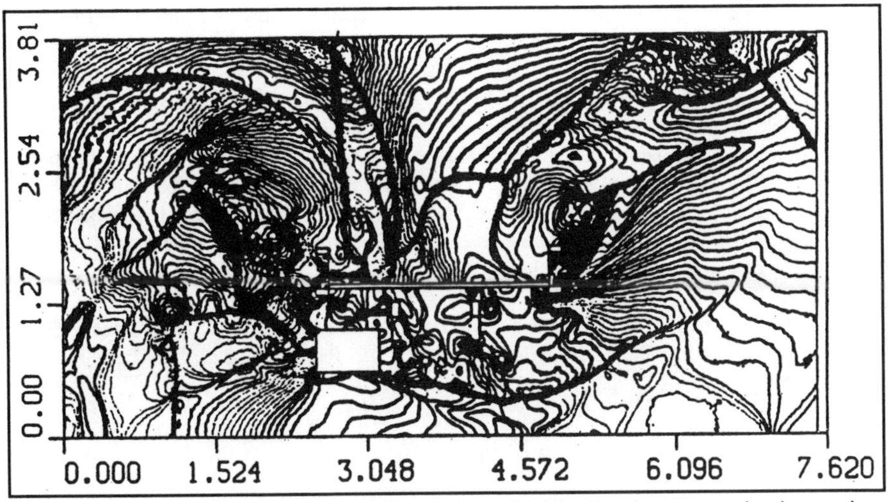

Figure 11. (Color Plate 12) Pressure contours for mine blast/truck cross-section interaction. t =2.32msec;Pmax = 3.69MPa; Pmin = 0.021MPa.

effective. This study presented the first results of the truck vulnerability analysis based on detailed numeric simulation of the mine blast-truck interaction. The simulations are performed with a new state-of-the-art code that has allowed detailed and accurate description of the complex flows over complex geometries. The new AUGUST code is based on adaptive, unstructured triangular grids. According to this technique, the computational grids dynamically evolve during the simulations, assuring a maximal resolution in the regions of the flow, with shock waves and contact discontinuities. The code was validated, and comparison of the experimental and simulation results testify to its high accuracy. Using the developed methodology of analysis, a base case of the mine blast-truck interaction was simulated. Computed results point to very high pressures that will develop in the area of an unprotected cargo bed. The presented work is a first step in a long-range program that will include analysis of the structural response as a result of blast load and comprehensive analysis of various blast hardening methods.

References

1. D. Mavriplis and A. Jameson, *Multigrid Solution of the Two-Dimensional Euler Equations on Unstructured Triangular Meshes* (document AIAA-87 0353, 1987).
2. J.D. Baum and R. Lohner, *Numerical Simulation of Shock-Elevated Box Interaction Using an Adaptive Finite Element Shock Capturing Scheme* (AIAA Preprint 89 0653, AIAA 27TH Aerospace Sciences Meeting, Reno, NV, 1989).

3. A. Jameson, T.J. Baker, and N.P. Weatherill, *Calculations of Inviscid Transonic Flow Over a Complete Aircraft* (AIAA Paper 86 0103, AIAA 24th Aerospace Sciences Meeting, Reno, NV, 1986).
4. B. van Leer, *J. Comp. Phys.* **32**, 101-136 (1979).
5. S. Eidelman, P. Collela, and R.P. Shreeve, *AIAA Journal* **22**, 10 (1984).
6. I. Lottati, S. Eidelman, and A. Drobot, A *Fast Unstructured Grid Second Order Godunov Solver (FUGGS)* (AIAA-90-0699, 28th Aerospace Sciences Meeting, Reno, NV, 1990).
7. I. Lottaff, S. Eldelman, and A. Drobot, *Solution of Euler's Equations on Adaptive Grids Using a Fast Unstructured Grid Second Order Godunov Solver* (Proceedings of the Free Lagarange Conference, Jackson Lake, WY, 1990).
8. P. Collela and P. Woodward, *J. Comp. Phys.* **54**, 174-201 (1984).
9. I.I. Glass and D.L. Zhang, *Interferometric Investigation of the Diffraction of Planar Shock Waves Over a Half-Diamond Cylinderin Air*, UTIAS Report No. 322, March 1988.

41

Synthesis of Metallo-tetrabenzporphyrins With High Third-Order Optical Nonlinearity for Military Laser Eye Protection

Reginald A. Willingham
John H. Cornell
David M. Alabran
John W. Cullen
Frank H. Bissett
Masato Nakashima
David E. Remy
Joseph F. Roach
U.S. Army
Natick Research, Development and Engineering Center
Natick, MA 01760-5020

Metallo-tetrabenzporphyrins and homologues possessing a high degree of third-order optical susceptibility, Chi(3), have been investigated as a part of an ongoing effort to develop laser eye protective devices for the soldier. The work has progressed through the synthesis, measurement of fundamental nonlinear optical properties, incorporation into a solid-state material, and evaluation of optical limiting capability. A series of metallo-tetrabenzporphyrins and a metallo-tetranaphthoporphyrin were prepared by reaction of the corresponding isoindole derivatives with an aldehyde and a metallic template ion, which is incorporated into the complex. These compounds were evaluated for third-order nonlinear optical susceptibility, Chi(3), using a frequency-doubled (532nm) Nd:YAG laser source and a degenerate four-wave mixing technique. The measured Chi(3) values were in the range 0.2 to 2.8x10^{-8} esu, which are among the highest reported to date.

The electronic effect of substituents on the meso-phenyl moieties was also investigated. A plot of Hammett sigma values against Chi(3) exhibited two linear relations of opposite slopes for positive and negative sigma values, with a minimum at the unsubstituted meso-phenyl complex. The authors discuss the significance of these results, with a view to designing porphyrin-based systems having higher third-order nonlinear optical activity values.

Introduction

A new threat to the soldier in the field has emerged through the potential use of tunable laser weapon systems. Laser range finders and target designators are already in use on the modern battlefield. Coherent radiation from laser sources, even at low energy levels, is capable of inflicting severe damage to the retina of the eye. One mode of protection against such laser threats is the use of nonlinear optical (NLO) media, which permit the transmission of visible light at low intensities but will reversibly block visible and near infrared radiation at intensities harmful to the eye. To be useful, the NLO media must have ultrafast response times.

The interaction of an electric field with a medium produces a separation of charges, resulting in a polarization in the medium. When the applied fields are weak, the induced polarization is linearly dependent on the applied field, and the propagation behavior is determined by the linear susceptibility of the material, Chi(1). At higher intensities, the polarization includes a nonlinear dependence, which becomes increasingly more important, and the susceptibility must be defined by additional terms such as Chi(2) and Chi(3). These terms represent second-order and third-order susceptibilities of the material and are determined by features at the molecular level, including electronic delocalization, molecular structure, and overall molecular packing. Chi(3) is a property of a bulk material. In addition, one can extract from Chi(3) an additional parameter, gamma, which is an intrinsic nonlinear property of the molecule.

Enhanced third-order susceptibilities can arise from centrosymmetric and noncentrosymmetric molecular structures possessing extended pi-electron delocalization, such as in the porphyrin ring systems. When these molecules are immobilized in an applied electrical or optical field, third-order nonlinear processes can occur, such as third harmonic generation (THG) and self-focusing. Also, a refractive index property change can result from third-order nonlinear processes. If the refractive index change is of sufficient magnitude and has an ultrafast response time, an optical limiting effect can be effected as required for eye protection from incident laser irradiation.

Measurements of Chi(3) on a material either in solution or dispersed in a solid polymeric phase contain both real and imaginary components. The real component is responsible for the nonlinear refractive index change. This brings about an intensity-dependent change of the phase of the laser beam. The imaginary component is responsible for nonlinear absorption, either by a two photon process or by excited state absorption.

This work focuses on the synthesis and measurement of third-order

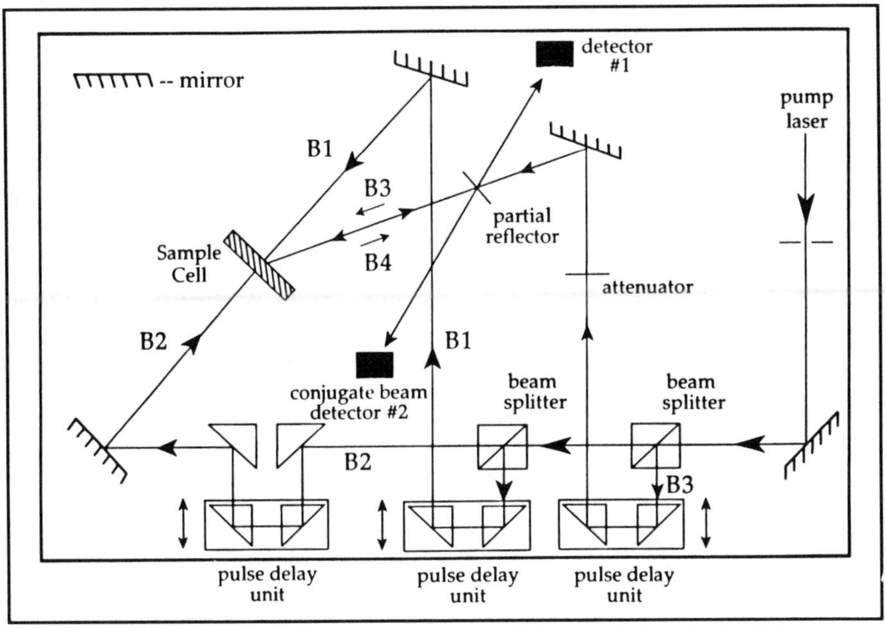

Figure 1. NLO measurement layout.

susceptibility, Chi(3), of a class of compounds called metallo-tetrabenzporphyrins (TBP) that have recently been shown to exhibit high Chi(3) values.[1,2] Both real and imaginary components of Chi(3) have been determined. In addition, preliminary experiments with these compounds incorporated into polymeric lenses has shown, for the first time, that optical limiting can be demonstrated in a device.

Experimental

The synthetic procedures for the preparation of the zinc tetrabenzporphyrins and the zinc tetra-1,2-naphthoporphyrin followed standard methodologies reported previously.[3,4]

The third-order NLO susceptibility, Chi(3), was measured by the degenerate four-wave mixing (DFWM) technique using tetrahydrofuran (THF) solutions of the porphyrins. The laser source is a Nd:YAG mode locked laser of 1064nm. It is frequency-doubled to give 17ps full width at half maximum (FWHM) single pulses, with repetition rates of 10 to 100Hz. The energy of each pulse was 25mJ, and the intensity was approximately 4MW/cm^2. Figure 1 shows the layout of the DFWM experiment.

The laser source used in the measurement is split into three separate beams, B1 through B3, which interact to produce a fourth beam, B4. Approximately 75 percent of the laser source beam is split into two pumping beams, B1 and B2, which are of equal intensities; the remaining portion of the laser source becomes a probing beam, B3. The delay units are adjusted so that B1 and B2 are colinear, and all three beams overlay in time and space in the sample cell (1 to 2mm path length). The pumping beams, B1 and B2, generate an electrical field that induces changes in the index of refraction of molecules in their paths. This refractive index change is maximum when B1 and B2 are in phase and zero when those beams are out of phase. The interaction creates a grating along the sample cell path. The probing beam, B3, interacts with this grating to generate a conjugate beam, B4, opposite in direction to B3. The intensity value (I) of the conjugate beam, B4, is used in equation (1) to determine Chi(3) of the solution relative to that of a standard.

$$\frac{\chi^{(3)}_{Sample}}{\chi^{(3)}_{cs2}} = \left(\frac{n_{Sample}}{n_{cs2}}\right)^2 \frac{d\alpha}{1 - e^{-d\alpha}}\, e^{\frac{d\alpha}{2}} \left(\frac{I_{Sample}}{I_{cs2}}\right)^{\frac{1}{2}} \frac{d_{cs2}}{d} \qquad (1)$$

In Equation 1 n is the index of refraction, d is the path length of the beam in the sample solution, d_{cs2} is the path length of the beam in CS_2, I is the intensity of B4, and α is the absorption coefficient of the solution at the laser wavelength.

The Chi(3) of the solution was calculated on the basis of the value for carbon disulfide as standard, which we take as 6.8×10^{-13} esu. The Chi(3) value for the solute is obtained from the plot of Chi(3) versus solute concentration.

$$\chi^{(3)}_{Solution} = \chi^{(3)}_{Solvent} \left(\frac{n^2 + 2}{3}\right)^4 \frac{\gamma^{(3)}_{Solute} A}{M} C \qquad (2)$$

In Equation 2, n is the index of refraction of solute, γ is the second hyperpolarizability, *A* is the Avogadro's Number, *M* is molecular weight, and *C* is concentration of a solute in g/mL. The index of refraction of the solute is determined by refractometric measurements of its solution.

Results and Discussion

A series of zinc tetrabenzporphyrins (Table 1, Ia to Ij) were prepared as outlined in Figure 2. By varying the nature of the aldehyde, several different

Table 1
Values of Chi(3) for Porphyrins

ZnTBP (I)	Chi (3) 10^{-8}esu
Ia. R = H, X = H	0.3
Ib. R = H, X = F	0.2
Ic. R = Me, X = H	1.5
Id. R = Ph, X = H	0.3
Ie. R = p - MePh, X = H	1.2
If. R = p - MeOPh, X = H	1.4
Ig. R = p - $N(Me)_2Ph$, X = H	2.8
Ih. R = p - $N(Me)_3Ph$, X = H	1.8
Ii. R = p - ClPh, X = H	0.56
Ij. R = m - FPh, X = H	1.3
ZnTNP (II)	
II. R = H	0.99

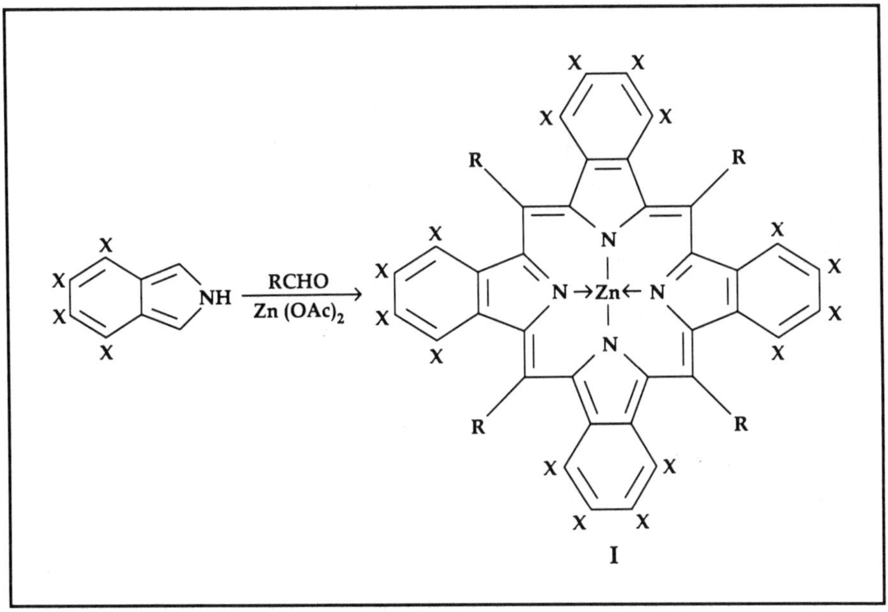

Figure 2. Preparation of Tetrabenzporphyrins.

structural variations were accommodated in the "meso" or R position on the porphyrin ring system. The macrocyclic compound series consisted of aliphatic and aryl substituents at the meso position. One example with substitu-

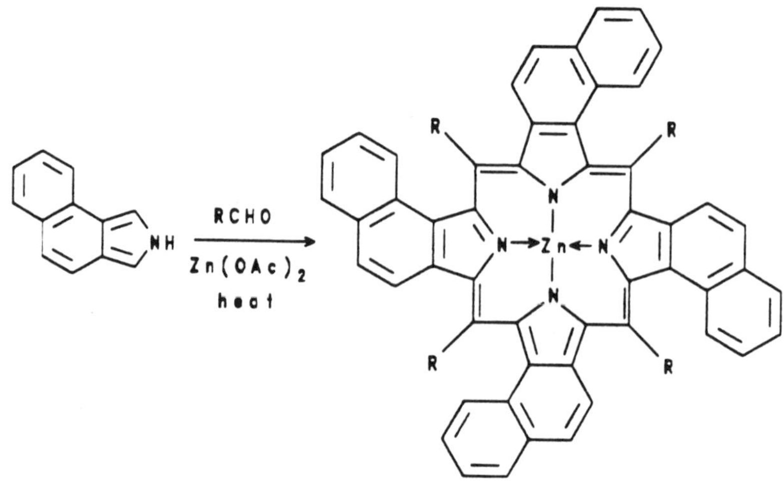

Figure 3. Preparation of Tetranaphthoporphyrin.

tion on the benzo- portion of the molecule is included (Ib). Macrocycle, Ib, has a fully fluorinated tetrabenzo periphery. In all cases, the metal coordinated at the center of the ring is zinc. Zinc tetra-1,2-naphthoporphyrin (II, R=H) was prepared as outlined in Figure 3.

The results of measurement of Chi(3) for the series of zinc tetrabenz- and a zinc tetranaphthoporphyrin are presented in Table 1. The effect on Chi(3) of electron delocalization over a larger ring system is shown by comparison of zinc tetrabenzporphyrin (Ia) with zinc tetra-1,2-naphthoporphyrin (II). The annelation of an additional aromatic ring to the benzporphyrin base structure has increased Chi(3) by a factor of three.

The introduction of electron donating groups into the meso-phenyl groups of the zinc tetrabenzporphyrin series led to a progressive rise in the measured Chi(3) value as the relative electron donating of the substituent increased. Interestingly, as the electropositive character of the substituent increased, the value of Chi(3) also increased. These findings are consistent with the fact that greater polarization, whether emanating from electron withdrawal or electron donation, enhances Chi(3).

Hammett linear free energy relationships have been traditionally developed to explain combined resonance and inductive effects on the reactivity of substituted aromatic systems. It is interesting to note, as shown in Figure 4, that a linear relationship is observed when Chi(3) values are plotted against the Hammett sigma constants. The plot shows a V-shape composed of two

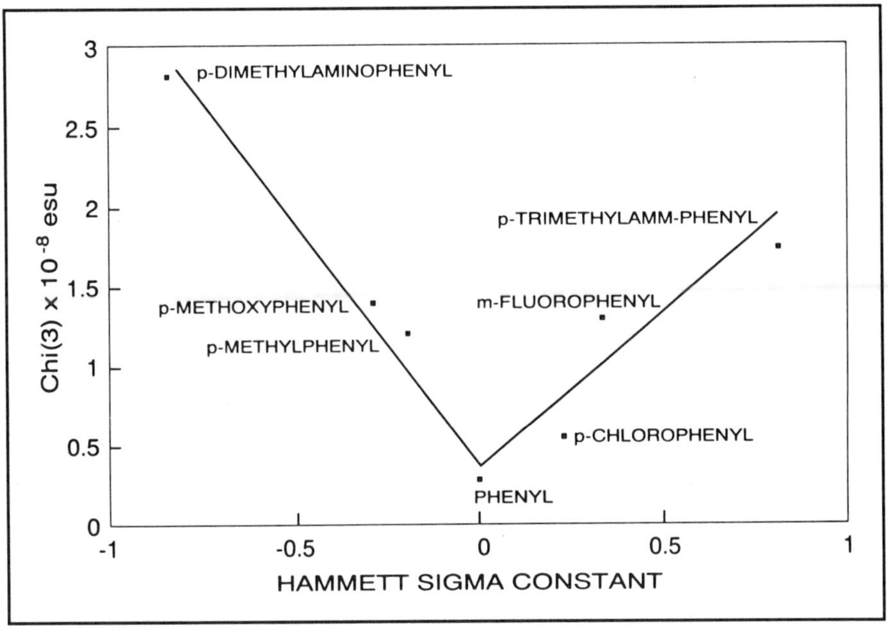

Figure 4. Hammett plot of Chi(3).

lines of opposite slope, depending on whether the substituent is electron withdrawing or donating. As expected, the greater the absolute value of the sigma constant, the greater is the polarization, and Chi(3) is increased proportionately.

Conjugated pi-electron systems such as in the porphyrin macrocycles are important to the manifest of the Chi(3) property. Apparently, the meso substituents contribute a resonance and inductive effect on the ring conjugated pi-electrons. A linear relationship is obtained when the meso-phenyl substituents' Hammett sigma constant values are plotted versus Chi(3). These Hammett sigma constant values are based on combined resonance and inductive contributions of the respective substituents. Steric constraints in the porphyrin ring dictate that the meso-phenyl substituents will be almost orthogonal to the plane of the pi-electron conjugated porphyrinic ring. Thus, the meso-substituents probably contribute a predominant inductive effect on the ring, with the resonance contribution being a lesser contributor.

Similar linear relationships have been proposed based on theoretic calculations of second-order susceptibility done by Goldfarb and Medrano for substituted benzthiazoles.[5] In addition, Meredith and Stevenson[6] calculated empirical substituent group parameters based on THG experiments and found values that were higher at both ends of the Hammett sigma scale.

Table 2
Chi(3) of TBP Solutions

Compound	Conc. Concentration g/l	Re ($\chi^{(3)}$) 10^{-14}esu	Im ($\chi^{(3)}$) 10^{-13}esu	α cm^{-1}	β cm/GW	σ 10^{-17}cm^{-2}
Ia	0.46	9.1	≤1.8	3.8	≤0.5	3
If	0.46	7.8	9	2.4	2.5	3

α - Ground state linear absorption coefficient.
β - Ground state two photon absorption coefficient.
σ - Excited state linear absorption coefficient.

Structure correlations that are predictive of nonlinear responses are at present based on an intuitive understanding of the phenomena and have yet to reach a rigorous state. The relationships observed contribute to methods that may be predictive of structures with enhanced values of Chi(3). In related work, we have started to investigate the effects of metal substituents on the TBP system. These results will not only be of interest in the search for compounds with high Chi(3) but will offer comparisons with published theoretic and experimental studies on metallo-phthalocyanines,[7] which are similar in structure to the benzporphyrins.

Nonlinear effect is limited not only to the induced change in the polarization but also in the absorptivity. In the mathematical derivation, Chi(3) has real and imaginary parts. The measure of the nonlinear absorptivity appears as the imaginary part of Chi(3). This nonlinear absorptivity is another useful property of these porphyrins in the optical limiting application. To determine the imaginary part of Chi(3), the transmission of laser as a function of laser intensity and fluence was measured, using picosecond and nanosecond pulses.[2] The results of the absorption experiments are summarized in Table 2. For a comparison, the real part of Chi(3) for the porphyrin solutions is listed also.

It was found that at certain fluence levels, the percent transmission did decrease from approximately 45 percent to 15 percent and 60 percent to 20 percent for compounds Ia and If, respectively. The decrease was observed from both picosecond and nanosecond laser pulses. It is inferred that it resulted from the absorption of laser by an excited state, which was populated by a one photon absorption and had a cross section larger than the ground state. Such an excited state has a lifetime of 15ns; the population became appreciable during these short laser exposures. The magnitude of this type of absorptivity is expressed as σ in Table 2. For both Ia and If, the values of σ are approximately equal. Besides the excited state absorption, an absorption due

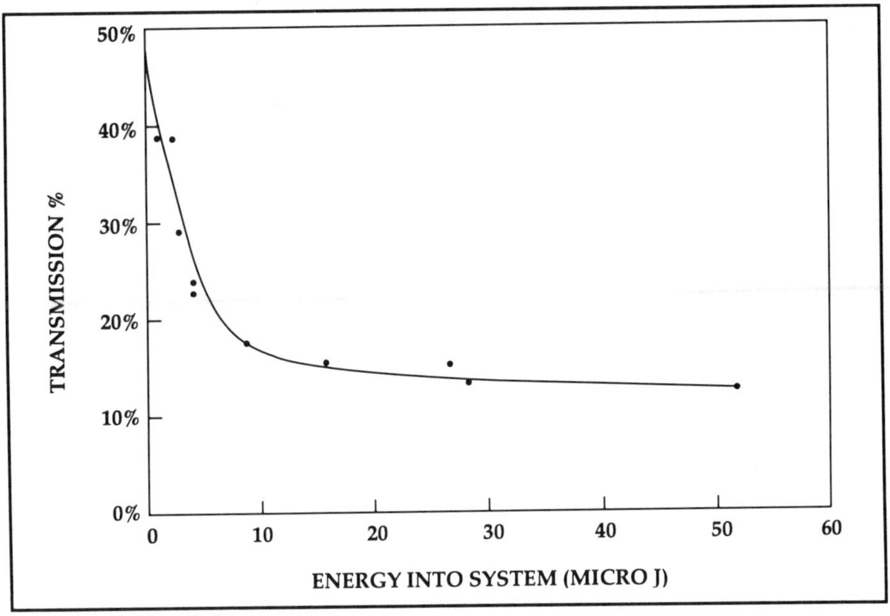

Figure 5. Optical limiting property.

to two-photon process was indicated. The two photon absorption cross-section is given as β in Table 2. It was of interest to note that If has a larger β than Ia. It is also noted that the magnitudes of Chi(3) obtained from these experiments were much smaller than those from DFWM. The values in Table 2 are expected to be underestimates due to the method of approximations used to determine these values.

Toward the development of optical limiting application devices, the tetrabenzporphyrin Ia was incorporated into the solid polymer matrix polycarbonate, and its optical limiting property was evaluated. Figure 5 shows that the optical limiting effect did occur. A 70 percent transmittance reduction was observed in these preliminary experiments. It is important to realize that this is the first demonstration of optical limiting in a prototype system. Thus, using a porphyrinic ring system, it is possible to engineer an optical limiting capability.

Summary

A series of zinc tetrabenzporphyrins and a zinc tetranaphthoporphyrin were synthesized for evaluation by a degenerative four-wave mixing technique. A high degree of third-order optical nonlinearity was found. The measured Chi(3) values were in the range of 0.2 to 2.8 x 10^{-8}esu, which are among the highest reported to date. The influence of the meso-substituent on the Chi(3) value was examined in relation to the Hammett σ constant. Metallo-porphyrins incorporating variations in the ring systems will be prepared and evaluated to explore the impact of molecular design on the magnitude of third-order optical nonlinearity.

Acknowledgments

The authors would like to thank Joel Carlson of Natick RD&E Center for the mass spectrometric characterization of all the porphyrins we have reported in this paper.

References

1. D.V.G.L.N. Rao *et al.*, *Appl. Phys. Lett.* **58**(12), 1241 (1991).
2. S. Guha *et al.*, *Optics Lett.* **17**(4), 264 (1992).
3. D.E. Remy, *Tetrahedron Lett.* **24**(14), 145 (1983).
4. D.E. Remy, G.C. Richard, and S.J. Weininger, U.S. Army Natick Technical Report NATICK/TR-87/029, AD A183957, 1987.
5. I.J. Goldfarb and J. Medrano, in *Nonlinear Optical Effects in Organic Polymers*, J. Messier *et al.*, Eds.(Kluwer Academic Publishers, New York, 1989), pp.93-99.
6. G.R. Meredith and S.H. Stevenson, *Ibid.*, pp.105-122.
7. James S. Shirk *et al.*, *Materials for Nonlinear Optics*, S.R. Marder, J.E. Sohn, and G.D. Stucky, Eds. (American Chemical Society, Washington, DC, 1991), p. 626.

41

Monolithic, Photoconductive Impulse-Generating Device

The Army has a need for impulse generators that can radiate pulses with narrow pulsewidths and picosecond rise times for synthetic aperture radars and smart high-power microwave jammers. The authors describe a novel, compact photoconductive wafer device that generates pulses with the desired characteristics. The impulse generator has a gallium arsenide photoconductive wafer that stores energy and forms a radial transmission line that is triggered by a mode-locked Nd:YAG laser. The impulse generator has successfully generated 5-kV impulses, about 500ps wide, into a 50-Ω load. Normally, such a device behaves as an rf capacitor with output pulse resembling that of a capacitive decay. A new mode of operation, however, has been observed with a newly designed device in which the turnoff is achieved in a time interval much smaller than the normal capacitive decay time. The conditions are described under which subnanosecond switch turnoff is achieved. A model is advanced that includes avalanche effects that account for the switch turnoff.

A. Kim
R. Zeto
R. Youmans
M. Weiner
U.S. Army Research Laboratory
Electronics and Power Sources Directorate
Fort Monmouth, NJ 07703-5601

L. Jasper
U.S. Army Research Laboratory
Signatures, Sensors, Signal and Information Processing Directorate
Adelphi, MD 20783-1197

Introduction

Picosecond switching is of interest to the Army for generating narrow-band pulses with picosecond rise times for synthetic aperture radars (SARs) and smart, high-power microwave (HPM) jammers. The photoconductive, semi-insulating gsallium arsenide (GaAs) switch is a candidate for generating the desired pulsewidths because of its high-voltage, holdoff capability, fast rise time, and large peak-power-handling capability. The low-jitter property

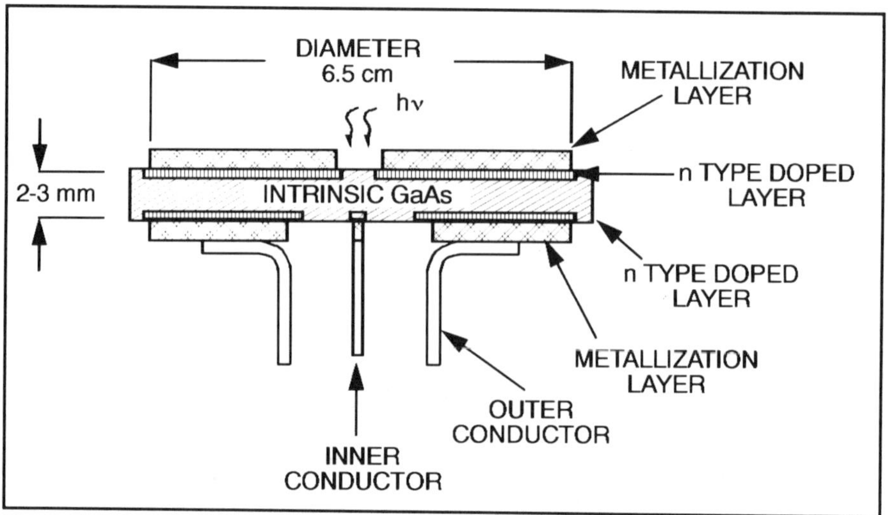

Figure 1. Schematic diagram of a monolithic, photoconductive GaAs impulse device.

of the switch, due to laser activation, gives the capability of phase arraying large numbers of switches for achieving gigawatt power levels at high repetition rates.

Recently a monolithic, photoconductive GaAs switch has been described.[1] As shown in Figure 1, the switch consists of a circular GaAs wafer, typically 7.5 cm in diameter, with metallized disc electrodes on each side. On the high-voltage side, an aperture at the center is etched out so as to provide access to the semiconductor for the light pulse, which, by producing carriers in the central region, causes this region to transition from an insulating state to a conducting one. On the opposite, grounded side, an annular region close to the center region was etched out, providing contact to an output coaxial transmission line, which was pressure fitted to this side of the wafer. Once the light was introduced, a centrally located conducting channel between the disk electrodes was created, so that the energy stored between the electrodes was delivered to the ouput coaxial line in the form of a narrow pulse. One may view the device as a radial transmission line that forms the energy-storing element. Once the light pulse produces the conducting channel, the radial line becomes connected to the output coaxial line to which the pulse energy is delivered. The light source has a much narrower pulsewidth (75ps) than the generated pulse so as not to have any effect on the circuit response.

An important property of the radial structure has to do with the impedance transformation[2,3] that occurs as the wave proceeds from the outer to the inner radius of the structure. At the outer perimeter, the radial impedance is small, whereas the impedance increases near the center, varying as $1/r$, where r is the radius. If Z_o and Z_i are the characteristic impedance levels at the outer and inner radii, then for small transformation ratios ($Z_i/Z_o < 2$), a simple approximation[4] for the peak output voltage V_{out} applies. For a matched load, with no loss in the switch, V_{out} is

$$V_{out} = (V_o/2)(Z_i/Z_o)^{1/2}, \tag{1}$$

where V_o is the charging voltage. For small transformation ratios, $(Z_i/Z_o)^{1/2}$ may be regarded as a gain factor (compared with the output of a uniform transmission line). For larger transformation ratios, however, more exact simulation program with integrated circuit emphasis (SPICE) simulation shows that the gain factor saturates very quickly at a value equal to about two, so that the output itself saturates at the value $V_{out} = V_o$.

As the impedance variation between device impedance at the inner radius (Z_i) and the load impedance (R_L) increases, the output waveform more and more resembles that of an rf capacitor, with the change accompanied by a widening of pulse but with an increase in the rise time. The resemblance to capacitor output becomes complete for a positive mismatch at the output, that is, $R_L > Z_i$, where R_L is the output impedance. In this case, the pulse does indeed exhibit a fast rise time followed by an R_LC decay, where C is the capacitance of the radial structure. Figure 2 shows the experimental output for the radial device shown in Figure 1, where Z_i is 6Ω and R_L is 50Ω, so that a positive mismatch prevails. The waveform in Figure 2 closely resembles that of a discharging capacitor. Results of the SPICE simulation agree with the experimental waveform. For most experiments, the positive mismatch is typical for the high-power GaAs wafer device. This is because it is very difficult to obtain 50-Ω output device impedance levels with the GaAs wafer device, unless one resorts to extremely small inner radii (<1mm). Such small radii become impractical because of voltage breakdown and problems associated with fabricating the transition to the output coaxial line.

Unfortunately, the relatively wide nanosecond pulsewidths obtained with the radial line limit the usefulness of the device for many applications where sub nanosecond pulses are desirable. In this investigation, we observe a new mode of operation for the monolithic, photoconductive GaAs wafer device, which yields subnanosecond pulsewidths. With this new mode, the current is turned off in times much shorter than the R_LC decay time. The device, biased to 6kV, produces pulses as narrow as 500ps and rise times as fast as 200ps, with an output voltage of 5kV. Narrower pulses at considerably higher voltages are foreseen.[2]

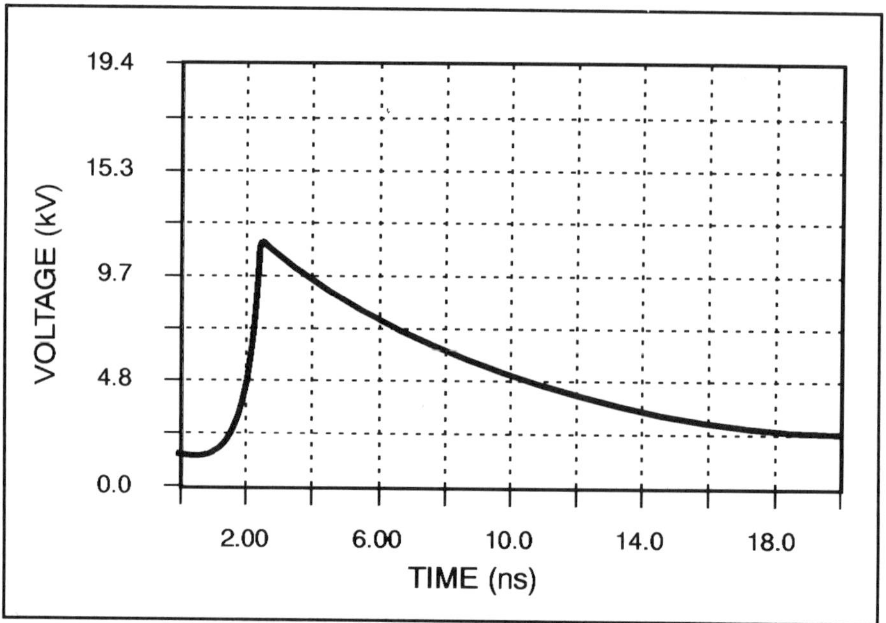

Figure 2. Output waveform obtained with a 3-mm-thick GaAs device. Switch is biased to 4kV and triggered by a mode-locked Nd:YAG laser. Output is delivered to 50-Ω load impedance.

Experimental Procedures and Results

The experimental procedures were similar to those reported previously,[1] for which a 50-Ω load was used. The laser light (either Nd:YAG laser or laser diode array) was conveyed to the GaAs wafer aperture by means of a fiber-optic bundle. The measurement sequence began with pulse charging of the radial line via a silicone controlled rectifier (SCR)-transformer circuit. Upon completion of the charging process, the laser was fired, and the generated pulse was delivered to a specially constructed, high-voltage coaxial line. The pulse was attenuated using large-bandwidth Barth attenuators, and the final pulse delivered to a Textronix SCD-5000 oscilloscope. For loads smaller than 50 Ω, a wideband current-viewing resistor was used to monitor the pulse.

Figures 3 and 4 show the output of a 2-mm-thick device when biased to 4kV and illuminated with a 75-ps-wide, 1-mJ pulse from an Nd:YAG laser. The inner and outer radii were 0.8 and 3.25cm, corresponding to characteristic impedance values of 4 and 1 Ω, respectively. Figures 3 and 4 are for 50- and 10-Ω loads, respectively. Unlike that of the 3-mm-thick device, the 2-mm device output exhibits periodic peaks, separated by intervals of about 1ns. The widths of the peaks themselves are less than 1ns wide. In Figure 3, which at 50Ω has the greater mismatch, the discharge more closely follows an R_LC

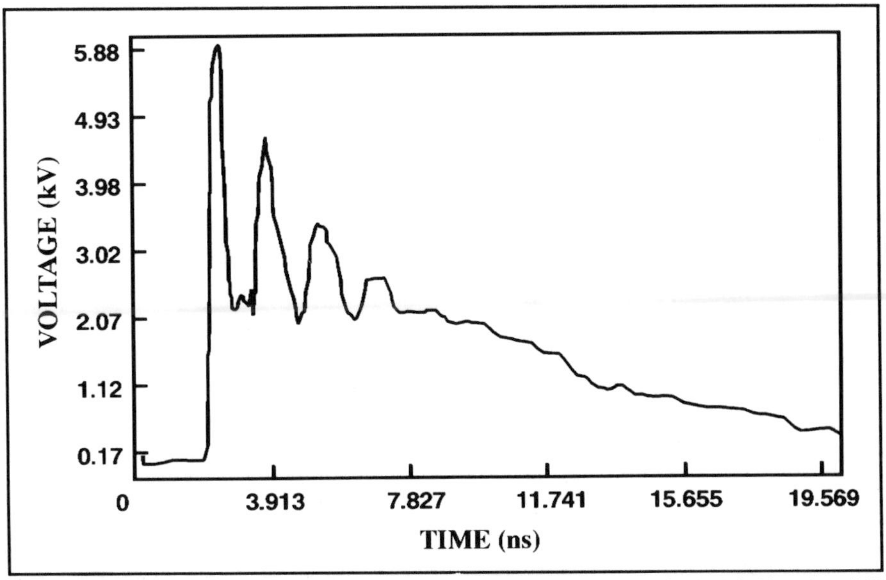

Figure 3. Output waveform obtained with a 2-mm-thick GaAs device. Switch is biased to 4kV and triggered by a mode-locked Nd:YAG laser. Output is delivered to 50-Ω load impedance.

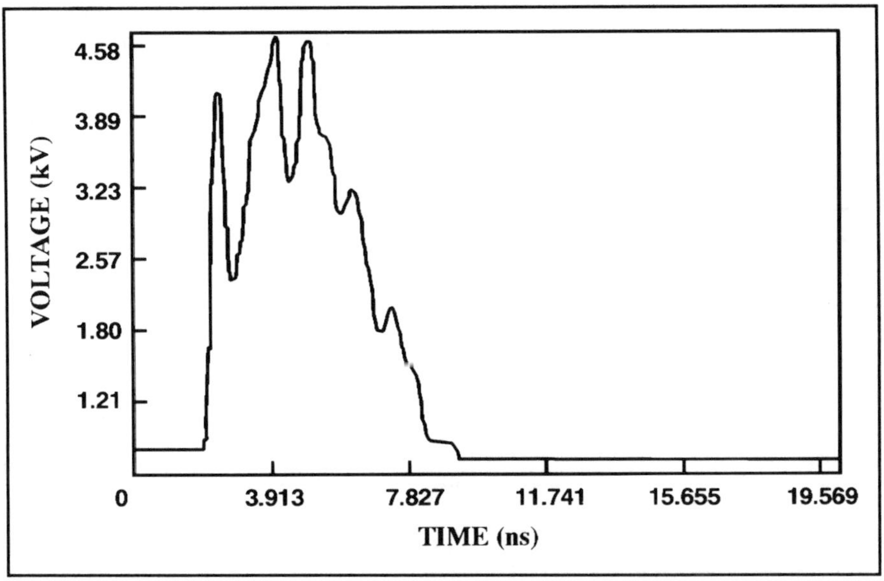

Figure 4. Output waveform obtained with a 2-mm-thick GaAs device. Switch is biased to 4kV and triggered by a mode-locked Nd:YAG laser. Output is delivered to 10-Ω load impedance.

decay, modulated by the periodic peaks. With the 10-Ω load, the pulsewidth narrowed and the risetime increased. Figures 5 and 6 show the same device illuminated with a laser diode (wavelength of 904nm), with a peak power of

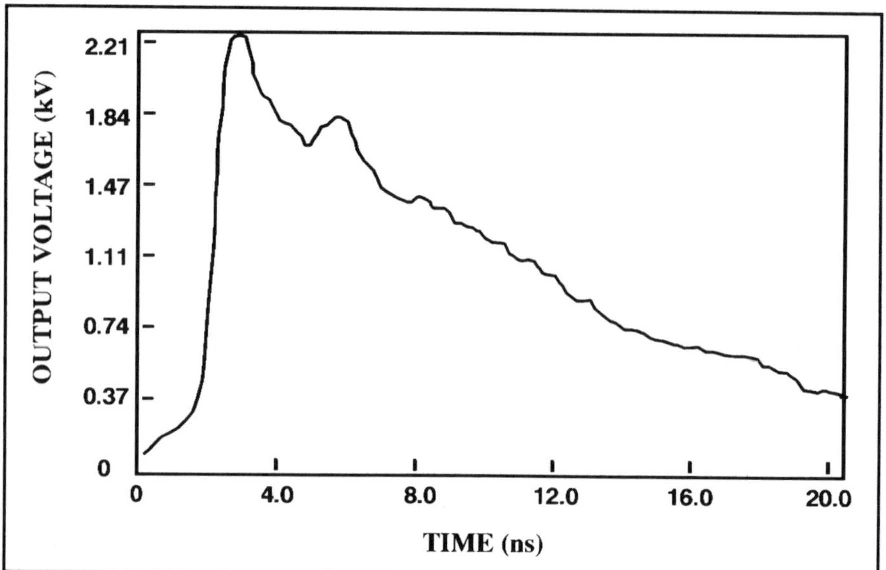

Figure 5. Output waveform obtained with a 2-mm-thick GaAs device. Switch is biased to 4kV and triggered by a laser diode. Output is delivered to 50-Ω load impedance.

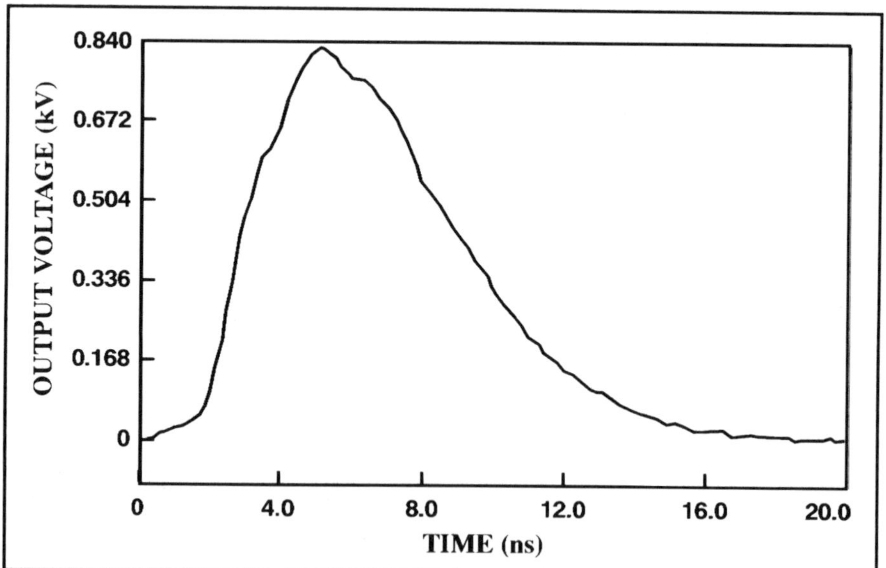

Figure 6. Output waveform obtained with a 2-mm-thick GaAs device. Switch is biased to 4kV and triggered by a laser diode. Output is delivered to 10-Ω load impedance.

100W and a rise time of 3ns. As before, the loads were 50Ω and 10Ω. Unlike those with the Nd:YAG laser, the laser-diode-triggered waveforms did not exhibit any periodic peaks. As before, the waveform for the 50-Ω load

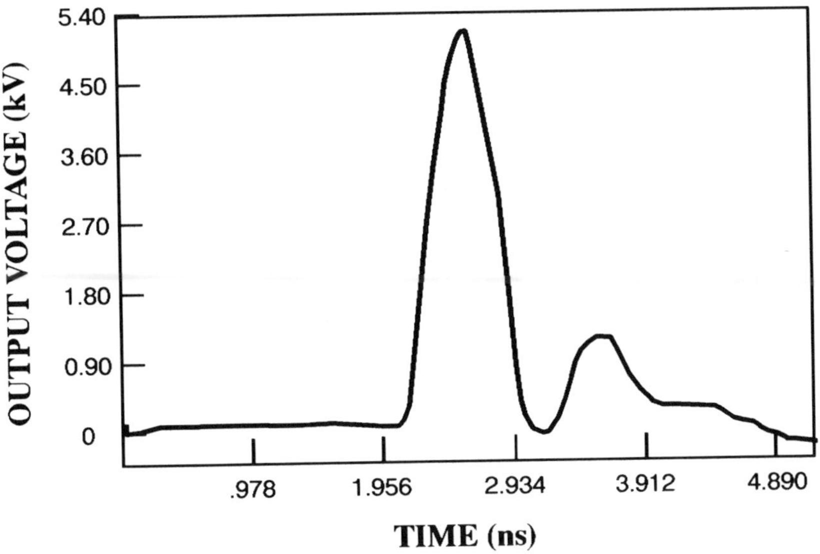

Figure 7. Output waveform obtained with a 2-mm-thick GaAs device. Switch is biased to 6kV and triggered by a mode-locked Nd:YAG laser. Output is delivered to 50-Ω load impedance.

exhibited a clear RLC decay. The risetime, however, was somewhat slower compared with that of the Nd:YAG laser. For the 10-Ω load, the waveform was still reminiscent of an R_LC waveform, although the pulsewidth was somewhat narrower and the rise time slower. SPICE simulations bear out this trend when more closely matched loads are used.

We then obtained results for a device similar to the previous one, except that the inner diameter was 0.2cm instead of 0.8cm. As a result, the estimated inner impedance was 16 Ω instead of 4 Ω. As before, the computer simulations predicted an R_LC-type discharge for the 50-Ω load. However, the actual experimental waveforms obtained were quite different. Figure 7 gives the pulse for a 6-kV bias voltage and subsequent Nd:YAG laser illumination. The pulsewidth obtained was extremely narrow, about 500ps, with a rise time of about 200ps. It is as though the first peak, shown in Figure 3, were allowed to pass through the device, while the remainder of the stored energy became trapped in the radial structure (the resistance in the input charging circuit eventually leaks off any trapped energy). As the bias voltage increases, the trend at higher bias voltages is to reduce the rise time and pulsewidth slightly, while increasing the relative output voltage. The waveform for laser diode excitation is shown in Figure 8. In this case, a single pulse also is obtained, but it is wider compared with that generated by the Nd:YAG laser. Nevertheless,

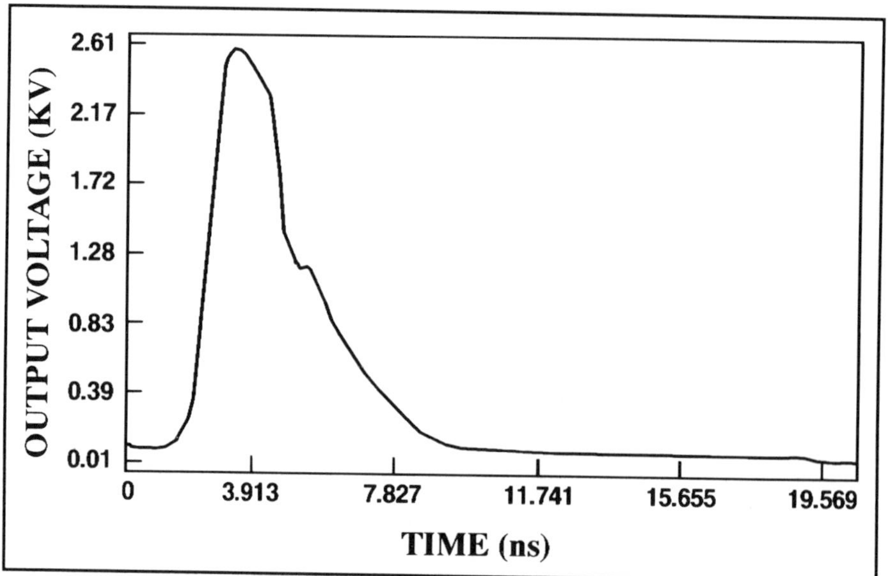

Figure 8. Output waveform obtained with a 2-mm-thick GaAs device. Switch is biased to 6kV and triggered by a laser diode. Output is delivered to 50-Ω load impedance.

the pulsewidth produced by the laser diode was narrower (by at least a factor of 3x) compared with the expected R_LC time constant.[3]

Discussion and Proposed Model

As mentioned previously, waveforms were obtained that were different from those predicted either by simple theory or SPICE simulations. In particular, the appearance of the output current spikes (Figures 3 and 4), or the single narrow pulses obtained in Figures 5 and 6, cannot be explained by the circuit theory for the radial line. The most glaring discrepancy occurs when the switch current is turned off, leaving energy trapped in the radial line.

To account for the experimental facts, we suggest that the observed current turnoff, with the resultant pulse narrowing, is related to the creation of carriers in the annular region of the semiconductor. If sufficient carriers are created in this region, then the pulse output path will be shorted out, and the energy will be trapped in the radial structure. When the device is turned on by the laser light illumination, an electric field (field strength determined by the bias voltage and the length of annular region) is established in the annular region, between the center electrode and the inner radii, and this field helps to create a transverse conducting path. For the thicker device (3mm) with the larger inner radius (0.8cm), the Nd:YAG laser fails to interrupt the current pulse (Figure 2). For the 2-mm thickness, however, the current interruption begins to appear (Figures 3 and 4). Because of the large annular region and the multiple wave reflections, however, the interruption is not complete, and oscillations of the peak current are observed. Mechanisms that might alternatively turn on and off the supply of carriers would include avalanche effects and rapid surface recombination. The initial carrier profile, determined by the laser pulse, may already partially bridge the gap between the inner and outer conductors, thus making it easy to short out the pulse for smaller inner gaps. One may speculate that if the closure of the annular region is caused by an avalanche, then the pulsewidth is determined by the delay time for the onset of the avalanche. For the same device (0.8-cm inner radius), it is interesting to note that with the laser diode, there are no current spikes. This is attributed to the smaller light pulse energy as well as the decreased light penetration of the 904-nm wavelength, which makes it more difficult to close the annular region. Under these conditions, the R_LC decay is observed without any current spikes.

The situation is far different when the inner radius is reduced from 0.8 to 0.2cm (maintaining the same center electrode radius of 1mm). For the Nd:YAG laser, a single pulse is observed, and we suggest that the annular region remains completely closed off because of the smaller spacing. This smaller spacing will result in a large field strength as well as a shorter plasma spreading path. The combination of these two factors may initiate an early onset of an avalanche and contribute to the generation of sub-nanosecond impulses. With the laser diode excitation, a single pulse is again achieved, but with a somewhat larger pulsewidth (2ns). The larger pulsewidth may be connected with the larger delay time for the onset of an avalanche. Such a long delay may be expected, because the laser diode produces fewer carriers to seed the avalanche initially.

Other possible mechanisms may account for the closure of the device, including breakdown in the output coaxial line. The observed current turnoff, however, occurs at electric fields as low as 5 to 10kV/cm in the coaxial line. The relatively small field, combined with the speed of the event, makes it unlikely that the source of the closure is the breakdown in the line.[4]

Conclusions

Narrow-pulse generation has been achieved in monolithic, GaAs switches fabricated in the form of a radial transmission line. Pulses as narrow as 500ps with 5-kV output into a 50-Ω load have been achieved. The large voltage enhancement (a 5-kV output voltage out of a 6-kV bias voltage) is attributed to the impedance transformation of the radial transmission line. The narrow-pulse operation is attributed to the plasma spreading into the annular region of the GaAs region. The presence of these carriers probably leads to avalanching, which disconnects the radial line from the output after a fraction of a nanosecond. Using this mode of operation, considerable improvements in peak voltage output, as well as further reduction in pulsewidth, are envisaged. In addition, extremely fast repetition rates may be possible because of the trapped energy in the transmission line, which would reduce subsequent charging times.

Acknowledgments

The authors acknowledge the work of J. Conrad and M. Wade for the switch fabrication.

Appendix: SPICE Simulation of Monolithic Photoconductive GaAs Pulser

A SPICE simulation may be used to obtain the response of the radial structure of the pulser. This appendix describes a general technique, which may then be particularized to the radial line. The technique involves first breaking up the transformer line into a number of segments of equal length, each of which has a constant impedance. The impedance value of each section follows the functional dependence of the transformer line with length, connecting the initial impedance to the final value. Once this is done, standard software, such as SPICE, may be used to find the transformer line response. In Figure 9, which demonstrates the technique, the impedance is a function of length $Z(\mathrm{x})$. Accordingly, the transformer line is divided into N sections (N = a positive integer), defining a set of lengths x_n given by

$$x_n = x_{\mathrm{i}} + n(x_f - x_{\mathrm{i}})/N \,, \qquad \text{(A-1)}$$

where x_{i} is the initial value of the device length, x_f is the final value, and n is an integer: 0, 1, 2, ..., N.

The length of each section is $(x_f - x_{\mathrm{i}})/N$, and the average impedance of each section is $Z_m = Z(x'_m)$, where

$$x'_m = (x_m + \mathrm{x}_{\mathrm{m-1}})/2 \qquad \text{(A-2)}$$

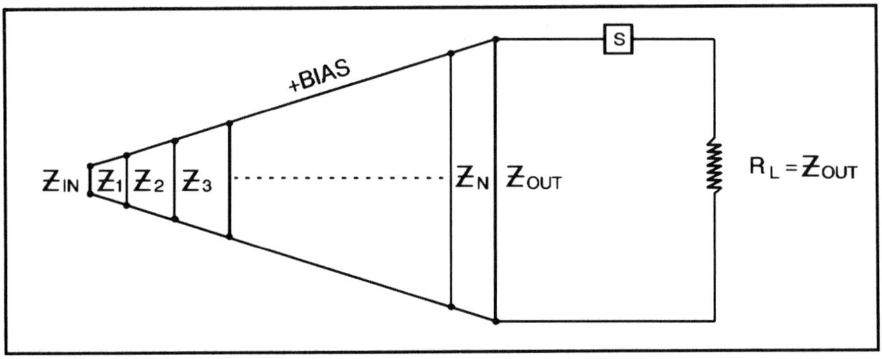

Figure 9. Division of transformer into a number of sections. Each section is approximated by a constant impedance, thus allowing for a simple computer simulation (such as SPICE).

and $m = 1, 2, \ldots, N$. The above technique may be applied to the radial line, where the radial characteristic impedance of the wave may be approximated by

$$Z(x) = (377d)/(2p\epsilon^{1/2}x) , \qquad \text{(A-3)}$$

where x is the radius, d is the separation between the circular plates, and ϵ is the relative dielectric constant. Equations A-1 and A-2 are then used to set up the SPICE simulation.

References

1. A. Kim *et al.*, *Appl. Phys. Lett.* **58**, 2812 (1991).
2. R. A. Petr *et al.*, *Rev. Sci. Instrum.* **59**, 132 (1988).
3. C. Bamber *et al.*, Particle Accelerators **23**, 255 (1988).
4. A. Kim, Ph.D. Thesis, Rutgers University, 1992, p. 47.

43

Review of Environmental Research Specific to Smart Weapons Operability Enhancement for the Battlefield Environment

James P. Welsh
U.S. Army Corps of Engineers
Cold Regions Research and
Engineering Laboratory
72 Lyme Road
Hanover, NH 03755-1290

Mike Hardaway
U.S. Army Corps of Engineers
Topographic Engineering Center
Fort Belvoir, VA 22060-5546

Wade West
U.S. Army Corps of Engineers
Waterways Experimentation
Station
3909 Halls Ferry Road
Vicksburg, MS 39180-6199

The Smart Weapons Operability Enhancement (SWOE) Program is developing, validating, and demonstrating an integrated, physics based, scene generation process to consider complex target and background environment interactions for varied battlefield conditions. The product of this program is an integrated process that will enhance the performance of future smart weapons systems for a global variety of battlefield environments. The validated capability combines methodologies and techniques that comprehensively treat environment interactions relevant to smart weapon performance. This capability is being developed and demonstrated for Department of Defense (DoD) wide user communities. The "weapon system environment" is number 11 on the DoD critical technologies list. "Synthetic Environments" is also a principal thrust area for DoD Science and Technology. Components of the environment have been quantitatively shown to impact the performance of electromagnetic and electro-optical weapon systems. The SWOE program has provided a focus for coordinated and cooperative investigations of critical environment factors and processes that significantly impact weapon performance. These studies have considered the broadest possible range of anticipated battlefield environmental conditions. The principal thrust has been to quantitatively define these factors and to provide the capabilities to measure, model, render, and extrapolate environmental impact on weapon performance.

Smart Weapons Operability Enhancement (SWOE) Scene Generation Process

The Smart Weapons Operability Enhancement (SWOE) scene generation process is an integration of measurement, information bases, models, and rendering capabilities. The Corps of Engineers (CoE) Tech Base effort, relevant to SWOE, is focused in four areas: measurement, information bases, modeling, and rendering.

Measurement and information bases. This area includes the following: what, where, when, and how to measure, obtain representative samples, and infer energy interactions between surface and volume scatters (including spatial scales of less than 0.5m); and construction of information bases for a global variety of battlefield environment conditions, particularly to develop site-to-site and area-to-area comparisons methods. Figure 1 summarizes the CoE effort for measurement and information bases.

An integral part of the three-dimensional (3-D) scene generation procedure, the information base contains all spatial and attribute data required to define the total landscape environment (land, water, atmosphere, sky, and so on). The content and structure of the information base are driven largely by the various requirements of the numeric models and the scene generation software.

The information base contains three kinds of data: digital terrain data (for example, topography, soil types, vegetation types); physical, thermal, and spectral terrain attribute data (for example, moisture content, emissivity, reflectance); and meteorologic (weather and atmospheric) data (for example, air temperature, visibility).

Digital terrain data are representations of portions of the earth's surface stored in computer-compatible formats. These data depict characteristics such as elevation, vegetation types, soil types, and other relevant environmental information. Digital terrain data used in the scene generation procedure are stored in raster and vector formats and managed by a geographic information system (GIS).

The physics-based thermal signature prediction models used in the scene generation procedure require as inputs complete, quantitative descriptions of the physical, thermal, and spectral attributes of each landscape feature. These data are most efficiently stored and retrieved in tubular format in a relational database management system (RDBMS). The RDBMS associates each stored numeric value with the corresponding landscape feature depicted in the GIS.

Meteorologic data are required throughout the scene generation process and have particular importance to the radiation field prediction models. Both surface weather and upper atmospheric profile data are required. These data are also stored in tabular format in the database management system (DBMS).

Models. The goal of the SWOE modeling effort is to assemble and integrate 3-D fundamental physics models of important environmental phenomena and objects (natural and manmade) for the infrared (IR) and millimeter wave (MMW) spectral bands. Specific objects are trees, with & without leaves; buildings; vehicles; roads; bridges; and so on. The boundary regions between adjacent objects, tree canopies, row crops, forest edges, and other textured surfaces are also important. Models of the energy budget are significantly affected by heterogeneity in the 3-D distribution of energy emitters and scatters. Figure 2 summarizes the CoE effort relevant to modeling.

The SWOE Interim Thermal Model is used to calculate the surface temperatures for a wide variety of surfaces, including vegetated and non-vegetated surfaces, bodies of water, and snow/ice-covered surfaces. The model results are valid for all seasons.

The thermal models in the package are driven by conventional weather data such as standard surface weather observations and radiosonde data. Default databases of seasonally dependent thermal properties are provided to cover a set of standard surfaces that are commonly encountered in scene simulation.

The SWOE thermal models package accommodates various vegetation effects. The effects of simple vegetation, such as grasses and crops, and forests can be included in the 1-D heat balance of soils. A separate 3-D model of the thermal balance for individual trees is included. Two geometric representations of trees, based on measurements of actual trees, have been included to calculate the temperature fields for trees.

The SWOE thermal models are driven by the radiation fields from the atmosphere. The atmospheric radiation budget is calculated using a modified version of LOWTRAN7, which is the standard atmosphere radiance and transmission computer code used by the DoD community.

The SWOE Radiance Models Software System contains two parallel computational paths, one for terrain and one for 3-D objects (currently individual trees and two military targets). The terrain radiance path is built around a new Fortran model called the Improved Background Radiance Model (IBRM).

Radiances for 3-D objects are computed with the Hardbody module of the SPIRITS code, a U.S. Government standard for aircraft. Both radiance models utilize the same basic algorithms and phenomena, which include:

Δ WES □ TEC ○ CRREL

TASK	FY 92	FY 93	FY 94
Measurements	Classification & Regression Based Winter Parameters ○ Standardize Snow Microstructure Methods ○ 3-D Scene Parameters Temperate ○Δ 35 GHz Measurements Temperate ○Δ	95 GHz Measurements Temperate & Desert ○Δ Forest Edge MMW Temperate ○Δ Slant Range IR Temperate & Desert ○Δ Forest Edge IR Temperate ○Δ	Winter Scene Sampling ○ Surface Energy Exchange in Winter ○ Surface Scatter Polarimetrics ○Δ
Evaluation	IR Metrics Ground to Ground ○Δ	MMW Metrics Air to Ground ○Δ IR Metrics Air to Ground ○Δ Correlation to Scenes ○Δ	MMW Metrics Ground to Ground ○Δ
Info Bases	Data Compression Δ Interface GIS & DBMS Δ Texture Reference Data Δ	Microstructure Info Base Snow Covered Area ○ Prototype 3-D Temperate, Desert & Tropic Info Bases ○Δ Transient Atmospheric Phenomena Info Base ○ Enhanced Data Base Procedures Δ	MMW Signature Data Base ○Δ IR Signature Data Base ○Δ Enhanced Data Compression Δ

Figure 1. Tech. base meausrement and information bases.

1. Radiances computed spectrally at 2 to 20cm^{-1} resolution, and band pass integrated (with optional filter function) only after atmospheric effects are added;

2. Radiance sources of thermal emission: the sun, the sky, and surrounding terrain;

3. Sky emission from broken clouds;

4. Solar shadowing;

5. Spectral directional emissivities for each material;

6. A spectral bidirectional reflectivity for each material;

7. Spectral atmospheric transmission and radiance (thermal and solar scatter) along all paths connecting the terrain, sun, sky, and sensor, utilizing the Air Force MODTRAN model (an upgrade to LOWTRAN7).

A separate model, SHADOW, automatically generates faceted shadows of the 3-D objects for inclusion within the scene.

The terrain is modeled with a set of textured polygons that overlay the topography grid. The polygon definitions and geometry are determined as part of the information base effort. The radiance models software computes a list of in-band radiance for each polygon, based in part on temperatures computed by the SWOE thermal models.

Trees and targets are described with a triangular geometry, typically with 3000 to 20,000 triangles per object. Tree geometries are based on trunk and branch measurements taken from sets of real tree measurements. Faceted leaves are generated using a fractal technique. The resulting geometry, plus a file with a separate temperatures for each triangle, are input to Hardbody for the tree radiance computations.

Targets require that the user have or prepare a set of computed target temperatures for the scene specified conditions. Hardbody then computes the radiances. Utility software is provided to convert thermal computations to the Hardbody format. Hardbody computations include facet-to-facet reflections in addition to the those listed previously. Hardbody also computes images for the objects as individuals.

Clouds are one of the more important modulators of the surface energy balance. The SWOE model package considers the influence of clouds for solar and infrared downwelling flux. During the thermal loading phase, also known as the model spin-up phase, a simple model is used to modulate the broadband downwelling flux in both spectral regions based on the geographical location of the scene, time, surface characteristics (slope and albedo), atmospheric conditions, cloud amount, and cloud type. This approach does not provide the radiant field information required at the time of the scene simulation. At scene simulation time, a modified version of LOWTRAN is used to calculate the spectrally dependent solar direct and diffuse and infrared downwelling flux. This information is used in the computation of reflections off of and between scene elements and absorption and scattering by atmospheric gases, aerosols, and clouds. Cloud shadows at the time of the scene simulation are generated by the Cloud Scene Simulation Model (CSSM). The CSSM uses a Successive Random Additions (SRA) fractal algorithm to generate the horizontal distribution of the clouds based on the cloud amount and type (stratiform, cirriform, or cumuliform). 1-D SRA and 2-D SRA algorithms are used to generate the upper and lower surface of the cloud while a 3-D SRA algorithm is used to modulate the liquid water density (LWD) information at each cloud grid point in the cloud volume. The mean LWD information as a function of cloud type and altitude has been obtained from an extensive cloud database. In the future, the LWD information will be used to determine cloud microphysical and optical properties for use in a model that will calculate the full 3-D cloud

△ WES □ TEC ○ CRREL

	Task	FY 92	FY 93	FY 94
Energy Budget	3-D Models	Geometric Mesh & Mechanics For 3-D Multiphase Calculations ○ Radiosity & Ray Tracing Methods For 3-D △ Low Resolution Mesoscale Weather Model △	3-D Multiphase Snow & Soil Model ○ Intermediate Resolution Mesoscale Weather Model △	3-D Model With Water Transport ○ Total Energy Flux Model △ High Resolution Mesoscale Weather Model △
	IR	First Generation Laser Scattering Model △	IR Scene Composit Models (Buildings, Cultural Features Etc.) ○△	
	MMW	Evaluation Of Lang 35GHz Snow & Vegetation Models ○ Active & Passive Signature Modeling Concepts ○△	Extension Of Lang Model To 95 GHz For Snow ○ First Gen Coupled 35 GHz Surface & Volume Scattering Models (Shi) ○ Interim Single Band Signature Model △	Extension Of Lang Model To 95 GHz For Vegetation ○ Integrate 35 GHz Lang & Coupled Models (Lang / Shi) ○ Multi-band Signature Model ○△
Texture	IR	Statistical IR Snow Relations ○ Second Generation Synthetic IR Texture Procedure △	Statistical IR Mixed Snow & Ground Model ○ First Gen Physics Based Texture Models ○△	Hybrid IR Snow & Frozen Ground ○ Refined Physics Based Texture Models △
	MMW	First Generation Synthetic MMW Texture Concepts ○△	Statistical MMW Snow Texture Model ○ First Generation MMW Vegetation & Soils Synthetic Texture Model △	Statistical MMW Mixed Snow & Ground Model ○ Enhanced MMW Vegetation & Soils Texture Model △
Boundary Layer			Model For Turbulent Heat Flux Exchange ○	Mesoscale Transient Boundary Layer Model ○

Figure 2. Tech. base modeling.

radiative interactions. The scene generated cloud characteristics are controlled by the Hurst and Lacunarity parameters in the SRA algorithm. Model default values controlled by the cloud type are used in the cloud simulation, but the user can modify these parameters. Cloud shadows are determined using a ray tracing technique, the 3-D cloud spatial distribution, and the solar azimuth and zenith angle or scene location and time of year and day.

Rendering. The rendering effort of the CoE laboratories is summarized in Figure 3. The SWOE rendering software provides the capability necessary to create 2-D visualization of 3-D objects and backgrounds. The software uses the depth buffer approach to resolve hidden surfaces. The output is a projection of the data contained in the information base onto a two-dimensional (2-D) image or pixel file. The input files contain physics models, initialization information, haze and lighting, and viewpoint . These inputs are used to generate the pixel data to create an image using the following processes:

- *Viewpoint manager* processes input initialization data and viewpoint

Δ WES ◻ TEC ○ CRREL

TASK		FY 92	FY 93	FY 94
Object Rendering	2-D Features	Standard Data Structures ◻	Non- Standard Data Structures ◻	
	3-D Solids	Initial Model Library ◻ Solid Model Insertion ◻	Expanded Model Library ◻ Solids Model Interactions ◻ Textured Ellipsoid ◻	Model Library Editing Functions ◻ Dynamic Models ◻
	Amorphous Objects	COMBIC Smokes ◻ Multiple Point Sources ◻	Other Obscurrents ◻ Clouds / Fog / Haze / Humidity ◻	Terrain / Wind Dynamics ◻
Techniques	Texture	Synthetic IR Texture Mapping ◻	Pixel Map Texturing ◻	Physics Based Texture ◻
	Rendering	Terrain Shadowing ◻ Anti-Aliasing ◻ Filtering / Smoothing ◻	Attribute Capability ◻ Integrated Z-Buffering / Ray Tracing / Textured Ellipsoid ◻ Solids Objects Shadowing ◻	Amorphous Object Shadowing ◻ Tracker Algorithm ◻ Radiosity Rendering ◻
	MMW	Technology Assessment ◻ Preliminary Geometry ◻	Interface Sensor Models ◻	Initial Phase / Brightness Soft ware ◻

Figure 3. Tech. base scene rendering

information to generate a sun vector, ambient and diffuse lighting parameters, bounding planes, and the world space to a viewpoint space transformation matrix, which is referred to as "viewpoint data."

• *World manager* uses viewpoint data and bounding planes to select root nodes in the data base for the terrain region. A node is defined as a subset of the information base that contains position data, information for level of detail (LOD) and field of view (FOV), materials properties, and so on, and pointers to the items associated with each node.

• *Node processor* traverses the node tree and uses the bounding planes to determine which nodes are in the FOV. The FOV test creates a list of active nodes, loads texture maps, and allocates nodes to the correct LOD.

• *Item processor* transforms sun vectors, calculates triangle face normals, eliminates back faces, determines polygon coloring and shading, and converts vertices to viewpoint space, clips to hither plane, and projects polygons to screen space.

- *Pretiler* clips polygon to screen space and creates triangles with incremental color, depth, and texture information.
- *Tiler* produces pixels for display for each triangle through the graphics processor and may modify color attributes of textured items.

The result of the rendering is a 2-D pixel space representation of the radiometric energy arriving at the aperture of a sensor for a specified viewing geometry.

Summary and Conclusions

The SWOE concept, as developed, depends on the research products of DoD technology base programs. The CoE technology base contributions to the SWOE program are significant. CoE has taken the lead by commitment to a focused tech base effort required for enhancement of future smart weapons system performance. This commitment has resulted in extremely high levels of coordination and cooperation between CoE laboratories as well as the work serving as a model for unprecedented levels of coordination and cooperation between all the armed forces in the SWOE program.

Mission Kill Metrics for Combat Damage Accounting

David C. Hardison
The John Hopkins University
Baltimore, MD

It has been determined that the metrics used for accounting combat damage are not as sophisticated as the current situation would demand. There is a need for improvement and wider use of "mission-damage" and "mission-kill" metrics. Metrics for item-level and force-level systems are proposed. Those proposed for item-level systems are mission-related and depend on the onset time and duration of the partial or total loss of ability to perform three battlefield functions: move, communicate/control, and operate. The metrics proposed for force-level systems are similar and pertain to the partial or total loss of ability to perform combat, combat support, and combat service support. The metrics are applicable to both conventional and non-lethal munitions.

The interest in better mission kill metrics stems from the recent development of "soft-kill" or "nonlethal" weapons, the need to be able to determine whether the best targets, or the most effective, were destroyed, and the need to know which of several targets should be attacked and with what munitions.

Current metrics can be criticized in the following areas: (1) mission-dependency; (2) their probabilistic character; (3) failure to account for time; and (4) limitations as to persons and items, rather than damage to a unit or force.

Metrics are needed for assessment of the amount of physical damage done when a particular munition is used to attack a particular target, thereby assessing the net battlefield implications of a system's loss or diminution of ability to move, communicate, or to operate.

A set of damage metrics has been established for item systems to be used in computations that can determine the net battlefield situation during combat and thereby propose a framework suited to consistent accounting.

Lightweight Exo-Atmospheric Projectile (LEAP)

Paul K. Baker
U.S. Army Space and Strategic Defense Command
Huntsville, AL 35807-3801

The Lightweight Exo-Atmospheric Projectile (LEAP) Program is a technology development effort to demonstrate lightweight component and integration technologies for kinetic energy kill vehicles. These technologies include lightweight, high-performance propulsion systems, advanced high-density electronics packaging, strapdown guidance, staring focal plane array based seekers, and advanced technology Inertial Measurement Units. The Army LEAP program is demonstrating these advanced technologies by integrating them into the smallest and lightest interceptor technology demonstrator being developed for the Strategic Defense Initiative Organization. An extensive test program has been developed to validate the performance of the LEAP and the issues associated with kill vehicle integration. This test program includes free-flight hover testing on the ground and exo-atmospheric flight testing.

Information provided by this work gives an update on the progress of the LEAP Program in the development and demonstration of these light weight technologies. The LEAP design and the results of the final two tests in the ground test program, a full-up strapdown test and a free-flight hover test, as well as future design enhancements and test plans, are all available.

Power Spectral Characterization of Fracture Surfaces of TNT, Composition B, and Octol

M. Yvonne D. Lanzerotti
James J. Pinto
U. S. Army Armament Research, Development and Engineering Center
Picatinny Arsenal, NJ 07806-5000

Allan Wolfe
New York City Technical College, Brooklyn, NY 11201

David J. Thomson
AT&T Bell Laboratories, Murray Hill, NJ 07974

Energetic materials are of significant interest for scientific and practical reasons in the extraction (mining) industry, space propulsion, and ordnance. The nature of fracture processes in such materials under high acceleration is of particular interest, especially in ordnance. This research is directed toward the development of new experimental and analysis techniques that allow us to characterize quantitatively and compare the fracture surfaces of different energetic materials, and to deduce specific fracture mechanisms. The techniques are widely applicable to other composite systems.

Topographical profiles spaced 1.0mm apart across the fracture surfaces of TNT, Composition B, and Octol have been obtained with a diamond stylus profilometer. Spatial power spectra (wavelengths of 1.0m to 1.0cm) have been calculated using a prolate spheroidal data window in the horizontal space domain prior to using a fast Fourier transform algorithm. The spatial power density of the fracture surface profiles is found in general to decrease with increasing spatial frequency over the region of interest, -1 to -1. Quasi-periodic peaks corresponding to the TNT grain size are observed in the TNT spatial power spectra; these peaks indicate the nonuniform grain size distribution in the TNT fracture surface. Quasi-periodic peaks corresponding to RDX and HMX particle sizes are observed in the Composition B and Octol spatial power spectra, respectively. These peaks indicate the inhomogeneous RDX and HMX grain size distribution in the Composition B and Octol fracture surfaces, respectively. The peaks in the TNT power spectra also indicate that much of the fracture in this material is occurring at grain boundaries: intergranular failure. Peaks in the Composition B and Octol spectra indicate that intergranular fracture often occurs between the TNT and RDX or HMX grains, respectively. Fractal analysis of the TNT, Composition B, and Octol power spectral slopes and amplitudes indicates the regions of deterministic intergranular failure and the regions of the nondeterministic trans-granular failure through TNT, RDX, or HMX grains. This non-deterministic (fractal) failure is chaotic and may indicate the origin of failure in the sample.

Imaging and Analysis of Combusting Liquid Propellant Sprays

Michael J. McQuaid
Avi Birk
Gary Bliesener
U.S. Army Research Laboratory
Weapons Technology Directorate
Aberdeen Proving Ground, MD 21005-5066

A series of experiments was conducted to elucidate the jet break-up, atomization, and combustion of liquid monopropellant sprays through the use of imaging and spectroscopic techniques. Hydroxylammonium nitrate (HAN)-based monopropellants and nitromethane were injected at 120 or 240 m/s through a circular orifice into 35 MPa, 500°C nitrogen. Spray dynamics were recorded via high speed cinematography. Breakthroughs in the imaging of combusting sprays were achieved through the use of novel experimental techniques, illumination of the spray with a pulsed laser sheet, and the seeding of propellants with lithium or sodium nitrate to enhance flame emission. Emission spectroscopy was used to identify combustion intermediates and their relative contribution to flame luminosity. The spectral data provided information necessary to process and analyze the cinematic records. The experimental series demonstrates the importance of particulate and vortex burning of HAN-based propellants at low gun pressures. The experiments with nitromethane show that dynamical behavior of the spray will change as temperature and pressure exceed supercritical values. These results are discussed in the context of understanding spray combustion phenomena in regenerative liquid propellant guns.

Battlefield Environmental Effects on Ultraviolet Atmospheric Sensing and Propagation

James B. Gillespie
David L. Rosen
U.S. Army Research Laboratory
Battlefield Environmental Directorate
White Sands Missile Range, NM 88002-5501

Traditionally Army electro-optic systems have operated in the infrared (or longer) wavelength regions in order to overcome atmospheric obscuration effects. The ultraviolet (UV) spectral region has largely been ignored. In recent years however, development of new sources and detectors has led to a renewed interest in the UV, particularly in remote sensing of chemical composition and in UV signatures, both solar blind and laser induced fluorescence. This became particularly apparent in the war with Iraq when biological weapons were possible. UV remote sensing is the only method to detect biological aerosols. The Army's war on drugs is another potential benefactor from the UV technology. UV remote sensing systems can potentially detect and locate illicit drug manufacturing facilities. To meet this new technology need, a research program has been implemented to quantify atmospheric effects on UV signatures and to quantify laser induced fluorescence properties. The program consists of an experimental effort to apply single particle levitation technology to the examination of fundamental fluorescence and scattering properties of aerosols and a modeling effort to simulate the effects that the atmosphere has on UV signatures and analysis of these signatures. We have developed an atmospheric propagation and lidar model, UVTRAN, to determine atmospheric transmission in the 200nm to 700nm spectral region.

We have also developed Factor Analysis - Rank Annihilation (FARA) codes (both time-resolved and spectrally-resolved techniques) to analyze complex laser induced fluorescence spectra for UV lidar spectral returns. Using the UVTRAN model with the FARA model we have simulated the effect of the battlefield atmosphere on laser induced fluorescence spectra and have used an overlap index method to quantify the applicability of UV remote sensor systems to army problems. Experimental results for the laboratory characterization of singly levitated particles are also available.

Camouflage Flicker: A Highly Observable Low Observable

Wendell R. Watkins
Samuel B. Crow
Fernando R. Palacios
Daniel R. Billingsley
U.S. Army Research Laboratory
Battlefield Environment Directorate
White Sands Missile Range, NM 88002-5501

Changes in meteorological conditions that may seem superficially innocuous can dramatically affect target signatures. Target surface characteristics are the key to camouflage protection. If simulations of the combined effects of the battlefield environment are not realistic, under design of target surfaces, such as camouflage flicker, will result. The design and testing of camouflage netting require application of more than a slowly varying irradiance level to determine signature characteristics. In the real world, typical meteorological and environmental conditions often produce large contrast changes in the infrared of under designed camouflage nets that are supposed to conceal the location of vital assets. The key to concealment is to match characteristics of the background. Use of low emissive paints and surfaces is a step in the right direction, but not a panacea. The recently developed Mobile Imaging Spectroscopy Laboratory has been used to assess current camouflage design, and the result was the discovery of camouflage flicker. The combination of solar loading changes, low windspeed, and moderate humidity causes the camouflage netting to act like a beacon. The low thermal mass of the netting, even with its low absorption coefficient surface, cannot match vegetative backgrounds that still retain evaporative cooling under low windspeed conditions and variable solar loading changes. Texturing the netting may help mitigate this camouflage flicker, but the solution must also address the future use of active imaging when altering the camouflage surface. If the camouflage is irradiated by a polarized laser source, the camouflage surface must also be characterized in terms of its enhanced backscatter. In addition, optical turbulence may reduce but not eliminate the backscatter, depending upon the atmospheric isoplanatic patch size and surface characteristics in terms of multiple scattering.

SUMMARY

River and Lake Ice Conditions as Interpreted frcm Microwave Imagery

Rae Melloh
Lawrence Gatto
U.S. Army Cold Regions Research and Engineering Laboratory
Hanover, NH 03755

Passive microwave (33.6GHz) and synthetic aperture radar (C-,Lr- and P-band) images of ice covered rivers and lakes can provide information on types and distribution of river and lake ice needed for studying river ice processes and dynamics, and monitoring ice during winter. Visible and IR remote sensing systems do not provide comparable data and are hindered by the darkness and cloud cover often encountered in northern regions. Our studies have resulted in several findings that will be useful for imagery analysts: active radar systems provide the capability to distinguish rough-textured from smooth-ice covers, zones of variable ice surface roughness within rough ice fields can be differentiated, C- and Lr-band data are more sensitive than P-band to the range of surface roughnesses encountered, and polarization synthesis and signatures provide a way to improve contrast of ice features on radar imagery. Ka-band passive microwave systems allow the differentiation of wet versus dry snow conditions; open water in ice covered rivers and lakes; and compacted areas in depth hoar snowpacks. Possible applications of these results include location of river bridging sites, dangerous thin-ice zones, packed-snow trails, and preferential water supply points.

High-Power Impulse Generation Using a Photoconductively Switched Radial Transmission Line

A. Kim
R. Zeto
R. Youmans
M. Weiner
U.S. Army Research Laboratory
Electronics and Power Sources Directorate
Fort Monmouth, NJ 07703-5601

L. Jasper
U.S. Army Research Laboratory
Signatures, Sensors, Signal and Information Processing Directorate
2800 Powder Mill Road
Adelphi, MD 20783-1197 I

The Army needs gigawatt impulse generators for applications such as sensors, communications, and electronic countermeasures. Our research is on impulse generation by the optical triggering of a gallium arsenide (GaAs) device that uses a tapered radial transmission line. The device has successfully generated nanosecond pulses with 40MW of peak power. The high-peak-power impulse-generating device was realized by housing a 1-cm-gap GaAs switch with two opposite gridded electrodes in the central region of a tapered radial transmission line structure. The tapered radial transmission line structure provided the large electrostatic energy storage capability as well as the impedance transformation, resulting in a voltage gain factor close to 1.7. A unique integration of the photoconductive GaAs switch into a nonuniform radial transmission line resulted in a pulser with high power capability, fast rise time, voltage gain, and compactness. Operation of this device into a 50-

load demonstrated 40-MW output power, 4-ns pulsewidth, a rise time of less than 250ps, and a fall time of less than 500ps.

Projectile Tracking Device Using GPS

George C. Wiles
John S. Eicke
Brian T. Mays
Andrew P. Ladas
U.S. Army Research Laboratory
Signatures, Sensors, Signal and Information Processing Directorate
2800 Powder Mill Road
Adelphi, MD 20783-1197

Significant increases in artillery accuracy and effectiveness can be achieved with the use of projectile tracking and trajectory modeling to predict the point of impact. As part of a U.S. Army cooperative program with the Human Engineering Laboratory and the Weapons Technology Directorate, the Signatures, Sensors, Signal and Information Processing Directorate is developing a Global Positioning System (GPS) based transponder capable of providing projectile tracking data. This effort includes development of both the projectile-mounted GPS transponder and the specialized receiver equipment needed to determine projectile position data.

The Signatures, Sensors, Signal and Information Processing Directorate has been performing the research into the design and implementation of the GPS transponder and the receiver equipment. Translator design issues included careful consideration of interference signals, antenna design requirements, minimization of components, and high-g factors in the design. The translator design was breadboarded and evaluated, and then two prototype projectile units were fabricated and fired in October 1991.

The translator receiver equipment was required to process the translated

GPS signals and provide position data. The correlation, tracking and position solution are performed in software. The short mission time (less than two minutes) made a random access memory (RAM)-based data storage system possible. This unique approach significantly reduced the complexity of the data recording system compared to existing systems. A breadboard of the translator receiver hardware has been fabricated and used to process the data at the field test.